The Evolving Continents: Understanding Processes of Continental Growth

The Geological Society of London
Books Editorial Committee

Chief Editor
BOB PANKHURST (UK)

Society Books Editors
JOHN GREGORY (UK)
JIM GRIFFITHS (UK)
JOHN HOWE (UK)
RICK LAW (USA)
PHIL LEAT (UK)
NICK ROBINS (UK)
RANDELL STEPHENSON (UK)

Society Books Advisors
MIKE BROWN (USA)
ERIC BUFFETAUT (FRANCE)
JONATHAN CRAIG (ITALY)
RETO GIERÉ (GERMANY)
TOM MCCANN (GERMANY)
DOUG STEAD (CANADA)
MAARTEN DE WIT (SOUTH AFRICA)

Geological Society books refereeing procedures

The Society makes every effort to ensure that the scientific and production quality of its books matches that of its journals. Since 1997, all book proposals have been refereed by specialist reviewers as well as by the Society's Books Editorial Committee. If the referees identify weaknesses in the proposal, these must be addressed before the proposal is accepted.

Once the book is accepted, the Society Book Editors ensure that the volume editors follow strict guidelines on refereeing and quality control. We insist that individual papers can only be accepted after satisfactory review by two independent referees. The questions on the review forms are similar to those for *Journal of the Geological Society*. The referees' forms and comments must be available to the Society's Book Editors on request.

Although many of the books result from meetings, the editors are expected to commission papers that were not presented at the meeting to ensure that the book provides a balanced coverage of the subject. Being accepted for presentation at the meeting does not guarantee inclusion in the book.

More information about submitting a proposal and producing a book for the Society can be found on its web site: www.geolsoc.org.uk.

It is recommended that reference to all or part of this book should be made in one of the following ways:

KUSKY, T. M., ZHAI, M.-G. & XIAO, W. (eds) 2010. *The Evolving Continents: Understanding Processes of Continental Growth*. Geological Society, London, Special Publications, **338**.

MARUYAMA, S., KAWAI, T. & WINDLEY, B. F. 2010. Ocean plate stratigraphy and its imbrication in an accretionary orogen: the Mona Complex, Anglesey–Lleyn, Wales, UK. *In*: KUSKY, T. M., ZHAI, M.-G. & XIAO, W. (eds) *The Evolving Continents: Understanding Processes of Continental Growth*. Geological Society, London, Special Publications, **338**, 55–75.

GEOLOGICAL SOCIETY SPECIAL PUBLICATION NO. 338

The Evolving Continents: Understanding Processes of Continental Growth

EDITED BY

T. M. KUSKY
China University of Geosciences, Wuhan

M.-G. ZHAI
Chinese Academy of Sciences, China

and

W. XIAO
Chinese Academy of Sciences, China

2010
Published by
The Geological Society
London

THE GEOLOGICAL SOCIETY

The Geological Society of London (GSL) was founded in 1807. It is the oldest national geological society in the world and the largest in Europe. It was incorporated under Royal Charter in 1825 and is Registered Charity 210161.

The Society is the UK national learned and professional society for geology with a worldwide Fellowship (FGS) of over 10 000. The Society has the power to confer Chartered status on suitably qualified Fellows, and about 2000 of the Fellowship carry the title (CGeol). Chartered Geologists may also obtain the equivalent European title, European Geologist (EurGeol). One fifth of the Society's fellowship resides outside the UK. To find out more about the Society, log on to www.geolsoc.org.uk.

The Geological Society Publishing House (Bath, UK) produces the Society's international journals and books, and acts as European distributor for selected publications of the American Association of Petroleum Geologists (AAPG), the Indonesian Petroleum Association (IPA), the Geological Society of America (GSA), the Society for Sedimentary Geology (SEPM) and the Geologists' Association (GA). Joint marketing agreements ensure that GSL Fellows may purchase these societies' publications at a discount. The Society's online bookshop (accessible from www.geolsoc.org.uk) offers secure book purchasing with your credit or debit card.

To find out about joining the Society and benefiting from substantial discounts on publications of GSL and other societies worldwide, consult www.geolsoc.org.uk, or contact the Fellowship Department at: The Geological Society, Burlington House, Piccadilly, London W1J 0BG: Tel. +44 (0)20 7434 9944; Fax +44 (0)20 7439 8975; E-mail: enquiries@geolsoc.org.uk.

For information about the Society's meetings, consult *Events* on www.geolsoc.org.uk. To find out more about the Society's Corporate Affiliates Scheme, write to enquiries@geolsoc.org.uk.

Published by The Geological Society from:
The Geological Society Publishing House, Unit 7, Brassmill Enterprise Centre, Brassmill Lane, Bath BA1 3JN, UK

(*Orders*: Tel. +44 (0)1225 445046, Fax +44 (0)1225 442836)
Online bookshop: www.geolsoc.org.uk/bookshop

The publishers make no representation, express or implied, with regard to the accuracy of the information contained in this book and cannot accept any legal responsibility for any errors or omissions that may be made.

© The Geological Society of London 2010. All rights reserved. No reproduction, copy or transmission of this publication may be made without written permission. No paragraph of this publication may be reproduced, copied or transmitted save with the provisions of The Copyright Licensing Agency Ltd, Saffron House, 6–10 Kirby Street, London EC1N 8TS, UK. Users registered with the Copyright Clearance Center, 222 Rosewood Drive, Danvers, MA 01923, USA: the item-fee code for this publication is 0305-8719/10/$15.00.

British Library Cataloguing in Publication Data

A catalogue record for this book is available from the British Library.
ISBN 978-1-86239-303-5

Typeset by Techset Composition Ltd, Salisbury, UK
Printed by MPG Books Ltd, Bodmin, UK

Distributors

North America
For trade and institutional orders:
The Geological Society, c/o AIDC, 82 Winter Sport Lane, Williston, VT 05495, USA
Orders: Tel. +1 800-972-9892
Fax +1 802-864-7626
E-mail: gsl.orders@aidcvt.com

For individual and corporate orders:
AAPG Bookstore, PO Box 979, Tulsa, OK 74101-0979, USA
Orders: Tel. +1 918-584-2555
Fax +1 918-560-2652
E-mail: bookstore@aapg.org
Website: http://bookstore.aapg.org

India
Affiliated East-West Press Private Ltd, Marketing Division, G-1/16 Ansari Road, Darya Ganj, New Delhi 110 002, India
Orders: Tel. +91 11 2327-9113/2326-4180
Fax +91 11 2326-0538
E-mail: affiliat@vsnl.com

Contents

Preface ... vii

KUSKY, T. M., ZHAI, M. & XIAO, W. The evolving continents: understanding processes of continental growth – introduction ... 1

Oceanic and island arc systems and continental growth

STERN, R. J. The anatomy and ontogeny of modern intra-oceanic arc systems ... 7

XIAO, W., HAN, C., YUAN, C., SUN, M., ZHAO, G. & SHAN, Y. Transitions among Mariana-, Japan-, Cordillera- and Alaska-type arc systems and their final juxtapositions leading to accretionary and collisional orogenesis ... 35

Tectonics of accretionary orogens and continental growth

MARUYAMA, S., KAWAI, T. & WINDLEY, B. F. Ocean plate stratigraphy and its imbrication in an accretionary orogen: the Mona Complex, Anglesey–Lleyn, Wales, UK ... 55

SANTOSH, M., MARUYAMA, S., KOMIYA, T. & YAMAMOTO, S. Orogens in the evolving Earth: from surface continents to 'lost continents' at the core–mantle boundary ... 77

WILDE, S. A., WU, F.-Y. & ZHAO, G. The Khanka Block, NE China, and its significance for the evolution of the Central Asian Orogenic Belt and continental accretion ... 117

Growth and stabilization of continental crust: collisions and intraplate processes

RAZAKAMANANA, T., WINDLEY, B. F. & ACKERMAND, D. Petrology, chemistry and phase relations of borosilicate phases in phlogopite diopsidites and granitic pegmatites from the Tranomaro belt, SE Madagascar; boron-fluid evolution ... 139

PENG, P. Reconstruction and interpretation of giant mafic dyke swarms: a case study of 1.78 Ga magmatism in the North China craton ... 163

Precambrian tectonics and the birth of continents

KRÖNER, A. The role of geochronology in understanding continental evolution ... 179

ROLLINSON, H., REID, C. & WINDLEY, B. Chromitites from the Fiskenæsset anorthositic complex, West Greenland: clues to late Archaean mantle processes ... 197

GARDE, A. A. & HOLLIS, J. A. A buried Palaeoproterozoic spreading ridge in the northern Nagssugtoqidian orogen, West Greenland ... 213

ZHAI, M., LI, T.-S., PENG, P., HU, B., LIU, F. & ZHANG, Y. Precambrian key tectonic events and evolution of the North China craton ... 235

OLIVEIRA, E. P., MCNAUGHTON, N. J. & ARMSTRONG, R. Mesoarchaean to Palaeoproterozoic growth of the northern segment of the Itabuna–Salvador–Curaçá orogen, São Francisco craton, Brazil ... 263

Active tectonics and geomorphology of continental collision and growth zones

PETTERSON, M. G. A review of the geology and tectonics of the Kohistan island arc, north Pakistan — 287

ALLEN, M. B. Roles of strike-slip faults during continental deformation: examples from the active Arabia–Eurasia collision — 329

SEARLE, M. P. & TRELOAR, P. J. Was Late Cretaceous–Paleocene obduction of ophiolite complexes the primary cause of crustal thickening and regional metamorphism in the Pakistan Himalaya? — 345

CUNNINGHAM, D. Tectonic setting and structural evolution of the Late Cenozoic Gobi Altai orogen — 361

OWEN, L. A. Landscape development of the Himalayan–Tibetan orogen: a review — 389

Index — 409

Preface

This Special Publication of the Geological Society of London, *The Evolving Continents: Understanding Processes of Continental Growth*, is dedicated to the long and spectacular career of Brian F. Windley, a pioneer in the application of uniformitarianism to Precambrian rocks, a leader in linking field geology with the geochemistry and geochronology of different orogenic units with global tectonic history, and an overall polymath who has had a deep influence on many fields of geological sciences.

Brian Windley has not only been a pioneer in science, but has been a mentor and teacher to many geologists who have become leaders in their fields, and in this volume, many of those scientists have contributed chapters that reflect the skill and knowledge that Brian instilled in his students and colleagues.

As an opener for this volume we asked Brian 'what has happened over the years'.

This was his reply.

> Following graduation at Liverpool in 1960, I went to Exeter University to study for a PhD under Ken Coe in an area in the *c*. 1.8 Ga Ketilidian orogenic belt in SW Greenland on a contract with GGU, the Geological Survey of Greenland, after which I was offered a job to undertake a reconnaissance of most of the Archaean craton of West Greenland, six delightful 3–4-month-long summers roaming free with a boat and a huge stack of aerial photographs. But GGU would only enable me to go to Greenland for the rest of my career (gneisses for ever), which I found a limiting thought, and so in 1968 I left and joined Leicester University in England from where the world became my geological backyard. But GGU decided to have its next base-camp in the Fiskenaesset region, where I continued in 1970 and 1972 to study the stratigraphy and structure of the anorthosite complex.
>
> In 1973 I went to southern India to find out how similar the Sittampundi complex was to the Fiskenaesset complex (remarkably so as it turned out). Subramaniam had described eclogites there in 1956, and so I collected some, but did nothing with them until recently when I passed them to Krishnan Sajeev in Bangalore, who found that they are indeed high-temperature eclogites. And we have found similar high-pressure rocks in the Scourian of Scotland and at Rodel in the Outer Hebrides. Interestingly, they all occur in high-grade layered cumulate complexes in which I was so interested in earlier years.
>
> Perhaps surprisingly, I have never done much work in the Scourian of northern Scotland. But in the late 1970s Hugh Rollinson and Jane Sills studied the granulites and layered mafic–ultramafic complexes for their PhDs. And this led to current work with Hugh on the Fiskenaesset chromites.
>
> As a result of a silly paper in *Nature* in 1970 comparing the new Apollo discovery of calcic anorthosites on the Moon with Archaean calcic anorthosites on Earth, I was invited by the great mineralogist Joe Smith to Chicago to make an investigation of the Fiskenaesset anorthositic complex in Greenland with his early, hand-operated ARL microprobe. So I commuted to his lab for 2 months a year for most of the 1970s when, for a week at a time, we alternated with one of us working on the probe in the daytime whilst the other slept, and then took the computer cards to the computer centre during the night.
>
> Although John Dewey had started the revolution into continental plate tectonics in 1969, by the late 1970s I saw that the filtering downwards of such ideas into the Precambrian was being hampered by a general lack of knowledge amongst the Precambrian community of modern collision tectonics, and so I started in 1980 a five-year project in the Himalayas of Pakistan with lots of students and staff such as Qasim Jan (mineralogy), Mike Coward (structure), Mike Petterson (Kohistan arc), Asif Khan (Chilas complex), Carol Pudsey (sedimentology) and Lewis Owen (terraces of the river Indus). To tackle the high mountains, glaciers and moraines of the Karakoram range in the next 5 years, two young mountaineering post-doc fellows, Mike Searle and Tony Rex, joined the team. With their mountaineering colleagues they organized the porters to carry the food in (and the rocks out) in the two-week stroll up the Baltoro River and glacier to the base of K2.
>
> After our horrific jeep accident above Skardu in 1981 with Mike Coward, Asif and Carol, I left the Himalayan–Karakoram scene and moved northwards into China to the North China craton with Jane Sills and Zhai Mingguo and then westwards to the Tien Shan range in Xinjiang with Mark Allen and Zhang Chi, and to the Chinese Altai with Guo Jinghui (with a second horrific jeep accident). The attraction of continuing a crustal section northwards led to more than a decade of summers in the Central Asian orogenic belt of Mongolia, but first the Cenozoic uplift had to be resolved with Dickson Cunningham starting in 1994; why were 4000 m mountains going up in the middle of Asia? I had met Dickson in Tierra del Fuego before Ian Dalziel's memorable and excellent GSA field trip to Antarctica in 1990. After much work in Mongolia especially with G. Badarch and Alfred Kröner (and more recently in Kyrgyzstan with Alfred and Dmitry Alexeiev), and after being nearly drowned in a river in Mongolia, I moved south to work with Xiao Wenjiao from Beijing on accretionary belts in the Qilian Shan (Shan means mountains in Chinese), the Kun Lun, Nei Mongol, and the Tien Shan, and currently in the Bei Shan. Currently, it is a pleasure to work with Lui Dunyi and Jian Ping from the SHRIMP laboratory in Beijing on zircon-related studies of ophiolites and associated rocks in China and Mongolia, and with Reimar Seltmann from CERCAM (Centre for Russian and EurAsian Mineral Studies, which belongs to the Natural History Museum, London) on granites and mineral deposits in Central Asia.

Going back a bit, a lifetime's interest in sapphirine-bearing rocks began in the mid-1960s when I came across Giesecke's 1809 type locality in Fiskenaesset harbour, after which Richard Herd came to Imperial College and Leicester, and joined the GGU team in 1970–1972 in the Fiskenaesset region to tackle all the minerals that the field geologists could hardly recognize. Then in the mid-1970s I found I was sitting in Chicago at a desk opposite Dietrich Ackermand from Kiel in Germany who was also visiting Joe Smith. So began a decade of sapphirine studies in 1982 with a NATO grant with Richard and Dietrich. This led to the first visit to Madagascar in 1986 with Dietrich to hunt for Lacroix's 1922–1923 report of sapphirine with anorthosite (sakenite); we never made it as we got badly stuck in a mosquito-infested Shagashak River on the plains of Horombe. But it was the interest in sapphirine that led to a study of the tectonics of Madagascar with Theodore Razakamanana, Alan Collins as a post-doc, and Alfred Kröner on zircon studies. And that was how Alfred came later to join the Mongolian project. Following a summer in 2006 with Richard Herd showing True North Gems how to find more rubies in sapphirine rocks throughout the Fiskenaesset region, my latest joy has been the opportunity to return to West Greenland in 2008 and 2009 with GEUS (the Geological Survey of Denmark and Greenland) in their re-evaluation of much of the Archaean craton.

Because the Pan-African Malagasy orogen extends northwards into Yemen it was natural to work with Martin Whitehouse and Mahfood Ba-bttat on the equivalent geology in Yemen in the early 1990s, which was great until our kidnap. Later, Martin, Vicky Pease and I continued across strike in Dhofur, Oman.

After so many years looking at the Central Asian orogenic belt in Mongolia, I thought I knew as much as anyone about how to interpret an old accretionary orogen. But after a year's visit to Shigenori Maruyama in Tokyo in 2001–2002 (one of the most stimulating years of my life) I realized that I knew very little indeed. We agreed that there was a poor understanding worldwide amongst the Precambrian community of the geology of Japan, arguably the best modern analogue for accretionary orogens back to Archaean greenstone belts. So, after much discussion, that led to the current ERAS project (Earth Accretionary Systems in Space and Time) of the International Lithosphere Programme led by Peter Cawood and Alfred Kröner. I had been taking students to Anglesey in North Wales for many years, but it was not until 2001 that I realized that the $c.$ 600 Ma Mona Complex is remarkably similar to the Mesozoic–Cenozoic accretionary orogen in Japan, and so Shige and I started a project with several Japanese staff and students, in particular Takahiro Kawai, to unravel the geology of Anglesey and Lleyn.

On arrival in Leicester in 1968, I started an undergraduate course on Earth Evolution which was essentially on continental plate tectonics through time, but found that there was no suitable textbook on that subject, and so I came to write in my spare time *The Evolving Continents*. Because I learnt so much that was useful for my research, there was little to hold me back writing more editions and updates.

In looking back, I feel it might have been better to do things in reverse order, and not start with some of the oldest rocks in the deep crust, and evolve into neotectonics in Mongolia and modern accretion in Japan. And what of the future? My only regret is that there is not enough time.

And finally, thanks to Judith for all that driving and for everything else.

Brian Windley, 30 June 2009

We hope you enjoy and benefit from this volume, as much as we have benefited from working with Brian over the years.

TIM KUSKY
MINGGUO ZHAI
WENJIAO XIAO

Brian Windley at Stornoway, Outer Hebrides, Scotland, 2009.

The evolving continents: understanding processes of continental growth – introduction

TIMOTHY M. KUSKY[1], MINGGUO ZHAI[2] & WENJIAO XIAO[2]

[1]*Ministry of Education, Three Gorges Geohazards Research Center and State Key Lab for Geological Processes and Mineral Resources, China University of Geosciences, Wuhan, Wuhan, China*

[2]*Institute of Geology and Geophysics, Chinese Academy of Sciences, Beijing, China*

*Corresponding author (e-mail: tkusky@gmail.com)

We have organized and edited this Special Publication of the Geological Society of London to honour the career of Brian F. Windley, who has been hugely influential in helping to achieve our current understanding of the evolution of the continental crust, and who has inspired many students and scientists to pursue studies on the evolution of the continents. Brian has studied processes of continental formation and evolution on most continents and of all ages, and has educated and inspired two generations of geologists to undertake careers in studies of continental evolution. The contributions in this volume represent only a small percentage of studies that Brian has influenced, yet the scope and significance of these papers are clear, and stand as a testimony to Brian's contributions to understanding processes of continental evolution, growth, and stabilization.

The volume is organized into six sections: oceanic and island arc systems and continental growth; tectonics of accretionary orogens and continental growth; growth and stabilization of continental crust: collisions and intraplate processes; Precambrian tectonics and the birth of continents; active tectonics and geomorphology of continental collision and growth zones.

The first section, oceanic and island arc systems and continental growth, begins with a paper by **Stern**, who summarizes the current state of knowledge about intra-oceanic arc systems from petrological, geophysical, and tectonic viewpoints and emphasizes that these systems have been the most important sites of juvenile continental crust formation for as long as plate tectonics has operated (the time of the start of plate tectonics is a matter of debate between some geologists). **Stern** describes the main components and zonation of intra-oceanic arc systems, including the trench, forearc, volcanic–magmatic arc, and back-arc, typically forming a system about 200 km wide, and strongly influenced by hydrous melting process in the underlying mantle. He then describes differences between the various stages of intra-oceanic arc systems, including juvenile arc lithosphere preserved in many forearcs, to mature arc systems where magmatism is concentrated along the magmatic–volcanic front. Mature intra-oceanic arc systems are typically extensional with volcanism and sea-floor spreading developing in the back-arc, and also show a transition in mantle types from serpentinized harzburgite beneath forearc sections, pyroxene-rich low-V_p mantle beneath the magmatic front, and lherzolite–harzburgite mantle beneath back-arc basins.

In the second paper in this section, **Xiao *et al.*** describe the major differences between the types of arc systems found in the circum-Pacific region, including Mariana-, Japan-, Cordillera-, and Alaska-type arcs, and compare these systems with accreted terranes in accretionary orogens. They show how arcs are complex systems that can be different along strike (such as in the Alaska-type systems), and change with time. They suggest that some unresolved issues in accretionary and collisional orogens may be related to geologists not appreciating some of the complexities in modern arc systems, and use examples from the Altaids and other systems to demonstrate their points.

Part 2 of the book, tectonics of accretionary orogens and continental growth, highlights a common thread of Brian Windley's multi-year efforts of trying to work out the framework and evolution of accretionary orogens. His studies followed through early studies on the Tien Shan, Altai, and Central Asian Orogenic Belt in Mongolia, the Solonker suture in China, the Qilian Shan, the western Kun Lun Mountains and the Bei Shan in China. His most recent work in Asia has all been done in collaboration with Xiao Wenjiao, and in the Mona Complex in Anglesey and Lleyn in Wales with Shigenori Maruyama. Appropriately, the first paper in this section is by **Maruyama *et al.*** They recognize three types of strongly imbricated oceanic plate stratigraphy in the Neoproterozoic accretionary orogen on the island of Anglesey and the Lleyn Peninsula.

From: KUSKY, T. M., ZHAI, M.-G. & XIAO, W. (eds) *The Evolving Continents: Understanding Processes of Continental Growth*. Geological Society, London, Special Publications, **338**, 1–6.
DOI: 10.1144/SP338.1   0305-8719/10/$15.00 © The Geological Society of London 2010.

These include an old section (at the structural top of the accretionary complex), a central section that was subjected to deep subduction and exhumed as blueschists, and the youngest section at the bottom of the structural sequence, which is an olistostrome-type deposit that formed by gravitational collapse of previously accreted material. **Maruyama et al.** use their structural observations to reconstruct the accretionary history of this orogen, providing a lesson to workers in other orogens world-wide about the use of recognizing and using ocean plate stratigraphy for deciphering the tectonics of accretionary orogens.

The second paper in this section is by **Santosh et al.** In this paper, the authors propose a new, untraditional classification of orogens that includes continental crustal material that is deeply subducted through tectonic erosion at trenches, subduction of young arc sequences, and continental collisions. Their classification includes: (1) deeply subducted material that is taken down to mantle depths and that never returns to the surface, termed ghost orogens; (2) orogens that are subducted to deep crustal levels, undergo melting and are recycled back to the surface, temporarily, called arrested orogens; (3) extant orogens, which are partly returned back to the surface after deep subduction; (4) concealed orogens, which have been deeply subducted and only the traces of which are represented on the surface by mantle xenoliths carried by younger magmas. **Santosh et al.** use this new classification to postulate material circulation within the Earth, throughout geological time, and note that this type of analysis leads to insight into understanding radiogenic heat generation in the mantle, and to models for the growth of continents through time.

The third paper in this section is by **Wilde et al.** In this paper the authors describe the geology and U–Pb geochronology of the Khanka Block, a poorly known part of the Central Asian Orogenic Belt. Granitic magmatism at 518 ± 7 Ma was followed by high-grade metamorphism at 500 Ma, suggesting a correlation with the Jiamusi block to the west. Younger magmatism at 112 ± 1 Ma in the Khanka Block is related to Pacific plate subduction and post-collisional extension. **Wilde et al.** suggest that the Khanka Block is a product of circum-Pacific accretion, and not a microcontinental block that was trapped by the northward collision of the North China craton with Siberia as part of the assembly of the main Central Asian Orogenic Belt.

Part 3 of the book, growth and stabilization of continental crust: collisions and intraplate processes, includes work related to Brian Windley's fascination with the $MgO-Al_2O_3-SiO_2-H_2O$ (MASH) system. In 1980, Brian organized a metamorphic group (with Dietrich Ackermand, Kiel and Richard Herd, Ottawa), whose aim was to calibrate natural mineral assemblages against equivalent, experimentally determined, assemblages in petrogenetic $P-T$ grids in the MASH system together with $P-T$ conditions calculated with standard geothermobarometric methods. For over 10 years they studied key assemblages, chemographic relationships, and specific mineralogical problems in rocks from the Limpopo belt in Zimbabwe; the Grenville belt in eastern Canada; Bahia, Brazil; Fiskenaesset, West Greenland; Madagascar; and Scotland. Their studies showed that it is possible to work out the complicated array of minerals, assemblages and reactions that are frozen into these refractory rock systems. In Limpopo rocks they defined 29 mineral reactions with 25 assemblages and used them to calculate an isothermal $P-T$ path without recourse to geothermobarometric methods. Appropriately, this section begins with a paper by **Razakamanana et al.**, on the petrology, chemistry, and phase relations of borosilicate phases in phlogopite diopsidites and granitic pegmatites from the Tranomaro belt, SE Madagascar, with special attention to the boron-fluid evolution. **Razakamanana et al.** discuss the role of boron-rich fluids in the evolution of Gondwana, including how the presence of sinhalite and serendibite associated with phlogopite lenses in metasedimentary diopsidites indicates an evaporitic origin from calc-silicate sediments. Other borosilicates are associated with shear zones that acted as conduits for the boron-rich fluids, derived from calc-silicate sedimentary protoliths. **Razakamanana et al.** use geothermometry and geobarometry of minerals from associated rocks to calculate that ambient pressures and temperatures changed in time from 7.5 to 4.0 kbar and from c. 800 °C to 700 °C. Their results confirm the important role of shear zones in channelling the fluid flow of boron-bearing fluids that were derived from crustal melt granites in the same shear zones, but that ultimately derived their boron from early metasediments.

The second paper in this section is by **Peng**, who uses the Taihang–Lvliang dyke swarm in the central North China craton as an example to show how these short-lived events are keys to the interpretation of continental evolution and tectonics, reconstruction of continental palaeogeographical regimes, and petrogenesis of the associated volcanism. **Peng et al.** relate this dyke swarm to the coeval Xiong'er volcanic province on the southern margin of the North China craton, and suggest that the dykes radiated from a triple junction rift centred on the Xiong'er volcanic province. The triple junction volcanism and radiating dyke swarms are related to the break-up of the North China craton at 1.78 Ga, which probably was influenced by the impact of a mantle plume at the base of the lithosphere. Similar volcanism and dyke swarms are located on other cratons, including the Uruguayan dykes on the Rio de Plata craton, the Avanavero dykes on the Guyana shield, the Crepori gabbro–dolerite sills and dykes in Australia (e.g. Harts Range volcanic rocks and sills and Eastern Creek volcanic rocks; Tewinga volcanic rocks; Mount Isa dykes; Hart doleritic sills) and possibly others (e.g. India: Dharwar dykes), and may relate to the break-up of the Columbia supercontinent.

Part 4 of the book is concerned with Precambrian tectonics and the birth of continents. This section highlights Brian's drive to apply the principles of uniformitarianism to help understand the evolution of ancient, complex high-grade terranes. Many of the secular research themes culminate in Brian's interest in the way the continents have evolved in the last 4 Ga. One of his main contributions has been in the form of innovative syntheses, using techniques such as U–Pb geochronology coupled with field mapping, to assess tectonic and crustal development of the Precambrian, and the wider issues of the growth and differentiation of the continental crust. This was initially done in collaboration with John Dewey, and all was brought together in his acclaimed book *The Evolving Continents* (Windley 1995). In his 1993 Hutton–Lyell commemoration paper to the Geological Society of London, Brian pointed out what uniformitarianism means today in terms of the operation of plate-tectonic processes since the start of the geological record.

Brian made the first proposal (with David Bridgwater) in 1971 that there are two main types of Archaean terrane representing different erosional levels, the greenstone–granite, and the then little known granulite–gneiss terranes. They also predicted that the oldest rocks were most likely to be found in deeply buried lower crustal rocks; later proved correct.

His work on the Archaean of West Greenland in the 1960s led to a detailed study with Joe Smith (1974) in Chicago of the silicate, oxide and sulphide chemistry of the anorthositic Fiskenaesset complex. They suggested that these rocks were tectonically intercalated with subduction-derived, tonalite-dominated continental rocks. Complementary geochemical studies were made by Barry Weaver and co-workers (1978) of Archaean complexes in the Limpopo belt of South Africa and the Scourian of Scotland. Joe and Brian first pointed out that Archaean tonalitic orthogneiss belts in the world most probably formed in Cordilleran–Andean-type continental margins later inter-thrust with oceanic crust. In 1980, Bob Newton, Joe Smith and Brian produced a new model to explain CO_2 vapour flux giving rise to carbonic metamorphism and formation of granulites. In a landmark paper, Norman Sleep and Brian proposed in 1982 that higher temperatures beneath Archaean mid-oceanic ridges resulted in more partial melting and at a greater depth than now, in turn resulting in an oceanic crust more than 20 km thick that was composed at least in part of tholeiitic lavas and anorthositic complexes. This model was widely supported by the geological community, and was used by Tim Kusky and co-workers to explain the thick sections of tectonically emplaced tholeiitic lavas in the Zimbabwean greenstone belts (Kusky & Kidd 1992; Kusky & Winsky 1995; Kusky 1998; Hoffman & Kusky 2004). With John Tarney *et al.* (1982), Brian suggested that tectonic underthrusting of Archaean oceanic crust into Archaean continental crust at Cordilleran-type margins led to crustal thickening and formation of granulites (followed by more thrust-controlled thickening in collisional environments.

Throughout much of Brian's career the mode of origin of many Precambrian orogens was poorly understood. For example, the role and implications of oceanic plateaux and Tibetan-type plateaux and their eroded and/or extended, collapsed equivalents were underestimated. Brian (Windley 1983) produced new ideas to explain the formation of four major Proterozoic orogens in terms of modern plate-tectonic processes: the Ketilidian in South Greenland, the Grenville in eastern North America, the Aravalli–Delhi in Rajasthan in NW India, and the Kola and Svecofennian orogens in the Baltic Shield. These studies showed that tectonic processes during the Proterozoic were not significantly different from those that operate today. With Japanese colleagues and Kevin Pickering, Brian produced a detailed comparative analysis of the similarities and differences between late Archaean island arcs and accretionary prisms and their close modern analogues in Japan (Taira *et al.* 1992).

The section begins with a review paper by **Kröner** about the role of geochronology in understanding continental evolution. **Kröner** highlights the importance of U–Pb dating for understanding processes of crustal growth and evolution, and discusses the merits of different techniques and how recent advancements such as the ability to perform *in situ* dating and to apply high-resolution ion microprobe and laser ablation inductively coupled plasma mass spectrometry to complex, multiple-deformed high-grade terranes has revolutionized models for crustal evolution that were previously based on just field and traditional geochronological and geochemical techniques. He shows how the combination of mineral ages with Sm–Nd, Lu–Hf and O isotopic systematics constrains magma sources and their evolution, and a picture is emerging that supports the beginning of modern-style plate tectonics in the Early Archaean.

Rollinson *et al.* contribute a paper on chromitites from the Fiskenaesset anorthositic complex, West

Greenland. The authors note that the chromitites in the Fiskenaesset complex have an unusual mineral assemblage, including highly calcic plagioclase, iron-rich aluminous chromites, and primary amphibole, and they relate its formation to partial melting of aluminous harzburgite in a mantle wedge above a subduction zone rather than in a continental layered intrusion. They propose that the aluminous mantle source of the parent magma was produced by the melting of a harzburgitic mantle refertilized by small-volume, aluminous slab melts. **Rollinson et al.** propose that this process ceased at the end of the Archaean because the dominant mechanism of crust generation changed such that the melt production shifted from the slab into the mantle wedge, thus explaining why highly calcic anorthosites are almost totally restricted to the Archaean.

Garde & Hollis describe an ophiolitic complex consisting of amphibolite-facies tholeiitic pillow lavas, chloritic shale, manganiferous banded iron formation (BIF), podded chert, jasper, andalusite–staurolite schist cut by numerous sills, and terrigineous sandstones in the northern Nagssugtoqidian orogen, West Greenland. By comparison with modern environments with similar rock associations such as the Resurrection ophiolite in southern Alaska, **Garde & Hollis** suggest that the ophiolite formed in a spreading centre undergoing burial in a forearc trench.

Zhai et al. follow with a paper on the Precambrian tectonic evolution of the North China craton. They outline the main crustal formation periods for the Precambrian evolution of the craton, beginning with the oldest crustal remnants forming at $c.$ 3.8 Ga, and the main crustal formation events occurring between 2.9 and 2.7 Ga. They suggest that by 2.5 Ga these microblocks amalgamated to form a coherent craton, which was cut by a major dyke swarm at 2.5 Ga. Volcanic and plutonic belts formed between 2.3 and 1.95 Ga, and Palaeoproterozoic mobile belts formed as intracontinental orogens between 1.9 and 1.85 Ga. **Zhai et al.** note that the strong metamorphism at $c.$ 1.8 Ga is not restricted to a central orogenic zone (termed the Trans-North China orogen) but instead is found nearly everywhere in the North China craton, so they argue that models calling for a simple collision between the western and eastern blocks of the craton at that time are not compatible with the data, including patterns of high-pressure and high-temperature or ultrahigh-temperature (UHT) metamorphism and uplift rates. **Zhai et al.** thus conclude that previous tectonic models for the North China craton need to be re-evaluated.

The final paper in this section is by **Oliveira et al.** on the the Itabuna–Salvador–Curaçá orogen, in the São Francisco craton, Brazil. They review the geology and present new U–Pb and Nd isotopic ages, along with major and trace element data to support a new tectonic model for the northern segment of the Itabuna–Salvador–Curaçá orogen in which oceanic and island arc sequences were accreted at 3.3 Ga to form the Mundo Novo greenstone belt, and then a second generation of accretion at 2.15–2.12 Ga formed the Rio Itapicuru and Rio Capim greenstone belts. Between 3.08 and 2.98 Ga, mafic crust experienced partial melting and formed the Retirolândia and Jacurici tonalite–trondhjemite–granodiorite belts of the Serrinha block. From 2.69 to 2.58 Ga an Andean-type arc with ocean crust remnants formed the Caraíba complex possibly at the Gavião block margin. Between 2110 and 2105 Ma, the Rio Itapicuru arc collided with the Retirolândia–Jacurici microcontinent, possibly involving slab breakoff. Oblique convergence between 2.09 and 2.07 Ga led to the collision of the Serrinha microcontinent with the Caraíba–Gavião superblock and reworked the Caraíba arc to granulite facies, locally at UHT conditions. At the same time, arc dacites spread over the Rio Itapicuru greenstone belt, and the 3.12–3.0 Ga Uauá terrane, crosscut by 2.58 Ga mafic dykes, extruded from south to north possibly together with the 2.15 Ga Rio Capim greenstone belt.

Part 5 of the book is on active tectonics and geomorphology of continental collision and growth zones. In 1979, Brian started British research in the Himalayas and Karakorum, partly because of the opening of the Karakorum Highway, and partly because at that time these classic orogenic belts were known only in reconnaissance outline. He also felt that a better understanding of Precambrian tectonics depends on a better knowledge of modern collisional orogenic belts. His NERC grant lasted for 10 years and brought to that region more than 20 students and staff, most notably Mike Coward, Mike Searle, and Qasim Jan. They produced new data on the lithology and structure, mineralogy, geochemistry, and isotopic history of the Pakistan Himalayas, and produced a comprehensive model for the igneous and tectonic development of the Kohistan arc–batholith in terms of a three-stage plutonic development (island arc, Andean-type batholith, post-collisional leucogranites), and for the mineral chemistry and tectonic environment of the Chilas complex that formed in the magma chamber of the island arc. In 1986, they produced a comprehensive synthesis of the tectonic evolution of the Himalayas (Coward et al. 1986; Searle & Windley 1986; Hoffman & Kusky 2004). With two mountaineering post-doctoral research fellows, Mike Searle and Tony Rex, Brian organized the remapping and study of structural and magmatic development of the Karakorum mountain range, which contains a mid-crustal gneissic block intruded by a mid-Cretaceous Andean-type granitic batholith, uplifted by thrusting and intruded by a crustal melt leucogranitic batholith in the Miocene.

The 7000 m Tien Shan and 4000 m Altai mountain ranges of Central Asia formed 2000 km and 2500 km respectively from the main deformation front in the Himalayas as a result of the post-collisional indentation of India into Asia. Brian's work in the Tien Shan with Mark Allen (Windley *et al.* 1990) led to new data and ideas on the Palaeozoic collision tectonics, the Mesozoic basin development, redeformation in the Late Cenozoic, and active tectonics of the region. They demonstrated for the first time that there are two Palaeozoic sutures in the Tien Shan, and that the Turfan basin has been downloaded by Cenozoic thrusts. The Altai is the northernmost mountain belt to be thrust and uplifted as a result of the India–Asia collision. Later, Dickson Cunningham and Brian carried on these ideas into the Altai in Mongolia, where transpressional restraining bends have controlled the uplift of flat-topped mountains in the Cenozoic (Cunningham *et al.* 1996).

In the first paper in this section, by **Petterson**, reviews more than 100 years of geological observations in Kohistan, including work he and Brian Windley were involved with during the past 30 years. The great bulk of the 30 000 km^2 Kohistan terrane represents growth and crustal accretion during the Cretaceous in an intra-oceanic island arc dating from *c.* 134 Ma to *c.* 90 Ma. This early period saw the extrusion and deposition of a *c.* 15–20 km thick arc sequence as well as the intrusion of the oldest parts of the Kohistan batholith, lower crustal plutons, crustal melting and the accretion of an ultramafic mantle–lower crust sequence. The crust had thickened sufficiently by *c.* 95 Ma to allow widespread granulite-facies metamorphism to take place within the lower arc. At around 90 Ma, Kohistan underwent a *c.* 5 Ma long high-intensity deformation caused by collision with Eurasia. Kohistan, now an Andean margin, was extended and further volcanic and plutonic series were emplaced. Collision with India at *c.* 55–45 Ma saw the rotation, upturning, underplating and wholesale preservation of the terrane. The Kohistan terrane represents a complete section through juvenile crust extracted from the mantle in a subduction-zone setting. The differences in composition between Kohistan and average continental crust indicate the substantial fractionation undergone by primitive arc crust to form mature continental crust.

The second paper in this section, by **Allen**, covers the roles of strike-slip faulting during continental deformation, with examples from the active Arabia–Eurasia collision. **Allen** notes that the active strike-slip faults in the Arabia–Eurasia collision zone play several roles, including acting as collision zone boundaries, accommodating tectonic escape, as strain partitioning structures, as shortening arrays, and as transfer zones. **Allen** notes how complex the roles of these strike-slip fault systems are in this active collision zone, and suggests that understanding their complexity can be used as a lesson for interpreting ancient orogens.

The third paper is by **Searle & Treloar**, and poses the question 'Was Late Cretaceous–Paleocene obduction of ophiolite complexes the primary cause of crustal thickening and regional metamorphism in the Pakistan Himalaya?' They note that the Pakistan Himalaya includes both ultrahigh-pressure coesite eclogite rocks and medium-pressure and -temperature kyanite–sillimanite-grade Barrovian metamorphic rocks that show that peak conditions were reached at about 47 Ma. ^{40}Ar–^{39}Ar hornblende cooling ages date post-peak metamorphism of both units through 500 °C by 40 Ma, some 20 Ma earlier than for metamorphic rocks in the central and eastern Himalaya. **Searle & Treloar** propose a new idea in which the earlier metamorphic and cooling ages of the Pakistan Barrovian metamorphic sequence are partially explained by Late Cretaceous to Early Paleocene crustal thickening linked to obduction of an ophiolite thrust sheet onto the leading edge of the Indian plate. Heating following on from this Paleocene crustal thickening explains peak Barrovian metamorphism within 5–10 Ma of subsequent obduction of Kohistan. Remnants of the ophiolite sheet, and underlying Tethyan sediments, are preserved in NW India and in western Pakistan but not in north Pakistan, suggesting that tectonic erosion has removed the ophiolites and other cover sequences from the Indian plate basement.

This section continues with a paper by **Cunningham** on the tectonic setting and structural evolution of the Late Cenozoic Gobi Altai orogen. This orogen is an intraplate, intracontinental transpressional orogen in southern Mongolia that formed in the Late Cenozoic as a distant response to the Indo-Eurasia collision. The basement consists of a series of Palaeozoic accreted terranes intruded by granitic rocks. The Quaternary faulting largely follows, and is controlled by Palaeozoic basement structural trends, Precambrian basement blocks, and stresses from the India–Asia collision. Some modifications of the fault pattern may have been induced from thermal weakening of the lower crust.

The final paper in this section is a review by **Owen** on the landscape development of the Himalayan–Tibetan orogen, focused on the dynamics of landscape development within the orogen. **Owen** describes many tectonic–climate–landform development links, including the influence of climate on surface uplift by denudational unloading; the limiting of topography by glaciation; localized uplift at syntaxes by enhanced fluvial and glacial erosion that, in turn, weaken the lithosphere, enhancing surface uplift and exhumation. He also documents climate-driven out-of-sequence thrusting and crustal channel flow; glacial damming leading to differential erosion and uplift; paraglaciation; and the influence of extreme events such as earthquakes, landslides, and floods as major formative processes. This contribution demonstrates how new technologies such as satellite remote sensing, global positioning system, and numerical modelling have led to recent advances in understanding landform development in active collision zones.

In summary, the papers in this volume attest to the remarkable and highly influential career of Brian Windley, and highlight some of the fundamental contributions he has made to understanding the evolution of the continental crust. Brian's career began with some of the oldest rocks on Earth in Greenland, and slowly evolved into neotectonics in central Asia, but his feet never left the foundation of the Precambrian. Brian has consistently applied principles of uniformitarianism to the analysis of old terranes, and his interdisciplinary approach to understanding continental tectonics and evolution from the oldest to the youngest terranes on Earth has shown how the same physical processes have operated throughout geological time. *Salut* to you Brian!

We would like to acknowledge the many reviewers of papers in this volume, including D. Cunningham, P. Mann, A. Polat, L. Ashwal, J. Connelly, T. Raharimahefa, M. Santosh, S. Liu, R. van Schmus, H. Halls, M. Sun, N. Pinter, L. Wang, L. Webb, R. Stern, O. Jagoutz, S. Wilde, B. Windley, H. Rollinson, and several anonymous reviewers.

References

COWARD, M. P., WINDLEY, B. F. ET AL. 1986. Collision tectonics in the NW Himalayas. *In*: COWARD, M. P. & RIES, A. C. (eds) *Collision Tectonics*. Geological Society, London, Special Publications, **19**, 203–219.

CUNNINGHAM, W. D., WINDLEY, B. F., DORJNAMJAA, D., BADAMGAROV, G. & SAANDAR, M. 1996. A structural transect across the Mongolian Western Altai: active transpressional mountain building in Central Asia. *Tectonics*, **15**, 142–156.

HOFFMAN, A. & KUSKY, T. M. 2004. The Belingwe greenstone belt: ensialic or oceanic? Chapter 15. *In*: KUSKY, T. M. (ed.) *Precambrian Ophiolites and Related Rocks*. Developments in Precambrian Geology, **13**, 487–537.

KUSKY, T. M. 1998. Tectonic setting and terrane accretion of the Archean Zimbabwe craton, *Geology*. **26**, 163–166.

KUSKY, T. M. & KIDD, W. S. F. 1992. Remnants of an Archean oceanic plateau, Belingwe greenstone belt, Zimbabwe. *Geology*, **20**, 43–46.

KUSKY, T. M. & WINSKY, P. A. 1995. Structural relationships along a greenstone/shallow water shelf contact, Belingwe greenstone belt, Zimbabwe. *Tectonics*, **14**, 448–471.

NEWTON, R. C., SMITH, J. V. & WINDLEY, B. F. 1980. Carbonic metamorphism, granulites and crustal growth. *Nature*, **288**, 45–50.

SEARLE, M. P. & WINDLEY, B. F. 1986. Thrust tectonics and the deep structure of the Pakistan Himalayas: comment. *Geology*, **14**, 441–442.

SLEEP, N. H. & WINDLEY, B. F. 1982. Archaean plate tectonics: constraints and inferences. *Journal of Geology*, **90**, 363–379.

TAIRA, A., PICKERING, K. T., WINDLEY, B. F. & SOH, W. 1992. Accretion of Japanese island arcs and implications for the origin of Archean greenstone belts. *Tectonics*, **11**, 1224–1244.

TARNEY, J., WEAVER, B. L. & WINDLEY, B. F. 1982. Geological and geochemical evolution of the Archaean continental crust. *Revista Brasileira de Geosciências*, **12**, 53–59.

WEAVER, B. L., TARNEY, J., WINDLEY, B. F., SUGAVANAM, E. B. & VENKATA RAO, V. 1978. Madras granulites: geochemistry and PT conditions of crystallisation. *In*: WINDLEY, B. F. & NAQVI, S. M. (eds) *Archaean Geochemistry*. Elsevier, Amsterdam, 177–204.

WINDLEY, B. F. 1983. A tectonic review of the Proterozoic. *Geological Society of America, Memoirs*, **161**, 1–10.

WINDLEY, B. F. 1993. Uniformitarianism today: plate tectonics is the key to the past. *Journal of Geological Society, London*, **150**, 7–19.

WINDLEY, B. F. 1995. *The Evolving Continents*. 3rd edn. John Wiley and Sons, New York.

WINDLEY, B. F. & BRIDGWATER, D. 1971. *The evolution of Archaean low- and high-grade terrains*. Geological Society of Australia, Special Publications, **3**, 33–46.

WINDLEY, B. F. & SMITH, J. V. 1974. The Fiskenaesset Complex, West Greenland. Pt. 2: General mineral chemistry from Qeqertarssuatsiaq. *Bulletin Grønlands Geologiske Undersøgelse*, **108**, 1–54.

WINDLEY, B. F., ALLEN, M. B., ZHANG, C., ZHAO, Z. H. & WANG, G. R. 1990. Paleozoic accretion and Cenozoic redeformation of the Chinese Tien Shan range, Central Asia. *Geology*, **18**, 128–131.

The anatomy and ontogeny of modern intra-oceanic arc systems

ROBERT J. STERN

Geosciences Department, University of Texas at Dallas, Richardson, TX 75083-0688, USA
(e-mail: rjstern@utdallas.edu)

Abstract: Intra-oceanic arc systems (IOASs) represent the oceanic endmember of arc–trench systems and have been the most important sites of juvenile continental crust formation for as long as plate tectonics has operated. IOASs' crustal profiles are wedge-shaped, with crust up to 20–35 km thick; a more useful definition is that IOASs occur as chains of small islands, generally just the tops of the largest volcanoes. A very small fraction of IOASs lie above sea level, but advancing marine technologies allow their most important features to be defined. Modern IOASs subduct old, dense oceanic lithosphere and so tend to be under extension. They consist of four parallel components: trench, forearc, volcanic–magmatic arc, and back-arc, occupying a ≥200 km zone along the leading edge of the overriding plate. These components form as a result of hydrous melting of the mantle and reflect the strongly asymmetric nature of subduction processes. Forearcs preserve infant arc lithosphere whereas magmatism in mature IOASs is concentrated along the volcanic–magmatic front. Mature IOASs often have minor rear-arc volcanism and, because most IOASs are strongly extensional, sea-floor spreading often forms back-arc basins. Sub-IOAS mantle is also asymmetric, with serpentinized harzburgite beneath the forearc, pyroxene-rich low-V_p mantle beneath the magmatic front, and lherzolite–harzburgite beneath back-arc basins. Because most IOASs are far removed from continents, they subduct oceanic lithosphere with thin sediments and have naked forearcs subject to tectonic erosion. IOASs evolve from broad zones of very high degrees of melting and sea-floor spreading during their first 5–10 Ma, with the volcanic–magmatic front retreating to its ultimate position *c.* 200 km from the trench.

Continental crust is basically a mosaic of orogenic belts, and these in turn are largely nests of island arcs, welded and melded together. Arcs are what is produced in the overriding plate of a convergent margin when subduction is sufficiently rapid (faster than *c.* 2 cm a^{-1}) for long enough that the subducted plate reaches magmagenetic depths (100–150 km) and causes the mantle to melt. This configuration must exist for long enough that melts generated in the overlying asthenospheric wedge not only reach the surface but also persist until a stable magmatic conduit system is established. When this happens, magma produced by hydrous fluxing of the mantle will rise towards the surface to form volcanoes and plutons of a magmatic arc. This magmatic locus and its often spectacular volcanoes are important parts of an arc, but neovolcanic zones make up a relatively minor part of any arc. Arcs can be built on continental or oceanic lithosphere. Here the focus is on those arcs that are built on oceanic crust, which may be called intra-oceanic arc systems (IOASs). These can be distinguished from arcs built on continental crust, also known as 'Andean-type arcs'.

IOASs are constructed on thin, mostly mafic crust, and consequently these magmas are not as contaminated by easily fusible felsic crust as are magmas from Andean-type margins, which are built on much thicker and more felsic continental crust (Fig. 1). IOASs are widely acknowledged as sites where thickened welts of juvenile (i.e. derived from melting of the mantle) crust is produced. In spite of this significance, IOASs are significantly less well studied than Andean-type arcs. The main reason for this is that the vast bulk of IOASs are submerged below sea level and difficult to study. Nevertheless, our understanding of IOASs has advanced greatly since these were first reviewed a quarter of a century ago by Hawkins *et al.* (1984); an overview of arcs in general was also prepared about that time (Hamilton 1988). New developments in marine technology [global positioning system (GPS), sonar swathmapping, deep-sea drilling, manned submersibles, remotely operated vehicles (ROVs) and autonomous underwater vehicles (AUVs)] are allowing the study of IOASs to advance rapidly, and we now have a good grasp of their most important features. The purpose of this paper is to summarize our understanding of IOASs for a broad geoscientific audience, with the hope that this understanding will allow geologists studying ancient crustal terranes to better identify fossil IOASs in orogens and cratons. This review draws heavily on 30 years of studying the Izu–Bonin–Mariana arc system, especially the Marianas, and examples are drawn heavily from this IOAS. The general tectonic and magmatic relationships observed there are mostly

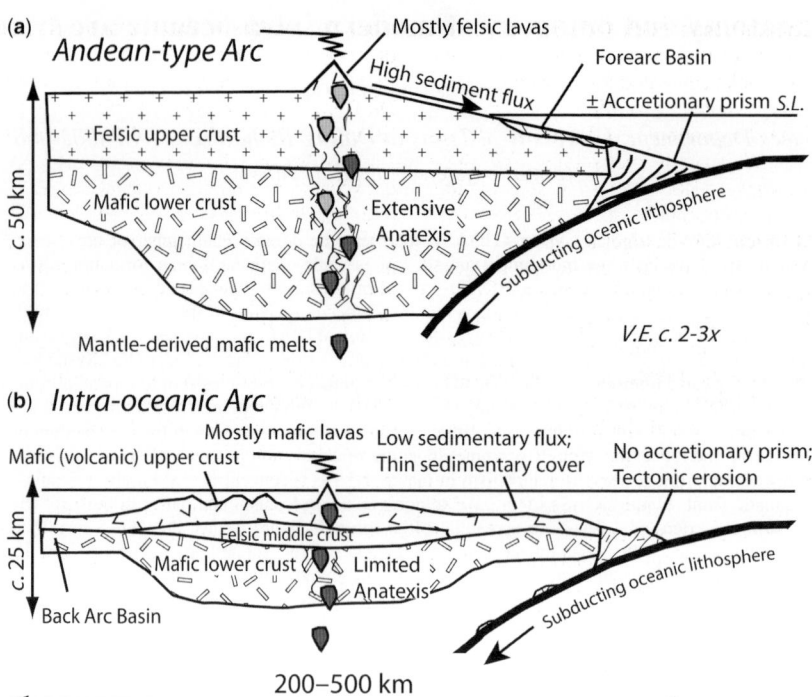

Fig. 1. Comparison of Andean-type arc (**a**) and intra-oceanic arc system (**b**), greatly simplified.

applicable to the general case of modern IOAS formation and evolution.

What is an intra-oceanic arc system?

Arc–trench systems comprise the lithosphere (crust and uppermost mantle) between the trench and the back-arc region. IOASs are one endmember in the spectrum of arc–trench systems, the opposite end of the spectrum from Andean-type arcs. This spectrum reflects differences in what the arc is constructed on and thus its crustal composition, thickness, and elevation. IOASs are built on oceanic crust, whereas Andean-type arcs are built on pre-existing continental crust. Correspondingly, IOAS crust is thinner and more mafic than that beneath Andean-type arcs, which is thicker and more felsic (Fig. 1). This reduces the opportunity for primitive, mantle-derived magmas to interact with the crust en route to the surface; consequently, IOAS lavas are less differentiated, more mafic, and less contaminated than are lavas erupted from Andean-type arcs. The more primitive nature of IOAS igneous rocks better preserves evidence of the subduction-modified mantle-derived melts that formed them than do igneous rocks from Andean-type arcs, and this 'window on subduction-zone processes' is a paramount reason that studies of IOAS arcs are advancing rapidly (although it must be noted that all arc lavas tend to be more fractionated than are mid-ocean ridge basalt (MORB) or 'hotspot' lavas).

IOASs are the fundamental building blocks that are assembled over geological time into orogens, cratons, and continents. If the continental crust can be likened to a brick wall, then an IOAS plays the role of a single brick. This analogy has limited utility, because bricks in a wall and IOASs look and behave very differently. First, bricks are homogeneous, whereas IOASs are heterogenous, with significant vertical and transverse compositional variations, both in the crust and in the mantle lithosphere. Second, bricks are annealed solid aggregates, whereas IOASs form from melts and continue through their lives to interact with mantle-derived melts as well as generating eutectic crustal melts. Continued magmatic additions to the base of IOAS occurs across a 'transparent Moho', with some mantle-derived melts permeating or traversing the crust whereas others are underplated to the base of the crust, at the same time that dense cumulates and residues 'delaminate' or sink back into the convecting asthenosphere. Third, bricks are rectangular parallelepipeds, or cuboids, whereas IOAS crust has the relative dimensions of flattened noodles: much wider (c. 250 km) than thick (15–35 km), much longer (hundreds to thousands of kilometres) than wide. Finally, bricks cemented into a wall are inert and immutable, whereas juvenile crust composed of accreted IOASs continues to deform as well as interact with and generate melts long after these are accreted. These melts, continuing deformation, and attendant metamorphism serve as the cement that welds different IOASs together.

IOASs are built on oceanic crust, but this crust generally forms when subduction begins, as discussed below. Crustal thicknesses are typically 20–35 km beneath the magmatic arc, thinning beneath the back-arc region and tapering trenchward. Detailed seismic studies of arcs are needed to actually measure crustal thicknesses, and these are relatively uncommon (the western Aleutian and Izu–Bonin–Mariana arcs are the IOASs with the best-imaged crustal structure; see Stern *et al.* 2003). However, the relative paucity of geophysical soundings of arcs does not seriously impede our ability to identify these among Earth's inventory of convergent margins. Because crustal thickness typically is reflected in elevation, intra-oceanic arcs are generally mostly below sea level. This serves as a useful (and simple) criterion for identifying intra-oceanic arcs: if much of the volcanic front of a given arc system lies below sea level, it can usefully be described as an intra-oceanic arc. Similarly, if the volcanic front of a given arc mostly lies above sea level, this is probably an Andean-type arc. This twofold classification differs from the threefold subdivision of de Ronde *et al.* (2003), who identified: (1) intra-oceanic arcs (those with oceanic crust on either side); (2) transitional or island arcs (those along the margins of island chains with a basement of young continental crust); (3) continental arcs (those developed along the margins of continents). De Ronde *et al.*'s (2003) intermediate category of 'island arc' is the principal difference, and although I agree that arcs built on crust that is transitional between true oceanic and continental crusts exist (e.g. the Philippines), assignment to this intermediate group requires more information about the nature of crustal substrate than is commonly available. Figure 2 shows the distribution of IOASs as they are identified here.

IOASs tend to form where relatively old oceanic crust, generally Cretaceous or older, is subducted. Such lithosphere is negatively buoyant, which causes the subducting sea floor to sink vertically as well as subduct down-dip, in turn causing the trench (and the associated arc system) to 'rollback' (Garfunkel *et al.* 1986; Hamilton 2007). Such a situation favours development of convergent plate margins within the oceanic realm and thus IOASs. In contrast, subduction of young oceanic lithosphere engenders a strongly compressional convergent margin so that the arc system migrates away from the ocean basin and onto any flanking continent. Subduction of older oceanic lithosphere favours development of an overall extensional strain regime in the hanging wall of the associated convergent margin and is an important reason why IOASs tend to be associated with back-arc basins, discussed further below.

It is essential to distinguish between the early stages in the formation of an IOAS and its subsequent evolution. The latter stages are usefully referred to as 'mature', and all of the IOASs that

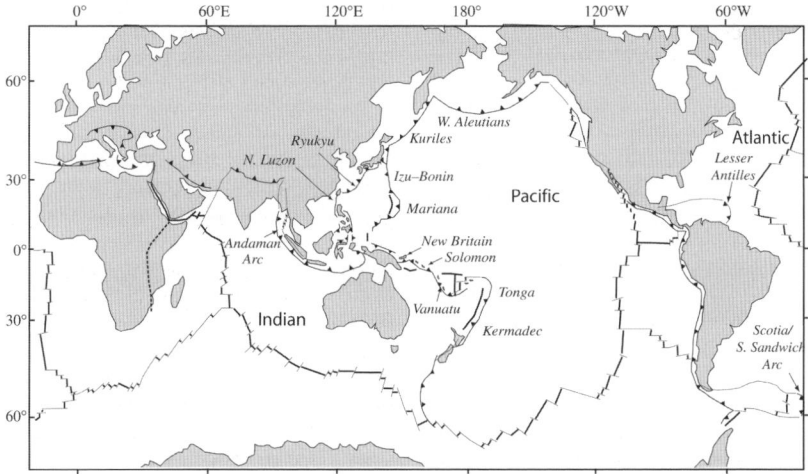

Fig. 2. Location of convergent plate margins and distribution of intra-oceanic arc systems.

are active today are in this stage. An IOAS changes somewhat once it becomes mature, but these changes are episodic as well as progressive. Episodic changes include the formation of back-arc basins and tectonic erosion of the forearc. Progressive changes include thickening of the crust beneath the volcanic–magmatic front, extent of serpentinization of subforearc mantle, and the thickness of sediments that are accumulated. The early stages in the life of an IOAS are called 'infant', 'nascent', or 'immature' and few of the characteristics of mature IOASs pertain to this stage. The infant arc stage is relatively brief, lasting $c.$ 5–10 Ma. Because none of the currently active IOASs are in this stage, our understanding of infant arcs is reconstructed from mostly early Cenozoic examples in the Western Pacific, especially the Izu–Bonin–Mariana arc. Except for the final section 'IOAS forearc structure preserves its early history', this review concentrates on mature IOASs.

How do we know about intra-oceanic arc systems?

Because arcs are large and heterogeneous geological entities, their study involves the full range of geoscientific perspectives: geochemistry, sedimentology, geophysics, geodynamics, structure, metamorphism, palaeontology, etc. Such studies of IOASs are more difficult in many ways than those of Andean-type arcs, simply because the former are largely below and the latter largely above sea level. Consequently, submarine geological studies are much more expensive than studies on land. It is only when studying the trench and deeper parts of the forearc (which are submerged for both Andean-type and intra-oceanic arc systems) that the same marine geological approaches must be used. Also, Andean-type arcs lie near many population centres and pose serious volcanic, landslide, and seismic hazards (also benefits such as geothermal energy and mineral deposits), whereas IOASs are isolated, sparsely populated, and any mineral deposits are difficult to exploit. Because traditional land-based geoscientific approaches can be used to study Andean-type arcs, and because many nations rightfully have concerns for the public good, studies of such systems and our understanding of them are relatively advanced. Some geoscientific work on IOASs can be done above sea level but this is limited to the upper slopes of the tallest volcanoes and isolated structural highs. Understanding IOASs requires using research vessels and marine geotechnology, which are expensive. Our understanding of IOASs naturally lags behind that of Andean-type margins but is advancing rapidly.

Intra-oceanic arc-itecture

At broad scale, IOASs consist of four components, as shown in Figure 3: trench, forearc, volcanic–magmatic arc, and back-arc. These same components are just as useful for subdividing the transverse structure of Andean-type arc-trench systems. IOASs are often associated with back-arc basins, where sea-floor spreading occurs, or with narrower rift zones. Other IOASs show no evidence of extension, but no modern IOAS is associated with back-arc shortening. Further details about these four components are provided below.

Trench

The trench marks where the two converging plates meet. Because one plate bends down to slide beneath the other, trenches are global bathymetric lows, several thousands of metres deeper than sea floor away from the trench. A trench can be filled with sediments or contain very little sediment, depending on how much is supplied to it. Sediment flux reflects proximity to continents. Trenches associated with Andean-type arcs can receive large volumes of sediments delivered by rivers or glaciers and thus are often filled. In contrast, very few of the trenches associated with IOASs contain significant sediment. The only IOAS trenches with significant sediment fill are found in the southernmost Lesser Antilles, where the Orinoco River delta lies at the southern terminus of this trench, the Aleutians, where sediment from Alaskan glaciers and rivers flows longitudinally westwards along the trench, and in the Andaman–Nicobar region, where the trench is fed by the Ganges–Brahmaputra river system. The sediment volume in the trench controls whether or not the forearc is associated with a significant accretionary prism (Clift & Vannucchi 2004; Scholl & von Huene 2007). Because most IOAS trenches are starved of sediment, accretionary prisms are not generally found in the inner trench

Fig. 3. (**a**) Schematic section through the upper 140 km of an intra-oceanic arc system (with an actively spreading back-arc basin), showing the principal crustal and upper mantle components and their interactions. It should be noted that the 'mantle wedge' (unlabelled) is that part of the asthenosphere beneath the magmatic front. The mantle between the asthenosphere and the trench is too cold to melt (modified after Stern *et al.* 2003). (**b**) Typical intra-oceanic forearc (modified after Reagan *et al.* (2010)). The exposed ophiolitic basement, thin sediments, absence of accretionary prism, and deep, empty trench should be noted (**c**) Crustal structure beneath the volcanic–magmatic front of a typical

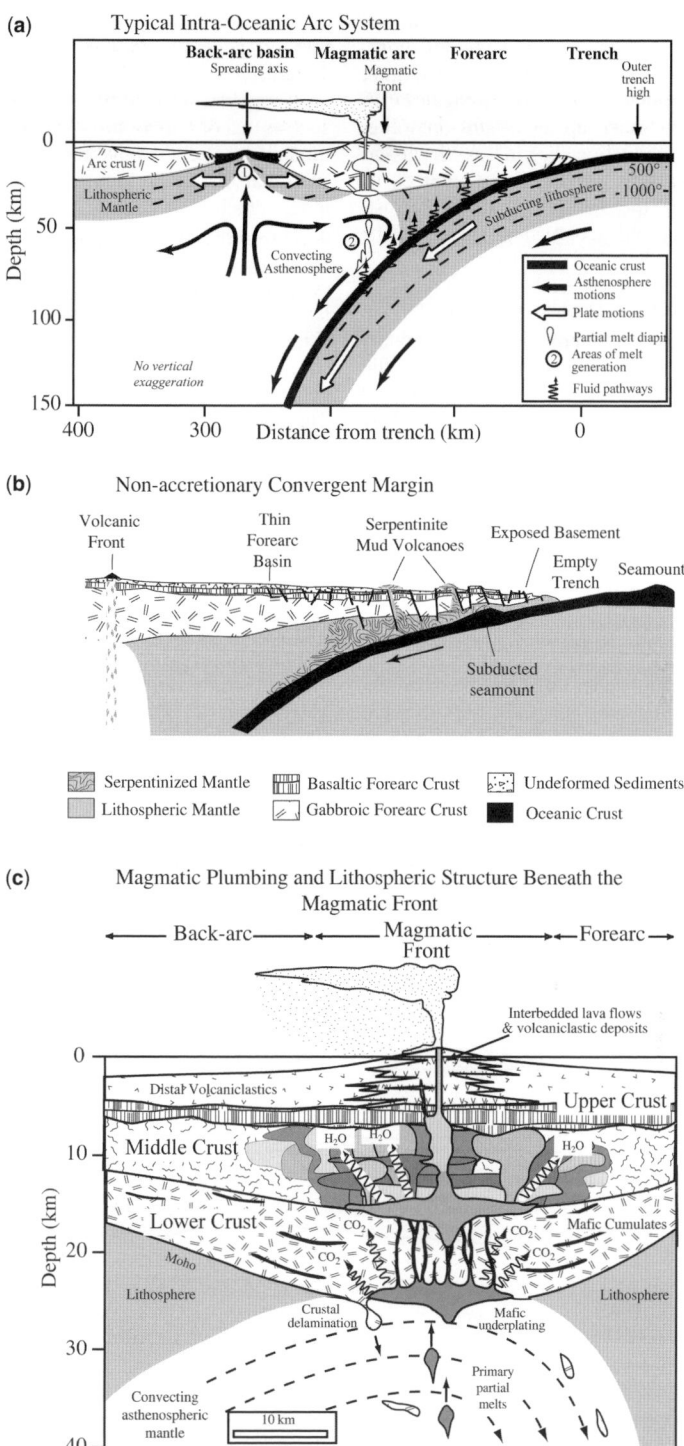

Fig. 3. (*Continued*) intraoceanic arc (modified after Stern 2003). The asthenosphere is shown extending up to the base of the crust; delamination or negative diapirism is shown, with blocks of the lower crust sinking into and being abraded by convecting mantle. Regions where degassing of CO_2 and H_2O is expected are also shown.

wall of IOASs, with the exceptions noted above. Instead, igneous basement, generally basalt, boninite, diabase, gabbro and serpentinized peridotite, is exposed in the inner trench wall (Reagan et al. 2010). These rocks make up an *in situ* ophiolite, produced when subduction began, as discussed below. Such exposures are important sources of information about the nature of forearc crust, along with drilling and geophysical studies (Clift & Vannucchi 2004; Scholl & von Huene 2007).

IOAS inner trench walls are highly fractured by continuing deformation and earthquakes associated with plate convergence. Subduction of seamounts and other bathymetric highs further fracture the inner trench wall. The cumulative effect is that much of the inner trench wall slope is defined by the angle of repose for fractured igneous rocks, especially at its base. The base of the inner trench wall consists of a talus prism of this material. This loose talus is carried into the subduction zone, resulting in significant 'tectonic erosion' of the outer forearc.

Forearc

The forearc is the lithosphere that lies between the trench and magmatic arc, generally 100–200 km wide. Its most important characteristic is a lack of recent igneous activity and remarkably low heat flow, even though forearc crust preserves strong evidence that igneous activity of unusual intensity occurred when subduction began. Morphologically, the forearc slopes gently towards the trench. It can be subdivided into a more stable inner forearc and a more deformed outer forearc. Tectonically stable forearcs, such as the Izu forearc, are also commonly deeply incised by submarine canyons. In contrast, actively deforming forearcs, such as the Mariana forearc, generally lack a well-developed canyon system.

Geophysical soundings of a typical IOAS forearc such as that of Izu–Bonin–Marianas reveal a predominantly mafic crust that tapers towards the trench, such that mantle peridotite is often exposed in the inner trench wall (Fisher & Engel 1969; Bloomer & Hawkins 1983; Pearce et al. 2000). P-wave velocities increase vertically downward from V_p consistent with fractured basaltic and boninitic lavas to V_p consistent with diabase and gabbro (Figs 4 & 5). This P-wave velocity structure is similar to that of oceanic crust or ophiolites, consistent with sampling of the IOAS inner trench walls. The significance of the ophiolitic nature of IOAS forearc crust is explored further in the final section 'IOAS forearc structure preserves its early history'.

The upper mantle beneath IOAS forearc crust is serpentinized, as demonstrated by low seismic velocities and the presence of serpentinite mud volcanoes in the outer forearc of at least one IOAS, the Marianas (Stern & Smoot 1998; Oakley et al. 2007). The extent of serpentinization seems to vary with distance from the trench; rocks are more serpentinized towards the trench, and less serpentinized away from it. This is shown by upper mantle P-wave velocities, which decrease significantly from V_p c. 8.1 km s^{-1} beneath the inner forearc to 6.4 km s^{-1} towards the trench. The greater serpentinization beneath the outer compared with the inner forearc makes the outer forearc weaker and easier to deform. Forearc mantle is composed of strongly depleted harzburgite, characterized by spinels with Cr-number $[= 100Cr/(Cr + Al)] > 50$ (Bloomer & Hawkins 1983; Parkinson & Pearce 1998; Okamura et al. 2006).

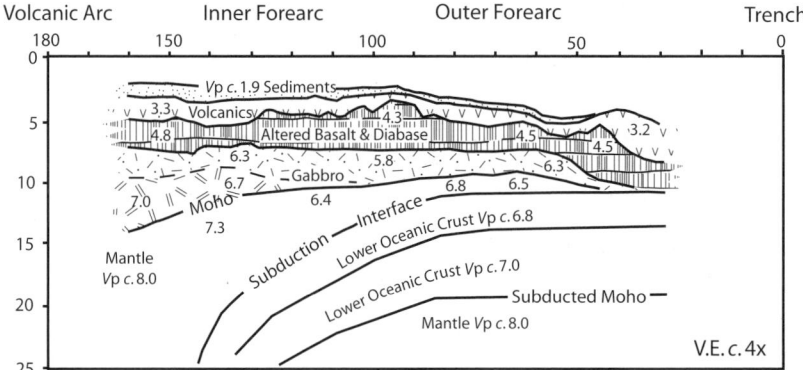

Fig. 4. Simplified P-wave velocity structure beneath Izu forearc (roughly east–west line at 30°50′E) modified after Kamimura et al. (2002), with interpreted lithologies. Forearc crustal thickness decreases from c. 11 km near the volcanic arc to 5 km or less near the trench. Also, V_p is significantly lower in uppermost mantle beneath the outer forearc (6.4–6.8 km s^{-1} v. 7.3–8.0 km s^{-1}), probably reflecting greater serpentinization beneath the outer forearc.

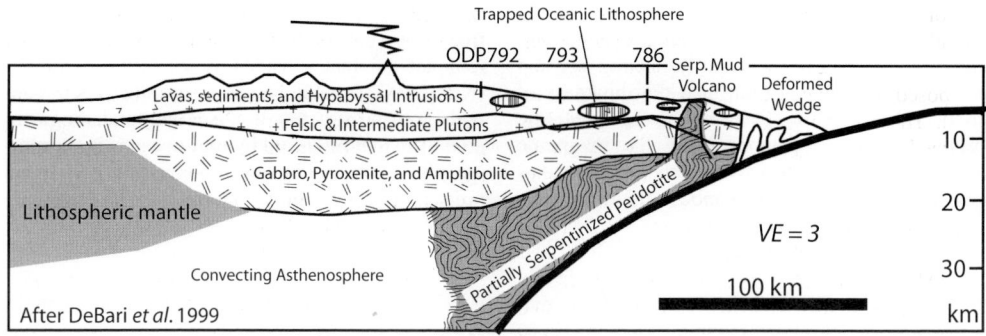

Fig. 5. Interpretation of the seismic structure of the crust and upper mantle of a typical intra-oceanic arc, modified after DeBari *et al.* (1999). This section is interpreted from the seismic refraction study of Suyehiro *et al.* (1996) for the Izu arc. The presence of a mid-crustal 'tonalite' layer should be noted.

Sediments are a relatively unimportant part of the forearc, but there are some. The relatively stable inner forearc may host a thin forearc basin, with a few hundred metres to few kilometres of sediment thickness. This paucity of sediment in IOAS forearcs is consistent with the fact that IOAS trenches are empty, and contrasts with the robust accretionary prisms and thick forearc basins characteristic of many Andean forearcs.

Volcanic–magmatic arc

The locus of continuing igneous activity in a mature IOAS defines the magmatic arc. This is recognized in modern IOASs as a linear or arcuate array of volcanoes that parallel the trench (England *et al.* 2004; Syracuse & Abers 2006). These are the largest and most productive volcanoes in a mature IOAS and are often the only feature that rises above sea level. These volcanoes are underlain by plutons and hypabyssal intrusions exposed by erosion in ancient, fossil IOASs. The trenchward limit of young igneous activity is referred to as the 'volcanic front' or 'magmatic front', and marks a steep gradient in heat flow, which is low towards the trench and high towards the back-arc region. The style of IOAS volcanism is fundamentally different from that of the other two great classes of oceanic volcanism, mid-ocean ridges and hotspots. Mid-ocean ridge volcanism is entirely volatile-poor tholeiitic basalt, which produces crust of nearly constant thickness (5–7 km) when magma fills the gap between two plates being pulled apart. Hotspot volcanism typically builds a linear chain of tholeiitic and/or alkalic lavas, with a progression of ages that increases in the direction of plate motion; generally only a volcano or two at one end of the hotspot chain are active. IOAS volcanoes, in contrast, remain relatively fixed with respect to the trench, so that the crust thickens at a rate of several hundred metres per million years as a result of the cumulative effects of magmatic addition at essentially the same place. Furthermore, IOAS magmas are relatively rich in silica and in volatiles, especially water, so that these eruptions are more violent and lavas are dominated by fragmental material, except in deep water, where the pressure suppresses violent degassing of magmas. This can sometimes be seen in the deposits of a growing IOAS volcano, as it grows from a base at 2–4 km below sea level to shallower water and then becomes an island. Eruptions from such a volcano are likely to change progressively with time from more effusive flows to increasingly fragmental as eruptions occur at increasingly lower P environments over time.

Hydrothermal activity and ore deposits associated with IOASs are also distinct from those associated with other oceanic igneous settings. Hydrothermal mineralization at mid-ocean ridges is controlled by the heating of seawater by hot rocks, which sets up a hydrothermal circulation that also leaches metals from the fractured basalts as it passes through these; these dissolved metals precipitate when hydrothermal fluids vent on the sea floor. Such circulation and leaching also occurs in association with IOAS submarine volcanoes, but in addition, the volatile-rich nature of IOAS magmas contributes directly to mineralization when these magmas degas (de Ronde *et al.* 2003; Baker *et al.* 2005). There are thus fundamental differences expected for mineral deposits that are related to mid-ocean ridges and IOASs.

IOAS arc volcanoes are built on a platform that varies in depth but that is characteristically well below sea level. The depth of this platform depends mostly on crustal thickness, and is shallower above thicker arc crust and deeper above thinner arc crust. Correspondingly, whether or not IOAS volcanoes rise above sea level depends not only on the size of the volcano (reflecting age and magma production rate) but also the thickness of the underlying crust. There is no characteristic spacing between volcanoes along the magmatic front (d'Ars et al. 1995). Where the sea floor has been imaged between volcanoes, there is little evidence of young lavas.

IOAS volcanoes erupt mostly basalts but these are distinct because they are characteristically vesicular, porphyritic, and fractionated. This is evidence that one or more magma reservoirs exist within the arc crust, as depicted in Figure 3c. Processing of magmas and crust in the middle crust is probably responsible for the formation of tonalitic middle crust. Thus IOAS crust thickens by addition of lavas on top as well as plutonism within the crust. IOAS crustal growth is discussed further below. Felsic magmas also erupt in IOASs, usually from large, submarine calderas. Such submarine calderas are now well documented in the Izu and Kermadec IOASs (Fiske et al. 2001; Graham et al. 2008). The origin of felsic melts in IOASs is controversial, and partly depends on the nature of lower arc crust, as discussed below.

IOAS magmatic arcs show strong asymmetries in the volume and composition of magmatic products. Melts along the volcanic–magmatic front reflect the highest degree of melting and much larger volcanoes compared with those at a greater distance from the trench.

Back-arc basins and intra-arc rifts

Convergent margins can show extension or contraction, or be strain-neutral. These strain regimes are most clearly manifested in back-arc regions. More often than not, IOASs are associated with active back-arc basins (BABs); such arcs include the Marianas (Mariana Trough; Fig. 7), Tonga–Kermadec (Lau–Havre Basin), New Britain (Manus Basin), Vanuatu (North Fiji Basin), Andaman–Nicobar (Andaman Sea), and South Scotia (East Scotia Basin), so this is shown as part of a typical IOAS in Figure 3a. Other IOASs are associated with extinct BABs, including the Lesser Antilles, Western Aleutians, and Izu–Bonin arcs. Active BABs form by sea-floor spreading, which can be fast (>10 cm a^{-1}) or slow ($1-2$ cm a^{-1}; Stern 2002; Martinez & Taylor 2003). Spreading results from extensional stresses that split the arc, mostly as a result of to trench rollback. Rifting to initiate a BAB can begin in the inner forearc, along the arc, or immediately behind the arc, as summarized in Figure 6. In any case, arc igneous activity may be extinguished temporarily as mantle-derived magmas are captured by the extension axis. The rifted part of the arc is progressively separated from the volcanic–magmatic front, forming a remnant arc that subsides as spreading continues (Karig 1972). BAB spreading produces sea floor with a crustal structure that is largely indistinguishable from that produced by mid-ocean ridges. Inter-arc rifts create significant basins within IOASs but no sea-floor spreading. These may or may not evolve into BABs; typical examples of inter-arc rifts are found in the Izu and Ryukyu arcs.

IOAS sediments

As mentioned before, IOASs are characterized by slow sedimentation rates and relatively thin sediments. Unless an IOAS lies near a continent, its trench will be empty and it will not have an accretionary prism. Significant sediment accumulations in IOASs occur only near the volcanic front, both on the back-arc side and forearc side. Sedimentation rates here will also reflect prevailing wind directions, which control the direction of volcanic ash dispersal and which flanks of subaerial volcanoes will be preferentially eroded by waves. Larger volcanoes are increasingly affected by flank collapse, which episodically sends tremendous volumes of sediment downslope.

Two sites of sedimentary basins in IOASs are worthy of further discussion: the forearc basin and the BAB immediately adjacent to the volcanic arc. Figure 8 shows typical examples of both, using a c. 350 km long multi-channel seismic reflection profile across the Mariana arc–trench system.

Most IOASs have an elongated basin parallel to the volcanic arc on the inner forearc. Such forearc basins are 50–80 km wide (Chapp et al. 2008). Sediments in the Mariana forearc basin have a maximum thickness of 1.5 s (two-way travel time; Fig. 8). Deep Sea Drilling Project (DSDP) Site 458 drilled into the distal edge of this forearc basin, penetrating c. 250 m of Oligocene–Pleistocene sediments and into Eocene basement (Shipboard Scientific Party 1982b). Physical property measurements indicate $V_p < 2.0$ km s^{-1} for these sediments, implying a maximum thickness of c. 1.5 km for this basin. These sediments have accumulated over the past 35 Ma, implying a maximum sedimentation rate of c. 43 m Ma^{-1}.

The other important sedimentary basin lies at the juncture between thin BAB crust and thicker crust beneath the volcanic–magmatic arc (Fig. 7). This results in a deep basin where arc-derived volcaniclastic sediments are deposited. In the case of the Mariana Trough profile shown in Figure 8, the sediments are <0.75 s thick. DSDP Site 455 drilled c. 100 m into a similar sediment pile just to the north of this profile (Shipboard Scientific Party

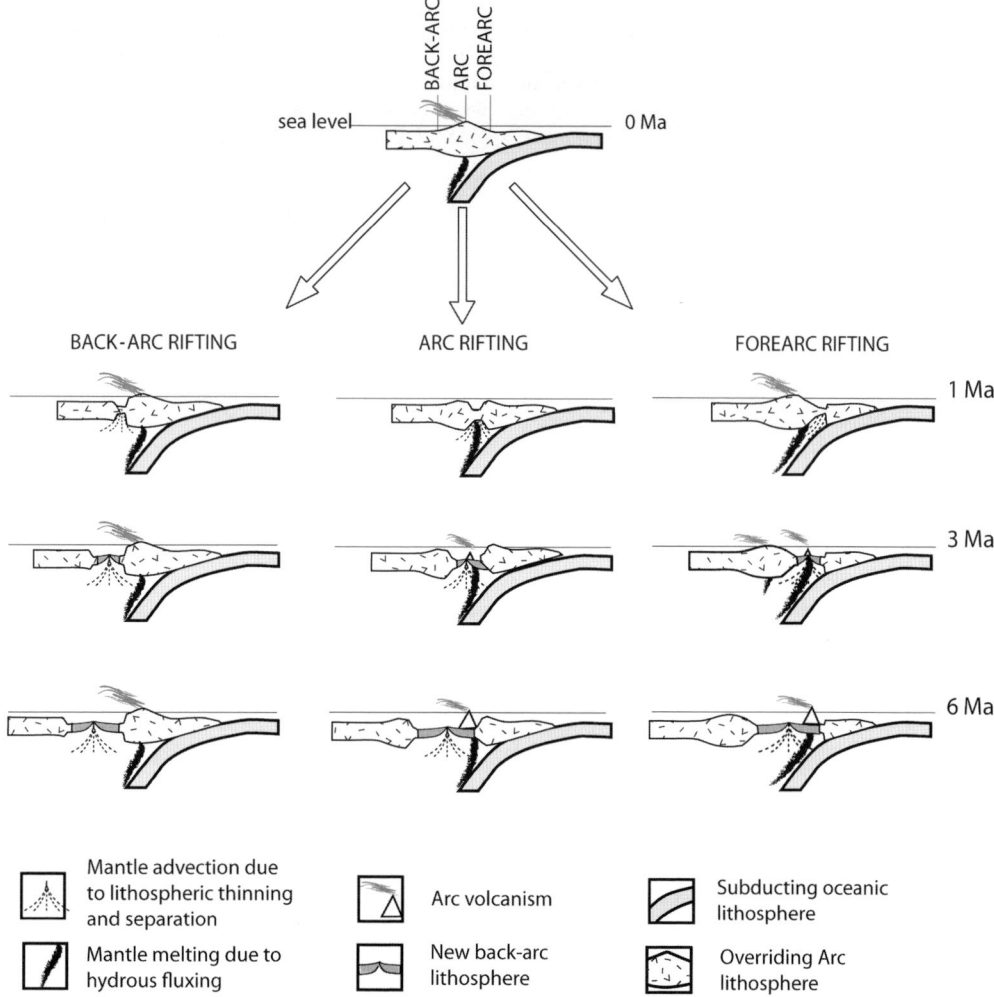

Fig. 6. Three ways in which an IOAS can rift to form a back-arc basin, modified after Martinez & Taylor (2006). The initial slab and arc lithospheric geometry is shown at the top. The three sets of panels depict the evolving geometry and interaction with arc melt sources (tied to the subducting slab) and back-arc melt generation (tied to mantle advection driven by separation of the overriding lithosphere). Left panels show BAB formation by rifting behind the active magmatic arc, in which case arc volcanism will not be interrupted and a narrow, thin remnant arc is produced. The middle panels show BAB formation by rifting the active arc, whereby the arc magmatic budget will initially be captured by the extension axis, shutting down the arc and forcing re-establishment of a new magmatic arc. The right panels shows BAB formation by forearc rifting, whereby the active magmatic arc will slowly migrate away from above the site of melt generation in the mantle and be extinguished, at the same time as a new magmatic arc forms above the site of mantle melt generation. In this case, a very broad and thick forearc is formed.

1982a) and DSDP Site 456 (about 20 km farther west, near the distal edge of the BAB sedimentary basin) penetrated 170 m of sediment and into BAB basaltic crust. Physical property measurements indicate V_p c. 2 km s^{-1} for DSDP 455 sediments, indicating a maximum thickness of c. 750 m. These sediments have accumulated over the past 7 Ma (since the Mariana Trough back-arc basin began to open), implying a maximum sedimentation rate of c. 100 m Ma^{-1}. The greater sedimentation rate for the back-arc basin adjacent to the arc relative to the forearc basin sedimentation rate reflects the fact that this basin is closer to the active arc than the forearc basin, which is separated from

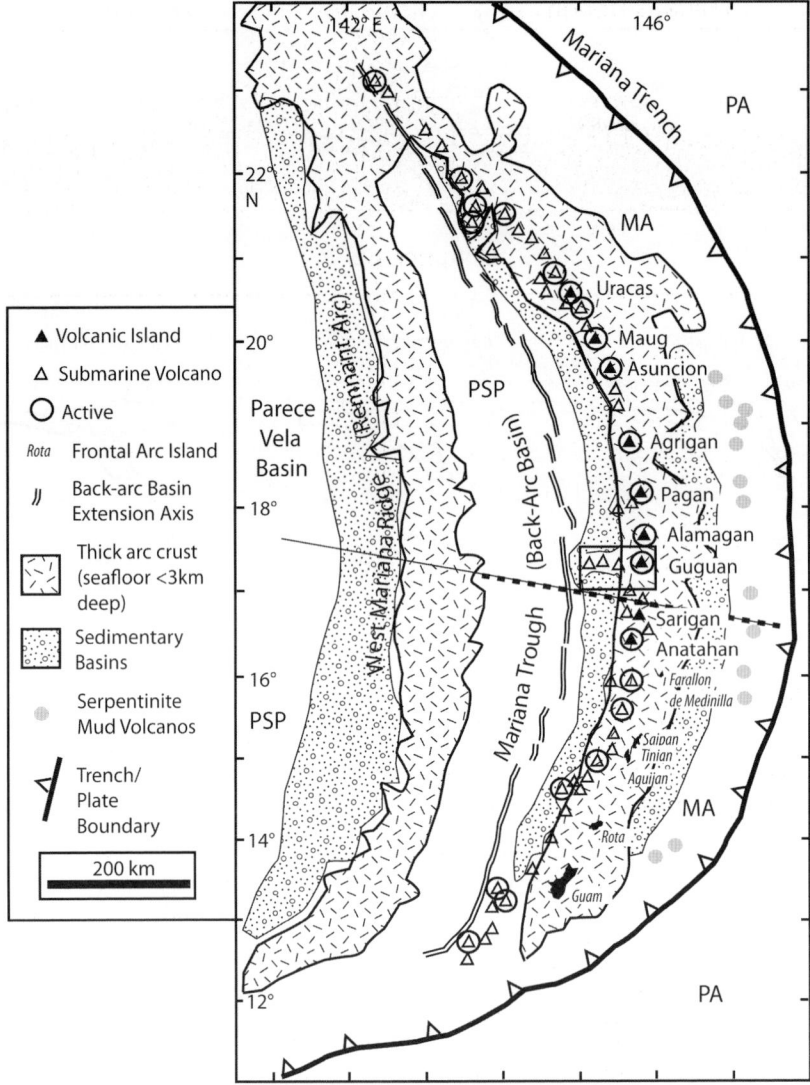

Fig. 7. Simplified tectonic map of the Mariana IOAS, showing the distribution of active arc volcanoes, back-arc basin spreading ridge (modified after Baker *et al.* 2008), frontal arc uplifted islands, remnant arc, major sedimentary basins, and serpentinite mud volcanoes (P. F. Fryer, pers. comm.). Location of boundaries between subducting Pacific plate (PA) and overriding Mariana plate (MA) and Philippine Sea plate (PSP) are also shown. Dashed line shows location of multichannel seismic profile shown in Figure 8; collinear fine continuous line shows position of lithospheric sounding profile in Figure 13. Box shows location of HMR-1 sonar image shown in Figure 9.

the active arc by the frontal arc structural high. Neither basin shows sedimentation rates that are particularly impressive.

Sediments in the Mariana forearc and back-arc basins are dominantly volcanogenic and derived from the active volcanic arc. These basins have sea floor that lies *c.* 3–4 km below sea level, mostly above the carbonate compensation depth so that carbonate sediments could be preserved, but biological productivity in this part of the ocean is low and biogenic sediments form a minor part of the fill of either basin. Forearc and back-arc basin sedimentation where biological productivity is higher (at high latitudes and equatorial regions) may have considerably larger proportions of biogenic components (Marsaglia 1995).

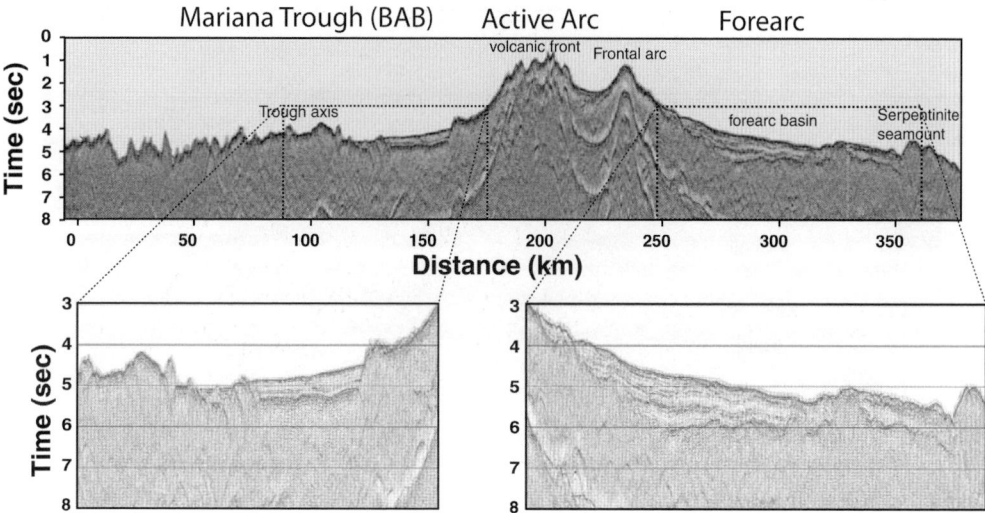

Fig. 8. Multi-channel reflection profile across the Mariana arc, from trench to back-arc basin, from Takahashi *et al.* (2008) with permission. Multiples should be noted. Location shown in Figure 7.

Volcanogenic sediments are transported from the active arc to forearc and back-arc basins probably by turbidity currents and by ash settling out of the water column. Many IOAS volcanoes are surrounded by submarine giant dunes, with wavelengths of up to 2 km, which define proximal sediment aprons that extend up to 60 km. The origin of these is poorly understood (Embley *et al.* 2006). Much other sediment must be transported by turbidites. Sonar backscatter images reveal well-developed, sinuous channels up to 100 km long, flowing downslope from the base of Mariana arc volcanoes towards the BAB to the west (Fig. 9). These may have evolved to transport turbidites produced by collapse of oversteepened arc volcano slopes, allowing mobilized sediments to move to depocentres in the eastern part of the Mariana Trough BAB. The relatively shallow depth of the BAB spreading ridge effectively partitions the two sides of any BAB into a sediment-starved portion that is distal to the active volcanic arc and an arc-proximal portion that is sediment-filled.

IOAS melts, lavas, and plutons

IOAS lavas are characteristically subalkaline and dominated by basalts and their fractionates. High-Mg andesites and rhyodacitic magmatic products also occur (Kelemen *et al.* 2003), with a minor amount of shoshonitic lavas in some cross-chains and along tectonically complex magmatic fronts. Boninites sometimes are formed when subduction zones begin. Tholeiitic basalts dominate BAB lavas, but the terms 'tholeiitic', 'calc-alkaline' and 'high-alumina basalt' all are used to describe the dominant IOAS volcanic front lavas. Arculus (2003) criticized especially the way that the term 'calc-alkaline' is commonly used, and proposed a new classification scheme for convergent margin igneous rocks. This scheme recognizes the importance of oxygen fugacity in controlling FeO^T/MgO v. silica variations in arc lavas as a complement to already established potassium–silica variations (high-Fe, medium-Fe, and low-Fe along with high-K, medium-K, and low-K suites).

The composition of these rocks reflect the unusual nature of subduction-related magmatic processes that make IOAS magmas. These begin with progressive metamorphism and perhaps melting of water-saturated subducted sediments, altered oceanic crust, and serpentinites as these are squeezed and heated in the subduction zone. Increasing pressure and heat closes fractures and pores, then dehydrates hydrous minerals. The result is a continuous release of water during progressive metamorphism to denser and drier minerals (Hacker *et al.* 2003). Kelemen *et al.* (2003) advocated an alternative interpretation whereby water-rich silicate melts from the subducted slab induce mantle melting. Subduction-derived fluids and melts rise into the base of the overlying mantle, corroding existing mantle minerals and forming new hydrous silicates. Hydrated mantle minerals release more water as induced convection drags these deeper. Released fluids percolate upwards into the hotter (asthenospheric) parts of the mantle wedge (Fig. 10), where they dramatically lower

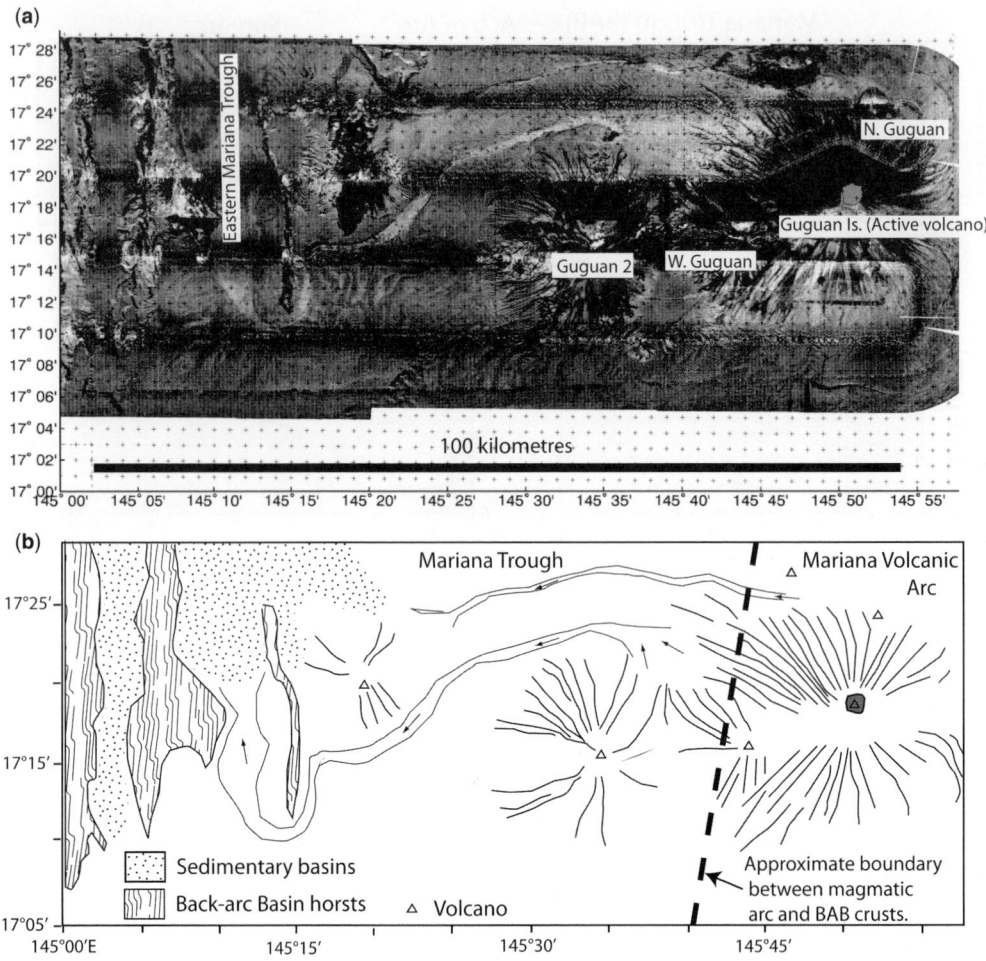

Fig. 9. Example of a large-scale sediment transport system associated with a mostly submarine IOAS. (**a**) HMR-1 sonar backscatter image of the Guguan cross-chain in the Mariana arc (Stern *et al.* 2006; location shown in Fig. 7). The volcanic–magmatic front is represented by Guguan island and submarine North Guguan volcanoes. The Guguan cross-chain extends due west from Guguan and trends perpendicular to the volcanic–magmatic front. The two long channels responsible for moving volcaniclastic sediments westward into deeper water on the eastern flank of the Mariana Trough BAB should be noted. (**b**) Interpretation sketch map of HMR-1 image in (a). See text for further discussion.

the temperature of peridotite and metasomatize the mantle source with fluid-mobile trace elements (K, Rb, U, Pb, etc.), giving the hybridized mantle source the distinctive trace element signature of arc melts, as discussed below. Melting as a result of decompression may also be important (Plank & Langmuir 1988; Conder *et al.* 2002).

The trenchward limit of asthenospheric flow in the core of the mantle wedge determines the location of the volcanic–magmatic front (Cagnioncle *et al.* 2007). Because fluid flux from the subducted slab decreases with increasing depth, the interaction of asthenosphere with the greatest fluid flux maximizes melt generation; because fluid flux diminishes with greater depth to the subducted slab and distance from the trench, melt generation also decreases rapidly in this direction. The characteristic depth to the subduction zone of *c.* 110 km below the volcanic–magmatic front probably reflects the minimum thickness of overriding crust and mantle that is required to allow asthenosphere to circulate.

Strong asymmetries are seen in IOAS magmatic products. Igneous activity in the forearc occurs only early in the life of an IOAS, as discussed below. Lavas erupted along the magmatic–volcanic front are dominated by tholeiitic and high-Al basalts and calc-alkaline andesites (and their differentiates;

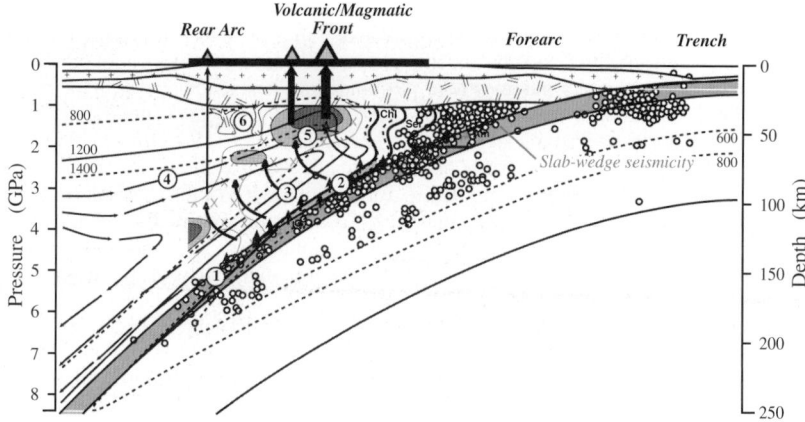

Fig. 10. Generation of melts in the mantle wedge beneath a convergent (intra-oceanic) margin, showing simplified processes modelled by Kimura & Stern (2008). 1, Dehydration and progressive metamorphism of subducted sediment, altered oceanic crust, and serpentinized lithosphere; 2, hydration–dehydration of mantle wedge sole and upward migration of supercritical hydrous fluid; 3, fluid transport and zone refining of mantle wedge; 4, induced asthenospheric convection; 5, open-system melting in the mantle wedge; 6, delamination of lower crust and refertilization of underlying asthenospheric mantle. This section is based on seismic tomography beneath NE Japan, Moho depth, crustal structure, and mantle wedge isotherms (dotted fine line with numbers; °C) and flow lines (fine black arrows). Hydrous mineral stability fields: Ser, serpentine; Am, amphibole; Zo, zoisite; Cld, chloritoid; Lw, lawsonite. Grey open circle, earthquake location; arrows indicate, from bottom to top, path of dehydrated fluid, first from subducted crust, sediments, and serpentinite; then from serpentinized wedge sole. Fluids traverse to the centre, hottest part of the mantle wedge, where shades and patterns map temperature and fluid or melt generation inferred from variations in V_s (Zhao et al. 1994). Vertical arrows from region of melt generation indicate basalt migration through crust to surface.

Kelemen et al. 2003). Rear-arc cross-chains are dominated by more enriched calc-alkaline and shoshonitic lavas. Back-arc basin basalts are tholeiitic, similar in most respects to MORB. These magmatic products are discussed in greater detail below.

Relative to other mantle-derived melts [MORB and ocean island basalt (OIB)], arc melts are enriched in silica, water, and large ion lithophile elements (LILE; e.g. K, U, Sr, and Pb). Figure 11 summarizes the distinctive trace element patterns of primitive (i.e. unfractionated) basalts from IOAS volcanic fronts. The trace element patterns observed for these lavas (and the derivatives of these melts, including fractionates and anatectic melts) show distinctive and similar trace element characteristics that testify to the processes shown in Figure 10, including high degrees of melting of hydrous metasomatized mantle in the spinel peridotite facies. First, the mantle melts more beneath a typical IOAS volcanic front than does the mantle associated with mid-ocean ridges or oceanic hotspots. This is shown by low Na_2O contents of primitive arc basalts (which vary inversely with extent of melting) and by low contents of incompatible trace elements that are not fluid-mobile [i.e. Ti, Zr, Nb, and heavy rare earth elements (HREE)]. For example, the dashed horizontal line approximates the abundances of fluid-immobile elements Zr to the HREE, which are c. 60% of those found in MORB. This implies that the mantle melts c. 70% more beneath IOAS volcanic fronts than it does

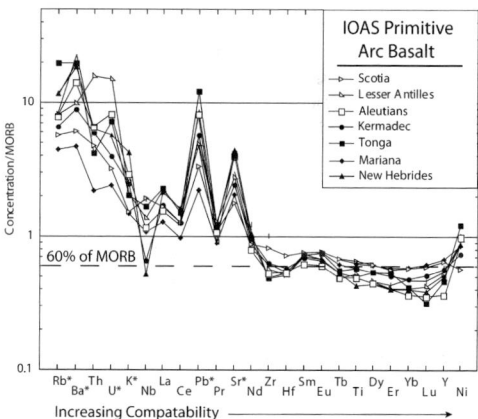

Fig. 11. Extended trace-element diagrams for average primitive IOAS active arc basalts. Concentrations are normalized to N-MORB (Hofmann 1988). Primitive arc basalts are remarkably similar, and consistently distinct from MORB. Elements with asterisks indicate fluid-mobile trace elements. Modified from Kelemen et al. (2003).

beneath mid-ocean ridges. Second, the greater melting beneath arcs reflects the important role that water plays in mantle melting and the much greater abundance of water in IOAS magmas relative to MORB or even hotspot basalts (Stern 2002). IOAS volcanic front lavas are typically coarsely porphyritic, consisting of phenocrysts of An-rich plagioclase and Fe-rich olivine (Fig. 12). This is best explained by high magmatic water contents suppressing the crystallization of plagioclase and rotating its solvus. IOAS volcanic front lavas are typically much more porphyritic than MORB or OIB, consistent with an interpretation that loss of magmatic water during ascent drove rapid crystallization. Third, elevated abundances of fluid-mobile elements such as LILE, shown as positive concentration spikes in Rb, Ba, U, K, Pb, and Sr (Fig. 11), are best explained as resulting from metasomatism of the mantle source region by hydrous fluids or melts. Finally, the flat HREE patterns shown by IOAS basalts are inconsistent with residual garnet in the mantle source region, instead suggesting melt generation in the spinel peridotite stability field. Spinel peridotite is stable at pressures corresponding to depths of 15–80 km, consistent with the expected position of the asthenosphere in the mantle wedge.

Primitive IOAS magma rises into the arc crust, where it may be temporarily stored in magma reservoirs, differentiate and mix with other magmas. The tremendous heat associated with the magmatic front also can lead to crustal anatexis and delamination to generate more felsic arc crust. The twin processes of magmatic fractionation and anatexis, responsible for producing IOAS felsic melts, occur in the crust.

To a certain extent, IOAS magmas reflect the arc's evolutionary stage as well as depth to the subducting slab. Early in the history of an IOAS, lavas are dominated by high-degree melts such as boninites and tholeiites (and their differentiates). It is not yet clear whether melts of the magmatic front evolve systematically as the arc ages. The Izu–Bonin–Mariana IOAS shows no systematic compositional evolution once the magmatic front stabilized c. 40 Ma ago (Straub 2003), whereas the Greater Antilles IOAS seems to have evolved from early tholeiites to later calc-alkaline and then shoshonitic lavas (Jolly et al. 2001). Systematic changes in melt composition may reflect changing upper plate stresses (lithospheric extension v. compression), or may be related to evolving crustal thickness (thicker crust encourages melt stagnation, enhancing fractionation and anatexis), or thermal conditioning (mature arcs with thick crust are

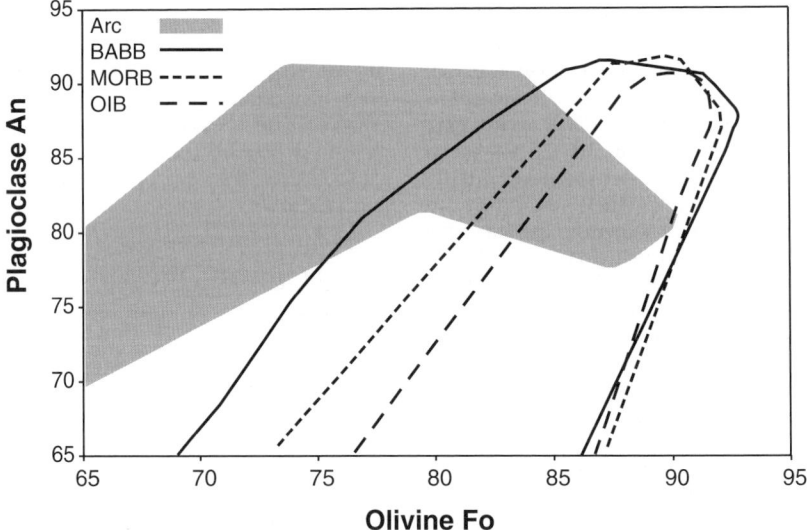

Fig. 12. Composition ranges for olivine–plagioclase assemblages for intra-oceanic arc and back-arc basin basalts (BABB) compared with those for OIB and MORB. Assemblages in basalts from the volcanic front show distinctively high-An (>85) feldspar in more evolved melts in equilibrium with Fo_{70-85} olivine. This is due to stabilization of high-An feldspar in hydrous melts. Anhydrous decompression melt assemblages show similar decreases in Plag An and Ol Fo with melt evolution. BABB assemblages are similar to those of MORB and OIB but partially overlap the arc range. Data for arc field compiled from Basaltic Volcanism Study Project (1981) and intra-oceanic arcs databases at PetDB; BABB from PetDB; MORB field from PetDB (Allan et al. 1987; Basaltic Volcanism Study Project 1981; Davis & Clague 1987); OIB field compiled from PetDB, Basaltic Volcanism Study Project (1981) and Garcia et al. (2000). Pairs of data points for fields: Arcs, 210; OIB, 178; MORB, 151; BABB, 112. Figure generated by E. Kohut.

more likely to experience anatexis around magmatic conduits and reservoirs). Systematic changes in the age of subducting lithosphere could also change IOAS active arc melt compositions, because subduction of very young (hot) oceanic crust (<20 Ma old) can melt to produce adakites (Defant & Drummond 1990). Because IOASs are concentrated where old, dense oceanic lithosphere is subducted, this mode of magmagenesis is not a significant aspect of modern IOASs.

IOAS igneous activity in mature systems is concentrated at the volcanic–magmatic front but can also occur farther away from the trench, along rear-arc or cross-chain volcanoes and in back-arc basins. These two magmatic regimes are very different. Rear-arc volcanoes are common in some IOASs such as the Izu, Bismarck, or Kurile arcs but not in others such as Tonga–Kermadec. Where cross-chains occur they comprise a few to several smaller volcanoes that extend away from a large volcanic front edifice at high angles. Rear-arc volcanoes erupt lavas that are more primitive (Mg-number >60) than volcanic front lavas. These rear-arc lavas are also less porphyritic and do not show the plagioclase–olivine compositional relationships characteristic of IOAS volcanic front lavas. These characteristics, along with the generally more enriched nature of cross-chain lavas, reflect diminished extent of melting as a result of lower fluid flux from the slab with greater distance from the trench. This is the basis for the K–h relationship, which describes the relative enrichment in K and other incompatible elements in arc lavas as a function of depth to the underlying subduction zone (Dickinson 1975; Kimura & Stern 2008).

The final locus of IOAS igneous activity occurs in BABs. Back-arc basin magmas are dominated by tholeiitic basalts that range from MORB-like to arc-like (Pearce & Stern 2006). BAB basalt (BABB) magmas are typically more hydrous than MORB (1–2% v. <0.5% water; Stern 2002). Compositions of BABBs result from four factors: (1) the composition of asthenosphere flowing into the region of melting beneath the BAB spreading ridge; (2) the composition and relative impact of a fluid- or melt-dominated subduction component; (3) how asthenosphere and subduction components interact; (4) the melting of water-rich mantle and the assimilation–crystallization history of the resulting hydrous magma. Trace element and water contents of BAB glasses indicate that decompression melting (similar to that beneath mid-ocean ridges) beneath back-arc basins is augmented by flux melting as a result of the abundance of hydrous fluids.

BAB spreading is generally asymmetric, with spreading axes often lying closer to the active volcanic arc than the remnant arc. BABs have relatively short lifespans compared with volcanic arcs, up to a few tens of million years. This is because as a BAB matures and widens the neovolcanic zone increases in distance from the trench, so that the water flux into the melting region decreases with time. Correspondingly, magma production diminishes over a BAB's life. Such a progression is observed for the 30–15 Ma old Parece Vela Basin, which progressed from magmatic spreading to amagmatic rifting before spreading stopped completely (Ohara 2006). These controls on melting and extension lead to the cessation of BAB spreading after a few tens of million years.

Because BAB basalts and gabbros have compositions and crustal structures that are similar to those of typical oceanic crust, but with trace element enrichments only found in convergent margins, they are often thought to be where ophiolites with 'supra-subduction zone' (SSZ) compositions form (Dilek 2003; Pearce 2003). Some ophiolites may originate in back-arc basins but most probably form in forearcs, where they are in a tectonic position that is much more likely to be emplaced on top of buoyant crust (obducted) if and when this arrives at a subduction zone (Stern 2004). The presence of boninites, which are common in forearcs but rare in BABs, may help resolve such uncertainty, but boninites are found only in some forearcs. Correspondingly, the absence of boninites in an SSZ ophiolite should not be taken as necessarily indicating that the ophiolite formed in a BAB instead of a forearc. Another way to distinguish BAB from forearc crust is that BAB crust forms at intervals after the magmatic arc, whereas forearc crust forms before the magmatic arc.

The nature of IOAS lower crust and upper mantle

The 'classic' Moho is strictly defined on the basis of seismic velocities to separate crust (V_p c. 6–7 km s^{-1}) from upper mantle ($V_p > 8$ km s^{-1}). This generalization does not hold for IOASs, where such Moho is rare beneath the forearcs and magmatic front, although the sub-BAB Moho is clear. The variable expression of Moho beneath IOAS magmatic front, forearc, and back-arc basins reflects different compositions and alteration histories. Simply put, forearc peridotites are dominated by harzburgite (Parkinson & Pearce 1998; Pearce et al. 2000) and are most serpentinized; peridotites associated with the active magmatic arc contain abundant pyroxenite and pyroxene-rich lithologies in addition to lherzolite and harzburgite; and back-arc basin peridotites are mostly lherzolites, similar in most regards to those associated with slow-spreading mid-ocean ridges.

Because the BAB Moho is the best defined, we discuss the composition of the upper mantle below this first. Until recently we had no samples of BAB peridotite to study but we now have had a chance to study peridotites from the northern Mariana Trough (Ohara *et al.* 2002) and the Parece Vela Basin (Ohara 2006). The Parece Vela Basin formed in Miocene time by fast to intermediate rates of sea-floor spreading and includes residual lherzolite and harzburgite, along with plagioclase-bearing harzburgite and dunite. These peridotites are similar to those of the Romanche Fracture Zone in the Mid-Atlantic Ridge and Indian and Arctic Ocean slow-spreading ridges. In contrast, the Mariana Trough is a typical slow-spreading ridge and it locally yields residual harzburgite that is also similar to that recovered from slow-spreading mid-ocean ridges. These two examples of BAB peridotites reveal different melt–peridotite interactions. The Parece Vela peridotites were affected by diffuse porous melt flow and pervasive melt–mantle interaction. Abundant dunites represent melt conduits, where melt and pyroxene reacted to form olivine. In contrast, the Mariana Trough peridotites are characterized by channelled melt–fluid flow and limited melt–mantle interaction.

Forearc peridotites (recovered from inner trench walls) are mostly harzburgites that are more depleted than abyssal peridotites from mid-ocean ridges. Forearc peridotites generally are the most depleted peridotites that form on Earth today (Bonatti & Michael 1989). This is shown by their low Al_2O_3 and elevated SiO_2, reflecting the fact that these are more orthopyroxene-rich than residues of fertile mantle peridotite (Herzburg 2004). Herzburg (2004) suggested on this basis that they were produced by melt–rock reaction. The extreme depletion of forearc peridotites is also shown by the composition of their spinels. Spinel is sensitive to melt depletion, which is reflected in $Cr/(Cr + Al)$. Spinel is also very resistant to alteration and generally preserves petrogenetic information in spite of significant serpentinization (Dick & Bullen 1984). Forearc peridotites have high-Cr-number $(Cr/(Cr + Al) > 0.4)$ spinel. These forearc peridotites are variably serpentinized, being more altered near the trench, and less altered towards the magmatic front. Serpentinization makes seismic recognition of the Moho difficult.

Multiple perspectives are needed to understand the mantle beneath IOAS magmatic arcs. One perspective comes from rare mantle xenoliths brought up in arc lavas. Arai & Ishimaru (2008) summarized this for West Pacific arc lavas. These are mostly from arcs built on continental crust but still provide an important glimpse of the mantle beneath the magmatic front. Almost all the peridotite xenoliths are spinel-bearing varieties without garnet or plagioclase. Arai & Ishimaru (2008) emphasized that mantle metasomatism is most pervasive and long-lived beneath the volcanic front, where magma transit is very long-lived. Metasomatism here reflects the importance of silica-rich, hydrous melts, and is likely to be manifested in the formation of secondary orthopyroxene at the expense of olivine. A large range of fertility and depletion is seen in arc xenoliths. Some are more fertile than the most fertile abyssal lherzolites, whereas others are more depleted than the most depleted abyssal peridotite. This lithological variation is probably due to the complex tectonic history of each as well as to across-arc variation in magma production conditions.

A second useful insight into the nature of the lower crust and upper mantle beneath the magmatic front comes from active source geophysical soundings of IOASs; investigations of the central Aleutian IOAS are particularly useful. Fliedner & Klemperer (2000) interpreted these results to indicate c. 30 km thick crust underlain by heterogeneous uppermost mantle ($V_p = 7.6$–8.2 km s^{-1}). Shillington *et al.* (2004) reinterpreted the same dataset to indicate that the crust was 20% thicker than concluded from the earlier study. They interpreted the regions with velocities of 7.3–7.7 km s^{-1} as lower crust made up of ultramafic–mafic cumulates and/or garnet granulites. They also found upper mantle velocities of 7.8–8.1 km s^{-1} at greater depth, significantly higher than that inferred from the Fliedner & Klemperer study. A similar situation is seen for the Izu–Bonin–Mariana IOAS. High-resolution seismic profiling (Kodaira *et al.* 2007; Takahashi *et al.* 2007) reveals an unusually low-velocity (7.2–7.6 km s^{-1}) material in the lower crust–upper mantle transition zone. This is interpreted at present as an uppermost mantle layer between the Moho above and reflectors in the upper mantle below. These velocities are best explained by a broad crust–mantle transition zone that is rich in pyroxene and/or amphibole.

P-wave velocity structure of the Mariana IOAS and complementary interpretation is shown in Figure 13 (Takahashi *et al.* 2008). Takahashi *et al.* (2008) found that the P-wave mantle velocity beneath the magmatic front is 7.7 km s^{-1}, markedly less than the normal upper mantle V_p of 8.0 km s^{-1}. Based on its continuity, Takahashi *et al.* (2008) identified the base of the 6.7–7.3 km s^{-1} layer as Moho. It should be noted that the sub-Moho reflectors lie beneath the Mariana arc magmatic front, the West Mariana Ridge (remnant arc), and the Mariana Trough spreading axis, in a part of the mantle that is characterized by low V_p (<8 km s^{-1}). These reflections might indicate rising magma bodies in the

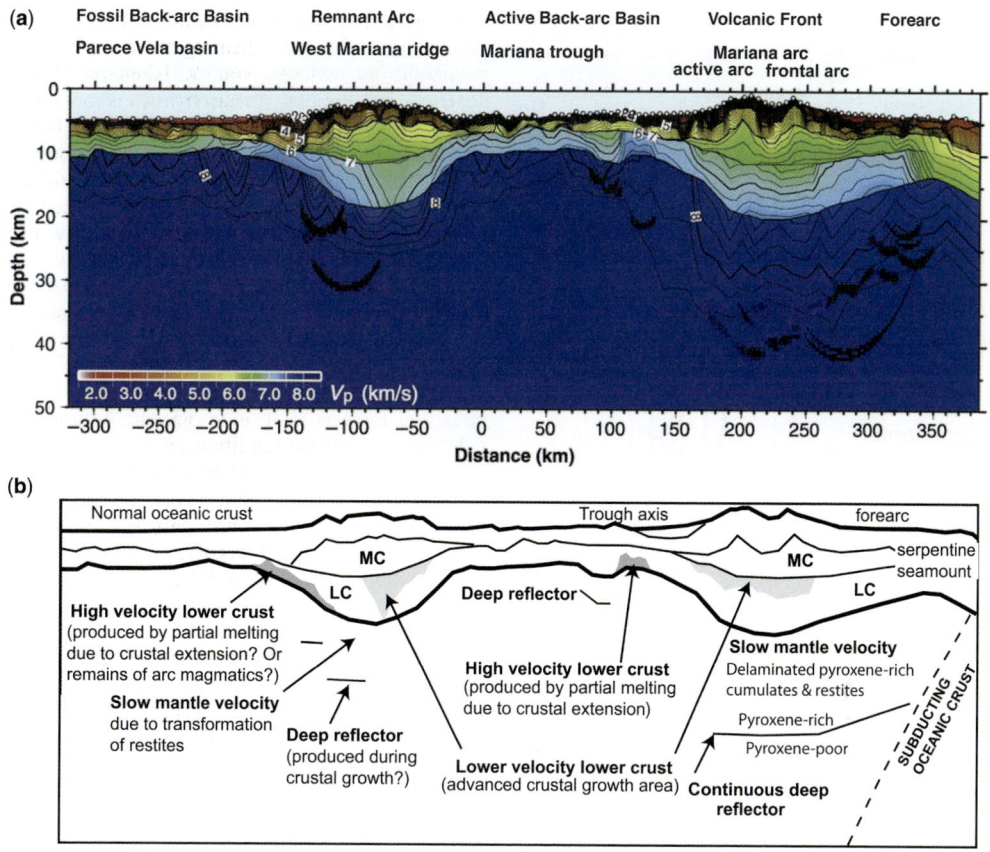

Fig. 13. (a) Velocity structure of the Mariana Arc (see Fig. 7 for location of profile; from Takahashi *et al.* (2008). Numerals indicate P-wave velocity (km s^{-1}). Thick black lines in blue area (inferred mantle) indicate positions of deep reflectors, perhaps marking boundaries in uppermost mantle separating upper regions with abundant pyroxene and pyroxenite from those below dominated by peridotite. (b) Schematic interpretation of seismic profile in (a) (modified after Takahashi *et al.* 2008).

upper mantle, or ghosts of dense lower crust that has sunk back into the mantle. The slow mantle region does not extend down to the upper mantle reflector. The 8.0 km s^{-1} velocity contour lies between the Moho and the deep reflector, which Takahashi *et al.* (2008) inferred has a velocity contrast of about 0.7 km s^{-1}. These observations are consistent with an interpretation that mantle peridotite dominates beneath the arc at depths >40 km, but the shallower mantle contains a significant proportion of pyroxenite and/or amphibolite.

The velocity structure of the Mariana IOAS is similar to that of other IOASs, at least those that have been studied in similar detail. Figure 14 compiles the P-wave velocity structure for 10 sites of three IOASs. These velocity profiles are beneath the active arc, except for one profile beneath the Mariana forearc. Comparison of the velocity structure beneath the Mariana active arc–forearc pair suggests that the forearc structure is c. 0.25–0.5 km s^{-1} slower than the active arc, but it is not yet known whether this is a systematic difference between forearc and active arc crustal structure. Most of the profiles show increases to V_p c. 7.8 km s^{-1} or more, interpreted to correspond to upper mantle and thus marking the Moho, at depths between 25 and 35 km. This is where IOASs are thickest, but is nevertheless significantly thinner than continental crust, which is typically c. 40 km thick. IOAS crust also differs from continental crust in being significantly faster (by 0.5–1.0 km s^{-1}) than typical continental crust. This indicates that IOAS crust is more mafic than continental crust.

A complementary perspective comes from fossil arcs in orogens, where deep crust and upper mantle

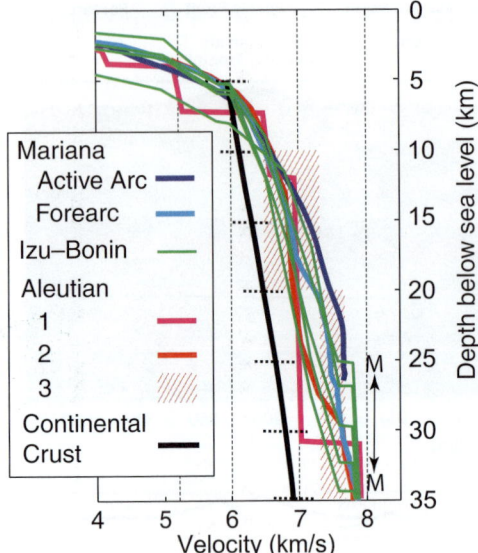

Fig. 14. Comparison of average P-wave velocities at 10 sites from three intra-oceanic arcs (Mariana, Izu–Bonin, and Aleutian), modified after Calvert *et al.* (2008). Profiles for Mariana are from Calvert *et al.* (2008); for Aleutian 1 after Holbrook *et al.* (1999); Aleutian 2 after Fliedner & Klemperer (2000); Aleutian 3 after Shillington *et al.* (2004); and five Izu–Bonin arc profiles are from Kodaira *et al.* (2007, fig. 8) at 162, 260, 307, 365, and 435 km. M, depth range of Moho. A noteworthy feature is the broad similarity of the 10 profiles, all of which are significantly faster (by 0.5–1.0 km s^{-1}) than typical continental crust (shown with 1 standard deviation, from Christiensen & Mooney 1995). All profiles are beneath active arcs except Mariana forearc, where upper crustal P-wave velocities are 240–360 m s^{-1} slower than beneath the active Mariana arc at equivalent depths.

are often exposed. The best studied examples are Mesozoic IOASs of Talkeetna (Alaska) and Kohistan (Pakistan). Reconstruction of the fossil (Jurassic) Talkeetna arc by Hacker *et al.* (2008) indicated *c.* 35 km crustal thickness. Hacker *et al.* (2008) reconstructed the following major layers, from top to bottom (Fig. 15): (1) a *c.* 7 km thick volcanic section (Clift *et al.* 2005); (2) intermediate plutonic rocks (tonalites and quartz diorites) at 5–12 km depth; (3) mafic metamorphosed gabbroic rocks from 12 to 35 km depth. These metagabbroic rocks include: (1) hornblende gabbronorites down to *c.* 25 km; (2) garnet diorites, first appearing at *c.* 25 km depth; (3) garnet gabbro, spinel-rich pyroxenite and orthogneiss, and hornblende gabbronorite at the base of the arc at *c.* 30–35 km depth (Fig. 15). The upper mantle beneath Talkeetna consists of abundant pyroxenites, consistent with inferences from geophysical studies of the Aleutians and Izu–Bonin–Mariana IOASs. The abundance of pyroxenite in the upper mantle beneath the active magmatic arc contrasts markedly with the more depleted, harzburgitic mantle beneath the forearc. The Border Ranges ultramafic complex fragments in the McHugh mélange complex were suggested by Kusky *et al.* (2007) to be the ophiolitic forearc to the Talkeetna IOAS

The upper mantle beneath the Cretaceous Kohistan palaeo-arc also contains abundant dunite, wehrlite, and Cr-rich pyroxenite in the inferred Moho transition zone [Jijal complex; (Garrido *et al.* 2006)]. Dhuime *et al.* (2007) reported that dunites, wehrlites and pyroxenites make up *c.* 50% of the Jijal complex. The origin of sub-arc pyroxenites and their relationship with the surrounding gabbros is controversial here as well. Are they cumulates from the fractionation of mafic magmas or do they result from melt– or fluid–rock interaction in the upper mantle? It is likely that sub-arc pyroxenites have multiple origins, some forming by reaction within the upper mantle, others forming in the lower crust and sinking back into the mantle as a result of delamination.

A key consideration for understanding how the lower crust and upper mantle beneath active magmatic arcs developed and behaves comes from advances in understanding the thermal structure of IOASs. Temperatures in IOAS lower crust, near the Moho beneath active magmatic arcs, approach or exceed 1000 °C. Because active arc volcanoes mark the surface projection of magma transport routes at depth, Moho temperatures probably also vary along strike of the arc, being higher beneath volcanoes and lower between them. In addition, temperatures in the crust beneath IOAS volcanic arcs may not vary smoothly with depth, but are strongly controlled by rheology. Figure 16b summarizes the rheological control on arc thermal structure, showing that a significantly lower thermal gradient exists through the upper, brittle part of the crust, where seawater is able to circulate through fractures and faults. At a certain depth in the middle crust corresponding to the brittle–ductile transition, such fractures and faults are annealed and a step in the geothermal gradient is likely to result from advective ascent and ponding of magmas. Regardless of this step in the geothermal gradient, temperatures of 800–1200 °C have been determined and inferred for IOAS Moho. Such a thermal structure implies significant remelting of mafic arc crust roots, depending on the composition of lower IOAS crust. If it is dry (meta-mafic granulite), solidus temperatures lie well below likely IOAS geotherms and remelting of IOAS arc crust is unlikely (Fig. 16c). On the other hand, if amphibole is an important component of IOAS arc crust, then the solidus should be significantly

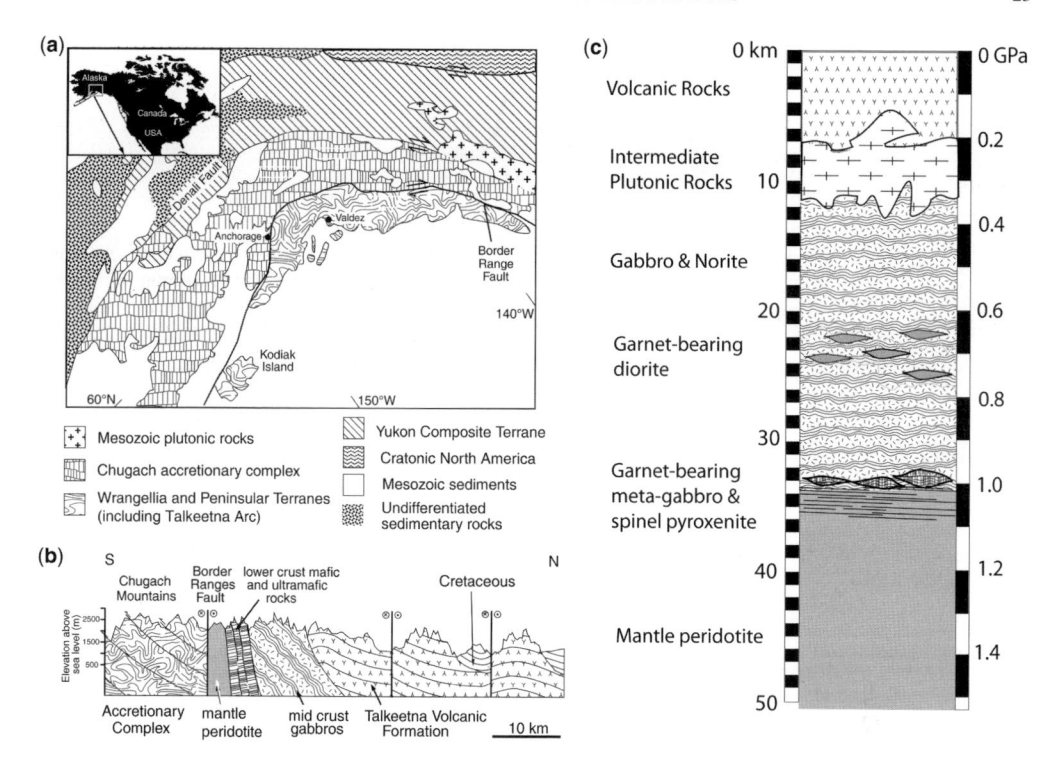

Fig. 15. (**a**) Tectonic map of Alaska and NW Canada showing the location of the Talkeetna arc and other exotic terranes accreted to cratonic North America. The study region is located within the Peninsular Terrane of south–central Alaska, sandwiched between Wrangellia to the north and the Cretaceous accretionary complexes exposed within the Chugach Terrane to the south (modified after Clift *et al.* 2005). Terrane boundaries are all faulted. The Talkeetna arc sequence is also associated with a fragmented forearc basement, a few million years older than the arc sequence (Kusky *et al.* 2007). (**b**) Schematic cross-section through the Talkeetna arc, showing the general northward dip of the arc, dissected by east–west strike-slip faults (modified after Clift *et al.* 2005). (**c**) Pseudostratigraphic column for the Talkeetna arc, modified after reconstruction of Hacker *et al.* (2008). K-feldspar-bearing plutons crystallized at 5–9 km, most hornblende gabbronorites crystallized at 17–24 km, garnet-bearing diorites crystallized at 22–24 km, and garnet gabbros crystallized at the base of the arc at *c.* 30 km depth.

lower than the geotherm (Fig. 16d) and remelting of large parts of the lower crust seems inevitable. Fractional crystallization coupled with anatexis in the lower arc crust seems likely, and dense residues and fractionates are likely to sink back into the upper mantle. In support of this idea, Brophy (2008) concluded from consideration of REE modelling that felsic magmas in IOASs could be produced either by fractional crystallization (with or without hornblende) or by amphibolite melting, and that fractional crystallization seemed to be the dominant mechanism.

The Talkeetna fossil arc again provides important insights into the fate of lower crust beneath IOAS magmatic fronts. Field studies show that the great volumes of mafic and ultramafic cumulates (gabbronorite and pyroxenite) expected to have been produced by fractionation of primitive, mantle-derived melts are missing from the lower crust. Behn & Kelemen (2006) argued that lower crustal $V_p > 7.4$ km s^{-1} in modern arcs indicates material that is gravitationally unstable relative to underlying asthenosphere; such material is likely to sink into the mantle. They also noted that the lower crust beneath modern arc magmatic fronts has $V_p < 7.4$ km s^{-1}, indicating that pyroxene-rich lower crust is generally not present. These observations support the conclusion that large volumes of lower arc crust founder rapidly on geological time scales, or that a significant proportion of high-V_p cumulates form beneath the Moho (Behn & Kelemen 2006).

Based on petrological modelling and V_p estimates for sub-Izu–Bonin–Mariana crust and mantle, Tatsumi *et al.* (2008) inferred that the uppermost mantle low-V_p layer was composed of mafic,

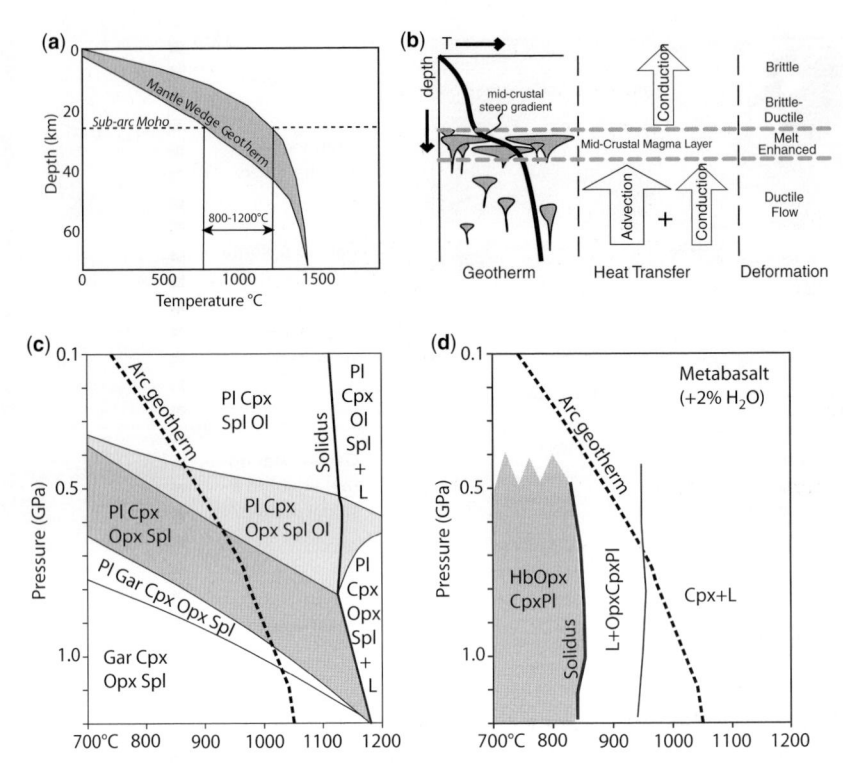

Fig. 16. Thermal structure of IOAS beneath volcanic–magmatic front. (**a**) Geothermal gradient in crust and mantle beneath the active volcanic–magmatic arc, simplified after Hacker et al. (2008). Mantle beneath the Moho of active magmatic arc is at 800–1200 °C and therefore will be weak. Dense, delaminated lower crustal materials could easily sink into this buoyant, weak mantle. (**b**) Schematic illustration of how a magma-rich layer could produce steep geothermal gradients in the middle crust. Advective heat transport below the level of magma emplacement is more efficient than conductive transport above. This forces steep geothermal gradients to form in the middle crust, separating low-temperature upper crust from higher temperature lower crust. Deformation below the magma layer is by penetrative ductile flow; deformation above the layer is focused in narrow shear zones (Shaw et al. 2005). (**c**) Phase relations for Talkeetna orthogneiss 1709P11, showing pressure-induced consumption of olivine from the igneous assemblage feldspar (fsp) + clinopyroxene (cpx) + orthopyroxene (opx) + spinel (spl) + olivine (ol) ± liquid (L) to produce cpx + opx + spl symplectite. A range of P–T paths compatible with this inferred transformation is shown (Hacker et al. 2008). (**d**) Simplified phase relationships for the hydrous synthetic low-K basalt with 2 wt% H_2O added compositions (Nakajima & Arima 1998). Solidus for anhydrous compositions is at c. 1150 °C, significantly hotter than typical arc crustal geotherm, but solidus for hydrous compositions is at c. 850 °C, well below the arc geotherm.

pyroxene-rich material that had sunk back into the mantle, not mantle peridotite. It appears that igneous activity beneath IOAS active magmatic arc evolves from mafic to intermediate as a result of continuous loss of mafic and ultramafic lower crustal components to the subarc mantle across a chemically transparent Moho (Tatsumi et al. 2008). This process will slow crustal thickening at the same time that increasingly felsic crust comes to make up the evolving arc crust, leading to a crustal thickening rate of c. 500 m Ma^{-1}. When IOAS crust thickens sufficiently to stabilize garnet in the lower crust (greater than c. 30 km thick), magma compositions will become more adakitic at the same time that delamination and crustal processing are likely to be enhanced [as a result of the greater density of garnet (3.6 g cm^{-3}) relative to pyroxene (3.2 g cm^{-3})]. This model is summarized in Figure 17.

IOAS forearc structure preserves early history

A final important point about IOASs concerns how they form, how the forearc is built and abandoned as a site of igneous activity, and how the active magmatic arc finally comes to the position that it subsequently occupies for as long as the IOAS exists. IOASs begin their evolution when subduction

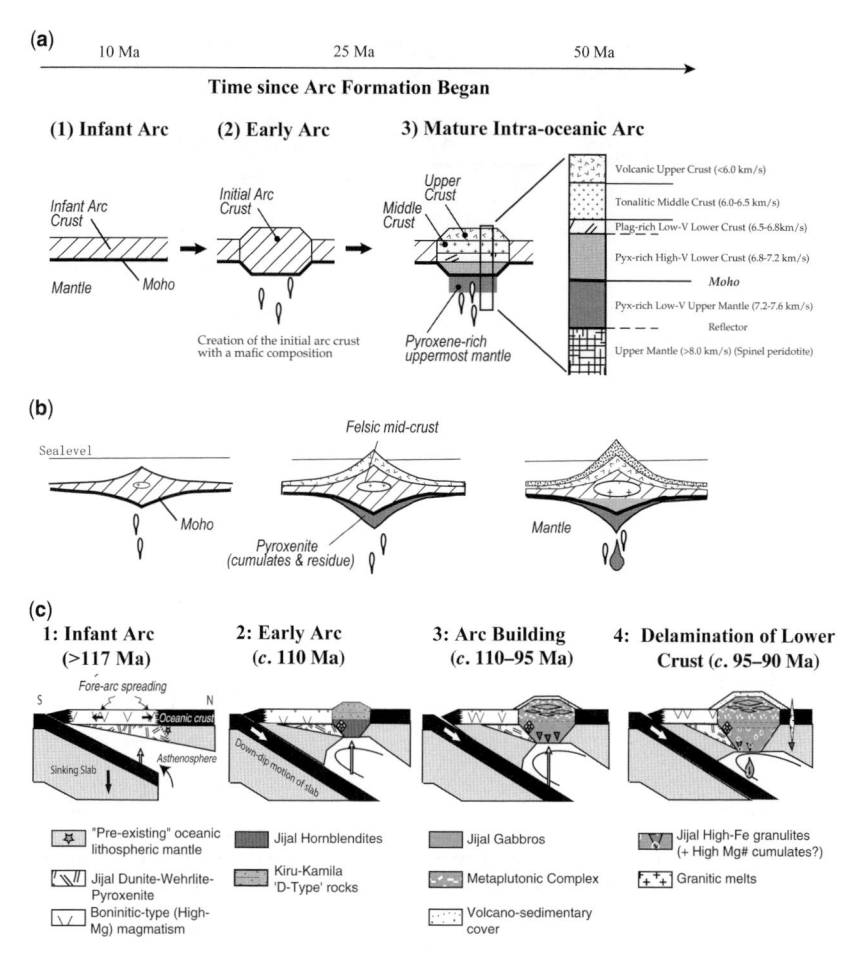

Fig. 17. Formation and refinement of juvenile arc crust. (**a**) Model of arc crust evolution from Tatsumi *et al.* (2008). (1) Incipient arc magmatism replaces the pre-existing oceanic crust to create (2) the initial mafic arc crust. (3) Continuing arc magmatism causes anatexis and differentiation of the arc crust, along with transformation of mafic crustal component into the mantle through the Moho, finally creating mature arc crust with an intermediate composition similar to the average continental crus. Generalized seismic velocity structure in the IBM arc after Suyehiro *et al.* (1996), Kodaira *et al.* (2007) and Takahashi *et al.* (2007). (**b**) Schematic illustration of progressive burial of early formed plutonic and volcanic rocks within a growing arc (modified after Kelemen *et al.* 2003). Pyroxenites near the base of the crust are denser than the underlying mantle, and temperatures are high near the Moho, so that these ultramafic cumulates may delaminate whenever their thickness exceeds some critical value (e.g. Jull & Kelemen 2001). Increasing pressure forms abundant garnet in Al-rich gabbroic rocks near the base of the section, and these too may delaminate. Intermediate to felsic plutonic rocks, and even volcanic rocks, may be buried to lower crustal depths, where they are partially melted. (**c**) Four-stage schematic evolution of the Kohistan arc during the period 117–90 Ma (after Dhiume *et al.* 2007). Four evolutionary stages span the history of the arc, from spontaneous initiation of subduction associated with extensive boninitic magmatism (stage 1) to arc building by tholeiitic basalts (stages 2 and 3) and a fourth stage of intracrustal differentiation. Stage 4 culminated in the intrusion of granitic rocks akin to continental crust in the upper level of the metaplutonic complex.

begins. Subduction zones can form either as a result of lithospheric collapse (spontaneous nucleation of subduction zone; SNSZ) or by forced convergence (induced nucleation of subduction zone; INSZ; Stern 2004). For SNSZ, IOASs start as broad zones of sea-floor spreading associated with subsidence of the adjacent lithosphere, whereas INSZ IOASs may be built on trapped crust, without construction of new arc substrate by large-scale sea-floor spreading. SNSZ results from gravitational instability of oceanic lithosphere adjacent to lithospheric zones of weakness such as fracture

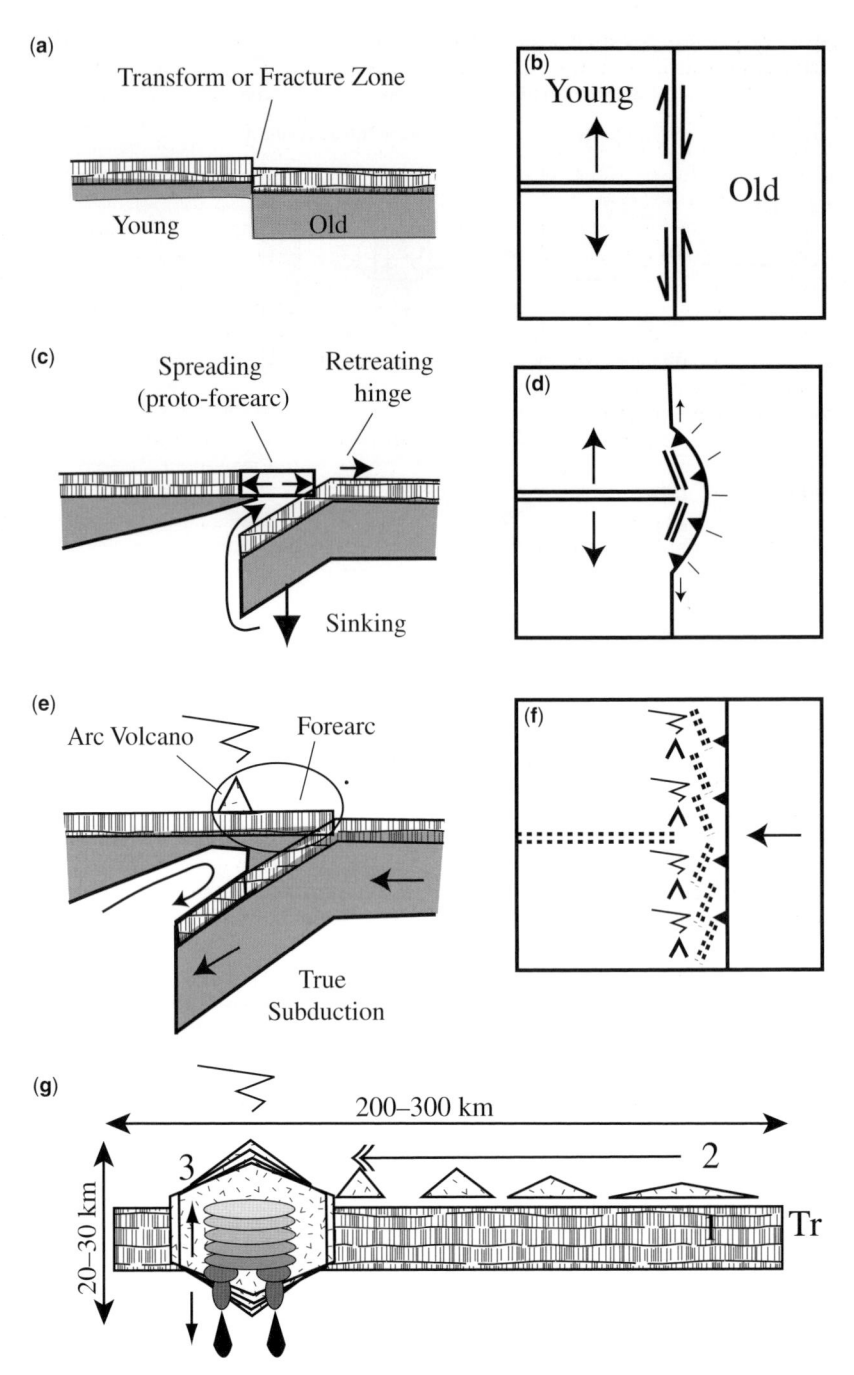

Fig. 18. Generation of forearc during subduction initiation and retreat of magmatic arc to the approximate position that it will subsequently occupy. (a–e) Subduction infancy model of Stern & Bloomer (1992), modified to show the third dimension. Left panels are sections perpendicular to the plate boundary (parallel to spreading ridge) and right panels are map views. (**a, b**) Initial configuration. Two lithospheres of differing density are juxtaposed across a transform fault or fracture zone. (**c, d**) Old, dense lithosphere sinks asymmetrically, with maximum subsidence nearest fault. Asthenosphere migrates over the sinking lithosphere and propagates in directions that are both orthogonal to the original trend of the transform or fracture zone as well as in both directions parallel to it. Strong extension in the region above the sinking lithosphere is accommodated by sea-floor spreading, forming infant arc crust of the proto-forearc.

zones (Fig. 18a, b). Because modern IOASs are concentrated in regions where the oceanic lithosphere is ancient, dense, and prone to collapse, it is likely that most IOASs form by SNSZ.

Lithospheric collapse to initiate SNSZ allows lighter asthenosphere to exchange places with denser, sinking lithosphere (Fig. 18c, d). This results in sea-floor spreading above the sinking lithosphere, and the combined result of decompression of rising asthenosphere and fluids released from the sinking slab is unusually high degrees of mantle melting. This produces very depleted tholeiites (low TiO_2 and K_2O) and subordinate boninites that make up the forearc crust. Felsic magmas are also produced during this 'infant arc' stage. Oceanic crust made up of these lavas and intrusions is underlain by genetically related ultradepleted harzburgitic mantle residues, and these rocks form what becomes the forearc. It is important to note that sea-floor spreading to form the forearc lasts only a few million years, while the subjacent lithosphere continues to sink and asthenospheric mantle flows above the sinking slab. This is the origin of most boninites and ophiolites. Stern & Bloomer (1992) argued that the Izu–Bonin–Mariana intra-oceanic arc system in the Western Pacific formed by collapse of old Pacific lithosphere east of a major transform, with much younger sea floor and hotter shallow mantle to the west. Stern & Bloomer (1992) estimated that during the first 10 Ma in the life of the Izu–Bonin–Mariana arc, melt generation occurred at a rate of $c.$ 120–180 $km^3\,km^{-1}\,Ma^{-1}$. This may be a significant underestimate, because recent geophysical studies of the Izu–Bonin–Mariana and Aleutian arcs infer a mean crustal growth rate of $c.$ 100 $km^3\,km^{-1}\,Ma^{-1}$ or more (Dimalanta et al. 2002; Jicha et al. 2006), orders of magnitude larger than mature arc eruption rates of $<10\,km^3\,km^{-1}\,Ma^{-1}$ (Kimura & Yoshida 2006; Carr et al. 2007).

Once down-dip motion of the slab (true subduction) begins, the asthenosphere is no longer able to flow into this region and the sub-forearc mantle rapidly cools. Slab-derived fluids migrating into the sub-forearc mantle no longer cause melting, only serpentinization. Igneous activity migrates away from the trench and eventually forearc igneous activity stops (Fig. 18e, f). A typical IOAS begins to mature when true subduction (i.e. down-dip motion of the sinking slab) begins and the subducted slab reaches $c.$ 130 km depth, focusing magmatism to begin building the magmatic arc and allowing the forearc lithosphere to cool and hydrate. The zone of infant arc crust produced by sea-floor spreading above the subsiding lithosphere becomes forearc as the IOAS matures and the locus of magmatism retreats from the trench. This makes up the forearc, lithosphere that is emplaced as an ophiolite when buoyant lithosphere is subducted beneath it (Pearce 2003). Thus the forearc preserves the early history of the arc.

Intra-oceanic arc systems and the growth of the continental crust

In contrast to continuing controversy about how the modern inventory of continental crust came to be and what it was in the past (Dewey & Windley 1981; Armstrong 1991; Windley 1993), a broad geoscientific consensus exists that continental crust today is mostly generated above subduction zones, with secondary sites associated with rifts, hotspot volcanism, and volcanic rifted margins. Subduction zones are also the most important sites of crust removal, by sediment subduction, subduction erosion, and deep subduction of continental crust. A comprehensive overview of how continental crust is created and destroyed by plate-tectonic processes has been given by Stern & Scholl (2010).

Continental crust mostly forms today and over the lifetime that plate tectonics has operated by magmatic additions from water-induced melting of mantle above subduction zones (Coats 1962; Dimalanta et al. 2002; Davidson & Arculus 2005). This produces juvenile IOAS arc crust, and magmatically adds to pre-existing continental crust at Andean-type arcs. Scholl & von Huene (2009) estimated that over the last tens to hundreds of million years, $c.$ 2.7 km^3 of juvenile continental crust has been generated by convergent margin igneous activity each year. This crust is mostly extracted from the mantle as mafic (basaltic or boninitic) magma, which must be further refined by anatectic remelting of the crust and delamination (foundering) to yield andesitic material approximating the

Fig. 18. (*Continued*) (**e, f**) Beginning of down-dip component motion in sinking lithosphere marks the beginning of true subduction. Strong extension above the sunken lithosphere ends, which also stops the advection of asthenosphere into this region, allowing it to cool and become forearc lithosphere. The locus of igneous activity retreats to the region where asthenospheric advection continues, forming a magmatic arc. (**g**) A detailed view of how magmatism in the arc evolves with time; 1, infant arc crust of the forearc ophiolite forms by sea-floor spreading during first $c.$ 5 Ma of subduction zone evolution, as shown in (c) and (d); 2, retreat of magmatic activity away from the trench (Tr) during the second $c.$ 5 Ma; 3, focusing of magmatic activity at the position of the active magmatic arc, resulting in crustal thickening and delamination.

bulk composition of continental crust (Rudnick & Gao 2003; Tatsumi 2005; Tatsumi *et al.* 2008). Although juvenile continental crust is mafic when first formed, it inherits from subduction zone processes the distinctive geochemical characteristics of subduction-related magmas: enrichment in silica and LILE (e.g. K, U, Sr and Pb) and depletion of high field strength incompatible elements (e.g. Nb, Ta and Ti). The similarity of trace element variations in arc rocks to that of bulk continental crust strengthens the inference that subduction-related magmatism is an important way to make continental crust today (Kelemen 1995; Rudnick & Gao 2003).

The distinctive chemical characteristics of IOAS magmas (Fig. 11) are likely to persist through later processing by anatexis within the crust and foundering of its base, so that juvenile mafic crust slowly matures into continental crust with andesitic bulk composition. The formation of continental crust at IOASs requires two stages: mantle melting to generate basalt and mafic crust and further refinement of this to yield granitic crust and refractory material, much of which founders back into the mantle (Hawkesworth & Kemp 2006). This is accomplished by distilling felsic magmas from mantle-derived juvenile crust via anatexis of amphibolites and fractional crystallization of mafic magmas. Much basaltic magma crystallizes as pyroxene-rich material near the Moho. Magmatic maturation of IOASs is also likely to occur in tandem with terrane accretion, whereby various juvenile crustal fragments coalesce to build progressively larger tracts of increasingly differentiated crust.

Conclusions

Intra-oceanic arc systems are enormous tracts of proto-continental crust, hundreds to thousands of kilometres long, several hundred kilometres wide, and up to 35 km thick. They tend to form where very old (>100 Ma) oceanic lithosphere is subducted, typically showing strong evidence for extension (normal faulting, intra-arc rifts, back-arc basins). Morphological asymmetries (trench, forearc, volcanic front, rear-arc volcanoes, and back-arc basin) reflect asymmetrical processes controlled by the dipping subduction zone. Crustal structure reflects this asymmetry, with IOAS crust thickening from the trench towards the magmatic front, and thinning again beneath back-arc basins. Distinctive compositions of igneous rocks and associated mantle also reflect the subduction-imposed asymmetries that reflect in part the evolution of the IOAS. Igneous rock compositions reflect decreasing water flux from the slab with distance from the trench, resulting in lower-degree melts and more enriched igneous rocks with this distance. This is the basis of the K–h relationship. In addition, ultrahigh-degree melting occurs early in the history of the IOAS to yield ultra-depleted tholeiites and some boninites to form what will become forearc crust. Magmatic products of a mature subduction zone manifest combination of a continuous flux of hydrous basalt from the mantle wedge, crustal magmatic process of fractionation and anatexis, and loss of dense pyroxene-rich cumulates from the base of the crust back to the mantle. The result of these processes is to produce crust that becomes increasingly felsic as it ages and slowly thickens. Mantle peridotites reflect these subduction-related asymmetries in melting and evolution: ultra-depleted harzburgites that are strongly serpentinized beneath the forearc, pyroxene-rich ultramafic rocks beneath the magmatic front, abyssal peridotite-like lherzolites and moderately depleted harzburgites beneath the back-arc basin. IOASs cannot be understood without understanding the remarkably different processes and magma production rates that accompany their early development. The characteristic dimensions and asymmetries of IOAS systems should be at least partly preserved in orogenic belts and discernible to careful observers.

This paper benefited from thoughtful reviews by M. Santosh and T. Kusky. Thanks go to E. Kohut for help with Fo–An relationships, and to A. Calvert for Figure 14. My research into the nature of intra-oceanic arc systems has been supported by the US National Science Foundation, most recently by OCE0405651. My studies of IOASs has benefited from many interactions with many colleagues in many ways over more than three decades. This is UTD Geosciences Department Contribution 1195.

References

ALLAN, J. F., BATIZA, R. & LONSDALE, P. F. 1987. Petrology and chemistry of lavas from seamounts flanking the East Pacific Rise Axis, 21° N: implications concerning the mantle source composition for both Seamount and Adjacent EPR lavas. *In*: KEATING, B., FRYER, P. F. & BATIZA, R. (eds) *Seamounts, Islands, and Atolls*. American Geophysical Union, Geophysical Monograph, **43**, 255–282.

ARAI, S. & ISHIMARU, S. 2008. Insights into petrologic characteristics of the lithosphere of mantle wedge beneath arcs through peridotite xenoliths. A review. *Journal of Petrology*, **49**, 665–695.

ARCULUS, R. J. 2003. Use and abuse of the terms calcalkaline and calcalkalic. *Journal of Petrology*, **44**, 929–935.

ARMSTRONG, R. L. 1991. The persistent myth of crustal growth. *Australian Journal of Earth Sciences*, **38**, 613–630.

BAKER, E. T., MASSOTH, G. J., NAKAMURA, K. I., EMBLEY, R. W., DE RONDE, C. E. J. & ARCULUS, R. J.

2005. Hydrothermal activity on near-arc sections of back-arc ridges: results from the Mariana Trough and Lau Basin. *Geochemistry, Geophysics, Geosystems*, **6**, doi:10.1029/2005GC000948.

BAKER, E. T., EMBLEY, R. W. ET AL. 2008. Hydrothermal activity and volcano distribution along the Mariana arc. *Journal of Geophysical Research, Solid Earth*, **113**, doi:10.1029/2007JB005423.

BASALTIC VOLCANISM STUDY PROJECT (BVSP) 1981. Ocean-floor basaltic volcanism. *In*: BVSP (eds) *Basaltic Volcanism on the Terrestrial Planets*. Pergamon, Oxford, 132–160.

BEHN, M. D. & KELEMEN, P. B. 2006. Stability of arc lower crust: insights from the Talkeetna arc section, south central Alaska, and the seismic structure of modern arcs. *Journal of Geophysical Research, Solid Earth*, **111**, doi:10.1029/2006JB004327.

BLOOMER, S. H. & HAWKINS, J. W. 1983. Gabbroic and ultramafic rocks from the Mariana Trench: an island arc ophiolite. *In*: HAYES, D. E. (ed.) *The Tectonic and Geologic Evolution of Southeast Asian Seas and Islands: Part 2*. American Geophysical Union, Geophysical Monograph, **27**, 294–317.

BONATTI, E. & MICHAEL, P. J. 1989. Mantle peridotites from continental rifts to ocean basins to subduction zones. *Earth and Planetary Science Letters*, **91**, 297–311.

BROPHY, J. G. 2008. A study of Rare Earth Element (REE) – SiO_2 variations in felsic liquids generated by basalt fractionation and amphibolite melting: a potential test for discriminating between the two different processes. *Contributions to Mineralogy and Petrology*, **156**, 337–357.

CAGNIONCLE, A. M., PARMENTIER, W. M. & ELKINS-TANTON, L. T. 2007. Effect of solid flow above a subducting slab on water distribution and melting at convergent plate boundaries, *Journal of Geophysical Research, Solid Earth*, **112**, doi:10.1029/2007/JB004934.

CALVERT, A. J., KLEMPERER, S. L., TAKAHASHI, N. & KERR, B. C. 2008. Three-dimensional crustal structure of the Mariana island arc from seismic tomography. *Journal of Geophysical Research, Solid Earth*, **113**, doi:10.1029/2007JB004939.

CARR, M. J., SAGINOR, I. ET AL. 2007. Element fluxes from the volcanic front of Nicaragua and Costa Rica. *Geochemistry, Geophysics, Geosystems*, **8**, Q06001, doi:10.1029/2006GC001396.

CHAPP, E., TAYLOR, B., OAKLEY, A. & MOORE, G. F. 2008. A seismic stratigraphic analysis of Mariana forearc basin evolution. *Geochemistry, Geophysics, Geosystems*, **9**, doi:10.1029/2008GC001998.

CHRISTIENSEN, N. J. & MOONEY, W. D. 1995. Seismic velocity structure and composition of the continental crust: a global view. *Journal of Geophysical Research, Solid Earth*, **100**, 9761–9788.

CLIFT, P. & VANNUCCHI, P. 2004. Controls on tectonic accretion versus erosion in subduction zones: implications for the origins and recycling of the continental crust. *Reviews of Geophysics*, **42**, doi:10.1029/2003RG000127.

CLIFT, P. D., DRAUT, A. E., KELEMEN, P. B., BLUSZTAJN, J. & GREENE, A. 2005. Stratigraphic and geochemical evolution of an oceanic arc upper crustal section: the Jurassic Talkeetna Volcanic Formation, south-central Alaska. *Geological Society of America Bulletin*, **117**, 902–925.

COATS, R. R. 1962. Magma type and crustal structure in the Aleutian Arc. *In*: MCDONALD, G. A. & KUNO, H. (eds) *Crust of the Pacific Basin*. American Geophysical Union, Geophysical Monograph, **6**, 92–109.

CONDER, J. A., WIENS, D. A. & MORRIS, J. 2002. On the decompression melting structure at volcanic arcs and back-arc spreading centers. *Geophysical Research Letters*, **29**, doi:10.1029/2002GL015390.

D'ARS, J. B., JAUPART, C. & SPARKS, R. S. J. 1995. Distribution of volcanoes in active margins. *Journal of Geophysical Research, Solid Earth*, **100**, 20 421–20 432.

DAVIDSON, J. P. & ARCULUS, R. J. 2005. The significance of Phanerozoic arc magmatism in generating continental crust. *In*: BROWN, M. & RUSHMER, T. (eds) *Evolution and Differentiation of the Continental Crust*. Cambridge University Press, Cambridge, 135–172.

DAVIS, A. S. & CLAGUE, D. A. 1987. Geochemistry, mineralogy, and petrogenesis of basalt from the Gorda Ridge. *Journal of Geophysical Research, Solid Earth*, **92**, 10 476–10 483.

DEBARI, S. M., TAYLOR, B., SPENCER, K. & FUJIOKA, K. 1999. A trapped Philippine Sea plate origin for MORB from the inner slope of the Izu–Bonin trench. *Earth and Planetary Science Letters*, **174**, 183–197.

DEFANT, M. J. & DRUMMOND, M. S. 1990. Derivation of some modern arc magmas by melting of young subducted lithosphere. *Nature*, **347**, 662–665.

DE RONDE, C. E. J., MASSOTH, G. J., BAKER, E. T. & LUPTON, J. E. 2003. Submarine hydrothermal venting related to volcanic arcs. *In*: SIMMONS, S. F. & GRAHAM, I. (eds) *Volcanic, Geothermal, and Ore-forming Fluids: Rulers and Witnesses of Processes within the Earth*. Society of Economic Geologists Special Publication, **10**, 91–110.

DEWEY, J. F. & WINDLEY, B. F. 1981. Growth and differentiation of the continental crust. *Philosophical Transactions of the Royal Society of London, Series A*, **301**, 189–206.

DHUIME, B., BOSCH, D., BODINIER, J.-L., GARRIDO, C. J., BRUGUIER, O., HUSSAIN, S. S. & DAWOOD, H. 2007. Multistage evolution of the Jijal ultramafic–mafic complex (Kohistan, N Pakistan): implications for building the roots of island arcs. *Earth and Planetary Science Letters*, **261**, 179–200.

DICK, H. J. B. & BULLEN, T. 1984. Chromian spinels as a petrogenetic indicator in abyssal and alpine-type peridotites and spatially associated lavas. *Contributions to Mineralogy and Petrology*, **86**, 54–76.

DICKINSON, W. R. 1975. Potash–depth (K–h) relations in continental margins and intra-oceanic magmatic arcs. *Geology*, **3**, 53–56.

DILEK, Y. 2003. Ophiolite concept and its evolution. *In*: DILEK, Y. & NEWCOMB, S. (eds) *Ophiolite Concept and the Evolution of Geological Thought*. Geological Society of America, Special Papers, **373**, 1–16.

DIMALANTA, C., TAIRA, A., YUMUL, G. P. JR, TOKUYAMA, H. & MOCHIZUKI, K. 2002. New rates of western Pacific island arc magmatism from seismic and gravity data. *Earth and Planetary Science Letters*, **202**, 105–115.

EMBLEY, R. W., CHADWICK, W. W., STERN, R. J., MERLE, S. G., BLOOMER, S. H., NAKAMURA, K. & TAMURA, Y. 2006. A synthesis of multibeam bathymetry and backscatter, and sidescan sonar of the Mariana submarine magmatic arc (abstract). *EOS Transactions, American Geophysical Union*, **87**, V41B-1723.

ENGLAND, P., ENGDAHL, R. & THATCHER, W. 2004. Systematic variation in the depths of slabs beneath arc volcanoes. *Geophysical Journal International*, **156**, 377–408.

FISHER, R. L. & ENGEL, C. G. 1969. Ultramafic and basaltic rocks dredged from the nearshore flank of the Tonga Trench. *Geological Society of America Bulletin*, **80**, 1136–1141.

FISKE, R. S., NAKA, J., IIZASA, K., YUASA, M. & KLAUS, A. 2001. Submarine silicic caldera at the front of the Izu–Bonin Arc, Japan: voluminous seafloor eruptions of rhyolite pumice. *Geological Society of America Bulletin*, **113**, 813–824.

FLIEDNER, M. M. & KLEMPERER, S. L. 2000. The transition from oceanic arc to continental arc in the crustal structure of the Aleutian Arc. *Earth and Planetary Science Letters*, **179**, 567–579.

GARCIA, M. O., PIETRUSKA, A. J., RHODES, J. M. & SWANSON, K. 2000. Magmatic processes during the prolonged Pu'u O'o eruption of Kilauea volcano, Hawaii. *Journal of Petrology*, **41**, 967–990.

GARFUNKEL, Z., ANDERSON, C. A. & SCHUBERT, G. 1986. Mantle circulation and the lateral migration of subducted slabs. *Journal of Geophysical Research, Solid Earth*, **91**, 7205–7223.

GARRIDO, C. J., BODINIER, J.-L. ET AL. 2006. Petrogenesis of mafic garnet granulite in the lower crust of the Kohistan paleo-arc complex (northern Pakistan): implications for intra-crustal differentiation of island arcs and generation of continental crust. *Journal of Petrology*, **47**, 1873–1914.

GRAHAM, I. J., REYES, A. G., WRIGHT, I. C., PECKETT, K. M., SMITH, I. E. M. & ARCULUS, R. J. 2008. Structure and petrology of newly discovered volcanic centers in the northern Kermadec–southern Tofua arc, South Pacific Ocean. *Journal of Geophysical Research, Solid Earth*, **113**, doi:10.1029/2007/JB005453.

HACKER, B. P., PEACOCK, S. M., ABERS, G. A. & HOLLOWAY, S. D. 2003. Subduction factory 2. Are intermediate-depth earthquakes in subducting slabs linked to metamorphic dehydration reactions? *Journal of Geophysical Research, Solid Earth*, **108**, doi:10.1029/2001JB001129.

HACKER, B. R., MEHL, L., KELEMEN, P. B., RIOUX, M., BEHN, M. D. & LUFFI, P. 2008. Reconstruction of the Talkeetna intraoceanic arc of Alaska through thermobarometry. *Journal of Geophysical Research, Solid Earth*, **113**, doi:10.1029/2007JB005208.

HAMILTON, W. 2007. Driving mechanism and 3-D circulation of plate tectonics. *In*: SEARS, J. W., HARMS, T. A. & EVENCHICK, C. A. (eds) *Whence the Mountains? Inquiries into the Evolution of Orogenic Systems: A Volume in Honor of Raymond A. Price*. Geological Society of America, Special Papers, **433**, 1–25.

HAMILTON, W. B. 1988. Plate tectonics and island arcs. *Geological Society of America Bulletin*, **100**, 1503–1527.

HAWKESWORTH, C. J. & KEMP, A. I. S. 2006. Evolution of the continental crust. *Nature*, **443**, 811–817.

HAWKINS, J. W., BLOOMER, S. H., EVANS, C. A. & MELCHIOR, J. T. 1984. Evolution of intra-oceanic arc–trench systems. *Tectonophysics*, **102**, 175–205.

HERZBURG, C. 2004. Geodynamic information in peridotite petrology. *Journal of Petrology*, **45**, 2507–2530.

HOFMANN, A. W. 1988. Chemical differentiation of the Earth: the relationship between mantle, continental crust, and oceanic crust. *Earth and Planetary Science Letters*, **90**, 297–314.

HOLBROOK, W. S., LIZARRALDE, D., MCGEARY, S., BANGS, N. & DEIBOLD, J. 1999. Structure and composition of the Aleutian island arc and implications for continental crustal growth. *Geology*, **27**, 31–34.

JICHA, B. R., SCHOLL, D. W., SINGER, B. S., YOGODZINSKI, G. M. & KAY, S. M. 2006. Revised age of Aleutian Island Arc formation implies high rate of magma production. *Geology*, **34**, 661–664.

JOLLY, W. T., LIDIAK, E. G., DICKIN, A. P. & WU, T.-W. 2001. Secular geochemistry of central Puerto Rican island arc lavas: constraints on Mesozoic tectonism in the eastern Greater Antilles. *Journal of Petrology*, **42**, 2197–2214.

JULL, M. & KELEMEN, P. B. 2001. On the conditions for lower crustal convective instability. *Journal of Geophysical Research, Solid Earth*, **106**, 6423–6446.

KAMIMURA, A., KASAHARA, J., SHINOHARA, M., HINO, R., SHIOBARA, H., FUJIE, G. & KANAZAWA, T. 2002. Crustal structure study at the Izu–Bonin subduction zone around 31°N: implications of serpentinized materials along the subduction plate boundary. *Physics of the Earth and Planetary Interiors*, **132**, 105–129.

KARIG, D. E. 1972. Remnant arcs. *Geological Society of America Bulletin*, **87**, 1057–1068.

KELEMEN, P. B. 1995. Genesis of high Mg# andesites and the continental crust. *Contributions to Mineralogy and Petrology*, **120**, 1–19.

KELEMEN, P. B., HANGHOJ, K. & GREENE, A. R. 2003. One view of the geochemistry of subduction-related magmatic arcs, with an emphasis on primitive andesite and lower crust. *In*: RUDNICK, R. L. & TUREKIAN, K. K. (eds) *Treatise of Geochemistry, 3*. Elsevier, Amsterdam, 593–659.

KIMURA, J.-I. & STERN, R. J. 2008. Neogene volcanism of the Japan island arc: The K–h relationship revisited. *In*: SPENCER, J. E. & TITLEY, S. R. (eds) *Circum-Pacific Tectonics, Geologic Evolution, and Ore Deposits*. Arizona Geological Society, Tucson, Digest, **22**, 187–202.

KIMURA, J.-I. & YOSHIDA, T. 2006. Contributions of slab fluid, mantle wedge and crust to the origin of Quaternary lavas in the NE Japan Arc. *Journal of Petrology*, **47**, 2185–2232.

KODAIRA, S., SATO, T., TAKAHASHI, N., ITO, A., TAMURA, Y., TATSUMI, Y. & KANEDA, Y. 2007. Seismological evidence for variable growth of crust along the Izu intra-oceanic arc. *Journal of Geophysical Research, Solid Earth*, **112**, doi:10.1029/2006JB004593.

KUSKY, T. M., GLASS, A. & TUCKER, R. 2007. Structure, Cr-chemistry, and age of the Border Ranges ultramafic complex: a suprasubduction zone ophiolite. *In*: RIDGEWAY, K. D., TROP, J. M., GLEN, J. M. G. & O'NEILL, M. O. (eds) *Tectonic Growth of a Collisional Continental Margin: Crustal Evolution of Southern Alaska*. Geological Society of America, Special Papers, **431**, 207–225.

MARSAGLIA, K. 1995. Interarc and backarc basins. *In*: BUSBY, C. & INGERSOLL, R. V. (eds) *Tectonics of Sedimentary Basins*. Blackwell, Oxford, 299–329.

MARTINEZ, F. & TAYLOR, B. 2003. Controls on back-arc crustal accretion: insights from the Lau, Manus and Mariana basins. *In*: LARTER, R. D. & LEAT, P. T. (eds) *Intra-Oceanic Subduction Systems: Tectonic and Magmatic Processes*. Geological Society, London, Special Publications, **219**, 19–54.

MARTINEZ, F. & TAYLOR, B. 2006. Modes of crustal accretion in back arc basins: inferences from the Lau Basin. *In*: CHRISTIE, D. M., FISHER, C. R., LEE, S.-M. & GIVENS, S. (eds) *Back-Arc Spreading Systems: Geological, Biological, Chemical and Physical Interactions*. American Geophysical Union, Geophysical Monograph, **166**, 5–30.

NAKAJIMA, K. & ARIMA, M. 1998. Melting experiments on hydrous low-K tholeiite: implications for the genesis of tonalitic crust in the Izu–Bonin–Mariana arc. *Island Arc*, **7**, 359–373.

OAKLEY, A. J., TAYLOR, B., FRYER, P., MOORE, G. F., GOODLIFFE, A. M. & MORGAN, J. K. 2007. Emplacement, growth, and gravitational deformation of serpentinite seamounts on the Mariana forearc. *Geophysical Journal International*, **170**, 615–634.

OHARA, Y. 2006. Mantle process beneath Philippine Sea back-arc spreading ridges: a synthesis of peridotite petrology and tectonics. *Island Arc*, **15**, 119–129.

OHARA, Y., STERN, R. J., ISHII, T., YURIMOTO, H. & YAMAZAKI, T. 2002. Peridotites from the Mariana Trough backarc basin. *Contributions to Mineralogy and Petrology*, **143**, 1–18.

OKAMURA, H., ARAI, S. & KIM, Y. U. 2006. Petrology of forearc peridotite from the Hahajima Seamount, the Izu–Bonin arc, with special reference to the chemical characteristics of chromian spinel. *Mineralogical Magazine*, **70**, 15–26.

PARKINSON, I. J. & PEARCE, J. A. 1998. Peridotites from the Izu–Bonin–Mariana forearc (ODP Leg 125): evidence for mantle melting and melt-mantle interaction in a supra-subduction zone setting. *Journal of Petrology*, **39**, 1577–1618.

PEARCE, J. A. 2003. Subduction zone ophiolites: the search for modern analogues. *In*: DILEK, Y. & NEWCOMB, S. (eds) *Ophiolite Concept and the Evolution of Geological Thought*. Geological Society of America, Special Papers, **373**, 269–294.

PEARCE, J. A. & STERN, R. J. 2006. The origin of back-arc basin magmas: trace element and isotope perspectives. *In*: CHRISTIE, D. M., FISHER, C. R., LEE, S.-M. & GIVENS, S. (eds) *Back-Arc Spreading Systems: Geological, Biological, Chemical and Physical Interactions*. American Geophysical Union, Geophysical Monograph, **166**, 63–86.

PEARCE, J. A., BARKER, P. F., EDWARDS, S. J., PARKINSON, I. J. & LEAT, P. T. 2000. Geochemistry and tectonic significance of peridotites from the South Sandwich arc–basin system, South Atlantic. *Contributions to Mineralogy and Petrology*, **139**, 36–53.

PLANK, T. & LANGMUIR, C. H. 1988. An evaluation of the global variations in the major element chemistry of arc basalts. *Earth and Planetary Science Letters*, **90**, 349–370.

REAGAN, M. K., ISHIZUKA, O. *ET AL*. 2010. Fore-arc basalts and subduction initiation in the Izu–Bonin–Mariana system. *Geochemistry, Geophysics, Geosystems*, **11**, Q03X12, doi: 10.1029/2009GC002871.

RUDNICK, R. A. & GAO, S. 2003. Composition of the continental crust. *In*: RUDNICK, R. L. & TUREKIAN, K. K. (eds) *Treatise on Geochemistry, 3*. Elsevier, Amsterdam, 1–64.

SCHOLL, D. W. & VON HUENE, R. 2007. Crustal recycling at modern subduction zones applied to the past – issues of growth and preservation of continental basement, mantle geochemistry, and supercontinent reconstruction. *In*: HATCHER, R. D., CARLSON, M. P., MCBRIDE, J. H. & CATALAN, J. M. (eds) *The 4D Framework of Continental Crust*. Geological Society of America, Memoirs, **200**, 9–32.

SCHOLL, D. W. & VON HUENE, R. 2009. Implications of estimated magmatic additions and recycling losses at the subduction zones of accretionary (non-collisional) and collisional (suturing) orogens. *In*: CAWOOD, P. & KRÖNER, A. (eds) *Earth Accretionary Orogens in Space and Time*. Geological Society, London, Special Publications, **318**, 105–125.

SHAW, C. A., HEIZLER, M. T. & KARLSTROM, K. E. 2005. ^{40}Ar/^{39}Ar thermochronologic record of 1.45–1.35 Ga intracontinental tectonism in the southern Rocky Mountains: interplay of conductive and advective heating with intracontinental deformation. *In*: KARLSTROM, K. E. & KELLER, G. R. (eds) *The Rocky Mountain Region – An Evolving Lithosphere: Tectonics, Geochemistry, and Geophysics*. American Geophysical Union, Geophysical Monograph, **154**, 163–184.

SHILLINGTON, D. J., VAN AVENDONK, H. J. A., HOLBROOK, W. S., KELEMEN, P. B. & HORNBACH, M. J. 2004. Composition and structure of the central Aleutian island arc from arc-parallel wide-angle seismic data. *Geochemistry, Geophysics, Geosystems*, **5**, doi:10.1029/2004GC000715.

SHIPBOARD SCIENTIFIC PARTY 1982*a*. Site 455: east side of the Mariana Trough. *In*: HUSSONG, D. M., UYEDA, S. *ET AL*. (eds) *Initial Reports of the Deep Sea Drilling Project, Leg 60*. US Government Printing Office, Washington, DC, 203–213.

SHIPBOARD SCIENTIFIC PARTY 1982*b*. Site 458: Mariana fore-Arc. *In*: HUSSONG, D. M., UYEDA, S. *ET AL*. (eds) *Initial Reports of the Deep Sea Drilling Project, Leg 60*. US Government Printing Office, Washington, DC, 263–307.

STERN, R. J. 2002. Subduction zones. *Reviews of Geophysics*, **40**, doi:10.1029/2001RG000108.

STERN, R. J. 2004. Subduction initiation: spontaneous and induced. *Earth and Planetary Science Letters*, **226**, 275–292.

STERN, R. J. & BLOOMER, S. H. 1992. Subduction zone infancy: examples from the Eocene Izu–Bonin–Mariana and Jurassic California. *Geological Society of America Bulletin*, **104**, 1621–1636.

STERN, R. J. & SCHOLL, D. W. 2010. Yin and Yang of continental crust creation and destruction by plate tectonics. *International Geology Review*, **52**, 1–31.

STERN, R. J. & SMOOT, N. C. 1998. A bathymetric overview of the Mariana forearc. *Island Arc*, **7**, 525–540.

STERN, R. J., FOUCH, M. J. & KLEMPERER, S. 2003. An overview of the Izu–Bonin–Mariana subduction factory. *In*: EILER, J. (ed.) *Inside the Subduction*

Factory. American Geophysical Union, Geophysical Monograph, **138**, 175–222.

STERN, R. J., KOHUT, E., BLOOMER, S. H., LEYBOURNE, M., FOUCH, M. & VERVOORT, J. 2006. Subduction factory processes beneath the Guguan Cross-chain, Mariana Arc: no role for sediments, are serpentinites important? *Contributions to Mineralogy and Petrology*, **151**, 202–221.

STRAUB, S. M. 2003. The evolution of the Izu–Bonin–Mariana volcanic arcs (NW Pacific) in terms of major element chemistry. *Geochemistry, Geophysics, Geosystems*, **4**, doi:10.1029/2002GC000357.

SUYEHIRO, K., TAKAHASHI, N. ET AL. 1996. Continental crust, crustal underplating, and low-Q upper mantle beneath an oceanic island arc. *Science*, **272**, 390–392.

SYRACUSE, E. M. & ABERS, G. A. 2006. Global compilation of variations in slab depth beneath arc volcanoes and implications. *Geochemistry, Geophysics, Geosystems*, **7**, doi:10.1029/2005GC001045.

TAKAHASHI, N., KODAIRA, S., KLEMPERER, S. L., TATSUMI, Y., KANEDA, Y. & SUYEHIRO, K. 2007. Crustal structure and evolution of the Mariana intra-oceanic island arc. *Geology*, **35**, 203–206.

TAKAHASHI, N., KODAIRA, S., TASUMI, Y., KANEDA, Y. & SUYEHIRO, K. 2008. Structure and growth of the Izu–Bonin–Mariana arc crust: 1. Seismic constraint on crust and mantle structure of the Mariana arc–back-arc system. *Journal of Geophysical Research, Solid Earth*, **113**, doi:10.1029/2007JB005120.

TATSUMI, Y. 2005. The subduction factory: how it operates in the evolving Earth. *GSA Today*, **15**, 4–10.

TATSUMI, Y., SHUKUNO, H., TANI, K., TAKAHASHI, N., KODAIRA, S. & KOGISO, T. 2008. Structure and growth of the Izu–Bonin–Mariana arc crust: 2. Role of crust–mantle transformation and the transparent Moho in arc crust evolution. *Journal of Geophysical Research, Solid Earth*, **113**, doi:10.1029/2007JB005121.

WINDLEY, B. F. 1993. Uniformitarianism today: plate tectonics is the key to the past. *Journal of the Geological Society, London*, **150**, 7–19.

ZHAO, D., HASEGAWA, A. & KANAMORI, H. 1994. Deep structure of Japan subduction zone as derived from local, regional, and teleseismic events. *Journal of Geophysical Research, Solid Earth*, **99**, 22 313–22 329.

Transitions among Mariana-, Japan-, Cordillera- and Alaska-type arc systems and their final juxtapositions leading to accretionary and collisional orogenesis

WENJIAO XIAO[1]*, CHUNMING HAN[1], CHAO YUAN[2], MIN SUN[3], GUOCHUN ZHAO[3] & YEHUA SHAN[2]

[1]*State Key Laboratory of Lithospheric Evolution, Institute of Geology and Geophysics, Chinese Academy of Sciences, Beijing 100029, China*

[2]*Guangzhou Institute of Geochemistry, Chinese Academy of Sciences, Guangzhou 510640, China*

[3]*Department of Earth Sciences, The University of Hong Kong, Hong Kong, China*

Corresponding author (e-mail: wj-xiao@mail.igcas.ac.cn)

Abstract: 'Arc system' is used here as a collective term for a variety of arcs that occur along continental margins or in oceanic plates; it includes associated units from adjacent plates. Four major arc systems (Mariana-, Japan-, Cordillera- and Alaska-type) can be distinguished along the Circum-Pacific region. Some Japan-type arc systems in ancient orogens (e.g. the Altaids) may have been largely regarded as microcontinents because they have so-called Precambrian basement. Often the Cordillera-type arc systems can be very complicated, and if they are rifted away from the host continent they become more difficult to recognize. Commonly these arc systems interact mutually and with continental marginal sequences, leading to complicated accretionary and collisional orogens. The alternation between Western Pacific archipelagos and the Eastern Pacific active margin is the stereotype of accretionary and collisional orogenesis. More importantly, these four main types of arc systems can be juxtaposed into a final orogenic collage, which is another main expression of accretionary orogenesis. Only some parts of accretionary and collisional orogens can be terminated by attachment of a continent-size craton such as Tarim or even India, and even so the accretionary and collisional processes may continue elsewhere along strike. The significance of the interactions among these arc systems and their final juxtaposition has not been fully appreciated in ancient orogens. The Altaids together with the Circum-Pacific orogens offers a good opportunity to study such accretionary–collisional orogenesis.

Orogens on Earth are subdivided into two types: collisional and accretionary (Windley 1995, 1998; Cawood *et al.* 2009). However, to distinguish between these two types of orogenesis is sometimes not easy, as each includes some aspects of the other. Collisional orogens have long been a major target of investigations, with the Tethys–Himalaya and Appalachian orogens as the classic examples, but even these collisional orogens also show a long history of accretion before final collision (Dewey 1969; Allègre *et al.* 1984; Chang *et al.* 1986, 1989; Bradley 1989; Grapes & Watanabe 1992; Van Staal 1994; Yin & Harrison 2000; Liou *et al.* 2004; Ratschbacher *et al.* 2004; Aikman *et al.* 2008; Santosh *et al.* 2009). In the mean time, accretionary orogens such as the North American Cordillera and the Altaids also have long been regarded as major sites with complicated accretionary and collisional geodynamic processes (Fig. 1; von Huene & Scholl 1991; Şengör *et al.* 1993; Kimura 1994; Kusky 1997; Condie 2000; Jahn *et al.* 2000; Van der Voo 2004; Kröner *et al.* 2007).

Accretionary orogens are increasingly noted for their considerable continental lateral enlargement and vertical growth with world-class metallogeny (Sillitoe 1974; Heinhorst *et al.* 2000; Jahn *et al.* 2000; Jenchuraeva 2001; Yakubchuk *et al.* 2001; Gray *et al.* 2002; Goldfarb *et al.* 2003; Xiao *et al.* 2008*a*). Therefore accretionary orogens actually have full records that can be studied to better understand the geodynamic evolution of mountain ranges on Earth. They have been the subject of many international efforts worldwide (Cawood & Buchan 2007; Condie 2007; Brown 2009; Cawood *et al.* 2009; Hall 2009), including international programmes such as the International Lithosphere Program (ILP), International Geological Correlation Programme (IGCP), and recently launched joint projects from European, American, and Asian

From: KUSKY, T. M., ZHAI, M.-G. & XIAO, W. (eds) *The Evolving Continents: Understanding Processes of Continental Growth.* Geological Society, London, Special Publications, **338**, 35–53.
DOI: 10.1144/SP338.3   0305-8719/10/$15.00 © The Geological Society of London 2010.

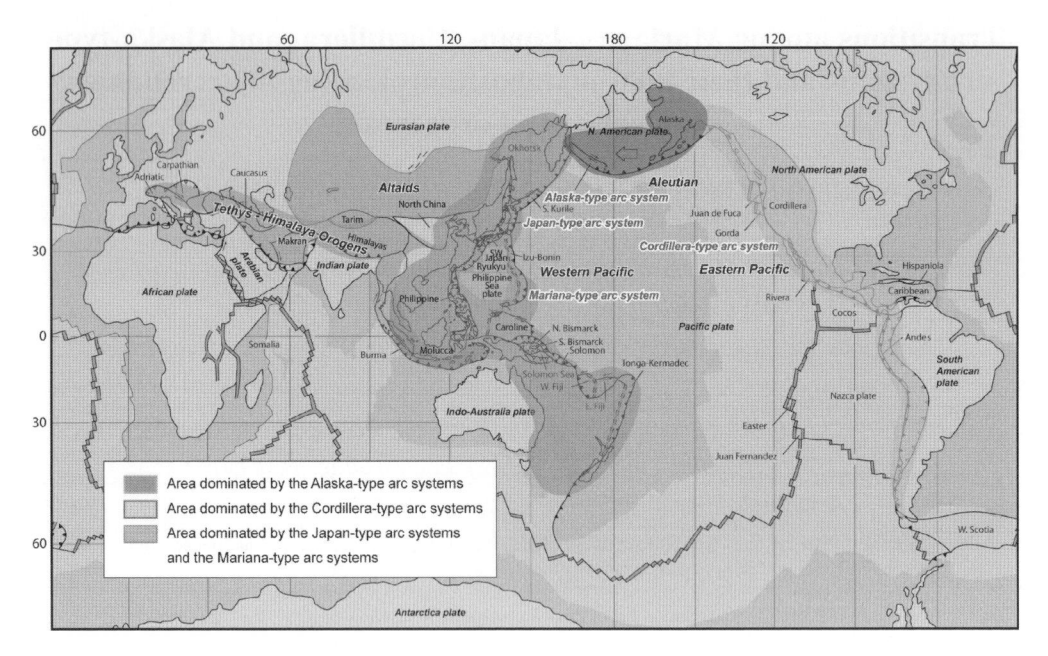

Fig. 1. Schematic tectonic map of orogens around the world and the circum-Pacific regions (Tagami & Hasebe 1999; Kröner et al. 2007; Rino et al. 2008).

countries (Hall & Spakman 2002; Pfänder et al. 2002; Tomurtogoo et al. 2005; Helo et al. 2006; Cawood & Buchan 2007; Condie 2007; Kröner et al. 2007; Windley et al. 2007; Brown 2009; Cawood et al. 2009; Hall 2009; Xiao et al. 2009a, c).

However, there is still controversy about the tectonic architecture and evolution of accretionary orogens, for which there are two major schools of thought: amalgamation of multiple terrane–ocean systems or accretion in terms of a single forearc (Ingersoll & Schweickert 1986; Mossakovsky et al. 1993; Şengör et al. 1993; Busby & Ingersoll 1995; Dickinson 1995; Şengör & Natal'in 1996a, b; Yakubchuk 2004, 2008). The focus of much controversy has centred on the recognition and interpretation of high-grade metamorphic rocks in some certain terranes. The proponents of multiple terrane–ocean systems have interpreted some high-grade gneiss–schist complexes as microcontinents (Chamberlain & Lambert 1985; Mossakovsky et al. 1993; Salnikova et al. 2001; Vaughan & Livermore 2005; Kozakov et al. 2007), and this model would necessitate some degree of collision (Mossakovsky et al. 1993). On the other hand, some high-grade gneiss–schist complexes have been interpreted as the basement or components of arcs and accretionary complexes, which emphasizes the role of accretion (Haeussler et al. 1995; Kozakov et al. 1999; Kuzmichev et al. 2001; Salnikova et al. 2001).

The Altaids of Central Asia and the Circum-Pacific including the Cordillera of Western North America are among the largest accretionary orogens (Fig. 1) that record considerable Phanerozoic continental growth (Coleman 1989; Şengör & Okurogullari 1991; Mascle et al. 1996; Condie 2000; Jahn et al. 2004; Van der Voo 2004). Both offer an opportunity to unravel the basic evolutionary history of accretionary orogens and they address the above controversy. This paper presents a comparative study of the palaeogeography and evolution of these orogens.

Typical arc systems

Arcs are tectonic belts formed by subduction of a plate of oceanic lithosphere beneath another oceanic or continental plate along a subduction zone with high seismic activity characterized by a high heat flow with active volcanoes bordered by a submarine trench (Windley 1995). As nearly all accretionary orogens are directly or indirectly related to subduction of oceanic plate(s), which generates different arcs, we define an arc system as a collective term for subduction-related tectonic settings along either continental or oceanic plate. We introduce this term because all accretionary orogens occur finally outboard of the margins of a host continent (Fig. 2; Coney et al. 1980; Cordey & Schiarizza 1993; Church et al. 1995; McClelland et al. 2000; Nance et al. 2002; Bradley et al. 2003; Stern 2004; Ring 2008) and accretionary orogenesis has a close relationship with the formation and

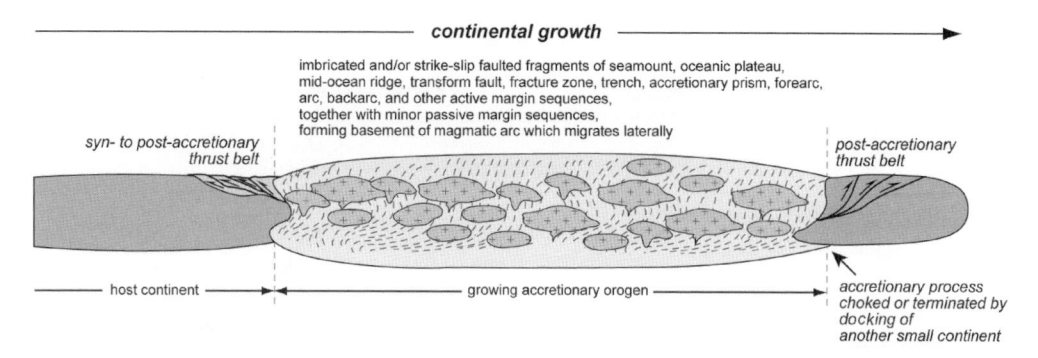

Fig. 2. A cross-sectional model of classic accretionary and collisional orogenesis (modified after Şengör 1992; Şengör et al. 1993).

evolution of supercontinents (Ilyin 1990; Dalziel et al. 2000; Zhao et al. 2002; Cawood & Buchan 2007; Murphy et al. 2010). Continental margins play a key role in accretionary orogenesis, and understanding of arc systems helps in unravelling the evolution of accretionary orogens.

Normally when arcs are studied the subduction zone and arc volcanic systems are mostly emphasized; however, the forearc and back-arc regimes are essential components of arc systems (Windley 1995). The back-arc regimes vary and sometimes have more remnant back-arcs and include the opposite side of passive margins and/or the forearc of another arc system (Bloomer et al. 1995; Hawkins 1995, 2003; Windley 1995). Other important features include the forearc accretionary complex and even granulite-facies rocks in the deep roots of mature arcs, and some important mineral deposits (Windley 1995). In the present-day circum-Pacific region, four major arc systems (Mariana-, Japan-, Cordillera- and Alaska-type) can be distinguished (Figs 3 & 4).

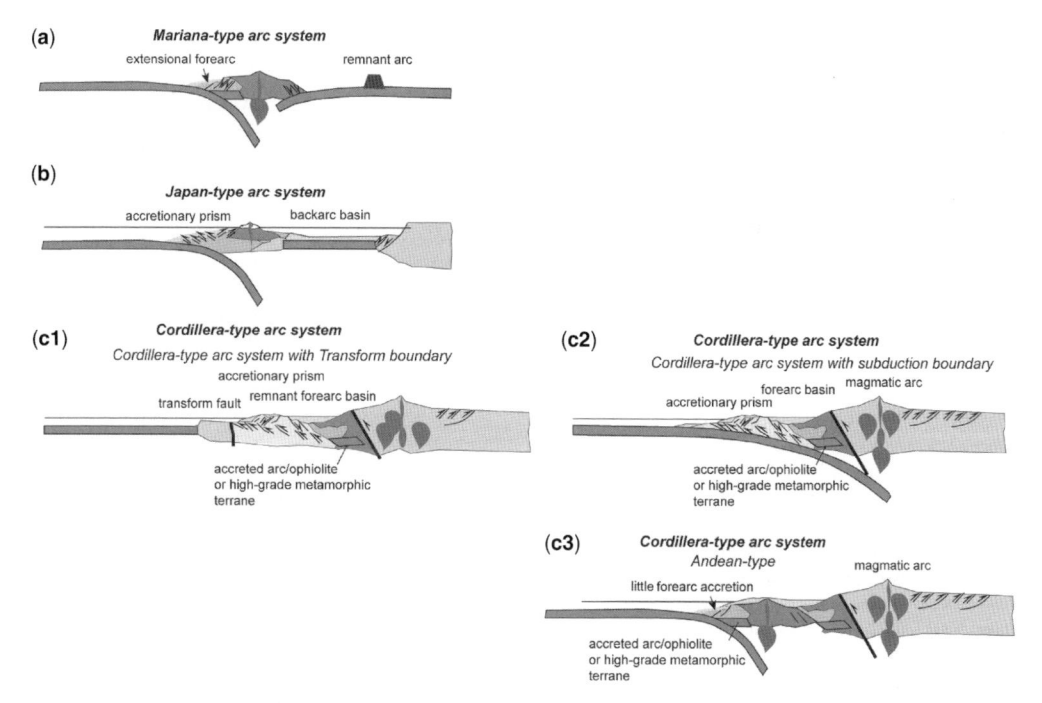

Fig. 3. Diagrams showing (**a**) Mariana-, (**b**) Japan-, and (**c**) Cordillera-types of arc systems related to accretionary and collisional orogens. It should be noted that the Cordillera-type arc system has three sub-types: (**c1**) Cordillera-type I with a transform boundary; (**c2**) Cordillera-type II with a subduction boundary; (**c3**) Cordillera-type III, the Andean-type, with little forearc accretion.

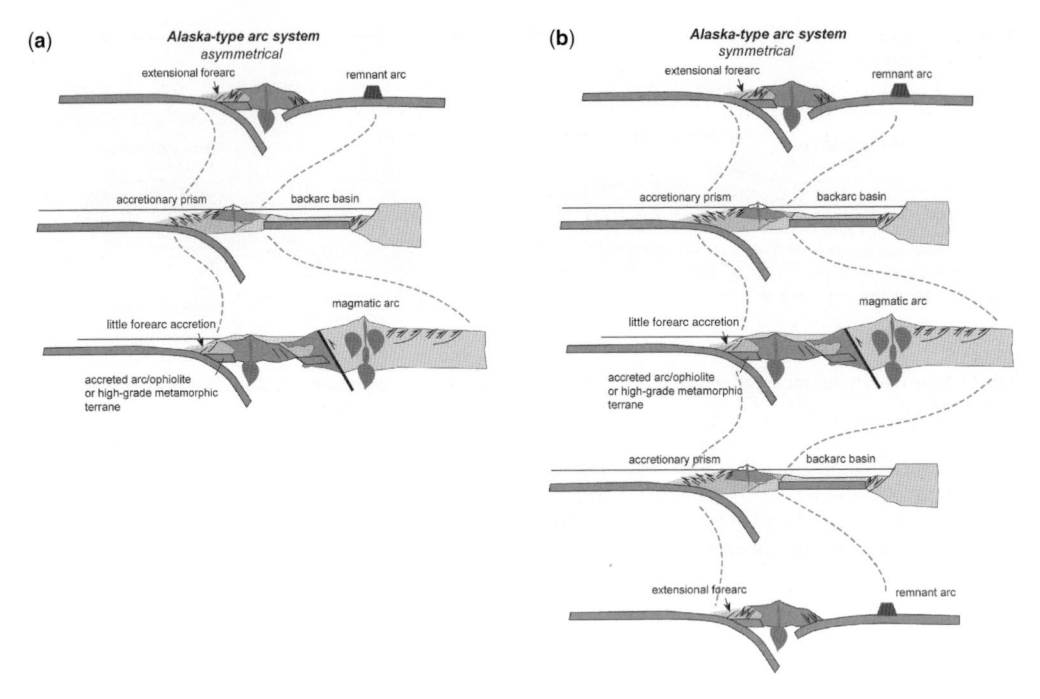

Fig. 4. Diagrams showing Alaska-type arc system related to accretionary and collisional orogens. It should be noted that the Alaska-type arc system has two cases, asymmetrical (**a**) and symmetrical (**b**), and that assemblage of these island arcs can be random.

Mariana-type arc system

The Mariana-type arc system, also termed the Izu–Bonin–Mariana arc system, consists of the Izu–Bonin arc in the north and the Mariana arc in the south (Figs 1 & 3a); it was generated by deep subduction of the Pacific Ocean plate beneath the Philippine Sea plate (Karig 1971; Miller *et al.* 2004). The Philippine Sea plate has moved northwestward and the Pacific slab has undergone possible rollback at the Mariana subduction zone (Deschamps & Fujiwara 2003).

This type of arc system is generated by subduction of one oceanic plate beneath another, characterized by an oceanic island arc, remnant arc(s) and back-arc basin(s) (Karig 1971) that are wide and separated by a remnant arc. In places there is only one back-arc basin. Partial melting of diapirs has produced voluminous mid-ocean ridge basalt (MORB)-type tholeiites, which have split the arc and formed the crust of a widening back-arc basin (Crawford *et al.* 1981; Deschamps & Fujiwara 2003; Gvirtzman & Stern 2004; Miller *et al.* 2004). Therefore, this type of arc system has no significant contribution of continental material during its formation.

Minor sediment accretion occurred along the Mariana-type arc system (Horine *et al.* 1990). The composition of dredged samples from the slopes of the Mariana trench (i.e. island arc tholeiites and boninites) suggests that there is little or no current sediment and/or there has been no oceanic crustal accretion in the Mariana system (Bloomer *et al.* 1995; Plank & Langmuir 1998; Fryer *et al.* 1999). A 9 km schematic section drawn by Hawkins (2003) shows that the rock types on the inner wall of the Mariana Trench are, from top to bottom, volcaniclastic rocks, serpentinites, serpentinized harzburgites, diabases–gabbros, basalts, boninites, quartz diorites, serpentinized harzburgites, basalts, gabbros, diabases, and serpentinized harzburgites. These lithologies are not the same as the typical ophiolitic sequences described in textbooks and illustrate the complications of Mariana-type arc systems. Blueschists also occur in the forearc, implying deep subduction and exhumation (Fryer *et al.* 1999). Usually these kinds of arc systems would generate ophiolites of suprasubduction-zone types (Hawkins 2003; Shervais *et al.* 2004). One can imagine how complicated the situation could be when such rocks of the Mariana-type arc system are juxtaposed into orogenic components in ancient orogens.

Japan-type arc system

The Japan-type arc system is generated by subduction of one oceanic plate beneath another, but with some pre-arc sialic basement (Windley

1995), characterized by an oceanic island arc and back-arc basin(s) (Fig. 3b).

The Japanese island arcs are regarded as one of the best examples of a subduction-related orogen along a continental margin on account of their extensive geological record (Kimura & Mukai 1991; Needham et al. 1991; Osozawa 1992; Taira et al. 1992a; Kimura & Onishi 1994; Faure et al. 1995; Isozaki 1997a; Isozaki & Ota 2001; Moore & Saffer 2001; Taira 2001; Aoya & Wallis 2003; Yoshikura & Hada 2004). Important features of the Japanese island arcs are that they now have an imbricated crustal thickness of 40 km, being underlain by intrusive plutonic arc rocks and cumulate mafic–ultramafic rocks (Taira et al. 1992b; Windley 1995; Taira 2001). The Japanese island arcs were rifted from the continental margin of East Asia (Figs 1 & 3), and have undergone semicontinuous accretion since the early Palaeozoic, and the Japan Sea (back-arc basin) developed only in the late Cenozoic (Filippov & Kemkin 2003). This type of island arc system, unlike the Mariana-type arc system, has components of old Precambrian rocks (Maruyama 1997). On its western side the high-grade Hida complex with Permo-Triassic eclogites may be derived from the Sulu belt of eastern China. The early geological history of the Japanese island arc system indicates break-up at about 750 Ma from the Proterozoic margin of China and formation of the proto-Pacific basin (Taira 2001). Arc growth has taken place since the early Palaeozoic along the continental margin of East Asia, as a result of subduction of the proto-Pacific ocean, and the Japan Sea back-arc basin formed at about 20 Ma (Hawkins 1995; Maruyama 1997; Taira 2001; Kawakami et al. 2004; Sato et al. 2004).

Huge accretionary prisms developed along the forearc of the Japan-type arc system in the Jurassic, Cretaceous, and Tertiary to present day (Cowan 1987; Kimura & Mukai 1991; Hasebe et al. 1993; Ogawa et al. 1994; Isozaki 1997a; Kanamatsu et al. 2001; Taira 2001; Miyazaki & Okumura 2002; Ujiie 2002; Yoshikura & Hada 2004). High-pressure metamorphic belts were exhumed along the forearc and within the accretionary prisms (Isozaki 1996, 1997a, b; Okamoto et al. 2000).

Cordillera-type arc system

The Cordillera-type arc system refers to the major western Pacific accretionary domain (Moores 1970, 1998; Coney et al. 1980; Draper et al. 1984; Cordey & Schiarizza 1993; Garver & Scott 1995; McClelland et al. 2000; Lagabrielle et al. 2007; Ring 2008). The Cordillera-type arc system has three sub-types (Figs 1 & 3). Cordillera system I is characterized by a transform boundary with only limited areas along strike being subducted by oceanic plates (Fig. 3c1) (Moores 1970, 1998; Cordey & Schiarizza 1993; Garver & Scott 1995; McClelland et al. 2000; Miller 2002; Underwood et al. 1999). Cordillera system II can have a long subduction boundary (Fig. 3c2), and Cordillera system III is referred to as a continental marginal arc generated by subduction of oceanic plates beneath it with little forearc accretion, similar to an Andean-type arc but without a back-arc basin (Fig. 3c3; Sillitoe 1974; Kay et al. 1988; Nelson 1996; Martin et al. 1999; Gutscher et al. 2000; Milkov et al. 2003; Reich et al. 2003). A basic characteristic is that there is a wide thrust belt with continent-directed thrust vergence in the retroarc setting, sometimes with a width of more than 1000 km (Moores 1998; McClelland et al. 2000; Miller 2002).

Different ages of accretionary prisms develop along the forearc of a Cordillera-type arc system *senso stricto* (Moores 1970; Bebout 1989; Ernst 1993; Isozaki & Blake 1994; Aalto et al. 1995; Tagami & Hasebe 1999). Along an Andean-type margin some parts of the trench can have no, or limited, accretionary material, as in the Andean Atacama Desert, where there are no rivers to bring material into the trench (Kay et al. 1988; Martin et al. 1999; Gutscher et al. 2000). High-pressure metamorphic belts are sometimes well developed, such as the well-known Franciscan complex (Bebout 1989; Ernst 1993; Isozaki & Blake 1994; Schemmann et al. 2008). Unlike the Japan-type and Mariana-type arc systems, this type of arc system can have oceanic, continental, and transitional geochemical properties (Kay et al. 1988; Windley 1995; Martin et al. 1999; Gutscher et al. 2000).

Alaska-type arc system

The Alaska-type arc system (Figs 1 & 4) is a special type with various degrees of combination of the Mariana-, Japan- and Cordillera-type arc systems (Barker & Sullivan 1989; Bradley & Kusky 1990; Sisson & Pavlis 1993; Haeussler et al. 1995; Himmelberg & Loney 1995; Goldfarb et al. 1997; Kusky & Bradley 1999; Lytwyn et al. 2000; Bradley et al. 2003; Nokleberg et al. 2005; Cole et al. 2006). The Alaskan orogen can be taken as a type example, the nature of which changes westwards from Cordillera-type in the east, via Japan-type in the middle, to Mariana-type in the west (Kusky & Bradley 1999; Kusky & Young 1999). A similar combined arc system can be found in the Sumatra–Banda region, where the Indo-Australian ocean plate is being subducted beneath a continental arc in the NW and a Japan-type arc in the SE (Honza et al. 2000; Hall 2002, 2009).

Different ages of accretionary prisms may develop with high-pressure metamorphic belts in this arc system type (Sample & Fisher 1986;

Bradley & Kusky 1990; Haeussler et al. 1995; Gutscher et al. 1998; Hansen & Dusel-Bacon 1998; Kusky & Bradley 1999; Kusky & Young 1999; Bradley et al. 2003). It can have oceanic, continental, and transitional geochemical properties along its Cordillera- and Japan-type arc segments and purely oceanic ones along its Mariana-type arc segment. A wide thrust belt with continent-directed thrust vergence can also develop in the retroarc setting (Clark et al. 1995).

Transitions between continental margins and arc systems

The four major types of arc systems are developed in the Circum-Pacific region (Figs 1, 3 & 4). There can be complicated types of accretion, but they are combined with these main types, and they all belong to the active margins of Asia and North America. Systematic investigations along these accretionary orogens show that these four types of arc systems can be mutually transitional. Two major transitional trends are discussed below.

Transition I: between the Mariana-, Japan- and Cordillera-type arc systems

The Mariana-type arc system involves an interaction between two oceanic plates, mostly in terms of subduction of one beneath the other. An intra-oceanic island arc will be generated with no contribution from any continental material.

Usually this kind of arc system extends for a relatively long distance, such as the Mariana–Izu–Bonin arc chain (Horine et al. 1990; Kanamatsu et al. 1996; Miller et al. 2004). When part of this long arc chain collides with a promontory or a rifted ribbon of the active margin of a host continent, a flip of subduction polarity can occur, forming a Cordillera-type arc system (Fig. 5). This is the geodynamic process that formed the Cordilleran active margin from the mid-Jurassic to the Neogene, when a mid-Jurassic intra-oceanic arc with west subduction polarity was accreted to the North American continent in the late Jurassic (Ingersoll & Schweickert 1986; Busby & Ingersoll 1995; Dickinson 1995).

The Japan-type arc system results from subduction of an oceanic plate beneath a thin continental fragment and it has a back-arc basin. As convergence continues, this arc system will collide with the host continent, closing the back-arc basin(s) and thus evolving into a Cordillera-type arc system (Fig. 5). This kind of geodynamic process is currently taking place along the Taiwan Island where an east-dipping subducted oceanic basin is colliding with the Luzon island arc (Fig. 1) (Huang et al. 2000; Chang et al. 2001; Lin 2002; Shyu et al. 2005).

In general, the subduction is beneath a host continent, which is characterized by a Cordillera-type arc system, and this can evolve with a high-angle slab when conditions are favourable, leading to extension in the back-arc, which can generate back-arc basins or marginal basins (Leggett et al. 1985; Wakita 1988; Kimura & Onishi 1994;

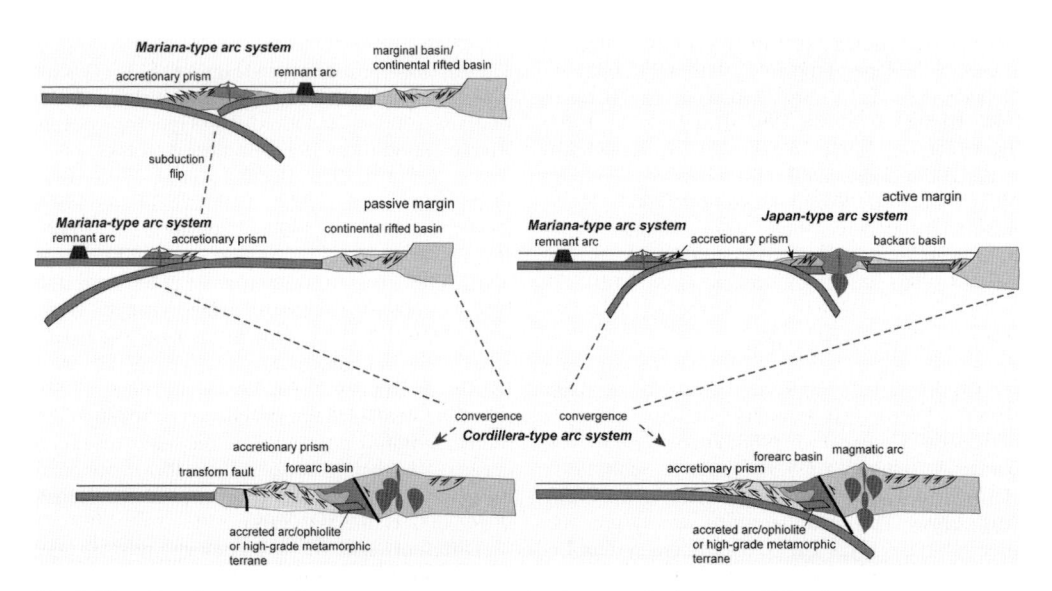

Fig. 5. Transitions between various types of arc systems related to accretionary and collisional orogens.

Okamoto et al. 2000; Taira 2001). Gradually a Japan-type arc system will be formed away from the host Cordillera-type arc system with a back-arc basin or basins.

Sometimes entrapment of an oceanic basin can generate back-arc basins with oceanic arcs that are characterized either by a Precambrian continental basement or by a purely Mariana-type arc system (Nokleberg et al. 2005). The former can be regarded as a transition from a Mariana- to a Japan-type arc system. A direct transition from a Japan- or Cordillera- to a Mariana-type is not common.

Transition II: between the Alaska-type and other types of arc systems

Transition I mentioned above can be best understood from its cross-sectional character. When looked at along the strike of arc systems, the role of the Alaska-type arc system is most important. The Alaska-type arc system combines the characters of the other three types, but in a more complicated manner (Kay 1978; Vrolijk et al. 1988; Holbrook et al. 1999). This type itself shows a transition along its trend (Fig. 4). In the asymmetrical case, an Alaskan-type arc system can have a core of the Cordillera-type arc system with each wing consisting of the Japan- and Mariana-type arc systems along strike (Fig. 4a). The order of the Japan- and Mariana-type arc systems can vary.

A complicated situation sometimes arises when the Alaska-type arc systems can have two transitional trends, as shown in the symmetrical case in Figure 4b, which means they can have double wings of intra-oceanic arcs with a core of Cordillera-type arc system. The order of the Japan- and Mariana-type arc systems in both wings can also vary. In this case, a four-fold complication would be expected.

In the SW Pacific, the Central Papua New Guinea arc system with continental crust basement has double wings of different intra-oceanic arc systems (Bloomer et al. 1995; Yamamoto et al. 2010).

Other transitions can be achieved by similar processes to Transition I mentioned above. For instance, the Alaskan-type arc system can evolve with each of the components of arc systems changing into another type. The tectonic evolution of the SW Pacific region in the last 50 Ma has been investigated in great detail (Hall 2002; Hall & Spakman 2002) and this scenario illustrates the complicated transitions among various kinds of arc systems including the Alaskan-type.

Discussion

The circum-Pacific accretionary orogens offer some modern examples for the study of accretionary processes. They can be roughly subdivided into the eastern (Cordillera-type arc system) and western (Japan- and Mariana-type arc systems) Pacific margins, with the Aleutians (Alaska-type arc system) as a special connection between them (Fig. 1).

Accretionary and collisional orogenesis: alternation of Circum-Pacific active margins

Systematic research has revealed that the Eastern Pacific active margin, represented by the Cordillera-type arc system in this paper, had archipelago palaeogeography since the Palaeozoic (Moores 1970; Unruh et al. 1995; Wakabayashi & Dilek 2000; Schemmann et al. 2008). A Japan-type island arc system developed near the North America craton, which is the host continent, with a back-arc basin between them (Fig. 6). The arc together with an accretionary complex migrated oceanward. This arc could have been enlarged on both sides by subduction of oceanic plates or accretion of other arcs or other terranes. Growing accretionary complexes may have also contributed to the growth of the arc system. In a later stage, this arc system would have become attached to the host continent (Fig. 6). With transitions between and amalgamation of these classic types of arc systems, the Eastern Pacific active margin has evolved into the present-day complicated accretionary stage. From this description, it is obvious that the Western Pacific archipelagos are the past of the Eastern Pacific.

The Western Pacific archipelagos involve continental ribbons and back-arc basins and nearly continuous trench systems from the Macquarie Ridge, via the Kermadec–Tonga Trench, to the Mariana Trench, and then to the Japan and Kurile Trench, which is connected with the Eastern Pacific via the Aleutian Trench in the Alaska-type arc system (Brueckmann 1989; Ballance 1993; Gutscher et al. 1999; Pankhurst et al. 1999; Ranero & von Huene 2003; Dilek et al. 2007). As the Indo-Australian plate is currently colliding with the Asian plate (Fig. 1), after its indentation into the archipelagos of SE Asia, a complicated scenario will develop with closure of the back-arc basins, giving rise to huge strike-slip fault systems (Van Staal et al. 1998), and then possibly will be evolving into a Cordillera-type arc system. Therefore, it seems that the Eastern Pacific active margin is the future of the Western Pacific archipelagos.

The interaction between the East Pacific Rise and the North America plate is causing formation of the Gulf of California, which is floored by oceanic crust (Fig. 1), and in future a continental ribbon (Baja California Range) will be translated northwards and will finally be rifted away from

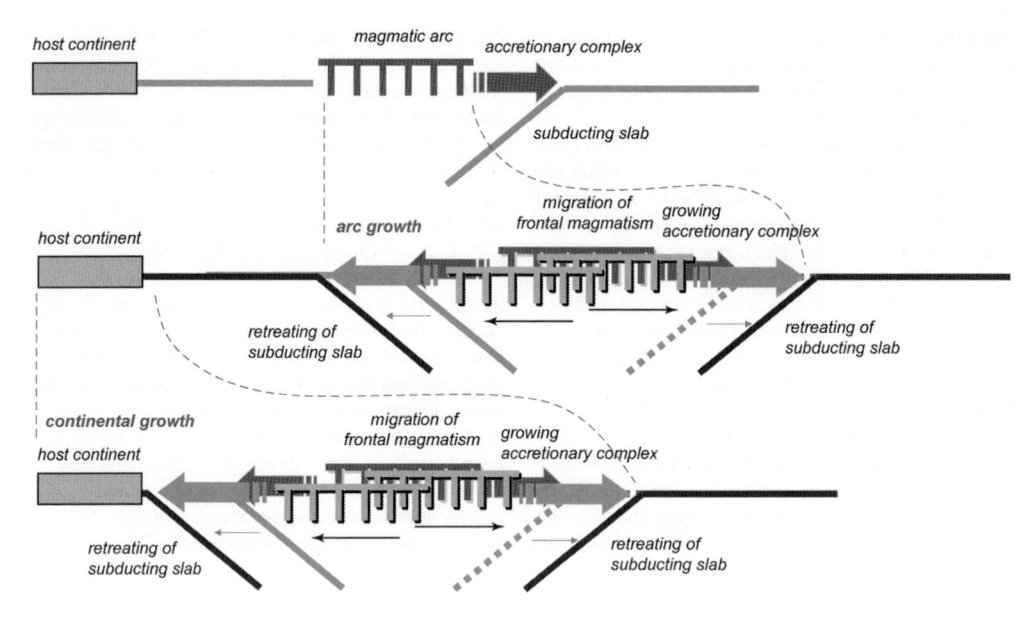

Fig. 6. Tectonic models of various types of arc systems related to accretionary and collisional orogens (modified after Şengör et al. 1993; Dickinson 1995).

the North American plate (Benoit et al. 2002; Zhang et al. 2009), the very beginnings of a new archipelago. Therefore the Western Pacific and Eastern Pacific scenarios can mutually change.

Juxtapositions of the typical arc systems and continental margin sequences

The four main types of arc systems can exist at different stages of palaeogeography. They can also coexist at one period of time or after a relatively long time one will evolve into another, but generally one or several types may coexist. The coexistence of these arc types provides a prime example of an archipelago palaeogeography of many accretionary orogens (Williams et al. 1989; Hsü et al. 1995; Scott & Gauthier 1996; Davis et al. 1998; Honza et al. 2000; Xiao et al. 2001, 2002a, b, 2003, 2009b, c). Some Japan-type arc systems in ancient orogens (e.g. the Altaids) may have been generally regarded as microcontinents because they have have so-called Precambrian basement. Often the Cordillera-type arc systems can be very complicated, and if they are rifted away from the host continent they become more difficult to recognize.

More importantly, these four main types of arc systems can be juxtaposed into a final orogenic collage, which is another main expression of accretionary orogenesis. One phase or several phases of accretionary orogenesis will finally amalgamate all these types of arc systems, together with continental margins such as passive margins, forming a wide and complicated orogenic collage, leading to arc growth and continental growth (Bloomer et al. 1995; Chaumillon & Mascle 1997; Ducea 2001; Badarch et al. 2002; Buslov et al. 2002).

Arc–arc collision is not uncommon in orogenesis. In the northern part of eastern Indonesia the Halmahera and Sangihe Arcs are the only intra-oceanic arcs in the world, which are currently colliding (Fig. 1; Hall & Spakman 2002; Hall 2009). The orogenic development of Inner Mongolia was characterized by two wide accretionary complexes that belonged to two Cordillera-type arc systems (Xiao et al. 2003). Amalgamation of active marginal sequences is also documented in the Caucasus (Khain 1975).

All these arc systems, as they have long chains, can be easily rotated and/or oroclinally bent. A good example is the Kazakhstan orocline, and this feature has been supported by palaeomagnetic data (Şengör et al. 1993; Şengör & Natal'in 1996a, b; Van der Voo 2004; Abrajevitch et al. 2008; Levashova et al. 2009). Other good examples are the western end of the Cantabrian orocline of Variscan Europe, and a ribbon continent that was buckled and accreted to the northwestern margin of North America (Van der Voo 2004).

Termination of accretionary and collisional orogenesis: the Altaids

The alternation between these typical arc systems and their interactions together with other marginal sequences are the main geodynamic processes of accretionary and collisional orogenesis, which is characterized by a long-lived history, a scenario not fully explained by the Wilson Cycle (Dalziel 1997; Dalziel et al. 2000; Lawver et al. 2003; Cawood & Buchan 2007). How then does a complicated accretionary and collisional orogen terminate? The Altaids (Central Asian Orogenic Belt) provides the type example.

The Altaids is one of the largest accretionary orogenic collages in the world, with the highest rate of Phanerozoic continental growth and significant metallogeny (Şengör & Okurogullari 1991; Şengör et al. 1993; Jahn et al. 2004; Wu et al. 2007; Utsunomiya & Jahn 2008). It is widely accepted that subduction-related development of the Altaids started in the late Precambrian (Khain et al. 2002; Kröner et al. 2007) and it gradually migrated southward (present coordinates), as recorded in the vast areas of Russia, Mongolia, China, and Kazakhstan and other Central Asian countries (Figs 7 & 8).

The late Palaeozoic to early Mesozoic accretion is best documented in two key areas, North Xinjiang in China and Kokchetav–Balkash in Kazakhstan in the west and Inner Mongolia in the east, together with neighbouring Mongolia, mainly called the 'Kazakhstan' and 'Tuva–Mongol' oroclines (Şengör et al. 1993; Xiao et al. 2003, 2004a, b, 2009c). The late Palaeozoic orogens of North Xinjiang and adjacent areas developed by continuous southward accretion along the wide southern active margin of Siberia with formation of an Alaska-type arc system (Kokchetav–North Tien Shan), some Japan-type arc systems (Altay, Chinese Central Tien Shan) and some Mariana-type arc systems (Balkash, West Junggar, and East Junggar) (Figs 8 & 9). Permian Alaskan-type zoned mafic–ultramafic complexes intruded along major faults in the Tien Shan (Xiao et al. 2008b). The final amalgamation of the passive margin of Tarim with the huge accretionary system to the north may have lasted to the end of the Permian to early or mid-Triassic (Xiao et al. 2003, 2004a, b, 2009c).

In Inner Mongolia of China and adjacent areas two wide accretionary wedges developed along the southern active margin of Southern Gobi–Mongolia and the northern active margin of the North China craton, which may have lasted to the Permian and mid-Triassic (Xiao et al. 2003, 2004a, b, 2009c; Zhang et al. 2007a, b). The final products of the long-lived accretionary processes in this part of the southern Altaids include a late Palaeozoic to Permian Cordillera-type arc system (northern

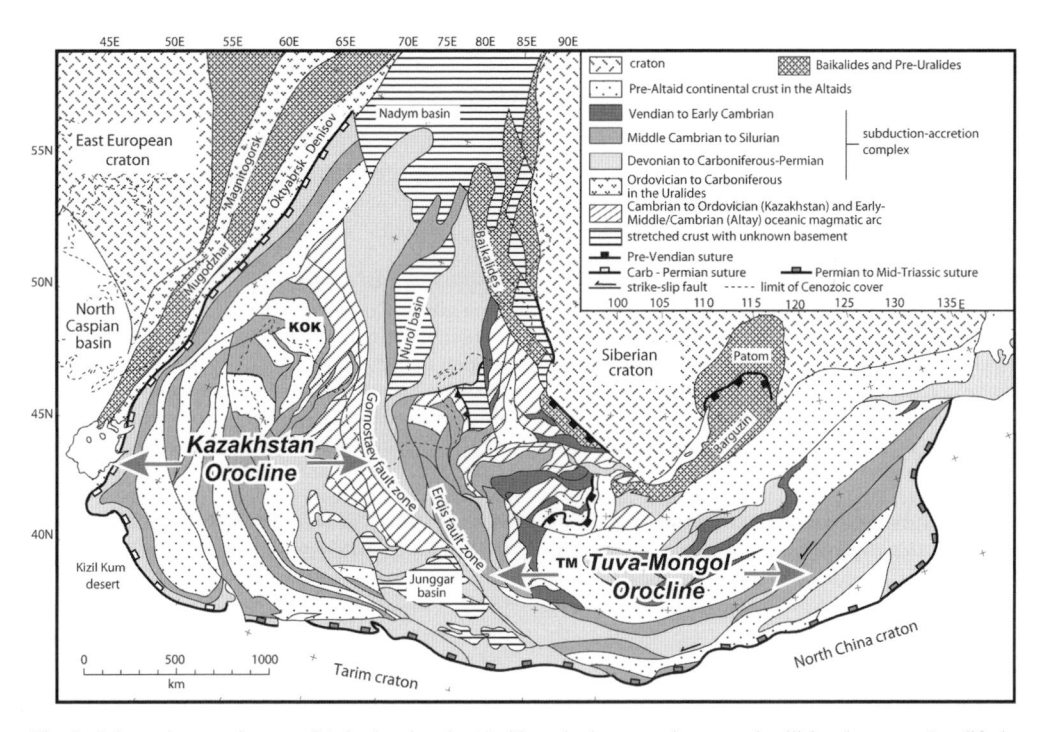

Fig. 7. Schematic tectonic map of Asia showing the Altaids and other accretionary and collisional orogens (modified after Şengör 1992; Şengör et al. 1993).

(a) Cambrian to mid-Ordovician

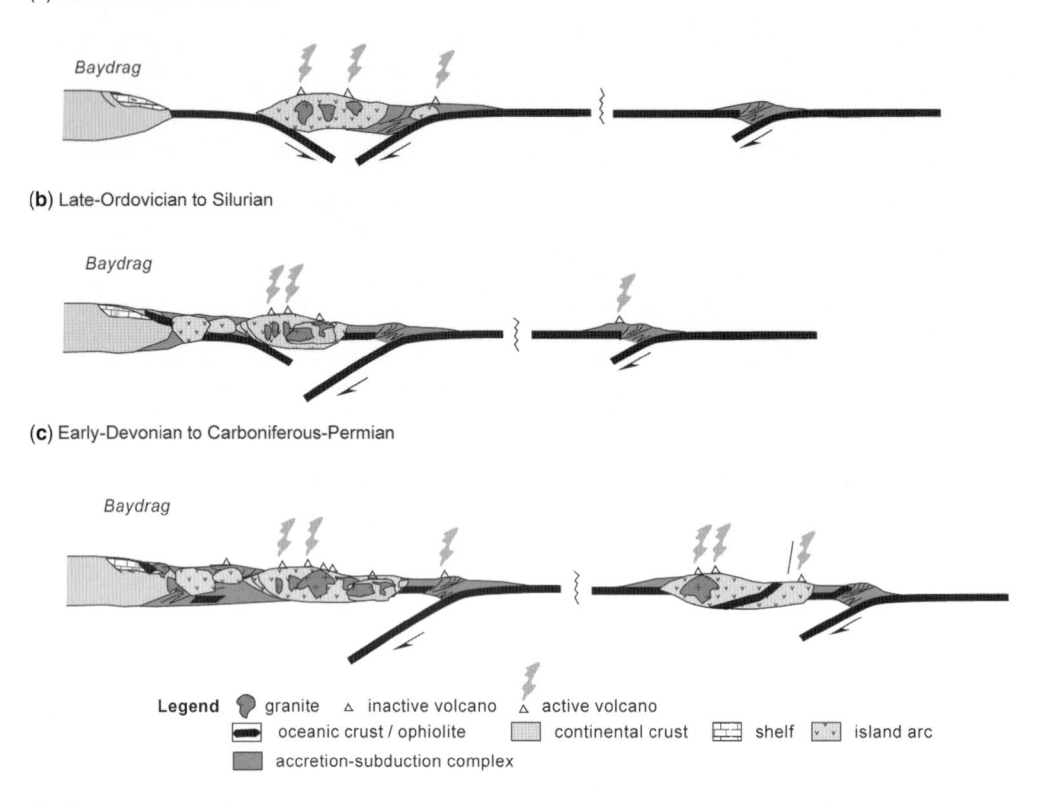

(b) Late-Ordovician to Silurian

(c) Early-Devonian to Carboniferous-Permian

Fig. 8. Tectonic evolutionary model of various types of arc systems related to the Altaids (modified after Şengör et al. 1993; Xiao et al. 2004b, 2009c; Windley et al. 2007).

part of the North China craton), a Japan-type arc system (Southern Gobi–Mongolia), and some Mariana-type arc systems (for instance, Bainaimiao), with late Palaeozoic to mid-Triassic accretionary wedges composed of radiolarian cherts, pillow lavas, and ophiolitic fragments, and high-pressure to ultrahigh-pressure metamorphic rocks (Wang & Fan 1997; Xiao et al. 2003, 2004a, b, 2009c; Zhang et al. 2007a, b).

The docking of the Tarim and North China cratons against the southern active margin of Siberia at the end of the Permian to the mid-Triassic resulted in the final closure of the Palaeoasian Ocean and terminated the accretionary orogenesis of the southern Altaids in this part of Central Asia. This complex geodynamic evolution led to the formation of giant metal deposits in Central Asia and to substantial continental growth.

Conclusions

'Arc system' is defined as a collective term for subduction-related tectonic environments distributed either along a continental margin or in an oceanic plate. Based on structural, plutonic and volcanic, sedimentary, metamorphic, and palaeogeographic data, the Mariana-, Japan-, Cordillera- and Alaska-type arc systems can be recognized. Commonly these arc systems interact mutually and with continental marginal sequences, leading to complicated accretionary and collisional orogens. Some Japan-type arc systems in ancient orogens (e.g. the Altaids) may have been generally regarded as microcontinents because they have so-called Precambrian basement. Often the Cordillera-type arc systems can be very complicated, and if they are rifted away from the host continent they become more difficult to recognize.

The alternation between Western Pacific archipelagos and the Eastern Pacific active margin is the stereotype of accretionary and collisional orogenesis. More importantly, these four main types of arc systems can be juxtaposed into a final orogenic collage, which is another main expression of accretionary orogenesis. This could not be terminated in one Wilson Cycle and thus is the reason

TRANSITIONS AMONG MARIANA-, JAPAN-, CORDILLERA- AND ALASKA-TYPE ARC SYSTEMS 45

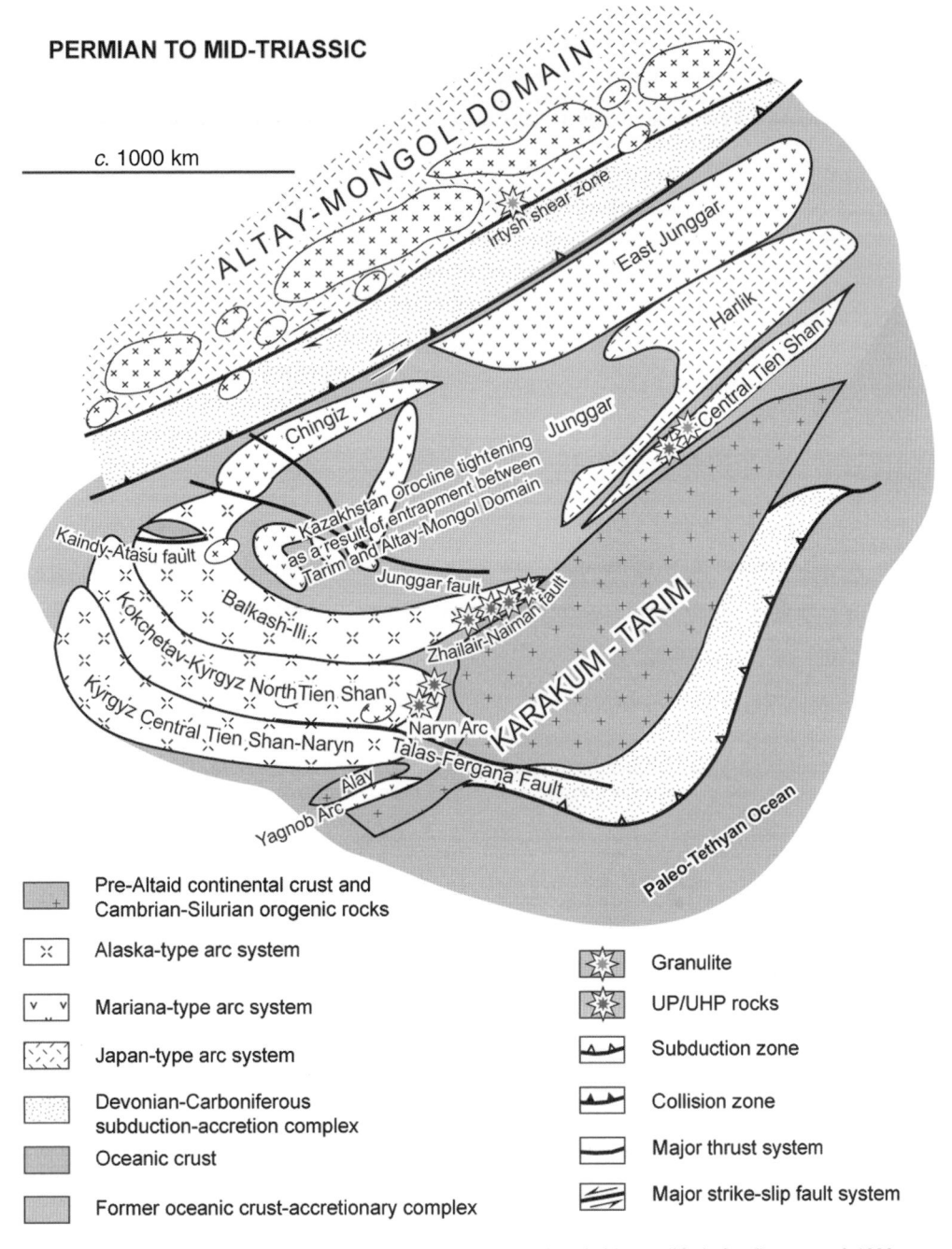

Fig. 9. Palaeogeography of various types of arc systems related to the Altaids (modified after Şengör *et al.* 1993; Xiao *et al.* 2009c).

why most accretionary orogenesis is long-lived. Only some segments of huge accretionary and collisional systems can be terminated by attachment of a subcontinent-sized cratonic block such as the Tarim or the Indian craton; nevertheless, the processes of accretionary and collisional orogenesis continue elsewhere. The Altaids is a prime example of a terminated accretionary–collisional orogen.

We dedicate this contribution to Brian Windley who has been working in Central Asia for more than two decades. We thank S. Sun, J. Li, B. Windley, A. Kröner and A. M. C. Şengör for advice and discussion, and T. Kusky for the critical reviews. This study was financially supported by the Chinese 973 Programme (2007CB411307), the National Natural Science Foundation of China (40725009), and the Hong Kong Research Grants Council (HKU7043/07P). This paper is a contribution to ILP (ERAs, Topo-Central Asia) and IGCP 480.

References

AALTO, K. R., MOLEY, K. & STONE, L. 1995. Neogene paleogeography and tectonics of northwestern California. In: FRITSCHE, A. E. (ed.) Cenozoic Paleogeography of the Western United States – II. Society of Economic Paleontologists and Mineralogists, Pacific Section, Los Angeles, CA, 162–180.

ABRAJEVITCH, A., VAN DER VOO, R., BAZHENOV, M. L., LEVASHOVA, N. M. & MCCAUSLAND, P. J. A. 2008. The role of the Kazakhstan orocline in the late Paleozoic amalgamation of Eurasia. Tectonophysics, 455, 61–76.

AIKMAN, A. B., HARRISON, T. M. & DING, L. 2008. Evidence for Early (>44 Ma) Himalayan crustal thickening, Tethyan Himalaya, southeastern Tibet. Earth and Planetary Science Letters, 274, 14–23, doi:10.1016/j.epsl.2008.06.038.

ALLÈGRE, C. J., COURTILLOT, V. ET AL. 1984. Structure and evolution of the Himalaya–Tibet orogenic belt. Nature, 307, 17–22.

AOYA, M. & WALLIS, S. R. 2003. Role of nappe boundaries in subduction-related regional deformation: spatial variation of meso- and microstructures in the Seba eclogite unit, the Sambagawa belt, SW Japan. Journal of Structural Geology, 25, 1097–1106.

BADARCH, G., CUNNINGHAM, W. D. & WINDLEY, B. F. 2002. A new terrane subdivision for Mongolia: implications for the Phanerozoic crustal growth of Central Asia. Journal of Asian Earth Sciences, 21, 87–110.

BALLANCE, P. F. 1993. The New Zealand Neogene forearc basins. In: BALLANCE PETER, F. (ed.) South Pacific Sedimentary Basins. Sedimentary Basins of the World. Elsevier, Amsterdam, 177–193.

BARKER, C. & SULLIVAN, G. 1989. Analysis of gases in individual fluid inclusions in minerals from the Kodiak accretionary complex, Alaska. In: BODNAR ROBERT, J. (ed.) Pan-American Conference on Research on Fluid Inclusions; Second Biennial Meeting; Program And Abstracts. Pacrofi, Virginia Polytechnic Institute and State University, Blacksburg, VA, 8.

BEBOUT, G. E. 1989. Mélange formation, hydrogeology, and metasomatism at 15–45 kilometer depths in an Early Cretaceous accretionary complex. In: DYMEK, R. F. & SHELTON, K. L. (eds) Geological Society of America, 1989 Annual Meeting. Geological Society of America, Abstracts with Programs, 339.

BENOIT, M., AGUILLÓN-ROBLES, A. ET AL. 2002. Geochemical diversity of Late Miocene volcanism in southern Baja California, Mexico: implication of mantle and crustal sources during the opening of an asthenospheric window. Journal of Geology, 110, 627–648.

BLOOMER, S., TAYLOR, B., MACLEOD, C. J., STERN, R. J., FRYER, P., HAWKINS, J. W. & JOHNSON, L. 1995. Early arc volcanism and the ophiolite problem: a perspective from ocean drilling in the western Pacific. In: NATLAND, J. & TAYLOR, B. (eds) Active Margins and Marginal Basins of the Western Pacific. American Geophysical Union, Geophysical Monograph, 88, 1–30.

BRADLEY, D. C. 1989. Taconic plate kinematics as revealed by foredeep stratigraphy, Appalachian orogen. Tectonics, 8, 1037–1049.

BRADLEY, D. C. & KUSKY, T. M. 1990. Kinematics of late faults along Turnagain Arm, Mesozoic accretionary complex, south–central Alaska. In: DOVER, J. H. & GALLOWAY, J. P. (eds) Geologic Studies in Alaska by the US Geological Survey, 1989. US Geological Survey Bulletin, 3–10.

BRADLEY, D. C., DUMOULIN, J. ET AL. 2003. Late Paleozoic orogeny in Alaska's Farewell terrane. Tectonophysics, 372, 23–40.

BROWN, M. 2009. Metamorphic patterns in orogenic systems and the geological record. In: CAWOOD, P. A. & KRÖNER, A. (eds) Earth Accretionary Systems in Space and Time. Geological Society, London, Special Publications, 318, 37–74.

BRUECKMANN, W. 1989. Porosity modeling and stress evaluation in the Barbados Ridge Accretionary Complex. Geologische Rundschau, 78, 197–205.

BUSBY, C. A. & INGERSOLL, R. V. (eds) 1995. Tectonics of Sedimentary Basins. Blackwell Science, Cambridge, MA.

BUSLOV, M. M., WATANABE, T., SAPHONOVA, I. Y., IWATA, K., TRAVIN, A. & AKIYAMA, M. 2002. A Vendian–Cambrian arc system of the Siberian continent in Gorny Altai (Russia, central Asia). Gondwana Research, 5, 781–800.

CAWOOD, P. A. & BUCHAN, C. 2007. Linking accretionary orogenesis with supercontinent assembly. Earth-Science Reviews, 82, 217–256.

CAWOOD, P. A., KRÖNER, A., COLLINS, W. J., KUSKY, T. M., MOONEY, W. D. & WINDLEY, B. F. 2009. Earth Accretionary orogens through Earth history. In: CAWOOD, P. A. & KRÖNER, A. (eds) Earth Accretionary Systems in Space and Time. Geological Society, London, Special Publications, 318, 1–37.

CHAMBERLAIN, V. E. & LAMBERT, R. S. J. 1985. Cordillera, a newly defined Canadian microcontinent. Nature, 314, 707–713.

CHANG, C. F., CHEN, N. ET AL. 1986. Preliminary conclusions of the Royal Society and Academia Sinica 1985 geotraverse of Tibet. Nature, 323, 501–507.

CHANG, C. F., PAN, Y. S. & SUN, Y. Y. 1989. The tectonic evolution of Qinghai–Tibet Plateau: a review. In: ŞENGÖR, A. M. C., YILMAZ, Y., OKAY, A. I. & GORUR, N. (eds) Tectonic Evolution of the Tethyan Region. NATO ASI Series, Series C, 259, 415–476.

CHANG, C. P., ANGELIER, J., HUANG, C. Y. & LIU, C. S. 2001. Structural evolution and significance of a mélange in a collision belt: the Lichi Mélange and the Taiwan arc–continent collision. Geological Magazine, 138, 633–651, doi:10.1017/S0016756801005970.

CHAUMILLON, E. & MASCLE, J. 1997. From foreland to forearc domains; new multichannel seismic reflection survey of the Mediterranean Ridge accretionary complex (eastern Mediterranean). *Marine Geology*, **138**, 237–259.

CHURCH, B. N., DOSTAL, J., OWEN, J. V. & PETTIPAS, A. R. 1995. Late Paleozoic gabbroic rocks of the Bridge River accretionary complex, southwestern British Columbia; geology and geochemistry. *Geologische Rundschau*, **84**, 710–719.

CLARK, M. B., BRANTLEY, S. L. & FISHER, D. M. 1995. Power-law vein-thickness distributions and positive feedback in vein growth. *Geology*, **23**, 975–978.

COLE, R. B., NELSON, S. W., LAYER, P. W. & OSWALD, P. J. 2006. Eocene volcanism above a depleted mantle slab window in southern Alaska. *Geological Society of America Bulletin*, **118**, 140–158.

COLEMAN, R. 1989. Continental growth of Northwest China. *Tectonics*, **8**, 621–635.

CONDIE, K. C. 2000. Episodic continental growth models: afterthoughts and extensions. *Tectonophysics*, **322**, 153–162.

CONDIE, K. C. 2007. Accretionary orogens in space and time. *In*: HATCHER, R. D. JR, CARLSON, M. P., MCBRIDE, J. H. & MARTÍNEZ CATALÁN, J. R. (eds) *4-D Framework of Continental Crust*. Geological Society of America, Memoirs, 200, 145–158.

CONEY, P. J., JONES, D. L. & MONGER, J. W. H. 1980. Cordilleran suspect terranes. *Nature*, **288**, 29–33.

CORDEY, F. & SCHIARIZZA, P. 1993. Long-lived Panthalassic remnant: the Bridge River accretionary complex, Canadian Cordillera. *Geology*, **21**, 263–266.

COWAN, D. S. 1987. Kinematic indicators in unmetamorphosed sandstone–mudstone mélange of the Shimanto accretionary prism, SW Japan. *In*: DICKINSON, W. R. (ed.) *Geological Society of America, 1987 Annual Meeting and Exposition*. Geological Society of America Abstracts with Programs, **19**, 629.

CRAWFORD, A. J., BECCALUVA, L. & SERRI, G. 1981. Tectono-magmatic evolution of the West Philippine–Mariana region and the origin of boninites. *Earth and Planetary Science Letters*, **54**, 346–356.

DALZIEL, I. W. D. 1997. Neoproterozoic–Paleozoic geography and tectonics; review, hypothesis, environmental speculation. *Geological Society of America Bulletin*, **109**, 16–42.

DALZIEL, I. W. D., LAWVER, L. A. & MURPHY, J. B. 2000. Plumes, orogenesis, and supercontinental fragmentation. *Earth and Planetary Science Letters*, **178**, 1–11.

DAVIS, J. S., ROESKE, S. M. & KARL, S. M. 1998. Late Cretaceous to early Tertiary transtension and strain partitioning in the Chugach accretionary complex, SE Alaska. *Journal of Structural Geology*, **20**, 639–654.

DESCHAMPS, A. & FUJIWARA, T. 2003. Asymmetric accretion along the slow-spreading Mariana Ridge. *Geochemistry, Geophysics, Geosystems*, **4**, 8622, doi:10.1029/2003GC000537.

DEWEY, J. F. 1969. Evolution of the Appalachian–Caledonian orogen. *Nature*, **222**, 124–129.

DICKINSON, W. R. 1995. Forearc basins. *In*: BUSBY, C. J. & INGERSOLL, R. V. (eds) *Tectonics of Sedimentary Basins*. Blackwell Science, Cambridge, MA, 221–261.

DILEK, Y., FURNES, H. & SHALLO, M. 2007. Suprasubduction zone ophiolite formation along the periphery of Mesozoic Gondwana. *Gondwana Research*, **11**, 453–475.

DRAPER, G., NAGLE, F., JOYCE, J. & PINDELL, J. 1984. Cretaceous accretionary complex of northern Hispaniola. *Geological Society of America, Abstracts with Programs*, **16**, 493.

DUCEA, M. 2001. The California Arc: thick granitic batholiths, eclogitic residues, lithospheric-scale thrusting, and magmatic flare-ups. *GSA Today*, **11**, 4–10.

ERNST, W. G. 1993. Metamorphism of Franciscan tectonostratigraphic assemblage, Pacheco Pass Area, east–central Diablo Range, California Coast Ranges. *Geological Society of America Bulletin*, **105**, 618–636.

FAURE, M., NATAL'IN, B. A., MONIE, P., VRUBLEVSKY, A. A., BORUKAIEV, C. & PRIKHODKO, V. 1995. Tectonic evolution of the Anuy metamorphic rocks (Sikhote Alin, Russia) and their place in the Mesozoic geodynamic framework of East Asia. *Tectonophysics*, **241**, 279–301.

FILIPPOV, A. N. & KEMKIN, I. V. 2003. Clastic rocks from Permian and Triassic cherty sequences in Sikhote Alin and Japan. *Lithology and Mineral Resources*, **38**, 36–47.

FRYER, P., WHEAT, C. G. & MOTTL, M. 1999. Mariana blueschist mud volcanism: Implications for conditions within the subduction zone. *Geology*, **27**, 103–106.

GARVER, J. I. & SCOTT, T. J. 1995. Trace elements in shale as indicators of crustal provenance and terrane accretion in the southern Canadian Cordillera. *Geological Society of America Bulletin*, **107**, 440–453.

GOLDFARB, R. J., GROVES, D. I. & GARDOLL, S. 1997. Metallogenic evolution of Alaska. *Economic Geology*, **9**, 4–34.

GOLDFARB, R. J., MAO, J. W., HART, C., WANG, D. H., ANDERSON, E. & WANG, Z. L. 2003. Tectonic and metallogenic evolution of the Altay Shan, northern Xinjiang Uygur Autonomous Region, northwestern China. *In*: MAO, J. W., GOLDFARB, R. J., SELTMANN, R., WANG, D. H., XIAO, W. J. & HART, C. (eds) *Tectonic Evolution and Metallogeny of the Chinese Altay and Tianshan*. IAGOD Guidebook Series 10. CERCAMS–NHM, London, 17–30.

GRAPES, R. & WATANABE, T. 1992. Metamorphism and uplift of Alpine schist in the Franz Josef–Fox Glacier area of the Southern Alps, New Zealand. *Journal of Metamorphic Geology*, **10**, 171–180.

GRAY, D. R., FOSTER, D. A. & BIERLEIN, F. P. 2002. Geodynamics and metallogeny of the Lachlan Orogen. *Australian Journal of Earth Sciences*, **49**, 1041–1056.

GUTSCHER, M. A., KUKOWSKI, N., MALAVIEILLE, J. & LALLEMAND, S. 1998. Episodic imbricate thrusting and underthrusting: analog experiments and mechanical analysis applied to the Alaskan accretionary wedge. *Journal of Geophysical Research, Solid Earth*, **103**, 10 161–10 176.

GUTSCHER, M. A., OLIVET, J. L., ASLANIAN, D., EISSEN, J. P. & MAURY, R. 1999. The 'lost Inca Plateau': cause of flat subduction beneath Peru. *Earth and Planetary Science Letters*, **171**, 335–341.

GUTSCHER, M.-A., SPAKMAN, W., BIJWAARD, H. & ENGDAHL, E. R. 2000. Geodynamics of flat subduction: seismicity and tomographic constraints from the Andean margin. *Tectonics*, **19**, 814–833.

GVIRTZMAN, Z. & STERN, R. J. 2004. Bathymetry of Mariana trench–arc system and formation of the Challenger Deep as a consequence of weak plate coupling. *Tectonics*, **23**, TC2011, doi:10.1029/2003TC001581.

HAEUSSLER, P. J., BRADLEY, D., GOLDFARB, R. J., SNEE, L. W. & TAYLOR, C. D. 1995. Link between ridge subduction and gold mineralization in southern Alaska. *Geology*, **23**, 995–998.

HALL, R. 2002. Cenozoic geological and plate tectonic evolution of SE Asia and the SW Pacific: computer-based reconstructions, model and animations. *Journal of Asian Earth Sciences*, **20**, 353–431.

HALL, R. 2009. The Eurasian SE Asian margin as a modern example of an accretionary orogen. *In*: CAWOOD, P. & KRÖNER, A. (eds) *Earth Accretionary Systems in Space and Time*. Geological Society, London, Special Publications, **318**, 351–372.

HALL, R. & SPAKMAN, W. 2002. Subducted slabs beneath the eastern Indonesia–Tonga region: insights from tomography. *Earth and Planetary Science Letters*, **201**, 321–336.

HANSEN, V. L. & DUSEL-BACON, C. 1998. Structural and kinematic evolution of the Yukon–Tanana upland tectonites, east–central Alaska: a record of late Paleozoic to Mesozoic crustal assembly. *Geological Society of America Bulletin*, **110**, 211–230.

HASEBE, N., TAGAMI, T. & NISHIMURA, S. 1993. The evidence of along-arc differential uplift of the Shimanto accretionary complex; fission track thermochronology of the Kumano acidic rocks, Southwest Japan. *Tectonophysics*, **224**, 327–335.

HAWKINS, J. W. 2003. Geology of supra-subduction – Implications for the origin of ophiolites. *In*: DILEK, Y. & NEWCOMB, S. (eds) *Ophiolite Concept and the Evolution of Geological Thought*. Geological Society of America, Special Papers, **373**, 227–268.

HAWKINS, J. W. J. 1995. The geology of Lau basin. *In*: TAYLOR, B. (ed.) *Backarc Basins: Tectonics and Magmatism*. Plenum, New York, 63–138.

HEINHORST, J., LEHMANN, B., ERMOLOV, P., SERYKH, V. & ZHURUTIN, S. 2000. Paleozoic crustal growth and metallogeny of Central Asia: evidence from magmatic–hydrothermal ore systems of Central Kazakhstan. *Tectonophysics*, **328**, 69–87.

HELO, C., HEGNER, E., KRÖNER, A., BADARCH, G., TOMURTOGOO, O., WINDLEY, B. F. & DULSKI, P. 2006. Geochemical signature of Paleozoic accretionary complexes of the Central Asian Orogenic Belt in South Mongolia: constraints on arc environments and crustal growth. *Chemical Geology*, **227**, 236–257.

HIMMELBERG, G. R. & LONEY, R. A. (eds) 1995. *Characteristics and Petrogenesis of Alaskan–type Ultramafic–Mafic Intrusions, Southeastern Alaska*. US Geological Survey, Professional Papers, 1564.

HOLBROOK, W. S., LIZARRALDE, D., MCGEARY, S., BANGS, N. & SDIEBOLD, J. 1999. Structure and composition of the Aleutian island arc and implications for continental crustal growth. *Geology*, **27**, 31–34.

HONZA, E., JOHN, J. & BANDA, R. M. 2000. An imbrication model for the Rajang accretionary complex in Sarawak, Borneo. *In*: METCALFE, I. & ALLEN, M. B. (eds) *Suture Zones of East and Southeast Asia*. Pergamon, Oxford.

HORINE, R. L., MOORE, G. F. & TAYLOR, B. 1990. Structure of the outer Izu–Bonin forearc from seismic-reflection profiling and gravity modeling. *In*: FRYER, B. P., PEARCE, J. A. & STOKKING, L. B. (eds) *Proceedings of the Ocean Drilling Program, Initial Reports, 125*. Ocean Drilling Program, College Station, TX, 81–94.

HSÜ, K. J., PAN, G. ET AL. 1995. Tectonic evolution of the Tibetan Plateau, a working hypothesis based on the archipelago model of orogenesis. *International Geology Review*, **37**, 473–508.

HUANG, C.-Y., YUAN, P. B., LIN, C.-W., WANG, T. K. & CHANG, C.-P. 2000. Geodynamic processes of Taiwan arc–continent collision and comparison with analogs in Timor, Papua New Guinea, Urals and Corsica. *Tectonophysics*, **325**, 1–21.

ILYIN, A. V. 1990. Proterozoic supercontinent, its latest Precambrian rifting, breakup, dispersal into smaller continents, and subsidence of their margins: evidence from Asia. *Geology*, **18**, 1231–1234.

INGERSOLL, R. & SCHWEICKERT, R. 1986. A plate-tectonic model for late Jurassic ophiolite genesis, Nevadan orogeny and forearc initiation, northern California. *Tectonics*, **5**, 901–912.

ISOZAKI, Y. 1996. Anatomy and genesis of a subduction-related orogen: a new view of geotectonic subdivision and evolution of the Japanese islands. *Island Arc*, **5**, 289–320.

ISOZAKI, Y. 1997a. Contrasting two types of orogen in Permo-Triassic Japan: Accretionary versus collisional. *Island Arc*, **6**, 2–24.

ISOZAKI, Y. 1997b. Jurassic accretion tectonics of Japan. *Island Arc*, **6**, 25–51.

ISOZAKI, Y. & BLAKE, M. C. 1994. Biostratigraphic constraints on formation and timing of accretion in a subduction complex – an example from the Franciscan Complex of northern California. *Journal of Geology*, **102**, 283–296.

ISOZAKI, Y. & OTA, A. 2001. Middle–Upper Permian (Maokouan–Wuchiapingian) boundary in mid-oceanic paleo-atoll limestone of Kamura and Akasaka, Japan. *Proceedings of the Japan Academy Series B, Physical and Biological Sciences*, **77**, 104–109.

JAHN, B.-M., WU, F.-Y. & CHEN, B. 2000. Granitoids of the Central Asian Orogenic Belt and continental growth in the Phanerozoic. *Transactions of the Royal Society of Edinburgh: Earth Sciences*, **91**, 181–193.

JAHN, B.-M., WINDLEY, B., NATAL'IN, B. & DOBRETSOV, N. 2004. Phanerozoic continental growth in Central Asia. *Journal of Asian Earth Sciences*, **23**, 599–603.

JENCHURAEVA, R. J. 2001. Paleozoic geodynamics, magmatism, and metallogeny of the Tien Shan. *In*: SELTMANN, S. & JENCHURAEVA, R. (eds) *Paleozoic Geodynamics and Gold Deposits in the Kyrgyz Tien Shan*. IAGOD Guidebook Series, **9**, 29–70.

KANAMATSU, T., HERRERO, B. E., TAIRA, A., SAITO, S., ASHI, J. & FURUMOTO, A. S. 1996. Magnetic fabric development in the Tertiary accretionary complex in the Boso and Miura peninsulas of central Japan. *Geophysical Research Letters*, **23**, 471–474.

KANAMATSU, T., HERRERO, B. E. & TAIRA, A. 2001. Magnetic fabrics of soft-sediment folded strata within a Neogene accretionary complex, the Miura Group, central Japan. *Earth and Planetary Science Letters*, **187**, 333–343.

KARIG, D. E. 1971. Origin and development of marginal basins in the western Pacific. *Journal of Geophysical Research*, **76**, 2542–2561.

KAWAKAMI, G., ARITA, K., OKADA, T. & ITAYA, T. 2004. Early exhumation of the collisional orogen and concurrent infill of foredeep basins in the Miocene Eurasian–Okhotsk Plate boundary, central Hokkaido, Japan: inferences from K–Ar dating of granitoid clasts. *Island Arc*, **13**, 359–369.

KAY, R. W. 1978. Aleutian magnesian andesites: melts from subducted Pacific oceanic crust. *Journal of Volcanology and Geothermal Research*, **4**, 117–132.

KAY, S. M., MAKSAEV, V., MOSCOSO, R., MPODOZIS, C., NASI, C. & GORDILLO, C. E. 1988. Tertiary Andean magmatism in Chile and Argentina between 288° and 338°: correlation of magmatic chemistry with a changing Benioff zone. *Journal of Southeast American Earth Sciences*, **1**, 21–39.

KHAIN, E. V., BIBIKOVA, E. V., KRÖNER, A., ZHURAVLEV, D. Z., SKLYAROV, E. V., FEDOTOVA, A. A. & KRAVCHENKO-BEREZHNOY, I. R. 2002. The most ancient ophiolite of Central Asian fold belt: U–Pb and Pb–Pb zircon ages for the Dunzhungur Complex, Eastern Sayan, Siberia, and geodynamic implications. *Earth and Planetary Science Letters*, **199**, 311–325.

KHAIN, V. E. 1975. Structure and main stages in the tectono-magmatic development of the Caucasus: an attempt at geodynamic interpretation. *American Journal of Science*, **275-A**, 131–156.

KIMURA, G. 1994. The latest Cretaceous–early Paleogene rapid growth of accretionary complex and exhumation of high-pressure series metamorphic rocks in northwestern Pacific margin. *Journal of Geophysical Research, B, Solid Earth and Planets*, **99**, 22 147–22 164.

KIMURA, G. & MUKAI, A. 1991. Underplated units in an accretionary complex; mélange of the Shimanto Belt of eastern Shikoku, Southwest Japan. *Tectonics*, **10**, 31–50.

KIMURA, G. & ONISHI, T. C. 1994. Change in relative convergence recorded in the mélange of the accretionary complex, the Shimanto Belt, Japan. Geological Society of America, Abstracts with Programs, **26**, 214.

KOZAKOV, I. K., KOTOV, A. B., SALIKOVA, E. B., KOVACH, V. P., KIRNOZOVA, T. I., BEREZHNAVA, N. G. & LYKIN, D. A. 1999. Metamorphic age of crystalline complex of the Tuva–Mongolia massif: U–Pb geochronology of granitoids. *Petrology*, **7**, 177–191.

KOZAKOV, I. K., SAL'NIKOVA, E. B., WANG, T., DIDENKO, A. N., PLOTKINA, Y. V. & PODKOVYROV, V. N. 2007. Early Precambrian crystalline complexes of the Central Asian microcontinent: Age, sources, tectonic position. *Stratigraphy and Geological Correlation*, **15**, 121–140.

KRÖNER, A., WINDLEY, B. F. ET AL. 2007. Accretionary growth and crust-formation in the central Asian Orogenic Belt and comparison with the Arabian–Nubian shield. *In*: HATCHER, R. D. JR, CARLSON, M. P., MCBRIDE, J. H. & MARTÍNEZ CATALAN, J. R. (eds) *4-D Framework of Continental Crust*. Geological Society of America, Memoirs, **200**, 181–209.

KUSKY, T. M. 1997. Are the ultramafic massifs of the Kenai Peninsula, Chugach Terrane (Alaska) remnants of an accreted oceanic plateau? Implications for continental growth. Geological Society of America, Abstracts with Programs, **29**, 246.

KUSKY, T. M. & BRADLEY, D. C. 1999. Kinematic analysis of mélange fabrics: examples and applications from the McHugh Complex, Kenai Peninsula, Alaska. *Journal of Structural Geology*, **21**, 1773–1796.

KUSKY, T. M. & YOUNG, C. P. 1999. Emplacement of the Resurrection Peninsula ophiolite in the southern Alaska forearc during a ridge–trench encounter. *Journal of Geophysical Research, Solid Earth*, **104**, 29 025–29 054.

KUZMICHEV, A. B., BIBIKOVA, E. V. & ZHURAVLEV, D. Z. 2001. Neoproterozoic (~800 Ma) orogeny in the Tuva–Mongolia Massif (Siberia): island arc–continent collision at the northeast Rodinia margin. *Precambrian Research*, **110**, 109–126.

LAGABRIELLE, Y., SUÁREZ, M. ET AL. 2007. Pliocene extensional tectonics in the Eastern Central Patagonian Cordillera: geochronological constraints and new field evidence. *Terra Nova*, **19**, 413–424.

LAWVER, L. A., DALZIEL, I. W. D., GAHAGAN, L. M., MARTIN, K. M. & CAMPBELL, D. A. 2003. *The PLATES 2003 Atlas of Plate Reconstructions (750 Ma to Present Day)*. PLATES Progress Report **280–0703**; UTIG Technical Report **190**.

LEGGETT, J., AOKI, Y. & TOBA, T. 1985. Transition from frontal accretion to underplating in a part of the Nankai Trough accretionary complex off Shikoku (SW Japan) and extensional features on the lower trench slope. *Marine and Petroleum Geology*, **2**, 131–146.

LEVASHOVA, N. M., VAN DER VOO, R., ABRAJEVITCH, A. V. & BAZHENOV, M. L. 2009. Paleomagnetism of mid-Paleozoic subduction-related volcanics from the Chingiz Range in NE Kazakhstan: the evolving paleogeography of the amalgamating Eurasian composite continent. *Geological Society of America Bulletin*, **121**, 555–573.

LIN, C.-H. 2002. Active continental subduction and crustal exhumation: the Taiwan orogeny. *Terra Nova*, **14**, 281–287.

LIOU, J. G., TSUJIMORI, T., ZHANG, R. Y., KATAYAMA, I. & MARUYAMA, S. 2004. Global UHP metamorphism and continental subduction/collision: the Himalayan model. *International Geology Review*, **46**, 1–27.

LYTWYN, J., LOCKHART, S., CASEY, J. & KUSKY, T. 2000. Geochemistry of near-trench intrusives associated with ridge subduction, Seldovia Quadrangle, southern Alaska. *Journal of Geophysical Research, Solid Earth*, **105**, 27957–27978.

MARTIN, M. W., KATO, T. T., RODRIGUEZ, C., GODOY, E., DUHART, P., MCDONOUGH, M. & CAMPOS, A. 1999. Evolution of the late Paleozoic accretionary complex and overlying forearc–magmatic arc, south central Chile (38°41°S); constraints for the tectonic setting along the southwestern margin of Gondwana. *Tectonics*, **18**, 582–605.

MARUYAMA, S. 1997. Pacific-type orogeny revisited: Miyashiro-type orogeny proposed. *Island Arc*, **6**, 91–120.

MASCLE, A., BIJU-DUAL, B., DE CLARENS, P. & MUNSCH, N. 1996. Growth of accretionary prisms: tectonic process from Caribbean examples. *In*: WEZEL, F. C. (ed.) *The Origin of Arcs*. Elsevier, Amsterdam, 375–400.

MCCLELLAND, W. C., TIKOFF, B. & MANDUCA, C. A. 2000. Two-phase evolution of accretionary margins: examples from the North American Cordillera. *Tectonophysics*, **326**, 37–55.

MILKOV, A. V., CLAYPOOL, G. E. ET AL. 2003. In situ methane concentrations at Hydrate Ridge, offshore Oregon; new constraints on the global gas hydrate inventory from an active margin. *Geology*, **31**, 833–836.

MILLER, K. C. 2002. Geophysical evidence for Miocene extension and mafic magmatic addition in the California Continental Borderland. *Geological Society of America Bulletin*, **114**, 497–512.

MILLER, M. S., KENNETT, B. L. N. & LISTER, G. S. 2004. Imaging changes in morphology, geometry, and physical properties of the subducting Pacific plate along the Izu–Bonin–Mariana arc. *Earth and Planetary Science Letters*, **224**, 363–370.

MIYAZAKI, K. & OKUMURA, K. 2002. Thermal modelling in shallow subduction; an application to low P/T metamorphism of the Cretaceous Shimanto accretionary complex, Japan. *Journal of Metamorphic Geology*, **20**, 441–452.

MOORE, J. C. & SAFFER, D. 2001. Updip limit of the seismogenic zone beneath the accretionary prism of southwest Japan: An effect of diagenetic to low-grade metamorphic processes and increasing effective stress. *Geology*, **29**, 183–186.

MOORES, E. 1970. Ultramafics and orogeny, with models of the US Cordillera and the Tethys. *Nature*, **228**, 837–842.

MOORES, E. M. 1998. Ophiolites, the Sierra Nevada, 'Cordillera', and orogeny along the Pacific and Caribbean margins of North and South America. *International Geology Review*, **40**, 40–54.

MOSSAKOVSKY, A. A., RUZHENTSEV, S. V., SAMYGIN, S. G. & KHERASKOVA, T. N. 1993. The Central Asian fold belt: geodynamic evolution and formation history. *Geotectonics*, **26**, 455–473.

MURPHY, J. B., NANCE, R. D. & CAWOOD, P. A. 2010. Contrasting modes of supercontinent formation and the conundrum of Pangea. *Gondwana Research*, **15**, 408–420.

NANCE, R. D., MURPHY, J. B. & KEPPIE, J. D. 2002. A Cordilleran model for the evolution of Avalonia. *Tectonophysics*, **352**, 11–31.

NEEDHAM, D. T., AGAR, S. M. & MACKENZIE, J. S. 1991. Controls on the structural development of an accretionary complex – the Shimanto Belt, southwest Japan. *Tectonophysics*, **191**, 433–433.

NELSON, E. P. 1996. Suprasubduction mineralization: metallo-tectonic terranes of the southernmost Andes. *In*: BEBOUT, G. E., SCHOLL, D. W., KIRBY, S. H. & PLATT, J. P. (eds) *Subduction from Top to Bottom*. American Geophysical Union, Geophysical Monograph, **96**, 315–330.

NOKLEBERG, W. J., BUNDTZEN, T. K. ET AL. 2005. Metallogenesis and Tectonics of the Russian Far East, Alaska and the Canadian Cordillera. US Geological Survey, Professional Papers, 1697.

OGAWA, Y., NISHIDA, Y. & MAKINO, M. 1994. A collision boundary imaged by magnetotellurics, Hidaka Mountains, central Hokkaido, Japan. *Journal of Geophysical Research-Solid Earth*, **99**, 22 373–22 388.

OKAMOTO, K., MARUYAMA, S. & ISOZAKI, Y. 2000. Accretionary complex origin of the Sanbagawa, high P/T metamorphic rocks, central Shikoku, Japan: layer-parallel shortening structure and greenstone geochemistry. *Journal of the Geological Society of Japan*, **106**, 70–86.

OSOZAWA, S. 1992. Double ridge subduction recorded in the Shimanto accretionary complex, Japan, and plate reconstruction. *Geology*, **20**, 939–942.

PANKHURST, R. J., WEAVER, S. D., HERVE, F. & LARRONDO, P. 1999. Mesozoic–Cenozoic evolution of the North Patagonian Batholith in Aysen, southern Chile. *Journal of the Geological Society, London*, **156**, 673–694.

PFÄNDER, J. A., JOCHUM, K. P., KOZAKOV, I., KRÖNER, A. & TODT, W. 2002. Coupled evolution of back-arc and island arc-like mafic crust in the late-Neoproterozoic Agardagh Tes-Chem ophiolite, Central Asia: evidence from trace element and Sr–Nd–Pb isotope data. *Contributions to Mineralogy and Petrology*, **143**, 154–174, doi:10.1007/s00410-001-0340-7.

PLANK, T. & LANGMUIR, C. H. 1998. The chemical composition of subducted sediment and its consequences for the crust and mantle. *Chemical Geology*, **145**, 325–394.

RANERO, C. R. & VON HUENE, R. 2003. Subduction erosion along the Middle America convergent margin. *Nature*, **404**, 748–752.

RATSCHBACHER, L., DINGELDEYA, C., MILLER, C., HACKER, B. R. & MCWILLIAMS, M. O. 2004. Formation, subduction, and exhumation of Penninic oceanic crust in the Eastern Alps: time constraints from $^{40}Ar/^{39}Ar$ geochronology. *Tectonophysics*, **394**, 155–170.

REICH, M., PARADA, M. A., PALACIOS, C., DIETRICH, A., SCHULTZ, F. & LEHMANN, B. 2003. Adakite-like signature of Late Miocene intrusions at the Los Pelambres giant porphyry copper deposit in the Andes of central Chile: metallogenic implications. *Mineralium Deposita*, **38**, 876–885.

RING, U. (ed.) 2008. *Deformation and Exhumation at Convergent Margins: The Franciscan Subduction Complex*. Geological Society of America, Special Papers, **445**.

RINO, S., KON, Y., SATO, W., MARUYAMA, S., SANTOSH, M. & ZHAO, D. 2008. The Grenvillian and Pan-African orogens: World's largest orogenies through geologic time, and their implications on the origin of superplume. *Gondwana Research*, **14**, 51–72.

SALNIKOVA, E. B., KOZAKOV, I. K. ET AL. 2001. Age of Palaeozoic granites and metamorphism in the Tuva–Mongolia massif of the Central Asian Mobile Belt: loss of a Precambrian microcontinent. *Precambrian Research*, **110**, 143–164.

SAMPLE, J. C. & FISHER, D. M. 1986. Duplex accretion and underplating in an ancient accretionary complex, Kodiak Islands, Alaska. *Geology*, **14**, 160–163.

Santosh, M., Maruyama, S. & Sato, K. 2009. Anatomy of a Cambrian suture in Gondwana: pacific-type orogeny in southern India? *Gondwana Research*, **16**, 321–341, doi:10.1016/j.gr.2008.12.012.

Sato, H., Iwasaki, T. *et al.* 2004. Formation and shortening deformation of a back-arc rift basin revealed by deep seismic profiling, central Japan. *Tectonophysics*, **388**, 47–58.

Schemmann, K., Unruh, J. R. & Moores, E. M. 2008. Kinematics of Franciscan Complex exhumation: new insights from the geology of Mount Diablo, California. *Geological Society of America Bulletin*, **120**, 543–555.

Scott, D. J. & Gauthier, G. 1996. Comparison of TIMS (U–Pb) and laser ablation microprobe ICP-MS (Pb) techniques for age determination of detrital zircons from Paleoproterozoic metasedimentary rocks from northeastern Laurentia, Canada, with tectonic implications. *Chemical Geology*, **131**, 127–142.

Şengör, A. M. C. 1992. The Paleo-Tethys suture: a line of demarcation between two fundamentally different architectural style in the structure of Asia. *Island Arc*, **1**, 78–91.

Şengör, A. M. C. & Natal'in, B. 1996a. Paleotectonics of Asia: fragments of a synthesis. *In*: Yin, A. & Harrison, M. (eds) *The Tectonic Evolution of Asia*. Cambridge University Press, Cambridge, 486–640.

Şengör, C. & Natal'in, B. 1996b. Turkic-type orogeny and its role in the making of the continental crust. *Annual Review of Earth and Planetary Sciences*, **24**, 263–337.

Şengör, A. M. C. & Okurogullari, A. H. 1991. The role of accretionary wedges in the growth of continents: Asiatic examples from Argand to plate tectonics. *Eclogae Geologicae Helvetiae*, **84**, 535–597.

Şengör, A. M. C., Natal'in, B. A. & Burtman, U. S. 1993. Evolution of the Altaid tectonic collage and Paleozoic crustal growth in Eurasia. *Nature*, **364**, 209–304.

Shervais, J. W., Kimbrough, D. L. *et al.* 2004. Multi-stage origin of the Coast Range ophiolite, California: implications for the life cycle of suprasubduction zone ophiolites. *International Geology Review*, **46**, 289–315.

Shyu, J. B. H., Sieha, T., Kerry, S. & Chen, Y.-G. 2005. Tandem suturing and disarticulation of the Taiwan orogen revealed by its neotectonic elements. *Earth and Planetary Science Letters*, **233**, 167–177.

Sillitoe, R. H. 1974. Tectonic segmentation of the Andes: implications for magmatism and metallogeny. *Nature*, **250**, 542–545.

Sisson, V. B. & Pavlis, T. L. 1993. Geologic consequences of plate reorganization – an example from the Eocene southern Alaska forearc. *Geology*, **21**, 913–916.

Stern, R. J. 2004. Subduction initiation: spontaneous and induced. *Earth and Planetary Science Letters*, **226**, 275–292.

Tagami, T. & Hasebe, N. 1999. Cordilleran-type orogeny and episodic growth of continents: Insights from the circum-Pacific continental margins. *Island Arc*, **8**, 206–217.

Taira, A. 2001. Tectonic evolution of the Japanese island arc system. *Annual Review of Earth and Planetary Sciences*, **29**, 109–134.

Taira, A., Byrne, T. & Ashi, J. 1992a. *Photographic Atlas of an Accretionary prism: Geological Structures of the Shimanto Belt, Japan*. University of Tokyo Press, Tokyo.

Taira, A., Pickering, K. T., Windley, B. F. & Soh, W. 1992b. Accretion of Japanese island arcs and implications for the origin of Archean greenstone belts. *Tectonics*, **11**, 1224–1244.

Tomurtogoo, O., Windley, B. F., Kroner, A., Badarch, G. & Liu, D. Y. 2005. Zircon age and occurrence of the Adaatsag ophiolite and Muron shear zone, central Mongolia: constraints on the evolution of the Mongol–Okhotsk ocean, suture and orogen. *Journal of the Geological Society, London*, **162**, 125–134.

Ujiie, K. 2002. Evolution and kinematics of an ancient décollement zone, mélange in the Shimanto accretionary complex of Okinawa Island, Ryukyu Arc. *Journal of Structural Geology*, **24**, 937–952.

Underwood, M. B., Shelton, K. L., McLaughlin, R. J., Laughland, M. M. & Solomon, R. M. 1999. Middle Miocene paleotemperature anomalies within the Franciscan Complex of Northern California; thermotectonic responses near the Mendocino triple junction. *Geological Society of America Bulletin*, **111**, 1448–1467.

Unruh, J. R., Loewen, B. A. & Moores, E. M. 1995. Progressive arcward contraction of a Mesozoic–Tertiary fore-arc basin, southwestern Sacramento Valley, California. *Geological Society of America Bulletin*, **107**, 38–53.

Utsunomiya, A. & Jahn, B. M. 2008. Preservation of oceanic plateaus in accretionary orogens: examples from Cretaceous Circum-Pacific region and Late Precambrian Gorny Altai, southern Siberia. *In*: Xiao, W. J., Zhai, M. G., Li, X. H. & Liu, F. (eds) *The Gondwana 13, Program & Abstracts*. Dali, China, 209.

Van der Voo, R. 2004. Paleomagnetism, oroclines and the growth of the continental crust. *GSA Today*, **14**, 4–9.

Van Staal, C. R. 1994. Brunswick subduction complex in the Canadian Appalachians: record of the Late Ordovician to Late Silurian collision between Laurentia and the Gander margin of Avalon. *Tectonics*, **13**, 946–961, doi:10.1029/93TC03604.

Van Staal, C. R., Dewey, J. F., Niocaill, C. M. & McKerrow, W. S. 1998. The Cambrian–Silurian tectonic evolution of the northern Appalachians and British Caledonides: history of a complex, west and southwest Pacific-type segment of Iapetus. *In*: Blundell, D. J. & Scott, A. C. (eds) *Lyell: the Past is the Key to the Present*. Geological Society, London, Special Publications, **143**, 197–242, doi: 10.1144/GSL.SP.1998.143.01.17.

Vaughan, A. P. M. & Livermore, R. A. 2005. Episodicity of Mesozoic terrane accretion along the Pacific margin of Gondwana: implications for superplume–plate interactions. *In*: Vaughan, A. P. M., Leat, P. T. & Pankhurst, R. J. (eds) *Terrane Processes at the*

Margins of Gondwana. Geological Society, London, Special Publications, **246**, 143–178.

VON HUENE, R. & SCHOLL, D. W. 1991. Observations at convergent margins concerning sediment subduction, subduction erosion, and the growth of continental crust. *Reviews of Geophysics*, **29**, 279–316.

VROLIJK, P., MYERS, G. & MOORE, J. C. 1988. Warm fluid migration along tectonic mélanges in the Kodiak accretionary complex, Alaska. *Journal of Geophysical Research, B, Solid Earth and Planets*, **93**, 10 313–10 324.

WAKABAYASHI, J. & DILEK, Y. 2000. Spatial and temporal relationships between ophiolites and their metamorphic soles: a test of models of forearc ophiolite genesis. *In*: DILEK, Y., MOORES, E., ELTHON, D. & NICOLAS, A. (eds) *Ophiolites and Oceanic Crust: New Insights from Field Studies and the Ocean Drilling Program*. Geological Society of America, Special Papers, 349, 53–64.

WAKITA, K. 1988. Early Cretaceous mélange in the Hida–Kanayama area, central Japan. *Bulletin of the Geological Survey of Japan*, **39**, 367–421.

WANG, Y. J. & FAN, Z. Y. 1997. Discovery of the Permian radiolarian fossils in the ophiolites north of the Xar Moron River, Inner Mongolia, and its geological implications. *Acta Palaeontologica Sinica*, **36**, 58–69.

WILLIAMS, P. R., SUPRIATNA, S., JOHNSTON, C. R., ALMOND, R. A. & SIMAMORA, W. H. 1989. A Late Cretaceous to early Tertiary accretionary complex in West Kalimantan. *Bulletin of the Geological Research and Development Centre, [Republic of Indonesia]*, **13**, 9–29.

WINDLEY, B. F. 1995. *The Evolving Continents*. Wiley, Chichester.

WINDLEY, B. F. 1998. Tectonic development of Early Precambrian orogens. *Origin and Evolution of Continents*. Memoir of National Institute of Polar Research, Special Issue, **53**, 8–28.

WINDLEY, B. F., ALEXEIEV, D., XIAO, W., KRÖNER, A. & BADARCH, G. 2007. Tectonic models for accretion of the Central Asian Orogenic belt. *Journal of the Geological Society, London*, **164**, 31–47.

WU, F. Y., YANG, J. H., LO, C. H., WILDE, S. A., SUN, D. Y. & JAHN, B. M. 2007. The Heilongjiang Group: A Jurassic accretionary complex in the Jiamusi Massif at the western Pacific margin of northeastern China. *Island Arc*, **16**, 156–172.

XIAO, W. J., SUN, S., LI, J. L. & CHEN, H. H. 2001. South China archipelago orogenesis. *In*: BRIEGEL, U. & XIAO, W. J. (eds) *Paradoxes in Geology*. Elsevier, Amsterdam, 15–37.

XIAO, W. J., WINDLEY, B. F., HAO, J. & LI, J. L. 2002a. Arc-ophiolite obduction in the Western Kunlun Range (China): implications for the Palaeozoic evolution of central Asia. *Journal of the Geological Society, London*, **159**, 517–528.

XIAO, W. J., WINDLEY, B. F., CHEN, H. L., ZHANG, G. C. & LI, J. L. 2002b. Carboniferous–Triassic subduction and accretion in the western Kunlun, China: implications for the collisional and accretionary tectonics of the northern Tibetan plateau. *Geology*, **30**, 295–298.

XIAO, W. J., WINDLEY, B. F., HAO, J. & ZHAI, M. G. 2003. Accretion leading to collision and the Permian Solonker suture, Inner Mongolia, China: termination of the Central Asian orogenic belt. *Tectonics*, **22**, 1069, doi: 10.1029/2002TC1484.

XIAO, W. J., ZHANG, L. C., QIN, K. Z., SUN, S. & LI, J. L. 2004a. Paleozoic accretionary and collisional tectonics of the Eastern Tianshan (China): implications for the continental growth of central Asia. *American Journal of Science*, **304**, 370–395.

XIAO, W. J., WINDLEY, B. F., BADARCH, G., SUN, S., LI, J. L., QIN, K. Z. & WANG, Z. H. 2004b. Palaeozoic accretionary and convergent tectonics of the southern Altaids: implications for the lateral growth of Central Asia. *Journal of the Geological Society, London*, **161**, 339–342.

XIAO, W. J., PIRAJNO, F. & SELTMANN, R. 2008a. Geodynamics and metallogeny of the Altaid orogen. *Journal of Asian Earth Sciences*, **32**, 77–81, doi:10.1016/j.jseaes.2007.10.003.

XIAO, W. J., HAN, C. M. ET AL. 2008b. Middle Cambrian to Permian subduction-related accretionary orogenesis of North Xinjiang, NW China: implications for the tectonic evolution of Central Asia. *Journal of Asian Earth Sciences*, **32**, 102–117, doi: 10.1016/j.jseas.2007.10.008.

XIAO, W. J., KRÖNER, A. & WINDLEY, B. F. 2009a. Geodynamic evolution of Central Asia in the Paleozoic and Mesozoic. *International Journal of Earth Sciences*, **98**, 1185–1188, doi: 10.1007/s00531-009-0418-4.

XIAO, W. J., WINDLEY, B. F., YONG, Y., YAN, Z., YUAN, C., LIU, C. Z. & LI, J. L. 2009b. Early Paleozoic to Devonian multiple-accretionary model of the Qilian Shan, NW China. *Journal of Asian Earth Sciences*, **35**, 323–333, doi:10.1016/j.jseaes.2008.10.001.

XIAO, W. J., WINDLEY, B. F. ET AL. 2009c. End-Permian to mid-Triassic termination of the accretionary processes of the southern Altaids: implications for the geodynamic evolution, Phanerozoic continental growth, and metallogeny of Central Asia. *International Journal of Earth Sciences*, **98**, 1189–1217, doi:10.1007/s00531-008-0407-z.

YAKUBCHUK, A. 2004. Architecture and mineral deposit settings of the Altaid orogenic collage: a revised model. *Journal of Asian Earth Sciences*, **23**, 761–779.

YAKUBCHUK, A. 2008. Re-deciphering the tectonic jigsaw puzzle of northern Eurasia. *Journal of Asian Earth Sciences*, **32**, 82–101.

YAKUBCHUK, A. S., SELTMANN, R., SHATOV, V. & COLE, A. 2001. The Altaids: tectonic evolution and metallogeny. *Society of Economic Geologists Newsletters*, **46**, 7–14.

YAMAMOTO, S., SENSHU, H., RINO, S., OMORI, S. & MARUYAMA, S. 2010. Granite subduction: arc subduction, tectonic erosion and sediment subduction. *Gondwana Research*, **15**, 443–453.

YIN, A. & HARRISON, T. M. 2000. Geological evolution of the Himalayan–Tibetan Orogen. *Annual Review of Earth and Planetary Sciences*, **28**, 211–280.

YOSHIKURA, S.-I. & HADA, S. 2004. Field Guide: Cretaceous to Holocene growth of continental margin recorded in the Shimanto Belt at Muroto Peninsula, Shikoku, Japan. *Gondwana Research*, **7**, 1409–1420.

ZHANG, S.-H., ZHAO, Y., SONG, B. & YANG, Y.-H. 2007*a*. Zircon SHRIMP U–Pb and *in-situ* Lu–Hf isotope analyses of a tuff from Western Beijing: evidence for missing Late Paleozoic arc volcano eruptions at the northern margin of the North China block. *Gondwana Research*, **12**, 157–165.

ZHANG, S.-H., ZHAO, Y., SONG, B., YANG, Z.-Y., HU, J.-M. & WU, H. 2007*b*. Carboniferous granitic plutons from the northern margin of the North China block: implications for a late Paleozoic active continental margin. *Journal of the Geological Society, London*, **164**, 451–463.

ZHANG, X., PAULSSEN, H., LEBEDEV, S. & MEIER, T. 2009. 3D shear velocity structure beneath the Gulf of California from Rayleigh wave dispersion. *Earth and Planetary Science Letters*, **279**, 255–262.

ZHAO, G., CAWOOD, P. A., WILDE, S. A. & SUN, M. 2002. Review of global 2.1–1.8 Ga orogens: implications for a pre-Rodinia supercontinent. *Earth-Science Reviews*, **59**, 125–162.

Ocean plate stratigraphy and its imbrication in an accretionary orogen: the Mona Complex, Anglesey–Lleyn, Wales, UK

S. MARUYAMA[1], T. KAWAI[1]* & B. F. WINDLEY[2]

[1]*Department of Earth and Planetary Sciences, Tokyo Institute of Technology, O-okayama 2-12-1, Meguro, Tokyo 152-8551, Japan*

[2]*Department of Geology, University of Leicester, Leicester LE1 7RH, UK*

*Corresponding author (e-mail: tkawai@geo.titech.ac.jp)

Abstract: We re-evaluate the Neoproterozoic, Pacific-type accretionary complex on Anglesey and in the Lleyn peninsula (Wales, UK), by reconstructing its ocean plate stratigraphy (OPS). Three types of distinctive OPS were successively emplaced downwards in an accretionary wedge: the oldest at the top formed when an ocean opened and closed from a ridge to a trench, the central OPS was subjected to deep subduction and exhumed as blueschists, and the youngest at the bottom is an olistostrome-type deposit that formed by secondary gravitational collapse of previously accreted material. The three types formed by successive eastward subduction of young oceanic lithosphere at the leading edge of Avalonia. The downward growth of the accretionary complex through time was almost coeval with exhumation of the blueschist unit at 550–560 Ma in the structural centre of the complex on Anglesey. From balanced sections we have reconstructed the ocean plate stratigraphy on Llanddwyn Island from which we calculate that about 8 km of lateral shortening of ocean floor took place during imbrication and accretion; this is comparable with the history of plate subduction around the Pacific Ocean. We also calculate that the age of the subducted lithosphere was very young, probably less than 10 Ma.

Ocean plate stratigraphy (OPS) is the fundamental, first-order structure of accretionary orogens that are forming today on the Pacific margins (Isozaki *et al.* 1990; Kusky *et al.* 1997a, b). It is a composite stratigraphic succession of the ocean floor reconstructed by means of microfossil biostratigraphy (mostly radiolaria, but also fusulinids and conodonts) from protolith sediments in accretionary wedges and mélanges. It consists of the following sediments in ascending order. At the base are cherts deposited under deep-sea pelagic conditions at or near a mid-oceanic ridge. The cherts may be overlain by limestones if the ridge rises above the carbonate compensation depth. The basal cherts are transported across the ocean as it opens, and are overlain by more and more cherts, as the pelagic ocean floor sediments are transported towards a subduction zone. When the pelagic sediments reach a hemipelagic environment on the offshore side of a trench typically near a continental margin, the oceanic cherts are succeeded by hemipelagic siliceous shales and mudstones that consist of radiolaria and fine-grained continental-derived detritus. Finally, when the sediments enter a trench, they are covered by voluminous clastic sediments in the form of shales, sandstones, conglomerates and turbidites that were deposited by turbidity currents that transport continental-derived detritus into the trench. The classic OPS section is represented by mid-ocean ridge basalt (MORB)–chert (limestone)–hemipelagic shale/mudstone–sandstone/conglomerate/turbidite (Matsuda & Isozaki 1991; Fig. 1). Moreover, when an oceanic plate passes over a plume, a seamount or volcanic island is created that is characterized by ocean island basalts overlain by reef limestones, given the appropriate latitude and carbonate compensation depth (Okamura 1991; Ota *et al.* 2007; Takayanagi *et al.* 2007). When cherts on the ocean floor pass over the plume head, they may be intruded by sills of alkali basalt, as illustrated in Figure 1. A new component to the OPS story was provided by Bradley *et al.* (2003), who pointed out that the subduction of a mid-oceanic ridge can explain the interbedding of trench sediments and pillow lavas, and the development of high-grade metamorphism of an accretionary wedge.

Although the OPS model is an idealized reconstruction, it is important to emphasize the fact that the idea that OPS represents a transition of sediment from a ridge to a trench has been extremely tightly constrained and supported by an enormous databank of material derived first from dredge and/or drill-hole material from ridges, the ocean floor and from trenches, and second from the presence of complete OPS in accretionary wedges either in still-evolving subduction–accretion complexes in the western and northern Pacific or in accreted

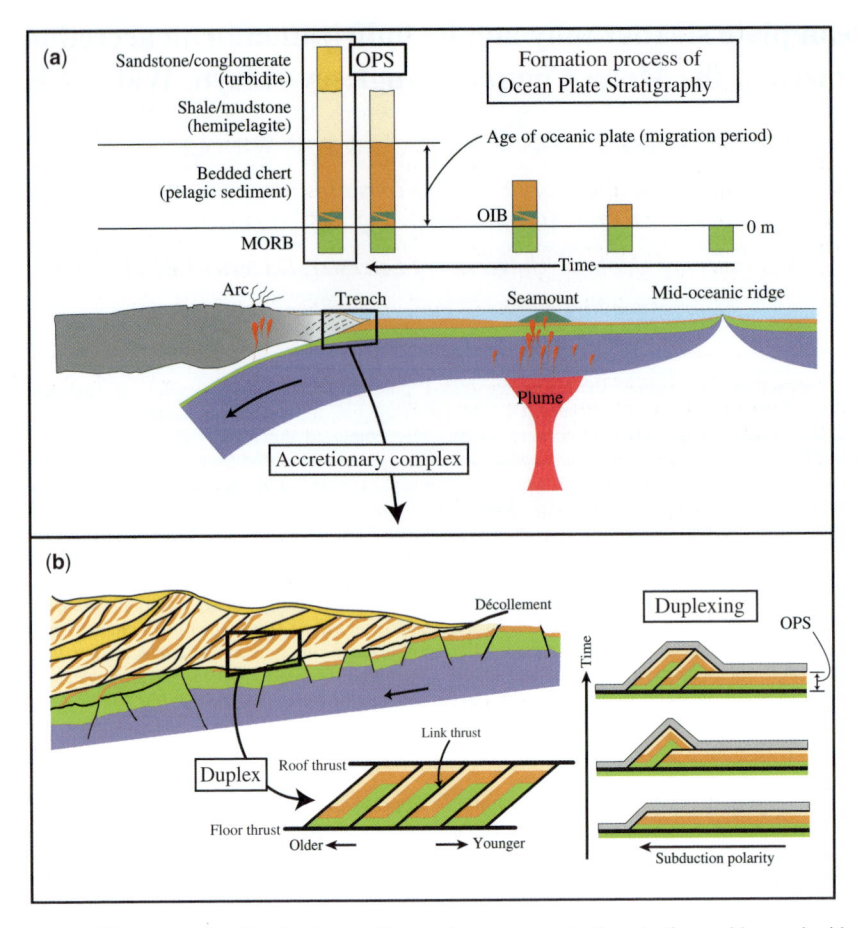

Fig. 1. Schematic illustration of an idealized travel history of ocean plate stratigraphy from mid-oceanic ridge to subduction zone (Isozaki *et al.* 1990). In (**a**) to the left of the mid-oceanic ridge the oceanic plate with its cover of deep-water bedded chert has passed over a hotspot plume creating a seamount with ocean island basalts; although common in many on-land accretionary complexes we do not see this in Anglesey–Lleyn. Farther left, older ocean floor has passed through deeper water where thicker pelagic cherts have been progressively deposited, and when they approach the trench (in b) they are overlain by hemipelagic mudstones. Finally, in the trench (b) turbidites are deposited, forming a cap to the OPS, and here already-accreted, water-saturated sediments may slump to form gravitational olistostrome deposits. (**b**) This shows the typical duplex structure of OPS of accretionary wedges, from which the subduction polarity can be determined.

material in Palaeozoic–Cenozoic accretionary and collisional orogens, the origin of that material having being well constrained by biostratigraphy and trace element geochemistry. Accordingly, it is well accepted that OPS represents a viable and indeed key plate-tectonic indicator (Wakita & Metcalfe 2005). OPS has proved useful as a means of determining the width of oceans, because the age range of chert deposited between a mid-ocean ridge and a trench documents the duration of oceanic plate migration (Xenophontos & Osozawa 2004). The OPS is present in all the subduction–accretion complexes across Japan from the Nankai Trough to the Japan Sea (Isozaki *et al.* 1990).

The classic or indeed type-section of OPS is at Inuyama in Central Japan; this consists of a series of stacked thrust sheets, for which the stratigaphy was first worked out by Matsuda & Isozaki (1991) using radiolaria. Each thrust sheet is composed of Early Triassic to Middle Jurassic lithological units (bottom cherts to top clastic deposits), all well defined by radiolaria into nine assemblage zones. The lower cherts and upper sandstones each have calculated original stratigraphic thicknesses of about 100 m. However, offscraping accretion gave rise first to a décollement in the hemipelagic mudstones, followed by in-sequence thrusting, then imbrication to produce up to six thrust sheets in a kilometre-scale duplex, and finally out-of-sequence thrusting, which increased the thickness of single chert and sandstone units by up to 2 km (Kimura & Hori 1993).

Syntheses of all the stratigraphic and structural data available from research, especially over the last 25 years, from the ocean floor and trenches integrated with comparable data in already accreted, onshore mélanges and accretionary wedges have given rise to the current, widely recognized ocean plate stratigraphy and to the mechanisms of development of the duplexes and thrust imbricates in accretionary wedges best exemplified by OPS, as shown in Figure 1. The duplex is a key structure that has been worked out and confirmed in many accretionary complexes in the western Pacific, where it occurs in the hanging wall of the trench against a still-open ocean, and thus its geometry can be used to indicate the subduction polarity (Isozaki et al. 1990; see Fig. 1); this geometry can be used in old accretionary orogens to indicate the direction of the original subduction, as we show in Anglesey. These sedimentary and structural relations in well-documented examples of modern OPS in the western and northern Pacific form the basis of our understanding of the Neoproterozoic OPS on Anglesey.

However, ocean plate stratigraphy is still poorly understood and demonstrated in pre-Mesozoic orogens. The few well-studied examples include early Palaeozoic subduction–accretion complexes in the Caledonides of the Southern Uplands of Scotland (Leggett et al. 1982) and Galway, western Ireland (Ryan & Dewey 2004), the Appalachians (Lash 1985), the Gorny Altai range in Siberia (Sennikov et al. 2004; Ota et al. 2007), northern Thailand (Hara et al. 2010), and the Lachlan fold belt of SE Australia (Miller & Gray 1996). Precambrian examples include the Palaeoproterozoic high-grade Lewisian complex of NW Scotland (Park et al. 2001), Archaean greenstone belts in the Slave Province of Canada (Kusky 1989, 1990, 1991), early Archaean greenstone belts at Cleaverville and Marble Bar in the Pilbara of Northern Australia (Kato et al. 1998; Kato & Nakamura 2003), and the early Archaean Isua belt of West Greenland (Komiya et al. 1999).

The Mona Complex in Anglesey, Wales is a fragment of a c. 600–500 Ma subduction–accretion orogen, but its ocean plate stratigraphy has been only briefly reported in relation to the overall tectonic evolution (Kawai et al. 2006, 2007). The aim of this paper is to present a detailed study of ocean plate stratigraphy in Anglesey, emphasizing the occurrence and interpretation of three types of OPS, and their delamination and imbrication during underplating and successive accretion.

The geological units on Anglesey

The Mona Complex (Fig. 2) was meticulously described by Greenly (1919), and interpreted as a subduction–accretion complex by Wood (1974), Shackleton (1975), Barber & Max (1979), Gibbons & Horák (1996) and Strachan et al. (2007). [For a detailed geological map see British Geological Survey (1980).] The complex continues in the adjacent Lleyn peninsula of Wales. Figure 2 shows an outline of the main tectonic belts of Anglesey and their relationships to those on Lleyn.

The most important starting materials of this Pacific-type subduction–accretion complex were the Gwna Group, the New Harbour Group, and the protoliths of the blueschists.

In the following, we summarize these geological units on Anglesey Island.

The Coedana granite–gneiss complex. In the centre of the island quartzo-feldspathic gneisses (up to sillimanite–kyanite grade) have a metamorphic U–Pb zircon age of 666 ± 7 Ma (Strachan et al. 2007), and within the gneisses is a post-tectonic granite (Coedana granite) that has a U–Pb zircon age of 613 ± 4 Ma (Tucker & Pharaoh 1991).

The Gwna Group. The Gwna Group, which crops out in about six areas on Anglesey, contains many lenses up to a few kilometres long of greenschist-grade, pillow-bearing metabasalt and deep-water chert and limestone in a matrix of chloritic greenschist to phyllite (Carney et al. 2000). On the northern coast a Gwna Group belt consists of a mélange made up of lenses largely of orthoquartzite and limestone (up to a kilometre long) in a matrix of chloritic schist (Wood 1974). On Llanddwyn Island (Figs 2 & 3) well-preserved basaltic pillow lavas are interbedded or intercalated with limestone, jaspery phyllite (mudstone), and bedded jasper (chert) (Greenly 1919).

The Blueschist Unit. The easternmost Gwna Group is bordered on its western side by a 20 km × 5 km Blueschist Unit of quartz-rich phengitic mica schist that contains lenses up to c. 5 km long of hornblende schist and glaucophane schist (Gibbons & Gyopari 1986), some of which contain lawsonite (Gibbons & Mann 1983). Phengites and crossites have ^{40}Ar/^{39}Ar mineral ages of 560–550 Ma (Dallmeyer & Gibbons 1987), and glaucophane schists have an ocean floor basalt chemical affinity (Thorpe 1972).

The New Harbour Group. The New Harbour Group, occurring in two belts, comprises a pelitic to psammitic mica schist with green biotite, chlorite and epidote that contains many lenses up to 2 km long of basaltic greenschist, some with

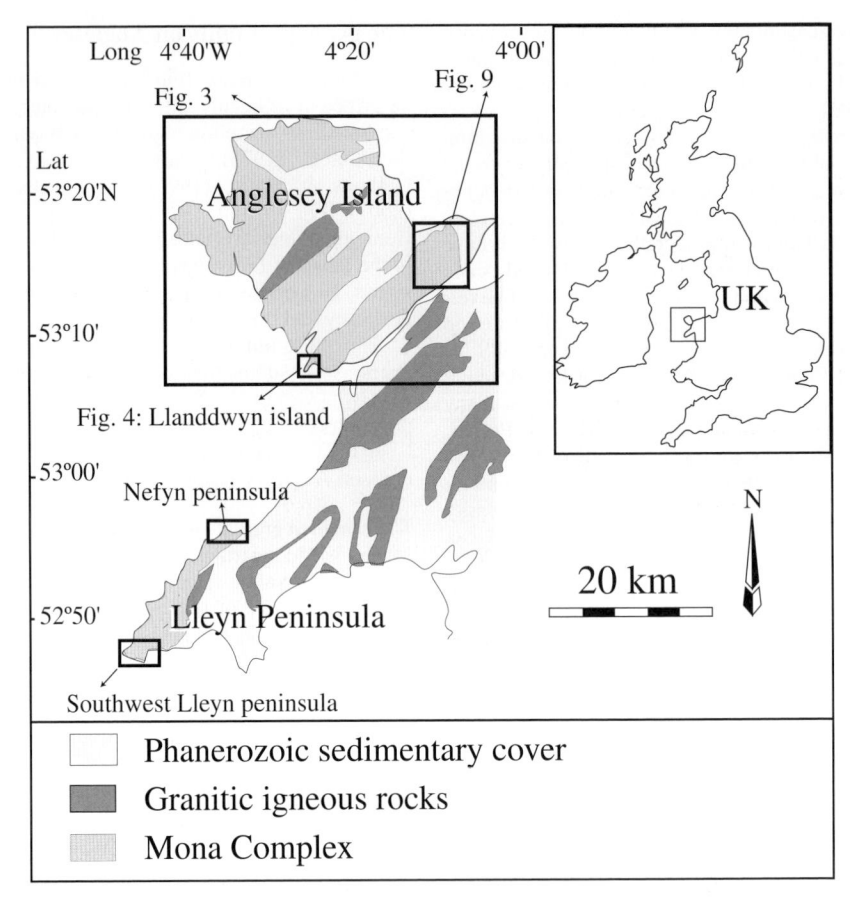

Fig. 2. Geological map of North Wales and the Lleyn peninsula showing the location of Figures 3, 4 and 9.

red chert. In addition, the western belt contains many lenses of serpentinite and gabbro up to 2 km long (Maltman 1977) commonly regarded as an ophiolitic remnant (Wood 1974; Thorpe 1978; Maltman 1979). Metabasalts from the Gwna Group have MORB chemical patterns, whereas metabasalts and meta-andesites from the New Harbour Group have suprasubduction-zone arc affinities (Thorpe 1993).

The South Stack Group. On Holy Island (Fig. 3), the South Stack Group, which mostly consists of thick quartzitic psammites interbedded with chloritic thin mafic pelites, contains detritus indicative of a quartzose to mixed recycled orogenic provenance (Phillips 1991), and accordingly most probably formed on a passive continental margin. The youngest detrital zircons of the South Stack Group have U/Pb secondary ionization mass spectrometry (SIMS) ages of 501 ± 10 Ma, indicating that this is the youngest group in Anglesey (Collins & Buchan 2004).

Although Anglesey is a flat island without any steep relief that might have made structural relationships easier to analyse, the available detailed geological maps, particularly those of Greenly (1919), suggest the common presence of low-angle thrusts, suggesting that the enveloping surface of the fundamental large-scale structure is subhorizontal, similar to that in many accretionary orogens worldwide (Maruyama 1997). Our recent analysis (Kawai *et al.* 2007) suggests that the overall structure of the island is, from structural top to bottom: (1) Coedana granite–gneiss complex; (2) Gwna Group; (3) blueschist and related metamorphic rocks; (4) olistostrome-type accretionary complex; (5) New Harbour Group; (6) South Stack Group. These units are all juxtaposed in a subhorizontal thrust–nappe pile. Throughout the island sedimentary cover rocks formed in extensional tectonic environments during Ordovician, Silurian, Devonian and Carboniferous times.

On Anglesey, Greenly (1919) understood the thrust repetition of beds belonging to what we now

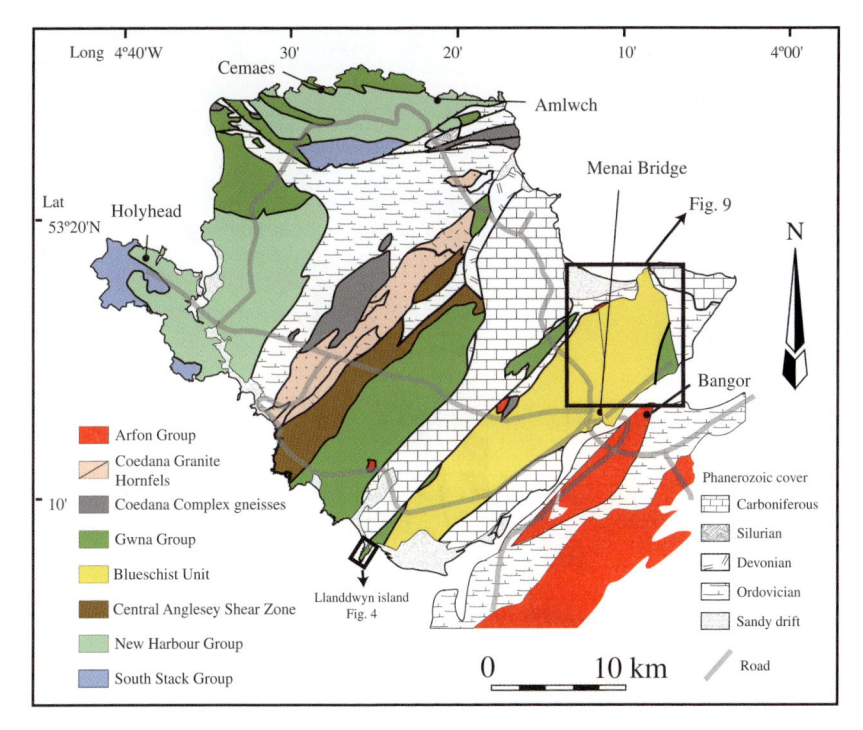

Fig. 3. Geological map of Anglesey (adapted from Greenly 1919; Shackleton 1975). Location of Figures 4 and 9 is indicated.

refer to as OPS. On p. 300, he stated that 2 km west of Cemaes village on the north coast (Fig. 3), in an area 'with a great development of jaspery phyllites [the hemipelagic mudstones] and spilitic lavas thrusting along planes nearly parallel to the old bedding is still dominant', and about 1 km west of Amlwch village farther east on the northern coast (Fig. 3), 'thrusting nearly parallel to the bedding, never permitting any one bed to range for more than a few yards, is still the dominating structure', with the result that 'jaspery phyllites and bedded jaspers become very numerous, 40 having been laid down upon the .0004 maps along 400 yards of coast, which, measured across the strike, is a proportion of one in every four yards'.

We now describe the three types of OPS in the order of their development.

Ridge–trench transition in OPS

The OPS of this type is best developed in the Gwna Group on Llanddwyn Island and SW Lleyn peninsula.

Llanddwyn Island

Reconstruction of OPS. Llanddwyn Island (300 m × 700 m) is off the SW coast of Anglesey (Figs 3 & 4). It is a fault-bound tectonic slice between the Blueschist Unit to the east and a coherent Carboniferous carbonate cover to the west, both sides being marked by high-angle faults.

It is dominantly composed of Gwna Group pillow basalts and subordinate pillow breccias or hyaloclastites, mafic mudstones, limestones, bedded cherts and in places turbiditic sediments that have a general strike to 030–045° and a vertical to overturned dip of 70–80° west (Fig. 4). Two close, parallel north–south-trending, late strike-slip faults (F_3) with a left-lateral sense of movement divide the island into eastern and western units. A 30 m thick brecciated clayey shear zone is developed between these faults (Fig. 4).

We have made a detailed geological map of a 5 m × 4 m area in western Llanddwyn Island to illustrate the imbricated OPS (Fig. 5a); Figure 5b and c shows vertical cross-sections. The imbricated OPS consists of repeated basalt, red chert, and mudstone and sandstone (Fig. 6a–e). Many faults are sub-conformable with bedding planes within the cherts and mudstones. In some places, such as on the limbs of isoclines (Fig. 5a), thrusts can be seen to be parallel to bedding, but elsewhere some thrusts can be observed only by the presence of finely sheared material on the bedding planes. The map shows three tight-isoclinal folds in bedded

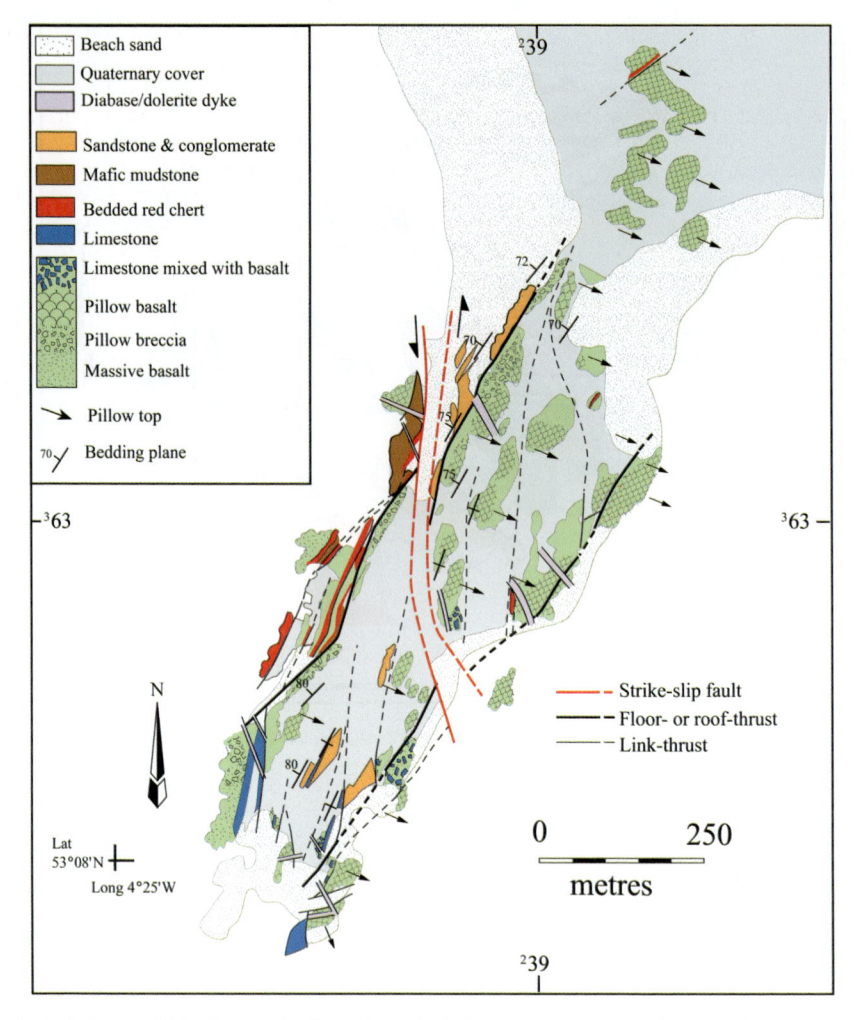

Fig. 4. Geological map of Llanddwyn Island showing principal stratigraphic units and structural data.

cherts; similar folds in bedded cherts are common in OPS sections of the Shimanto Group in Japan. Figure 5d shows the observed OPS along the line D–D'. This type of imbricated OPS structure can also be observed on a map scale (Figs 4, 7 & 8).

A large-scale textbook-like duplex is well preserved on Llanddwyn Island (Figs 4 & 7). The eastern unit is subdivided by several bedding-parallel link thrusts (F_1) into six separate horses separated by two roof- and floor-thrusts (F_2). The eastern and western units are separated by two F_3 close and parallel strike-slip faults. The western unit has 17 horses separated by two major roof- and floor-thrusts (F_2) (Fig. 7). In spite of extensive tectonic modification by layer-parallel shortening, the stratigraphic top of each horse can be easily determined by the primary shapes of undeformed pillows (Fig. 6a). This indicates that all the units face downwards to the SE (Fig. 4). In each horse, the stratigraphic bottom is a MORB-like, pillow-bearing or massive basaltic flow. Pillows are 30–60 cm in diameter and are cemented by a glassy matrix. Each pillow is well identified by its chilled margin, which is a few millimetres to 1 cm thick. In places pillows are connected to form an elongate tube or channel. Under the microscope the pillow lava is fine-grained and intersertal ophitic with swallow-tail augites and plagioclase laths. The matrix consists of red jasper or limestone.

In unit I the horses, numbered from 1 to 6, in general comprise 100 m thick (maximum) MORB-like pillow lava (with red chert in the inter-pillow spaces; Fig. 6g) that is covered by the 20 m thick chlorite-rich sandstone.

In unit II the horses, numbered from 7 to 17, in general comprise 100 m thick, MORB-like pillow lava flows, overlain by 40 m thick pillow breccias with a matrix of mafic mudstone containing several lenses of limestone (Fig. 6h), in turn succeeded by 40 m thick pelagic carbonates, overlain by a *c.* 50 m thick sandstone. A special situation exists in eastern Llanddwyn, where unit II basalts with inter-pillows of red chert are succeeded by basalts with inter-pillows of limestone, overlain by 2 m of grey limestone, and 5 m of pink manganiferous rhodochrosite limestone, which we interpret as pelagic limestone deposited at or near a thick, high-standing ridge.

In unit III the horses, numbered from 18 to 23, are characterized by a 2 m thick red bedded chert capped by 18 m thick hemipelagic mafic mudstone and a 20 cm sandstone (Figs 5 & 6b–e).

In general, the sedimentological changes reflect the gradual change in environment from eruption of vesicle-absent lavas in the deep ocean at or near a ridge, through deposition of pelagic chert or limestone in an ocean remote from a continent, via hemipelagic mudstone either approaching or on the lip of a trench, to a deep trench flooded by an arkosic–quartzitic sandstone to conglomerate in the form of a cap turbidite. This is OPS (Matsuda & Isozaki 1991) that contains a record of the travel history from the birth of an oceanic plate to its demise at a trench.

Sense of subduction and duplexing history. The sense of subduction can be estimated from the structural pattern of duplexes and the sense of shear along thrusts shown in the inset diagram of Figure 1 and in Figure 7b. To make a duplex defined by a top- and a floor-thrust by underplating subducted oceanic crust with a specific OPS, the sense of subduction must be from left to right in Figure 7b.

To explain the lithological changes of the OPS from unit I to unit III, a break of accretion is necessary, indicated by a sudden change in lithology in the OPS. Judging from the deformation by the latest F_3 strike-slip faults, the sequence of accreted units must have been I, followed by II and finally by III (Figs 8 and 9). Possible time gaps between the accretion of the three units might account for tectonic erosion that has destroyed some already-formed accreted material. Such tectonic erosion is a common process at modern consuming plate boundaries (Kopp *et al.* 2006).

To reconstruct the original succession of each horse in each unit, the minimum lateral shortening must be estimated (Fig. 8). In Figure 8, the exact shape of each horse is oriented to calculate the minimum shortening. The results indicate 3000 m, 3200 m and 1600 m of shortening for units I, II and III, respectively (Fig. 8). This is a surprising amount of shortening, because it means that the (only) *c.* 300 m thick accretionary complex on Llanddwyn Island preserves an amount of subduction that would be responsible for over 7800 m of subducted lithosphere. However, it is not surprising, in comparison with the current rate of subduction around the Pacific Ocean and the amount of subaerial subduction–accretion complexes (e.g. Kopp *et al.* 2006).

The depth of accretion in this part of Anglesey seems to have been shallow, because of the very low (sub-greenschist, zeolite facies) metamorphic grade associated with the very low deformation; it could have been caused either by offscraping or underplating of <10 km of lithosphere. Accreted MORB crust, the thickness of which never exceeds 150 m, is common in circum-Pacific regions such as California and Japan, and reflects the mechanically weak zone that develops at about 100–200 m depth during hydrothermal alteration (Kimura *et al.* 1996).

SW Lleyn

Another variation of ridge–trench OPS crops out on the southern coast of the Lleyn peninsula (Fig. 10b, for more detail see Kawai *et al.* 2007, fig. 7). More than 10 slices of accretionary complex are cut by several high-angle normal faults, and by layer-parallel thrusts. The reconstructed OPS is shown in Figure 11; a 20 m thick pillow basalt is overlain by a 2 m thick bedded pelagic dolomitic carbonate, which in turn is overlain conformably by 5 m of red bedded chert; there is no arkosic–quartzite turbidite at the top of this succession. The dolomitic carbonate and bedded chert beds make a spectacular lithology. These beds constitute the topmost geological units on Anglesey above the Blueschist Unit.

Subducted OPS

This type of OPS, metamorphosed under blueschist-facies conditions, is exposed on Anglesey and in the Lleyn peninsula.

Anglesey

Greenly (1919) called the eastern part of the Blueschist Unit on Anglesey 'Gwna greenschist' and the western part 'mica-schist', although both parts underwent the same greenschist–blueschist transition and the same extensive penetrative deformation (Kawai *et al.* 2006). The eastern and western schists belong to the eastern and western limbs, respectively, of the exhumation-related antiform

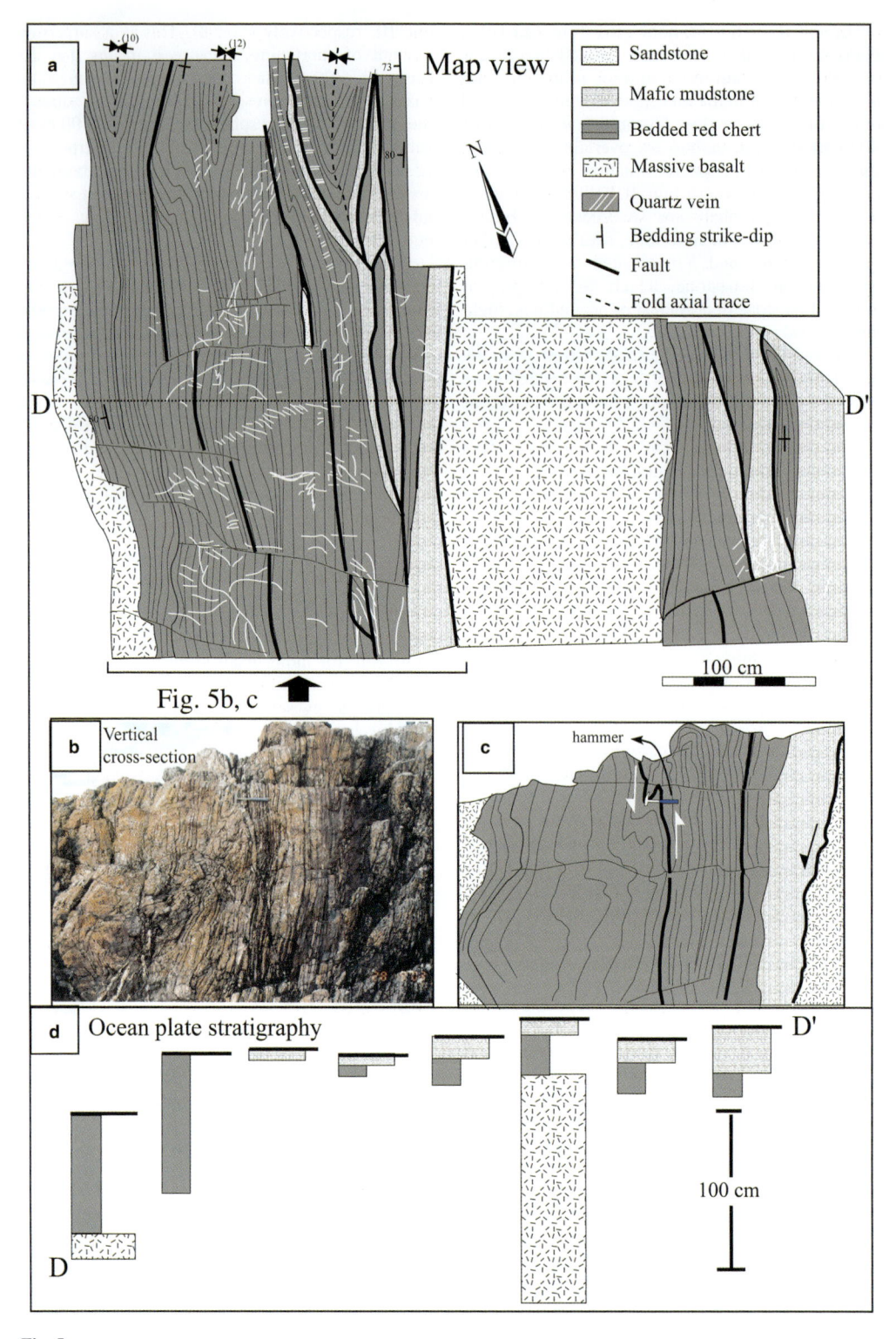

Fig. 5.

outlined by Kawai *et al.* (2007, fig. 3). The detailed maps by Greenly (1919) show innumerable slabs of quartzite or quartzitic schist (up to 30 m thick), of limestone up to several metres to tens of metres thick, and of greenstone (blueschist or hornblende schist, up to 1 km thick) particularly in the eastern area, and clearly show their relative relationships. Although the exposure is insufficient to examine many of the boundaries between these rocks, we have confirmed in a few excellent exposures that some quartzites or quartzitic schists up to 30 m thick are situated on the western side of several greenstones. This might be taken to suggest that the original facing direction of these rocks was to the west. The west–east distribution between the quartzite and greenstone, respectively, reflects the gently east-dipping fault-bound overturned parts of this accretionary unit.

In spite of extensive deformation and metamorphism of the blueschist to albite–epidote amphibolite facies, parts of the OPS can be reconstructed even at these higher grades of metamorphism, where the quartzite–greenstone slabs occur in a matrix of mica schist or mafic chloritic schist, the protolith of which we interpret as trench-fill pelitic turbidite. The important feature of this type of OPS is the enormous volume of late clastic sediment that engulfed the slabs of quartzite–greenstone, limestone and quartzite in the trench. The structures within this accretionary complex often make it difficult to decipher the subduction polarity. However, the quartzite–greenstone relationship suggests that the subduction polarity was to the east, which is consistent with that exhibited on Llanddwyn Island.

Northern Lleyn Peninsula

A unit of blueschist-facies metamorphic rocks (which include glaucophane–phengite schists; Gibbons 1981) crops out on the Nefyn peninsula on the northern coast of the Lleyn peninsula (Fig. 10a; for more detail see Kawai *et al.* 2007, fig. 4). About 2.5 km along strike to the west of Nefyn is the Porth Dinllaen peninsula, on the western side of which is another mafic schist, but without glaucophane.

The protoliths of the blueschist-facies metamorphic rocks on Nefyn (Fig. 10a) are 80 m thick (tectonic thickness) basic greenschist (basalt), chert and carbonate up to a few metres thick, and a 20 m thick mafic mudstone (phengite schist). Pillow basalts are flattened and deformed extensively, and contain high-pressure minerals such as barroisite, crossite, lawsonite and phengite, as well as actinolite, epidote, chlorite, albite, titanite, carbonate and quartz. Even so, the thin chilled margins, the vesicles cemented by carbonate and quartz, and the pale green, less competent glassy matrix are clearly distinguishable from the dark green cores of the pillows.

The schist unit on the western side of Porth Dinllaen peninsula consists of abundant chlorite, and it contains many lenses up to 2 m long, which record a crude stratigraphy from lower lenses of red bedded chert, a 2 m thick bed of red chert, central lenses of chert and limestone, to upper lenses of quartzite; this schist with its lenses is overlain by a 50 m thick succession of well-bedded turbiditic sandstone. We consider that the overall tectonostratigraphic sequence in this area of northern Lleyn corresponds to an OPS, which we interpret as a 20 m thick basal pillow basalt originally capped by 5–10 m thick bedded chert, limestone and quartzite, in turn covered and engulfed by a 20 m thick, arc-derived volcaniclastic mafic mudstone, and finally by a 50 m thick well-bedded sandstone turbidite deposited in a trench (Fig. 11). Such quartzites associated with mafic mudstones might be derived from outbursts of siliceous volcanism from the arc.

SW Lleyn peninsula

Along the southwestern coast of the Lleyn peninsula, a subhorizontal chloritic schist unit without glaucophane crops out with underlying and overlying accretionary units (Fig. 10b). Kawai *et al.* (2007, fig. 5) established (in agreement with Gibbons 1983) that this schist unit in SW Lleyn lies structurally below the Gwna Group, and thus occupies the same structural position as the Blueschist Unit on Anglesey and the blueschist-bearing schist unit on Nefyn peninsula. The OPS protoliths of the schist unit consist of lenses of basal metabasite (predominantly basaltic tuff) overlain by sandstone, hemipelagic mudstone, and rare thin chert and limestone in a widespread matrix of slaty mudstone or chlorite schist with a vertical cleavage, which we interpret as a thick cap of fine-grained clastic sediment.

Fig. 5. (*Continued*) (**a**) Detailed map of a 5 m × 4 m area of Unit II on Llanddwyn Island [NGR 386 629]. Looking north. (**b**) Photograph of a vertical face, looking north, at the lower edge of the area in (a), showing the stratigraphy of the western unit of OPS. (**c**) Sketch section across the lower edge of (a) drawn from (b). The major fold outlined indicates a left-lateral motion, which is consistent with underthrusting of the western unit of OPS eastwards below the eastern unit. (**d**) Observed OPS along the line D–D′ marked in (a).

Fig. 6.

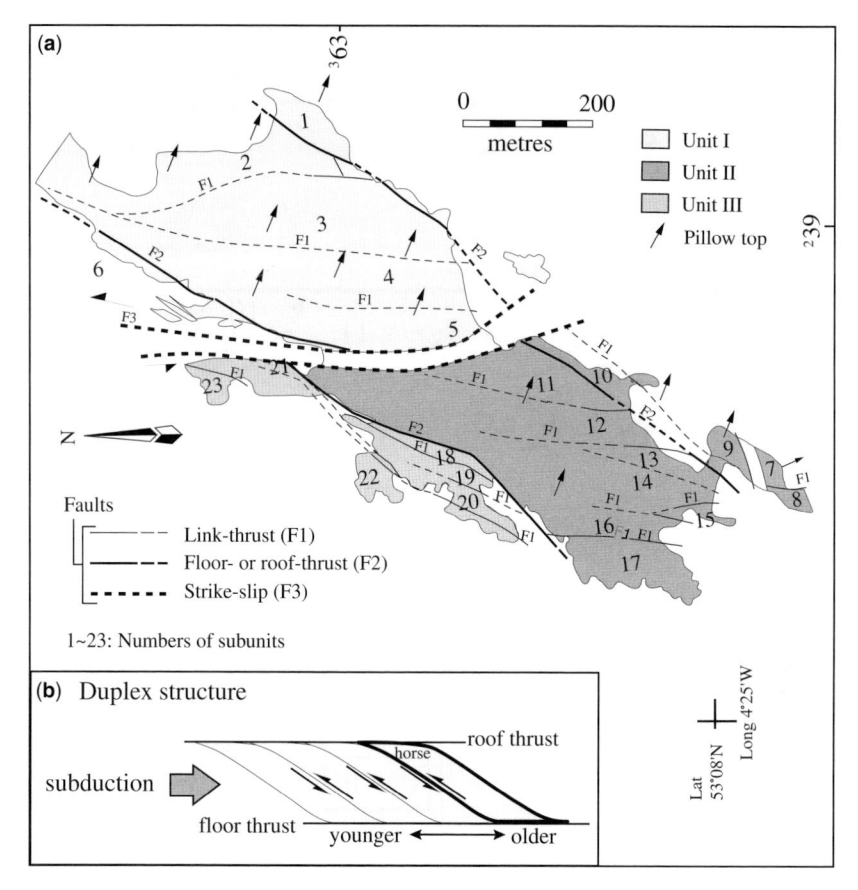

Fig. 7. (**a**) Map showing the distribution of 23 duplex horses (subunits) on Llanddwyn Island. (**b**) Schematic Illustration demonstrating the sense of subduction that can be estimated from the structural geometry of duplexes and the sense of shear along the thrusts.

Olistostrome-type OPS

On the southwestern end of the Lleyn peninsula the above chloritic schist unit is separated by a low-angle phyllonite-bearing thrust from an underlying unmetamorphosed, steeply dipping accreted unit, which we interpret as an olistostrome-type mélange (Fig. 10b; for more detail see Kawai et al. 2007, fig. 5).

The olistotrome contains many lenses of quartzite, red chert, dolomitic limestone and basaltic greenschist in a matrix of mafic mudstone; most lenses are about 5–30 cm across, some are up to 6 m long, and the largest is a spectacular 17 m × 25 m lens of white quartzite in dark mudstone illustrated in photographs by Roberts (1979, fig. 14) and also shown in Figure 6h.

The olistostrome is underlain by a unit with well-preserved OPS that passes downwards from a 35 m thick, well-bedded sandstone that is underlain conformably by 1 m of mudstone on top of a 7 m thick red bedded chert below which is 1 m of sheared basalt. This basalt is cut off below by a normal fault that defines the base of this unit of OPS (Fig. 11).

Fig. 6. (*Continued*) Photographs of OPS lithologies on Llanddwyn Island. (**a**) Pillow basalt at the base of Unit I on the NE coast. Pillows face upward to the right. (**b**) Pillow basalt and chert of Unit III. (**c**) Bedded red chert in Unit III shows the folding. (**d**) Mafic mudstone of top of Unit III. (**e**) A sandstone bed between two red, bedded cherts in Unit III on the west coast. Looking south. Grading of the sandstone to the left (eastward) should be noted. (**f**) Lenses of limestone in a matrix of deformed chlorite-rich mafic mudstone. (**g**) Pillow lavas (dark grey) with interpillow chert (red). (**h**) Quartzite clasts in a mudstone matrix in the olistostrome-type accretionary complex shown in Figure 10.

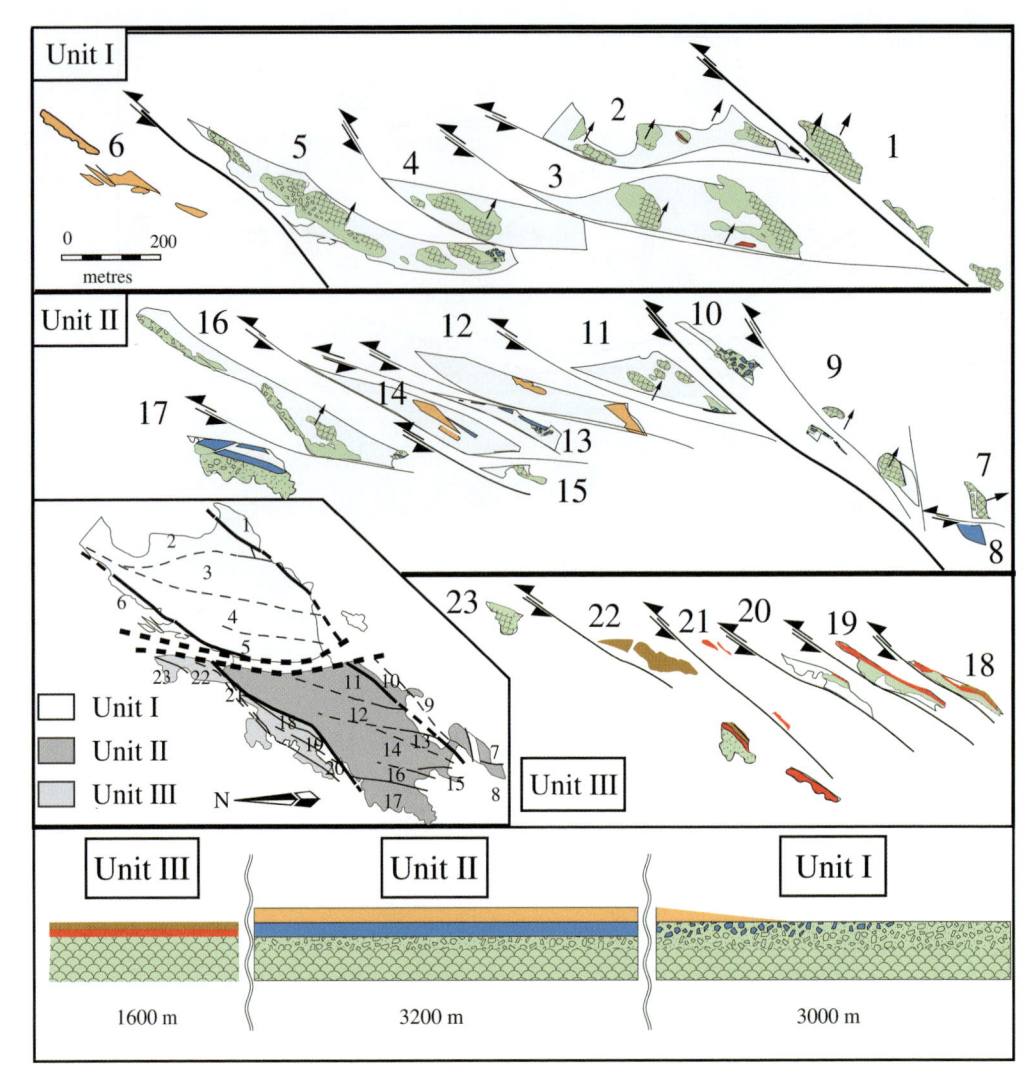

Fig. 8. The ocean plate stratigraphy of the three units on Llanddwyn Island. The figure shows a balanced section of the growth history of the three units with an inset of the general structure from Figure 7a, and at the bottom are the original reconstructed stratigraphies of the three units. Lithologies have the same legend as in Figure 4.

Structural relationships between the three types of ocean plate stratigraphy

The three units of the accretionary complex in Anglesey have different and characteristic OPS. Here we first summarize their character, then compare them, and discuss their mutual structural relationships.

Structural top

The uppermost segment of OPS in Anglesey–Lleyn is characterized by a ridge–trench transition type of OPS on Llanddwyn Island and the SW Lleyn peninsula. We have subdivided this structural top segment on Llanddwyn Island into Units I, II and III (Fig. 11), which represent the upper, middle and lower parts. Unit I is characterized by basal basalt overlain by only sandstone. Unit II has basal basalt and volcanic breccia overlain by a thick limestone and sandstone; this is suggestive of deposition in a shallow marine environment such as an intraoceanic seamount. Unit III consists of basal pillow lavas overlain by thin chert succeeded by mudstone and thin limestone; we interpret this stratigraphy as ocean-floor type.

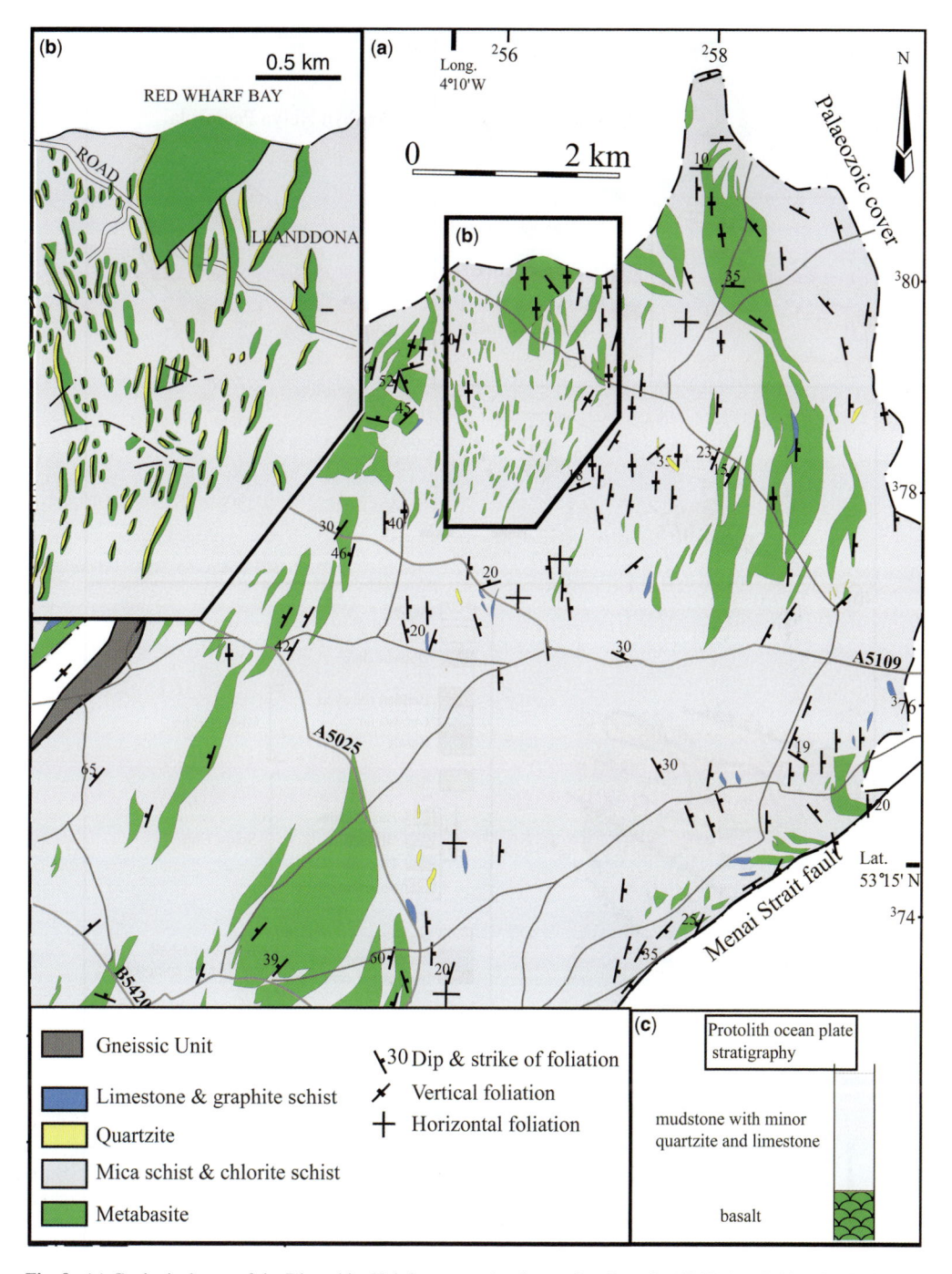

Fig. 9. (**a**) Geological map of the Blueschist Unit in eastern Anglesey after Greenly (1919) (see British Geological Survey 1980). Metabasite lenses with a mineralogy of glaucophane schist or retrogressed hornblende schist are locally bordered on their western sides by quartzite or quartzitic schist. Lenses of the metabasites, and of limestones and quartzites, occur in a matrix of mica schist and chlorite schist. (**b**) Enlarged view of Llanddona area. (**c**) Protolith ocean plate stratigraphy of the Blueschist Unit.

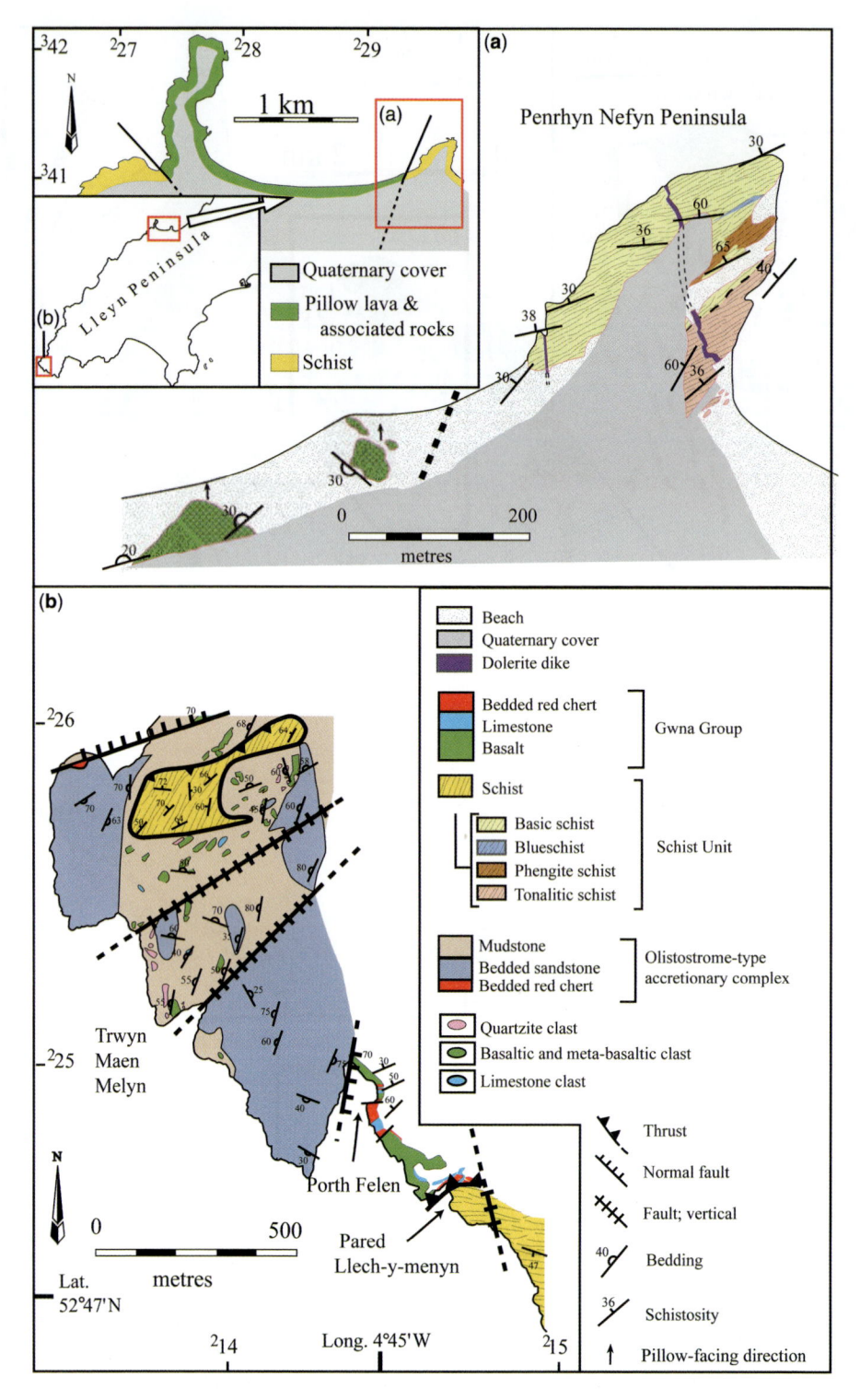

Fig. 10. (a) Geological map of Penrhyn Nefyn peninsula. (b) Geological map of SW Lleyn peninsula. Inset shows the locations of the Nefyn peninsula and SW Lleyn.

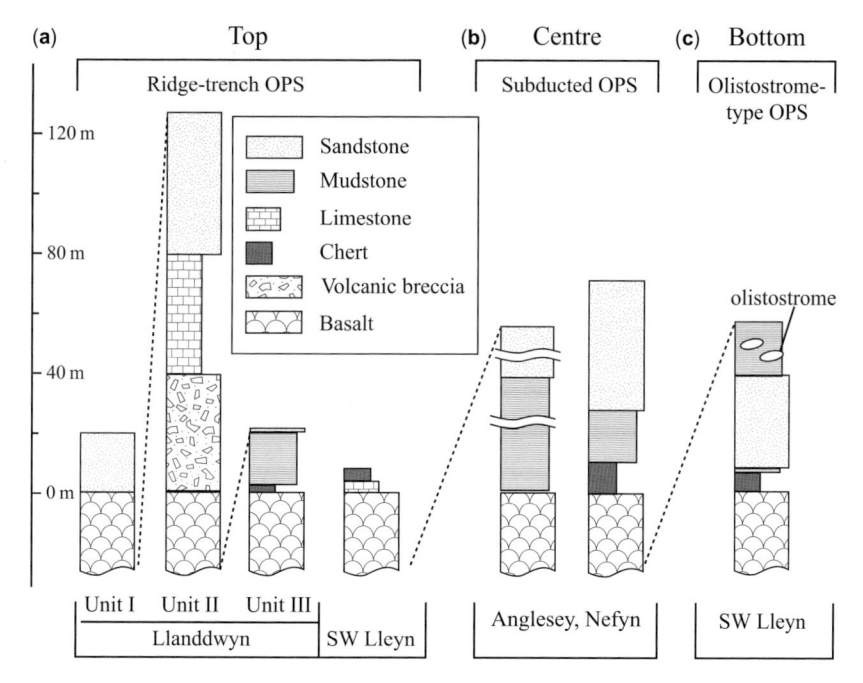

Fig. 11. Stratigraphic columns to demonstrate the three types of OPS. (**a**) The topmost and oldest ridge–trench transition-type OPS of the Gwna Group on Llanddwyn Island and in SW Lleyn. (**b**) The central Blueschist Unit, showing its protolith stratigraphy preserved in chloritic schists and their inclusions on the western side of Porth Dinllaen peninsula on the north coast of the Lleyn peninsula. (**c**) The lowermost and youngest unit in SW Lleyn, which contains an olistostrome deposit above a unit of OPS.

Although the structural relationship between the Gwna Group, which is characterized by ridge–trench transitional OPS, and the neighbouring Blueschist Unit is unknown on Llanddwyn, the equivalent unit clearly rests structurally on the schist or Blueschist Unit on the southern coast of the Lleyn peninsula (Fig. 10b).

Structural centre

This central type of OPS is the Blueschist Unit, which on Anglesey is characterized by protoliths with an upwards stratigraphy of basalt, minor limestone and quartzo-feldspathic sandstone (quartzite), and predominant hemipelagic to clastic, mafic to siliceous mudstone, which has deluged or engulfed the earlier parts of the stratigraphy (Figs 9 & 11). The schist units in Lleyn have similar OPS, particularly in so far as they each have a late development of thick clastic sediment, which we interpret as trench-fill (Fig. 11).

Structural bottom

The structurally lowermost unit is an olistostrome-type accretionary complex (which crops out only on SW Lleyn), which has a matrix of black mafic mudstone that contains many lenses consisting of different parts of the OPS, such as pillow basalt, red bedded chert, dolomitic carbonate, and huge blocks of quartzitic sandstone (Fig. 11). These relations suggest this is an olistostrome-type OPS in which already-formed accretionary material had collapsed from the inner wall of a trench to form a gravitational deposit.

Discussion and implications

Presence of a Pacific-type accretionary complex

The term mélange was first coined in Anglesey by Greenly (1919); the typical exposures are on the northern coast and are probably equivalent to the olistostrome type of accretionary complex underneath the Blueschist Unit.

Since Leggett *et al.* (1979) first reported the presence of an accretionary complex in Britain in the Southern Uplands of Scotland, no-one has used OPS in the UK to unravel an accretionary complex by combining stratigraphy and thrust imbrication. Although the best accretionary complex in the UK is on Anglesey, its OPS has not been described, and thus the use of OPS in understanding its tectonic history has not been appreciated. Using established methods of working out the accretionary geology of Pacific-type orogens such as Japan and Alaska, we propose the following history of subduction for Anglesey from the Neoproterozoic to the Cambrian.

Evidence for the successive underplating or off-scraping of subducting oceanic lithosphere to form an accretionary complex on the hanging wall is best recorded on Llanddwyn Island. The process formed a 300 m thick accretionary package. Only the top 100–150 m of the subducted plate was accreted, because most, perhaps as much as 98%, of the magmatic oceanic crust was subducted into the mantle. Although the island is only 700 m long, reconstruction of the balanced sections of the three types of OPS on Llanddwyn Island suggests that the subducted oceanic lithosphere was at least 7800 m long, and probably far longer, as suggested by the presence of three types of OPS.

We estimate that the age of subducted lithosphere was very young, probably less than 10 Ma, if the sedimentation rate was the same as in the Phanerozoic (1 m equivalent to 1 Ma); that is, the same as for the Gwna Group on Llanddwyn. This topic will be further discussed below.

The eastward subduction polarity developed an accretionary complex downward with time; hence the structural top is the oldest. The structural top was the Gwna Group, which developed at or not far from a ridge in an intra-oceanic environment, like that in the present-day western Pacific. As the ridge basalts moved across the ocean they were covered by pelagic cherts or limestones, which near the trench were succeeded by hemipelagic mafic mudstones, which were overlain in the trench by granitic clastic cap turbidites derived from a hanging-wall arc or continental mass to complete the first stage of OPS. With time, some accreted OPS, which contained large volumes of trench turbidites, was subducted to the pressure–temperature conditions of blueschist- and epidote amphibolite-facies metamorphism; return to the surface was most probably as a result of subduction of a mid-ocean ridge (Kawai et al. 2007). Finally, some sediment that had been accreted to the hanging wall of the trench was incorporated in debris flows and recycled into olistostromes to give rise to the third type of OPS, which was accreted to form the lowest structural unit of the accretionary complex.

Comparisons with present-day OPS

The Ocean Drilling Program (ODP), which started drilling of extant ocean floor in 1966, has now completed more than 2000 drill sites in the world's oceans. The OPS of the Pacific Ocean has been well established by the ODP. One of the representative OPS sections is shown in Figure 12, in which the OPS is drawn from the East Pacific Rise through mid-Pacific seamount chains and the oldest (c. 200 Ma) part of the Pacific Ocean, to a modern accretionary complex on the inner wall of the Mariana Trench (Johnson et al. 1991).

The world's largest segment of oceanic lithosphere, the Pacific plate, was born at the ridge of the East Pacific Rise, which is not yet covered by deep-sea sediments because it is so young. The measured thickness of pelagic sediment reaches 1200 m on c. 60 Ma old oceanic crust in the mid-Pacific, and finally it is up to 2000 m thick off the Mariana Trench, where the age of the plate is as much as 200 Ma. The petrography and geochronology of pelagic clay strongly indicates that the major portion of the clay was derived from aeolian dust that originated from desert areas in mid-latitudes (e.g. Mahoney et al. 1998; Mahoney 2005; Stancin et al. 2006). The pelagic clay, which is a mixture of aeolian dust and organic matter, accumulates on the ocean floor at a rate depending on biological activity and the production rate of the aeolian dust; the rate is about 2 km in 200 Ma; that is, 10 m per million years. On considering the compaction rate during diagenesis of the older on-land rocks, this rate would reduce by a few metres per million years.

In Japan trench–forearc accretionary wedges are commonly capped by crustal-derived clastic deposits that are interbedded with arc-derived, mafic or even siliceous volcaniclastic sediments (Ogawa & Taniguchi 1988; Stowe et al. 1998). On Anglesey–Lleyn very comparable trench-fill clastic turbiditic sediments are commonly interbedded with mafic mudstones that probably formed as mafic volcaniclastic sediments derived from nearby arc volcanoes, and were later transformed into chlorite schists during accretion.

Red bedded chert is up to 5 m thick in the SW Lleyn peninsula and 2 m on Llanddwyn Island; this suggests that the subducted plate was as old as a few million years, to <10 Ma, although the thickness of bedded chert may be difficult to assess because of folding and layer-parallel faulting (see Fig. 7).

Near-trench, hemipelagic sediments accumulate on the Pacific plate, although not off Mariana, because of the absence of a nearby large continent. In contrast, off SW Japan, where the outer rise has been drilled (Klein et al. 1980), hemipelagic sediments dominated by mudstone have been well characterized petrographically and interpreted as the episodic voluminous overflow of sediment from the Asian continent across and into the trench (Mahoney et al. 1998; Mahoney 2005). An extremely thick cap of turbidite has long been recognized in deep trenches, as in the Japanese Nankai Trough,

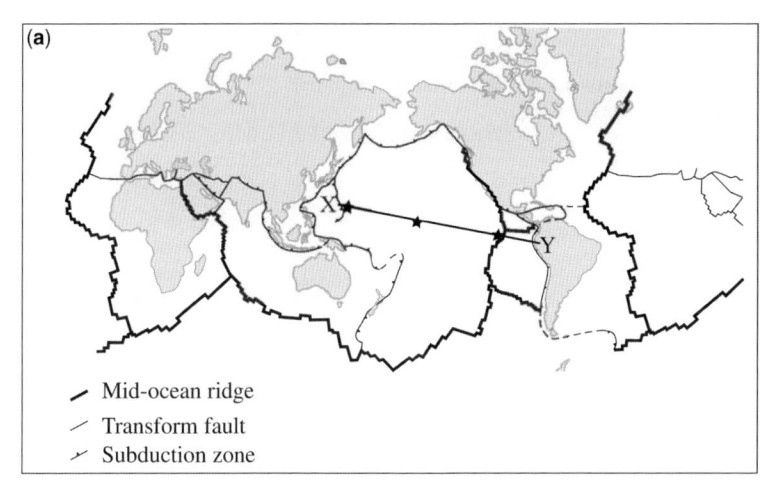

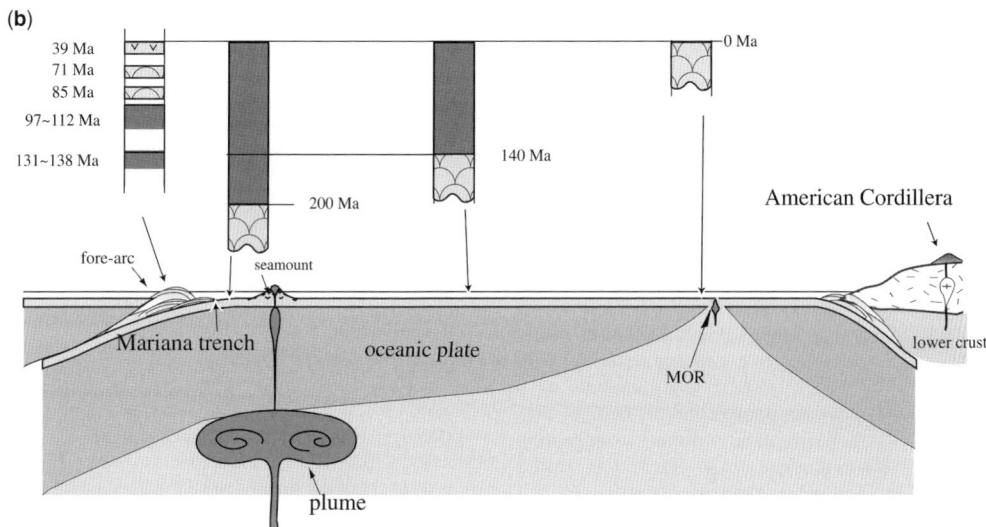

Fig. 12. (a) Map showing the main plate boundaries in the world. (b) A representative section, marked as XY in (a), across the Pacific Ocean showing the various types of ocean plate stratigraphy with their times of formation from the East Pacific Rise.

trenches around Indonesia and other circum-Pacific trenches (e.g. Curtis & Echols 1980; Betzler *et al.* 1991; Taira & Niituma 2007). Moreover, we emphasize the common presence in Japanese trench-dominated accretionary wedges of arc-derived volcaniclastic sediments within continental crust-derived turbiditic clastic cap sediments, as in the Miocene–Pliocene Miura Group near Tokyo (Ogawa & Taniguchi 1988; Stowe *et al.* 1998).

In a Pacific-type trench oceanic crust is commonly delaminated with the result that ultramafic rocks, gabbros and sheeted dykes are subducted, and only basalts and overlying sediments are accreted (Kimura & Ludden 1995). During early formation of an accretionary wedge in the trench basalts and sediments are typically intensely imbricated by bedding-parallel thrusts, and this thrust repetition can be most easily demonstrated by duplication of the OPS, as in the Shimanto Group in Japan (Okamoto *et al.* 2000), and in the Permian–Cretaceous accretionary orogen of Alaska where 'the ocean plate stratigraphy is repeated hundreds of times' [Kusky & Bradley 1999; according to Kusky (pers. comm.) about 330 times]. The Nankai Trough off Japan is one of the best sites for the study of the early stages of development of an accretionary wedge. The results of Deep Sea Drilling Project (DSDP) Legs 31, 87 and ODP Leg 131 revealed the lithologies of the Nankai accretionary wedge, showing that the stratigraphy consists of lower cherts and upper turbidites (Taira et al. 1992), and multichannel seismic reflection data demonstrated the structure of the wedge in the form of basal décollement, thrust ramps and duplexes (Moore et al. 1990).

Evidence for Mesozoic–Cenozoic ridge–trench transition is documented in Japan by microfossils (Matsuda & Isozaki 1991; Hori & Wakita 2004), bedding-parallel structure and greenstone geochemistry (Okamoto et al. 2000), thrust structure and sedimentary facies that are comparable with those in the Nankai Trough (Taira et al. 1988), and by stratigraphy and age (Xenophontos & Osozawa 2004). Mesozoic–Cenozoic ridge–trench transition is demonstrated in Alaska by progressive deformation (Kusky et al. 1997a, b) and geochemical data (Kusky & Young 1999), in Sakhalin (Kimura et al. 1992), Indonesia (Wakita 2000) and SE Asia (Wakita & Metcalfe 2005) by ocean plate stratigraphy and structure, and in accreted complexes in Troodos, Cyprus (Robertson & Hudson 1974) and the Himalayan Yarlung–Tsangpo suture (Ziabrev et al. 2004). The first prescient description and appreciation of OPS was by Steinmann (1925), who compared Mesozoic pelagic chert–hemipelagic clay–clastic sandstone successions in the Alps and Apennines with those in Pacific trenches, and pointed out their presence in the Variscan and Caledonian orogens of Europe.

The OPS of the Gwna Group on Anglesey is characterized by a ridge–trench transition either with mafic mudstone as a cap sediment or with continental-derived sandstone–conglomerate as a cap turbidite that commonly contains beds of arc-derived volcaniclastic mafic sediments, suggesting that the sedimentological environment was similar to that of the present-day western Pacific.

Growth pattern of the Anglesey–Lleyn accretionary complex

Our data presented above demonstrate that the growth pattern of the Anglesey–Lleyn accretionary complex was downward with time, and the large-scale structure is today subhorizontal (Kawai et al. 2007). The 667 ± 7 Ma (Strachan et al. 2007) Coedana granite–gneiss complex is the structural top unit and hence should be the oldest. Below it is the Gwna Group with a ridge–trench transition OPS, below that is the 560–550 Ma (Dallmeyer & Gibbons 1987) Blueschist Unit with its distinctive, trench-thick OPS, and the structural bottom unit is the olistostrome-type OPS, which is the youngest. Below the Gwna Group is the accretionary New Harbour Group, and structurally below that is the youngest South Stack Group with its 501 ± 10 Ma detrital zircons (Collins & Buchan 2004), which was the last package to accrete to the accretionary collage of the Mona Complex. The recognition of OPS in this Pacific-type accretionary complex on Anglesey–Lleyn is the key to reconstructing its structural development.

We thank M. Wood for useful information and discussions on the geology of Anglesey. The field survey and collection of rock samples used in this study were carried out with M. Watanabe, who is gratefully acknowledged. We thank T. Kusky for comments and editing this manuscript.

References

BARBER, A. J. & MAX, M. D. 1979. A new look at the Mona Complex, Anglesey, North Wales. *Journal of the Geological Society, London*, **136**, 407–432.

BETZLER, C., NEDERBRAGT, A. J. & NICHOLS, G. J. 1991. Significance of turbidites at site 767 (Celebes Sea) and site 768 (Sulu Sea). *In*: SILVER, E., RANGIN, C. ET AL. (eds) *Proceedings of the Ocean Drilling Program, Scientific Results, 124*. Ocean Drilling Program, College Station, TX, 431–446.

BRADLEY, D., KUSKY, T. ET AL. 2003. Geologic signature of early Tertiary ridge subduction in Alaska. *In*: SISSONS, V. B., ROESKE, S. M. & PAVLIS, T. L. (eds) *Geology of a Transpressional Orogen Developed during Ridge–Trench Interaction along the North Pacific Margin*. Geological Society of America, Special Papers, **371**, 19–49.

BRITISH GEOLOGICAL SURVEY 1980. *Geological Map, England and Wales*. Special Sheet, Solid with Drift edition, 1:50 000. British Geological Survey, Keyworth, Nottingham.

CARNEY, J. N., HORÁK, J. M., PHARAOH, T. C., GIBBONS, W., WILSON, D., BARCLAY, W. J. & BEVINS, R. E. 2000. *Precambrian Rocks of England and Wales*. Geological Conservation Review Series **20**.

COLLINS, A. S. & BUCHAN, C. 2004. Provenance and age constraints of the South Stack Group, Anglesey, UK: U–Pb SIMS detrital zircon data. *Journal of the Geological Society, London*, **161**, 743–746.

CURTIS, D. M. & ECHOLS, D. J. 1980. Lithofacies of the Shikoku and Parece Vela Basins. *In*: KLEIN, G. D., KOBAYASHI, K. ET AL. (eds) *Initial Reports of the Deep Sea Drilling Project, 58*. US Government Printing Office, Washington, DC, 701–709.

DALLMEYER, R. D. & GIBBONS, W. 1987. The age of blueschist metamorphism in Anglesey, North Wales: evidence from $^{40}Ar/^{39}Ar$ mineral dates of the Penmynydd schists. *Journal of the Geological Society, London*, **144**, 843–852.

GIBBONS, W. 1981. Glaucophanitic amphibole in a Monian shear zone on the mainland of North Wales. *Journal of the Geological Society, London*, **138**, 139–143.

GIBBONS, W. 1983. The Monian 'Penmyndd zone of metamorphism' in Llŷn, North Wales. *Geological Journal*, **18**, 21–41.

GIBBONS, W. & MANN, A. 1983. Pre-Mesozoic lawsonite in Anglesey, northern Wales: preservation of ancient blueschists. *Geology*, **11**, 3–6.

GIBBONS, W. & GYOPARI, M. 1986. A greenschist protolith for blueschist in Anglesey, U.K. *In*: EVANS, B. W. & BROWN, E. H. (eds) *Blueschists and Eclogites*. Geological Society of America, Memoirs, **164**, 217–228.

GIBBONS, W. & HORÁK, J. M. 1996. The evolution of the Neoproterozoic Avalonian subduction system: evidence from the British Isles. *In*: NANCE, R. D. & THOMPSON, M. D. (eds) *Avalonian and Related Peri-Gondwana Terranes of the Circum North Atlantic*. Geological Society of America, Special Papers, **304**, 269–280.

GREENLY, G. 1919. *The Geology of Anglesey*. Memoir of the Geology Survey of Great Britain. HMSO, London.

HARA, H., WAKITA, K. ET AL. 2010. Nature of accretion related to Paleo-Tethys subduction recorded in northern Thailand: constraints from mélange kinematics and illite crystallinity. *Gondwana Research*, **16**, 310–320.

HORI, N. & WAKITA, K. 2004. Reconstructed oceanic plate stratigraphy of the Ino Formation in the Ino district, Kochi prefecture, central Shikoku, Japan. *Journal of Asian Earth Sciences*, **24**, 185–197.

ISOZAKI, Y., MARUYAMA, S. & FURUOKA, F. 1990. Accreted oceanic materials in Japan. *Tectonophysics*, **181**, 179–205.

JOHNSON, L. E., FRYER, P. & TAYLOR, B. 1991. New evidence for crustal accretion in the outer Mariana fore-arc: Cretaceous radiolarian cherts and mid-ocean ridge basalt-like lavas. *Geology*, **19**, 811–814.

KATO, Y. & NAKAMURA, K. 2003. Origin and global tectonic significance of Early Archean cherts from the Marble Bar greenstone belt, Pilbara craton, Western Australia. *Precambrian Research*, **125**, 191–243.

KATO, Y., OHTA, I., TSUNEMATSU, T., WATANABE, Y., ISOZAKI, Y., MARUYAMA, S. & IMAI, N. 1998. Rare-earth element variations in mid-Archean banded iron formations: implications for chemistry of ocean and continent, and plate tectonics. *Geochimica et Cosmochimica Acta*, **62**, 3475–3497.

KAWAI, T., WINDLEY, B. F., TERABAYASHI, M., YAMAMOTO, H., MARUYAMA, S. & ISOZAKI, Y. 2006. Mineral isograds and zones of the Anglesey blueschist belt, UK: implications for the metamorphic development of a subduction–accretion complex. *Journal of Metamorphic Geology*, **24**, 591–602.

KAWAI, T., WINDLEY, B. F., TERABAYASHI, M., YAMAMOTO, H., MARUYAMA, S. & ISOZAKI, Y. 2007. Geotectonic framework of the blueschist unit on Anglesey–Lleyn, UK, and its role in the development of a Neoproterozoic accretionary orogeny. *Precambrian Research*, **153**, 11–28.

KIMURA, K. & HORI, R. 1993. Offscraping accretion of Jurassic chert–clastic complexes in the Mino–Tamba belt, central Japan. *Journal of Structural Geology*, **15**, 145–161.

KIMURA, G. & LUDDEN, J. 1995. Peeling oceanic crust in subduction zones. *Geology*, **23**, 217–220.

KIMURA, G., RODZDESTVENSKIY, V. S., OKUMURA, K., MELINIKOV, O. & OKAMURA, M. 1992. Mode of mixture of oceanic fragments and terrigenous trench fill in an accretionary complex: example from southern Sakhalin. *Tectonophysics*, **202**, 361–374.

KIMURA, G., MARUYAMA, S., ISOZAKI, Y. & TERABAYASHI, M. 1996. Well-preserved underplating structure of the jadeitized Franciscan complex, Pacheco Pass, California. *Geology*, **24**, 75–78.

KLEIN, G. DE V., KOBAYASGHU, K. ET AL. 1980. *Initial Reports of the Deep Sea Drilling Project, 58*. US Government Printing Office, Washington, DC, 101–145.

KOMIYA, T., MARUYAMA, S., MASUDA, T., NOHDA, S., HAYASHI, M. & OKAMOTO, K. 1999. Plate tectonics at 3.8–3.7 Ga: field evidence from the Isua accretionary complex, Southern West Greenland. *Journal of Geology*, **107**, 515–554.

KOPP, H., FLUEH, E. R., PETERSEN, C. J., WEINREBE, W., WITTWER, A. & MERAMEX SCIENTISTS 2006. The Java margin revisited: evidence for subduction erosion off Java. *Earth and Planetary Science Letters*, **242**, 130–142.

KUSKY, T. M. 1989. Accretion of the Archean Slave Province. *Geology*, **17**, 63–67.

KUSKY, T. M. 1990. Evidence for Archean ocean opening and closing in the southern Slave Province. *Tectonics*, **9**, 1533–1563.

KUSKY, T. M. 1991, Structural development of an Archean orogen, western Point Lake, Northwest Territories. *Tectonics*, **10**, 820–841.

KUSKY, T. M. & BRADLEY, D. C. 1999. Kinematic analysis of mélange fabrics: examples and applications from the McHugh Complex, Kenai Peninsula, Alaska. *Journal of Structural Geology*, **21**, 1773–1796.

KUSKY, T. M. & YOUNG, C. P. 1999. Emplacement of the Resurrection Peninsula ophiolite in the southern Alaska a forearc during a ridge–trench encounter. *Journal of Geophysical Research*, **104**, 29 025–29 054.

KUSKY, T. M., BRADLEY, D. C. & HAEUSSLER, P. 1997a. Progressive deformation of the Chugach accretionary complex, Alaska, during a Paleogene ridge–trench encounter. *Journal of Structural Geology*, **19**, 139–157.

KUSKY, T. M., BRADLEY, D. C., HAEUSSLER, P. & KARL, S. 1997b. Controls on accretion of flysch and mélange belts at convergent margins: evidence from the Chugach bay thrust and Iceworm mélange, Chugach Terrane, Alaska. *Tectonics*, **16**, 855–879.

LASH, G. G. 1985. Accretion-related deformation of an ancient (early Paleozoic) trench-fill deposit, central Appalachian orogen. *Geological Society of America Bulletin*, **96**, 1167–1178.

LEGGETT, J. K., MCKERROW, W. S. & EALES, M. H. 1979. The Southern Uplands of Scotland: a Lower Palaeozoic accretionary prism. *Journal of the Geological Society, London*, **136**, 755–770.

LEGGETT, J. K., MCKERROW, W. S. & CASEY, D. M. 1982. The anatomy of a Lower Palaeozoic accretionary forearc: the Southern Uplands of Scotland. *In*: LEGGETT, J. K. (ed.) *Trench–Forearc Geology: Sedimentation and Tectonics on Modern and Ancient Active Plate Margins*. Geological Society, London, Special Publications, **10**, 495–520.

MAHONEY, J. B. 2005. Nd and Sr isotopic signatures of fine-grained clastic sediments: a case study of western Pacific marginal basins. *Sedimentary Geology*, **182**, 183–199.

MAHONEY, J. B., HOOPER, R. L. & MICHAEL, G. 1998. Resolving compositional variations in fine-grained clastic sediments: a comparison of isotopic and mineralogical sediment characteristics, Shikoku Basin, Philippine Sea. *In*: SCHIEBER, J., ZIMMERLE, W. & SETHI, P. (eds) *Mudstones and Shales*. Schweizerbart, Stuttgart, 177–194.

MALTMAN, A. J. 1977. Serpentinites and related rocks of Anglesey. *Geological Journal*, **12**, 113–128.

MALTMAN, A. J. 1979. Tectonic emplacement of ophiolitic rocks in the Precambrian Mona Complex of Anglesey, North Wales. *Nature*, **277**, 327.

MARUYAMA, S. 1997. Pacific-type orogeny revisited: Miyashiro-type orogeny proposed. *Island Arc*, **6**, 91–120.

MATSUDA, T. & ISOZAKI, Y. 1991. Well-documented travel history of Mesozoic pelagic chert in Japan: from remote ocean to subduction zone. *Tectonics*, **10**, 475–499.

MILLER, J. McL. & GRAY, D. R. 1996. Structural signature of sediment accretion in a Palaeozoic accretionary complex, southeastern Australia. *Journal of Structural Geology*, **18**, 1245–1258.

MOORE, G. F., SHIPLEY, T. H. ET AL. 1990. Structure of the Nankai Trough accretionary zone from multichannel seismic reflection data. *Journal of Geophysical Research*, **95**, 8753–8765.

OGAWA, Y. & TANIGUCHI, H. 1988. Geology and tectonics of the Miura–Boso peninsulas and the adjacent area. *Marine Geology*, **12**, 147–168.

OKAMOTO, K., MARUYAMA, S. & ISOZAKI, Y. 2000. Accretionary complex origin of the Sanbagawa, high P/T metamorphic rocks, Central Shikoku, Japan – layer-parallel shortening structure and greenstone geochemistry. *Journal of the Geological Society of Japan*, **106**, 70–86.

OKAMURA, Y. 1991. Large-scale mélange formation due to seamount subduction: an example from the Mesozoic accretionary complex in central Japan. *Journal of Geology*, **99**, 661–674.

OTA, T., UTSUNOMIYA, A. ET AL. 2007. Geology of the Gorny Altai subduction–accretion complex, southern Siberia: tectonic evolution of an Ediacaran–Cambrian intra-oceanic arc–trench system. *Journal of Asian Earth Sciences*, **30**, 666–695.

PARK, R. G., TARNEY, J. & CONNELLY, J. N. 2001. The Loch Maree Group: Palaeoproterozoic subduction–accretion complex in the Lewisian of NW Scotland. *Precambrian Research*, **105**, 205–226.

PHILLIPS, E. 1991. The lithostratigraphy, sedimentology and tectonic setting of the Monian Supergroup, western Anglesey, North Wales. *Journal of the Geological Society, London*, **148**, 1079–1090.

ROBERTS, B. 1979. *The Geology of Snowdonia and Llŷn: an Outline and Field Guide*. Adam Hilger, Bristol.

ROBERTSON, A. H. F. & HUDSON, J. D. 1974. Pelagic sediments in the Cretaceous and Tertiary history of the Troodos massif, Cyprus. *In*: International Association of Sedimentologists, Special Publications, **1**, 403–436.

RYAN, P. D. & DEWEY, J. F. 2004. The South Connemara Group reinterpreted: a subduction–accretion complex in the Caledonides of Galway Bay, western Ireland. *Journal of Geodynamics*, **37**, 513–529.

SENNIKOV, N. V., OBUT, O. T., IWATA, K., KHLEBNIKOVA, T. V. & ERMIKOV, V. D. 2004. Lithological markers and bio-indicators of deep-water environments during Paleozoic siliceous sedimentation (Gorny Altai segment of the Paleo-Asian ocean). *Gondwana Research*, **7**, 843–852.

SHACKLETON, R. M. 1975. Precambrian rocks of Wales. *In*: HARRIS, A. L. (ed.) *A Correlation of Precambrian Rocks in the British Isles*. Geological Society, London, Special Publications, **6**, 76–82.

STANCIN, A. M., GLEASON, J. D., REA, D. K., OWEN, R. M., MOORE, T. C., BLUM, J. D. & HOVAN, S. A. 2006. Radiogenic isotopic mapping of late Cenozoic eolian and hemipelagic sediment distribution in the east–central Pacific. *Earth and Planetary Science Letters*, **248**, 840–850.

STEINMANN, G. 1925. Gibt es fossile Tiefseeablagerungen von erdgeschichtlicher Bedeutung. *Geologische Rundschau*, **16**, 435–468. [Also *International Journal of Earth Sciences*, **91**, 18–42, in English].

STOWE, D. A. V., TAIRA, A., OGAWA, Y., SOH, W., TANIGUCHI, H. & PICKERING, K. T. 1998. Volcaniclastic sediments, process interaction and depositional setting of the Mio-Pliocene Miura Group, SE Japan. *Sedimentary Geology*, **115**, 351–381.

STRACHAN, R. A., COLLINS, A. S., BUCHAN, C., NANCE, R. D., MURPHY, J. B. & D'LEMOS, R. S. 2007. Terrane analysis along a Neoproterozoic active margin of Gondwana: insights from U–Pb zircon geochronology. *Journal of the Geological Society, London*, **164**, 57–60.

TAIRA, A. & NIITSUMA, N. 2007. Turbidite sedimentation in the Nankai trough as interpreted from magnetic fabric, grain size, and detrital modal analyses. *In*: *Initial Reports of the Deep Sea Drilling Project*, **87**. US Government Printing Office, Washington, DC, 611–709.

TAIRA, A., KATTO, J., TASHIRO, M., OKAMURA, M. & KODAMA, K. 1988. The Shimanto belt in Shikoku, Japan – evolution of a Cretaceous to Miocene accretionary prism. *Modern Geology*, **12**, 5–46.

TAIRA, A., HILL, I. ET AL. 1992. Sediment deformation and hydrogeology of the Nankai Trough accretionary prism: synthesis of shipboard results of OPD Leg 131. *Earth and Planetary Science Letters*, **109**, 431–450.

TAKAYANAGI, H., IRYU, Y. ET AL. 2007. Carbonate deposits on submerged seamounts in the northwestern Pacific Ocean. *Island Arc*, **16**, 394–419.

THORPE, R. S. 1972. Ocean floor basalt affinity of Precambrian glaucophane schist from Anglesey. *Nature, Physical Science*, **240**, 164–166.

THORPE, R. S. 1978. Tectonic emplacement of ophiolitic rocks in the Precambrian Mona Complex of Anglesey. *Nature*, **275**, 57–58.

THORPE, R. S. 1993. Geochemistry and eruptive environment of metavolcanic rocks from the Mona Complex of Anglesey, North Wales, U.K. *Geological Magazine*, **130**, 85–91.

TUCKER, R. D. & PHARAOH, T. C. 1991. U–Pb zircon ages for Late Precambrian igneous rocks in southern Britain. *Journal of the Geological Society, London*, **148**, 435–445.

WAKITA, K. 2000. Cretaceous accretionary–collision complexes in central Indonesia. *Journal of Asian Earth Sciences*, **18**, 739–749.

WAKITA, K. & METCALFE, I. 2005. Ocean plate stratigraphy in East and South and Southeast Asia. *Journal of Asian Earth Sciences*, **24**, 679–702.

WOOD, D. S. 1974. Ophiolites, mélanges, blueschists and ignimbrites: early Caledonian subduction in Wales? *In*: DOTT, R. H. & SHAVER, R. H. (eds) *Modern and*

Ancient Geosynclinal Sedimentation. Society of Economic Paleontologists and Mineralogists, Special Publications, **19**, 334–344.

XENOPHONTOS, C. & OSOZAWA, S. 2004. Travel time of accreted igneous assemblages in western Pacific orogenic belts and their associated sedimentary rocks. *Tectonophysics*, **393**, 241–261.

ZIABREV, S. V., AITCHISON, J. C., ABRAJEVITCH, A. V., BADENGZHU, DAVIS, A. M. & LUO, H. 2004. Bainang terrane, Yarlung–Tsangpo suture, southern Tibet (Xizang, China): a record of intra-Neotethyan subduction–accretion processes preserved on the roof of the world. *Journal of the Geological Society, London*, **161**, 523–538.

Orogens in the evolving Earth: from surface continents to 'lost continents' at the core–mantle boundary

M. SANTOSH[1,2]*, SHIGENORI MARUYAMA[3], TSUYOSHI KOMIYA[3] & SHINJI YAMAMOTO[3]

[1]*Faculty of Science, Kochi University, Akebono-cho, Kochi 780-8520, Japan*

[2]*Department of Earth and Atmospheric Sciences, Center for Environmental Sciences, Saint Louis University, St. Louis, MO 63108, USA*

[3]*Department of Earth and Planetary Sciences, Tokyo Institute of Technology, Tokyo 152-8551, Japan*

*Corresponding author (e-mail: santosh@cc.kochi-u.ac.jp)

Abstract: Orogens and their posthumous traces are the basic elements that can be used to understand the material circulation within the Earth. Although information preserved in the rocks on the surface ranging in age from 4.4 Ga to the present has been used to characterize orogens, it is important to understand orogens on a whole-Earth scale to evaluate global material circulation through time. In this paper, we synthesize the general concepts and characteristics of orogens and orogenic belts. The collision type and accretionary type constitute the two end-member types of orogens, both sharing similar structural features of subhorizontal disposition, bounded above and below by paired faults. Their exhumation generally occurs in two steps: first by wedge extrusion to form a sandwich structure with subhorizontal boundaries, which is followed by domal uplift of all the units. In the accretionary type, oceanic lithosphere subducts under the continental margin, and in the collision type, buoyant continents collide with each other. Of the various types of subduction and collision processes, arc–arc collision orogeny may have been widespread in the Archaean, although most of the intra-oceanic arc crust must have been destroyed and dragged down to the Archaean core–mantle boundary (CMB). Here we propose a broad twofold classification of orogens and their subducted remnants, based on (1) their thermal history and (2) temporal constraints. Based on their thermal history, orogens are grouped into three types: cold orogens, hot orogens and ultra-hot orogens. Two extreme situations, which are anomalous and unlikely to occur on Earth, termed here super-cold and super-hot orogens, are also proposed. We discuss the characteristics of each of these subtypes. Based on temporal constraints, we group orogens into Modern and Ancient, where in both cases regional metamorphic belts occupy the orogenic core. In both groups, the overlying and underlying units of the regional metamorphic belts are weakly metamorphosed or unmetamorphosed, and are either accretionary complex in origin (Pacific type) or continental basement and cover (collision type). Major structures are subhorizontal with oceanward vergence of deformation, for both types. Orogens in the Modern Earth are grouped into four sub-categories: (1) deeply subducted orogens that are taken down to mantle depths and never return to the surface, termed here 'ghost orogens'; (2) those that are subducted to deep crustal levels, undergo melting and are recycled back to the surface, forming resurrected and temporarily 'arrested orogens'; (3) 'extant orogens', which are partly returned to the surface after deep subduction; (4) 'concealed orogens', which have been deeply subducted and only the traces of which are represented on the surface by mantle xenoliths carried by younger magmas. The preservation of orogens on the surface of the Earth occurred through an unusual return process from their natural course of total destruction, a phenomenon that operated more efficiently in the Phanerozoic through exhumation from ultra-deep domains against the slab-pull force of the plate, aided by fluids derived by dehydration of subducted lithosphere. Orogens at present represented on the surface of the Earth constitute only a fraction of the total volume formed in Earth history. Traces of the deeply subducted 'lost orogens' are sometimes returned to the surface in the form of melt or mantle xenoliths through combined processes of plume and plate tectonics. From a synthesis of the processes associated with the various categories of orogens proposed in this study, we trace the time-dependent transformations of orogens in relation to the history of the evolving Earth.

An orogen in general terms is a belt of rocks (Suess 1875; Miyashiro 1973) composed of regional metamorphic units at the core enveloped by unmetamorphosed rocks including an accretionary complex or coherent passive margin sediments. Orogens also include ophiolitic rocks, with or without large contemporaneous igneous suites such as tonalite–trondhjemite–granodiorite (TTG) batholiths (Windley 1995; Maruyama 1997; Kusky et al. 2004). All of these units have more or less similar geological ages based on which they are grouped together, although the various units may be separated in space and time. Orogens can be used as the basis to explain the geological and tectonic history of a region (Dewey & Bird 1970; Windley 1995; Maruyama 1997; Maruyama et al. 2004).

In this paper, we synthesize the general concepts and characteristics of orogens and orogenic belts. We propose a new classification for orogens to understand material circulation in the solid Earth. The preservation of orogens on the surface of the Earth is interpreted as a miraculous retrieval process from their course of total destruction, a process that operated more efficiently in the Phanerozoic through exhumation from ultra-deep domains against the slab-pull force of the plate. This is probably due to the highly lubricant role played by fluids derived by dehydration of subducted lithosphere. We argue that the natural fate of orogens is their total demise, and that substantial volumes of orogens were destroyed and the material was deeply subducted during Earth history and never returned to the surface. Thus, orogens at present represented on the surface of the Earth constitute only a fraction of the total volume. However, small traces of the deeply subducted 'lost orogens' are sometimes returned to the surface in the form of melt or mantle xenoliths through combined processes of plume and plate tectonics. We evaluate these aspects in an attempt to trace the time-dependent transformations of orogens in relation to the history of the evolving Earth.

We would like to clarify that the term 'orogen' is used in this paper frequently in a sense different from that of the conventional usage. In modern geology, orogen refers to a region that has been formed as a result of uplift associated with collision, and the generalized and widely accepted classification (e.g. Windley 1995) considers an Andean-type orogen to result from the collision and subduction of an oceanic plate beneath a continental plate, an Alpine–Himalayan-type orogen to result from the collision of two continental plates, and the Tethyan type to be generated from the sweeping up of smaller arcs and continental fragments during the subduction of an oceanic slab. Rogers & Santosh (2004) grouped orogens into three types: (1) intercratonic, formed by the closure of ocean basins; (2) intracratonic, developed within continents where there was no pre-existing oceanic crust; (3) confined orogens formed by closure of small oceanic basins. Windley (1995) broadly grouped orogens into collisional and acccretional types. The question arises whether there is a need to propose a new classification scheme for orogens. Although various categories of orogens have been well defined and are in common use in the literature, we consider that the history of orogens has not yet been addressed on a whole-Earth scale. Plate tectonics is a surface manifestation of the motion of Earth's lithosphere, and it contributes to the generation of new continental crust that is horizontally transported and eventually destroyed at subduction zones prior to orogenic 'suturing'. The subducted material accumulates at 660 km depth, being transformed from a curtain-like sheet to a large blob that drops vertically to the core–mantle boundary (CMB) (Maruyama et al. 1994). Recent models propose that these blobs accumulate as slab 'graveyards' and provide the fuel for the generation of superplumes by the consumption of recycled mid-ocean ridge basalt (MORB) (see Maruyama et al. 2007a, b). The superplume rises and penetrates the mantle transition zone to finally appear as hotspots. Superplumes also supply the fuel materials (high T and enriched basaltic component) in the upper mantle to drive plate tectonics. Thus, global material circulation may work episodically on a whole-mantle scale as a consequence of material fractionation at both the uppermost regions and the base. Through time, continents gradually grow on the top of the surface, and 'anti-continents' must have been generated simultaneously on the CMB (Maruyama et al. 2007a). To understand the mechanism of global material circulation through time, geologists use information preserved in the rocks on the surface ranging in age from 4.4 Ga to the present. Orogens in space and time are the potential source of this information, and it is important to understand their characteristics on a whole-Earth scale to characterize global material circulation from the surface to the bottom of the mantle. It is with this objective, and based on recent concepts of plate, plume and anti-plate tectonics, that we attempt to propose a new classification scheme in this paper, although some of the terms that we employ to classify orogens might be considered debatable.

General characteristics of orogens

Orogeny includes a collage of processes, such as: (1) magmatism, which generates continental crust; (2) rejuvenation and recrystallization by metamorphism where metamorphic belts occupy the orogenic core; (3) deformation to produce major structures of orogenic belts; (4) sedimentation where mountain building occurs, through the transportation of large volumes of sedimentary material. The proposal of Miyashiro (1961) on 'paired metamorphic belts' in fact represents the parallel juxtaposition of two synchronous orogens of different thermal characters and represents a milestone in thinking of how these processes may be related at different levels and in different tectonic settings. Several recent studies have addressed the characteristics of hot orogens, mostly comprising granulite-facies terranes. Collins (2002) proposed that granulite terranes were too hot to have formed during continental collision and that along with the high-grade metamorphic terranes that typify continental crust, most formed in accretionary orogens during tectonic switching, when prolonged lithospheric extension was interrupted by intermittent, transient contraction. Brown (2006) identified that crustal melt flow is an integral part of large hot orogens and pointed out that granites commonly occur associated with major transtensional fault systems or in extensional detachments close to the base of the upper continental crust. Beaumont et al. (2006) numerically modelled crustal-scale channel flow in hot orogens taking the Himalayan–Tibetan tectonics as an example, and showed that gravitationally driven channel flows of low-viscosity, melt-weakened, middle crust can explain both outward growth of the Tibetan Plateau and ductile extrusion of the regional metamorphic belts as the core of the Greater Himalayan Sequence.

In the following sections, we synthesize the general characteristics of orogens in the plate-tectonic perspective.

Accretionary-type and collision-type orogens

Orogeny has long been considered on the basis of the concept of geosynclines (e.g. Dickinson 1971). It has been held that the formation of orogens is initiated through extensional tectonics during the breakup of continents leading to the accumulation of voluminous sediments as geosynclines, which are subsequently transferred to build mountain belts when the regional stress field changes from extension to compression, followed by post-orogenic volcano-plutonism. Thus, the space–time relationship and tectonic history from geosynclines to mountain belts have been the focus of geological studies for many decades. After the advent of the plate-tectonic theory, orogeny has been considered from a different point of view and the formation of orogens along consuming plate boundaries came into focus with a broad two-fold classification into Cordilleran versus collision types. The global examples for these were considered to be Western North America for the former, and the Alpine and Himalayan belts for the latter (Dewey et al. 1973; Dickinson 1981; England & Thompson 1984).

The introduction of the term core complex (Coney & Harms 1984) and the proposal linking it with extensional tectonics shifted the focus of the formation of orogens from consuming plate boundaries to those of extension. However, detailed documentation of recent orogens formed along zones of extension has not yet been carried out, apart from a few cases of core complexes in post-collisional extensional settings (Lister & Davis 1989; Matte 1991).

Seismological and other investigations along spreading centres on the ocean floor, for instance in the Atlantic, Indian and Pacific Oceans and the Philippine Sea, led to the discovery of mega-mullion and caldera structures (Cann et al. 1997; Ohara et al. 2001). A mega-mullion exposes the Moho boundary of the oceanic crust and mantle over a scale of several hundred kilometres, indicating a regional-scale metamorphism with simultaneous on-axis and off-axis extensive magmatism at the divergent plate boundary (Umino et al. 2008). Thus, it can be envisaged that under the African rift valley, where extension is continuing, similar extensive regional metamorphism and related deformation must prevail.

The two major types of orogens, the accretionary type (also known as the Pacific type; see Maruyama et al. 1996; Santosh et al. 2009c; see also Cawood et al. 2009, for a recent synthesis on accretionary-type orogens) and collision type, are illustrated in Figure 1a and b. The major components are different for the two types of orogens. In the case of collision-type orogens (Fig. 1a), protoliths can be formed along the passive continental margins and include coarse-grained shallow marine deposits such as conglomerates, sandstone, mudstone and evaporites. Sedimentary facies changes oceanward from coarse-grained to fine-grained mudstones. Platform carbonate is a characteristic rock type, including rhythmically interlayered clastic sediments. Hence the carbonate lithology is remarkably different between the collision type and the accretionary type. Volcanic rocks are frequently associated with rifting and have a characteristic bimodal nature, commonly manifested by alkaline to strongly alkaline basalts with or without carbonatites and kimberlites, together with felsic volcanic rocks such as phonolites and rhyodacites. Moreover, no TTG batholiths occur landwards, indicating that there is no crustal growth by continent collision, but instead only reorganization or rejuvenation of the rock materials by deformation and/or metamorphism. During the migration of continents to the consuming plate boundary, such as in the case of collision of India with Asia, accretionary-type orogeny may have prevailed to form continental

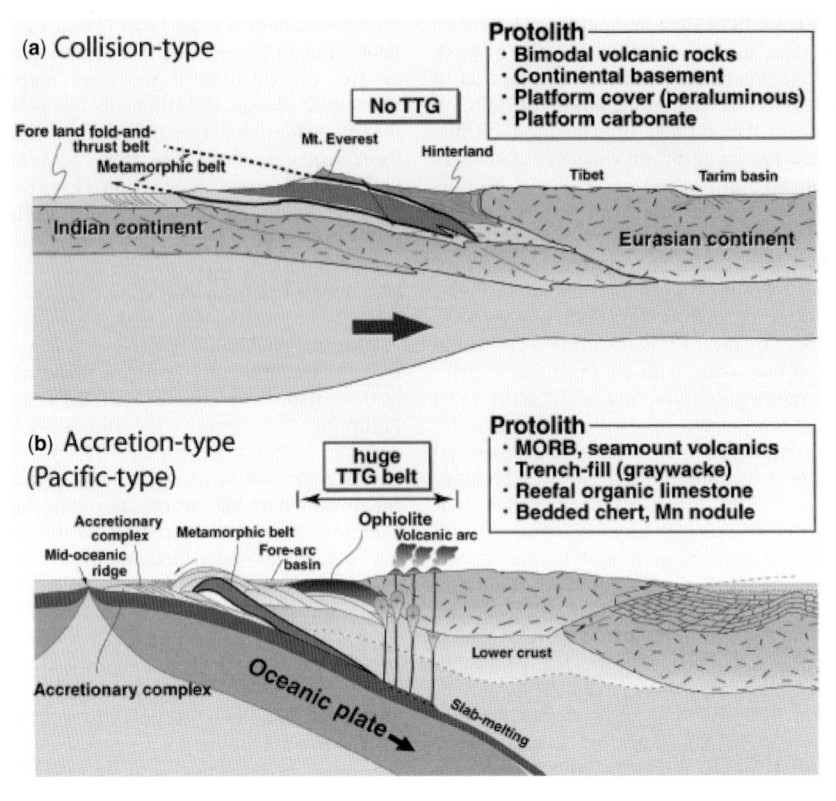

Fig. 1. (**a**, **b**) Schematic illustrations of collision-type and accretion (Pacific)-type orogens (after Maruyama et al. 1996).

margins (see Santosh et al. 2009a). The final collision of continents and subduction of continental margin to a maximum depth of 200–300 km results in the recrystallization of the passive margin package of rock units under ultrahigh-pressure (UHP) to high-pressure (HP) conditions. These rocks are finally returned to the surface to become juxtaposed with a collision-type orogen between an accretionary (Pacific)-type orogen and unsubducted continent. Thus, two distinct boundaries are defined. The central part of the collisional orogen is occupied by an orogenic core of regional metamorphic belts, such as in the case of the Himalaya and the well-studied Caledonides, Hercynides and Alpides. In these regional metamorphic orogens, the grade of metamorphism has long been considered as kyanite–sillimanite intermediate-pressure type (Miyashiro et al. 1982), and has been regarded as characteristic for the collision type. However, the discovery of the ubiquitous occurrence of HP to UHP minerals and rocks from major collisional orogens (e.g. Chopin 2003; Zhang et al. 2009, and references therein) has drastically changed the prevailing concepts on the progressive nature of collision metamorphism. The discovery of coesite- and diamond-bearing eclogites and related HP to UHP minerals clearly indicates that the progressive metamorphism belongs to the same facies series as that of the accretionary type. The strong overprinting by Barrovian metamorphism in these cases may be the result of late-stage hydration during the exhumation of the rocks from mantle depths to the surface (Maruyama et al. 2004).

The accretionary-type orogen (Fig. 1b) includes an accretionary complex comprising oceanic materials such as MORB, seamounts, ocean island basalt (OIB), and the carbonates that cap them, together with deep-sea sediments. All of these are finally capped by trench turbidites (Matsuda & Isozaki 1991). These materials are subsequently incorporated to generate the accretionary complex. Some of them are tectonically removed to the mantle depths and regionally metamorphosed under low-T–high-P conditions of usually up to 700–800 °C and at less than 60 km depth, returned to the surface and inserted within a shallow-level accretionary complex. Simultaneously, a huge batholith belt with felsic volcanic rock components forms above on the continental side. Forearc basins are developed between these regions, displaying a normal stratigraphic sequence. Among the lithological components of the accretionary-type orogens, the volcano-plutonic sequences are usually within 200–300 km width, the forearc basin deposits are within or slightly more than 100 km wide, and the regional metamorphic belts a few tens of kilometres wide and less than 2 km thick, with unmetamorphosed accretionary complex below and oceanward. The total width is thus around 400 km, with the regional metamorphic belts at the centre, and all of the units formed within a time span of 100–200 Ma (Fig. 1b), as in the case of Japan (Maruyama & Seno 1986; Isozaki 1997).

A recent synthesis by Cawood et al. (2009) provided a detailed evaluation of accretionary-type orogens. The formation of accretionary-type orogens is linked to intra-oceanic and continental margin convergent plate boundaries and includes the suprasubduction-zone forearc, magmatic arc and back-arc components. Cawood et al. (2009) grouped accretionary orogens into retreating and advancing types. The modern western Pacific constitutes an example of a retreating orogen with a characteristic back-arc basin. Advancing orogens such as the Andes develop foreland fold and thrust belts and also show crustal thickening. Cawood et al. (2009) noted that accretionary orogens have been active throughout Earth history and were responsible for the major growth of the continental lithosphere through the addition of juvenile magmatic products. These orogens also mark major sites of consumption and reworking of continental crust through time by sediment subduction and subduction erosion. According to the model of Cawood et al., the net growth of the continental crust from the Archaean is effectively zero because the rates of crustal growth and destruction are roughly equal.

Structure and size

In both accretion-type and collision type orogens, the major structures are similar; that is, subhorizontal and bounded above and below by paired faults (Maruyama et al. 1996). A normal fault on the top and a reverse fault on the bottom bound the regional metamorphic belts. Asymmetrical structure indicating an orogenic root occurs usually landward or in the direction of subduction. Vergence indicates the polarity of subduction and usually shows a landward dip. There is a marked difference in size between the two major types of orogens. The thickness of regional metamorphic belts in the accretionary type is usually less than 2 km, and the width is around at least 100 km (much more in the case of Alaska; for example, see Kusky et al. 2004) and lateral extent over 1000 km. On the other hand, collision orogens contain relatively thicker regional metamorphic belts (up to 5 km), with widths of 200–300 km and lateral continuity of over a few thousand kilometres depending on the size of the collided continental mass. In both orogen types, olistrosomes are rather common, formed by the collapse of gravitationally unstable units (Kaneko 1997; Maruyama et al. 1997). They form either during the rifting of collision-type orogens or during the exhumation of both collision and accretionary types. Alpine-type peridotites are also associated with both orogen types. Ophiolitic rocks rest on the regional metamorphic belts, or occur as fragments incorporated within an accretionary complex, but are less common in the collision type. Traditionally, in Alpine orogens, both accretionary type and collision type are mixed, so some ophiolitic rocks and minor pelagic sediments are metamorphosed under high-P–low-T conditions and are regarded as products of progressive collisional metamorphism (Maruyama et al. 1996). At the same time, continental protoliths are also regionally metamorphosed. Moreover, resting on the top of the Alpine regional metamorphic belts are unmetamorphosed ophiolitic sequences, suggesting the subduction of both oceanic lithosphere and a small volume of continental passive margin sediments, and their return to the surface to be juxtaposed with the oceanic crust.

Orogenic history through a Wilson cycle

Although the Japanese Islands at presently are within the Pacific Ocean, the major part was originally formed along the continental margin of Asia. They were separated during Miocene times (20–15 Ma) with the opening of the Japan Sea, resulting in the apparent isolated occurrence of Japanese archipelagos in the ocean (Maruyama et al. 1997). The Pacific crust off the Marianas is the oldest, with an age range of up to 200 Ma (Hilde et al. 1977). However, the birth of the Pacific Ocean dates back to 600 Ma, as documented by the oldest passive margin along North America, South America, Antarctica, eastern Australia and East Asia (Maruyama et al. 1989, 1997). Since its birth, the Pacific Ocean expanded to become a super-ocean at some time around 450 Ma. The tectonic style then switched from passive margin to active margin, initiating subduction and accretionary-type orogeny (Nakagawa et al. 2009). Episodic formation of orogenic belts around the Pacific margins has been recorded, occurring approximately once every 100 Ma (Maruyama & Seno 1986). Thus Japan is underlain by c. 450 Ma, 320 Ma, 250 Ma, (?170 Ma), 100 Ma and presumably 20 Ma aged orogenic belts (Maruyama 1997). It can be predicted that in future, such as 50 Ma after the present, the Australian continent will collide and amalgamate with the future Asian continents. Thereafter, 250 Ma from the present, North America will collide with the Asian margin to form the future supercontinent Amasia (Hoffman 1992, 1999; Maruyama et al. 2007a), although there is debate as to whether or not the Pacific will close to form Amasia or the Atlantic will close again to form Pangaea Ultima. In summary, orogenic history through a Wilson cycle indicates that the accretionary-type orogenic belts will be finally transformed to collision-type orogens.

Episodic nature of accretionary-type orogens

During the subduction process in accretionary-type orogens, the regional metamorphic belts are not exhumed in a continuous manner. Instead, they are episodically brought up as a result of the episodic subduction of mid-ocean ridges (Sisson & Pavlis 1993; Kusky et al. 1997; Bradley et al. 1993). Thus, the formation of the c. 100 Ma Sambagawa belt (orogen) in SW Japan is related to the subduction of the Pacific–Kula oceanic spreading centres. A gradual change of the subduction angle from deep to shallow will extrude the deep subducted accretionary wedge to the surface, thus promoting the exhumation of regional metamorphic belts (see Maruyama et al. 1996). At the same time, hot plate subduction causes extensive formation of felsic melts to underplate the hanging wall and to form a wide batholith belt. The Cretaceous example from Japan clearly documents the discontinuous development of orogenic belts, rather than a steady-state formation of accretionary complex, continental crust, deformation and sedimentation. Thus it can be envisaged that one orogenic belt corresponds to one subduction, particularly ridge subduction. The repetition of five Pacific (accretionary)-type orogenic belts from the structural top to the structural bottom with a subhorizontal relationship for the last 500 Ma in Japan suggests the annihilation of the oceanic lithosphere such as the Farallon, Kula, Pacific, Philippine Sea and some other unnamed plates. All of these must have subducted underneath the East Asian margin in the past 500 Ma period.

The Tethyan region, on other hand, is different from the Pacific margins. Consuming plate boundaries are restricted to the northern margin of the Indian and Tethyan oceans, dating back to the Permian (Metcalfe 2006). Continuous separation of continental fragments from the northern margin of Gondwana formed the passive margin sediments, and the migrated continental fragments caused the episodic growth of the southern margin of Central to Western Asia such as India, Saudi Arabia and many fragmental continents in the Tethyan orogenic belts (Santosh et al. 2009a). The extensive subduction of oceanic lithosphere before the final collision of continents is notable, suggesting the common occurrence of accretionary-type orogenic belts in the Tethyan realm. For example, to the north of the Himalayan belts, lawsonite–glaucophane schist belts are underlain by the Tethyan ophiolite, both of Cretaceous age, and continue laterally over few thousand kilometres, although sporadically, running parallel to the orogen-parallel Cretaceous batholith belt to the north (Windley 1983). This 200–300 km wide orogen-parallel rock unit should be defined as an accretionary-type orogen.

Thus, the accretionary-type orogens have dominated along the Pacific Ocean margin over the past 500 Ma of Earth history and ultimately all of these with be docked by continental collision in future. On the other hand, frequently repeated orogeny of both accretionary type and collision type occurred in the Tethyan domain. The future supercontinent Amasia will be characterized on the eastern side by the accretionary type-dominated orogens, and on the western side by the Tethyan-type orogens until the final completion of Amasia is achieved by about 250 Ma from the present.

Exhumation of orogens

To understand the exhumation mechanism of regional metamorphic belts, the modern analogue for continent subduction and ridge subduction is critically important. One of the typical examples for the former is the Timor–Tanimbar region, where a non-volcanic forearc occurs in the eastern part of the Indonesian region, to the north of Australia (Kaneko et al. 2007). The Australian continent was rifted away from Antarctica at c. 60 Ma ago, migrated northward at c. 7–8 cm a^{-1} and was underthrust below the Indonesian arc at c. 20–30 Ma ago (Kaneko et al. 2007). Volcanism ended in the late Miocene, and the subduction of passive margin deposits on the northern flank of Australia extended underneath the volcanic front. The exhumed regional metamorphic belts reached the surface from west Timor at about 20 Ma ago and propagated episodically to the east over 1500 km. At the eastern end, Lavabo Island, close to Tanimbar Island, the topmost portion of the blueschist-bearing regional metamorphic belts has just reached sea level. The extensive development of high-angle normal faults and related mud volcanoes is common in the Timor islands (which is a type locality for mud volcanoes; Charlton et al. 2002). This region is also characterized by the highest speed of mountain building, equivalent to the Himalayan upheaval (Kaneko et al. 1997). Quaternary reef carbonates have been raised to 1000–2000 m elevation (see Charlton et al. 2002). Flat-lying orogenic core sandwiched by overlying ophiolitic rocks and underlying passive margin sediments are all cut by high-angle secondary normal faults and the central portion is selectively lifted, currently witnessing the second stage of domal uplift. The protoliths of blueschist eclogites belong to the collision-type regional metamorphic belts, originally formed during an unknown continental rifting event, presumably within the Indonesian region, and perhaps derived from Gondwana. There is unusual and frequent occurrence of seismic activity under Timor. The earthquakes in this region are distributed perpendicular to the suggested Benioff plane. Osada & Abe (1981) speculated that these earthquakes indicate slab break-off, indicating that the dense older oceanic lithosphere is being released from the buoyant Australian continental lithosphere. If this is true, the stress caused by buoyant return by the slab break-off would push up the deep-seated high-temperature and ductile regionally recrystallized UHP–HP units, which will then move along the Benioff thrust to the forearc region of the Timor–Tanimbar regions. The reason why the UHP units never rise vertically is the presence of a brittle mantle wedge, which acts as a metal-like hard barrier, diverting the highly ductile and buoyant felsic material oceanward along a small and thin flow channel. The underlying low-temperature mass of continents and continental lithosphere also behaves as a brittle block. The temperature at the bottom of the lithosphere is around 800 °C, but that in the central domain is only around 300 °C. The top surface has a higher temperature because of the heating by wedge corner flow in the mantle wedge brought about by arc magmatism. Thus, the continental lithosphere is relatively cold and brittle. The surface of the passive continental margin, originally formed during Permian time and deposited on the northern flank of the Australian continent, migrated northward and finally was subducted. The subduction-zone metamorphism presumably occurred under high-P/T geothermal gradients including the formation of blueschist–eclogites, heated by the wedge corner flow (Ishikawa et al. 2007). The ductile material was removed to the surface along the Benioff thrust, passing over to a much lower temperature domain of feebly to unmetamorphosed passive margin sediments, and was subsequently hydrated at mid-crustal levels to be in equilibrium with Barrovian hydration mineral assemblages. Subsequent to the completion of slab break-off doming started, and attained the current stage of mountain building seen in western Timor. On the other hand, the eastern portion is still in the stage of exhumation and has not yet reached the second stage of doming. However, the orogen has already nearly reached sea level, indicating that wedge extrusion must have preceded final slab break-off. The successive subduction of continental lithosphere driven by a heavy anchor of oceanic lithosphere caused the UHP to HP regional metamorphism (Parkinson et al. 2004). Presumably, the change of subduction angle from deep to shallow could have been a major trigger to return the passive margin sediments to the surface. Maruyama et al. (1996) defined this mechanism as wedge extrusion (Fig. 2).

The foregoing example indicates that exhumation occurs in two steps: first by wedge extrusion to form a sandwich structure with subhorizontal boundaries, which is followed by domal uplift of all the units. The so-called mountain-building stage is restricted to the second stage. The large pressure gap between regional metamorphic belts and the surrounding units denotes formation by tectonic juxtaposition of the thin slice of regional metamorphic belt without any mountain building.

The classic concepts of collision orogeny and the mechanism of exhumation refer to uplift and erosion by buoyancy, exposing intermediate-pressure regional metamorphic rocks characterized by clockwise $P-T$ paths (e.g. England & Thompson 1984). The Himalayan metamorphic belt has long been regarded as a world standard, and a similar model was also extended to collisional metamorphic belts in other parts of the world. If the model of England & Thompson (1984) is valid, the Himalayan orogenic core must be underlain by progressively higher $P-T$ rocks until mantle peridotite is reached at depth. However, geologists who worked in the Himalayan regions for over a century have clearly documented that the flat-lying regional metamorphic belts here are underlain by low-grade to unmetamorphosed platform cover sediments, implying that the classic concepts do not hold good (Searle et al. 1987). A significant breakthrough came when Chopin (1984) identified an earlier episode of blueschist- to eclogite-facies metamorphism from the highest grade rocks in the Alps. He discovered coesite in eclogite that occupied the orogenic core. Moreover, from the Pakistan Himalaya, O'Brien et al. (2001) discovered coesite eclogites, and later Sachan et al. (2004) made a similar discovery from Ladak, Indian Himalaya. Recently, possible eclogites were also reported in the Chinese and Nepal Himalaya (Parkinson & Kohn 2002). In the case of the Himalayas, Kaneko et al. (2003) reported a 48 Ma zircon concordia age for the progressive blueschist–eclogite metamorphism and subsequent hydration during the mid-Miocene at mid-crustal level, indicating an extremely slow exhumation rate as a narrow channel flow along the Benioff thrust. The domal uplift started after 9 Ma in the Himalayan region, with no mountain building prior to this, in spite of

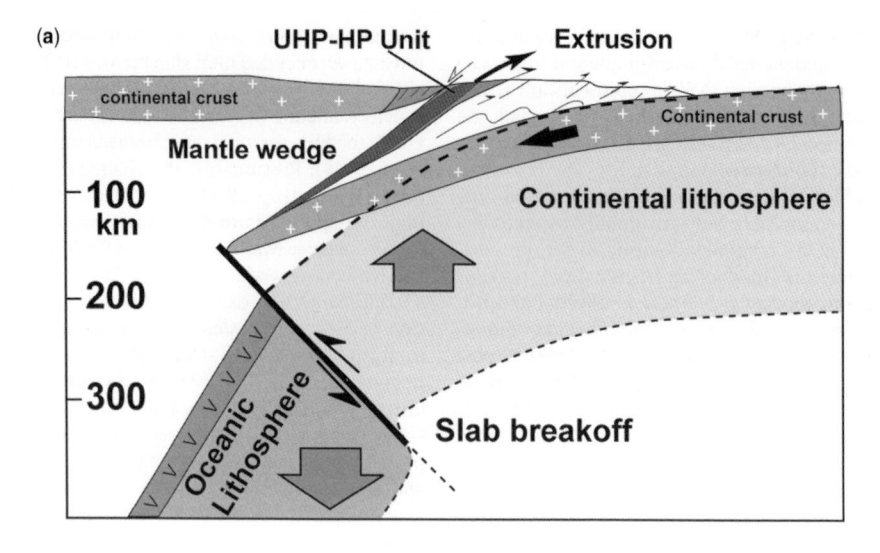

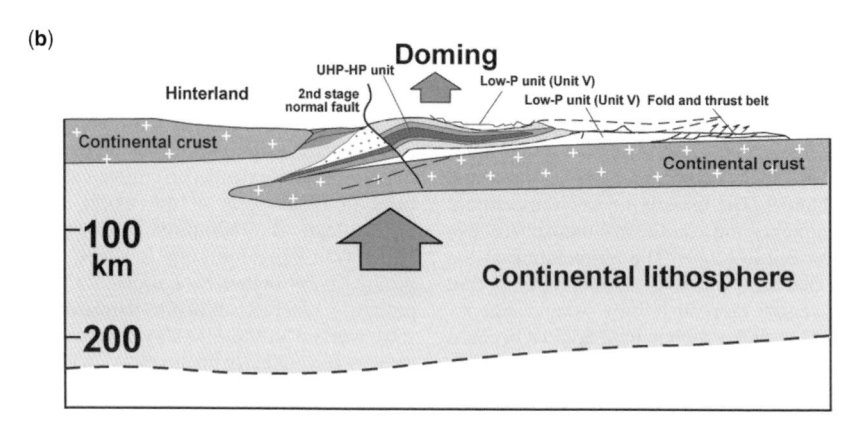

Fig. 2. Schematic illustrations of exhumation mechanisms: (**a**) tectonic extrusion; (**b**) domal uplift (after Maruyama *et al.* 1996).

the tectonic juxtaposition of UHP–HP regional metamorphic rocks to the mid-crustal level. These discoveries further support the wedge extrusion model of exhumation discussed in the previous section.

Further support for the wedge extrusion model comes from the progressive $P-T-t$ path defined through detailed thermodynamic modelling and inclusion mineralogy, such as, for example, from the diamond-bearing UHP metamorphic rocks of Kokchetav (e.g. Katayama *et al.* 2001; Masago *et al.* 2009). The estimated progressive geotherm is anti-clockwise with a kink point at about 13 kbar and 700 °C, followed by an exhumation path characterized by isothermal decompression (Masago & Omori 2004) and extensive hydration at mid-crustal level. Katayama *et al.* (2001) identified zoned zircon with index minerals such as quartz–coesite, graphite–diamond and albite–jadeite, systematically tracing the $P-T-t$ path. The results indicate rapid subduction, c. 100 km Ma^{-1}, but ultraslow exhumation from mantle depths as great as 200 km to the mid-crustal level, at a rate of 1 mm a^{-1}. This exhumation rate is one to two orders of magnitude slower than the next step of the domal uplift process. The current domal uplift in the Timor region and the Himalayas is estimated to be at around 1 cm a^{-1}. On the other hand, Rubatto & Hermann (2001) proposed ultrafast exhumation by measuring zoned zircons, which indicate 3–4 cm a^{-1}, faster than the subduction of European continent underneath Africa. Similar studies followed for the Himalaya by Leech et al. (2005). Baldwin et al. (2004) measured young zircon from eclogites in Papua New Guinea, and proposed that these represent the world's youngest eclogites, with an age of 4.3 Ma; this region is an extension of the Timor–Tanimbar region further to the east. These are famous core complex regions, including post-orogenic batholiths, and the entire island is occupied by plutons with the same age. Although Baldwin et al. proposed an ultrarapid exhumation model against the ultraslow uplift models, because the study measured the post-orogenic stage (equivalent to the Kokchetav massif), the logic may not hold good for the ultrarapid exhumation model.

Tectonic erosion

Sedimentary erosion is the process by which rain and wind remove weathered surface rocks into the ocean. Tectonic erosion denotes the destruction of the leading edge of the overriding plate by the subducting lithosphere. If the coupling between the two plates is too strong, then tectonic erosion is expected, such as in the case of subduction in Chile (Von Huene & Lallemand 1990). An opposite scenario occurs in the case of loosely coupled plates, such as in the Mariana Trench, where gravitational collapse of the hanging wall causes the tectonic erosion to transfer olistostromal deposits to the Mariana Trench. The olisostromes are transported to the deep mantle together with the subducting Pacific plate. For an intermediate degree of coupling, an accretionary complex can be formed only when voluminous clastic sediments are supplied. Continental crust including granites, high-grade gneiss or low-grade metamorphic rocks, and even sedimentary rocks originally derived from different rock types on the Earth's surface, will all be removed from the surface to the deep mantle through tectonic erosion.

Geologists have long believed that once a continent is formed on the surface of the Earth, it will never subduct into the mantle, but will remain on the surface, thus contributing to a one-way growth of continental crust through geological time (e.g. Macdonald 1961; Moorbath 1976; Condie 1986). This traditional concept has been revised recently after the discovery of evidence for tectonic erosion, which is now known to be common around the southern Pacific region (Scholl et al. 1980; Armstrong 1991; von Huene & Ranero 2003). The first recognition of this phenomenon was in the Peru Trench, where, near the trench line, in the landward wall of the trench, deep drilling has revealed high-grade gneiss, which forms the basement of South American craton, in spite of subduction over 500 Ma. The absence of a Phanerozoic accretionary complex indicates that tectonic erosion must have occurred some time in the Phanerozoic with the lack of formation of an accretionary complex since then. Although a minor accretionary complex may be present, this is not sufficient in volume to account for the last 500 Ma of subduction. In the case of the Japan Trench off Hokkaido and NE Japan, very small amounts of post-Miocene accretionary complex could have been formed since the Miocene, but Cretaceous granites and a Jurassic accretionary complex with north–south and west–east strike trends indicate that the older basement rocks are truncated by the Japan Trench, offering important clues for large-scale tectonic erosion during the time of formation of the Japan Sea (Maruyama 1997). On the other hand, a voluminous post-Miocene accretionary complex was formed along the Nankai Trough off SW Japan by the subduction of the Philippine Sea plate. Moreover, the subducting oceanic slab at 100 Ma as documented by the distribution of the Sambagawa high-pressure metamorphic belt from the surface to the bottom of the Moho, and the present-day Benioff plane of the Philippine Sea plate (Park et al. 2002) clearly demonstrate the downward growth of the accretionary complex, ranging in thickness from 20 to 30 km, through the supply of huge amounts of clastic wedges before and/or during the opening of the Japan Sea. However, the 450 Ma age estimated for the Benioff plane from the presence of high-P metamorphic rocks immediately above the 100 Ma Sambagawa rocks (Isozaki & Maruyama 1991) indicates that there was no growth of accretionary wedge from 450 to 100 Ma. Moreover, the occurrence of 450 Ma granites in the oldest orogen in Japan suggests possible tectonic erosion underneath, with a missing 30 km thick continental crust for the period 450–100 Ma. Therefore, orogenic growth along the Pacific margin does not support continuous growth oceanward, and sometimes may support even a lack of growth of continental margin. The quantitative evaluation of the role of tectonic erosion to erase the older Pacific orogeny remains an important problem to be investigated in detail (von Huene et al. 2000, 2004).

Role of arc collision-type orogeny

In the foregoing sections, we have discussed only the two end-member orogens, the accretion type and the collision type. In the former case, oceanic lithosphere subducts under the continental margin, and in the latter case huge continents collide with other continents. In nature, there are several types of subduction and collision, such as arc–arc collision and arc–continent collision, as well as the subduction and collision of huge oceanic plateaux. In the oceanic domain, arc–arc collision is observed rather commonly in the triangular western Pacific region. Here, the largest oceanic plate, the Pacific plate, subducts westward to form the Izu–Mariana intraoceanic arc, and to the south the Pacific plate subducts below the oceanic domains and the Tonga–Kermadec arc is formed. To the north of the Indo-Australian plate, the Java–Sumatra Trench is at present characterized by the subducting oceanic portion of the Indian Ocean as well as the subduction of Australian continent underneath the oceanic regions. Thus, continuing continent–intra-oceanic arc collision is seen in the Timor–Tanimbar regions (Kaneko et al. 2007). The triangular western Pacific region, on the other hand, witnesses double-sided subduction from the south and from the east (see Maruyama et al. 2007a; Santosh et al. 2009a). A number of oceanic microplates are present in this region, and the consuming plate motions of these microplates produce a number of intra-oceanic arcs. Arc–arc collision in the Philippine Islands, arc–continent collision in Taiwan, arc–arc collision in the Celebes Islands and several other arc collisions are under way in the western Pacific region. Moreover, the Izu–Mariana arc has been long consumed to deform the Honshu orogenic belts. The oroclinal bending to the north of Izu clearly supports the idea of a buoyant arc–arc collision. The Tanazawa metamorphic rocks are exposed along the consuming boundary between the two arcs. Collision started at about 18 Ma and the thickness of the continental crust of Izu reaches 30 km (Taira et al. 1998). Farther to the south, the thickness decreases to 20 km. Therefore, 20–30 km thick arc crust must have subducted underneath Honshu for the last 18 Ma. The rate of consumption ranged from 2 to 5 cm a^{-1}, and the total accreted mass can be calculated to be thicker than 100 km under the Honshu arc. However, seismological data on the thickness of continental crust show only 30–35 km normal thickness compared with the rest of the Japan arc (Seno & Maruyama 1984; Maruyama et al. 1996), strongly indicating that almost all of the crust must have been subducted. Arc subduction is also observed to the south of Kyushu, where four arcs are currently subducting without any strong deformation on the hanging wall (Yamamoto et al. 2009). These are the Kyushu–Palau arc (dead arcs after 18 Ma), Amami plateau (late Cretaceous, as the granites and andesites dredged so far are of late Cretaceous age), and Daito ridge and Okino–Daito ridge, both arcs of late Eocene age. All these are moving northwestward at 6 cm a^{-1}. The geological features of the hanging wall do not indicate any strong resistance or accretion of the buoyant arc materials. Instead, all the arcs so far show that tangential or oblique subduction proceeded to subduct the material into the deep mantle without any accretion. On the other hand, in the Celebes, where large ophiolite bodies occur, parallel collision of two arcs indicates collision and amalgamation without subduction between the two arcs (Hamilton 1979). To the north, the Celebes ophiolites continue within two arms to the ocean floor between the two arcs, and have not yet been ultimately subducted, with an intervening ocean present between the two arcs. In the Celebes, only the central portion of the two arcs is connected, with exhumed ultrahigh-pressure metamorphic rocks as well as unmetamorphosed ophiolites (Kadarusman et al. 2004). In summary, the arc–arc collision in the western Pacific shows that most of the intra-oceanic arcs subduct without leaving any trace on the surface. The growth of continental crust in the oceanic domain is apparently less effective, although the orogenic style seems to be similar to the continent collision type. In this case, subducted orogens (ghost orogens; see below) related to arcs may be common, and those arcs must have moved down to the CMB to accumulate sialic rocks on the bottom of the mantle through time. Such ghost orogens must have been widespread in the Archaean Earth, because no large continents were present at that time on the planet, and the Earth was covered by a number of micro-oceanic plates (over 300) (Sleep & Windley 1982; Kusky & Polat 1999; Komiya et al. 2002a). Therefore, arc–arc collision orogeny must have been widespread in the Archaean, although most of the intra-oceanic arc crust must have been destroyed by subduction as ghost orogens down to the Archaean CMB. Rino et al. (2008) estimated the volume of such material and compared it with the total continental crust on the Earth, and computed 6–7 times more volume of continental crust in the Archaean. In spite of the double production rate following the fast rate of subduction because of higher temperatures in the Archaean Earth, the observed continental crust of Archaean age represents only about 20% of that originally present. The remaining 80% of continental crust was formed during the Proterozoic and Phanerozoic, predominantly during the Proterozoic. The drastic depletion of Archaean continental crust therefore suggests extensive subduction of juvenile arcs into the mantle.

A new classification of orogens

Traditionally, the term orogen has been applied to describe a mountain belt composed of different types of rocks or rock strata forming a complex of variable size, typically tens to hundreds of kilometres wide and several thousand kilometres long, which is later fragmented during younger geological time by various processes (Suess 1875; Miyashiro

1961). Orogens commonly form along consuming plate boundaries, at which there is either continent collision or Pacific-type subduction, as discussed above. Whereas the term orogens generally refers to those exposed on the surface of the Earth, in the present work we extend the term to include the regionally metamorphosed units that have undergone deep subduction and were never returned to the surface. Although some of these remain within the crust, others have gone to the bottom of the mantle as 'lost continents'. Large parts of most old orogens have been destroyed, ending up in the deep mantle together with subducted oceanic crust through geological time. Presumably only a small fraction of these have returned to the surface, either as xenoliths or in on-land orogenic belts.

Here we make a broad two-fold classification of orogens, based on (1) their thermal history and (2) temporal constraints. These, as well as their subtypes, are briefly discussed below.

Classification based on the thermal history of orogens

Based on their thermal history, we make an attempt to characterize the various orogens in a $P-T$ diagram (Fig. 3) where orogens are classed in three groups: cold orogens, hot orogens and ultra-hot orogens. The regions of two extreme situations, which are anomalous and unlikely to occur on Earth, termed here super-cold and super-hot orogens (except perhaps in the rare case of Archaean komatiites, some of which may fall in the field of super-hot orogens), are also shown in the figure. The upper temperature limit of hot orogens may range up to 750–800 °C, with a migmatite zone appearing towards the higher temperature region. The documentation of the widespread occurrence of ultrahigh-temperature (UHT) granulites in various terranes formed under extreme thermal conditions of 900–1100 °C and pressures of 8–12 kbar (e.g. Brown 2007; Santosh et al. 2007a, b; Kelsey 2008) indicates that these orogens constitute an important category. Because they fall beyond the realm of hot orogens, we define their field as ultra-hot orogens. More than 40 localities of UHT rocks are now known worldwide, mostly from within Precambrian terranes, with only a few Phanerozoic examples (see Kelsey 2008), signifying the global nature of ultra-hot orogens.

The oceanic counterpart of the paired metamorphic belts of Miyashiro (1961) ranges in metamorphic grade from zeolite through greenschist and/or blueschist up to eclogite facies. In general, the temperatures of these rocks are less than 600 °C, albeit the pressures are high, ranging up to 10–15 kbar. The P/T ratio is markedly different from that for the low-pressure counterpart of the paired belts. Although recrystallization is generally limited at temperatures below 600 °C, these rocks exhibit a recrystallized nature with strong schistosity, indicating penetrative fabric formation under extensive shearing, even under relatively low-temperature conditions. These rocks, previously considered as high-P/low-T regional metamorphic belts, are designated as cold orogens in our study.

UHP rocks were first described from the Alps (Chopin 1984), followed by several recent reports from various parts of the world (see Brown 2007, and references therein), particularly from Phanerozoic orogenic belts (Zhang et al. 2009). Recent studies indicate that the upper pressure limit of many UHP belts exceeded 30 kbar, as against the earlier concept by Miyashiro (1961) of 15 kbar. Another issue concerning UHP rocks is their relationship with the surrounding host gneisses. Detailed mineral inclusion work on zircons from the country rocks clearly demonstrates that the whole package of the regional metamorphic belt in which these rocks are enclosed was once metamorphosed under UHP–HP conditions and retrogressed during their exhumation (Maruyama & Parkinson 2000; Katayama et al. 2001; Zhang et al. 2009). Usually a 20–30 Ma difference exists between the UHP–HP metamorphism and the subsequent Barrovian hydration events (Katayama et al. 2001). The temperature maximum was extended from the earlier 600 °C to 800–900 °C, although this was achieved only at extreme high-pressure conditions (Kokchetav, Masago 2000; Masago et al. 2009; Dabie–Sulu, Liu et al. 2004; see Zhang et al. 2009, for a recent review; Norwegian Caledonides, Cuthbert et al. 2000; Brueckner et al. 2002). Thus, these rocks would fall in the transition between hot and ultra-hot orogens. The Barrovian hydration can occur at high temperatures also, as shown by some of the examples of granulite-facies retrogressed assemblages enveloping UHP rocks (e.g. Kaneko 1997; Kaneko et al. 2003).

According to our classification in Figure 3, cold orogens and ultra-hot orogens define two end-members. The former generally defines temperatures <600 °C, as originally envisaged by Miyashiro (1961) for the oceanic side of the paired metamorphic belts, whereas the latter lies above 900 °C within the UHT granulite field. Normal hot orogens fall between these two extremes, although a zone of considerable overlap exists. In the $P-T$ field in Figure 3, there is a wide 'blank zone' in the data plots of various orogens, where none of the categories above are represented. This region realistically represents the subsolidus rocks that dominate the Earth's interior and are distinctly different from the unrealistic extreme zones designated as the super-cold and super-hot regions.

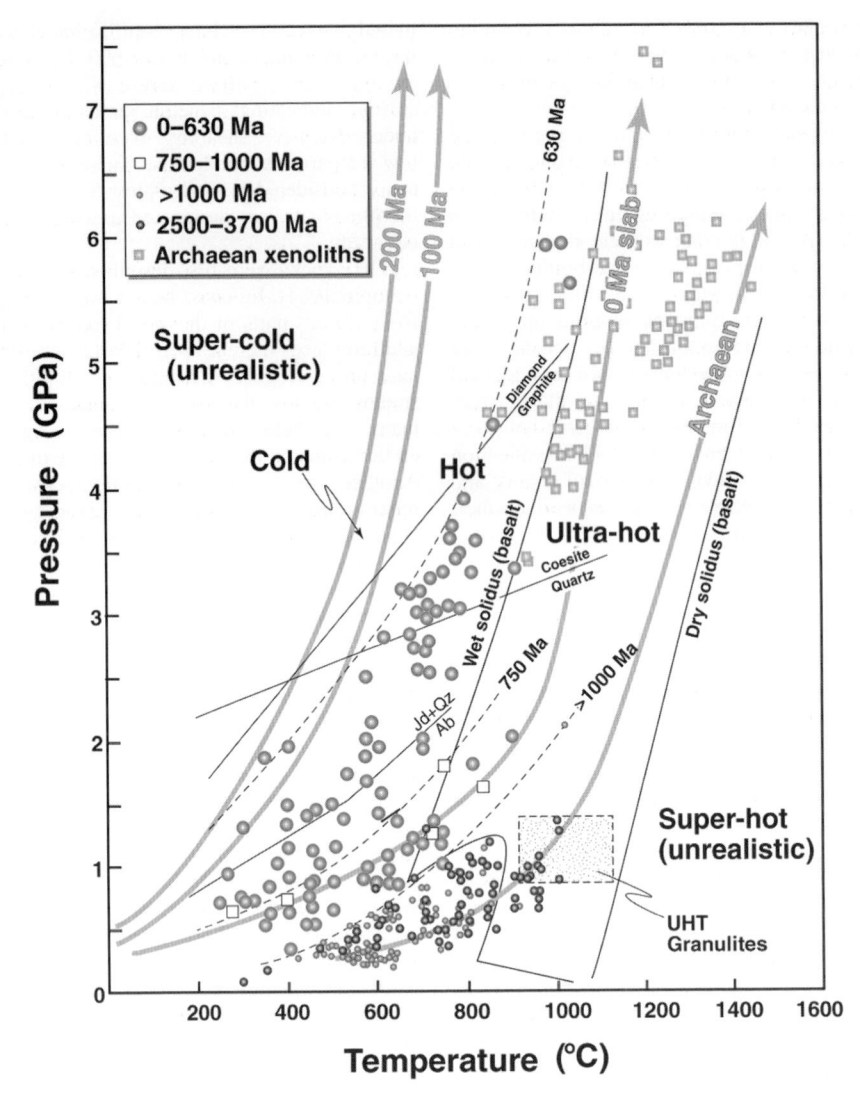

Fig. 3. Classification of orogens shown in $P-T$ space. Also included are the geotherms for 0 Ma, 100 Ma and 200 Ma slabs, as well as the dry and wet solidi of basalt (Peacock et al. 1994; Peacock 1996; Vielzeuf & Schmidt 2001; Maruyama & Okamoto 2007). Also plotted in the figure are $P-T$ data for metamorphic rocks from various orogens belonging to different timespans in Earth history (Maruyama et al. 1996; Hayashi et al. 2001) and the mean geothermal gradients in each case are also shown. The fields of Archaean xenoliths (Jacob 2004) as well as those of UHT granulites (after Kelsey 2008; Santosh & Omori 2008a, b) are also shown. Based on thermal history, orogens can be grouped into cold orogens, hot orogens, and ultra-hot orogens. Two extreme situations, which are anomalous and unlikely to occur in the Earth, are also shown as super-cold and super-hot orogens. Among the orogenic belts on the Earth's surface, the highest pressure recorded is from Kokchetav, where diamond–coesite-bearing 540 Ma collisional orogenic belts formed under $P-T$ conditions of c. 7 GPa and 1000 °C (Katayama et al. 2001; Maruyama et al. 2004). Orogenic belts characterized by low geothermal gradients are formed along subduction zones. Examples for the lowest T/P regional metamorphic belts include the Franciscan (Ernst 1984; Maruyama 1988) and Middle Guatemala lawsonite–glaucophane schist (Tsujimori et al. 2005). Rocks below this region are unrealistic, although the calculated geotherm on the top of oldest slab at 200 Ma covers this unrealistic region. This might suggest a possible presence of such orogens, but they have already been dragged down to the bottom of the mantle. The highest temperature side is defined by the dry solidus of MORB. Beyond this line MORB is melted, and c. 150 °C above the dry liquidus 100% melting occurs. In the solid mantle, temperatures higher than that of dry basalt liquidus are unrealistic. Between the dry basalt solidus and the coldest geotherm represented by the 200 Ma slab, a wide region is present, which includes the super-hot (unrealistic), ultra-hot, hot and cold orogens as defined in the text. The wet solidus of MORB is also shown,

A major enigma is the near-absence of these 'realistic' rocks on the surface, except as mantle xenoliths (e.g. Griffin *et al.* 1999, 2003). This problem is further addressed below.

In the modern Earth, the $P-T$ realm of ridge subduction or subduction zones covers more than half of the surface, yet the rocks belonging to the 'blank zone' do not appear on the surface of the Earth. This is possibly because all of these rocks have undergone deep subduction; not only the slab peridotite but also the trapped sediments, seamount volcanoes, and all associated rocks, which have sunk below the bottom of the mantle and were never returned to the surface (see Maruyama *et al.* 2007*a*). In the Archaean, subduction-zone geotherms must have been on the high-temperature side, and similar to the case above, the deeply buried rocks never returned to the surface. However, mantle xenoliths of eclogites transported by Phanerozoic kimberlites or lamproites (e.g. Schulze *et al.* 2000) clearly show subducted Archaean oceanic crust or were metamorphosed at high P and T conditions shown in Figure 3. This explains well the presumed high Archaean subduction-zone geotherm.

The boundary between the 'blank' region and the region of common metamorphic belts on the Earth's surface roughly corresponds to the wet solidus of basalts. Beyond the wet solidus, if water is absent, then the rocks are generally dry and hence denser; therefore they sink deep into the mantle. This might be one of the reasons for the absence of these rocks on the surface of the Earth, except as xenoliths in magmas, as mentioned above.

In Figure 3, the subduction-zone geotherms passing through the 'blank' zone might have extended further up (to higher pressures corresponding to the bottom of the mantle). For the Archaean, these geotherms would be shifted towards the higher temperature side. However, the limit is defined by the dry solidus of basalt, beyond which melting occurs and magma would be present at all depths. Therefore, we designate this region as the super-hot region, a zone that is unrealistic on the modern Earth. The other end-member is the low-temperature limit, which we term the super-cold region, defined by the subduction-zone geotherm characterized by the oldest age of oceanic plate (*c.* 200 Ma). The $P-T$ region below the curve for the coldest subduction-zone geotherm cannot be achieved in the solid Earth's interior.

The boundary between the wet solidus and the Archaean geotherm corresponds to the ultra-hot region and lies above the amphibole dehydration-melting curve. The amphibole dehydration-melting curve incorporates a solidus backbend (Wolf & Wyllie 1994), because amphibole-bearing MORB may have no free water, and the H_2O is mostly structurally bound within the mineral. Once partial melting of the amphibole begins, free water appears; hence the curve is strongly bent, convex to higher P and T. If free water is available, UHT rocks cannot be formed (see Santosh & Omori 2008*a*, *b*). Instead, if a non-aqueous fluid is available, such as CO_2, UHT rocks will be stabilized. In the Archaean tectonic setting, H_2O-rich fluids were available probably only in near-surface regions, and the highly inflected geothermal gradients prevailing in the Archaean precluded hydrous fluids from becoming a dominant factor until fairly mature stages of Earth evolution. During the Phanerozoic, H_2O-dominated subducting slab-derived fluids became a significant factor down to depths of *c.* 660 km. The propagating H_2O 'front' in deeper mantle may correspond to the apparent increase in pressure through geological time (Santosh & Omori 2008*a*, *b*; Santosh *et al.* 2009*b*; Omori & Santosh 2009). Thus in the early history of the Earth, rocks were probably dehydrated at relatively shallow levels. However, water circulation became dominant in the Phanerozoic, penetrating even to the bottom of the upper mantle (Maruyama *et al.* 1996; Maruyama & Okamoto 2007). The rocks in the 'blank' zone, not exposed on the surface of the Earth, have been subducted deep into the mantle. They belong to the category of hot orogens, although they are not exposed on the surface. Thus, they constitute the largest group

Fig. 3. (*Continued*) which divides the region into a lower temperature side (hot) and a higher temperature side (ultrahot). The $P-T$ conditions of most of the orogenic belts at the Earth's surface are plotted on the lower temperature side, suggesting that the orogenic belts that formed above the wet solidus must have all subducted into deep mantle, corresponding to ghost orogens. A relatively small domain beyond the wet solidus and above the amphibole dehydration curve constitutes the ultra-hot region represented at the surface by UHT metamorphic rocks (e.g. Kelsey 2008; Santosh & Omori 2008*a*, *b*). The Archaean eclogite xenoliths define a region towards the high $P-T$ side in the figure. They represent basalt that was subducted into deep mantle in Archaean times and brought to the surface by kimberlite in the form of mantle xenoliths (termed concealed orogens in this paper). In most cases, the pressure conditions recorded from Archaean xenoliths are less than 10 GPa, although some recent studies (e.g. Moore & Gurney 1985) have reported the presence of majorite in the rocks, indicating much greater depths for their origin (up to pressures of 20 GPa). Although these rocks have almost completely recrystallized at lower pressures (<10 GPa), indicating the critical role of fluids, their derivation from a much higher $P-T$ source, even down from the bottom of the mantle at 2900 km depth, cannot be excluded.

of hot orogens on the Earth because all of the oceanic crust, occupying two-thirds of the surface of the Earth, belongs to this group, albeit more or less hydrated. Our model predicts that the subducted material accumulates at the CMB, with thickness locally ranging from 250 to 500 km as estimated from mantle tomography and seismic reflection work (Zhao 2004; Zhao et al. 2007). These are the so-called stagnant slab graveyards (see Maruyama et al. 2007a, for a detailed description).

Classification based on temporal constraints

Based on temporal constraints, we group orogens into Modern and Ancient. We give below a brief account of these two categories. Some examples have been discussed in more detail because of their importance in understanding the various processes addressed in this paper, whereas other cases are treated only briefly.

Modern orogens. The Modern orogens include those of accretionary type (Cretaceous of Japan) and collision type (Himalaya and Alps). Structurally, these orogens have many similarities and differences (see Maruyama et al. 1996). In both cases, regional metamorphic belts occupy the orogenic core. The accretionary type is characterized by a large orogenic welt towards the continental side, composed of a large TTG batholith; this feature is absent in the collision type, with only minor late-stage leucogranites, as in the case of the Himalayas. In both orogen types, overlying and underlying units of the regional metamorphic belts are weakly metamorphosed or unmetamorphosed, and are either accretionary complex in origin (Pacific type) or continental basement and cover (collision type). Major structures are subhorizontal with oceanward vergence of deformation, for both types. The size of regional metamorphic belts tends to be smaller in the accretionary type, by up to one order of magnitude. In both cases, regional metamorphic belts occur as thin slabs or sheets, with an aspect ratio of 1:10:100 (thickness:width:length/extension). However, the dimensions in the Precambrian are different, as will be discussed below.

We group orogens in the Modern Earth into four categories: (1) deep-subducted orogens that are taken down to mantle depths and never return to the surface, termed here 'ghost orogens'; (2) those that are subducted to deep crustal levels, undergo melting and are recycled back to the surface, forming resurrected but temporarily 'arrested orogens'; (3) 'extant orogens', which are partly returned to the surface after deep subduction; (4) 'concealed orogens', which have been deeply subducted and the traces of which are represented on the surface by mantle xenoliths carried by younger magmas.

These four sub-categories are schematically illustrated in Figure 4 and briefly considered below. Our usage of the term 'orogen' to include the subducted material in various tectonic environments may be debatable, but as mentioned in the introduction, we use this term in a different sense to address the material circulation on a whole-Earth scale.

Ghost orogens. Orogens of both collisional and accretionary types are grouped in this category. These orogens have been deeply subducted and most of them have been dragged down along with the down-going slab and are now components in the D'' layer incorporated within slab graveyards (Maruyama et al. 2007a). Recent studies along active margins such as the Circum-Pacific Trench clearly indicate the subduction of felsic trench turbidites and their transportation down to deep mantle (Scholl et al. 1980). The tectonic erosion at trenches, such as those off NE Japan and the Peru Trench off South America, provides further evidence for the transport of felsic materials to deep mantle (e.g. Bangs et al. 2009). It is also now known that immature arcs in general subduct into deep mantle, as displayed well in the case of SW Japan including the Izu–Mariana Arc, Kyushu–Palau Arc, Amami Plateau (Cretaceous arc), and Okino–Daito and Daito ridges (both arcs), which are all subducting along the Nankai Trough without accretion to the hanging wall (Maruyama et al. 2007a; Yamamoto et al. 2009). Whereas these are the modern analogues along the Circum-Pacific consuming plate boundaries, it is possible that a similar mechanism might have operated since plate tectonics started through the history of the Earth, presumably from 4.2 Ga. These processes must have led to the accumulation of a vast amount of subducted felsic material at the bottom of the mantle, which forms the 'ghost orogens' defined in this study and which is seldom returned to the surface. Although some of this material may be recycled as magmas originating from the deep mantle (e.g. Hofmann & White 1982), other material remains as 'lost continents' at the bottom of the CMB, forming a slab graveyard.

Our model of 'orogens' being recycled goes against the existing definitions and concepts on orogeny and orogens and may appear confusing. We imply that the material making up the rocks that were 'orogens' was recycled and not the orogens themselves.

Arrested (resurrected) orogens. In the framework of plate tectonics, it has been speculated that almost all orogens were formed at consuming plate boundaries, either by oceanic slab subduction (accretionary type) or continental or arc collision (collision type). However, the concept of

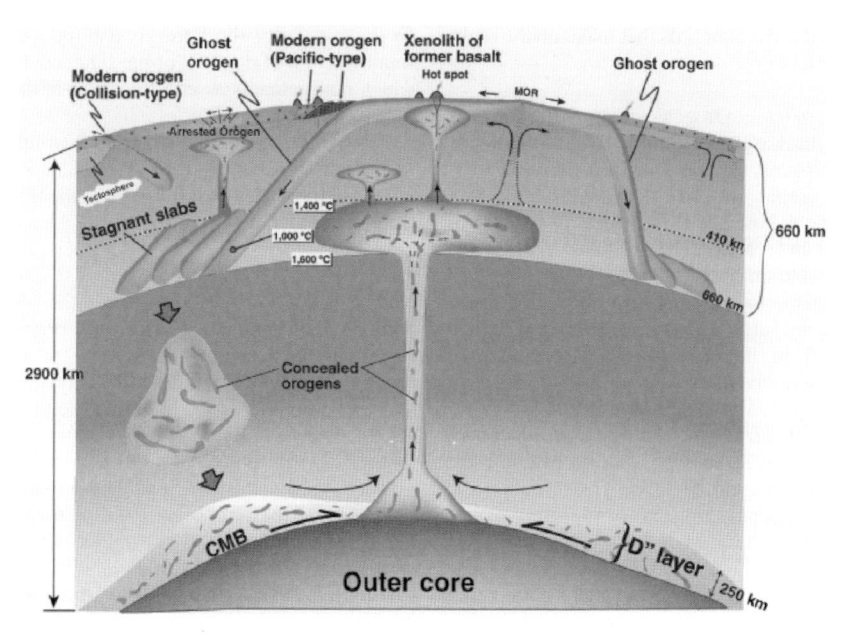

Fig. 4. Schematic section of the Earth illustrating the various types of orogens defined in this paper. The Pacific-type modern orogenic belts formed by subduction of oceanic lithosphere along continental margin, as exemplified by the type locality of Japan. The collisional-type orogenic belts include the Himalaya and the Alps. The Pacific-type orogen is composed dominantly of accretionary complex, and a mixture of oceanic material surrounded by predominant trench turbidites. Subducting oceanic slabs with or without trench sediments recrystallize along deep subduction zones and move down to the mantle transition zone or even to the CMB. These recrystallized rocks do not return to the surface, and we name them ghost orogens. However, traces of these may be returned to the surface either as melts of hotspot magma such as in Hawaii or as small mantle xenoliths with an eclogitic assemblage, which we term concealed orogens. The concealed orogens can be derived either from the D″ layer or from the mantle transition zone between 410 and 660 km. When continents rift as a result of a rising plume underneath, the lower crust is recrystallized or migmatized, and is normally not exhumed to the surface, remaining as resurrected but arrested orogens along passive continental margins. Listric normal faults extensively develop in these rift zones, such as in Africa. The metamorphic core complex formed in these regions may locally expose these arrested orogens. Under mid-oceanic ridges, usually at depths greater than 2500 m, extensive caldera development is found, such as mega-mullions, where similar arrested orogens may be present. Sedimentary rocks are generally absent, and the lower crust is completely recrystallized to form amphibolites or basic granulites. The figure also shows the stagnant slabs along the deep subduction zones from 410 to 660 km depth. Strongly deformed ghost orogens may be present in these zones. The recrystallized MORB with minor trench turbidites may catastrophically collapse and sink to the CMB. After a long period of heating by the underlying liquid outer core, the recycled MORB with trench turbidites can be partially melted to release dense melts on the bottom of the mantle, and the restite may move to the surface within a superplume.

metamorphic core complexes, such as the Miocene core complex in North America (Coney & Harms 1984) and Variscan fold belts in Europe (Echtler & Malavieille 1990), illustrates cases of post-collision orogenic extension. Moreover, along the African rift valley, extensive continental extension together with the formation of felsic magma suggests that widespread regional metamorphism under high geothermal gradients might be continuing in the lower crust in this region. The listric normal faults in this region would extensively modify the large-scale structure of the continental crust to promote the formation of the current core complexes. With time, rifting develops towards the oceanic side, such as the Red Sea or Atlantic Ocean. In these passive continental margins, such as those off New York, thick shelf turbidites (thickness over 5–10 km) occur, which totally cover the possible regional metamorphic units containing felsic plutonic and volcanic rocks. Thus, these orogens will not be exposed, and we define them as resurrected but temporarily 'arrested' orogens to indicate that they remain isolated in the middle to deep crust. Sometimes these orogens are modified by later collisional events with the consumption of the intervening ocean, and these have been referred to by some workers as collisional orogens (e.g. the Himalayas). Here again, our definition does not imply that the 'orogens' themselves are wholly resurrected and recycled back to the surface, but it is the materials that make up the rocks that are brought up.

Extant orogens. Orogens at present on the surface of the Earth are commonly designated as exhumed orogens, and in simple terms they represent the deeper structures of an orogenic belt that are now exposed at the surface as a result of weathering and erosion, or in other words, the deeper levels of an orogenic belt. We use the term extant orogens to describe the orogenic belts exposed on the surface of the Earth ranging in age from 4.0 Ga to the Recent. Examples for the Recent include the Cascades in North America and the Timor–Tanimbar region in Indonesia (Maruyama *et al.* 1996; Baldwin *et al.* 2004; Kaneko *et al.* 2007). It is possible to evaluate the ratio of ghost orogens to those preserved on the surface now (extant orogens) using the surface record of accretionary wedges and an estimate of subducted slabs. Such calculations demonstrate that the subducted oceanic lithosphere underneath SW Japan alone must have been *c.* 20 000 km during the last 180 Ma (Engebertson *et al.* 1992; Lithgow-Bertelloni & Richards 1998). Recent detailed seismic profiles (Replumaz *et al.* 2004) have revealed the deep structure of SW Japan and confirmed the lateral downward extension of the Sambagawa metamorphic belt. Two regionally metamorphosed and subsequently exhumed belts in the time span of the last 100 Ma have been confirmed from these studies. It is also believed that older, *c.* 200 Ma exhumed belts may be present at depth, particularly in the Kii Peninsula, where the belts may have exhumed almost to the surface, although they have not yet been exposed. If we assume two possible exhumed regional metamorphic belts, it is possible to estimate the exhumed versus subducted orogen, considering the thickness of the Sambagawa belt and the 200 Ma belts to be about 2 km. The length of the Sambagawa belt is *c.* 100 km, and assuming the same length for the 20 Ma belt, a total of 200 km long lithosphere remains as extant orogens. On the other hand, the total amount of subducted lithosphere for the last 100 Ma is about 10 000 km long. The accumulated total volume of clastic wedges is not clearly known, but a large part of these remained near the surface, and were not deeply subducted. Large volumes of oceanic lithosphere were subjected to regional metamorphism and transferred to blueschist to eclogite facies underneath Japan; all of these must have been subducted, and probably reached the deep mantle and were never returned to the surface. Considerable amounts of sedimentary rocks must have also been subducted along with the oceanic lithosphere. Thus, it is clear that the extant orogens, those currently represented on the surface of the Earth, constitute only a very small fraction of the bulk of the subducted material, which now remains as ghost orogens at the CMB.

Concealed orogens. Mantle xenoliths often include the traces of deeply subducted material. For example, Cenozoic alkaline basalts on the Colorado Plateau of central North America carry xenoliths of 100 Ma lawsonite eclogites (Usui *et al.* 2003), which were transported horizontally over a distance of 200 km from the subduction zone in California and subsequently brought to the surface trapped in magmas. There are also many modern examples of basalts carrying mantle xenoliths, such as those in Hawaii (e.g. Sen *et al.* 1993). Chesley *et al.* (1991) reported Re–Os systematics of mantle xenoliths from the Labait volcano of the East African Rift on the eastern boundary of the Archaean Tanzanian craton. The harzburgitic xenoliths here yielded ages of 2.8–2.0 Ga. The chronology of eclogitized basalts (garnetiferous basaltic rocks) occurring as xenoliths in alkaline basalts has been poorly studied. These xenoliths may provide important clues to the older history of the Earth, and their $P-T$ history may yield vital information on the style of Archaean subduction. In our study, we consider that these xenoliths represent traces of orogens that have been deeply subducted in the earlier history of the Earth, and therefore term them concealed orogens. Mantle xenoliths, although only minor and fragmentary in nature, are significant, as they indicate the source and apparent depth of magma. Because of the high speed of transportation of the magmas, volatilization occurs, and volatiles transport elements and heat. Thermometric studies based on the stability of mineral assemblages in mantle xenoliths indicate high temperatures, often exceeding 1000 °C, and sometimes up to 1400 °C (e.g. Chesley *et al.* 1991) classifying them as imprints of ultra-hot orogens. We should note here that this classification does not include all xenoliths by definition, such as a gneissic xenolith in a granitic pluton.

Discussion. The accretionary-type or collision-type extant orogens are preserved well on the surface of the Earth. They represent orogens that were subducted and metamorphosed along the ancient consuming plate boundaries. Against the downward movement of the subducting slab, part of the deep-seated and recrystallized metamorphic units has returned to the surface to form extant orogens. The next category of orogens represents formation along the subducting lithosphere, and their protoliths originally formed on the surface. Dehydration reactions gradually modify their mineralogy and these orogens move down to the bottom of the mantle and become stagnant at the mantle transition zone from 410 to 660 km in the temperature range of 1000–1600 °C. These orogens cannot normally return to the surface and they gradually accumulate at the CMB to grow another continent at the bottom of the mantle. These ghost orogens constitute the 'lost continents' as designated in this study, and would build up an anti-crust at the bottom of the mantle (see Maruyama *et al.* 2007*a*). When a rising superplume or plumes that gather fragments of some of the deeply buried old orogens break out through the surface of the Earth, traces of these fragments are delivered to the surface in the form of mantle xenoliths in strongly alkaline rocks. Although they are very small in size and population, these xenoliths are the fossils of the concealed orogens. Examples of these are now seen in Hawaii and other volcanic islands in the Pacific, Indian and Atlantic Ocean regions, and also in the continents, such as the 100 Ma lawsonite eclogites trapped in Cenozoic alkaline basalts of the Colorado Plateau in central North America described by Usui *et al.* (2003). The metamorphism of these eclogites occurred along the western margin of North America, similar to the Franciscan jadeite–glaucophane schist units, and then migrated in the upper mantle from the subduction zone of California to central North America before being delivered to the surface by alkaline magmas. Although mantle xenoliths are minor and provide only fragmentary evidence, we consider them as the vestiges of former large orogens and thus designate them as concealed orogens. Finally, the category of arrested orogens forms during continental break-up in the mid- to lower crust. Large-scale post-collision extension generates voluminous felsic magmas and widespread regional metamorphism under high geothermal gradients. The orogens thus formed are, however, covered by a thick cover sequence and are not generally exposed on the surface.

We show in Figure 5 a $P-T$ diagram to illustrate the thermal architecture of the Earth's interior in relation to the formation of various orogens. The measured $P-T$ conditions of regional metamorphic rocks exposed on the surface define the minimum and maximum temperature limits bounded by the 200 Ma cold slab on the lower $P-T$ side and the dry solidus of basalt on the higher $P-T$ side. It should be noted that the T maximum of the regional metamorphic belts exposed on the surface of the Earth lies below the dry solidus of MORB. The $P-T$ conditions of UHT granulite-facies rocks range from 900 to 1150 °C and 8 to 12 kbar

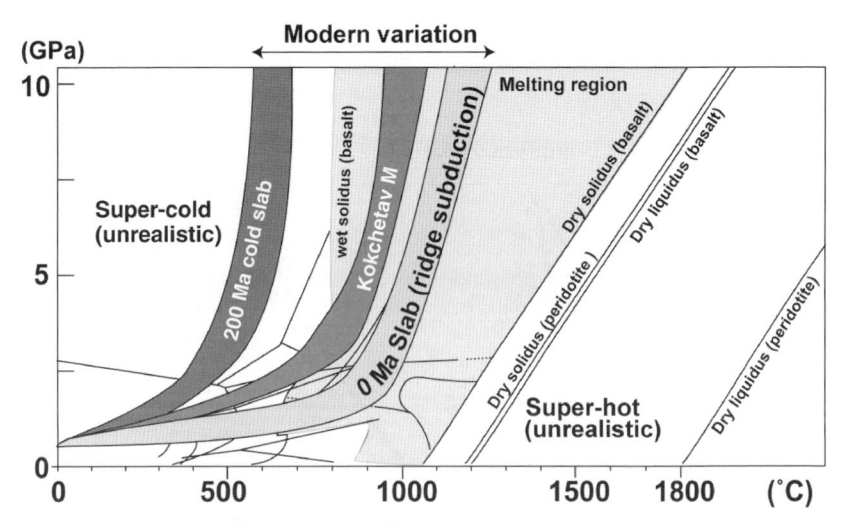

Fig. 5. $P-T$ diagram to illustrate the thermal architecture of the solid Earth interior in relation to the formation of various orogens. The T maximum of the regional metamorphic belts exposed at the surface of the Earth lies below the dry solidus of MORB. Also shown are the simulated subduction-zone geotherms with different ages of slabs. The 200 Ma Pacific slab on the Mariana Trench gives the lowest-temperature $P-T$ conditions in the present-day solid Earth. The 0 Ma subduction-zone geotherm lies below the dry MORB solidus, defining a curve close to the temperature maximum of regional metamorphic belts. Two forbidden or unrealistic zones can be defined, one below the 200 Ma slab geotherm and the other above the dry solidus of MORB. The zone between the 200 Ma subduction-zone geotherm and the low-temperature minimum of regional metamorphic rocks would define the $P-T$ conditions for a subducting orogen. References: 200 Ma slab and 0 Ma slab Peacock (1996), Hyndman *et al.* (1997) and Maruyama & Okamoto (2007); wet and dry solidus of basalt Peacock *et al.* (1994) and Vielzeuf & Schmidt (2001); liquidus and solidus of peridotite Takahashi (1986).

(Santosh et al. 2007a, b, 2008; Kelsey 2008) and would be placed in a narrow region above the amphibole dehydration-melting curve (see Santosh & Omori 2008b). If water is available, the lower crust would undergo partial melting at temperatures below the UHT conditions. Also shown in the figure are the simulated subduction-zone geotherms with different ages of slabs. The oldest slab on the Earth now is the 200 Ma Pacific slab at the Mariana Trench (Hilde et al. 1977). This curve gives the lowest temperature $P-T$ conditions in the present-day solid Earth. On the other hand, the 0 Ma subduction-zone geotherm defines a curve close to the temperature maximum of regional metamorphic belts. Two forbidden or unrealistic zones can be defined, one below the 200 Ma slab geotherm, and the other above the dry solidus of MORB. The $P-T$ conditions between the 0 Ma subduction-zone geotherm and dry MORB solidus may be achieved by a rising plume from deep mantle. The zone between the 200 Ma subduction-zone geotherm and the low-temperature minimum of regional metamorphic rocks would define the $P-T$ conditions for a subducting orogen.

The general $P-T$ conditions for mantle xenoliths are shown in Figure 6, along with subduction-zone geotherms for the Archaean and 200 Ma. The mantle xenoliths range from low-temperature lawsonite eclogites to high- and ultrahigh-temperature categories as shown in the figure. The Archaean xenoliths define the highest $P-T$ group. A number of studies have been published on the $P-T$ estimates from mantle xenoliths based on mineral phase equilibria considerations, as follows. Chesley et al. (1991) reported temperatures of 990–1400 °C from mantle xenoliths in the East African Rift. Qi et al. (1995) studied the petrology of mantle peridotite xenoliths from SE China occurring within Cenozoic basalts and reported $P-T$ conditions of 770–1250 °C and 10–27 kbar. Girod et al. (1981) obtained $P-T$ conditions of 1000–1100 °C and 20–25 kbar from lherzolite xenoliths in Quaternary alkali olivine basalts and nephelinites from the Hoggar area in southern Algeria. Griffin et al. (1984) studied xenoliths of spinel lherzolite, wehrlite and garnet websterite in basanite tuff from Gnotuk Maars, Victoria, Australia and reported $P-T$ conditions of 900–1100 °C and 11–16 kbar. Henjes-Kunst & Altherr (1992) reported $P-T$ conditions of 950–1065 °C and 23–28 kbar from peak assemblages and 770–860 °C and 9–11.5 kbar from late-stage assemblages in pyroxenite and peridotite xenoliths trapped within the Quaternary volcanic field in Marsabit, northern Kenya. McGuire (1988) identified three groups of mantle xenoliths from the margin of the Red Sea in Harrat al Kishb of western Saudi Arabia and reported $P-T$ conditions of 900–980 °C and 13–19 kbar (peridotites), 1050–1070 °C (Al-augite spinel pyroxenites) and 1000–1030 °C and 13.8–16.5 kbar (garnet–spinel websterites). Witt-Eickschen & Kramm (1997) obtained $P-T$

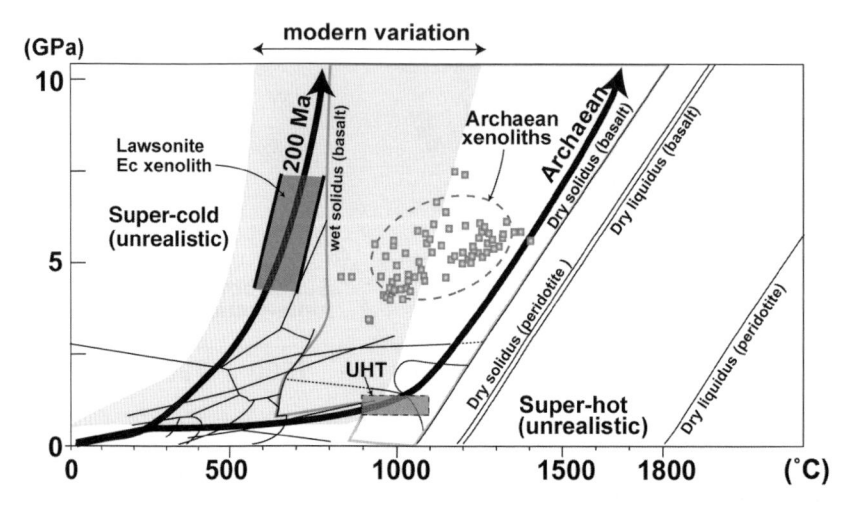

Fig. 6. $P-T$ conditions for mantle xenoliths along with subduction-zone geotherms in the Archaean and a 200 Ma old oceanic plate. The mantle xenoliths range from low-temperature lawsonite eclogites to high- and ultrahigh-temperature categories. The Archaean xenoliths define the lowest $P-T$ group, whereas lawsonite eclogite xenoliths in Colorado Plateau of western USA are estimated as the highest $P-T$ field (Usui et al. 2003). The Archaean eclogite, garnet-clinopyroxenite xenoliths contain garnet and pyroxene, possibly indicating a residue after partial melting of subducted oceanic crust at the subduction zone, consistent with the field hotter than the wet solidus of basalt. References for stability fields are as in Figure 5.

conditions of up to 1030 °C and 15 kbar from xenoliths of spinel lherzolite, harzburgite and wehrlite in alkali olivine basalt from Rhon in Central Europe. Adalberto et al. (2005) reported P–T conditions of 850–1170 °C and 10–25 kbar from mantle-derived ultramafic xenoliths in alkaline magmas in NE Mexico. Although the mineral assemblages in the mantle xenoliths have undergone re-equilibration during their capture and ascent in hot magmas, the temperature ranges obtained in most cases indicate that they are the traces of ultrahot orogens. The P–T diagram in Figure 6 illustrates the case of hot v. cold orogens. As the figure shows, eclogites have a wide temperature stability region from 1400 °C to 400 °C and include UHT eclogite, through HT eclogite, to blueschist eclogite, with decreasing temperature. The Archaean high geothermal gradient can produce only UHT eclogites, which are now observed in mantle xenoliths. UHP rocks within one orogenic belt sometimes comprise eclogite with pressure exceeding 70 kbar, moving down to epidote–amphibolite- to blueschist-facies rocks within a thickness of a few kilometres. The wedge extrusion model can best explain the mode of occurrence of various types of eclogites with the UHP belt; these typically occur as a thin film or tectonic slice within overlying and underlying low-pressure units.

It is now recognized that many eclogite xenoliths from kimberlites have Archaean ages. For example, Jagoutz et al. (1984) obtained a 2.7 ± 0.1 Ga Sm–Nd isochron age for Roberts Victor eclogite. Pearson et al. (1995) reported that whole-rock analyses of eclogites from Udachnaya kimberlite defined a Re–Os isochron age of 2.9 ± 0.4 Ga. Jacob & Foley (1999) showed a Pb–Pb isochron age for Udachnaya eclogite of 2.57 ± 0.2 Ga. These chronological data for eclogite xenoliths involve the Archaean subduction and emplacement of eclogitic components in the deep mantle. A number of studies have been published on the pressure–temperature estimates from eclogite xenoliths based on mineral phase equilibria considerations. For orthopyroxene-bearing eclogites pressures can be calculated using several geobarometers applied for peridotites. However, orthopyroxene-bearing eclogites are not common. Although no appropriate geobarometers exist for bimineralic (garnet–clinopyroxene) eclogites, the occurrence of coesite (>30 kbar) and diamond (>50 kbar) clearly represents high-pressure assemblages. Equilibrium temperatures are usually estimated using the Fe^{2+}–Mg exchange between garnet and clinopyroxene at an assumed pressure, often at a pressure of 50 kbar. The most widely used geothermometer is that by Ellis & Green (1979) and Ai (1994). For instance, the temperatures of equilibration for Roberts Victor eclogites range between 1000 and 1250 °C (Basu et al. 1986) and for Udachnaya eclogites range from 950 to 1300 °C (Sovolev et al. 1994). A compilation of equilibration temperature of eclogite for various kimberlite pipes has been given by Jacob (2004). The equilibration temperature of eclogite generally varies between c. 800 and 1380 °C. The garnet–clinopyroxene pairs in eclogitic diamond have also been used to calculate temperatures of equilibration based on the Fe^{2+}–Mg exchange thermometer between garnet and clinopyroxene. Appleyard et al. (2004) reported equilibration temperatures from 1138 to 1179 °C at an assumed pressure of 50 kbar for garnet–clinopyroxene pairs in diamond from the Finsch kimberlite pipe. Gurney et al. (1985) reported equilibration temperatures up to 1200 °C from coexisting garnet–clinopyroxene pairs from Orapa kimberlite. Premier eclogitic diamonds exhibit much higher temperatures between 1200 and 1400 °C (Gurney et al. 1985). Argyle eclogitic diamond show the widest range in temperatures, and also the highest equilibration temperatures, >1400 °C (Jaques et al. 1989).

We attempt a comparison of the Phanerozoic orogens and Archaean orogens along a cross-section of the Earth in Figure 7, where the double-layered mantle convection in the Archaean (left figure) and episodic whole mantle convection in the Phanerozoic (right figure) are illustrated (after Komiya et al. 2002a). The contrast in size and number of the plates is also shown. The right-hand figure shows that the Phanerozoic Earth is characterized by episodic whole-mantle convection. Subducted orogens are for a time stagnant at the mantle transition zone (410–660 km) and then catastrophically drop to the bottom of the mantle. In this case, significant amounts of continental material such as intra-oceanic arc and trench turbidite sediments are all metamorphosed with the down-going slab and finally accumulate at the CMB, forming stratified 'anti-continents' on the so-called D″ layer (see Maruyama et al. 2007a). Several such accumulated layers have been seismologically identified (Zhao 2004). A small portion of these anti-continents, either from the D″ layer or from those orogens floating at the mantle transition zone, may finally be delivered to the surface by plumes or superplumes.

The left-hand part in Figure 7 shows Archaean orogens along a cross-section of the Earth. Mantle convection in the Archaean was double layered, with the upper mantle separated from the lower mantle and no exchange of materials by mixing until a catastrophic collapse occurs as a result of cooling of the upper mantle and heating by radiogenic elements in the lower mantle (Komiya et al. 2004; Rino et al. 2004). Such mantle overturn has been considered to have occurred at 2.7 Ga and

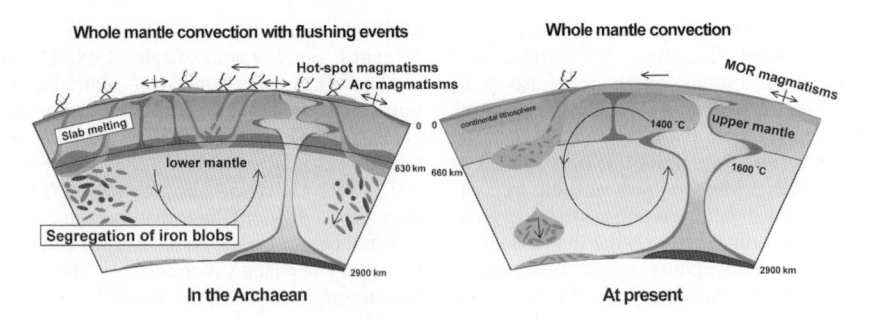

Fig. 7. Schematic illustrations of double-layered mantle convection in the Archaean (left) and episodic whole-mantle convection in the Phanerozoic (right). In the Archaean, because of the 150–200 K higher mantle temperature, a double-layered mantle convection predominated (Komiya 2004, 2007). The number of plates was much larger, but they were of small size (*c.* 700 km across). On the other hand, the Phanerozoic Earth is covered by a smaller number of plates of larger size (*c.* 3000 km across). During the transition from Archaean to Phanerozoic type, a catastrophic mantle overturn occurred at 2.7–2.8 Ga (Condie 1998). In the Archaean, orogenic belts developed within oceanic regions as island arcs (Komiya 2004). Collision of arcs led to the growth of primitive continents. Arc–arc collision gradually developed the primordial continent and finally formed the so-called continents such as Africa during the Proterozoic (right). In the Archaean, subducted slab has been commonly partially melted to yield TTG arc crust. The restite after melting may have accumulated at the bottom of the upper mantle (Maruyama *et al.* 2007*a*). If MORB crust segregated at 630 km depth, it might have preferentially sunk into the CMB. In the Archaean, slab melting may have been common because of the higher mantle temperature. Continental crust formed by slab melting may be predominant in the oceanic region, but most of such arc crust may have been subducted into the deep mantle (Maruyama *et al.* 2007*a*). Segregated MORB restite may have been removed to the bottom of the lower mantle, transporting iron-rich residue to the CMB (Komiya 2004).

presumably at 1.9 Ga (Sleep & Windley 1982; Komiya *et al.* 2000*a, b*). In the Archaean, the plates were less than 600–700 km across, much smaller than modern plates, which have an average width of 3000 km. For this reason, a number of intra-oceanic island arcs were present on the surface of the Earth during the Archaean, and collision of these led to the gradual growth of the proto-continents. However, the majority of them must have been subducted into the mantle transition zone in the upper mantle. Therefore we consider that most of the ancient orogens must have been subducted to grow the 'anti-continents' in the deep mantle, among which some orogens were returned to the surface along consuming plate boundaries such as those represented by the Archaean belt-type regional metamorphic units, which are smaller in size than the Phanerozoic ones, reflecting the smaller size of the plates in the Archaean. Geothermal gradients in the Archaean were slightly higher (Grove & Parman 2004) and hence the temperature conditions of Archaean orogenic belts never exceeded 1100 °C and pressures were generally lower than 10 kbar. Most of the subducted orogens were melted to generate arc or primordial continents. The restite after the partial melting must have sunk into the deep mantle. Although quartzites and platform sediments were formed in the Archaean, the volume of sedimentary rocks formed in the Archaean is much less than in the Phanerozoic probably because of the sea-level highstand.

Ancient orogens. The beginning of plate tectonics and the formation of granitic continental rocks are key issues to understand the history of orogeny in the Earth because both characterize the fundamental elements of global tectonics. Debate surrounding the timing of initiation of plate tectonics arose immediately after its establishment in late 1960s (Dewey & Spall 1975). Most geologists easily accepted that plate tectonics could date back to the Mesozoic because it accounts for the drift of African and South American continents. However, it is still controversial whether plate tectonics was in operation even in the Precambrian, and especially in the Archaean (e.g. Stern 2005; Condie & Kroner 2008). It is generally considered that the three features favouring plate tectonics are absent in the Precambrian geological history: ophiolites, Franciscan-type mélange and blueschists. All of them characterize divergent or consuming plate boundary processes on the modern Earth. However, Kusky (2004) assembled strong evidence for the occurrence of 2.5 Ga ophiolites and related rocks in China, Canada and other regions, which suggest that the features considered as diagnostic for plate tectonics are present in the older Earth history also. In this section, we summarize the evidence for plate tectonics in the Archaean from the available data, and compare the Precambrian tectonics with the modern equivalents. We also look at the salient differences between the Precambrian and modern orogens.

Komiya et al. (1999) proposed two geological criteria for plate tectonics in the Archaean: an accretionary complex and an oceanic plate stratigraphy. In addition, duplex structure is also regarded as an essential feature to characterize ancient accretionary complex. The oceanic plate stratigraphy comprises, in ascending order, abyssal basaltic lava flows, deep-sea or pelagic sediments, and terrigenous sediments. The abyssal basaltic lava flows and pelagic sediments contain no terrigenous materials, indicating that they were deposited very far from a continent or island arc; that is, in the open ocean. On the other hand, the terrigenous sediments contain mudstone, sandstone and conglomerate, deposited in the continental margin. As a result, the presence of these sediments indicates a lateral change of depositional environment from open sea to continental margin, as well as the presence of open ocean. The presence of ocean is very important for plate tectonics because it promoted the efficient cooling of the surface of the Earth and produced the rigid lithosphere (Komiya et al. 1999). On the Venusian surface, for example, the surface temperature reaches about 400 °C because of the lack of ocean water, and hence no rigid lithosphere is formed (Komiya et al. 1999). Formation of an accretionary complex necessarily means that a subduction process has occurred, and the presence of an accretionary complex indicates significant lateral movement of rigid lithosphere and its subduction. The horizontal passive movement of a rigid plate results in the presence of multiple plates and also the linearity of plate boundaries (Turcotte & Schubert 1982). Thus, the discovery of an accretionary complex, comprising the oceanic plate stratigraphy and duplex structure, proves the existence of plate tectonics including rigid plates, significant lateral movement of the plate, multiple plates surrounded by plate boundaries, and linearity of plate boundaries. Komiya et al. (1999) discovered evidence for an accretionary complex in the early Archaean Isua supracrustal belt.

Kusky & Polat (1999) suggested that Archaean granite–greenstone terranes represent juvenile continental crust formed in a variety of plate-tectonic settings and most of them acquired their first-order structural and metamorphic characteristics at convergent plate margins, where large accretionary wedges grew through offscraping and accretion of oceanic plateaux, oceanic crustal fragments, juvenile island arcs, rifted continental margins, and pelagic and terrigenous sediments similar to the Phanerozoic examples. Dehydration of the subducting slabs hydrated the mantle wedges below the new arcs and generated magmas similar to the sanukitoid suite in the mantle wedge. TTG magmatic suites were generated by melting of hot young subducted slabs. Eventual collision of these juvenile orogens with other continental blocks formed anatectic granites. Kusky & Polat (1999) noted that most preserved granite–greenstone terranes have been tectonically stable since the Archaean, and form the cratonic interiors of many continents. The foregoing discussion therefore suggests that plate tectonics date back to, at least, the early Archaean on the Earth.

The absence of blueschists in the Archaean was long considered as counter-evidence for the operation of Archaean plate tectonics. However, this lack of evidence does not necessarily mean the absence of plate tectonics in the Archaean because the Archaean geotherm at the subduction zone was markedly different from the modern equivalent. Archaean subduction-related metamorphic terranes belong to the intermediate- or low-pressure type metamorphic series (Hayashi et al. 2001; Komiya et al. 2002a). In addition, Hayashi et al. (2001) compiled evidence for subduction-related metamorphism in the Archaean, and divided 111 examples of the Archaean metamorphic terranes into two groups as 'metamorphic belts' and 'metamorphic areas'. In the category of 'belts', the length is three times larger than the width, whereas this ratio is less than three in the case of 'areas'. Many Archaean metamorphic terranes belong to the 'metamorphic area' type, whereas some are classified into 'the metamorphic belts' but have much shorter length than the modern equivalents. In contrast, many of the Phanerozoic metamorphic belts extend for several thousand kilometres along plate boundaries. This difference is attributed to the shorter plate boundaries during the Archaean, resulting from the presence of a number of microplates. The sizes of plates or number of microplates depends on the temperature of the upper mantle, as mentioned below (Hargraves 1986). The geochemistry of mid-oceanic basalts in the Archaean shows that the mid-Archaean mantle was about 150 °C hotter than the modern mantle (Ohta et al. 1996; Komiya et al. 2002a, 2004), which resulted in about 300 small plates on the Earth, compared with about 20 large plates on the modern Earth (Sleep & Windley 1982). Generally speaking, the age of a subducting plate exerts significant control on the geotherm at the subduction zone (e.g. Peacock 1996). A comparison of the $P-T$ estimates of the Archaean subduction-related metamorphisms with the Phanerozoic equivalents indicates that the geothermal gradient of the Archaean subduction zones was more than 200 °C higher than that in the Phanerozoic (Komiya et al. 2002a). Although the Phanerozoic geotherm ranges from low to high P/T types, depending on the age of the subducted slab (e.g. Maruyama & Okamoto 2007), the Archaean geotherm does not extend to the high P/T side, and is restricted to the low and intermediate P/T type (Komiya et al. 2002a). The variation of the subduction geotherm is explained by the difference of average lifespan of the subducting slab, which is due to the difference of mantle temperature. The lifespans of the oceanic lithosphere range from 0 to 200 Ma in the Phanerozoic, whereas the thermal history of the mantle indicates that the average lifespan in the Archaean was about 15 Ma (Komiya 2004). As a result, in the early Precambrian, younger oceanic plates subducted, whereas in the Phanerozoic, not only young but also much older lithosphere subducts under the continental crust and island arc. Defant & Drummond (1990) showed that subduction of young oceanic plate (<25 Ma) necessarily involves slab melting to form granitic magma. The age variation of subducting oceanic plates in the Archaean resulted commonly in slab melting and formation of granitoids as a result of the high geothermal gradient at the subduction zone. In summary, the lack of blueschists in the Archaean geology results not from non-operation of plate tectonics but from the youthful nature of the subducting slab. On the other hand, the presence of blueschists in the late Proterozoic indicates that the subduction geotherm was cooled to within the stability field of blueschists. The higher temperature of the mantle produced thin oceanic lithosphere, up to c. 30 km thick, with a thick (>15 km) basaltic oceanic crust in the Archaean. Slab melting produces granitic magma and a complementary garnet-bearing heavy residue, which exerts significant negative buoyancy and serves as a driving force for plate tectonics. Therefore, subducted oceanic materials commonly underwent slab melting in the Archaean, consistent with the presence of siliceous garnet-pyroxenite xenoliths, the so-called eclogite xenoliths, in kimberlite magmas (Rollinson 1997). Secular change of the potential temperature of the upper mantle indicates that the style of plate tectonics switched from an early stage characterized by slab melting (Precambrian-type) to a subsequent stage with or without slab melting (Phanerozoic-type) in the late Proterozoic, consistent with the presence of blueschists in the late Proterozoic (Maruyama et al. 1996) and ophiolite with thin basaltic crust (Moores 1993). Older ophiolite suites of 2.5 Ga have also been documented (Kusky et al. 2004). The thickness of oceanic crust on the mantle peridotite may generally prevent the common emplacement of peridotite on continental crust, accounting for the scarcity of ophiolites in the Archaean, although there are exceptions, with older ophiolite fragments in wedges reported from several regions (see Kusky et al. 2004, and papers therein). In summary, plate tectonics was already in operation even in the Archaean, but the higher temperature of the mantle caused short lifespans of the oceanic plates, different structure of oceanic plate, thin oceanic lithosphere with thick basaltic crust, and common slab melting. As a result, the subduction geotherm in the Archaean was much higher (by c. 200 °C) than that in the Phanerozoic, resulting in the absence of blueschists.

In the Archaean, the amount of continental crust was much smaller (e.g. Rino et al. 2004, 2008), and plate boundaries were much shorter, as mentioned above. These features prevented the formation of large continents, and many small island arcs were common on the Earth. Because the supply of terrigenous materials through rivers into trenches was limited, only thin terrigenous sediments were deposited on the oceanic plate at the trench, as evident from the relatively small volume of terrigenous sediments compared with more voluminous chert and basaltic lava flows in the Archaean accretionary complexes. More common slab melting generated several granitic batholiths, and formed granite–greenstone terranes (Kusky & Polat 1999). In addition, lack of large continents explains why sedimentary basins on the continents and passive margin sediments were rare in the early Archaean, and why they were not produced until the Late Archaean.

Tectonic framework of cold and hot orogens

Cold and hot orogens

The cold orogens generally are parallel to the consuming plate boundaries where water-dominated fluids are available to recrystallize the rocks through extensive shearing along the Benioff plane (Hasegawa et al. 1991, 1994). In the absence of circulating fluids, rocks do not recrystallize at relatively low-temperature conditions. For example, near the Moho depth in North America and other continents, there is very little recrystallization even though granulite-facies conditions prevail (Valley et al. 2003), which might be due to the absence of migrating fluids. This inference is also supported by the preservation of dry granulite-facies rocks in the Archaean on the Earth, reflecting the scarcity of water circulation, except along consuming plate boundaries.

Subduction along consuming plate boundaries induces secondary mantle convection and generates a volcanic front behind, parallel to the trench. The curtain-like mantle upwelling here creates hot orogens, classically termed paired metamorphic belts by Miyashiro (1961). However, the thermal maximum here does not exceed 800 °C because water-dominated fluid is available from the mantle wedge. This leads to the generation of granitic melts and migmatites, buffering the temperatures below 800 °C. Collision-type orogenic belts suggest that the temperature never exceeded 800 °C, except in the case of rare UHT rocks (see Santosh & Omori 2008b).

Considering the distribution of the divergent and convergent plate boundaries and hotspots, and the position of the two superplumes, the Pacific and African, as well as the major back-arc basins on the Earth (see Maruyama et al. 2007a), it is possible to speculate on the locations of hot, ultra-hot and cold orogens based on heat-flow measurements and S-wave whole-mantle tomography (Grand 2002). Thus, we evaluate various tectonic settings such as those for the African rift valley, Tibet Himalaya, the western Mediterranean, the Indian Ocean, NE Japan, and back-arc basins, with illustrative schematic cross-sections for some of these. We briefly discuss below the possible tectonic scenarios associated with these cases. Our consideration of these examples is for the purpose of discussion only, and does not mean that the global examples are limited only to these cases.

Hot orogens

African rift valley. Figure 8 schematically illustrates the possible scenario beneath the African rift valley to address the case where H_2O is nearly absent, and instead a CO_2-rich fluid is available, leading to the formation of ultra-hot orogens. This is particularly relevant in cases where a decarbonating tectosphere supplies abundant CO_2, such as those speculated to exist under the South Africa along the rift valley where mantle-derived CO_2 from decarbonating tectosphere is predicted by seismic tomographic images (Grand 2002; Zhao 2004; Zhao et al. 2007; Santosh & Omori 2008a). The tectosphere (carbonated subcontinental lithosphere) is buoyant, chemically distinct and is enriched in volatiles, predominantly CO_2, as well as in incompatible elements (Jordan 1988). Burke et al. (2003), based on a compilation of African alkaline igneous rocks and carbonatites, showed that the majority of these are concentrated within known or inferred Proterozoic suture zones. They suggested that the deformed alkaline rocks and carbonatites (DARCs) mark the places where vanished oceans have opened and then closed. Burke et al. (2003) also postulated that DARCs taken into the mantle lithosphere to c. 100 km depths at collision could provide source material for later alkaline magmatism. Thus, if continental rifts follow old sutures as indicated by the study of Burke et al. (2003), then we suggest that these regions would mark potential sites for a CO_2 source. Remnants of the Archaean tectosphere are still present, and are identified seismologically from the markedly high velocity of seismic waves, which indicates lower temperature (Zhao 2004; Zhao et al. 2007), and by a distinct compositional difference compared with oceanic mantle, as recorded from petrological studies of mantle xenoliths (O'Reilly et al. 1991). Santosh & Omori (2008b) compiled the distribution of carbonated subcontinental lithosphere (tectosphere) worldwide based on published information on S-wave seismic tomography. The carbonated mantle underlies the continental regions with pre-2.0 Ga orogenic belts, the largest of such regions being in North America

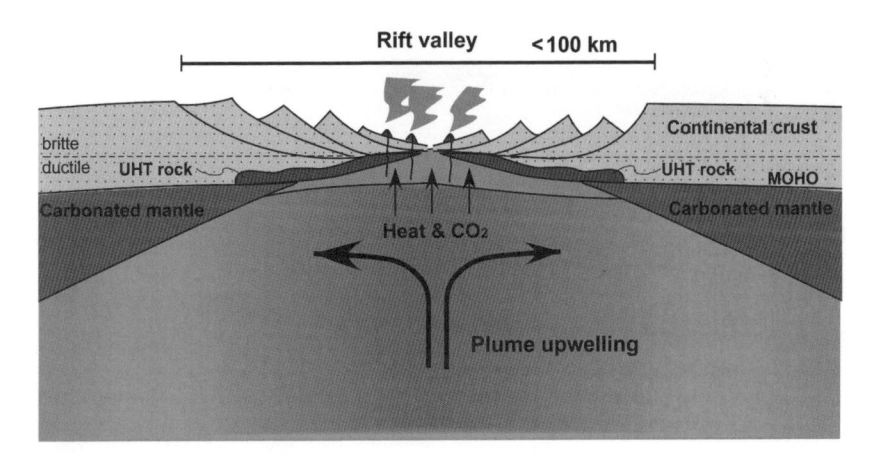

Fig. 8. Schematic section to illustrate extensional orogeny at a divergent plate boundary of a continental rift as exemplified in the present-day African rift valley (after Santosh & Omori 2008b). When a superplume rises, the continental crust is domed up to 2–3 km elevation and 2000 km width. Rifting starts above the centre of plume head by extensive development of listric normal faults. When the rising plume strikes the subcontinental carbonated mantle, small amounts of melts from the plume and the higher temperature recrystallize the lower crust to yield UHT rocks, which temporarily remain as 'arrested orogens'.

(about 3000 km across), where oceanic domain is totally absent. The tectosphere is generally confined beneath continents, with no clear examples from oceanic mantle. The presence of a tectosphere beneath subcontinental mantle in many regions is also inferred from the occurrence of carbonate minerals in mantle xenoliths. Santosh & Omori (2008b) illustrated from orogenic belts during the Archaean, Proterozoic and Phanerozoic that a close correspondence exists between the distribution of tectosphere and the overlying edge of the Archaean continental crust. Decarbonation associated with plate-tectonic processes drives the CO_2 out of the subcontinental mantle and into the atmosphere through magmatic conduits. Phanerozoic subduction of hydrated oceanic lithosphere tends to stagnate in the mantle transition layer and liberate free H_2O into the tectosphere. It is known that the water infiltration can induce decarbonation or partial melting in the mantle. Once decarbonated, the tectosphere transforms to a normal subcontinental mantle. A typical case is the North China Craton, where a flat-lying slab has been clearly detected underneath Beijing, about 2000 km away from the Japan Trench (Zhao et al. 2007).

Tibet Himalaya. Post-orogenic extension following continent–continent collision leads to the formation of hot orogens. For example, post-orogenic high-T migmatization has been noted in the Tibet Himalaya (Fig. 9) after the collision and subduction of the Indian continental crust (Kaneko 1997;

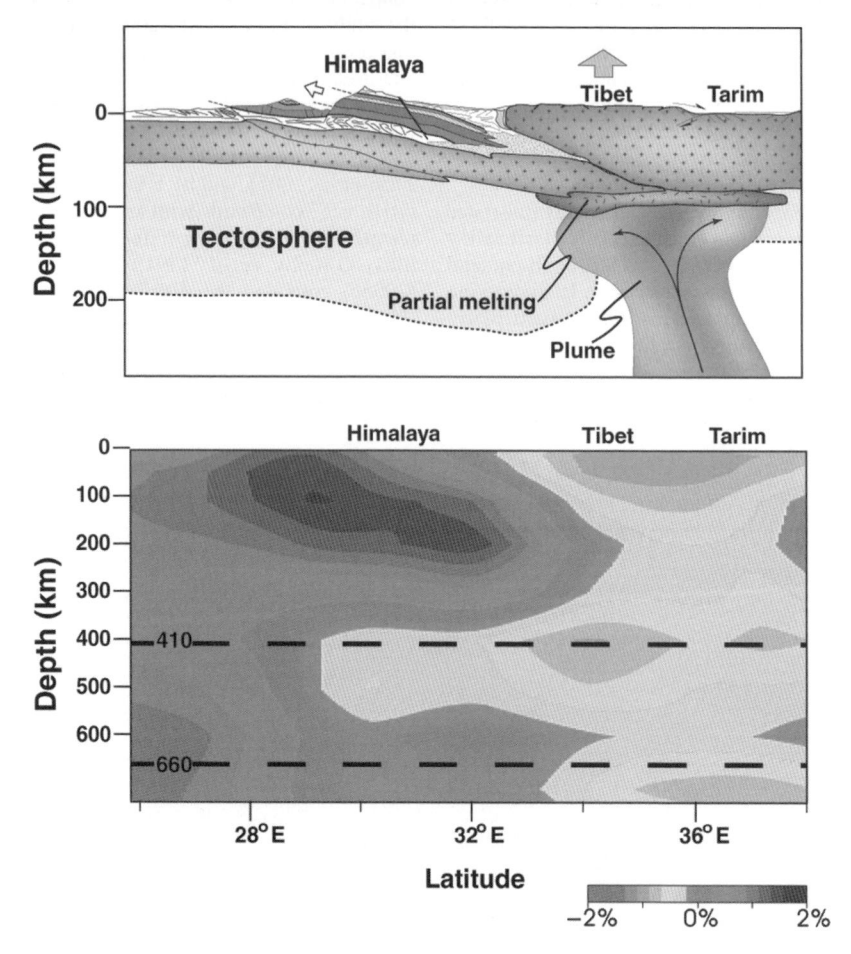

Fig. 9. The upper figure shows a schematic cross-section of the Himalaya. The lower figure shows a vertical cross-section of the P-wave tomography along a NE–SW profile passing through North India and the Tibetan Plateau (after Huang & Zhao 2006). Red and blue denote low and high velocities, respectively. The velocity perturbation scale is shown at the bottom. The two dashed lines show the 410 and 660 km discontinuities. For reference to colour, please consult the online version of the article.

Kaneko et al. 2003). It is well established that the Himalayan orogenic system is still undergoing convergence (e.g. Goscombe et al. 2006; Sol et al. 2007; Goscombe & Gray 2009). However, evidence also shows that, starting around 9 Ma, the region has started to dome upwards, forming a series of high-angle normal faults (Kaneko 1997). Seismological transects suggest large-scale strike-slip faulting and the presence of partial melting in the lower crust (Allègre et al. 1984). The regional metamorphic belt characterized mostly by quartzo-feldspathic lithologies is cut at the top by sub-horizontal normal faults, and at the bottom by reverse faults (Kaneko 1997; see also Goscombe et al. 2006, for a different model). A rigid raft of tectosphere was created by the Himalayan orogenic belts after 50 Ma (see Santosh & Omori 2008b). The Indian continent, supported by a buoyant tectosphere, has indented into Asia over a distance of 2000 km. Santosh & Omori (2008b) postulated that the oceanic lithosphere, which subducted before the collision, was delaminated and cut off from the Indian tectosphere, dropping into the midmantle. This has led to the upwelling of higher temperature asthenospheric mantle to heat the Tibetan lower crust, thereby generating felsic partial melts and leading to a hot orogen. If the underlying mantle of Tibet was carbonated, then UHT rocks must have commonly occurred in this region, leading to an ultra-hot orogen. However, the subcontinental lithosphere of Tibet is geologically young, and must have been hydrated and decarbonated by the subduction of oceanic crust before the collision. Therefore no carbonated mantle is present beneath Tibet. Thus, instead of generating dry UHT metamorphic rocks, the collision and subsequent injection of the asthenospheric mantle led to partial melting as seen from the extensive migmatite formation in this region (see Goscombe et al. 2006).

Another example is the collision of Europe against Africa, where the western Mediterranean Sea (Fig. 10) is underlain by extended and attenuated continental crust (see Windley 1995). The whole region is now in the stage of post-orogenic extension with magmatism in an extensive extensional tectonic regime. Recrystallized lower continental crust, as predicted in our models, is present below the Mediterranean Sea.

Indian Ocean. In the oceanic region, large-scale caldera ('mega-mullion') formation has been described (Tuckhole et al. 1998), indicating tilting along listric normal faults to expose the Moho. Mega-mullions are sea-floor edifices first described from the Mid-Atlantic Ridge with two distinctive characteristics: a domelike or turtleback shape extending over a diameter of some 15–30 km, and conspicuous grooves or corrugations (mullions) that formed as part of the faulting process and that are parallel to the direction of fault slip over the domed surface. A number of mega-mullions have been recognized where strong tectonic extension occurs on slow- to intermediate-spreading mid-ocean ridges in the Indian and Atlantic Oceans (Tuckhole et al. 1998), although these have not yet been reported in the Pacific Ocean, probably because of the fast spreading rate there. It has been suggested that oceanic core complexes (OCC) or mega-mullion structures are a common tectonic feature on slow-spreading ridge flanks. The core of a mega-mullion is an abnormally elevated (0.5–2.0 km above the surrounding sea floor), dome-like structure comprising unroofed, deformed lower-crustal and upper mantle rocks (MacLeod et al. 2002). Their surfaces display corrugations and striations parallel to the plate spreading direction (Searle et al. 2003). They are characterized by positive mantle Bouguer anomalies, and the sampled structures show outcrops of serpentinized peridotite and gabbro (Escartin et al. 2003). The recrystallization process operating in the mega-mullion structures has not yet been studied in detail. However, on-land ophiolites provide an example

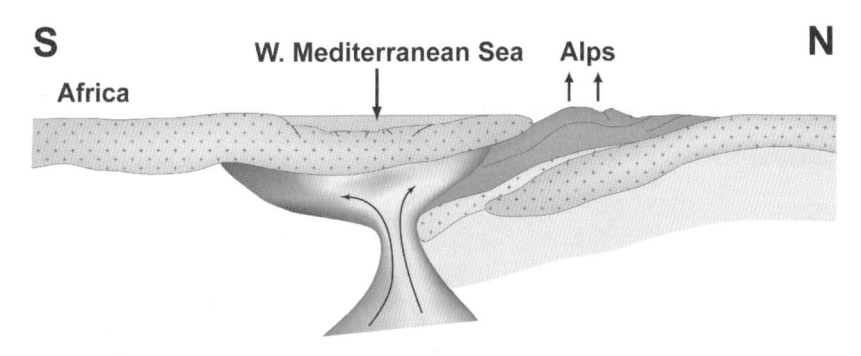

Fig. 10. Schematic cross-section of the Western Mediterranean.

for studying this process (Dilek et al. 2000, and papers therein). A continuous section from uppermost mantle through the mantle transition zone to the lower crust is exposed in SW Japan, providing on-land examples of recrystallization resulting from circulation of a water-rich fluid from the bottom to the lower crust at high geothermal gradients. The high geothermal gradients here generate granulite-facies rocks, and thus this structure corresponds to a hot orogen. The highest grade does not exceed 800 °C. Another case is Macquarie Island to the south of Australia and close to the mid-oceanic ridge, where 10.9 Ma ophiolite predominated over peridotite (Daczko et al. 2005), which constitutes an example of a hot orogen. Although the situation in these cases is similar to that of the African rift valley system, in the latter case, the highest-grade portion is never exposed at the surface, and the orogen evolves into a passive margin sequence buried by thick deltaic sediments. The development of hot orogens in calderas within divergent settings in the mid-ocean is similar to that in the African rift valley, although the predominant volatile in this case is H_2O, and therefore these are not ultra-hot orogens.

It must be noted here that although we have discussed only the case of the Indian Ocean in this section, a similar argument can be extended for the case of the Atlantic and Pacific Oceans with their impressive rift chains of volcanoes.

Back-arc basins. Mega-mullions have been recently discovered in the Cenozoic back-arc basin of the Philippine Sea plate, where the large Godzilla mullion constitutes the best example (Ohara et al. 2001; Ohara 2006). Giant mega-mullion structures have also been described from the Parece Vela back-arc basin in the northwestern Pacific, which are nearly an order of magnitude larger than the similar structures in the slow-spreading Mid-Atlantic Ridge. The elevated mantle Bouguer anomaly and the occurrence of serpentinized peridotites and gabbros suggest that the mega-mullions are exposing oceanic crust and upper mantle. These regions also constitute examples of hot orogens. Here it can be debated whether back-arc basins can be called orogens, although they result from a collision outboard and may be captured as an ophiolite during a Tethyan-, Pacific- or Himalayan-type orogeny. Therefore, this category is rather tentative.

Cold orogens

NE Japan. Seismic tomographic data from the Japan Trench to China (e.g. Zhao et al. 2007; Zhao & Ohtani 2009) shows that from the east, c. 100 Ma old Pacific plate is subducting to 660 km depth and delivering surface water to the mantle wedge (Hacker & Peacock 1995). In response to the deep subduction, a secondary mantle convection is generated starting from 410 km depth and rising parallel to the subducting Pacific plate for more than 500 km to intersect the Moho plane at about 100 km from the trench and generate a volcanic front (Fig. 11; Tatsumi 1989; Tatsumi & Eggins 1995; Poli & Schmidt 2002). The dehydration reaction of hydrous wadsleyite (beta phase olivine) to yield olivine $+ H_2O$ (magma) would further promote the secondary convection (Inoue et al. 1995; Komabayashi et al. 2004). At depths of about 200 km, another dehydration front from the centre of the Pacific slab enhances the secondary convection and the rising curtain-flow. This is further aided by a third dehydration reaction from the MORB crust under the volcanic front. A few plumes may be formed by the first and second dehydration reactions and migrate to the surface through secondary convection. The traditionally explained hotspots in Korea (Sleep 1992; Zhao et al. 2007) may be a reflection of subduction-related wet plumes generated through curtain-like mantle convection to form isolated plumes. On the corner of the mantle wedge, extensive earthquakes frequently occur at depths of 50–60 km and are predominant within the subducted MORB crust where blueschist to amphibole–eclogite transition reactions occur, scrapping off the oceanic crust to attach to the hanging wall (Ota et al. 2004). Simultaneously, the huge amounts of water derived by the dehydration reactions mentioned above play a crucial role in generating earthquakes and tectonic modifications. The released water moves upward along the Benioff thrust to hydrate a small corner of the mantle wedge, and some of the fluid is removed to the surface along the décollement zone above the subducting Pacific plate. The remaining part of the fluid may move downward together with the Pacific slab. A highly serpentinized thin film saturated with pore fluids moves down along the Pacific slab, and earthquakes are generally absent in this zone because of interstitial fluid saturation (Seno et al. 2001). These hydrated thin films with antigorite and other hydrous silicates would deliver water-rich fluids to promote mantle convection and magma generation (Schmidt & Poli 1998). Within the Pacific slab, a double seismic zone has been identified from shallow depths to 180 km depth (Hasegawa et al. 1994; Zhang et al. 2004). Both upper and lower seismic planes occur in slab peridotites. The source of water may be related to hydration along extensional normal faults outside the trench as a result of bending of the Pacific slab. The propagating hydration front formed by slab bending may transport surface water to 40 km depth along the cracks developed in the lower crust and uppermost mantle, which

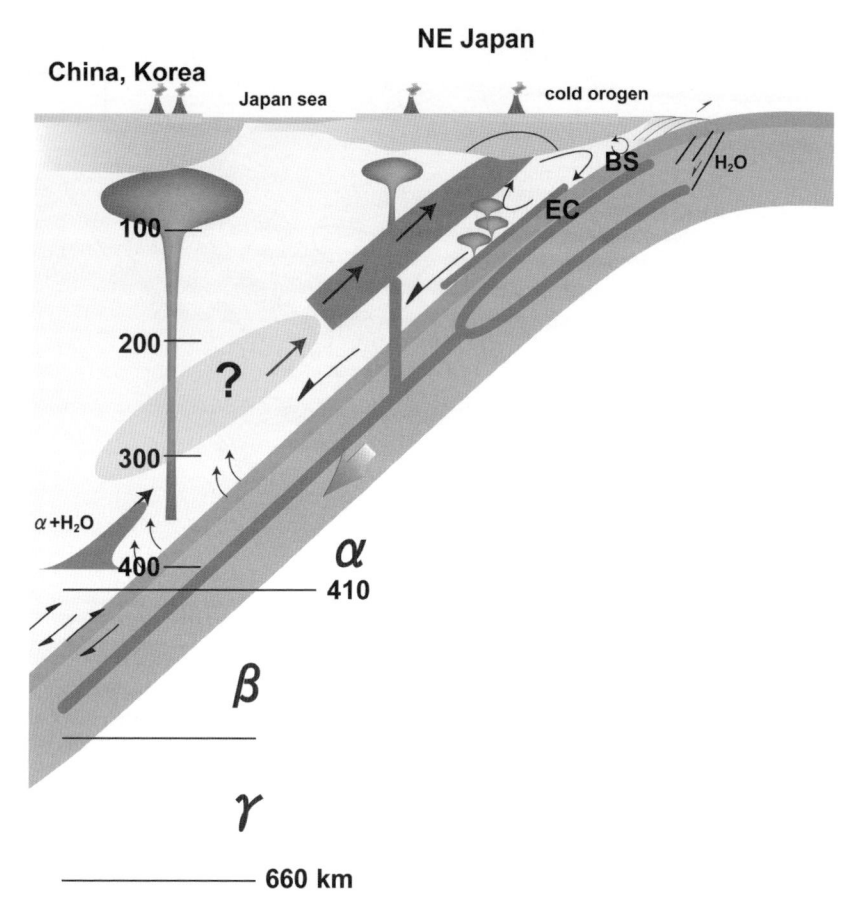

Fig. 11. Schematic cross-section from NE Japan through Korea to China. Slab earthquakes continue sporadically even down to 660 km underneath North Korea to China. Stagnant slabs extend horizontally to the west of Beijing. Intermittent earthquakes are recorded at depths of 400 and 500 km. In general, the distribution of earthquakes along the lower seismic plane is heterogeneous. This indicates that the hydration into the deep Pacific slab is sporadic. Tomographic images by Lei & Zhao (2005) clearly identify the root of the mantle plume in North Korea and China starting above 410 km depth, suggesting the breakdown reaction of hydrous wadsleyite at this depth. Fluid-dominated strongly alkaline basalts may be generated by this process. Lei & Zhao (2005) newly defined a major mantle wedge that covers the region from Japan to China. (Note the scale difference between this and the mantle wedge in NW Japan, the latter being only about 200 km wide.) The lower crust under North Korea to China may have been recrystallized by the rising hydrous plume. BS, blueschist; EC, eclogite.

are otherwise composed of anhydrous gabbroic rocks (Tatsumi & Eggins 1995). Below a depth of 180 km, seismicity is not continuous, although it is detected to 660 km depth (Fukao 1992; Zhao et al. 2007), nearly to the bottom of the upper mantle, suggesting that dehydration reactions occur within the whole upper mantle, but only along subduction zones. Thus, subduction can effectively transport surface water into the mantle transition zones (Ohtani et al. 2004; Komiya & Maruyama 2007; Maruyama & Okamoto 2007). Figure 8d thus explains the detailed process of water transportation and generation of cold orogens. Under the volcanic front, a hot orogen can be formed, but not an ultra-hot orogen, because of the dominance of water. Along the Benioff plane, a cold orogen can be formed by the circulating H_2O-rich fluids, in spite of the low temperatures in this region.

Figure 12 shows the tectonic scenario in NE Japan to a depth of more than 100 km with a large mantle wedge, with details of the hot orogen and cold orogen, the adherence of the oceanic crust to the hanging wall by dehydration reactions, and the nature of exhumation. As shown in the figure, the rising convection soon turns downward and hydration is initiated by water released through the dehydration reaction from the subducting lithosphere. At about 50 km depth, there is a neutral

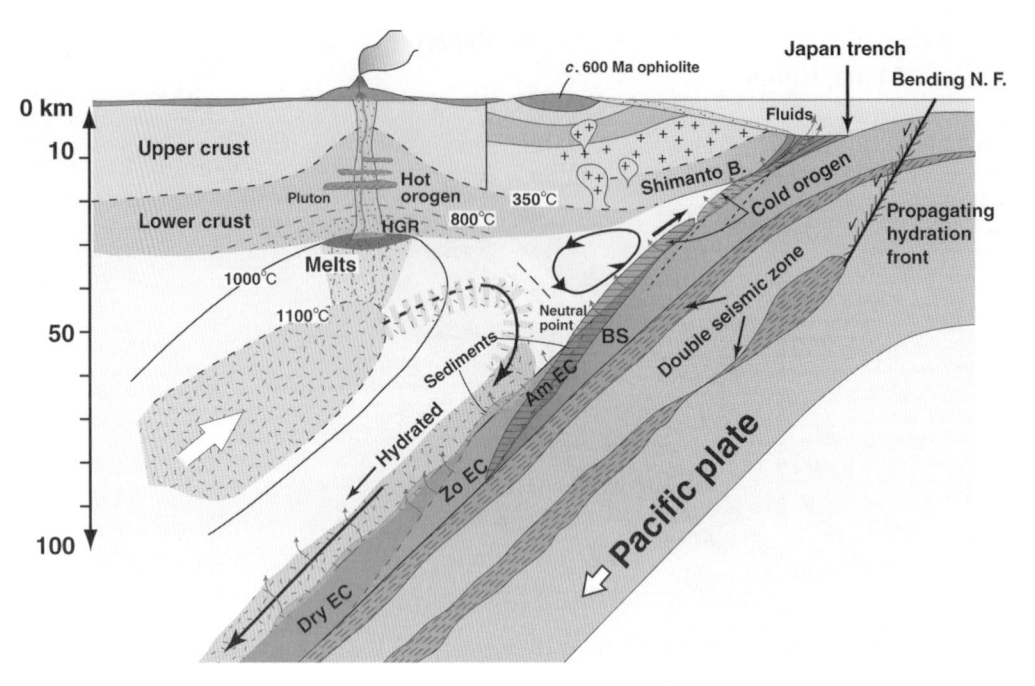

Fig. 12. Detailed schematic cross-section of NE Japan showing the continuing processes. The Pacific slab of c. 60 km thickness is subducting with the top 6 km composed of MORB, possibly hydrated. A double seismic zone has been precisely detected in this region (Hasegawa et al. 1978). The double seismic zones represent the dehydration of hydrous slab releasing water-dominated fluid upward. The lower seismic plane is supplied by water from the Pacific slab bending outward from the trench. Surface water penetrates into the centre of the slab along the normal faults (N.F.). The continuous dehydration on the double seismic zone merges to produce one seismic plane near the centre of the Pacific slab at about 180 km depth. A fluid-dominated plume occurs at 180 km depth and upward migration results in the lowering of viscosity in the mantle wedge. The hydrated mantle plume is removed obliquely towards the Moho of the volcanic front. Recent experiments indicate the temperature of the rising plume to be about 1100 °C, releasing basaltic to andesitic melts at Moho depths and creating new continental crust off NE Japan. The hydrous plume heats the lower crust, leading to recrystallization of the presumably mafic lower crust to generate a hot orogen. The rising wet mantle wedge turns oceanward after releasing andesitic melts to heat the cold Pacific slab and then moves downward with the slab by thin mantle flow. Under the forearc, a small triangular region is present, which has been cooled continuously by the cold Pacific subduction. Seismologically, the 50–60 km zone is identified as the focus of earthquakes centred in the subducted oceanic crust, suggesting large amounts of fluid release. The expected dehydration reaction is the blueschist–amphibole to eclogite transition. From this point, up to the shallow depths near the trench, earthquakes predominate along the top of the oceanic crust, indicating a plate boundary earthquake style. A number of horst and graben structures are developed in the outer wall of the trench, with vertical displacement 5 up to 500 m and widths of a few kilometres. These grabens are filled by trench turbidites with granitic composition. The sediments must have been dragged down to depth, indicating the subduction of granitic material instead of the formation of an accretionary wedge. Post-Miocene accretionary complexes are almost absent in NE Japan, whereas they are extensively developed along the Nankai Trough of SW Japan, with a width of over 100 km. The boundary above and below at 50–60 km roughly corresponds to plate boundary-type to within-plate-type earthquakes. This suggests that metasediments may play a key role in the various styles of earthquakes, with a more slippery plane above that is absent below. The penetration of the décollement zone into oceanic crust below 50–60 km suggests underplating of the Pacific MORB crust into the hanging wall of the mantle. It should be noted that neutral points of mantle convection are present at this depth. The small corner beneath the forearc region might be characterized by anti-clockwise convection. On the other hand, clockwise mantle convection predominates under the volcanic front. Along the anticlockwise rotation beneath the forearc is a narrow channel of regional metamorphic belts, which are exhumed as a tectonic slice to the surface in spite of the downgoing Pacific MORB crust. Presumably, the dehydrated water is channelized along the Benioff thrust and thus transported to the trench–slope break to form a cold seep. Bio-communities colonize the cold seep, and methane and other nutrients are concentrated there. We summarize the orogenic belts in this region as follows. Pacific-type cold orogens can be formed along the subducting slab from the trench to 50–60 km depth. The top surface of the Pacific slab below this neutral point is a ghost orogen, which moves down to the deep mantle. A high-temperature orogen (hot orogen) is present underneath the volcanic front. Fluid released from the subducting Pacific slab causes the metasomatic recrystallization of a Pacific-type orogen and serpentinization of the mantle wedge. We therefore envisage a metasomatic and metamorphic factory under the forearc region. On the other hand, a magmatic factory is present under the volcanic front, and is driven by the fluid released from the Pacific slab. Am EC, amphibole eclogite; BS, blueschist; Dry EC, dry eclogite; HGR, High grade granulite; Zo EC, Zoisite eclogite.

point of convection in the small corner wedge. An anticlockwise convection cell develops, which would transport the cold orogen to the surface as a thin channel flow, similar to the process of wedge extrusion described by Maruyama *et al.* (1994). The top portion of the channel is truncated by a normal fault, and the bottom boundary is a thrust parallel to the top boundary. These paired faults transport the thin slices of cold orogens to the surface.

The role of fluids

Our study speculates that the nature and style of orogens are dictated not only by temperature and pressure, but also by the critical role played by fluids. The origin of ultra-hot orogens and cold orogens can thus be demonstrated based on the contrast in fluid regimes. In Figure 13 we show a schematic cross-section of the Earth to illustrate the availability and distribution of fluids in the

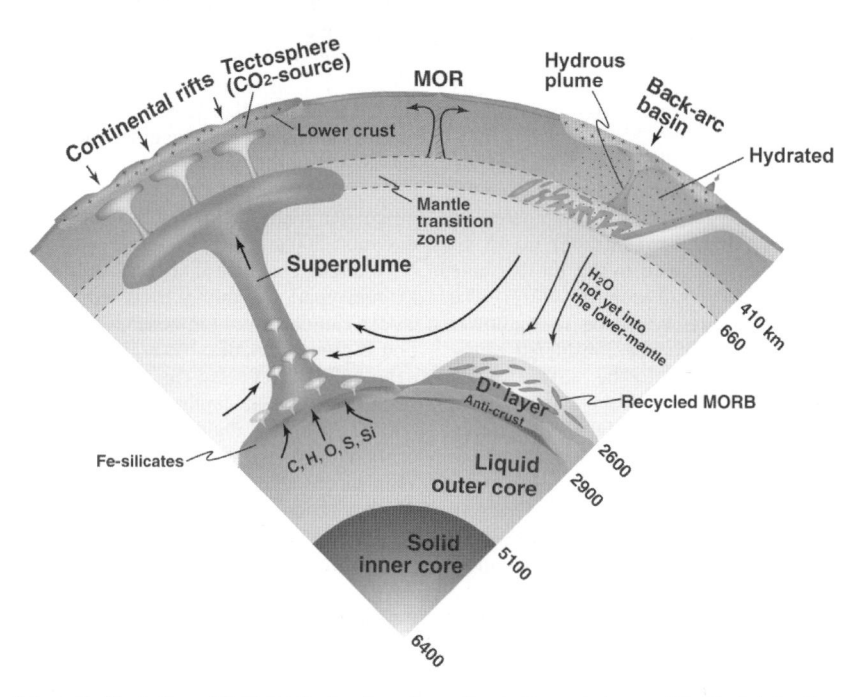

Fig. 13. Schematic illustration of fluid distribution from the surface to the central core of the Earth (after Santosh *et al.* 2009*b*). C–O–H–S fluids may constitute up to 10% of the metallic core, and may have been stored during the formation of the Earth at about 4.56 Ga. These light elements not only promote compositional convection in the outer core to drive the dynamo, but also cause chemical buoyancy of superplumes. The recycled MORB starts from the trench and moves through the mantle transition zone and finally down to the CMB. This material is then heated by the outer core, leading to partial melting. The dense iron-rich melts accumulate at the bottom of the D″ layer. The remaining restite MORB with a dominant andesitic composition rises to form a superplume. (It must be mentioned here that alternate views exist for the D″ layer, and some workers consider this to represent material that floated upward from the crystallization of silicates at the inner core–outer core boundary, or from the outer core crystallizing at the D″ layer). The subducted slab from the trench is hydrous, and is heated by the surrounding mantle, which releases the water in the mantle wedge and enhances the viscosity. The hydrous regions above stagnant slabs and subduction zones can be active because of the presence of released water-dominated fluid, which may cause back-arc opening. The vertically rising superplume enters the upper mantle, transforms to horizontal flow and branches out into several hotspots. These hotspots cause the rifting of continents, such as the present-day African rift valley, and deliver the mantle fluid to the surface. Surface CO_2 was selectively transported into the mantle in the Hadean to the Archaean. After the Neoproterozoic, surface water started to be transported into the mantle transition zones (410–660 km). From the Hadean through the Archaean and to the Proterozoic, water must have been recycled only in the near-surface region. CO_2 may have been removed from the primordial atmosphere into the mantle in the early Archaean, and later returned to the surface periodically with plume activity. For the major part of the Earth's fluid history, fluid transport was mostly one way, from the outer core to the surface. The return flow of water started only after 750 Ma, although it has not yet entered the lower mantle. Prior to water subduction, CO_2 subduction occurred in the Hadean to the Archaean, and thereafter returned to the surface through periodic magmatic and plume activity. In the Archaean, the upper mantle was characterized by CO_2 bound within magnesite in the peridotite. However, since the Neoproterozoic, the mantle transition zone became enriched in water through continuous subduction.

solid Earth (after Santosh et al. 2009b). It can be seen that fluid circulation is present only in a very restricted area, mostly along the plate boundaries, particularly along subduction zones where there is whole upper mantle-scale fluid circulation. Here the fluid is water-dominated, with the mantle transition zone from 410 to 660 km serving as a huge water tank. About five times the volume of all the surface ocean water can be stored in the dense hydrous silicates in this zone. However, the circulating fluid is perhaps less than 1% of the surface water (Maruyama & Okamoto 2007). For all the other tectonic settings, even if the temperature is very high (800 °C, Moho depth), fluids are generally absent and hence there is no recrystallization of rocks.

In the divergent zones, fluids are mostly mantle derived and/or recycled water. The volatiles in the lower mantle are considered to be dominated by CO_2 together with light elements (C, O, H, S) which were originally trapped in the Earth's core (Javoy 1997; Frezzotti & Peccerillo 2007). The volatiles and light elements escape to shallower levels and a superplume can be considered as the pipe that connects the core to the surface of the Earth. These regions are possible candidates to yield ultra-hot orogens, particularly where a carbonated tectosphere is available (see Santosh & Omori 2008a, b). If high-temperature material (superplume) or the rising curtain flow in divergent zones strikes the tectosphere, then even small amounts of melts that form would be extremely enriched in CO_2; typical examples of such melts are carbonatites, kimberlites or lamproites. Therefore, the lower crust can be considered to be enriched in CO_2 rather than water, and hot as well as ultra-hot orogens can be formed through recrystallization under extremely high-temperature conditions (>900 °C) without melting. Also, the curtain flow under mid-ocean ridges delivers small amounts of water-rich fluid, probably from the mantle transition zone. The route of the curtain-like convection under mid-ocean ridges has recently been traced by P-wave whole-mantle tomography (see Maruyama et al. 2007a). Small amounts of volatiles may have been derived from the liquid outer core through a superplume, although this has been debated (Maruyama et al. 2007b).

If we consider the zone of fluid circulation worldwide, it can be inferred to occur along restricted zones, perhaps covering less than 1% of the total area. Subduction zones, where back-arc basin formation occurs, are also related to the circulation of water, although in this case, the water does not penetrate into the lower mantle, because of the high temperatures (about 1600 °C) at 660 km, which inhibit the stability field of hydrous silicates in the lower mantle. It can be predicted that in future, perhaps after 2 Ga, the lower mantle may hold a substantial amount of water (see Maruyama & Okamoto 2007).

In summary, water plays a critical role in generating cold orogens. The oceanic regions of the globe may not have ultra-hot orogens because of the lack of tectosphere. Similarly, in simple collision–subduction regimes, ultra-hot orogens do not develop. If a collision zone was formed related to a tectosphere, then ultra-hot orogens can be generated. The Indian subcontinent, supported by a buoyant tectosphere, has indented into Asia over a distance of 2000 km, but the hanging wall is Phanerozoic with no tectosphere, a situation similar to that in the Alps, and therefore ultra-hot orogens are absent. On the other hand, the sense of subduction during the Grenvillian and Pan-African orogenies was conducive to generating ultra-hot orogens in many regions of the Earth (see Santosh & Omori 2008a).

We also conclude that recrystallization as well as rejuvenation leading to resetting of isotopic clocks can occur only in zones where fluid circulation is present. In all other regions, no changes are expected, as reflected in the predominance of Archaean crust at depths where temperatures are 800 °C in many regions of the world, in most cases without any recrystallization or age resetting. This means that hot orogens might dominate at depth, although most of them have been subducted deep into the mantle, and form slab graveyards (Maruyama et al. 2007a).

Time dependence of orogens

UHP rocks are common in the Phanerozoic, but are rare in the Archaean, a feature that has been correlated to changing subduction-zone geotherms and secular cooling of the geotherm, among other factors, as discussed by many workers (e.g. Brown 2007, and references therein; Santosh et al. 2009b). We address here the question of why UHP rocks formed in the Archaean have all gone down to the mantle, and why only the Phanerozoic UHP rocks have been brought to the surface. We propose that this enigma can be explained by a highly fluid-dependent tectonic process. Examples of Phanerozoic UHP rocks extruded from depths of 200–300 km to the surface exist (see Maruyama et al. 1994; Brown 2007; Zhang et al. 2009), but not for the Archaean. According to our interpretation, although UHP metamorphic rocks formed during the Archaean, these were not brought up because of the absence of hydrous fluids as a lubricant to return these rocks to the surface. For example, 3 Ga eclogites brought up as xenoliths in Cretaceous kimberlites have been described from Africa (Valley et al. 1998; see also the field of Archaean xenoliths in Fig. 3). These UHT eclogites represent UHP rocks formed in the Archaean deep mantle by subduction, although they were never returned to the surface, except as xenoliths. In contrast, 100 Ma lawsonite eclogites trapped within Cenozoic alkaline basalts in central North America (Usui et al. 2003) were transported horizontally for a distance of over 2000 km from the subduction zone in California and subsequently brought to the surface by magmas (Fig. 14). This example also illustrates the changing

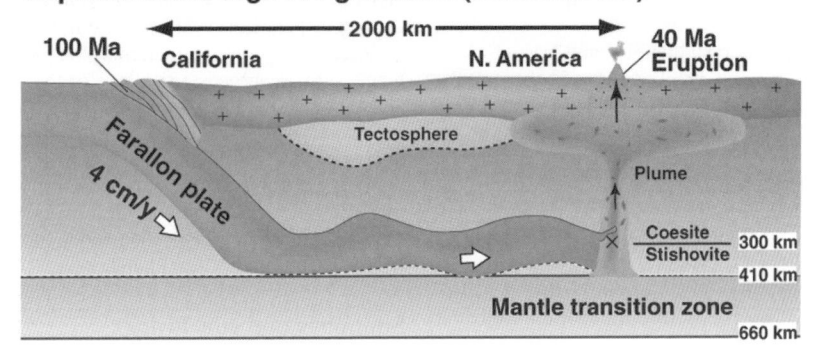

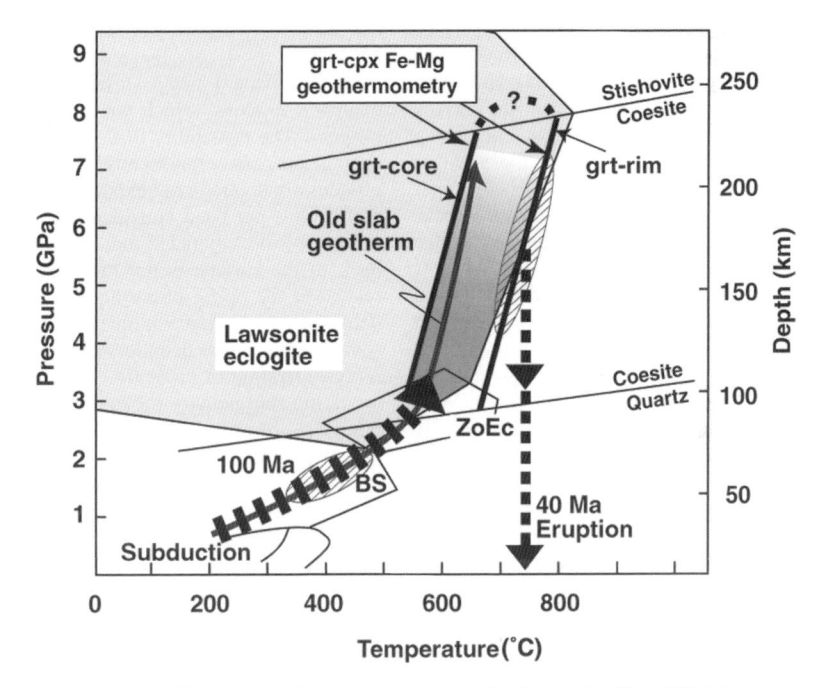

Fig. 14. Upper figure: history of Four Corners lawsonite eclogites in North America. The 100 Ma lawsonite eclogites must have been formed in California by Pacific-type orogeny. The overlying MORB crust was cooled, recrystallized as lawsonite eclogites and subducted in the upper mantle; it then travelled from California eastward for over 2000 km in the stability field of coesite and finally was caught up by a rising plume that probably originated from 410 km depth, or even lower, and brought to the surface by strongly alkaline basalts at 40 Ma (Usui et al. 2003). The average speed from California to the Four Corners region in the upper mantle was 4 cm a^{-1}. Lower figure: P–T–t history of lawsonite eclogites in the Four Corners region (after Usui et al. 2003). Rapid subduction from the trench along a cold geotherm with an anti-clockwise path produced lawsonite eclogites from 600 to 750 °C and up to 8 GPa. They were then returned quickly to the surface. The temperature estimate is based on grt–cpx Fe–Mg exchange geothermometry, and the pressure condition is given by the presence of coesite in garnet. However, these eclogites may have been subjected to pressure in the stishovite field. Stishovite is not a quenchable mineral, and hence it is difficult to demonstrate its original presence.

subduction-zone geotherm (e.g. Brown 2007; see also geotherms in Fig. 6 and the position of the 100 Ma lawsonite eclogites and Archaean xenoliths shown in $P-T$ space). These occurrences offer another important message: the lawsonite eclogite with coesite must have travelled almost 200 km in the mantle, but still retains a Cretaceous age memory. This is possibly because no fluid was available after the eclogite formation, until the Cenozoic trap by alkaline magma, so that there was only slight recrystallization and partial overprint of the Cenozoic event. Zircons in these rocks still preserve the memory of the Tertiary event (Usui *et al.* 2003) despite the high (>1000 °C) temperatures in the mantle. The lack of fluid is the dominant factor here, indicating the critical role exerted by fluids.

Water has a distinctly different behaviour as compared with CO_2. Water works as an effective lubricant. Regional metamorphic belts, whether Archaean or Phanerozoic, are most often truncated at the top and bottom by paired faults (e.g. Maruyama *et al.* 1994). The syntectonic slice between these faults has selectively moved to the surface through wedge extrusion at subduction zones since the Archaean (Maruyama & Liou 2005). However, water-rich fluid was available only near the surface in the high geothermal gradient prevalent during the Archaean. On the other hand, in the younger Earth, water became available at the whole upper mantle scale, so that much of the deeper rocks could be extended through the process of wedge extrusion.

Buoyant continental subduction can be achieved only through slab-pull force. Continents can subduct to some critical depth, which is the neutral point at which a balance is maintained of the buoyancy of continental crust against the slab-pull force. With time, the subducted slab becomes mechanically released, or slab break-off occurs. Subsequently, the deeply subducted continental shelf deposits along with the basement would return to the surface by squeezing-out subhorizontally through the wedge extrusion process, bounded by the paired faults described above. It should be noted that the extrusion is not vertical, but wedge-type, with accelerated extrusion if a hydrous lubricant is available along the paired faults.

In the Phanerozoic scenario (Fig. 15a), water-dominated fluids carried by subducting oceanic slabs became available down to 660 km depth in the modern Earth, as inferred from the distribution of seismicity (Hasegawa *et al.* 1991, 1994; Zhao 2004; Zhao *et al.* 2007). Water derived by dehydration may be concentrated at the bottom of the basement and the top of the subducted slab against the hanging wall. The tectonic slice separated by these two weak zones, or water channels, may be transported to the surface by the wedge extrusion mechanism. This propagating water front in the deep mantle may correspond to the apparent increase in pressure through geological time. Thus we conclude that although UHP rocks are widespread throughout geological time, those formed in the Archaean seldom returned to the surface, and remained as eclogitic MORB crust in the mantle, and only traces of these were transferred to the surface as xenoliths such as those in kimberlites.

Archaean regional metamorphic belts can be formed by a process similar to that described above, as shown in Figure 15b. However, circulating water was probably not available at depths equivalent to >10 kbar (mantle depths). This may explain the observation that Archaean regional metamorphic rocks seldom show pressures higher than 10 kbar even though their mode of occurrence is identical to that of the Phanerozoic.

Summary and conclusions

The nature and origin of orogens in various geological settings have been studied for several decades. However, most of these studies focused on orogens on the Earth's surface, or the so-called exhumed orogens, which we term here the extant orogens. The possibility that a large portion of the ancient orogens remains now as ghost or arrested orogens that are not exposed on the Earth's surface has not been considered so far. The exact volume of the subducted material and that brought back to the surface is not known. Recent studies related to the origin of earthquakes have indicated dehydration reaction within or around subducting slabs, and this helps in understanding the continuing recrystallization of orogenic welts, with or without overlying sedimentary rocks deposited at the plate boundary. Such analyses suggest that most of the orogens move down to the deep mantle and are seldom brought back to the surface.

In this paper, we have summarized the $P-T$ conditions of orogen-forming units and also the time-dependent evolution of orogens. We propose a classification of orogens based on their distribution in $P-T$ space. Isotope geochemists have long discussed the possibility of global material circulation from the surface to the bottom of the CMB. We illustrate that remnants of deeply subducted orogens are brought up through plumes or superplumes originating from the CMB. We also propose that large volumes of subducted orogens must have accumulated on the CMB as 'lost continents' constituting the major component in the D'' layer.

A consideration of the cooling history of the Earth indicates that this might have started from subduction zones, which made it possible for

(a)

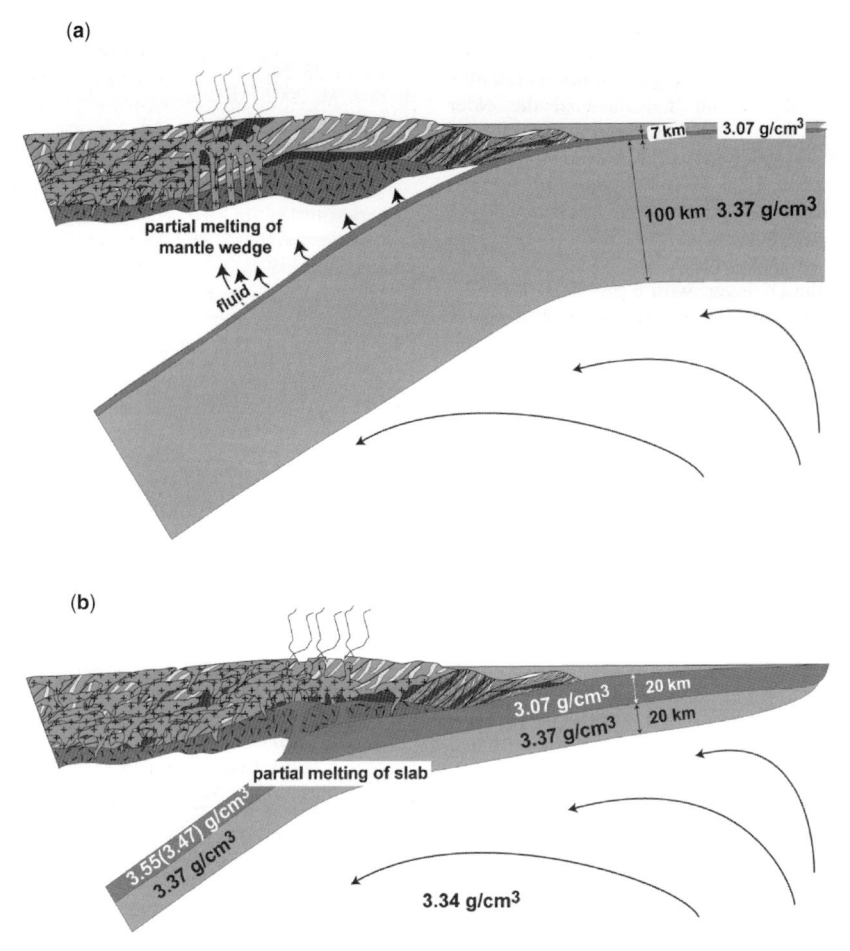

(b)

Fig. 15. Contrasting aspects of the lithosphere, Archaean (**b**) versus Phanerozoic (**a**) (after Komiya *et al.* 2002; Komiya 2004). In the Archaean, a thin (40 km) lithosphere, composed of an overlying MORB (20 km) and a mantle lithosphere (20 km), subducts into the mantle to release slab melts with TTG composition. Subduction of a thin and buoyant Archaean lithosphere into mantle peridotite (density 3.34 g cm^{-3}) seems to be difficult because of the lighter nature of amphibolites (density 3.07 g cm^{-3}) as compared with the the surrounding mantle. However, after partial melting of the MORB crust, the remaining restite is transformed to garnet pyroxenite (density 3.55–3.47 g cm^{-3}), which is heavier than the surrounding mantle. Therefore, slab melting dominated subduction must have been more active in the Archaean compared with the Phanerozoic because the Phanerozoic slab (density 3.37 g cm^{-3}) is denser than the surrounding mantle (3.34 g cm^{-3}). In the Phanerozoic, thick lithosphere (about 100 km) capped by a thin (6–7 km) MORB subducts and dehydrates, releasing water down to 660 km. The hanging wall of the overriding lithosphere is the accretionary complex formed by subduction of the oceanic slab, which grows oceanward through time with subsequent penetration of felsic magma by the melting of the wet mantle wedge.

material to be returned to the surface even from very great depths, but only after about 600 Ma. Global material circulation must have occurred even in the Archaean albeit only at shallow levels. The critical depth for transport of material back to the surface evolved to increasingly deep levels with time. Whereas the maximum depth corresponded to pressures less than 10 kbar in the Archaean, it increased to pressures of 60–70 kbar in the Phanerozoic; an increase of almost seven times, as a result of the increased stability field of hydrous silicates through cooling brought about by water subduction to depth. This allowed fragments of orogens to be brought back to the surface from deeper levels. An evaluation of the total volume of continents on the Earth shows that almost 50% is constituted of Precambrian orogens, whereas the remaining 50% is made up of Phanerozoic orogens formed during the last 540 Ma. Considering that the Phanerozoic timespan represents only about one-seventh of the total Earth history, the volume of Phanerozoic orogens on the surface is considerably higher than that of the Precambrian orogens. This would further support the notion that most of the older orogens were deeply subducted and never returned to the surface in the early Earth history, whereas fluid lubrication aided the exhumation of Phanerozoic orogens. Ghost orogens and arrested orogens may thus constitute more than 90% of the total volume of orogens. These are now major components in the D" layer, with a probable thickness of 200–250 km, almost 10 times more than that of the present-day continental crust.

We thank T. Kusky for inviting us to contribute to this volume, and for his constructive comments and suggestions on the manuscript, which greatly helped in improving our presentation. The manuscript also benefited from reviews by J. J. W. Rogers and an anonymous referee.

References

ADALBERTO, T.-C., ALONSO, R.-F.-J., FERNANDO, V.-T. & PEDRO, R.-S. 2005. Mantle xenoliths and their host magmas in the Eastern Alkaline Province, Northeast Mexico. *International Geology Review*, **47**, 1260–1286.

AI, Y. 1994. A revision of the garnet–clinopyroxene Fe^{2+}–Mg exchange geothermometer. *Contributions to Mineralogy and Petrology*, **115**, 467–473.

ALLÈGRE, C. J., COURTILLOT, V. ET AL. 1984. Structure and evolution of the Himalaya–Tibet orogenic belt. *Nature*, **307**, 17–22.

APPLEYARD, C. M., VILJOEN, K. S. & DOBBE, R. 2004. A study of eclogitic diamonds and their inclusions from the Finsch kimberlite pipe, South Africa. *Lithos*, **77**, 317–332.

ARMSTRONG, R. L. 1991. The persistent myth of crustal growth. *Australian Journal of Earth Sciences*, **38**, 613–630.

BALDWIN, S. L., MONTELEONE, B. D., WEBB, L. E., FITZGERALD, P. G., GROVE, M. & JUNE, H. E. 2004. Pliocene eclogite exhumation at plate tectonic rates in eastern Papua New Guinea. *Nature*, **431**, 263–267.

BANGS, N. L., GULICK, S. P. S. & SHIPLEY, T. H. 2009. Seamount subduction erosion in the Nankai Trough and its potential impact on the seismogenic zone. *Geology*, **34**, 701–704.

BASU, A. R., ONGLEY, J. S. & MACGREGOR, I. D. 1986. Eclogites, pyroxene geotherm, and layered mantle convection. *Science*, **233**, 1303–1305.

BEAUMONT, C., NGUYEN, M. H., JAMIESON, R. A. & ELLIS, S. 2006. Crustal flow models in large, hot orogens. *In*: LAW, R. D., SEARLE, M. P. & GODIN, L. (eds) *Modelling Channel Flow and Ductile Extrusion Processes*. Geological Society, London, Special Publications, **268**, 91–145.

BRADLEY, D. C., HAEUSSLER, P. J. & KUSKY, T. M. 1993. Timing of early Tertiary ridge subduction in southern Alaska. *In*: TILL, A. (ed.) *Geologic Studies in Alaska by U.S. Geological Survey during 1990*. US Geological Survey Bulletin, **2068**, 163–177.

BROWN, M. 2006. Crust melt flow in large hot orogens. *Geological Society of America, Abstracts with Programs*, **38**, 342.

BROWN, M. 2007. Metamorphic conditions in orogenic belts: a record of secular change. *International Geology Review*, **49**, 193–234.

BRUECKNER, H. K., CARSWELL, D. A. & GRIFFIN, W. L. 2002. Paleozoic diamonds within a Precambrian peridotite lens in UHP gneisses of the Norwegian Caledonides. *Earth and Planetary Science Letters*, **203**, 805–816.

BURKE, K., ASHWAL, L. D. & WEBB, S. J. 2003. New way to map old sutures using deformed alkaline rocks and carbonatites. *Geology*, **31**, 391–394.

CANN, J. R., BLACKMAN, D. K. ET AL. 1997. Corrugated slip surfaces formed at ridge–transform intersections on the Mid-Atlantic Ridge. *Nature*, **385**, 329–332.

CAWOOD, P. A., KRONER, A., COLLINS, W. J., KUSKY, T. M., MOONEY, W. D. & WINDLEY, B. F. 2009. Accetionary orogens through Earth history. *In*: CAWOOD, P. A. & KRONER, A. (eds) *Earth Accretionary Systems in Space and Time*. Geological Society, London, Special Publications, **318**, 1–36.

CHARLTON, T. R., BARBER, A. J. ET AL. 2002. The Permian of Timor: stratigraphy, palaeontology and palaeogeography. *Journal of Asian Earth Sciences*, **20**, 719–774.

CHESLEY, J. T., RUDNICK, R. L. & LEE, C.-T. 1991. Re–Os systematics of mantle xenoliths from the East African Rift: age, structure, and history of the Tanzanian craton. *Geochimica et Cosmochimica Acta*, **63**, 1203–1217.

CHOPIN, C. 1984. Coesite and pure pyrope in high-grade blueschists of the Western Alps: a first record and some consequences. *Contributions to Mineralogy and Petrology*, **86**, 107–118.

CHOPIN, C. 2003. Ultrahigh-pressure metamorphism: tracing continental crust into the mantle. *Earth and Planetary Science Letters*, **212**, 1–14.

COLLINS, W. J. 2002. Hot orogens, tectonic switching, and creation of continental crust. *Geology*, **30**, 535–538.

CONDIE, K. C. 1986. Origin and early growth rate of continents. *Precambrian Research*, **32**, 261–278.

CONDIE, K. C. 1998. Episodic continental growth and supercontinents: a mantle avalanche connection? *Earth and Planetary Science Letters*, **163**, 97–108.

CONDIE, K. C. & KRONER, A. 2008. When did plate tectonics begin? Evidence from the geological record. *Geological Society of America Bulletin*, **440**, 281–294.

CONEY, P. J. & HARMS, T. A. 1984. Cordilleran metamorphic core complexes: Cenozoic extensional relics of Mesozoic compression. *Geology*, **12**, 550–554.

CUTHBERT, S. J., CARSWELL, D. A., KROGH-RAVNA, E. J. & WAIN, A. 2000. Eclogites and eclogites in the Western Gneiss Region, Norwegian Caledonides. *Lithos*, **52**, 165–195.

DACZKO, N. R., MOSHER, S., COFFIN, M. F. & MECKEL, T. A. 2005. Tectonic implications of fault-scarp-derived volcaniclastic deposits on Macquarie Island: sedimentation at a fossil ridge–transform intersection? *Geological Society of America Bulletin*, **117**, 18–31.

DEFANT, M. J. & DRUMMOND, M. S. 1990. Dehydration of some modern arc magmas by melting of young subducted lithosphere. *Nature*, **347**, 662–665.

Dewey, J. F. & Bird, J. M. 1970. Mountain belts and new global tectonics. *Journal of Geophysical Research*, **75**, 2625–2647.

Dewey, J. F. & Spall, H. 1975. Pre-Mesozoic plate tectonics: how far back in Earth history can the Wilson Cycle be extended? *Geology*, **3**, 422–424.

Dewey, J. F., Pitman, W. C., Ryan, W. B. F. & Boin, J. 1973. Plate tectonics and evolution of the Alpine system. *Geological Society of America Bulletin*, **84**, 3137–3184.

Dickinson, W. R. 1971. Plate tectonics in geologic history. *Science*, **174**, 107–113.

Dickinson, W. R. 1981. Plate tectonics and the continental margin of California. *In*: Ernst, W. G. (ed.) *The Geotectonic Development of California, Vol. 1*. Prentice–Hall, Englewood Cliffs, NJ, 1–28.

Dilek, Y., Moores, E. M., Elthon, D. & Nicolas, A. (eds) 2000. *Ophiolites and Oceanic Crust: New Insights from Field Studies and the Ocean Drilling Program*. Geological Society of America, Special Papers, **349**.

Echtler, H. & Malavieille, J. 1990. Extensional tectonics, basement uplift and Stephano-Permian collapse in a late Variscan metamorphic core complex (Montagne Noire, Southern Massif Central). *Tectonophysics*, **177**, 125–138.

Ellis, D. J. & Green, D. H. 1979. An experimental study of the effect of Ca upon garnet–clinopyroxene Fe–Mg exchange equilibria. *Contributions to Mineralogy and Petrology*, **71**, 13–22.

Engebertson, D. C., Kelley, K. P., Cashman, H. J. & Richards, M. A. 1992. 180 million years of subduction. *GSA Today*, **2**, 93–100.

England, P. C. & Thompson, J. B. 1984. Pressure–temperature–time paths of regional metamorphism. I. Heat transfer during the evolution of regions of thickened continental crust. *Journal of Petrology*, **25**, 894–928.

Ernst, W. G. 1984. Californian blueschists, subduction, and the significance of tectono-stratigraphic terranes. *Geology*, **12**, 436–440.

Escartin, J., Mevel, C., MacLeod, C. J. & McCaig, A. M. 2003. Constraints on deformation conditions and the origin of oceanic detachments: the Mid Atlantic Ridge core complex at 15°45'N. *Geochemistry, Geophysics, Geosystems*, **4**, doi:10.1029/2002GC000472.

Frezzotti, M. L. & Peccerillo, A. 2007. Diamond-bearing COHS fluids in the mantle beneath Hawaii. *Earth and Planetary Science Letters*, **262**, 273–283.

Fukao, Y. 1992. Seismic tomogram of the Earth's mantle: Geodynamic implications. *Science*, **258**, 625–630.

Girod, M., Dautria, J. M. & Giovanni, R. D. 1981. A first insight into the constitution of the upper mantle under the Hoggar area (southern Algeria): the lherzolite xenoliths in alkali basalts. *Contributions to Mineralogy and Petrology*, **77**, 66–73.

Goscombe, B. D. & Gray, D. R. 2009. Metamorphic response in orogens of different obliquity, scale and geometry. *Gondwana Research*, **15**, 151–167.

Goscombe, B., Gray, D. & Hand, M. 2006. Crustal architecture of the Himalayan metamorphic front in eastern Nepal. *Gondwana Research*, **10**, 232–255.

Grand, S. P. 2002. Mantle shear-wave tomography and the fate of subducted slabs. *Philosophical Transactions of the Royal Society of London, Series A*, **360**, 2475–2491.

Griffin, W. L., Wass, S. Y. & Hollis, J. D. 1984. Ultramafic xenoliths from Bullenmerri and Gnotuk Maars, Victoria, Australia: petrology of a subcontinental crust–mantle transition. *Journal of Petrology*, **25**, 53–87.

Griffin, W. L., O'Reilly, S. Y. & Ryan, C. G. 1999. The composition and origin of subcontinental lithospheric mantle. *In*: Fei, Y., Bertka, C. M. & Mysen, B. O. (eds) *Mantle Petrology: Field Observations and High-Pressure Experimentation: A Tribute to Francis R. (Joe) Boyd*. Geochemical Society, Special Publications, **6**, 13–45.

Griffin, W. L., O'Reilly, S. Y. et al. 2003. The origin and evolution of Archaean lithospheric mantle. *Precambrian Research*, **127**, 19–41.

Grove, T. L. & Parman, S. W. 2004. Thermal evolution of the Earth as recorded by komatiites. *Earth and Planetary Science Letters*, **219**, 173–187.

Gurney, J. J., Harris, J. W., Richard, R. S. & Moore, R. O. 1985. Inclusions in Premier Mine diamonds. *Transactions of the Geological Society of South Africa*, **88**, 301–310.

Hacker, B. R. & Peacock, S. M. 1995. Creation, preservation and exhumation of UHPM. *In*: Coleman, R. G. & Wang, X. (eds) *Ultrahigh-Pressure Metamorphism*. Cambridge University Press, Cambridge, 159–182.

Hamilton, W. B. 1979. *Tectonics of the Indonesian Region*. US Geological Survey, Professional Papers, **1078**.

Hargraves, R. B. 1986. Faster spreading or greater ridge length in the Archaean? *Geology*, **14**, 750–752.

Hasegawa, A., Umino, N. & Tagai, A. 1978. Double-planed deep seismic zone and upper-mantle structure in the Northeastern Japan Arc. *Geophysical Journal International*, **53**, 281–296.

Hasegawa, A., Zhao, D., Hori, S., Yamamoto, A. & Horiuchi, S. 1991. Deep structure of the northeastern Japan arc and its relationship to seismic and volcanic activity. *Nature*, **352**, 683–689.

Hasegawa, A., Horiuchi, S. & Umino, N. 1994. Seismic structure of the northeastern Japan convergent margin: a synthesis. *Journal of Geophysical Research*, **99**, 22 295–22 311.

Hayashi, M., Komiya, T., Nakamura, Y. & Maruyama, S. 2001. Archaean regional metamorphism of the Isua supracrustal belt, southern West Greenland: implications for a driving force of Archaean plate tectonics. *International Geology Review*, **42**, 1055–1115.

Henjes-Kunst, F. & Altherr, R. 1992. Metamorphic petrology of xenoliths from Kenya, Northern Tanzania and implications for geotherms and lithospheric structures. *Journal of Petrology*, **33**, 1125–1156.

Hilde, T. W., Uyeda, S. & Kroenke, L. 1977. Evolution of the western Pacific and its margins. *Tectonophysics*, **38**, 145–146.

Hoffman, P. F. 1992. Rodinia, Gondwanaland, Pangea and Amasia: alternating kinematic scenarios of supercontinental fusion (abstract). *EOS (Trans. American Geophysical Union)*, **73**, No. 14, Supplement, 282.

HOFFMAN, P. F. 1999. The break-up of Rodinia, birth of Gondwana, true polar wander and the snowball Earth. *Journal of African Earth Sciences*, **28**, 17–33.

HOFMANN, A. W. & WHITE, W. M. 1982. Mantle plumes from ancient oceanic crust. *Earth and Planetary Science Letters*, **57**, 421–436.

HUANG, J. & ZHAO, D. 2006. High-resolution mantle tomography of China and surrounding regions. *Journal of Geophysical Research*, **111**, B09305, doi:10.1029/2005JB004066.

HYNDMAN, R. D., YAMANO, M. & OLESKEVICH, D. A. 1997. The seismogenic zone of subduction thrust faults. *Island Arc*, **6**, 244–260.

INOUE, T., YURIMOTO, Y. & KUDOH, T. 1995. Hydrous modified spinel, $Mg_{1.75}SiH_{0.5}O_4$: a new water reservoir in the mantle transition region. *Geophysical Research Letters*, **22**, 117–120.

ISHIKAWA, A., KANEKO, Y., KADARUSMAN, A. & OTA, T. 2007. Multiple generations of forearc mafic–ultramafic rocks in the Timor–Tanimbar ophiolite, eastern Indonesia. *Gondwana Research*, **11**, 200–217.

ISOZAKI, Y. 1997. Jurassic accretion tectonics of Japan. *Island Arc*, **6**, 25–51.

ISOZAKI, Y. & MARUYAMA, S. 1991. Studies on orogeny based on plate tectonics in Japan and a new geotectonic subdivision of the Japanese islands. *Journal of Geography*, **100**, 697–761 [in Japanese].

JACOB, D. E. 2004. Nature and origin of eclogite xenoliths from kimberlite. *Lithos*, **77**, 295–316.

JACOB, D. E. & FOLEY, S. F. 1999. Evidence for Archaean ocean crust with low high field strength element signature from diamondiferous eclogite xenoliths. *Lithos*, **48**, 317–336.

JAGOUTZ, E., DAWSON, J. B., HOERNES, S., SPETTEL, B. & WANKE, H. 1984. Anorthositic oceanic crust in the Archaean Earth. *Abstracts, 15th Lunar and Planetary Science Conference*. Lunar and Planetary Science Institute, Huston, TX, 395–396.

JAQUES, A. L., HALL, A. E. ET AL. 1989. Composition of crystalline inclusions and C-isotopic composition of Argyle and Ellendale diamonds. *In*: ROSS, J., JAQUES, A. L., FERGUSON, J., GREEN, D. H., O'REILLY, S. Y., DANCHIN, R. V. & JANSE, A. J. A. (eds) *Kimberlites and Related Rocks, Vol. 2: Their Mantle/Crust Setting, Diamonds and Diamond Exploration*. Geological Society of America, Special Papers, **14**, 966–989.

JAVOY, M. 1997. The major volatile elements of the Earth: their origin, behavior, and fate. *Geophysical Research Letters*, **24**, 177–180.

JORDAN, T. 1988. Structure and formation of the continental tectosphere. *Journal of Petrology, Special Lithosphere Issue*, 11–37.

KADARUSMAN, A., MIYASHITA, S., MARUYAMA, S., PARKINSON, C. D. & ISHIKAWA, A. 2004. Petrology, geochemistry and paleogeographic reconstruction of the East Sulawesi Ophiolite, Indonesia. *Tectonophysics*, **392**, 55–83.

KANEKO, Y. 1997. Two-step exhumation model of the Himalayan metamorphic belt, Central Nepal. *Journal of the Geological Society of Japan*, **103**, 203–226.

KANEKO, Y., KATAYAMA, I. ET AL. 2003. Timing of Himalayan ultrahigh-pressure metamorphism: sinking rate and subduction angle of the Indian continental crust beneath Asia. *Journal of Metamorphic Geology*, **21**, 589–599.

KANEKO, Y., MARUYAMA, S. ET AL. 2007. On-going orogeny in the outer-arc of the Timor–Tanimbar region, eastern Indonesia. *Gondwana Research*, **11**, 218–233.

KATAYAMA, I., MARUYAMA, S., PARKINSON, C. D., TERADA, K. & SANO, Y. 2001. Ion microprobe U–Pb zircon geochronology of peak and retrograde stages of ultrahigh-pressure metamorphic rocks from the Kokchetav massif, northern Kazakhstan. *Earth and Planetary Science Letters*, **188**, 185–198.

KELSEY, D. E. 2008. On ultrahigh temperature crustal metamorphism. *Gondwana Research*, **13**, 1–29.

KOMABAYASHI, T., OMORI, S. & MARUYAMA, S. 2004. Petrogenetic grid in the system $MgO-SiO_2-H_2O$ up to 30 GPa, 1600 °C: applications to hydrous peridotite subducting into the Earth's deep interior. *Journal of Geophysical Research*, **109**, doi:10.1029/2003JB002651.

KOMIYA, T. 2004. Material circulation model including chemical differentiation within the mantle and secular variation of temperature and composition of the mantle. *Physics of the Earth and Planetary Interiors*, **146**, 333–367.

KOMIYA, T. 2007. Material circulation through time – Chemical differentiation within the mantle and secular variation of temperature and composition of the mantle. *In*: YUEN, D. A., MARUYAMA, S., KARATO, S. & WINDLEY, B. F. (eds) *Superplumes: Beyond Plate Tectonics*. Springer, New York, 187–234.

KOMIYA, T. & MARUYAMA, S. 2007. A very hydrous mantle under the western Pacific region: implications for formation of marginal basins and style of Archaean plate tectonics. *Gondwana Research*, **11**, 132–147.

KOMIYA, T., MARUYAMA, S., NOHDA, S., MASUDA, T., HAYASHI, M. & OKAMOTO, S. 1999. Plate tectonics at 3.8–3.7 Ga; Field evidence from the Isua accretionary complex, southern West Greenland. *Journal of Geology*, **107**, 515–554.

KOMIYA, T., HAYASHI, M., MARUYAMA, S. & YURIMOTO, H. 2002a. Intermediate P/T type Archaean metamorphism of the Isua supracrustal belt: implications for secular change of geothermal gradients at subduction zones and for Archaean plate tectonics. *American Journal of Science*, **302**, 806–826.

KOMIYA, T., MARUYAMA, S., HIRATA, T. & YURIMOTO, H. 2002b. Petrology and geochemistry of MORB and OIB in the mid-Archaean North Pole region, Pilbara craton, Western Australia: implications for the composition and temperature of the upper mantle at 3.5 Ga. *International Geology Review*, **44**, 988–1016.

KOMIYA, T., MARUYAMA, S., HIRATA, T., YURIMOTO, H. & NOHDA, S. 2004. Geochemistry of the oldest MORB and OIB in the Isua supracrustal belt (3.8 Ga), southern West Greenland: implications for the composition and temperature of early Archaean upper mantle. *Island Arc*, **13**, 47–72.

KUSKY, T. M. (ed.) 2004. *Precambrian Ophiolites and Related Rocks*. Developments in Precambrian Geology, **13**.

KUSKY, T. M. & POLAT, A. 1999. Growth of granite–greenstone terranes at convergent margins, and stabilization of Archaean cratons. *Tectonophysics*, **305**, 43–73.

KUSKY, T. M., BRADLEY, D. C. & HAEUSSLER, P. 1997. Progressive deformation of the Chugach accretionary complex, Alaska, during a Paleogene ridge–trench encounter. *Journal of Structural Geology*, **19**, 139–157.

KUSKY, T. M., GANLEY, R., LYTWYN, J. & POLAT, A. 2004. The Resurrection Peninsula ophiolite, mélange and accreted flysch belts of southern Alaska as an analog for trench–forearc systems in Precambrian orogens. *In*: KUSKY, T. M. (ed.) *Precambrian Ophiolites and Related Rocks*. Developments in Precambrian Geology, **13**, 627–674.

LEECH, M. L., SINGH, S., JAIN, A. K., KLEMPERER, S. L. & MANICKAVASAGAM, R. M. 2005. The onset of India–Asia continental collision: early, steep subduction required by the timing of UHP metamorphism in the western Himalaya. *Earth and Planetary Science Letters*, **234**, 83–97.

LEI, J. & ZHAO, D. 2005. P-wave tomography and origin of the Changbai intraplate volcano in Northeast Asia. *Tectonophysics*, **397**, 281–295.

LISTER, G. S. & DAVIS, G. A. 1989. The origin of metamorphic core complexes and detachment faults formed during Tertiary continental extension in the northern Colorado River region. *U.S.A. Journal of Structural Geology*, **11**, 65–94.

LITHGOW-BERTELLONI, C. & RICHARDS, M. A. 1998. The dynamics of Cenozoic and Mesozoic plate motions. *Reviews of Geophysics*, **36**, 27–78.

LIU, F., XU, X., LIOU, J. G. & SONG, B. 2004. SHRIMP U–Pb ages of ultrahigh-pressure and retrograde metamorphism of gneisses, south-western Sulu terrane, eastern China. *Journal of Metamorphic Geology*, **22**, 315–326.

MACDONALD, J. F. 1961. Surface heat flow from a differentiated Earth. *Journal of Geophysical Research*, **66**, 2489–2493.

MACLEOD, C. J., ESCARTIN, J. ET AL. 2002. First direct evidence for oceanic detachment faulting: the Mid-Atlantic Ridge, 14°45′N. *Geology*, **30**, 879–882.

MARUYAMA, S. 1997. Pacific-type orogeny revisited; Miyashiro-type orogeny proposed. *Island Arc*, **6**, 91–120.

MARUYAMA, S. 1988. Petrology of Franciscan metabasites along the jadeite–glaucophane-type facies series, Cazadero, California. *Journal of Petrology*, **29**, 1–37.

MARUYAMA, S. & LIOU, J. G. 2005. From Snowball to Phanerozoic Earth. *International Geology Review*, **47**, 775–791.

MARUYAMA, S. & OKAMOTO, K. 2007. Water transportation from the subducting slab into the mantle transition zone. *Gondwana Research*, **11**, 148–165.

MARUYAMA, S. & PARKINSON, C. D. 2000. Overview of the geology, petrology and tectonic framework of the HP–UHPM Kokchetav Massif, Kazakhstan. *Island Arc*, **9**, 441–442.

MARUYAMA, S. & SENO, T. 1986. Orogeny and relative plate motions: example of the Japanese Islands. *Tectonophysics*, **127**, 305–329.

MARUYAMA, S., LIOU, J. G. & SENO, T. 1989. Mesozoic and Cenozoic evolution of Asia. *In*: BEN-AVRAHAM, Z. (ed.) *The Evolution of the Pacific Ocean Margins*. Oxford University Press, New York, 75–99.

MARUYAMA, S., LIOU, J. G. & ZHANG, R. 1994. Tectonic evolution of the ultrahigh-pressure (UHP) and high-pressure (HP) metamorphic belts from central China. *Island Arc*, **3**, 112–121.

MARUYAMA, S., LIOU, J. G. & TERABAYASHI, M. 1996. Blueschists and eclogites of the world and their exhumation. *International Geology Review*, **38**, 485–594.

MARUYAMA, S., ISOZAKI, Y., KIMURA, G. & TERABAYASHI, M. 1997. Paleogeographic maps of the Japanese Islands: Plate tectonic synthesis from 750 Ma to the present. *Island Arc*, **6**, 121–142.

MARUYAMA, S., MASAGO, H., KATAYAMA, I., IWASE, Y. & TORIUMI, M. 2004. A revolutionary new interpretation of a regional metamorphism, its exhumation, and consequent mountain building. *Journal of Geography*, **113**, 727–768 [in Japanese].

MARUYAMA, S., SANTOSH, M. & ZHAO, D. 2007a. Superplume, supercontinent, and post-perovskite: mantle dynamics and anti-plate tectonics on the Core–Mantle Boundary. *Gondwana Research*, **11**, 7–37.

MARUYAMA, S., YUEN, D. A. & WINDLEY, B. F. 2007b. Dynamics of plumes and superplumes through time. *In*: YUEN, D. A., MARUYAMA, S., KARATO, S. & WINDLEY, B. F. (eds) *Superplumes: Beyond Plate Tectonics*. Springer, New York, 441–502.

MASAGO, H. 2000. Metamorphic petrology of the Barchi-Kol metabasites, western Kokchetav ultrahigh-pressure–high-pressure massif, northern Kazakhstan. *Island Arc*, **9**, 358–378.

MASAGO, H. & OMORI, S. 2004. Application of thermodynamics forward-modeling to estimation of a metamorphic P–T path. *Journal of Geography*, **113**, 647–663 [in Japanese].

MASAGO, H., OMORI, S. & MARUYAMA, S. 2009. Counterclockwise prograde P–T path in collisional orogeny and water subduction at the Precambrian–Cambrian boundary: The ultrahigh-pressure pelitic schist in the Kokchetav massif, northern Kazakhstan. *Gondwana Research*, **15**, 137–150.

MATSUDA, T. & ISOZAKI, Y. 1991. Well-documented travel history of Mesozoic pelagic chert in Japan: from remote ocean to subduction zone. *Tectonics*, **10**, 475–499.

MATTE, P. 1991. Accretionary history and crustal evolution of the Variscan belt in Europe. *Tectonophysics*, **196**, 309–337.

MCGUIRE, A. V. 1988. Petrology of the mantle xenoliths from Harrat al Kishb: the mantle beneath western Saudi Arabia. *Journal of Petrology*, **29**, 73–92.

METCALFE, I. 2006. Palaeozoic and Mesozoic tectonic evolution and palaeogeography of East Asian crustal fragments: the Korean Peninsula in context. *Gondwana Research*, **9**, 24–46.

MIYASHIRO, A. 1961. Evolution of metamorphic belts. *Journal of Petrology*, **2**, 277–311.

MIYASHIRO, A. 1973. *Metamorphism and Metamorphic Belts*. Allen & Unwin, London.

MIYASHIRO, A., AKI, K. & SENGOR, A. M. 1982. *Orogeny*. Wiley, Chichester.

MOORBATH, S. 1976. Age and isotope constraints for the evolution of Archaean crust. *In*: WINDLEY, B. F. (ed.) *The Early History of the Earth*. Wiley, New York, 351–360.

MOORE, R. O. & GURNEY, J. J. 1985. Pyroxene solid solution in garnets included in diamond. *Nature*, **318**, 553–555.

MOORES, E. M. 1993. Neoproterozoic oceanic crustal thinning, emergence of continents and origin of the Phanerozoic ecosystem: a model. *Geology*, **21**, 5–8.

NAKAGAWA, M., SANTOSH, M. & MARUYAMA, S. 2009. Distribution and mineral assemblages of bedded manganese deposits in Shikoku, Southwest Japan: implications for accretion tectonics. *Gondwana Research*, **16**, 609–621, doi:10.1016/j.gr.2009.05.003.

O'BRIEN, P. J., ZOTOV, N., LAW, R., KHAN, M. A. & JAN, M. Q. 2001. Coesite in Himalayan eclogite and implications for models of India–Asia collision. *Geology*, **29**, 435–438.

OHARA, Y. 2006. Mantle process beneath Philippine sea back-arc spreading ridges: a synthesis of peridotite petrology and tectonics. *Island Arc*, **15**, 119–129.

OHARA, Y., YOSHIDA, T., KATO, Y. & KASUGA, S. 2001. Giant megamullion in the Parece Vela backarc basin. *Marine Geophysical Researches*, **22**, 47–61.

OHTA, H., MARUYAMA, S., TAKAHASHI, E., WATANABE, Y. & KATO, Y. 1996. Field occurrence, geochemistry and petrogenesis of the Archaean mid-oceanic ridge basalts (AMORBs) of the Cleaverville area, Pilbara craton, Western Australia. *Lithos*, **37**, 199–221.

OHTANI, Y., LITASOV, K., HOSOYA, T., KUBO, T. & KONDO, T. 2004. Water transport into the deep mantle and formation of a hydrous transition zone. *Physics of the Earth and Planetary Interiors*, **143–144**, 255–269.

OMORI, S. & SANTOSH, M. 2008. Metamorphic decarbonation in the Neoproterozoic and its environmental implication. *Gondwana Research*, **14**, 97–104.

O'REILLY, S. Y., GRIFFIN, W. L. & RYAN, C. G. 1991. Residence of trace elements in metasomatized spinel lherzolite xenoliths: a proton-microprobe study. *Contributions to Mineralogy and Petrology*, **109**, 98–113.

OSADA, M. & ABE, K. 1981. Mechanism and tectonic implications of the great Banda Sea earthquake of November 4 1963. *Physics of the Earth and Planetary Interiors*, **25**, 129–139.

OTA, T., TERABAYASHI, M. & KATAYAMA, I. 2004. Thermobaric structure and metamorphic evolution of the Iratsu eclogite body in the Sanbagawa belt, central Shikoku, Japan. *Lithos*, **73**, 95–126.

PARK, J.-P., TSURU, T. ET AL. 2002. A deep strong reflector of the Nankai accretionary wedge from multichannel seismic data: implications for underplating and interseismic shear stress release. *Journal of Geophysical Research*, **107**, 3–16.

PARKINSON, C. D. & KOHN, M. J. 2002. A first record of eclogite from Nepal and consequences for the tectonic evolution of the Greater Himalayan Sequence. *Eos Transactions, American Geophysical Union*, **82**, F1302.

PARKINSON, C. D., MIYAZAKI, K., WAKITA, K., BARBER, A. J. & CARSWELL, D. A. 2004. An overview and tectonic synthesis of the pre-Tertiary very-high-pressure metamorphic and associated rocks of Java, Sulawesi and Kalimantan, Indonesia. *Island Arc*, **7**, 184–200.

PEACOCK, S. M. 1996. Thermal and petrologic structure of subduction zones. In: BEBOUT, G. E., SCHOLL, D. W., KIRBY, S. H. & PLATT, J. P. (eds) *Subduction: Top to Bottom*. American Geophysical Union, Geophysical Monograph, **96**, 19–133.

PEACOCK, S. M., RUSHMER, T. & THOMPSON, A. B. 1994. Partial melting of subducting oceanic crust. *Earth and Planetary Science Letters*, **121**, 227–244.

PEARSON, D. G., SYNDER, G. A., SHIREY, S. B., TAYLOR, L. A., CARLSON, R. W. & SOBOLEV, N. V. 1995. Archaean Re–Os age for Siberian eclogites and constraints on Archaean tectonics. *Nature*, **374**, 711–713.

POLI, S. & SCHMIDT, M. W. 2002. Petrology of subducted slabs. *Annual Review of Earth and Planetary Sciences*, **30**, 207–235.

QI, Q. U., TAYLOR, L. A. & ZHOU, X. 1995. Petrology and geochemistry of mantle peridotite xenoliths from SE China. *Journal of Petrology*, **36**, 55–79.

REPLUMAZ, A., KARASON, H., VAN DER HIRST, R. D., BESSE, J. & TAPPONNIER, P. 2004. 4-D evolution of SE Asia's mantle from geological reconstructions and seismic tomography. *Earth and Planetary Science Letters*, **221**, 103–115.

RINO, S., KOMIYA, T., WINDLEY, B. F., KATAYAMA, S., MOTOKI, A. & HIRATA, T. 2004. Major episodic increases of continental crust growth determined from zircon ages of river sands; implication for mantle overturns in the Early Precambrian. *Physics of the Earth and Planetary Interiors*, **146**, 369–394.

RINO, S., KON, Y., SATO, W., MARUYAMA, S., SANTOSH, M. & ZHAO, D. 2008. The Grenvillian and Pan-African orogens: World's largest orogenies through geologic time, and their implications on the origin of superplume. *Gondwana Research*, **14**, 51–72.

ROGERS, J. J. W. & SANTOSH, M. 2004. *Continents and Supercontinents*. Oxford University Press, New York.

ROLLINSON, H. 1997. Eclogite xenoliths in West African kimberlites as residues from Archaean granitoid crust formation. *Nature*, **389**, 173–176.

RUBATTO, D. & HERMANN, J. 2001. Exhumation as fast as subduction? *Geology*, **29**, 21–36.

SACHAN, H. K., MUKHERJEE, B. K., OGASAWARA, Y., MARUYAMA, S., ISHIDA, H. & YOSHIOKA, N. 2004. Discovery of coesite from Indus Suture Zone (ISZ), Ladakh, India. *European Journal of Mineralogy*, **16**, 235–245.

SANTOSH, M. & OMORI, S. 2008a. CO_2 flushing: a plate tectonic perspective. *Gondwana Research*, **13**, 86–102.

SANTOSH, M. & OMORI, S. 2008b. CO_2 windows from mantle to atmosphere: Models on ultrahigh-temperature metamorphism and speculations on the link with melting of snowball Earth. *Gondwana Research*, **14**, doi:10.1016/j.gr.2007.11.001.

SANTOSH, M., TSUNOGAE, T., LI, J. H. & LIU, S. J. 2007a. Discovery of sapphirine-bearing Mg–Al granulites in the North China Craton: implications for Paleoproterozoic ultrahigh-temperature metamorphism. *Gondwana Research*, **11**, 263–285.

SANTOSH, M., WILDE, S. A. & LI, J. H. 2007b. Timing of Paleoproterozoic ultrahigh-temperature metamorphism in the North China Craton: evidence from SHRIMP U–Pb zircon geochronology. *Precambrian Research*, **159**, 178–196.

SANTOSH, M., TSUNOGAE, T., OHYAMA, H., SATO, K., LI, J. H. & LIU, S. J. 2008. Carbonic metamorphism at ultrahigh-temperatures: evidence from North China Craton. *Earth and Planetary Science Letters*, **266**, 149–165.

Santosh, M., Maruyama, S. & Yamamoto, S. 2009a. The making and breaking of supercontinents: some speculations based on superplumes, superdownwelling and the role of tectosphere. *Gondwana Research*, **15**, 324–341.

Santosh, M., Maruyama, S. & Omori, S. 2009b. A fluid factory in the Solid Earth. *Lithosphere*, **1**, 29–33.

Santosh, M., Maruyama, S. & Sato, K. 2009. Anatomy of a Cambrian suture in Gondwana: Pacific type orogeny in southern India? *Gondwana Research*, **16**, 321–341, doi:10.1016/j.gr.2008.12.012.

Schulze, D. J., Valley, J. W. & Spicuzza, M. J. 2000. Coesite eclogites from the Roberts Victor kimberlite, South Africa. *Lithos*, **54**, 23–32.

Schmidt, M. W. & Poli, S. 1998. Experimentally based water budgets for dehydrating slabs and consequences for arc magma generation. *Earth and Planetary Science Letters*, **163**, 361–379.

Scholl, D. W., Von Huene, R., Vallier, T. L. & Howell, D. G. 1980. Sedimentary masses and concepts about tectonic processes at underthrust ocean margins. *Geology*, **8**, 564–568.

Searle, M. P., Windley, B. F. et al. 1987. The closing of Tethys and the tectonics of the Himalaya. *Geological Society of America Bulletin*, **98**, 678–701.

Searle, R. C., Cannat, M., Fujioka, K., Mevel, C., Fujimoto, H., Bralee, A. & Parson, L. 2003. FUJI Dome: a large detachment fault near 64°E on the very slow-spreading southwest Indian Ridge. *Geochemistry, Geophysics, Geosystems*, **4**, doi:10.1029/2003GC000519.

Sen, G., Frey, F. A., Shimizu, N. & Leeman, W. P. 1993. Evolution of the lithosphere beneath Oahu, Hawaii: rare earth element abundances in mantle xenoliths. *Earth and Planetary Science Letters*, **119**, 53–69.

Seno, T. & Maruyama, S. 1984. Paleogoegraphic reconstruction and origin of the Philippine Sea. *Tectonophysics*, **102**, 53–84.

Seno, T., Zhao, D., Kobayashi, Y. & Nakamura, M. 2001. Dehydration of serpentinized slab mantle: seismic evidence from southwest Japan. *Earth Planets Space*, **53**, 861–871.

Sisson, V. B. & Pavlis, T. L. 1993. Geologic consequences of plate reorganization: An example from the Eocene southern Alaska fore arc. *Geology*, **21**, 913–916.

Sleep, N. H. 1992. Hot spot volcanism and mantle plumes. *Annual Review of Earth and Planetary Sciences*, **20**, 19–43.

Sleep, N. H. & Windley, B. F. 1982. Archaean plate tectonics: Constraints and inferences. *Journal of Geology*, **90**, 363–379.

Sol, S., Meltzer, A. et al. 2007. Geodynamics of the southeastern Tibetan Plateau from seismic anisotropy and geodesy. *Geology*, **35**, 563–566.

Sovolev, V. N., Taylor, L. A. & Snyder, G. A. 1994. Diamondiferous eclogites from the Udachnaya kimberlite pipe, Yakutia. *International Geology Review*, **36**, 42–64.

Stern, R. J. 2005. Evidence from ophiolites, blueschists, and ultrahigh-pressure metamorphic terranes that the modern episode of subduction tectonics began in Neoproterozoic time. *Geology*, **33**, 557–560.

Suess, E. 1875. *The Origin of the Alps*. Braumüller, Vienna [in German].

Taira, A., Saito, S. et al. 1998. Nature and growth rate of the Northern Izu–Bonin (Ogasawara) arc crust and their implications for the continental formation. *Island Arc*, **7**, 395–407.

Takahashi, E. 1986. Melting of a dry peridotite KLB-1 up to 14 GPa: implications on the origin of peridotitic upper mantle. *Journal of Geophysical Research*, **11**, 148–165.

Tatsumi, Y. 1989. Migration of fluid phases and genesis of basalt magmas in subduction zones. *Journal of Geophysical Research*, **94**, 4697–4707.

Tatsumi, Y. & Eggins, S. 1995. *Subduction Zone Magmatism*. Blackwell, Oxford.

Tsujimori, T., Liou, J. G. & Coleman, R. G. 2005. Coexisting retrograde jadeite and omphacite in a jadeite-bearing lawsonite eclogite from the Motagua Fault Zone, Guatemala. *American Mineralogist*, **90**, 836–842.

Tuckhole, B. E., Lin, J. & Kleinrock, M. C. 1998. Megamullions and mullion structure defining oceanic metamorphic core complexes on the Mid-Atlantic Ridge. *Journal of Geophysical Research*, **103**, 9857–9866.

Turcotte, D. L. & Schubert, G. 1982. *Geodynamics: Applications of Continuum Physics to Geological Problems*. Wiley, New York.

Umino, S., Geshi, N., Kumagai, H. & Kishimoto, K. 2008. Do off-ridge volcanoes on the East Pacific Rise originate from the Moho transition zone? *Journal of Geography*, **117**, 190–219 [in Japanese].

Usui, T., Nakamura, E., Kobayashi, N., Maruyama, S. & Helmstaedt, H. 2003. Fate of the subducted Farallon plate inferred from eclogite xenoliths in the Colorado Plateau. *Geology*, **31**, 589–592.

Valley, J. W., Kinny, P. D., Schulze, D. J. & Spicuzza, M. J. 1998. Zircon megacrysts from kimberlite: oxygen isotope variability among mantle melts. *Contributions to Mineralogy and Petrology*, **133**, 1–11.

Valley, J. W., Bohlen, S. R., Essene, E. J. & Lamb, W. 2003. Metamorphism in the Adirondacks: II. The role of fluids. *Journal of Petrology*, **31**, 555–596.

Vielzeuf, D. & Schmidt, M. W. 2001. Melting relations in hydrous systems revisited: application to metapelites, metagreywackes and metabasalts. *Contributions to Mineralogy and Petrology*, **141**, 251–267.

Von Huene, R. & Lallemand, S. 1990. Tectonic erosion along the Japan and Peru convergent margins. *Geological Society of America Bulletin*, **102**, 704–720.

Von Huene, R. & Ranero, C. R. 2003. Subduction erosion and basal friction along the sediment-starved convergent margin off Antofagasta, Chile. *Journal of Geophysical Research*, **108**, 2079, doi:10.1029/2001JB001569.

Von Huene, R., Ranero, C. R., Weinrebe, W. & Hinz, R. 2000. Quaternary convergent margin tectonics of Costa Rica, segmentation of the Cocos Plate and central American volcanism. *Tectonics*, **19**, 314–334.

Von Huene, R., Ranero, C. R. & Vannucchi, P. 2004. Generic model of subduction erosion. *Geology*, **32**, 913–916.

Windley, B. F. 1983. Metamorphism and tectonics of the Himalaya. *Journal of the Geological Society, London*, **140**, 849–865.

WINDLEY, B. F. 1995. *The Evolving Continents*. 3rd edn. Wiley, Chichester.

WITT-EICKSCHEN, G. & KRAMM, U. 1997. Mantle-upwelling and metasomatism beneath Central Europe: geochemical and isotopic constraints from mantle xenoliths from the Rhon (Germany). *Journal of Petrology*, **40**, 479–493.

WOLF, M. B. & WILLIE, P. J. 1994. Dehydration-melting of amphibolite at 10 kbar: the effects of temperature and time. *Contributions to Mineralogy and Petrology*, **115**, 369–383.

YAMAMOTO, S., SENSHU, H., RINO, S., OMORI, S. & MARUYAMA, S. 2009. Granite subduction: arc subduction, tectonic erosion and sediment subduction. *Gondwana Research*, **15**, 443–453.

ZHANG, H., THURBER, C. H., SHELLY, D., IDE, S., BEROZA, G. C. & HASEGAWA, A. 2004. High-resolution subducting-slab structure beneath northern Honshu, Japan, revealed by double-difference tomography. *Geology*, **32**, 361–364.

ZHANG, R. Y., LIOU, J. G. & ERNST, W. G. 2009. The Dabie–Sulu continental collision zone: a comprehensive review. *Gondwana Research*, **16**, 1–26.

ZHAO, D. 2004. Global tomographic images of mantle plumes and subducting slabs. *Physics of the Earth and Planetary Interiors*, **146**, 3–34.

ZHAO, D. & OHTANI, E. 2009. Deep slab subduction and dehydration and their geodynamic consequences: evidence from seismology and mineral physics. *Gondwana Research*, **16**, 401–413, doi:10.1016/j.gr.2009.01.005.

ZHAO, D., MARUYAMA, S. & OMORI, S. 2007. Mantle dynamics of Western Pacific and East Asia: insight from seismic tomography and mineral physics. *Gondwana Research*, **11**, 120–131.

The Khanka Block, NE China, and its significance for the evolution of the Central Asian Orogenic Belt and continental accretion

SIMON A. WILDE[1]*, FU-YUAN WU[2] & GUOCHUN ZHAO[3]

[1]*The Institute for Geoscience Research, Department of Applied Geology, Curtin University of Technology, PO Box U1987, Perth, WA 6845, Australia*

[2]*Institute of Geology and Geophysics, Chinese Academy of Sciences, Qijiahuozi, PO Box 9825, Beijing 100029, China*

[3]*Department of Earth Sciences, University of Hong Kong, Pokfulam Road, Hong Kong, China*

*Corresponding author (e-mail: s.wilde@curtin.edu.au)

Abstract: Sensitive high-resolution ion microprobe and inductively coupled plasma mass spectrometry U–Pb dating of zircons from granitoids and paragneiss in the Chinese segment of the Khanka Block reveals that granite magmatism occurred at 518 ± 7 Ma and was followed shortly after by high-grade metamorphism at *c*. 500 Ma (timing ranging from 491 ± 4 Ma in medium-grained granitoid, through 499 ± 10 Ma in porphyritic granite, to 501 ± 8 Ma in paragneiss). Such a scenario has previously been established on similar lithologies in the Jiamusi Block to the west, with identical ages. This suggests that the Khanka and Jiamusi blocks form part of a single terrane and that the Dunhua–Mishan Fault, which was previously considered to separate two unique terranes, cannot be a terrane boundary fault. Previous suggestions of a link between the Khanka Block and the Hida Block in Japan are not supported following a comparison of the new zircon data with published ages for the Japanese terranes. A granitoid with an age of 112 ± 1 Ma in the Khanka Block probably records the effect of Pacific plate subduction, as such ages are common further south in the extreme eastern part of the North China Craton, where they have been related to post-collisional extension and lithospheric thinning in the Jiaodong Peninsula. The presence of such young granitoids, and the previous dating of blueschist-facies metamorphism as late Early Jurassic in the Heilongjiang Complex of the Jiamusi Block, supports the view that the current location of the Jiamusi–Khanka terrane is a product of circum-Pacific accretion rather than it being a microcontinental block that was trapped by the northward collision of the North China Craton with Siberia as part of the assembly of the main Central Asian Orogenic Belt.

The Central Asian Orogenic Belt (CAOB) is a major zone of largely juvenile crust that extends from the Urals and Kazakhstan in the west to the Sea of Japan in the east, separating the Siberian Craton to the north from the Chinese Tarim and North China cratons to the south (Fig. 1 inset). It is generally considered to have formed between 500 and 100 Ma by accretion of arc complexes, with the emplacement of large volumes of granitic magma in the Palaeozoic to Mesozoic (Şengör *et al.* 1993; Chen *et al.* 2000; Jahn *et al.* 2000*a*). The area has also been referred to as the Altaid Tectonic Collage by Şengör *et al.* (1993). The identification of extensive areas of juvenile crust in the CAOB is especially significant (Şengör *et al.* 1993; Wu *et al.* 2000; Jahn *et al.* 2000*b*), as it has challenged conventional views regarding the timing of formation of continental crust.

In the southeastern part of the CAOB, Şengör and Natal'in (1996) separated the 'Manchurides', which lie immediately north of the North China Craton (NCC), from the Altaid accreted terranes that extend further north to the Siberian Craton. The junction between the 'Manchurides' and Altaids is represented by the Solonker Suture Zone (Fig. 1) and has been considered to mark the boundary between the Chinese and Siberian blocks by many workers (Tang 1990; Şengör *et al.* 1993), although the timing of collision has long been debated. Recent studies have favoured final closure of the intervening Palaeo-Asian Ocean in the Late Palaeozoic to Early Jurassic (Xiao *et al.* 2003; Wu *et al.* 2007*a*).

Geological overview

Within the CAOB, several microcontinental blocks have been recognized, although it is unclear if all of these truly predate accretion of the belt (Wu *et al.* 2007*b*). In the southeastern CAOB, up to five

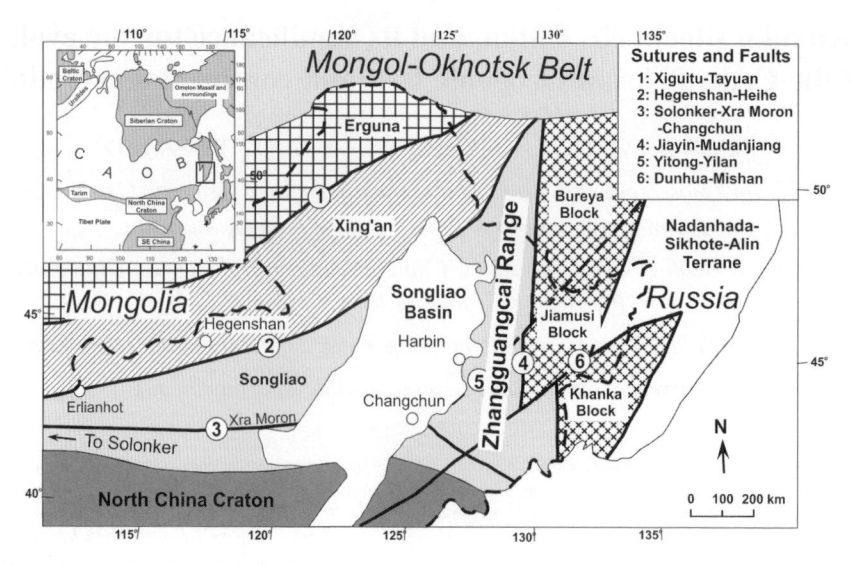

Fig. 1. Geological setting of the Khanka Block with respect to adjacent terranes. The bold dashed line marks international borders. Inset shows the regional setting of the Central Asian Orogenic Belt (CAOB), with the rectangle outlining the area of the Khanka Block.

possible blocks have been identified which, from NW to SE, are named the Erguna, Xing'an, Songliao, Jiamusi and Khanka blocks (Fig. 1); their main features are briefly summarized below.

The Erguna Block is considered to be the eastern extension of the Central Mongolian microcontinent (Wu *et al.* 2007*a*), although its geological evolution is largely unknown, as the area is heavily forested and has been little studied. The block is believed to be characterized by Proterozoic to Palaeozoic volcano-sedimentary strata and granites (HBGMR 1993). Recent zircon U–Pb isotopic dating indicates that many of the gneisses are deformed Early Palaeozoic (506–547 Ma) intrusions (Miao *et al.* 2007), and most granites in the area were emplaced during the early Jurassic, with a few formed in the early Palaeozoic (Ge *et al.* 2005*a*, 2007*a, b*; Sui *et al.* 2007). Limited Nd isotopic data obtained from the granites indicate a crustal formation age of 1680–1060 Ma (HBGMR 1993), which is much older than the Xing'an and Songliao Blocks (Wu *et al.* 2003). This conclusion is also supported by recent zircon Hf isotopic studies (Ge *et al.* 2007*a*; Sui *et al.* 2007).

The Xing'an Block consists of voluminous Palaeozoic to Mesozoic granites and related volcanic rocks, possibly overlying Precambrian basement. Palaeozoic granitoids and sedimentary rocks are locally present, but are largely covered by the Mesozoic volcanic rocks (Yan *et al.* 1989; Zhang & Tang 1989; Wu *et al.* 2002, 2003; Ge *et al.* 2005*b*, 2007*b*; Wang *et al.* 2006; Zhang *et al.* 2008). Some amphibolite- to greenschist-facies metamorphic rocks have also been found locally, but their age and extent is largely unknown, although some are Mesozoic orthogneisses (Miao *et al.* 2003).

The Songliao Block consists of a 'basement' of accreted arc complexes (Wu *et al.* 2000; Gao *et al.* 2007; Pei *et al.* 2007), intruded in the east by Mesozoic granites of the Zhangguangcai Range and overlain by the Mesozoic to Cenozoic Songliao Basin (Wu *et al.* 2000, 2001*a*; Fig. 1). Local amphibolite-facies metamorphic rocks are present, including the Dongfengshan Group in the north and the Hulan Group in the south (JBGMR 1988; HBGMR 1993; Wu *et al.* 2007*a*), although Phanerozoic granites are dominant.

The Jiamusi Block extends northward to the Russian border (Fig. 1), beyond which it is referred to as the Bureya Block. In China, it consists of three components: the Mashan Complex, Heilongjiang Complex and various granitoids (Fig. 2). The oldest component is the Mashan Complex, which is a c. 500 Ma granulite-facies metamorphic terrane, considered to be a Late Pan-African terrane possibly derived from Gondwana (Wilde *et al.* 1997, 2000). The Heilongjiang Complex occurs only in the western part of the Jiamusi Block, where it is structurally interleaved with the Mashan Complex. It is also a metamorphic complex, but at blueschist-facies (Cao *et al.* 1992; Wu *et al.* 2007*b*). It has previously been considered to define a Palaeozoic suture between the Jiamusi and Songliao Blocks (Cao *et al.* 1992); however, recent work indicates

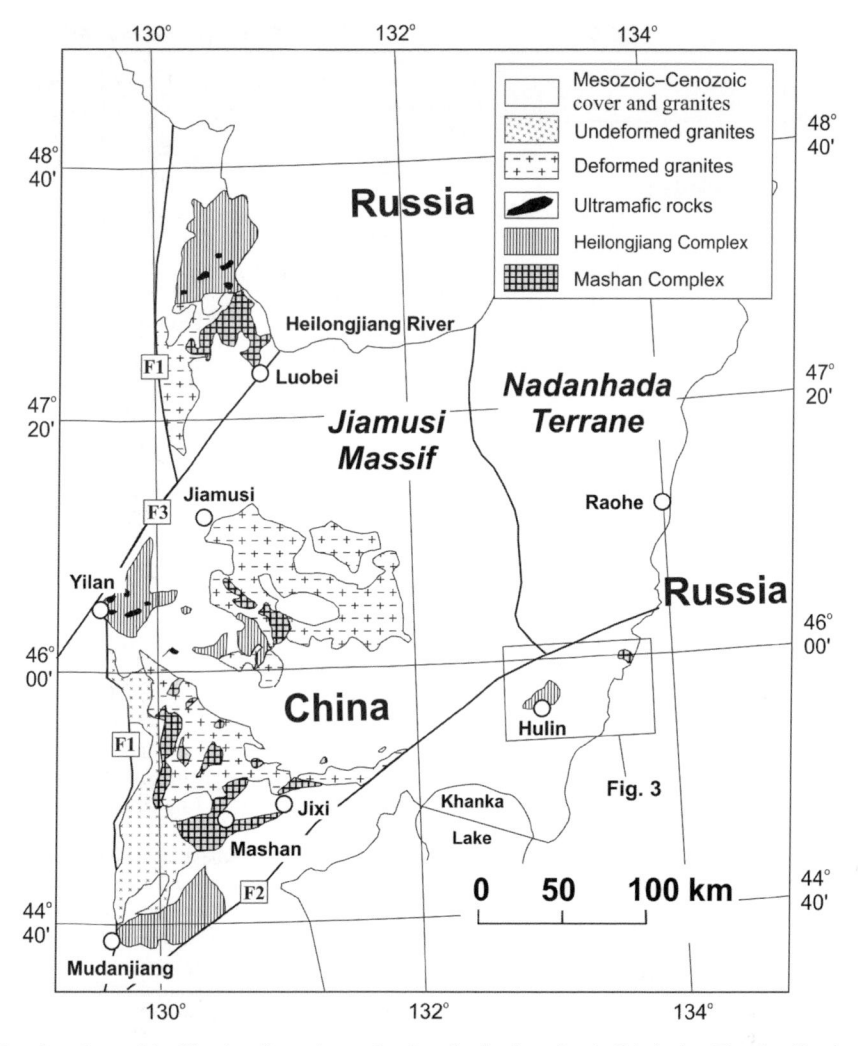

Fig. 2. Local geology of the Khanka–Jiamusi area showing distribution of main lithologies. The classification of the Khanka Block rocks at Hulin and to the NE is based on regional mapping by the Heilongjiang Bureau of Geology and Mineral Resources (HBGMR 1993).

that suturing took place in the Mesozoic at c. 185 Ma and may have been related to Pacific Ocean subduction (Wu et al. 2007b). The granitoids in the Jiamusi Block are of two types: deformed Late Pan-African granitoids associated with the Mashan Complex (Wilde et al. 2003) and pervasive undeformed Permian granites that intrude the Mashan Complex and make up much of the Jiamusi Block (Wilde et al. 2000; Wu et al. 2001b). In addition, Permian (268–288 Ma) volcanic rocks have also been identified in the eastern part of the massif (Meng et al. 2008).

The Khanka Block is the most southerly terrane in the CAOB (Shao & Tang 1995) and is largely contained in Far East Russia, with only a small segment cropping out in NE China (Fig. 1). According to Jia et al. (2004), the block is composed of Precambrian metamorphosed basement rocks overlain by carbonates of Cambrian age. In addition, Ordovician and Silurian volcanic rocks and clastic sediments and Siluro-Devonian granites are exposed along the eastern margin of the block, possibly representing a magmatic arc. One view is that the Khanka Block is equivalent to the Southern Kitakami magmatic arc and Hida Block in NW Japan (Natal'in 1993; Şengör & Natal'in 1996). Although the rocks are poorly described from China, field mapping has suggested a relationship to the

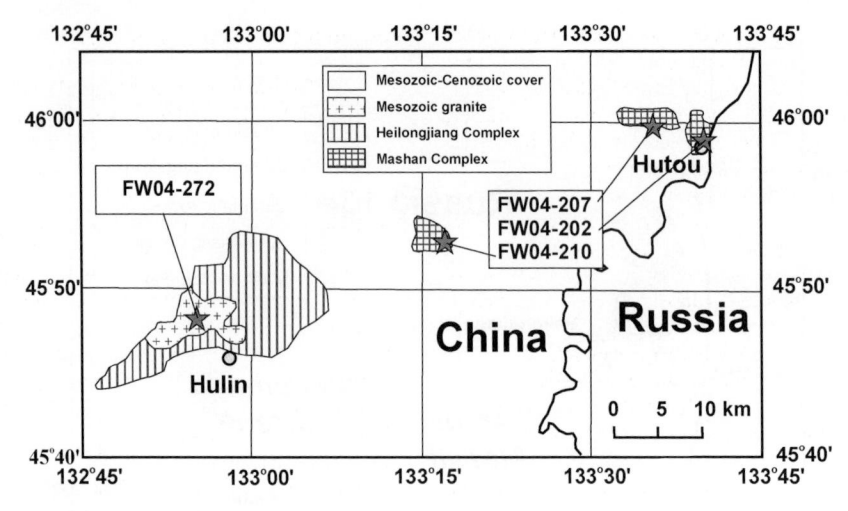

Fig. 3. Geology of the Chinese segment of the Khanka Block, showing sample locations. Classification of rocks as in Figure 2.

Jiamusi Block, with high-grade gneisses equated with the Mashan Complex and lower-grade assemblages considered to be equivalent to the Heilongjiang Complex (HBGMR 1993; Fig. 3). Mesozoic granites are also described as cutting so-called Heilongjiang 'group' gneisses (HBGMR 1993) in the Khanka Block, similar to the relationship in the Jiamusi Block. The Dunhua–Mishan Fault (Fig. 1) is commonly taken to mark the boundary between these two blocks, although Jia et al. (2004) have argued against this, considering that an $^{40}Ar-^{39}Ar$ crossite age of 154.7 ± 0.7 Ma shows that the fault is too young, believing instead that the two blocks collided at 283.5 ± 0.4 Ma, based on $^{40}Ar-^{39}Ar$ dating of hornblende from the Bamiantong belt, which they considered to mark the suture. With respect to the NCC, Jia et al. (2004) considered that the contact with the Khanka Block is marked by the Yanbian Fold Belt. The Khanka Block segment of this fold belt consists of Late Proterozoic metamorphic basement rocks associated with interleaved Permian volcanic rocks and littoral to fluvial sediments (Jia et al. 2004), taken to represent an active continental margin (Yin & Nie 1993; Shao & Tang 1995). However, Zhang et al. (2004, 2008) proposed that the collision between the Khanka Block and the NCC took place during the Jurassic, based on the extensive development of granites at this time.

To clarify some of the issues raised above with respect to the presence of microcontinental blocks, their amalgamation and their role in the development of the CAOB, we undertook a geochronological study of the key rock units exposed in the Chinese segment of the Khanka Block.

Characteristics of the Khanka Block

The Chinese part of the Khanka Block is just a small portion of the terrane and is rather poorly exposed (Fig. 3). The Russian segment to the east and north appears to be better documented, but few papers are available in English. Kojima (1989) outlined two additional terranes in China and Far East Russia, adjacent to the Khanka Block, that appear relevant to the discussion: the Nadanhada terrane to the north (in China) and the Western Sikhote-Alin terrane to the east (in Russia) (Figs 1 & 2). However, the lithologies and faunal sequences are essentially similar in both and they are commonly regarded as a single entity (Ishiwitari & Tsujimori 2003). Sedimentation within the Nadanhada–Western Sikhote-Alin terrane commenced in the Carboniferous and there is no direct evidence that older rocks are present, although the presence of older basement is suggested in the literature. One of the main features of the terrane is the presence of blocks or olistoliths of Carboniferous to Middle Jurassic marine sediments within a post-Middle Jurassic clastic matrix (Kojima 1989), typical characteristics of Pacific accretionary complexes in the region (Kojima 1989; Ishiwitari & Tsujimori 2003). Accretion occurred in the Late Jurassic to Early Cretaceous, as these complexes are overlain by Early Cretaceous volcanic rocks (Kojima 1989). The composite Nadanhada–Sikhote-Alin terrane is considered to be faulted against the Khanka Block.

Khanchuk et al. (1996) referred to two Precambrian terranes within the Russian part of the Khanka Block, separated by a younger east–west accretionary terrane. In the north, the Matveevka terrane is considered to be either Archaean or Proterozoic in age and composed of diopside-, forsterite-, dolomite- and graphite-bearing marble, biotite–sillimanite and garnet–biotite–cordierite gneiss, hypersthene–magnetite and fayalite quartzite, with some biotite–amphibole schist and amphibolite. The Nakhimovka terrane in the south of the Khanka Block is believed to be of similar age and is composed of quartzo-feldspathic biotite-gneiss and biotite–amphibole gneiss, with lenses of marble, and biotite–, diopside–, and muscovite–graphite schist. It should be noted that there are no reliable published ages to confirm that these terranes are indeed Precambrian in age. The intervening Kabarga terrane is considered to be Neoproterozoic to Palaeozoic in age and consists of mica schist, quartzite, chert, banded iron formation, shale and carbonate rocks. There is also a serpentine-rich mélange in the Spassk zone near Khanka Lake (see below).

Kojima (1989) reported the occurrence of Cambrian sediments, including limestone, overlying Precambrian basement gneisses in the Russian part of the Khanka Block. The age of the gneisses is based on early radiometric data (method not stated) in the range of 830–400 Ma (Mishkina & Lelikov 1982; quoted by Kojima 1989) that was interpreted to date an early metamorphic event. However, the presence of Cambrian sediments appears significant, as they would predate the early Palaeozoic igneous and metamorphic events determined in this study. Unfortunately, there is insufficient information available to resolve this issue.

More recent work in the Russian Primorye has identified ophiolites in both the Khanka Block and the Sikhote-Alin terrane to the east (Ishiwitari & Tsujimori 2003). In the former, the Khanka Ophiolite lies within the Spassk zone that extends ESE from the southern tip of Khanka Lake. The ophiolite consists of serpentinized harzburgite, pyroxenite, gabbro and basalt, emplaced over Early Cambrian limestone and shale and overlain by Middle Cambrian conglomerate (Ishiwitari & Tsujimori 2003). There are also major ophiolitic complexes further east in the Sikhote-Alin terrane and these are aligned NNE–SSW, parallel to the eastern margin of the Khanka Block. Available geochronology on one of these, the Sergeevka metagabbro body, includes multi-grain U–Pb zircon ages of 528 ± 3 Ma for gneissose metagabbro, 504 ± 3 Ma for gneissose diorite, and 493 ± 12 Ma for granite (Khanchuk et al. 1996). Interestingly, although these rocks were grouped with the ophiolites by Ishiwitari & Tsujimori (2003), Khanchuk et al. (1996) considered that they represent the margin of the Khanka Block. This latter interpretation is more in harmony with the results presented in this paper, as the ages are virtually identical.

Sample locations and descriptions

The Khanka Block is named after Khanka Lake, which straddles the border between China and Far East Russia (Fig. 2). As stated above, it is mostly exposed in Russia, with only a small segment extending into eastern Heilongjiang Province. According to the last regional geological survey (HBGMR 1993), the Precambrian rocks in this area can be classified into two groups, and these have been named using the terminology adopted in the Jiamusi Block to the west: the Mashan Group (or Complex) exposed around the town of Hutou and the Heilongjiang Group (or Complex) cropping out in Hulin County (Fig. 3). The 'Mashan Group' rocks consist of sillimanite-, garnet- and cordierite-bearing high-grade metamorphic paragneiss, hypersthene-bearing mafic granulite, marble, graphitic schist and fine-grained granitic gneiss. The 'Heilongjiang Group' mostly consists of strongly foliated schist, with a typical mineral association of garnet + muscovite + biotite + quartz + albite.

For this study, three samples were selected for sensitive high-resolution ion microprobe (SHRIMP) zircon U–Pb dating. Sample FW04-202 was collected on the outskirts of Hutou Town (45°58′20″N, 133°39′55″E) and is a deformed porphyritic granite, characterized by large phenocrysts of K-feldspar. The mineral assemblage is quartz, plagioclase, K-feldspar, biotite and garnet, with accessory apatite and zircon. The deformation fabric is weakly developed, although some gneissic enclaves of fine-grained granitoid are locally present; we sampled away from these xenoliths.

Sample FW04-207 was collected from a small hill to the west of Hutou (45°59′52″N, 133°35′39″E). It is a medium-grained deformed granite that shows similar features to the previous sample, except that K-feldspar phenocrysts are only locally present and the rock also contains xenoliths of marble. The sample is composed of quartz, plagioclase, K-feldspar, biotite and garnet, with accessory apatite and zircon and some secondary epidote.

Sample FW04-210 is a fine-grained gneiss collected c. 35 km WSW of Hutou (45°52′37″N, 133°16′55″E) from an area mapped as Liumao 'Formation' of the Mashan 'Group'. The outcrop is intensely deformed and composed of marble and fine-grained gneiss, cut by later granitic veins. Sample FW04-210 has a mineral assemblage of clinopyroxene + hornblende + quartz + K-feldspar, with accessory apatite, titanite and zircon, and was considered in the field to be of sedimentary origin. The conversion of clinopyroxene to hornblende is common in rocks from this outcrop and has been interpreted to result from later metamorphism

(HBGMR 1993), or at least some subsequent thermal disturbance.

In addition, sample FW04-272 was collected from a Mesozoic granitoid that intrudes the so-called 'Heilongjiang Group', c. 5 km NNW of Hulin (45°48′00″N, 132°55′16″E; Fig. 3) to place a lower limit on the formation age of the metamorphic rocks. The rock is an undeformed granodiorite composed of quartz, plagioclase, K-feldspar and biotite, with accessory apatite and zircon.

Analytical techniques

Zircon crystals were extracted by lightly crushing the samples in a Tema ring mill and then concentrated by the use of heavy liquids. For all samples except FW04-272, single crystals were hand picked and mounted, along with several pieces of the Curtin University standard (CZ3; with a conventionally measured ^{206}Pb/^{238}U age of 564 Ma), onto double-sided adhesive tape and enclosed in epoxy resin discs. The discs were polished, so as to effectively section the zircons in half, and then gold coated. Th–U–Pb analyses were performed using a SHRIMP II ion microprobe at Curtin University, following standard operating procedures described by Nelson (1997) and Williams (1998). Spot size ranged between 20 and 30 μm and each analysis spot was rastered over 100 μm for 2 min to remove any common Pb on the surface or contamination from the gold coating. The mass resolution used to measure Pb/Pb and Pb/U isotopic ratios was c. 5000 (1%) during each analytical session and the Pb/U ratios were normalized to those measured on the standard zircon (CZ3; ^{206}Pb/^{238}U = 0.0914). The uncertainty associated with the measurement of Pb/U isotopic ratios for the standard, at 1 SD, was <1.5% during each analytical run. The measured ^{204}Pb values in the unknowns were similar to those recorded for the standard zircon and so common lead corrections were made assuming an isotopic composition of Broken Hill lead, as the common lead is considered to be mainly associated with surface contamination in the gold coat (Nelson 1997). All ages have been calculated using the U and Th decay constants recommended by Steiger & Jäger (1977). Data reduction was performed using both the Krill 007 (P. D. Kinny, Curtin University) and Squid/Isoplot (Ludwig 2001a, b) programs, applying the ^{204}Pb correction. Uncertainties on single analyses are based mainly on counting statistics and are quoted at the 1σ level. Errors on pooled analyses are quoted at 2σ or 95% confidence.

Zircons from sample FW04-272 were analysed using the laser-ablation inductively coupled plasma mass spectrometry (LA-ICPMS) technique at the State Key Laboratory of Continental Dynamics, Northwest University in Xi'an, using a 193 nm laser attached to the ICPMS system and zircon 91500 as the standard. The spot diameter was 30 μm. The raw count rates for ^{29}Si, ^{204}Pb, ^{206}Pb, ^{207}Pb, ^{208}Pb, ^{232}Th and ^{238}U were collected for age determination and the detailed experimental procedures have been described by Yuan et al. (2004). U, Th and Pb concentrations were calibrated using ^{29}Si as an internal standard and NIST 610 as an external standard. ^{207}Pb/^{206}Pb and ^{206}Pb/^{238}U ratios were calculated using the GLITTER 4.0 program (van Achterbergh et al. 2001). Following Ballard et al. (2002), measured ^{207}Pb/^{206}Pb, ^{206}Pb/^{238}U and ^{208}Pb/^{232}Th ratios in zircon 91500 were averaged over the course of the analytical session and used to calculate correction factors. These correction factors were then applied to correct for both instrumental mass bias and depth-dependent elemental and isotopic fractionation. The common Pb correction used the method described by Andersen (2002), as the signal intensity of ^{204}Pb is much lower than the other Pb isotopes and there is a large isobaric interference from Hg. The age calculations and concordia plots were made using Isoplot (ver. 3.0) (Ludwig 2003).

Results

SHRIMP analyses

The zircons from porphyritic granite sample FW04-202 are pale pink, elongate prismatic crystals with pyramidal to slightly rounded terminations and average length to width ratios of 2.5:1. Many crystals show oscillatory zoning in transmitted polarized light, whereas those with rounded terminations either lack oscillatory zoning or show patchy zoning and record younger ages (see below) from the unzoned portions. Rare inclusions of apatite are present and some grains contain irregular cracks. A total of 20 analyses were made from 20 zircons along with 12 analyses of the CZ3 standard that recorded a 1σ variation in Pb/U isotopic ratios of 1.12% over the analytical session. The data are presented in Table 1 and on a concordia diagram in Figure 4a. There are two grains that record older ^{206}Pb/^{238}U ages of 667 ± 14 Ma and 630 ± 10 Ma, both being elongate zircons with slightly rounded terminations and showing oscillatory zoning. The main zircon population is largely concordant, but with some spread along concordia, the total dataset defining a weighted mean ^{206}Pb/^{238}U age of 507 ± 19 Ma (MSWD = 23.4). However, further processing of the data allows identification of two sub-populations that do not overlap within analytical error. The older population (n = 9) records a weighted mean $^{206}Pb/^{238}U$ age of 518 ± 7 Ma (MSWD = 0.34) and the younger population (n = 7) a weighted mean $^{206}Pb/^{238}U$ age of 499 ± 10 Ma (MSWD = 1.94) (Fig. 4a). There are also two younger analyses, 439 ± 7 and 466 ± 8 Ma, that were omitted from the calculations, both recording high U and Th concentrations (Table 1). Similar ages were recorded from deformed granitoids in the Jiamusi Massif (Wilde et al. 2003) and were interpreted to reflect the effect of high-grade metamorphism on a slightly older igneous population. Importantly, the ages of 518 ± 7 Ma and 499 ± 10 Ma are virtually identical to those recorded from deformed granitoids in the Jiamusi Block.

Medium-grained granite sample FW04-207 contains zircons that are colourless to pale pink, stubby to elongate crystals with length to width ratios ranging from 1:1 to 3:1. Thin cracks parallel to the long axis are common in the more elongate grains. Most grains contain at least one inclusion of apatite or other unidentified minerals. A total of 17 zircons were analysed during the same analytical session as sample FW04-202 above. The data are presented in Table 1 and on a concordia diagram in Figure 4b. All zircon data form a tight cluster on concordia and record a weighted mean $^{206}Pb/^{238}U$ age of 491 ± 4 Ma (MSWD = 0.44) (Fig. 4b). This age is interpreted to record the time of high-grade metamorphism that affected the granite, with no remaining evidence of the time of crystallization of the protolith.

The fine-grained gneiss sample (FW04-210) contains zircons that are a mixture of colourless stubby and pale pink elongate prismatic grains; the latter have length to width ratios ranging up to 2.5:1, but with somewhat rounded terminations. The elongate grains show oscillatory zoning in transmitted polarized light, whereas the stubby grains do not: both contain minute inclusions and some larger elongate apatite crystals. A total of 23 analyses were made on 20 grains, along with 13 analyses of the CZ3 standard that recorded an uncertainty in Pb/U isotopic ratios of 0.91% during the analytical session. The data are presented in Table 1 and shown on a concordia diagram in Figure 4c. There is a broad spread of data along concordia which requires careful interpretation. The preferred interpretation is that the three older concordant to slightly discordant zircons, with ages of 942 ± 19, 771 ± 12 and 609 ± 11 Ma, are detrital grains incorporated in the original sedimentary rock and that survived the effects of subsequent metamorphism. The bulk of the zircons (n = 14) cluster on concordia and give a weighted mean $^{206}Pb/^{238}U$ age of 504 ± 8 Ma (MSWD = 3.02) (Fig. 4c). This is interpreted to record the timing of the high-grade metamorphic event in the paragneiss and is consistent not only with the interpretation given for samples FW04-202 and FW04-207, but also with the timing of granulite-facies metamorphism in the Jiamusi Block to the west. The remaining zircons reveal a spread down concordia (Fig. 4c) that culminates in three analyses that record a weighted mean $^{206}Pb/^{238}U$ age of 258 ± 5 Ma (MSWD = 0.93). The analyses that plot between this group and the c. 505 Ma zircon population are likely to be the result of disturbance of the latter by the younger event. The age of 258 ± 5 Ma for the young population includes analyses from two discrete zircon rims (210-3 and 210-16b) and is identical within error to the age of unfoliated Permian granitoids in the Jiamusi Block to the west (Wilde et al. 2001). Although no such granitoids were observed in close proximity to the site at which sample FW04-210 was collected, the overall degree of exposure is poor (Fig. 3) and such an interpretation appears feasible, especially given the close similarity in all datasets to events previously recognized in the Jiamusi Block.

ICP-MS analyses

Granodiorite sample FW04-272 contains colourless to pale pink elongate zircons with length to width ratios up to 2:1. A total of 23 analyses of 23 zircons were made by LA-ICPMS and the data are presented in Table 2 and on a concordia diagram in Figure 4d. One nearly concordant grain has a $^{206}Pb/^{238}U$ age of 203 ± 1 Ma and is older than the main population of 18 concordant to nearly concordant analyses that record a weighted mean $^{206}Pb/^{238}U$ age of 112 ± 1 Ma (MSWD = 6.3). The former probably represents the age of an inherited xenocryst, whereas the bulk of the zircons record the time of crystallization of the granodiorite. Although ages of c. 110 Ma have not been published from this or immediately adjacent areas, similar ages have recently been obtained from granitic dykes cutting the Heilongjiang Group near Mudanjiang in the Jiamusi Block (J. B. Zhou, pers. comm.) and from granites intruded into the Nadanhada terrane (Cheng et al. 2006). Such ages are also common for Mesozoic volcanic rocks in NE China and farther south at the northern margin of the NCC (e.g. Zhang et al. 2004; Pei et al. 2008).

Discussion

Major geological events recorded in the basement rocks of the Khanka Block

The U–Pb zircon results presented above reveal that the basement rocks of the Chinese part of the Khanka Block formed in the Phanerozoic and not

Table 1. SHRIMP II Th–U–Pb data for samples from the Khanka Block, NE China

Spot	U ppm	Th ppm	Th/U	Pb ppm	$^{204}Pb/^{206}Pb$	f^{206} %	$^{208}Pb*/^{232}Th$	+/−	$^{208}Pb*/^{206}Pb*$	+/−	$^{207}Pb*/^{206}Pb*$	+/−	$^{207}Pb*/^{235}U$
FW04-202													
202-1	1625	222	0.14	123	0.00002	0.03	0.0259	0.0007	0.0441	0.0010	0.0591	0.0005	0.65
202-2	2744	1995	0.73	241	0.00074	1.18	0.0254	0.0004	0.2442	0.0016	0.0573	0.0006	0.60
202-3	1395	375	0.27	109	0.00020	0.32	0.0082	0.0004	0.0267	0.0011	0.0584	0.0006	0.67
202-4a	1498	290	0.19	119	0.00037	0.60	0.0248	0.0007	0.0591	0.0014	0.0575	0.0007	0.64
202-4b	563	233	0.41	49	0.00023	0.37	0.0254	0.0006	0.1257	0.0022	0.0570	0.0010	0.66
202-5	3486	360	0.10	232	0.00015	0.24	0.0207	0.0005	0.0303	0.0006	0.0573	0.0003	0.56
202-6	1204	116	0.10	121	0.00003	0.05	0.0317	0.0009	0.0284	0.0007	0.0636	0.0004	0.94
202-7	1092	162	0.15	87	0.00006	0.09	0.0250	0.0008	0.0444	0.0011	0.0581	0.0006	0.67
202-8	1692	243	0.14	126	0.00038	0.60	0.0244	0.0008	0.0459	0.0014	0.0586	0.0006	0.62
202-9	1795	188	0.10	136	0.00029	0.46	0.0180	0.0009	0.0235	0.0011	0.0567	0.0005	0.63
202-10	1674	107	0.06	135	0.00111	1.77	0.0341	0.0027	0.0270	0.0021	0.0578	0.0010	0.64
202-11	965	455	0.47	85	0.00003	0.05	0.0263	0.0005	0.1452	0.0012	0.0583	0.0006	0.68
202-12	2684	866	0.32	286	0.00000	0.00	0.0475	0.0008	0.1496	0.0005	0.0585	0.0002	0.83
202-13	584	312	0.53	52	0.00008	0.13	0.0261	0.0005	0.1669	0.0019	0.0579	0.0008	0.67
202-14	1920	1432	0.75	182	0.00009	0.14	0.0261	0.0004	0.2276	0.0010	0.0575	0.0004	0.68
202-15	1591	679	0.43	133	0.00063	1.00	0.0226	0.0005	0.1213	0.0017	0.0584	0.0008	0.64
202-16	1448	670	0.46	127	0.00016	0.26	0.0260	0.0005	0.1436	0.0012	0.0578	0.0005	0.67
202-17	382	140	0.37	32	0.00011	0.17	0.0257	0.0008	0.1127	0.0027	0.0569	0.0012	0.66
202-18	788	254	0.32	65	0.00000	0.00	0.0262	0.0005	0.1021	0.0008	0.0575	0.0005	0.66
202-19	896	153	0.17	74	0.00050	0.80	0.0273	0.0013	0.0559	0.0024	0.0582	0.0011	0.67
FW04-207													
207-1	227	141	0.62	19	0.00018	0.29	0.0239	0.0007	0.1885	0.0041	0.0564	0.0017	0.61
207-2	124	152	1.23	12	0.00008	0.12	0.0247	0.0007	0.3760	0.0071	0.0570	0.0027	0.63
207-3	113	108	0.96	10	0.00012	0.19	0.0242	0.0007	0.2935	0.0071	0.0553	0.0029	0.60
207-4	145	180	1.24	14	0.00004	0.07	0.0250	0.0006	0.3928	0.0070	0.0599	0.0027	0.65
207-5	337	257	0.76	29	0.00002	0.04	0.0241	0.0005	0.2368	0.0034	0.0576	0.0014	0.62
207-6	115	98	0.85	11	0.00059	0.95	0.0236	0.0009	0.2557	0.0086	0.0520	0.0036	0.56
207-7	106	100	0.94	10	0.00063	1.01	0.0230	0.0010	0.2695	0.0104	0.0515	0.0043	0.57
207-8	142	192	1.35	14	0.00033	0.52	0.0237	0.0006	0.4037	0.0070	0.0559	0.0026	0.61
207-9	162	115	0.71	15	0.00000	0.00	0.0260	0.0006	0.2221	0.0027	0.0596	0.0010	0.68
207-10	174	131	0.75	16	0.00013	0.21	0.0249	0.0007	0.2304	0.0044	0.0564	0.0018	0.63
207-11	232	130	0.56	19	0.00016	0.25	0.0234	0.0008	0.1689	0.0049	0.0545	0.0021	0.58
207-12	211	134	0.63	19	0.00010	0.16	0.0254	0.0008	0.1992	0.0049	0.0590	0.0021	0.66
207-13	186	211	1.13	18	0.00024	0.38	0.0238	0.0006	0.3409	0.0059	0.0561	0.0023	0.61
207-14	107	112	1.05	10	0.00039	0.63	0.0245	0.0008	0.3184	0.0080	0.0567	0.0032	0.63
207-15	103	111	1.08	10	0.00022	0.36	0.0246	0.0008	0.3384	0.0094	0.0600	0.0038	0.65
207-16	100	87	0.87	9	0.00049	0.79	0.0233	0.0010	0.2545	0.0091	0.0502	0.0037	0.55
207-17	238	182	0.76	21	0.00024	0.39	0.0238	0.0006	0.2259	0.0045	0.0530	0.0018	0.59
FW04-210													
210-1	1676	417	0.25	129	0.00000	0.00	0.0250	0.0003	0.0793	0.0004	0.0569	0.0002	0.62
210-2	566	82	0.14	41	0.00002	0.03	0.0234	0.0007	0.0445	0.0013	0.0575	0.0007	0.60
210-3	2062	268	0.13	78	0.00002	0.03	0.0124	0.0002	0.0400	0.0005	0.0517	0.0003	0.29
210-4	211	157	0.75	19	0.00008	0.12	0.0255	0.0005	0.2310	0.0036	0.0571	0.0014	0.65
210-5	959	472	0.49	83	0.00006	0.10	0.0251	0.0004	0.1492	0.0010	0.0577	0.0004	0.66
210-6	3970	547	0.14	307	0.00006	0.09	0.0253	0.0004	0.0428	0.0003	0.0575	0.0002	0.65
210-7	381	123	0.32	30	0.00004	0.07	0.0253	0.0005	0.1031	0.0017	0.0573	0.0008	0.63
210-8	935	136	0.15	74	0.00002	0.02	0.0212	0.0005	0.0366	0.0007	0.0580	0.0004	0.67
210-9	975	432	0.44	81	0.00003	0.05	0.0251	0.0003	0.1383	0.0009	0.0578	0.0004	0.64
210-10a	1536	327	0.21	127	0.00002	0.03	0.0264	0.0004	0.0661	0.0005	0.0579	0.0003	0.68
210-10b	726	98	0.13	32	0.00003	0.04	0.0168	0.0005	0.0488	0.0015	0.0522	0.0007	0.33
210-10c	543	122	0.22	32	0.00010	0.16	0.0260	0.0007	0.1001	0.0022	0.0550	0.0010	0.44
210-11	1123	389	0.35	93	0.00033	0.54	0.0240	0.0004	0.1019	0.0013	0.0580	0.0006	0.65
210-12	840	109	0.13	32	0.00001	0.02	0.0131	0.0005	0.0417	0.0015	0.0520	0.0007	0.29

+/-	$^{206}Pb^*/^{238}U$	+/-	% Conc	$^{208}Pb^*/^{232}Th$	+/-	$^{207}Pb^*/^{206}Pb^*$	+/-	$^{207}Pb^*/^{235}U$	+/-	$^{206}Pb^*/^{238}U$	+/-
0.01	0.0800	0.0013	87	516	14	573	19	510	8	496	8
0.01	0.0755	0.0012	93	506	9	502	25	475	8	469	7
0.01	0.0829	0.0014	94	165	7	546	21	519	8	513	8
0.01	0.0810	0.0013	98	494	14	511	25	504	8	502	8
0.02	0.0837	0.0014	106	507	12	490	38	513	10	518	8
0.01	0.0704	0.0012	87	413	11	502	13	449	7	439	7
0.02	0.1069	0.0018	90	630	18	730	14	672	9	655	10
0.01	0.0835	0.0014	97	498	15	535	22	520	8	517	8
0.01	0.0764	0.0013	86	487	16	551	24	488	8	474	8
0.01	0.0802	0.0013	104	360	18	479	21	494	8	497	8
0.02	0.0807	0.0013	96	678	53	521	37	504	10	500	8
0.01	0.0853	0.0014	98	524	10	540	21	530	8	527	8
0.01	0.1025	0.0017	115	938	15	549	7	612	8	629	10
0.02	0.0835	0.0014	98	521	11	526	31	518	9	517	8
0.01	0.0855	0.0014	103	520	9	512	16	526	8	529	8
0.01	0.0797	0.0013	91	453	10	543	28	503	9	494	8
0.01	0.0839	0.0014	99	519	9	523	20	520	8	519	8
0.02	0.0837	0.0014	106	513	15	488	48	513	12	518	8
0.01	0.0828	0.0014	101	523	10	509	18	512	8	513	8
0.02	0.0835	0.0014	96	545	25	538	43	521	11	517	8
0.02	0.0786	0.0013	104	478	13	470	67	485	14	488	8
0.03	0.0808	0.0014	102	493	13	490	104	499	21	501	9
0.03	0.0789	0.0014	115	484	15	425	115	479	22	490	8
0.03	0.0790	0.0014	82	499	13	600	98	510	20	490	8
0.02	0.0776	0.0013	93	482	11	516	53	488	12	482	8
0.04	0.0788	0.0014	172	472	18	284	157	454	27	489	8
0.05	0.0800	0.0015	187	459	20	265	192	457	32	496	9
0.03	0.0794	0.0014	110	473	12	448	103	485	20	492	8
0.02	0.0827	0.0014	87	518	11	590	35	527	10	512	8
0.02	0.0811	0.0014	108	497	13	468	71	497	15	503	8
0.03	0.0777	0.0013	123	468	16	392	86	467	16	482	8
0.03	0.0806	0.0014	88	507	15	567	76	512	16	500	8
0.03	0.0792	0.0014	107	476	12	458	91	486	18	492	8
0.04	0.0804	0.0014	104	488	15	480	125	495	24	498	9
0.04	0.0784	0.0014	81	491	16	604	137	508	27	487	9
0.04	0.0798	0.0015	240	465	19	206	168	447	29	495	9
0.02	0.0805	0.0014	152	476	12	329	77	470	15	499	8
0.01	0.0786	0.0009	100	500	6	486	9	487	5	488	6
0.01	0.0759	0.0009	92	468	15	511	25	479	7	472	6
0.00	0.0402	0.0005	94	249	5	270	15	256	3	254	3
0.02	0.0825	0.0011	104	510	10	494	56	508	12	511	6
0.01	0.0828	0.0010	99	501	7	519	17	514	6	513	6
0.01	0.0814	0.0010	98	505	7	513	7	506	5	504	6
0.01	0.0792	0.0010	98	505	10	504	31	493	8	491	6
0.01	0.0842	0.0010	99	425	9	529	15	522	6	521	6
0.01	0.0803	0.0010	96	501	7	521	15	502	6	498	6
0.01	0.0852	0.0010	100	527	7	527	11	527	6	527	6
0.01	0.0464	0.0006	99	337	11	295	32	292	5	292	3
0.01	0.0580	0.0007	88	518	13	412	40	370	7	364	4
0.01	0.0816	0.0010	96	480	8	529	22	510	7	506	6
0.01	0.0405	0.0005	90	262	10	286	32	259	5	256	3

(Continued)

Table 1. *Continued*

Spot	U ppm	Th ppm	Th/U	Pb ppm	$^{204}Pb/^{206}Pb$	f^{206} %	$^{208}Pb*/^{232}Th$	+/−	$^{208}Pb*/^{206}Pb*$	+/−	$^{207}Pb*/^{206}Pb*$	+/−	$^{207}Pb*/^{235}U$
210-13	952	166	0.17	47	0.00000	0.00	0.0240	0.0004	0.0836	0.0007	0.0544	0.0004	0.38
210-14	65	31	0.48	9	0.00001	0.01	0.0395	0.0013	0.1490	0.0044	0.0646	0.0020	1.13
210-15	209	69	0.33	21	0.00002	0.03	0.0311	0.0008	0.1023	0.0024	0.0652	0.0011	0.90
210-16a	310	257	0.83	29	0.00001	0.01	0.0259	0.0004	0.2627	0.0023	0.0573	0.0008	0.65
210-16b	1058	137	0.13	41	0.00027	0.44	0.0132	0.0006	0.0420	0.0018	0.0517	0.0008	0.29
210-17	546	60	0.11	79	0.00000	0.01	0.0473	0.0011	0.0338	0.0007	0.0711	0.0005	1.50
210-18	546	237	0.43	46	0.00002	0.04	0.0253	0.0004	0.1330	0.0014	0.0591	0.0006	0.67
210-19	197	113	0.57	17	0.00003	0.05	0.0249	0.0006	0.1774	0.0035	0.0584	0.0014	0.65
210-20	822	361	0.44	69	0.00001	0.01	0.0269	0.0004	0.1467	0.0011	0.0577	0.0005	0.64

*Corrected using measured ^{204}Pb.
f^{206}% is (common ^{206}Pb/total ^{206}Pb) × 100.
% Conc = % concordance defined as [($^{206}Pb/^{238}U$ age)/($^{207}Pb/^{206}Pb$ age)] × 100.

in the Precambrian as previously considered (Jia et al. 2004). Igneous zircon with distinct oscillatory zoning in porphyritic granite sample FW04-202 from Hutou records a $^{206}Pb/^{238}U$ age of 518 ± 7 Ma. Two igneous xenocrysts with $^{206}Pb/^{238}U$ ages of 667 ± 14 and 630 ± 10 Ma are present and indicate the presence of some late Neoproterozoic material at depth. A metamorphosed sedimentary rock (sample FW04-210), from an outcrop midway between Hutou and Hulin (Fig. 3), contains detrital igneous zircons with $^{206}Pb/^{238}U$ ages of 942 ± 19, 771 ± 12 and 609 ± 11 Ma, indicating

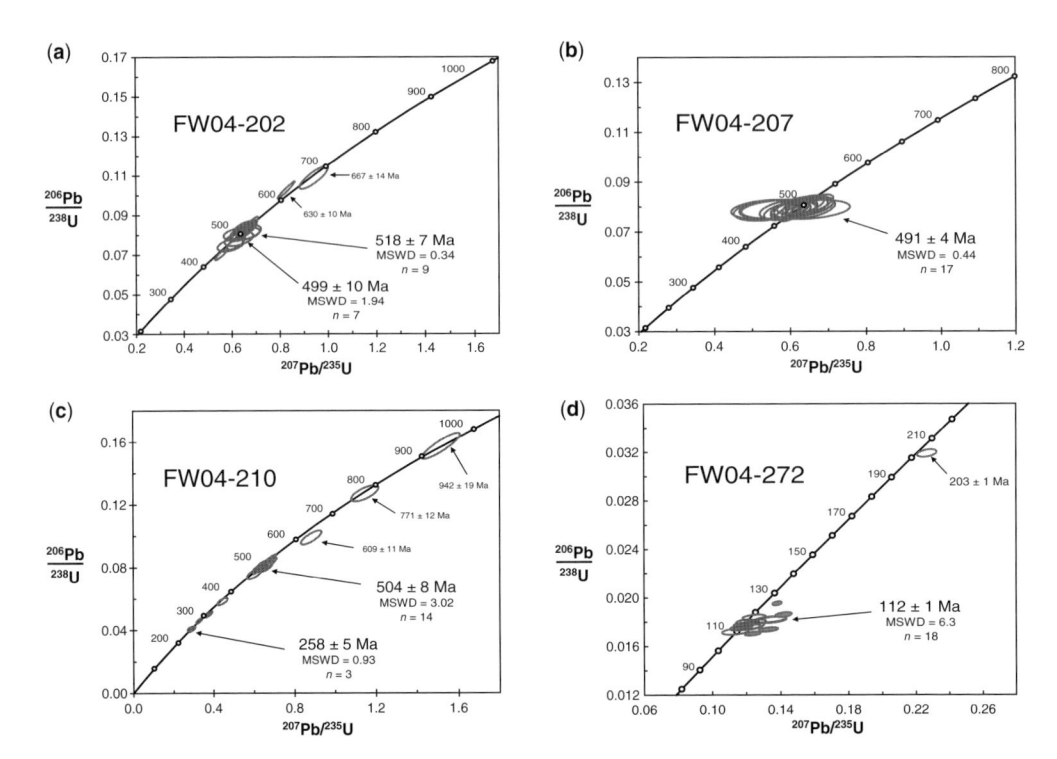

Fig. 4. U–Pb zircon data for rocks from the Khanka Block: (**a**) sample FW04-202; (**b**) sample FW04-207; (**c**) FW04-210; (**d**) FW04-272. (a–c) are SHRIMP II data; (d) is based on ICP-MS data.

+/−	$^{206}Pb^*/^{238}U$	+/−	% Conc	Age								
				$^{208}Pb^*/^{232}Th$	+/−	$^{207}Pb^*/^{206}Pb^*$	+/−	$^{207}Pb^*/^{235}U$	+/−	$^{206}Pb^*/^{238}U$	+/−
0.01	0.0501	0.0006	81	479	7	387	17	324	4	315	4
0.04	0.1271	0.0018	101	783	26	761	64	769	19	771	10
0.02	0.1000	0.0013	79	619	17	780	36	651	11	614	7
0.01	0.0819	0.0010	101	517	8	502	32	506	8	507	6
0.01	0.0408	0.0005	95	266	12	270	37	259	5	258	3
0.02	0.1528	0.0019	96	935	22	960	13	930	9	917	10
0.01	0.0825	0.0010	90	505	8	569	23	522	7	511	6
0.02	0.0802	0.0010	91	498	12	545	54	506	12	497	6
0.01	0.0805	0.0010	96	536	8	517	17	502	6	499	6

that such material was exposed at the surface in the catchment area at the time the sediment was initially deposited.

A medium-grained deformed granitoid from near Hutou (sample FW04-207) contains a single zircon population that records a weighted mean $^{206}Pb/^{238}U$ age of 491 ± 4 Ma. This matches, within error, the age of unzoned zircon domains in porphyritic granite sample FW04-202 with a $^{206}Pb/^{238}U$ age of 499 ± 10 Ma. Furthermore, the main zircon population in paragneiss sample FW04-210 defines a weighted mean $^{206}Pb/^{238}U$ age of 504 ± 8 Ma. All three results overlap within error and are interpreted to record the time of high-grade metamorphism in the area.

Importantly, overgrowth rims and a single zircon grain in paragneiss sample FW04-210 record a younger $^{206}Pb/^{238}U$ age of 258 ± 5 Ma. No known rocks of this age have been reported from the Khanka Block, but it is noted that such Permian ages are common in the adjacent Jiamusi Block to the west, where they record the emplacement age of a suite of undeformed granitoids (Wilde et al. 2000; Wu et al. 2001b).

The youngest rock dated in this study is a granodiorite from Hulin (Fig. 3) that has a weighted mean $^{206}Pb/^{238}U$ zircon age of 112 ± 1 Ma. A single xenocryst with a $^{206}Pb/^{238}U$ age of 203 ± 1 Ma is also present. The latter age has also been recorded from granites at Yanbian in the NCC and from the Nadanhada terrane (Zhang 2002; Cheng et al. 2006).

Relationship of the Khanka Block to Japanese terranes

As mentioned above, one view is that the Khanka Block is equivalent to the Southern Kitakami magmatic arc and Hida Block in NW Japan (Natal'in 1993; Sengör & Natal'in 1996). A major problem in establishing such links is the inability to unequivocally match the geology of Japan with terranes to the west (Ishiwitari & Tsujimori 2003; Oh & Kusky 2009). The Southern Kitakami portion of the Higo terrane has recently been redefined (Tazawa 2004) as an early Ordovician to late Devonian accretionary complex, with younger portions of the Higo terrane now grouped as part of the Hida–Abukuma nappe (Tazawa 2004; Osanai et al. 2006). Although U–Pb ages as old as 2155 Ma have been recorded by SHRIMP dating of zircon cores (Sakashima et al. 2003) and ages between 550 and 450 Ma obtained by CHIME (chemical Th–U–total lead isochron method, by electron microprobe) zircon dating (Suzuki et al. 1998), these are from paragneisses and thus record the age of detrital zircon components. The main metamorphic event in the area occurred at 260–230 Ma (see Osanai et al. 2006) and is thus dissimilar to the c. 500 Ma metamorphism in the Khanka Block.

With regard to the Hida Block and its possible connections to the Asian mainland, there are a range of postulates, including a link with South China (Faure & Charvet 1987), North China (e.g. Kojima 1989; Arakawa & Shinmura 1995), the CAOB (e.g. Arakawa et al. 2000; Jahn et al. 2000c), or with various blocks in Korea (e.g. Hiroi 1981; Tsujimori et al. 2006), including extension, via the Imjingang Belt, across to the Dabie–Sulu zone separating the North and South China cratons (e.g. Oh 2006). It has also been considered a separate microcontinent that collided with the NCC in the late Palaeozoic (Ernst et al. 1988; Maruyama et al. 1989).

The Hida Block has long been regarded as an exotic terrane, forming the nucleus of Japan, and

Table 2. *ICP-MS Th–U–Pb zircon data for sample FW04-272 from the Khanka Block, NE China*

Spot	U ppm	Th ppm	Th/U	$^{208}Pb*/^{232}Th$		$^{207}Pb*/^{206}Pb*$	+/−	$^{207}Pb*/^{235}U$	+/−
272-01	220	157	0.71	0.0054	0.0001	0.0482	0.0015	0.12	0.00
272-02	452	224	0.50	0.0055	0.0001	0.0496	0.0011	0.12	0.00
272-03	512	261	0.51	0.0056	0.0000	0.0543	0.0023	0.13	0.01
272-04	2501	4353	1.74	0.0068	0.0001	0.0663	0.0016	0.18	0.00
272-05	844	522	0.62	0.0057	0.0000	0.0536	0.0024	0.13	0.01
272-06	364	149	0.41	0.0058	0.0000	0.0494	0.0018	0.12	0.00
272-07	250	139	0.56	0.0055	0.0001	0.0506	0.0019	0.12	0.00
272-08	462	301	0.65	0.0062	0.0001	0.0496	0.0012	0.12	0.00
272-09	282	137	0.49	0.0054	0.0001	0.0481	0.0023	0.11	0.01
272-10	348	141	0.40	0.0059	0.0001	0.0484	0.0013	0.12	0.00
272-11	195	128	0.66	0.0050	0.0001	0.0581	0.0032	0.13	0.01
272-12	150	116	0.77	0.0060	0.0001	0.0506	0.0022	0.12	0.01
272-13	714	343	0.48	0.0056	0.0001	0.0490	0.0012	0.12	0.00
272-14	482	157	0.33	0.0056	0.0000	0.0506	0.0018	0.12	0.00
272-15	382	334	0.87	0.0063	0.0001	0.0556	0.0014	0.14	0.00
272-16	566	279	0.49	0.0053	0.0000	0.0530	0.0014	0.12	0.00
272-17	551	112	0.20	0.0072	0.0001	0.0516	0.0010	0.23	0.00
272-18	404	89	0.22	0.0055	0.0001	0.0484	0.0018	0.12	0.00
272-19	814	515	0.63	0.0054	0.0000	0.0559	0.0016	0.13	0.00
272-20	292	128	0.44	0.0055	0.0001	0.0501	0.0010	0.12	0.00
272-21	480	202	0.42	0.0056	0.0000	0.0480	0.0013	0.12	0.00
272-22	581	291	0.50	0.0057	0.0001	0.0494	0.0010	0.12	0.00
272-23	659	302	0.46	0.0061	0.0001	0.0515	0.0008	0.14	0.00

*Data corrected using the method of Andersen (2002) – as required.
% Conc = % concordance defined as $[(^{206}Pb/^{238}U$ age$)/(^{207}Pb/^{206}Pb$ age$)] \times 100$.

consists of three tectonic units: the Hida gneiss region, the Unazuki Belt and the Hida Marginal Belt (Hiroi 1981). The Oki metamorphic belt on Oki-Dongo island to the west has previously been considered a part of the Hida Block (e.g. Suzuki & Adachi 1994), although important differences were found in the monazites and zircons used in CHIME dating (Suzuki & Adachi 1994) and, more recently, Dallmeyer & Takasu (1998) have shown that metamorphic conditions and exhumation styles were different, although they were coeval in the two areas. Thus, only the Hida gneiss should possibly be compared with the Khanka Block.

The Hida gneisses are dominantly quartzo-feldspathic, with units of marble, amphibolite and minor pelite. The presence of andalusite, sillimanite and cordierite in the metapelites indicates low-pressure metamorphic conditions. The rocks have long been considered Precambrian in age, although this is not supported by the limited geochronological data that are available. Studies indicate that metamorphism in the Hida gneisses took place at c. 350 Ma (granulite facies) and at 240–220 Ma (medium-pressure amphibolite facies) and that igneous events occurred at 420–410 Ma, 340–320 Ma and 230–180 Ma (see Arakawa et al. 2000); none of these match activity recorded from the Chinese segment of the Khanka Block in this study. Sr–Nd data from the paragneisses obtained by Tanaka in 1992 (quoted by Arakawa et al. 2000) and Kagami et al. (2006) identified two types of gneisses; those with T_{DM} model ages of 2.2–1.4 Ga and those with 0.74–0.55 Ga ages. This reflects the provenance and does not assist in determining the age of sedimentation. With respect to older ages, these have been obtained only from detrital zircon grains in the metasediments and include SHRIMP U–Pb zircon core ages extending back to c. 3420 and c. 2560 Ma, although the data are discordant. There are also more concordant Proterozoic populations at 1840 Ma, 1130 Ma and 580 Ma, and Phanerozoic populations at 400 Ma, 360 Ma, 285 Ma and 250 Ma (Sano et al. 2000). The Archaean and Palaeoproterozoic ages are in accord with ages obtained from the NCC, but these, and the Palaeozoic zircons, do not match the results obtained in this study of the Khanka Block (Table 3). We therefore conclude that a genetic link between the Hida and Khanka Blocks cannot be substantiated.

$^{206}Pb^*/^{238}U$	+/−	% Conc	Age $^{208}Pb^*/^{232}Th$	+/−	$^{207}Pb^*/^{206}Pb^*$	+/−	$^{207}Pb^*/^{235}U$	+/−	$^{206}Pb^*/^{238}U$	+/−
0.0178	0.0002	103	109	2	111	54	114	3	114	1
0.0175	0.0001	63	111	1	176	35	115	2	112	1
0.0180	0.0002	57	114	1	382	99	128	5	115	1
0.0201	0.0002	15	137	1	816	33	171	3	128	1
0.0181	0.0002	32	114	1	354	105	127	5	115	1
0.0182	0.0002	70	116	1	166	88	119	4	116	1
0.0174	0.0002	50	111	2	222	65	116	4	111	1
0.0178	0.0001	63	125	1	178	37	116	2	114	1
0.0170	0.0002	104	109	1	105	105	109	5	109	1
0.0174	0.0001	94	118	2	118	45	112	3	111	1
0.0161	0.0001	36	101	1	532	123	123	6	103	1
0.0175	0.0002	50	120	2	223	77	117	5	112	1
0.0175	0.0001	75	113	1	148	41	113	2	112	1
0.0178	0.0002	50	113	1	224	85	119	4	114	1
0.0185	0.0001	33	126	1	437	38	134	3	118	1
0.0169	0.0001	47	107	1	330	61	118	3	108	1
0.0319	0.0002	75	145	3	268	27	208	3	203	1
0.0174	0.0002	92	111	1	121	84	111	4	111	1
0.0172	0.0001	32	108	1	448	66	126	3	110	1
0.0178	0.0001	88	111	1	199	31	118	2	114	1
0.0176	0.0001	112	113	1	101	63	112	3	113	1
0.0176	0.0001	68	115	1	165	31	115	2	112	1
0.0194	0.0001	47	122	1	262	19	131	2	124	1

Table 3. *Comparison of available SHRIMP U–Pb zircon ages for the Hida and Khanka blocks*

Hida Block[1]		Khanka Block[2]	
Zircon type	Age	Age	Zircon type
Igneous	3420–2560		
Metamorphic	1840		
Metamorphic	1130		
		950	Detrital (igneous)
		770	Detrital (igneous)
		650	Inherited (igneous)
		600	Inherited (igneous)
Metamorphic	580		
		520	Igneous
		500–490	Metamorphic
Metamorphic	400		
Metamorphic	360		
Metamorphic	285		
Metamorphic	250		
		260	Igneous/Metamorphic
		110	Igneous

[1]Sano *et al.* (2000).
[2]This paper.

Relationship of the Khanka Block to the Jiamusi Block

As there appears to be little in the currently available data to support a link between the Khanka Block and the Japanese terranes, it is pertinent to examine the possible association with the nearest adjacent terrane that has also been considered exotic to the region (Wu *et al.* 2007*b*), the Jiamusi Block (Fig. 1). This block is fault-bounded to the east, SE and west, whereas the northern boundary is marked by the Heilongjiang (or Amur) River (Fig. 2), with a possible northward extension into the Bureya Block on the Russian side of the border (Wilde *et al.* 2000). As outlined above, the Jiamusi Block consists of three components; the Mashan Complex, Heilongjiang Complex, and various deformed and undeformed granitoids (Fig. 2). The Mashan Complex is composed of granitic gneisses tectonically interleaved with khondalitic metasedimentary rocks that include graphite- and sillimanite-bearing schists and gneisses and diopside- and olivine-bearing marbles, which have been metamorphosed to amphibolite or granulite facies. The Heilongjiang Complex consists of granitic gneiss, two-mica schist, muscovite–albite schist, blueschist, greenschist, marble and ultramafic rocks that have undergone greenschist- to blueschist-facies metamorphism, accompanied by intense deformation. As alluded to above, two stages of granitoids have been identified in the Jiamusi Block: metamorphosed and deformed granitoids associated with the Mashan Complex (Wilde *et al.* 2003) and weakly to undeformed granites intruded in the Permian (Wilde *et al.* 1997; Wu *et al.* 2000).

The depositional age of the khondalites in the Mashan Complex has not been established. However, detrital zircons as old as 1700 Ma have been recorded from sillimanite gneiss, whereas *c.* 1900 Ma and *c.* 1100 Ma detrital zircon populations from a metasedimentary enclave in a garnet granite (Wilde *et al.* 2000) indicate that the sedimentary protolith was deposited some time after 1100 Ma. The youngest possible age of sedimentation is constrained by the time at which the rocks underwent granulite-facies metamorphism, which occurred at *c.* 500 Ma (Wilde *et al.* 2000, 2001). An identical metamorphic age was also recorded from orthogneisses in the area, including the formation of anatectic granite (Wilde *et al.* 1997, 2000). Furthermore, geochronological evidence confirms that the deformed granitoids associated with the Mashan Complex were also metamorphosed at this time (Wilde *et al.* 1997, 2003), indicating tectonic juxtaposition of all components of the Mashan Complex prior to the attainment of peak metamorphism at *c.* 500 Ma.

Although the depositional age of the Heilongjiang Complex cannot be determined because of a lack of suitable lithologies, recent work has clarified the time of blueschist-facies metamorphism. Li *et al.* (1999) obtained $^{40}Ar/^{39}Ar$ ages of 175.3 ± 0.9 Ma and 166 ± 1.2 Ma from two muscovite samples from a mica schist, whereas hornblende from amphibolite gave an age of 167.1 ± 1.5 Ma, both samples being obtained from the Mudanjiang area (Fig. 2). Wu *et al.* (2007*b*) dated three samples of mica schist from the Yilan area, which gave phengite $^{40}Ar/^{39}Ar$ ages of 173.6 ± 0.5 Ma, 175.3 ± 0.4 Ma and 174.8 ± 0.5 Ma. Thus blueschist-facies metamorphism in the Heilongjiang Complex took place in the Jurassic, not in the Neoproterozoic–Palaeozoic as previously believed (Zhang 1992).

The emplacement age of deformed granitoids in the Jiamusi Block ranges from 523 ± 8 Ma to 515 ± 4 Ma (Wilde *et al.* 2003). These rocks underwent high-grade metamorphism at *c.* 500 Ma, which is identical to the event recognized in the more intensely deformed orthogneisses intimately associated with rocks of the Mashan Complex (Wilde *et al.* 1997, 2000). The undeformed granites in the Jiamusi Block were mainly emplaced between 258 and 254 Ma, with some slightly earlier plutons emplaced at *c.* 270 Ma (Wilde *et al.* 1997; Wu *et al.* 2001*b*).

It becomes clear from the above review that both the Mashan Complex and deformed granitoids show remarkable similarities in age and rock type to the rocks present in the Chinese part of the Khanka Block. For porphyritic granite sample FW04-202, the magmatic age of 518 ± 7 Ma is identical to that obtained (515 ± 8 Ma) from a similar porphyritic granite (sample 00-SAW-204), collected *c.* 30 km NW of Jiamusi (Wilde *et al.* 2003). Furthermore, metamorphism of the Jiamusi sample was interpreted to have occurred at 497 ± 5 Ma, compared with 499 ± 10 Ma for the Khanka Block sample in the present study. Likewise, the age of 491 ± 7 Ma obtained from the more strongly deformed Khanka Block granitoid sample FW04-207 is identical, within error, to metamorphic ages obtained from Jiamusi Block orthogneisses and deformed granitoids, which range from 507 ± 12 Ma to 498 ± 11 Ma (Wilde *et al.* 1997, 2000, 2003).

With respect to the paragneisses, the clinopyroxene + hornblende + quartz + K-feldspar assemblage of sample FW04-210 from the Khanka Block is mineralogically similar, and has undergone similar metamorphic conditions, to components of the Mashan Complex. It is interpreted to have been metamorphosed at 505 ± 6 Ma, identical to the metamorphic age obtained from Mashan Complex paragneisses in the Jiamusi Massif, which range from 502 ± 10 to 490 ± 15 Ma, defining a major event at c. 500 Ma (Wilde et al. 1997, 2000, 2003; Wu et al. 2001b). Interestingly, one sample of sillimanite gneiss from the Mashan Complex (97-SAW-042) at Xi Mashan contains a range of detrital zircons that have survived the high-grade metamorphism and record concordant to slightly discordant ages of 700, 900, 1050, 1300 and 1600 Ma. Furthermore, a garnet granulite (97-SAW-034) from the complex at Xi Mashan contains detrital zircons with ages of 650, 750, 950, 1050, 1100 and 1300 Ma (Wilde et al. 2000), although some are more discordant than in the sillimanite gneiss. Sample FW04-202 from the Khanka Block contains two inherited zircons with ages of 667 ± 14 Ma and 630 ± 10 Ma and sample FW04-210 contains detrital zircons with ages of $942 \pm 19, 771 \pm 12$ and 609 ± 11 Ma; comparable in age with those in the Jiamusi Block paragneisses (Wilde et al. 2000), although no zircon older than 1000 Ma has been recorded.

The younger intrusive granodiorite (sample FW04-272) from the Khanka Block has an age of 112 ± 1 Ma. Such ages have not been recorded from the Jiamusi Block, but are present in much of NE China (Zhang et al. 2004). As indicated above, this age post-dates blueschist-facies metamorphism in the Heilongjiang Complex and has implications for the overall evolution of the terrane (discussed below). The age of post-tectonic granitoids in the Jiamusi Block is tightly constrained between 270 and 254 Ma (Wu et al. 2001b). Interestingly, concordant zircons of this age (258 ± 5 Ma) are recorded from Khanka gneiss sample FW04-210 (Fig. 4c). Although no granitoids were observed nearby in the field, the most likely explanation is that these zircons grew as a result of the effects of subsurface emplacement of Permian granites in the area.

The new ages obtained from granitoids and gneiss in the Khanka Block thus show strong affinity with similar rocks in the Jiamusi Block in terms of their magmatic crystallization age, timing of metamorphism and detrital zircon signature. Our results thus support the view of the geologists who mapped the area for the Heilongjiang Bureau of Geology and Mineral Resources (HBGMR 1993) that the Khanka and Jiamusi blocks contain the same lithologies. We therefore propose that they should be regarded as the same terrane, with the implication that the Dunhua–Mishan Fault (Fig. 1) that separates the blocks is not a major terrane boundary.

Assembly of the combined Khanka–Jiamusi Block with the CAOB

If the Jiamusi and Khanka blocks do indeed represent a single terrane, the question arises as to their relationship with adjacent terranes in the CAOB to the west, generally attributed to north–south closure of oceans between the North China and Siberia cratons, and with the accretionary terranes to the east, generally attributed to Pacific plate subduction.

If the microcontinental blocks in the CAOB represent exotic fragments derived from Gondwana (Wilde et al. 2000), it is important to establish the sequence of events that led to their amalgamation. It has not been precisely established when the Erguna and Xing'an blocks (Fig. 1) collided along the Xiguitu–Tayuan suture, although large ophiolitic complexes have been identified along the southern boundary of the former (Li 1991). The presence of Early Palaeozoic rocks in the northern part of the Xing'an Block, especially Ordovician island-arc associations and related blueschists, indicates that collision most probably occurred in the Ordovician (Yan et al. 1989; Zhang & Tang 1989). According to Ye et al. (1994), the arc complexes in the Songliao Block were accreted to the Xing'an–Erguna composite block along the Hegenshan–Heihe suture (Fig. 1) in the late Devonian to early Carboniferous. Subsequently, the Zhangguangcai Range evolved to the east in the Jurassic, possibly as a result of Pacific Ocean subduction from the east (Wu et al. 2007b). Finally, extension led to the formation of the Songliao Basin in the Late Mesozoic (Wu et al. 2001a). The collision between this composite Songliao–Xing'an–Erguna block (now the eastern part of the Central Mongolian microcontinent) and the NCC was originally believed to have taken place in the Late Triassic (Zonenshain et al. 1985, 1990), but recent palaeomagnetic data favour amalgamation before the Late Permian. Suturing between the Siberian and North China cratons probably commenced in the Late Permian to Early Triassic (Zonenshain et al. 1985; Dobretsov et al. 1990, 1995; Sengör et al. 1993; Sengör & Natal'in 1996; Xiao et al. 2003), with the combined North China–Mongolian Block finally colliding with the Siberian Craton upon closure of the Mongol–Okhotsk Ocean in the Late Jurassic (Zhao et al. 1990; Xu et al. 1997; Zorin 1999; Kravchinsky et al. 2002) or possibly Early Cretaceous (Tomurtogoo et al. 2005).

However, just how the Jiamusi Block fits into this scenario is a matter for debate. One view is that it too was located in a peri-Gondwana position, drifting northward, along with the North China, South China and Tarim blocks, to dock with the Siberian Craton some time between the Late Permian and Late Jurassic (Wilde et al. 2000). This implies that the Jiamusi Block forms an integral part of the CAOB. More recently, Wu et al. (2007b) have proposed an alternative model,

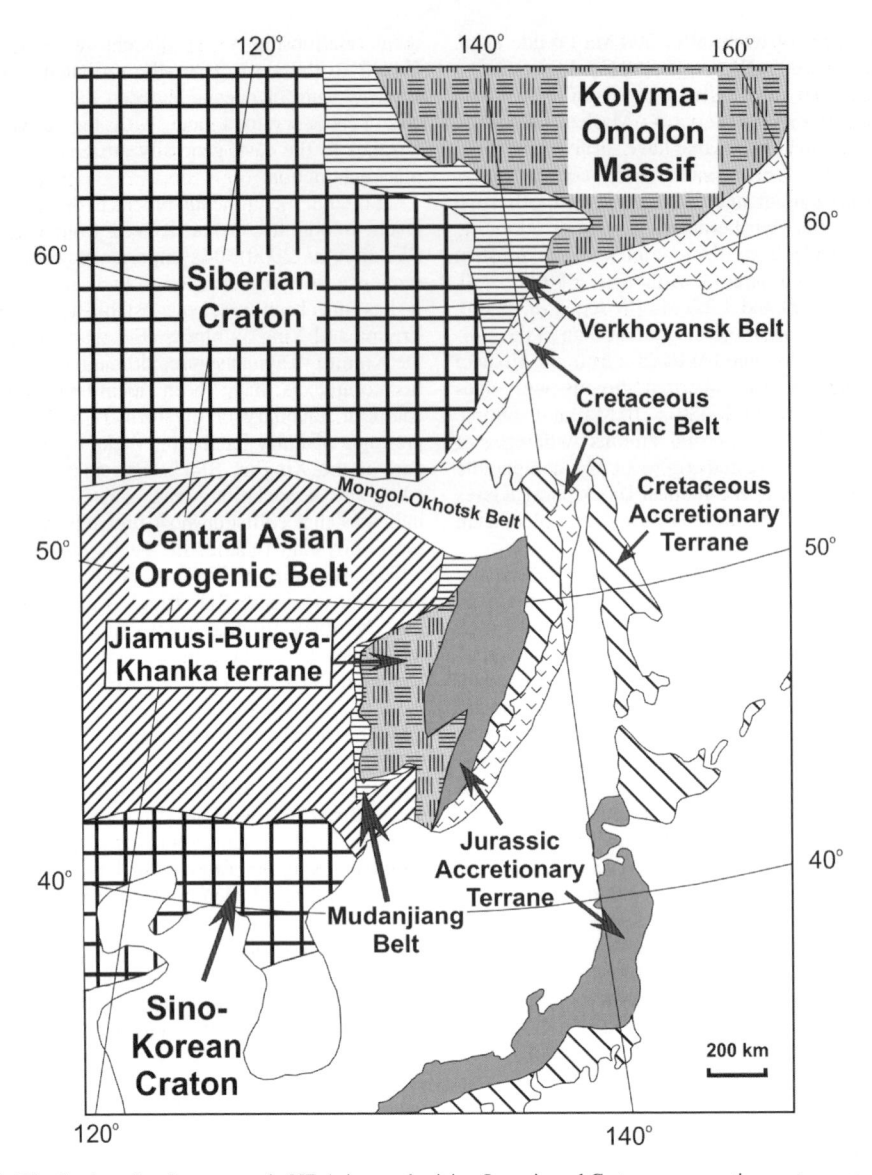

Fig. 5. Distribution of major terranes in NE Asia, emphasizing Jurassic and Cretaceous accretionary terranes. Interpreted to show the contiguity of the Jiamusi–Bureya–Khanka blocks; modified from Wu et al. (2007b).

based on new argon data, in which the Jiamusi Block was actually transported westward to dock with the CAOB in the Jurassic, the boundary being defined by the Heilongjiang Complex. These are not mutually exclusive hypotheses, as the earliest known component of the Jiamusi Block (the Mashan Complex) has a Pan-African Gondwanan signature. The real question is the route it took to reach its present position; did it simply drift northward, or did it somehow follow a separate path from the other blocks?

An earlier view that the Khanka Block (and, from this study, the Jiamusi Block as well) also collided with the NCC during the Late Permian–Early Jurassic (Jia et al. 2004) is refuted by recent geochronology undertaken in the area, as the granitoids in the intervening Dashanzui–Antu–Kaishantun belt are Late Jurassic and therefore much younger (Zhang et al. 2004). However, this belt is one of a series of Jurassic–Cretaceous accretion complexes related to Pacific plate subduction that are widely distributed along the eastern Asian continental margin (Fig. 5) (Natal'in & Borukayev 1991; Natal'in 1993; Sengör & Natal'in 1996). From this time, the amalgamation of such terranes led to the continued growth of the Eurasian continent (Faure & Natal'in 1992; Jia *et al.* 2004; Zhang *et al.* 2004).

Based on current knowledge, it thus appears likely that the Jiamusi–Khanka Block does not represent a microcontinental block trapped between the North China and Siberian cratons during the main assembly of the CAOB. Instead, c. 180 Ma blueschist to greenschist metamorphism of the accretionary Heilongjiang Complex along the western margin of the Jiamusi Block indicates accretion during the Jurassic (Wu *et al.* 2007b). Work is currently in progress on the Bureya Block and this will be critical to evaluating whether this is part of the same crustal entity; preliminary results (Wilde *et al.* 2008) do suggest a continuation of the 500 Ma metamorphic terrane northward beyond the Heilongjiang (Amur) River. The location of the Jiamusi–Khanka(–Bureya) Block (Fig. 5) prior to this event is unknown. It may possibly have been located farther south along the Gondwana margin than the other Chinese blocks, thus taking longer to drift northward. Alternatively, it might have a more direct relationship to the Siberian Craton, where high-grade metamorphism in the Baikal region (Salnikova *et al.* 1998) is virtually coeval with that affecting the Mashan Complex. Such a linkage was previously considered possible by Wilde *et al.* (1997). Further detailed work in both north China and Russia is required to determine the extent of Late Pan-African events in the region.

Conclusions

(1) The Chinese segment of the Khanka Block is poorly exposed, with rocks cropping out mainly around the towns of Hulin and Hutou (Fig. 3). They consist of metamorphosed basement rocks overlain by Cambrian to Silurian carbonates, volcanic rocks and clastic sediments, locally intruded by later granitoids. The basement rocks were previously considered to be Precambrian in age, but our U–Pb dating indicates that the rocks were formed in the Phanerozoic.

(2) Oscillatory zoned igneous zircons from a deformed porphyritic granite obtained near Hutou (sample FW04-202) define a weighted mean $^{206}Pb/^{238}U$ age of 518 ± 7 Ma, which is taken to record the crystallization of the granite. It is identical, within error, to ages of similar granitoids dated from the adjacent Jiamusi Block to the west (Wilde *et al.* 2003).

(3) Unzoned zircon domains in sample FW04-202 record a $^{206}Pb/^{238}U$ age of 499 ± 10 Ma; this is identical within error to the age of deformed medium-grained granitoid sample FW04-207, obtained c. 7 km west of Hutou, which contains a single zircon population recording a weighted mean $^{206}Pb/^{238}U$ age of 491 ± 4 Ma. In addition, the main zircon population in paragneiss sample FW04-210, located midway between Hutou and Hulin (Fig. 3), defines a weighted mean $^{206}Pb/^{238}U$ age of 504 ± 8 Ma. These results are interpreted to record the time of high-grade metamorphism in the area and are identical to the age of granulite-facies metamorphism recorded in the Jiamusi Block to the west (Wilde *et al.* 1997, 2000, 2001). Thus, both the timing of magmatism and metamorphism are identical in the Khanka and Jiamusi blocks, further supporting the view that they form part of a single terrane (see also HBGMR 1993).

(4) Inherited zircons in porphyritic granite sample FW04-202 and detrital igneous zircons in paragneiss sample FW04-210 record ages ranging from 609 to 942 Ma, indicating that Neoproterozoic rocks are present in the basement and in the source area of the protolith to the gneiss. However, there is no evidence that such rocks are exposed at the surface in the Chinese segment of the Khanka Block. Inherited and detrital zircons of this age, and even older, are also reported from the Jiamusi Block (Wilde *et al.* 1997, 2001), supporting the link between the two areas.

(5) A granodiorite from Hulin (Fig. 3) has a younger weighted mean $^{206}Pb/^{238}U$ zircon age of 112 ± 1 Ma. Such ages are common for diorite and lamprophyre dykes in the Jiaodong Peninsula farther south in the North China Craton (Zhang *et al.* 2003), but have not previously been published from either the Khanka or the Jiamusi Block. Such young ages suggest the influence of Pacific plate subduction in magma genesis (see Wu *et al.* 2005) and are probably related to post-collisional events following accretion of the combined Jiamusi–Khanka blocks in the late Early Jurassic (Wu *et al.* 2007b).

(6) The data presented in this study from the Chinese portion of the Khanka Block support the view that it shares a common history with the Jiamusi Block to the west. One implication of this is that the Dunhua–Mishan Fault, traditionally considered to mark the boundary between the two blocks, cannot be a major terrane boundary. Furthermore, a comparison of our new U–Pb zircon data with published information on the Japanese terranes fails to establish any link between the Khanka and Hida blocks, as previously postulated by several workers.

(7) It has recently been proposed that the Jiamusi Block amalgamated with the CAOB in the late Early Jurassic as a result of Pacific plate subduction,

based on the timing of blueschist-facies metamorphism in the Heilongjiang Complex (Wu et al. 2007b), with ^{40}Ar/^{39}Ar dating of biotite and phengite recording ages of 184–174 Ma for this event. Although the early sequences in the Jiamusi Block (Mashan Complex) probably formed part of a peri-Gondwana terrane (Wilde et al. 1997), collision with the Asian continental margin did not occur until the Early Jurassic. If this scenario is correct, then the combined Jiamusi–Khanka terrane is related to circum-Pacific accretion and does not form a microcontinental block that was trapped between the Siberia and North China cratons during the main assembly of the Central Asian Orogenic Belt.

We thank A. Frew for assistance with the SHRIMP II data collection. This is The Institute for Geoscience Research (TIGeR) Publication 200.

References

ANDERSEN, T. 2002. Correction of common lead in U–Pb analyses that do not report ^{204}Pb. *Chemical Geology*, **192**, 59–79.

ARAKAWA, Y. & SHINMURA, T. 1995. Nd–Sr isotopic and geochemical characteristics of two contrasting types of calc-alkaline plutons in the Hida belt, Japan. *Chemical Geology*, **124**, 217–232.

ARAKAWA, Y., SAITO, Y. & AMAKAWA, H. 2000. Crustal development of the Hida belt, Japan: evidence from Nd–Sr isotopic and chemical characteristics of igneous and metamorphic rocks. *Tectonophysics*, **328**, 183–204.

BALLARD, J. R., PALIN, J. M., WILLIAMS, I. S., CAMPBELL, I. H. & FAUNES, A. 2002. Two ages of porphyry intrusion resolved for the super-giant Chuquicamata copper deposit of northern Chile by ELA-ICP-MS and SHRIMP. *Geology*, **29**, 383–386.

CAO, X., DANG, Z. X., ZHANG, X. Z., JIANG, J. S. & WANG, H. D. 1992. *Jiamusi Composite Terranes*. Jilin Publishing House of Science and Technology, Changchun [in Chinese, with English and Russian abstracts].

CHEN, B., JAHN, B. M., WILDE, S. & XU, B. 2000. Two contrasting Palaeozoic magmatic belts in northern Inner Mongolia, China: petrogenesis and tectonic implications. *Tectonophysics*, **328**, 157–182.

CHENG, R. Y., WU, F. Y., GE, W. C., SUN, D. Y. & YANG, J. H. 2006. Emplacement age of the Raohe Complex in eastern Heilongjiang Province and the tectonic evolution of the eastern part of Northeastern China. *Acta Petrologica Sinica*, **22**, 353–376 [in Chinese with English abstract].

DALLMEYER, R. D. & TAKASU, A. 1998. ^{40}Ar/^{39}Ar mineral ages from the Oki metamorphic complex, Oki-Dogo, SW Japan: implications for regional correlations. *Journal of Asian Earth Sciences*, **16**, 437–448.

DOBRETSOV, N. L., DOOK, V. L. & KITSUL, V. I. 1990. Geotectonic evolution of the Siberian platform during the Precambrian and comparison with the lower Precambrian complexes of eastern Asia. *Journal of Southeastern Asian Earth Sciences*, **4**, 259–266.

DOBRETSOV, N. L., BERZIN, N. A. & BUSLOV, M. 1995. Opening and tectonic evolution of the Paleo-Asian Ocean. *International Geology Review*, **37**, 335–360.

ERNST, W. G., CAO, R. & JIANG, J. 1988. Reconnaissance study of Precambrian metamorphic rocks, northeastern Sino-Korean shield, People's Republic of China. *Geological Society America Bulletin*, **100**, 692–701.

FAURE, M. & CHARVET, J. 1987. Late Permian/early Triassic orogeny in Japan: piling up of nappes, transverse lineation and continental subduction of the Honshu block. *Earth and Planetary Science Letters*, **84**, 295–308.

FAURE, M. & NATAL'IN, B. 1992. Geodynamic evolution of the Eurasian margin in Mesozoic times. *Tectonophysics*, **208**, 397–411.

GAO, F. H., XU, W. L., YANG, D. B., PEI, F. P., LIU, X. M. & HU, Z. C. 2007. LA-ICP-MS zircon U–Pb dating from granitoids in southern basement of Songliao basin: constraints on ages of the basin basement. *Science in China (D)*, **50**, 995–1004.

GE, W. C., WU, F. Y., ZHOU, C. Y. & ABDEL RAHAMAN, A. A. 2005a. Emplacement age of the Tahe granite and its constraints on the tectonic nature of the Eguna block in the northern part of the Great Xing'an Range. *Chinese Science Bulletin*, **50**, 2097–2105.

GE, W. C., WU, F. Y., ZHOU, C. Y. & ZHANG, J. H. 2005b. Zircon U–Pb ages and its significance of the Mesozoic granites in the Wulanhaote region, central Da Hinggan Mountain. *Acta Petrologica Sinica*, **21**, 749–762 [in Chinese with English abstract].

GE, W. C., SUI, Z. M., WU, F. Y., ZHANG, J. H., XU, X. C. & CHENG, R. Y. 2007a. Zircon U–Pb ages, Hf isotopic characteristics and their implications of the Early Palaeozoic granites in the northwestern Da Hinggan Mts, northeastern China. *Acta Petrologica Sinica*, **23**, 423–440 [in Chinese with English abstract].

GE, W. C., WU, F. Y., ZHOU, C. Y. & ZHANG, J. H. 2007b. Porphyry Cu–Mo deposits in the eastern Xing'an–Mongolian Orogenic Belt: mineralization ages and their geodynamic implications. *Chinese Science Bulletin*, **52**, 3416–3427.

HBGMR (HEILONGJIANG BUREAU OF GEOLOGY AND MINERAL RESOURCES) 1993. *Regional geology of Heilongjiang Province*. Geological Publishing House, Beijing [in Chinese with English abstract].

HIROI, Y. 1981. Subdivision of the Hida metamorphic complex, central Japan, and its bearing on the geology of the far east in pre-Sea of Japan time. *Tectonophysics*, **76**, 317–333.

ISHIWITARI, A. & TSUJIMORI, T. 2003. Palaeozoic ophiolites and blueschists in Japan and Russian Primorye in the tectonic framework of East Asia: a synthesis. *Island Arc*, **12**, 190–206.

JAHN, B.-M., WU, F. Y. & CHEN, B. 2000a. Granitoids of the Central Asian Orogenic Belt and continental growth in the Phanerozoic. *Transactions of the Royal Society of Edinburgh: Earth Sciences*, **91**, 181–193.

JAHN, B.-M., WU, F. Y. & CHEN, B. 2000b. Massive granitoid generation in Central Asia: Nd isotope evidence and implication for continental growth in the Phanerozoic. *Episodes*, **23**, 82–92.

JAHN, B.-M., WU, F. Y. & HONG, D. W. 2000c. Important crustal growth in the Phanerozoic: isotopic evidence of granitoids from East–Central Asia. *Proceedings of the Indian Academy of Sciences (Earth and Planetary Sciences),* **109**, 5–20.

JBGMR (JILIN BUREAU OF GEOLOGY AND MINERAL RESOURCES) 1988. *Regional Geology of Jilin Province.* Geological Publishing House, Beijing [in Chinese with English summary].

JIA, D. C., HU, R. Z., LU, Y. & QIU, X. L. 2004. Collision belt between the Khanka block and the North China block in the Yanbian Region, Northeast China. *Journal of Asian Earth Sciences,* **23**, 211–219.

KAGAMI, H., KAWANO, Y. ET AL. 2006. Provenance of Palaeozoic–Mesozoic sedimentary rocks in the Inner Zone of Southwest Japan: an evaluation based on Nd model ages. *Gondwana Research,* **9**, 142–151.

KHANCHUK, A. I., RATKIN, V., RYAZANTSEVA, M. D., GOLOZUBOV, V. V. & GONOKHOVA, N. G. 1996. *Geology and Mineral Deposits of Primorsky Krai (Territory).* Dalnauka, Vladivostok [in Russian].

KOJIMA, S. 1989. Mesozoic terrane accretion in northeast China, Sikhote-Alin and Japan regions. *Palaeogeography, Palaeoclimatology, Palaeoecology,* **69**, 213–232.

KRAVCHINSKY, V. A., COGNE, J. P., HARBERT, W. P. & KUZMIN, M. I. 2002. Evolution of the Mongol–Okhotsk Ocean as constrained by new palaeomagnetic data from the Mongol–Okhotsk suture zone, Siberia. *Geophysical Journal International,* **148**, 34–57.

KWON, S., SAJEEV, K., MITRA, G., PARK, Y., KIM, S.-W. & RYU, I.-C. 2009. Evidence for Permo-Triassic collision in Far East Asia: The Korean collisional orogen. *Earth and Planetary Science Letters,* **279**, 340–349, doi: 10.1016/j.epsl.2009.01.016.

LI, J. Y., NIU, B. G., SONG, B., XU, W. X., ZHANG, Y. H. & ZHAO, Z. R. 1999. *Crustal Formation and Evolution of Northern Changbai Mountains, Northeast China.* Geological Publishing House, Beijing [in Chinese with English abstract].

LI, R.-S. 1991. Xinlin ophiolite. *Heilongjiang Geology,* **2**, 19–32 [in Chinese with English abstract].

LUDWIG, K. R. 2001a. *Squid 1.02: a user's manual.* Berkeley Geochronology Center, Special Publications, **2**.

LUDWIG, K. R. 2001b. *User's manual for Isoplot/Ex version 2.05.* Berkeley Geochronology Center, Special Publications, **1a**.

LUDWIG, K. R. 2003, *ISOPLOT 3.0 – A Geochronological Toolkit for Microsoft Excel.* Berkeley Geochronology Center Special Publication, **4**, 1–70.

MARUYAMA, S., LIOU, J. G. & SENO, T. 1989. Mesozoic and Cenozoic evolution of Asia. *In*: BEN AVRAHAM, Z. (ed.) *The Evolution of the Pacific Ocean Margins.* Oxford Monographs in Geology and Geophysics, **8**, 75–99.

MENG, E., XU, W. L., YANG, D. B., PEI, F. P., JI, W. Q., YU, Y. & ZHANG, X. Z. 2008. Permian volcanisms in eastern and southeastern margins of the Jiamusi Massif, northeastern China: zircon U–Pb chronology, geochemistry and its tectonic implications. *Chinese Science Bulletin,* **53**, 1231–1245.

MIAO, L. C., FAN, W. M. ET AL. 2003. Zircon SHRIMP geochronology of the Xinkailing-Kele complex in the northwestern Lesser Xing'an Range, and its geological implications. *Chinese Science Bulletin,* **49**, 2201–2209.

MIAO, L. C., LIU, D. Y., ZHANG, F. Q., FAN, W. M., SHI, Y. R. & XIE, H. Q. 2007. Zircon SHRIMP U–Pb ages of the 'Xinghuadukou Group' in Hanjiayuanzi and Xinlin areas and the 'Zhalantun Group' in Inner Mongolia, Da Hinggan Mountains. *Chinese Science Bulletin,* **52**, 1112–1134.

NATAL'IN, B. A. 1993. History and modes of Mesozoic accretion in southeastern Russia. *Island Arc,* **2**, 15–34.

NATAL'IN, B. A. & BORUKAYEV, C. B. 1991. Mesozoic sutures in the southern Far East of USSR. *Geotectonics,* **25**, 64–74.

NELSON, D. R. 1997. *Compilation of geochronology data, 1996.* Geological Survey of Western Australia Record, **1997/2**.

OH, C. W. 2006. A new concept on tectonic correlation between Korea, China and Japan: histories from the late Proterozoic to Cretaceous. *Gondwana Research,* **9**, 47–61.

OSANAI, Y., OWADA, M., KAMEI, A., HAMAMOTO, T., KAGAMI, H., TOYOSHIMA, T., NAKANO, N. & NAM, T. M. 2006. The Higo metamorphic complex in Kyushu, Japan as the fragment of Permo-Triassic metamorphic complexes in East Asia. *Gondwana Research,* **9**, 152–166.

PEI, F. P., XU, W. L., YANG, D. B., ZHAO, Q. G., LIU, X. M. & HU, Z. C. 2007. Zircon U–Pb geochronology of basement metamorphic rocks in the Songliao basin. *Chinese Science Bulletin,* **52**, 942–948.

PEI, F. P., XU, W. L., YANG, D. B., JI, W. Q., YU, Y. & ZHANG, X. Z. 2008. Mesozoic volcanic rocks in the Southern Songliao Basin: zircon U–Pb ages and their constraints on the nature of basin basement. *Journal of China University (Earth Sciences),* **33**, 603–617.

SAKASHIMA, T., TERADA, K., TAKESHITA, T. & SANO, Y. 2003. Large-scale displacement along the Median Tectonic Line, Japan: evidence from SHRIMP zircon U–Pb dating of granites and gneisses from the South Kitakami and paleo-Ryoke belts. *Journal of Asian Earth Sciences,* **21**, 1019–1039.

SALNIKOVA, E. B., SERGEEV, S. A., KOTOV, A. B., YAKOVLEVA, S. Z., STEIGER, R. H., REZNITSKIY, L. Z. & VASILEV, E. P. 1998. U–Pb zircon dating of granulite metamorphism in the Sludyanskiy Complex, eastern Siberia. *Gondwana Research,* **1**, 195–205.

SANO, Y., HIDAKA, H., TERADA, K., SHIMIZU, H. & SUZUKI, M. 2000. Ion microprobe U–Pb zircon geochronology of the Hida gneiss: finding of the oldest minerals in Japan. *Geochemical Journal,* **34**, 135–153.

ŞENGÖR, A. M. C. & NATAL'IN, B. A. 1996. Turkic-type orogeny and its role in the making of the continental crust. *Annual Review of Earth and Planetary Sciences,* **24**, 263–337.

ŞENGÖR, A. M. C., NATAL'IN, B. A. & BURTMAN, V. S. 1993. Evolution of the Altaid tectonic collage and Palaeozoic crustal growth in Eurasia. *Nature,* **364**, 299–307.

SHAO, J.-A. & TANG, K.-D. 1995. *Terranes in Northeast China and Evolution of Northeast Asia Continental Margin.* Seismic Press, Beijing [in Chinese].

STEIGER, R. H. & JÄGER, E. 1977. Subcommission on geochronology: convention on the use of decay constants in geo- and cosmochronology. *Earth and Planetary Science Letters,* **36**, 359–362.

Sui, Z. M., Ge, W. C., Wu, F. Y., Zhang, J. H., Xu, X. C. & Chang, R. Y. 2007. Zircon U–Pb ages, geochemistry and its petrogenesis of Jurassic granites in northwestern part of the Da Hinggan Mts. *Acta Petrologica Sinica*, **23**, 461–480 [in Chinese with English abstract].

Suzuki, K. & Adachi, M. 1994. Middle Precambrian detrital monazite and zircon from the Hida gneiss on Oki-Dogo Island, Japan: their origin and implications for the correlation of basement gneiss of Southwest Japan and Korea. *Tectonophysics*, **235**, 277–292.

Suzuki, K., Adachi, M., Takagi, H. & Osanai, Y. 1998. CHIME monazite age of the Higo metamorphic rocks. *In*: *105th Annual Meeting, Geological Society of Japan*, 214 [in Japanese].

Tang, K.-D. 1990. Tectonic development of Palaeozoic fold belts at the north margin of the Sino-Korean craton. *Tectonics*, **9**, 249–260.

Tazawa, J. 2004. The strike-slip model: a synthesis on the origin and tectonic evolution of the Japanese Islands. *Journal of the Geological Society of Japan*, **110**, 503–517 [in Japanese with English abstract].

Tomurtogoo, O. B., Windley, B. F., Kroner, A., Badarch, G. & Liu, D. Y. 2005. Zircon age and occurrence of the Adaatsag ophiolite and Muron shear zone, central Mongolia: constraints on the evolution of the Mongol–Okhotsk ocean, suture and orogen. *Journal of the Geological Society, London*, **162**, 125–134.

Tsujimori, T., Liou, J. G., Ernst, W. G. & Itaya, T. 2006. Triassic paragonite- and garnet-bearing epidote-amphibolite from the Hida Mountains, Japan. *Gondwana Research*, **9**, 167–175.

van Achterbergh, E., Ryan, C. G., Jackson, S. E. & Griffin, W. L. 2001. Data reduction software for LA-ICP-MS. *In*: Sylvester, P. (ed.) *Laser Ablation-ICPMS in the Earth Sciences*. Mineralogical Association of Canada, Short Course, **29**, 239–243.

Wang, F., Zhou, X. H., Zhang, L. C., Ying, J. F., Zhang, Y. T. & Wu, F. Y. 2006. Timing of volcanism succession of the Great Xing'an Range, Northeastern Asia, and its tectonic significance. *Earth and Planetary Science Letters*, **251**, 179–198.

Wilde, S. A., Dorsett-Bain, H. L. & Liu, J. 1997. The identification of a Late Pan-African granulite facies event in Northeastern China: SHRIMP U–Pb zircon dating of the Mashan Group at Liu Mao, Heilongjiang Province, China. *In*: Qian, X. L., You, Z. D. & Halls, H. C. (eds) *Proceedings of the 30th International Geological Congress. Precambrian Geology and Metamorphic Petrology*, **17**, 59–74.

Wilde, S. A., Zhang, X. Z. & Wu, F. Y. 2000. Extension of a newly-identified 500 Ma metamorphic terrain in North East China: further U–Pb SHRIMP dating of the Mashan Complex, Heilongjiang Province, China. *Tectonophysics*, **328**, 115–130.

Wilde, S. A., Wu, F. Y. & Zhang, X. Z. 2001. The Mashan Complex: SHRIMP U–Pb zircon evidence for a Late Pan-African metamorphic event in NE China and its implications for global continental reconstructions. *Geochimica*, **30**, 35–50 [in Chinese, with English abstract].

Wilde, S. A., Wu, F. Y. & Zhang, X. Z. 2003. Late Pan-African magmatism in Northeastern China: SHRIMP U–Pb zircon evidence from granitoids in the Jiamusi Massif. *Precambrian Research*, **122**, 311–327.

Wilde, S. A., Wu, F. Y., Zhao, G. C. & Sklyarov, E. 2008. The Jiamusi, Khanka and Bureya blocks; a contiguous crustal entity accreted to the Central Asian Orogenic Belt in the Early Jurassic? *In*: *33rd International Geological Congress, Oslo. Pre-Mesozoic Accretionary Tectonics in Central Asia*, Abstracts CD-ROM, ASI-06.

Williams, I. S. 1998. U-Th-Pb geochronology by ion microprobe. *In*: McKibben, M. A. (ed.) *Understanding Mineralizing Processes*. Society of Economic Geologists, Reviews in Economic Geology, **7**, 1–35.

Wu, F. Y., Jahn, B.-M., Wilde, S. A. & Sun, D. Y. 2000. Phanerozoic continental crustal growth: Sr–Nd isotopic evidence from the granites in northeastern China. *Tectonophysics*, **328**, 87–113.

Wu, F. Y., Sun, D. Y., Li, H. M. & Wang, X. L. 2001a. The nature of basement beneath the Songliao Basin in NE China: geochemical and isotopic constraints. *Physics and Chemistry of the Earth (Part A)*, **26**, 793–803.

Wu, F. Y., Wilde, S. & Sun, D. Y. 2001b. Zircon SHRIMP ages of gneissic granites in Jiamusi Massif, northeastern China. *Acta Petrologica Sinica*, **17**, 443–452 [in Chinese with English abstract].

Wu, F. Y., Sun, D. Y., Li, H. M., Jahn, B. M. & Wilde, S. A. 2002. A-type granites in Northeastern China: age and geochemical constraints on their petrogenesis. *Chemical Geology*, **187**, 143–173.

Wu, F. Y., Jahn, B. M. *et al*. 2003. Highly fractionated I-type granites in NE China (I): geochronology and petrogenesis. *Lithos*, **66**, 241–273.

Wu, F. Y., Lin, J. Q., Wilde, S. A., Zhang, X. O. & Yang, J. H. 2005. Nature and significance of the Early Cretaceous giant igneous event in eastern China. *Earth and Planetary Science Letters*, **233**, 103–119.

Wu, F. Y., Zhao, G. C., Sun, D. Y., Wilde, S. A. & Zhang, G. L. 2007a. The Hulan Group: its role in the evolution of the Central Asian Orogenic Belt of NE China. *Journal of Asian Earth Sciences*, **30**, 542–556.

Wu, F. Y., Yang, J. H., Lo, C. H., Wilde, S. A., Sun, D. Y. & Jahn, B.-M. 2007b. The Heilongjiang Group: a Jurassic accretionary complex in the Jiamusi Massif at the western Pacific margin of northeastern China. *Island Arc*, **16**, 156–172.

Xiao, W. J., Windley, B., Hao, J. & Zhai, M. G. 2003. Accretion leading to collision and the Permian Solonker suture, Inner Mongolia, China: termination of the Central Asian Orogenic Belt. *Tectonics*, **22**, 1069, doi: 2002TC001484.

Xu, X., Harbert, W., Dril, S. & Kravchinksy, V. 1997. New paleomagnetic data from the Mongol–Okhotsk collision zone, Chita region, south–central Russia: implications for Palaeozoic paleogeography of the Mongol–Okhotsk Ocean. *Tectonophysics*, **269**, 113–129.

Yan, J. Y., Tang, K. D., Bai, J. W. & Mo, Y. C. 1989. High pressure metamorphic rocks and their tectonic environment in northeastern China. *Journal of Southeast Asian Earth Sciences*, **3**, 303–313.

Ye, M., Zhang, S. H. & Wu, F. Y. 1994. The classification of the Paleozoic tectonic units in the area crossed by

Manzhouli–Suifenghe geoscience transect. *Journal of Changchun University of Earth Sciences*, **24**, 241–245 [in Chinese with English abstract].

YIN, A. & NIE, S. 1993. An indentation model for the North and South China collision and the development of the Tan-Lu and Nonam fault systems, eastern Asia. *Tectonics*, **12**, 801–813.

YUAN, H. L., GAO, S., LIU, X. M., LI, H. M., GUNTHER, D. & WU, F. Y. 2004. Accurate U–Pb age and trace element determinations of zircon by laser ablation–inductively coupled plasma Mass spectrometry. *Geostandards Newsletter*, **28**, 353–370.

ZHANG, J. H., GE, W. C., WU, F. Y., WILDE, S. A., YANG, J. H. & LIU, X. M. 2008. Large-scale Early Cretaceous volcanic events in the northern Great Xing'an Range, Northeastern China. *Lithos*, **102**, 138–157.

ZHANG, X. O., CAWOOD, P. A., WILDE, S. A. & LIU, R. Q. 2003. Timing of mineralization at the Cangshan gold deposit, northwest Jiaodong Peninsula, China: a SHRIMP U–Pb study. *Mineralium Deposita*, **38**, 141–153.

ZHANG, X. Z. 1992. Heilongjiang mélange: the evidence of Caledonian suture zone of the Jiamusi massif. *Journal of Changchun University of Earth Sciences, Special Issue: Doctoral Thesis*, **22**, 94–101 [in Chinese with English abstract].

ZHANG, Y. B. 2002. *Geochronological framework of the granitic magmatism in Yanbian area*. Doctoral dissertation, Jilin University [in Chinese with English summary].

ZHANG, Y. B., WU, F. Y., WILDE, S. A., ZHAI, M. G., LU, X. P. & SUN, D. Y. 2004. Zircon U–Pb ages and tectonic implications of 'Early Paleozoic' granitoids at Yanbian, Jilin province, NE China. *Island Arc*, **13**, 484–505.

ZHANG, Y. B., WU, F. Y., ZHAI, M. G., LU, X. P. & ZHANG, H. F. 2008. Geochronology and tectonic implications of the Seluohe Group in the northern margin of the North China Craton. *International Geology Review*, **50**, 135–153.

ZHANG, Y. P. & TANG, K. D. 1989. Pre-Jurassic tectonic evolution of intercontinental region and the suture zone between the North China and Siberian platforms. *Journal of Southeast Asian Earth Sciences*, **3**, 47–55.

ZHAO, X. X., COE, R. S., ZHOU, Y. X., WU, H. R. & WANG, J. 1990. New paleomagnetic results from northern China: Collision and suturing with Siberia and Kazakhstan. *Tectonophysics*, **181**, 43–81.

ZONENSHAIN, L. P., KUZMIN, M. I. & KONONOV, M. V. 1985. Absolute reconstructions of the Palaeozoic oceans. *Earth and Planetary Science Letters*, **74**, 103–116.

ZONENSHAIN, L. P., KUZMIN, M. I. & NATAPOV, L. M. (eds) 1990. *Geology of the USSR: A Plate Tectonic Synthesis*. American Geophysical Union, Geodynamics Series, **21**.

ZORIN, Y. A. 1999. Geodynamics of the western part of the Mongolia–Okhotsk collisional belt, Trans-Baikal region (Russia) and Mongolia. *Tectonophysics*, **306**, 33–56.

Petrology, chemistry and phase relations of borosilicate phases in phlogopite diopsidites and granitic pegmatites from the Tranomaro belt, SE Madagascar; boron-fluid evolution

THÉODORE RAZAKAMANANA[1], BRIAN F. WINDLEY[2] & DIETRICH ACKERMAND[3]

[1]*Département des Sciences de la Terre, Université de Toliara 601, Toliara, Madagascar*
[2]*Department of Geology, University of Leicester, Leicester LE1 7RH, UK*
[3]*Institut für Geowissenschaften, Christian-Albrechts-Universität, 24098, Kiel, Germany*
*Corresponding author (e-mail: brian.windley@btinternet.com)

Abstract: The boron-bearing minerals grandidierite, werdingite, serendibite and sinhalite are common in high-grade rocks of the Tranomaro belt in southeastern Madagascar. The mutual occurrence of these phases allows a new understanding of the role of boron-rich fluids in the crustal evolution of Gondwana, and we provide critical borosilicate data to constrain that development. We distinguish two types of grandidierite depending on their B_2O_3 and Al_2O_3 contents and on their relations with associated borosilicate phases. (1) At Vohibola the presence of sinhalite and serendibite associated with phlogopite lenses in metasedimentary diopsidites indicates an evaporitic origin from calc-silicate sediments. (2) At Cape Andrahomana borosilicates are associated with pegmatites and granites that were emplaced along shear zones on the boundary of the Tranomaro belt. The shear zones acted as conduits for boron-bearing fluids and for granitic partial melts, which had derived their boron from calc-silicate sedimentary protoliths. Using geothermometry and geobarometry of minerals from associated rocks, we calculate that ambient pressures and temperatures changed in time from 7.5 to 4.0 kbar and from c. 800 °C to 700 °C. Our results confirm the important role of shear zones in channelling the fluid flow of boron-bearing fluids that were derived from crustal melt granites in the same shear zones, but that ultimately derived their boron from early metasediments. We provide new information on the mineralogy, phase assemblages and paragenetic history of multiple borosilicates.

The Precambrian basement of Madagascar forms part of the East African orogen (Stern 1994; Shackleton 1996; de Wit *et al.* 2001; Kusky *et al.* 2003), which developed by collision between the Dharwar and Congo cratons belonging to Eastern and Western Gondwana, respectively (Lardeaux *et al.* 1999; Collins 2000, 2006). In Madagascar that collision zone is marked by the Betsimisaraka suture (Collins *et al.* 2000; Kröner *et al.* 2000; Collins & Windley 2002; Raharimahefa & Kusky 2006, 2009). This collisional geology occurs north of the NW–SE-trending Ranotsata shear zone (Fig. 1). The Betsimisaraka suture extends eastwards (present coordinates) into the Palghat–Cauvery suture zone of southern India (Santosh & Sajeev 2006; Collins *et al.* 2007).

South of the Ranotsara shear zone the basement of southern Madagascar (Fig. 2) (which includes the Tranomaro belt) is characterized by paragneisses, diopsidites with phlogopite mineralization, marbles, quartzites, granulite- to amphibolite-facies metamorphism, crustal melt granites and shear zones (Ackermand *et al.* 1989; Windley *et al.* 1994; Martelat *et al.* 1997, 2000; Lardeaux *et al.* 1999; Pili *et al.* 1999; Markl *et al.* 2000).

The first deformation in southern Madagascar gave rise at 650–600 Ma to recumbent isoclinal folds and regional shallow-dipping foliation (Martelat *et al.* 1997, 2000) that was associated with an ultrahigh-temperature (UHT), granulite-grade, peak metamorphism at 8–11 kbar and 950–1000 °C (Jöns 2006, quoted by Raith *et al.* 2008). The second deformation gave rise at c. 530–500 Ma to many north–south-trending steep shear zones (Fig. 1 shows only the major shear zones, not the many minor ones) that were responsible for retrogression of the granulites during clockwise, near-isothermal decompression from 1000 °C at 10 kbar to 750 °C at 3 kbar (Ackermand *et al.* 1989) or from 850 °C to c. 500 °C, and down to 5 kbar (Raith *et al.* 2008). Major shear zones are >350 km long and 20–35 km wide, and minor shear zones are <140 km long and 7 km wide; the major shears were conduits for mantle-derived CO_2, and the minor shears for crustal-derived, H_2O-rich fluids (Pili *et al.* 1997a, b, 1999). The shear zones separate blocks with progressively deeper crustal levels from 3–5 kbar in the east (the region of Tranomaro) to 8–11 kbar in the west of Madagascar (Nicollet 1990a; Martelat *et al.* 1997; Pili *et al.*

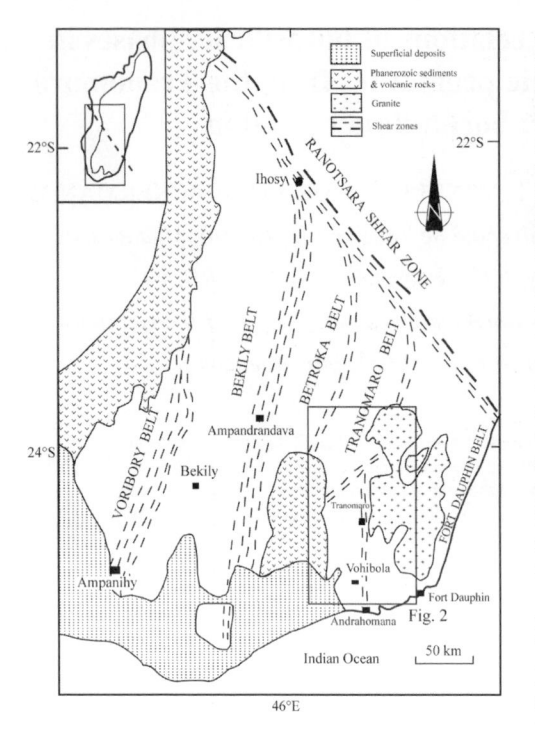

Fig. 1. Map of southern Madagascar south of the Ranotsara shear zone showing location of the grandidierite-bearing localities at Vohibola and Cape Andrahomana (on the coast), modified after Razakamanana (1999) and Martelat *et al.* (2000). Only the main shear belts are marked (after Pili *et al.* 1999), showing that the grandidierite localities are situated in or against a major north–south-trending shear zone. The shear zones separate five tectonic belts; Vohibola and Andrahomana belong to the Tranomaro belt. The location of Figure 2 is indicated.

1997a, b, 1999). Zircon [U–Pb sensitive high-resolution ion microprobe (SHRIMP)] and monazite (U–Th–Pb) dating suggests that the decompressive $P-T$ trajectory took place between 560 and 530 Ma (Paquette *et al.* 1994; Kröner *et al.* 1996; Ito *et al.* 1997; Martelat *et al.* 2000; Jöns 2006). The shear zones contain abundant tourmaline-bearing granitic intrusions, pegmatites, phlogopite and uranothorianite mineralization, which resulted from crustal-scale fluid ingress (Moine *et al.* 1985; Pierdzig 1992; Pili *et al.* 1997a, b). Much fluid circulation was restricted to the vicinity (<100 m) of granitic plutons (Pili *et al.* 1999); according to Markl *et al.* (2000) this granitic activity was associated with cordierite-producing metamorphism at 540–520 Ma at 4 ± 1 kbar and 690 ± 40 °C.

Diopsidite is a key lithology in the evolution of the borosilicate phases in southern Madagascar. We follow the common definition of diopsidite as containing more than 90% diopside; but usually it is much more, up to nearly 100%. It occurs in several forms. Moreau (1963) recognized two types of diopsidite: that closely associated with marble, and that intercalated within alumino-silicate gneisses. Moine *et al.* (1985) added a third, occurring as reaction zones up to a few metres wide along the contact between marbles and granitic to syenitic bodies. We have observed diopsidite layers up to 50–100 m wide in hypersthene–diopside–hornblende gneisses, and diopsidite lenses from 10 cm to 10 m wide within marbles that reach 15 m wide. Some diopsidite layers in gneisses contain spinel marble lenses of up to 10 m width, and between these there may be a 30 cm wide reaction zone of calcite–olivine rock. Some coarse diopsidite layers in gneisses contain disseminated plagioclase grains up to 15 cm long, and their textures suggest they were derived from coarse metagabbros. Such diopsidites are traversed by calcite veins up to 30 cm wide.

For this paper we have selected two representative occurrences of borosilicate-bearing rocks at Vohibola and Andrahomana in the Tranomaro belt. Key mineral assemblages contain one or more of four diagnostic boron-bearing phases: grandidierite, sinhalite, serendibite, werdingite. Our aim is to define the main rocks and constituent assemblages, and discuss their implications for the provenance of the boron-bearing protoliths and fluids.

General geology of the Tranomaro belt

Tranomaro is one of five north–south-trending, high-grade Neoproterozoic gneissic belts (Fig. 1) south of the Ranotsara shear zone (Pili *et al.* 1999; Martelat *et al.* 2000). The Tranomaro belt, which is about 250 km long and 50–60 km wide, is characterized by paragneiss, diopsidite, marble, quartzite and minor orthogneiss, complex fold interference patterns, and north–south-trending shear zones (Lacroix 1922–1923; Pili *et al.* 1999; Razakamanana 1999). Most abundant are paragneissses with various combinations of biotite, diopside, garnet, cordierite, sillimanite, K-feldspar and orthopyroxene. The Tranomaro belt is bordered by shear zones of up to 5–10 km width, and within the belt many minor shear zones are up to 5 m wide. Near the village of Tranomaro there is a 3–5 km wide, >100 km long shear zone (Martelat *et al.* 2000).

Throughout southern Madagascar paragneisses commonly contain diopsidite layers, up to many tens of metres wide, which locally contain pegmatitic lenses ('les poches' of early French mine workers, which reach 30 m × 6 m × 15 m; Lacroix 1941) that contain phlogopite (crystals reach 2 m long in museum samples), diopside, calcite,

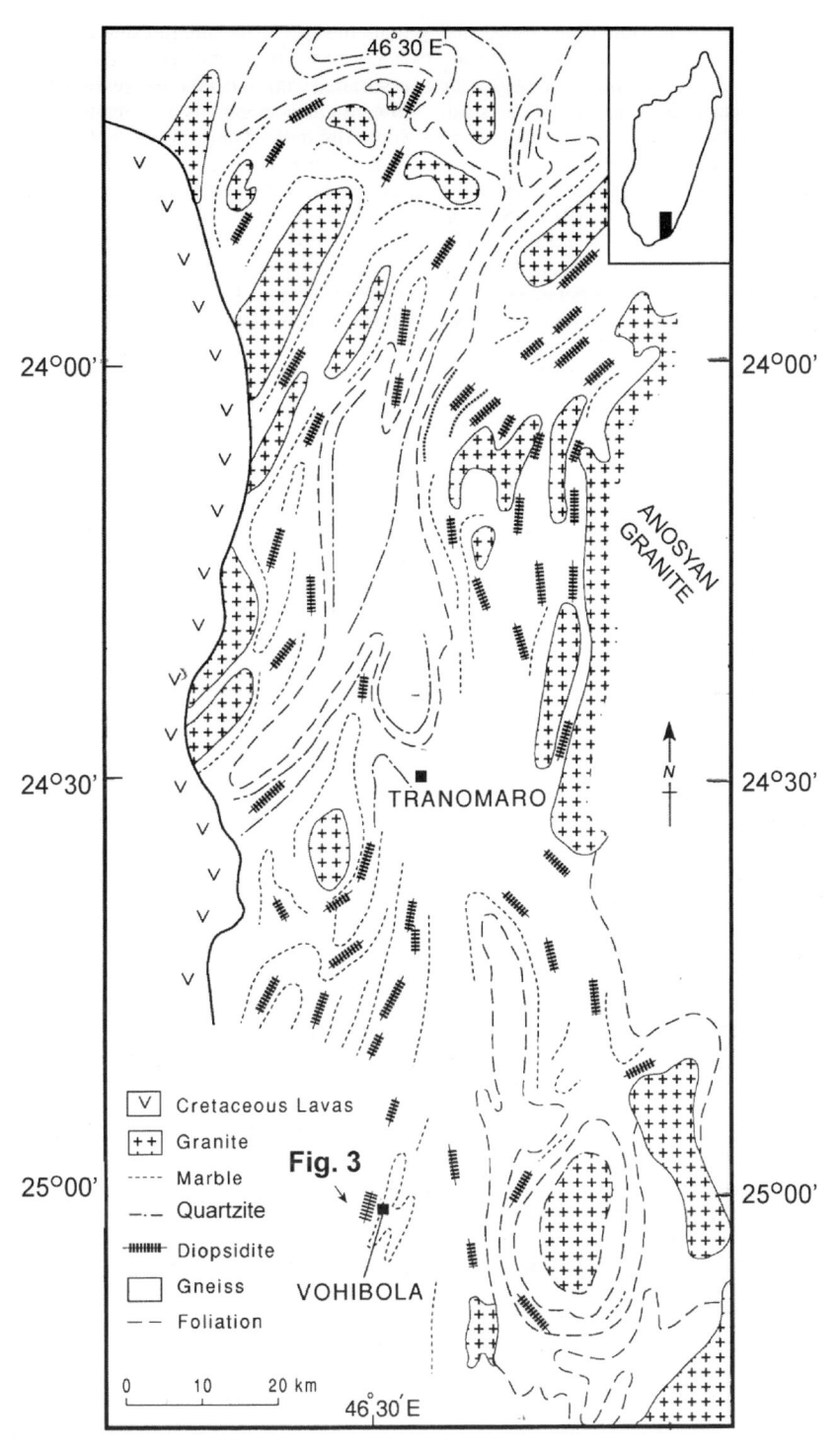

Fig. 2. Simplified geological map of the southern Tranomaro belt with the location of Vohibola and Tranomaro towns. The map (minus shear zones) shows that paragneisses contain abundant layers of diopsidite, marble and quartzite. In addition to the main granites shown, there are innumerable small bodies of crustal-melt granites throughout the belt. Figure 3 is located at the diopsidite near Vohibola.

apatite, scapolite, anorthite, anhydrite, spinel and fluorite (Pierdzig 1992). These largest lenses are located at former phlogopite mines or prospects. The Vohibola occurrence (Fig. 3) is a small example of this relationship.

From the Tranomaro belt diopsidite has a monazite age of 500–600 Ma (Moine *et al.* 1985) and a U–Pb zircon age of 565 ± 15 Ma (Paquette *et al.* 1994). Andriamorofahatra *et al.* (1990) gave a U–Pb date of 561 ± 12 Ma for monazites and zircons from a granodiorite dyke contemporaneous with granulite-facies metamorphism. Zircon in diopside-bearing calcite veins that cross marbles has a U–Pb age of 523 ± 5 Ma at 850–650 °C and 6–4 kbar (Paquette *et al.* 1994; de Grave *et al.* 2002). Berger *et al.* (2006) used the (U + Th)/Pb electron probe method to date a 5 cm monazite crystal from a pegmatite in a granulite-facies granite (charnockite) near Tranomaro town (Fig. 1). The monazite has a core age of 554 ± 34 Ma and a rim age of 492 ± 15 Ma. Berger *et al.* (2006) related the former to a period of garnet-grade (peak) granulite-facies metamorphism, and the latter to low-pressure (retrograde) cordierite metamorphism associated with sinistral transpressional shear zones; in consequence, garnets are commonly rimmed by cordierite in the Tranomaro belt. The garnet to cordierite breakdown took place during exhumation and isothermal decompression in a transpressive shear zone regime (Martelat *et al.* 1999).

Throughout southern Madagascar sapphirine occurs in layers enriched in Mg and Al typically associated with phlogopite-rich rocks and diopsidites (Lacroix 1941), and gem corundum deposits occur in veins in impure marbles, pyroxenites and calc-silicate gneisses in an assemblage of K-feldspar–sapphire–F-apatite–calcite–phlogopite that formed at 500 °C and 2 kbar (Rakotondrazafy *et al.* 2008). In the Betroka belt to the west of the Tranomaro belt there are sapphirine–phlogopite enrichments at Vohidava (Razakamanana *et al.* 2006), and some gneisses contain layers enriched in kornerupine and tourmaline (Megerlin 1968; Ackermand *et al.* 1991*b*). Shear zones played an important role in channelling boron-rich fluids giving rise to borosilicate phases such as grandidierite, kornerupine, serendibite, sinhalite and tourmaline (Razakamanana *et al.* 2001).

In the following sections grandidierite-bearing rocks at two key localities, Vohibola and Andrahomana, will be described and interpreted in turn.

Grandidierite–sinhalite–serendibite-bearing diopsidites from Vohibola

Field relationships

Grandidierite at Vohibola was first reported by Behier (1960) from a diopsidite that occurs as a layer in paragneiss.

There are two types of metasedimentary gneisses in the Vohibola region: (1) cordierite–garnet–sillimanite–spinel–diopside–orthoclase gneiss; (2) hypershene–diopside–hornblende with or without biotite and K-feldspar. This second type contains abundant lenses and layers, up to 50–100 m wide and several kilometres long, of coarse-grained spinel diopsidite, which contain many pegmatitic lenses or pockets that consist of phlogopite, diopside, calcite, apatite, scapolite, anorthite, anhydrite, spinel and fluorite. Figure 3 illustrates how the borosilicate phases grandidierite, serendibite and sinhalite with diopside are concentrated in fracture fillings up to a few centimetres wide along the margins of the phlogopite-rich lenses. The gneisses and diopsidite lenses are traversed by isolated granitic veins (Fig. 3); Behier (1960) reported grandidierite-bearing granites in the Tranomaro belt. From our observations we conclude that the

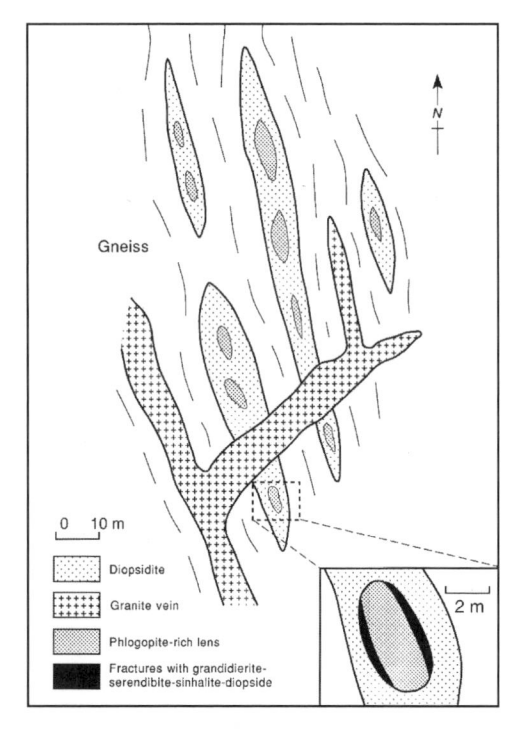

Fig. 3. Field relationships at the Vohibola phlogopite mine (modified after Ackermand *et al.* 1989) showing diopsidite lenses within regional diopside-bearing paragneisses. The diopsidites enclose phlogopite-rich lenses along the contacts of which there are fractures (black in inset) that contain the borosilicates grandidierite, serendibite and sinhalite, plus diopside.

grandidierite-bearing diopsidites predate the grandidierite-bearing granites and pegmatites.

Other rocks of sedimentary origin at Vohibola include the following. Marble layers of 100 m length and quartzite layers 10 m width occur in orthoclase–cordierite–garnet–sillimanite–spinel–diopside gneiss, and wollastonitite occurs in lenses of 10 m length in the marble and in diopsidite. Hypersthene–diopside–hornblende gneiss grades into a two-pyroxene pyroxenite that contains lenses and layers up to 10 m wide and several kilometres long of diopsidite and minor lenses of scapolitite. Dolomitic marbles near Tranomaro village are characterized by the presence of clinohumite, humite and chondrondite (Pradeepkumar & Krishnanath 2000). Hibonite–corundum deposits are associated with marbles throughout the Tranomaro region (Rakotondrazafy et al. 1996), and metasomatic sapphire deposits occur in marbles at the contact of granites (Rakotondrazafy et al. 2008).

Petrography

The boron-bearing phases grandidierite, serendibite and sinhalite (Table 1) are concentrated together with diopside, phlogopite, spinel, rare plagioclase, chlorite, dolomitic carbonate, illite and kaolin in fractures along the contact of the phlogopite-rich lenses with their host diopsidites (see Fig 3). At the centre of the phlogopite-rich lenses grandidierite occurs in drusy cavities (several centimetres in size) with deep green spinel, the grain size of which ranges up to 1 cm. At the margins of the cavities the mode of occurrence of spinel differs in three types of micro-domains dependent on its position, grain size and contact relations with the grandidierite, as follows.

(1) The innermost micro-domain is characterized by the presence of large porphyroblasts of spinel in contact with grandidierite. In the centre of this domain centimetre-size cavities are filled with euhedral spinel; this is probably an indication of fracture-filling. Relics of phlogopite occur with grandidierite where diopside has disappeared.

(2) An intermediate micro-domain has similar spinel, but the grain size is much smaller than in domain (1). Phlogopite grains are smaller, and grandidierite is associated with sinhalite mantled by calcite.

(3) An outer micro-domain is marked by rare, small spinels and by scarce grandidierite. Serendibite occurs where there is less phlogopite and calcite. Serendibite surrounds grandidierite, and elongate phlogopite grains with a grain size up to 5 cm are typical of this outer micro-domain.

The object of this description is to demonstrate the fact that borosilicate phases do not occur in granitic pegmatite dykes (as in Andrahomana; see below), and that their genesis is more related to the movement of boron-bearing fluids within and marginal to the phlogopite-rich lenses and to the contacts with their host metasedimentary diopsidites.

Diopside, which is usually coarse-grained and euhedral and has no pleochroism, commonly coexists with phlogopite. Triple junctions are formed between sinhalite, serendibite and diopside within a mosaic granoblastic texture that indicates that the rock formed at a high temperature.

Grandidierite, in crystals up to 3 cm long, shows no pleochroism and is usually colourless, but in polarized light it shows pleochroism from colourless to grey. Crystals are porphyroblastic to poikiloblastic (Fig. 4a), and in patches it may contain many small grains of phlogopite, sinhalite and Fe–Ti-oxides (Fig. 4b) (Ackermand et al. 1989, 1991a; Razafiniparany et al. 1989).

Serendibite, a rare boron silicate, was first described in Madagascar in calc-silicate paragneiss from Ihosy and Ianapera (Nicollet 1990b), and in phlogopite diopsidite from Vohibola (Razafiniparany et al. 1989; Ackermand et al. 1991a). In our rocks serendibite is often in stable contact with grandidierite (Fig. 4b), and along microfractures and cleavages it contains many inclusions filled with chlorite and calcite. Serendibite is colourless to yellow green, anhedral, and occurs in single grains and aggregates. It overgrows sinhalite and spinel.

Sinhalite is a rare borate, stable in high-grade skarns and gneisses (Werding & Schreyer 1996), and is rare in the Tranomaro rocks (Razafiniparany et al. 1989; Ackermand et al. 1991a). In our samples (e.g. M56 23-1B and M56 23-7) sinhalite occurs in unstable relationships with grandidierite, serendibite, spinel, calcite, diopside and rare pargasitic amphibole. It is orthorhombic and has an olivine-type structure; the grains are anhedral to subhedral, polarized and colourless to orange–yellow. Sinhalite grains are locally mantled by calcite and spinel in a corona-like texture, and frequently it forms pale-coloured inclusions in darker grandidierite as in Figure 4a, or in included patches with opaque oxides within grandidierite as in Figure 4b. In patches sinhalite appears to be replaced (by c. 20–30%) by a fine-grained mineral optically similar to calcite or hydrocarbonate.

Phlogopite is common and accompanied by diopside. It is mostly coarse-grained, strongly bent and often oriented, giving the rocks their prominent schistosity. It occurs as medium-grained, anhedral to sub-idioblastic flakes. It shows characteristic weak pleochroism, the pleochroic scheme being $x =$ colourless, $y = z =$ light brown. Some grains show crosscutting relationships with earlier phases indicating their post-kinematic crystallization. In some

Table 1. *Locality, rock type (mineral assemblage), texture and modal composition of rocks with boron-bearing minerals from the Tranomaro belt*

Sample	Di	Phl	Spl	Grd	Chl	Dsp	Carb	Ser	Sinh	Amp	Scp	Texture
Vohibola mine												
Rock type 2 (R2)												
Grd–Spl–Phl diopsidite												
M56 23-5	oooo	x	oo	x			x	x		x		Granoblastic
M56 23-7	ooo	o	oo	o			x		x			Granoblastic
M56 23-9	oooo			x						o		Granoblastic
M56 23-13	ooo			o								Granoblastic
M56 23-17	ooo	oo	o	x			o					Granoblastic, corona
M56 23-18	oo	o	o	o			o			ooo	oo	Granoblastic
Rock type 3 (R3)												
Grd phlogopitite												
M56 23-6	oo	oooo	oo	o						x		Granoblastic
Rock type 4 (R4)												
Grd–Srd–Sinh spinellite												
M56 23-1B		oo	oooo	o			o	x	x			Porphyritic, poikiloblastic
Andrahomana Cape												
Rock type 1 (R1)												
Grd–Spl–Phl pegmatite												
M55-1		oooo	o	ooo	oo	o	x					Porphyritic, poikiloblastic

Amp, amphibole; Carb, carbonate; Chl, chlorite; Crn, corundum; Di, diopside; Dsp, diaspore; Grd, grandidierite; Phl, phlogopite; Scp, scapolite; Ser, serendibite; Sinh, sinhalite; Spl, spinel; Crn, corundum. Symbols: x, <1 vol.%; o, 1–10 vol.%; oo, 10–20 vol.%; ooo, 20–30 vol.%; oooo, >30 vol.%. Reference example, M56; outcrop, 23; selected locality, 5.

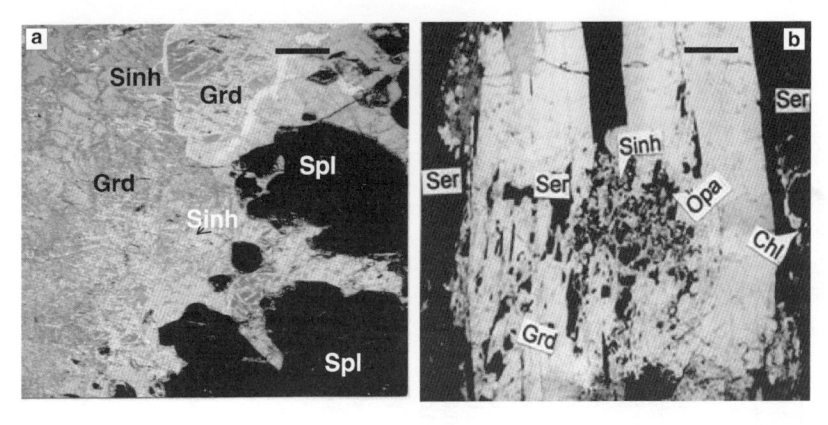

Fig. 4. Photomicrographs of textural relations in diopsidites from Vohibola (×3.2). (**a**) Large grandidierites contain inclusions of pale-coloured sinhalite and of black spinel. (**b**) Large poikiloblasts of grandidierite are in stable contact with large grains of serendibite (black), and contain included patches filled with sinhalite and opaque oxides.

patches phlogopite is corroded and deformed, when in contact with undeformed grandidierite. Fine-grained post-tectonic phlogopite is often chloritized on grain boundaries with grandidierite.

Spinel occurs as green subhedral to anhedral blasts in all mineral associations and locally may form up to 70 vol.% or more of the rock. It has various grain shapes: fine- to coarse-grained and xenoblastic. Sometimes, spinel forms inclusions in grandidierite (Fig. 4a), or is rimmed by calcite and sinhalite.

Amphibole (hornblende) is present in small amounts and is medium-grained. It shows a weak pleochroism, and it defines a lineation on foliation planes with phlogopite and diopside. It may be rimmed by basic plagioclase.

Chlorite occurs with serendibite, spinel, and grandidierite, and in patches with phlogopite, sericite and fine-grained corundum.

Opaque minerals are frequent. In places an unknown radiating opaque phase, probably a boron-bearing mineral, is abundant as a breakdown product of serendibite and sinhalite.

Corundum occurs in very small grains detectable only with the microprobe.

Mineral chemistry

Mineral analyses made in this study were undertaken with wavelength-dispersive microprobes in Kiel and Leicester. The operating conditions were 15 kV accelerating voltage and 15 nA beam current, which allow a beam diameter smaller than of 1.0 μm. Representative mineral analyses are given in Tables 2–4.

Diopside has almost no compositional change from grain to grain and a weak zoning from grain core to rim. The stoichiometry of diopside is calculated on the basis of six oxygens and the cation proportions of all analyses indicate the homogeneity of this phase, although some analyses show deviations. X_{Mg} varies between 0.94 and 0.97 with a statistical mean value (of more than 100 analyses) of 0.96, and TiO_2 differs from 0.05 to 0.34 wt%. In sample M56/23-12 the content of TiO_2 is unusually up to 2 wt%, suggesting a thermal influence. Al_2O_3 ranges between 0.87 and 2.27 wt% with a statistical mean content of 1.96 wt%, but diopsides from the outer micro-domain have contents up to 2.78 wt%. In contact with grandidierite, compositional variations record an increase of Ti and a decrease of Fe from core to rim; this suggests that iron fluctuation was able to impede the progress of the heating process. A plot of Al v. Ti has a positive slope, probably suggesting a progressive substitution between these two elements during chemical equilibration; this could be influenced by pressure and temperature variations from the inner to outer micro-domains.

Grandidierite has the theoretical structural formula [(Mg,Fe)Al$_3$SiBO$_9$] (McKie 1965), calculated on nine oxygens per formula, but in practice the electron microprobe cannot directly determine the light element boron. Boron and therefore B_2O_3 can be calculated from the difference between 100 wt% and the anhydrous total, because grandidierite contains no H_2O or CO_2. The grains are chemically almost homogeneous from core to rim with an X_{Mg} between 0.95 and 0.97, and are remarkably Mg-rich compared with other published analyses. MnO is negligible (<0.01 wt%) and TiO_2 and Cr_2O_3 are commonly between 0.01 and 0.04 wt%. AlO_2 decreases when SiO_2 increases; a similar relationship is observed between Al_2O_3

Table 2. *Representative electron microprobe analyses of grandidierite from Vohibola and Cape Andrahomana*

	Vohibola mine							Cape Andrahomana		
Rock type:	R4		R3		R2			R1		
Reference:	23-1B		23-6	23-9		23-13		55-B		
Analysis no.:	20/2 rim	20/2 core	14/1 rim	1/6 rim	1/7 rim	10/2 rim	10/2 core	2/rim	4/core	23/rim	24/core
SiO$_2$	21.19	20.98	21.62	21.01	20.78	21.82	20.90	20.91	20.84	21.12	21.23
TiO$_2$	0.00	0.00	0.00	0.05	0.00	0.01	0.02	0.01	0.01	0.00	0.04
Al$_2$O$_3$	53.06	52.98	51.83	54.12	54.22	53.58	53.82	53.37	53.47	52.87	52.57
Cr$_2$O$_3$	0.00	0.01	0.01	0.09	0.00	0.03	0.00	0.00	0.01	0.00	0.02
FeO*	1.40	1.19	0.96	1.14	0.95	1.02	1.24	3.48	3.43	3.44	3.71
MnO	0.04	0.05	0.00	0.01	0.04	0.01	0.00	0.04	0.02	0.05	0.06
MgO	13.35	13.31	13.66	13.52	13.56	13.87	13.75	12.04	12.26	11.95	12.04
CaO	0.00	0.04	0.03	0.02	0.00	0.11	0.01	0.00	0.01	0.00	0.00
K$_2$O	0.00	0.00	0.02	0.00	0.00	0.02	0.01	0.01	0.00	0.00	0.00
Na$_2$O	0.00	0.01	0.01	0.00	0.02	0.01	0.00	0.00	0.01	0.00	0.00
Total	89.04	88.57	88.14	90.01	89.80	90.51	89.75	89.86	90.06	89.43	89.67
Oxide basis[†]											
Si	7.5	7.5	7.5	7.5	7.5	7.5	7.5	7.5	7.5	7.5	7.5
Ti	1.01	1.01	1.04	0.97	0.98	1.02	0.99	1.00	0.99	1.01	1.02
Al	–	–	–	0.007	–	0.001	0.003	–	–	–	0.002
Cr	2.98	2.99	2.94	3.01	3.02	2.96	3.00	3.00	3.00	2.99	2.97
Fe	–	0.002	0.002	0.013	–	0.004	–	–	–	–	0.001
Mn	0.06	0.05	0.04	0.05	0.04	0.04	0.05	0.14	0.14	0.14	0.15
Mg	0.001	0.002	–	0.001	0.002	0.001	–	0.001	0.001	0.001	0.002
Ca	0.95	0.95	0.98	0.95	0.96	0.97	0.97	0.86	0.87	0.85	0.86
K	–	0.002	0.001	0.001	–	0.007	0.001	–	0.001	–	–
Na	–	–	0.001	–	–	0.001	0.001	0.001	–	–	–
	–	0.001	0.001	–	0.001	0.001	–	–	0.001	–	–
Total	5.00	5.00	4.99	5.01	5.01	5.00	5.01	5.00	5.01	4.99	5.00
X$_{Mg}$	0.94	0.95	0.96	0.95	0.96	0.94	0.95	0.86	0.86	0.86	0.85

*Total Fe as FeO.
[†] Oxide basis without Bor.

Table 3. *Representative electron microprobe analyses of serendibite and sinhalite from the Vohibola mine*

Rock type:	R1		R2		R2	
Phase:	Serendibite		Serendibite		Sinhalite	
Reference:	23-1B		23-17		12	18
Analysis no.:	22/5 rim	22/8 core	1	2	20/1 rim	20/8 core
SiO$_2$	22.20	23.19	25.81	25.43	0.04	0.00
TiO$_2$	0.06	0.10	0.02	0.07	0.02	0.02
Al$_2$O$_3$	32.96	33.05	32.44	32.96	42.25	42.46
Cr$_2$O$_3$	0.02	0.02	0.01	0.00	0.01	0.01
FeO	3.81	4.17	4.33	4.13	2.90	3.11
MnO	0.00	0.02	0.03	0.04	0.02	0.03
MgO	13.99	14.08	14.44	14.08	29.84	29.63
CaO	15.78	16.31	15.70	16.11	0.02	0.00
K$_2$O	0.07	0.04	0.00	0.02	0.00	0.02
Na$_2$O	0.00	0.01	0.40	0.32	0.00	0.00
Total	88.89	90.99	93.18	93.16	75.10	75.28
Oxide basis	37	37	37	37	5	5
Si	5.71	5.84	6.31	6.22	0.002	0.000
Ti	0.012	0.019	0.004	0.013	0.001	0.001
Al	9.99	9.81	9.34	9.500	2.05	2.05
Cr	0.004	0.004	0.002	0.000	0.000	0.000
Fe^{2+}	0.820	0.878	0.885	0.845	0.100	0.107
Mn	0.000	0.004	0.006	0.008	0.001	0.001
Mg	5.37	5.28	5.26	5.13	1.83	1.812
Ca	4.35	4.39	4.11	4.22	0.001	–
K	0.023	0.013	–	0.006	–	0.001
Na	–	0.005	0.190	0.152	–	–
Total	26.28	26.25	26.11	26.10	3.98	3.97
X_{Mg}	0.87	0.86	0.86	0.86	0.94	0.95

and B$_2$O$_3$ (Fig. 5). In comparison with compositions of grandidierite from Andrahomana, Vohibola grandidierite shows a negative slope of B$_2$O$_3$ v. Al$_2$O$_3$; this clearly signifies a zoning of the Al and B from the inner to the outer micro-domains. Alkalis in grandidierite from Vohibola are higher than those in grandidierites from other localities. Grandidierites have an X_{Mg} between 0.95 to 0.0.96; in comparison, in other occurrences they have an X_{Mg} less than 0.7, indicating that they formed under hydrothermal conditions at low pressure and high temperature (Olesch & Seifert 1976; Werding & Schreyer 1978, 1984, 1996). Heide (1992) showed that Mg-rich grandidierite is stable at 4 kbar and 300–600 °C and becomes unstable at medium pressure (5–10 kbar) depending on temperature and water activity.

Serendibite [Ca$_2$(Mg,Fe)$_3$(Al,Fe)$_{4.5}$B$_{1.5}$Si$_3$O$_{23}$], belonging to the aenigmatite group like sapphirine, shows a very small compositional range. Nevertheless, there is substitution between B and Al, but B usually occupies preferentially the tetrahedral site (van der Veer et al. 1993). The substitution between Mg and Al in serendibite is shown in Figure 6. TiO$_2$ is nearly undetectable or has values under 0.05 wt%; MnO contents are even lower, and Na$_2$O is <0.1 wt%. The homogeneity of this phase is shown by its X_{Mg} values, which are 0.87 ± 0.01.

Sinhalite [(Mg,Fe)AlBO$_4$], a phase that forms on the margin of or as an inclusion in grandidierite, shows no large compositional changes from grain to grain, which suggests that the sinhalite from Vohibola has a composition very close to its ideal formula (Hayward et al. 1994). Calculation of the stoichiometric formula on the basis four oxygens and determination of the B$_2$O$_3$ contents by the difference of total wt% from 100 wt% gives a B$_2$O$_3$ content that ranges from 24.5 to 26.7 wt%. Aluminium is around 42 wt%. No minor elements were detected. B occupies tetrahedron-sites with Al, and the negative slope between Mg and Al in Figure 6 suggests that Mg and Al mutually

Table 4. Representative electron microprobe analyses of oxides and boron-free silicates from Vohibola and Cape Andrahomana

	Vohibola mine										Cape Andrahomana			
Rock type:	R4	R3	R2	R3	R2	R4	R2		R4	R2	R4			R1	
Phase:	Di	Di	Di	Amp	Amp	Spl	Spl		Phl	Phl	Chl	Spl	Crn	Phl	Chl
Reference:	23-1B	23-6	23-17	23-6	23-17	23-1B	23-15		23-1B	23-6	23-18	23-1B	55-M		
Analysis no.:	1/15 core	11	5	10	1	20/1 core	20/11 rim	5/5 rim	10/8 rim	20/2 core	5	6	107 rim	41	11	22	28	1	26	29

SiO$_2$	55.00	53.68	53.24	43.12	42.13	0.00	0.00	0.09	0.04	42.23	42.22	42.13	39.51	19.72	0.00	0.00	0.58	40.33	41.25	33.91
TiO$_2$	0.04	0.15	0.03	0.40	0.1	0.12	0.00	0.02	0.02	0.21	0.50	0.51	0.23	0.02	0.01	0.02	0.04	1.79	1.73	0.03
Al$_2$O$_3$	0.41	1.17	1.59	14.06	16.97	67.81	67.7	68.37	68.61	13.57	14.00	14.54	15.17	48.03	63.22	63.56	84.51	12.65	12.73	33.22
Cr$_2$O$_3$	0.02	0.01	0.00	0.02	0.00	0.03	0.03	0.00	0.00	0.23	0.02	0.06	0.00	0.00	0.00	0.00	0.76	0.00	0.00	0.00
FeO*	2.82	1.15	1.26	3.46	3.61	9.01	10.41	11.04	9.01	2.01	2.71	3.02	2.49	6.34	23.93	23.37	0.0	4.10	3.47	7.25
MnO	0.11	0.01	0.01	0.06	0.06	0.09	0.10	0.02	0.05	0.05	0.03	0.00	0.01	0.09	0.20	0.15	0.3	0.07	0.02	0.06
MgO	17.43	17.61	16.92	18.44	17.15	22.96	21.97	20.59	23.09	26.66	26.18	25.87	25.72	14.76	12.81	13.05	0.03	24.54	23.97	12.37
CaO	24.22	26.18	26.62	13.23	13.72	0.01	0.00	0.04	0.00	0.02	0.06	0.00	0.02	0.07	0.02	0.00	0.02	0.01	0.01	0.20
K$_2$O	0.00	0.01	0.00	1.44	0.51	0.00	0.00	0.01	0.00	10.46	8.56	8.82	10.74	0.02	0.00	0.00	0.00	9.12	9.59	0.17
Na$_2$O	0.21	0.05	0.12	2.70	2.69	0.02	0.00	0.09	0.04	0.10	0.05	0.03	0.06	0.04	0.00	0.00	0.00	0.67	0.58	0.02
Total	100.26	100.02	99.79	96.93	96.43	100.05	100.48	100.37	100.86	95.54	94.33	94.98	93.95	89.09	100.19	100.15	86.24	93.28	93.35	87.23

Oxide basis
	6	6	6	46	46	4	4	4	4	22	22	22	22	28	4	4	3	22	22	28
Si	1.99	1.94	1.95	12.30	12.06	–	–	0.002	0.001	5.92	5.87	5.85	5.67	3.63	–	–	0.012	5.83	5.93	6.27
Ti	0.001	0.004	0.001	0.086	0.022	0.002	–	0	0	0.022	0.052	0.053	0.025	0.003	–	–	0.001	0.195	0.187	0.004
Al	0.018	0.05	0.068	4.737	5.736	1.97	1.97	1.99	1.98	2.24	2.30	2.38	2.56	10.4	1.97	1.98	1.97	2.15	2.16	7.24
Cr	–	–	–	0.005	0.000	0.001	0.001	–	–	0.025	0.002	0.007	–	–	–	–	–	–	–	–
Fe^{2+}	0.086	0.035	0.038	0.825	0.864	0.186	0.215	0.228	0.184	0.235	0.315	0.35	0.299	0.977	0.53	0.516	0.013	0.495	0.417	1.12
Mn	0.003	–	–	0.014	0.014	0.002	0.002	–	0.001	0.006	0.004	–	0.001	0.014	0.005	0.003	0	0.009	0.002	0.009
Mg	0.943	0.934	0.921	7.841	7.320	0.845	0.827	0.780	0.84	5.56	5.43	5.35	5.50	4.06	0.506	0.514	0.009	5.29	5.14	3.41
Ca	0.942	1.00	1.04	4.04	4.21	–	–	0.001	0	0.003	0.009	–	0.003	0.014	0.001	–	0.001	0.002	0.002	0.04
Na	–	–	–00	1.493	1.494	0.001	–	0.004	0.002	0.027	0.013	0.008	0.017	0.014	–	–	0.001	1.681	1.76	0.04
K	0.015	0.004	0.008	0.524	0.186	–	–	–	–	1.86	1.52	1.56	1.93	0.006	–	–	–	0.188	0.162	0.007
Total	4.00	3.985	4.02	31.86	31.90	3.01	3.02	3.01	3.01	15.99	15.51	15.552	15.99	19.15	3.01	3.01	2.00	15.83	15.76	18.14
X_{Mg}	0.92	0.96	0.96	0.90	0.89	0.82	0.79	0.77	0.82	0.96	0.95	0.94	0.95	0.80	0.50	0.50	0.41	0.91	0.93	0.75

*Total Fe as FeO.
c, core; r, rim.

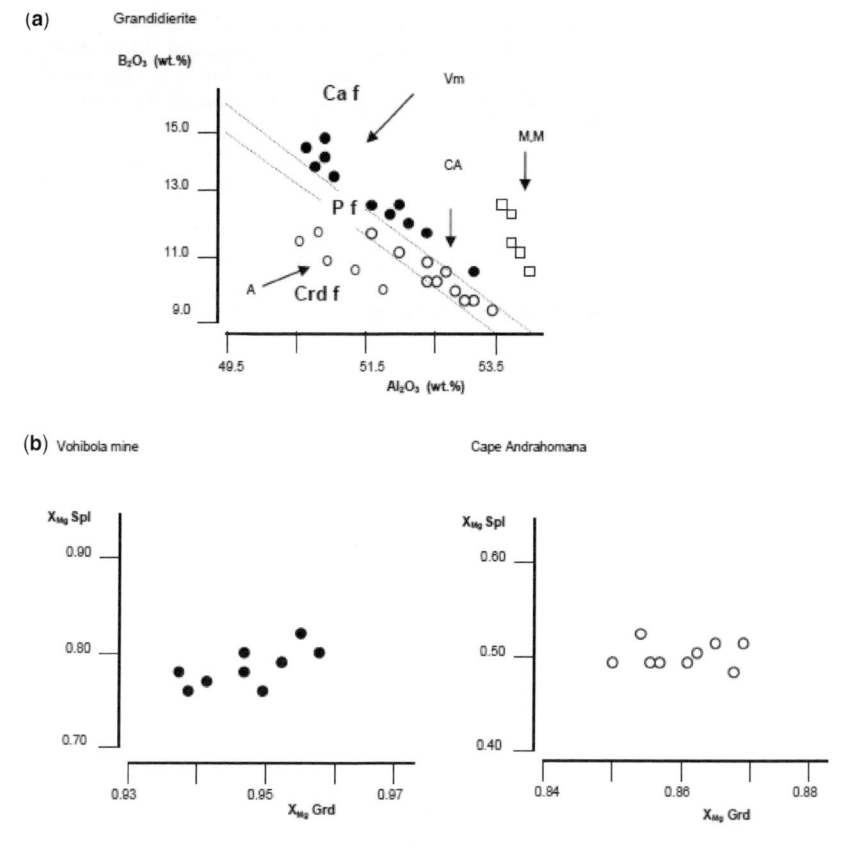

Fig. 5. Compositional variations of grandidierite and spinel from Vohibola and Andrahomana. (**a**) Al$_2$O$_3$ v. B$_2$O$_3$ (wt%) in grandidierite. Vm, Vohibola (●); CA, Cape Andrahomana (○); for comparison data are given from Mchinji in Malawi (M. M., Haslam 1980) and Anteto, 40 km south of Ihosy (A, Razakamanana 1999); Ca f, calc-silicate field; P f, pegmatite field; Crd f, cordieritite field. (**b**) Spinel X_{Mg} v. grandidierite X_{Mg} for Vohibola (●) and Andrahomana (○).

exchanged in an inverse sense during the crystallization process. X_{Mg} values usually vary between 0.94 and 0.96, but become lower to 0.8 when connected with 48.03 wt% of Al$_2$O$_3$ and 14.8 wt% of MgO. This can indicate that there was locally an increase of iron during the replacement of a boron mineral by sinhalite.

Phlogopite is the most magnesium-rich phase in the rocks with X_{Mg} oscillating between 0.95 and 0.96. TiO$_2$ varies between 0.15 and 0.52 wt%, but in sample M65-XV it reaches 1.61 wt% with an X_{Mg} of 0.97.

Spinel is zoned, as indicated by changes in the Mg/Fe ratio from an X_{Mg} of 0.69 in the core to 0.83 in the rim (rim analyses shown in Fig. 5b). It may contain a few anhedral corundum grains, as in sample M56/23-15. Na$_2$O and CaO are detectable at 0.13 wt% and 0.31 wt%, respectively.

Chlorite has TiO$_2$ contents up to 0.79 wt% and its X_{Mg} is always close to 0.97.

Phase relations and textural interpretations

In summary, X_{Mg} shows the following relationships between the Mg- and Fe-bearing phases:

(1) phlogopite > grandidierite > spinel (sample M65 X);

(2) phlogopite > grandidierite > diopside > sinhalite > serendibite > chlorite > spinel (M56 23-1B);

(3) phlogopite > grandidierite > diopside > hornblende > spinel (sample M56 23-6);

(4) grandidierite > diopside > hornblende (sample M56 23-9);

(5) diopside > grandidierite (samples M56 23-12 and M56 23-13);

(6) grandidierite > spinel (M56 23-15 and 23-17).

Textural observations suggest that early phases were diopside, phlogopite, sinhalite and spinel (Fig. 4a), and late phases were grandidierite and

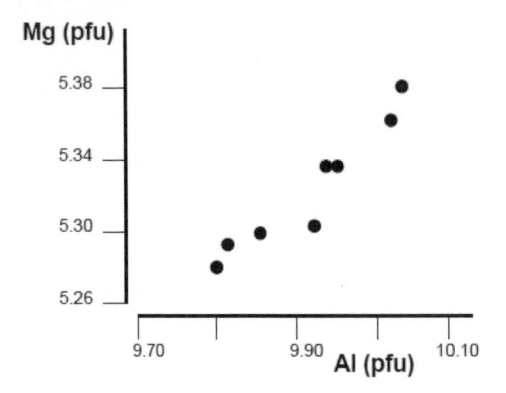

Serendibite

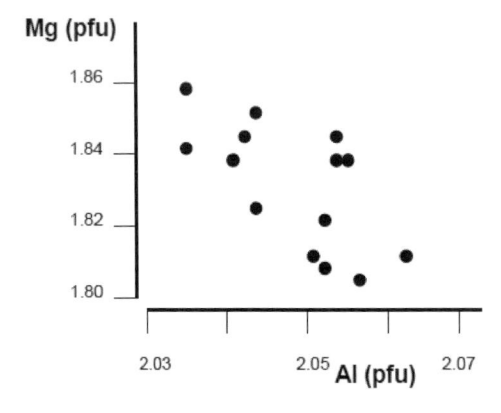

Mg (pfu)

Fig. 6. Variation diagrams of Mg v. Al per formula unit (p.f.u.) for serendibite and sinhalite from Vohibola, showing positive and negative relationships, respectively.

serendibite (Fig. 4b). The latest phases along cleavages in grandidierite were phlogopite, calcite, sinhalite and fine-grained diaspore. Diopside was stable during all stages of metamorphism. There are sharp and stable contacts between grandidierite–serendibite (Fig. 4b), sinhalite–diopside–grandidierite and phlogopite–grandidierite.

The components of these phases (K_2O, Na_2O, CaO, FeO, MgO, Al_2O_3, SiO_3, B_2O_3 and H_2O) occur in different proportions. This means that the reactions can be studied only in the system KNC(FM)AS(BH). However, in general, the alkali elements do not play an important role in all the rocks, therefore these two elements can be neglected. Thus, the system can be written (CFM)ASHB or in a simple combination CMAS. The chemical compositions of the silicates and spinel show a negligible content of Fe_2O_3. B_2O_3 must have played an important role in stabilizing some phases, as discussed below. K_2O as a major component is only found in phlogopite; CaO only in diopside, calcite, serendibite and rarely in plagioclase; and Na_2O in plagioclase. Thus, the main components are reduced to MgO, FeO, Al_2O_3 and SiO_2. There are very few experimental data for these mineral associations. Werding & Schreyer (1978) studied grandidierite experimentally in the system MASBH, but no data for sinhalite and serendibite are available. The Vohibola parageneses are similar to those in the Adirondack Mountains, USA, where grandidierite occurs as a secondary mineral in serendibite-bearing calc-silicate marbles (Grew et al. 1990; Grew 1996).

On the basis of the observed assemblages and textural relations between the Tranomaro minerals, three crystallization stages are deduced: (1) a coarsening development giving rise to porphyroblasts and poikiloblasts; (2) formation of symplectites; (3) corona formation.

Coarsening development of grandidierite. Between pale green diopsidite and the borosilicate concentrations is a 1 cm wide leucocratic rim consisting of diopside and an unknown alteration mineral. Microscopically, many parageneses are observed, but seven main reactions are critical, as follows.

Phlogopite occurs as a relict in grandidierite without diopside, and diopsidite is commonly invaded by melt products along fractures where grandidierite is developed. This suggests the following reaction:

$$Di + Phl + fluid = Grd. \quad (R1)$$

Spinel–phlogopite dominates in small pockets in grandidierite. Within a coarse matrix of spinel and phlogopite are the reactants grandidierite and sinhalite. Prismatic blades of grandidierite contain inclusions of phlogopite, diopside and spinel. These relationships suggest the following reaction:

$$Phl + Spl1 + fluid = Grd + Sinh. \quad (R2)$$

This reaction is valid for textural relations and reactions in Figure 7a–d; in particular those in Figure 7a.

Formation of symplectites. Where grandidierite is in contact with diopside there is often a symplectite of sinhalite–calcite; this gives rise to the reaction

$$Spl + Di + fluid = Grd + Sinh + Cc \quad (R3)$$

(see Fig. 7a, which shows that the dashed tie line Spl–Di is broken and Grd–Sinh–C is stable).

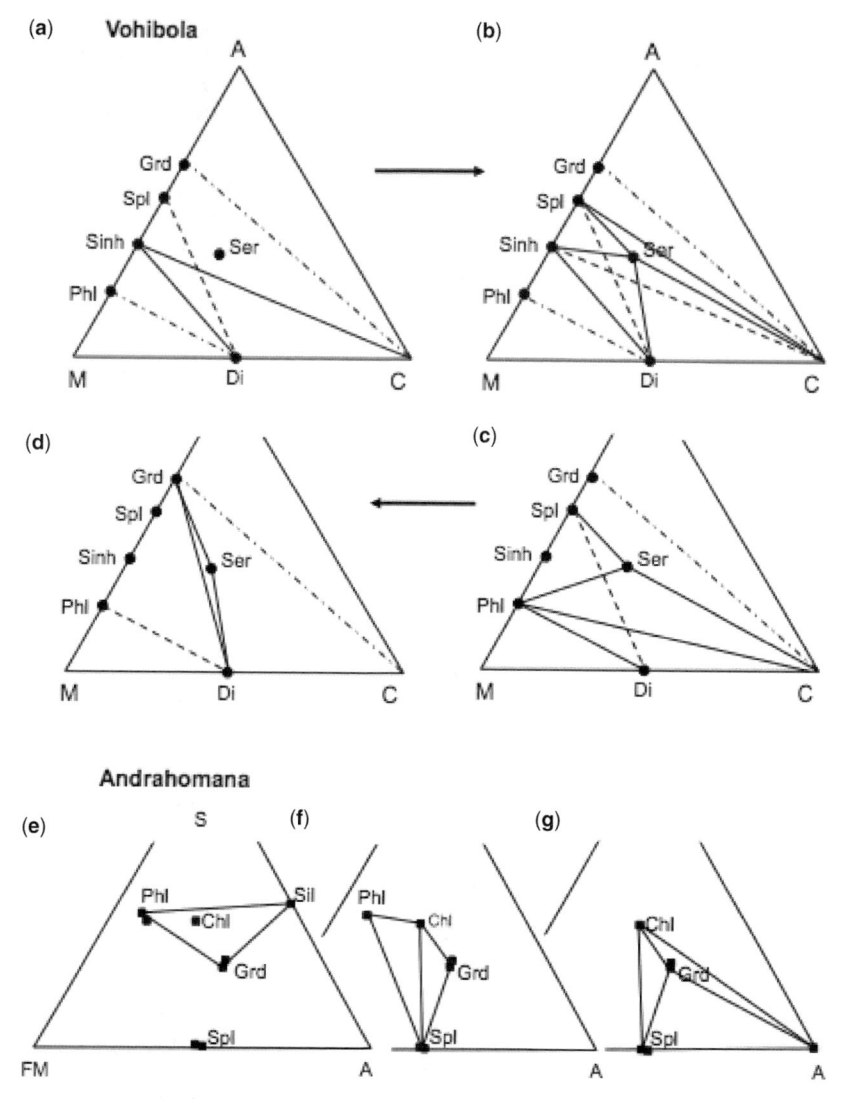

Fig. 7. ACM diagrams to illustrate the sequence of observed textural relations at Vohibola, and ASFM diagrams to illustrate the textural relations seen at Andrahomana. In (a)–(d) the dashed tie lines enclose fields (e.g. Grd–C(calcite)–Di–Phl) representing the whole-rock chemical compositions of assemblages with boron-bearing phases. The triangles (e)–(g) show observed textural relations that are compatible with the three reactions given in the text.

In patches with reactants, there are blades of grandidierite within symplectitic aggregates of phlogopite and calcite; this feature points to the following reaction:

$$\text{Spl} + \text{Di} + \text{fluid} = \text{Grd} + \text{Cc} + \text{Phl}. \quad (R4)$$

Formation of serendibite. Within spinel-rich areas, in a paragenesis comparable with reaction (R2), there are crystals of serendibite intergrown with a fine-grained symplectite of calcite–phlogopite.

Although there is no diopside in direct contact with these reactants, this implies the following reaction between the diopside and spinel:

$$\text{Spl} + \text{Di} + \text{fluid} = \text{Ser} + \text{Cc} + \text{Phl} \quad (R5)$$

(see Fig. 7c).

Degradation of serendibite. Very small grains of serendibite are complexly mantled by calcite, sinhalite and often with an opaque oxide. However, serendibite and sinhalite are not stable under heating conditions, and thus give rise to a reaction that may be related to fluid circulation during faulting or shearing:

$$Ser + O_2 + CO_2 = Sinh + Cc + Fe\text{-}Ox. \quad (R6)$$

(This reaction cannot be shown in ACM space of Fig. 7.)

Degradation of sinhalite. In spinel-rich rocks, spinel appears as a porphyroblast or corona. Small grains of sinhalite are rimmed by spinel. This texture shows that the sinhalite, as a Mg–Al borate, could be earlier than the matrix spinel and points to the reaction

$$Sinh = Spl\ 2 + fluid + vapour. \quad (R7)$$

(This reaction cannot be shown in ACM space of Fig. 7.)

Sinhalite grains sometimes occur within spinel 2, and corundum is hydrated to diaspore and opaque oxides. These relations suggest that an oxidation reaction developed as follows:

$$Sinh = Spl\ 2 + Crn + Opa + fluid\ (H_2O + O_2) \quad (R8)$$

(see Fig. 7c, which shows Sp has broken down to produce Sp and Crn (in A) on one side and opa (in M) on the other.

Metamorphic P–T conditions

Rocks that surrounded the grandidierite-bearing rocks are garnet–cordierite–biotite paragneisses. Several geothermobarometric models are currently used to estimate the $P–T$ conditions of granulites. We have estimated $P–T$ with the Opx–Cpx geothermobarometer of Lindsley (1983), which uses the ternary diagram Ca–Mg–Fe with a pressure from less than 1 atm to 15 kbar. For these conditions we calculate that $P = 5$ kbar and $T = 800–900\ °C$. The estimate with a garnet–cordierite geobarometer is $P = 5 \pm 1$ kbar and $750 \pm 50\ °C$.

Synthesis and phase relationships demonstrate that a serendibite–spinel–sinhalite–diopside paragenesis is stable under conditions of $P = 2–5$ kbar and $T\ c.\ 600\ °C$ (Werding et al. 1990). Grandidierite precipitates in low-temperature and low-pressure regions where $T = 150–250\ °C$ and $P = 3–4$ kbar (Cartwright & Buick 1994). Heide (1992) published stability conditions for Mg-grandidierite at $P = 4$ kbar and $T = 300–500\ °C$, but it became unstable at a medium pressure (5–10 kbar) depending on temperature and water activity. Experimental data by Werding & Schreyer (1996) and Schreyer & Werding (1997) indicate that grandidierite, whether rich in boron or not, has an upper stability of $c.\ 10$ kbar at $c.\ 950\ °C$. Grew (1996) pointed out that the Vohibola grandidierite assemblage is very similar to that in the Adirondacks, USA, which is stable at a pressure of 7–8 kbar.

Using other geobarothermometric mineral pairs (e.g. garnet–cordierite; spinel–cordierite; garnet–biotite; orthopyroxene–clinopyroxene) from neighbouring rocks, $P–T$ values vary from 4 to 7.5 kbar and 700 to 800 °C (Ackermand et al. 1989; Nicollet 1990b; Razakamanana 1999; Pierdzig 1992; Rakotondrazafy 1992). Experimentally, Olesch & Seifert (1976) synthesized grandidierite at a high temperature with $X_{Mg} = 1.0$ at 780 °C and $X_{Mg} = 0.9$ at 700 °C at low pressure. Black (1970) estimated a minimum temperature of 600 °C from the aureole of a granodiorite body in New Zealand. Grandidierite-bearing pegmatite formed during a late phase of metamorphism and deformation at $T < 750\ °C$ and $P = 4–5$ kbar in a Rogaland shear zone, Norway (Huijsmans et al. 1982). Grandidierite can be a prograde high-temperature mineral (Schau et al. 1986; Carson et al. 1995). On the other hand, Vry & Cartwright (1994) suggested that grandidierite can appear under low-temperature and low-pressure conditions ($T = 150–250\ °C$ and $P = 3–4$ kbar). Using hypothetical geobarothermometric data compiled by Grew (1996), we confirm the $P–T$ conditions previously estimated by many workers for Tranomaro rocks (e.g. Moine et al. 1985, 1997; Nicollet 1990b).

Grandidierite–spinel–phlogopite pegmatite at Andrahomana

Field relations

Lacroix (1902, 1904) discovered the type locality of grandidierite on the cliffs at Andrahomana on the southern coast of Madagascar. Lacroix (1922–1923) stated that the grandidierite occurs in a pegmatite and aplite as crystals up to 8 cm long that enclose poikilitically all other phases in the rock: quartz, microcline, almandine, pleonaste, andalusite and biotite. The pegmatite occurs at an old pegmatite mine near Andrahomana in a biotite–garnet–cordierite gneiss that has many pegmatitic and granitic veins (Behier 1960; Besairie 1973; Grew 1996; our field observations), and Behier (1961) reported tourmaline at this outcrop. The occurrence is clearly located within a major shear zone (Fig. 1). Now it is practically impossible to find any fresh grandidierite-bearing rocks at Andrahomana. We based our study on a sample of grandidierite–phlogopite pegmatite, in which the grandidierite is accompanied by diopside, spinel and phlogopite (von Knorring et al. 1969).

Petrography

Our coarse-grained sample contains deep green–blue grandidierite crystals up to 1 cm long in an aplitic matrix of phlogopite–spinel–tourmaline–aluminium silicate–K-feldspar. Microscopically, the grandidierite is porphyroblastic to poikiloblastic. Phlogopite forms anhedral to sub-euhedral microcracked flakes. Grandidierite–phlogopite symplectites are frequent, and phlogopite is patchily chloritized. Coarse spinel is poikilitic and contains inclusions of grandidierite and phlogopite, and spinel occurs within the grandidierite. Chlorite has two generations: chlorite I is often intergrown with spinel that mantles grandidierite; chlorite II occurs as a corrosion product of phlogopite. Rare poikiloblastic phases are quartz, microcline, garnet and werdingite (Grew 1996). Spinel surrounds grandidierite and corundum, which is patchily altered to diaspore.

Mineral chemistry

Mineral analyses are given in Tables 2 and 4.

Grandidierite has a calculated borate content (B_2O_3) between 8.5 and 12.5 wt%, but this scatter depends on the analyses of the other elements. The average X_{Mg} value is 0.86, and a plot of Al_2O_3 v. B_2O_3 contents shows a negative slope, with Al_2O_3 increasing with decrease in B_2O_3 (Fig. 5a). This trend points to a substitution of Al for B, which probably happened during melt passage; this substitution is clearly overlapped by substitution of Si for Al. Both substitutions permit the following substitution equation: (Si, B) = (Al, Al). There is only a slight compositional change from grain to grain and from grain core to rim. TiO_2 and Cr_2O_3 contents are between 0.01 and 0.04 wt%, and CaO, Na_2O and K_2O are negligible. Grandidierite inclusions in spinel contain TiO_2 up to 0.04 wt%.

Spinel is a Mg-hercynite and the composition changes from grain to grain but shows no zoning from grain core to rim. Stoichiometric calculations on the basis of four oxygens and three cations show only a small amount of Fe^{3+}. X_{Mg} ranges from 0.48 to 0.52 (Fig. 5b) and indicates that the spinel is a Mg-hercynite. A plot of FeO v. MnO (MnO variation from 0.15 to 0.22 wt%) contents shows an inverse relation of these elements. TiO_2 is very low, between 0.01 and 0.03 wt%.

Phlogopite shows a weak compositional change from grain to grain. The analyses show a small amount of alkalis with 1.8 instead of 2.0 calculated with 22 oxygens per formula unit. X_{Mg} is constant at 0.92. TiO_2 is also constant at 1.77 wt%; only phlogopite as an inclusion in spinel and grandidierite has lower TiO_2; nevertheless, adjacent grandidierite contains up to 2.0 wt% TiO_2. Ti varies inversely with X_{Mg}. Using Ti for calculation of precipitation temperature (geothermometer of Lal 1991), the calculated temperature is close to 835 °C. Cl and F contents (not shown in Table 4) are 0.05 wt% and 0.50–1.97 wt%, respectively; these values are confirmed by phlogopite compositions in similar rocks (de la Roche 1963; Joo' 1971; Moine et al. 1996). BaO varies between 0.03 and 0.07 wt%.

Chlorites, formed by alteration of various phases, have different compositions, especially shown by their MgO–FeO relations. Chlorite in spinel and grandidierite has lower MgO (12 wt% on average) than in phlogopite (20 wt% on average). The structural formula of chlorite is based on 28 oxygens, and X_{Mg} values are close to 0.75. Alkali (calc-alkali) elements are on average similar to those of chlorites from other localities (0.02 wt% Na_2O, 0.17 wt% K_2O and 0.20 wt% CaO), but chlorite intimately associated with diaspore contains higher alkali contents.

The above X_{Mg} values indicate the following sequence: phlogopite > chlorite (in Phl) > grandidierite > spinel > chlorite (in Spl). The mutual distribution between $(X_{Mg})^{Spl}$ and $(X_{Mg})^{Grd}$, shown in Figure 5b, suggests that the Fe–Mg distribution between coexisting hercynite spinel and grandidierite is temperature dependent.

Textural interpretation and phase relations

We use the FMAS system to plot all the analysed phases, but this system can give only a preliminary view of the phase relations, because other oxides such as water and boron oxide have to be taken into account. From textural observations of small phlogopite and sillimanite grains in grandidierite porphyroblasts the following three reactions are suggested, which we correlate with the three ASFM diagrams in Figure 7e–g.

$$Phl + Sill + fluid\ 1 = Grd + fluid\ 2 \quad (1)$$

(Fig. 7e). Fluid 2 removed the elements necessary for the crystallization of grandidierite. Grandidierite and phlogopite relicts in spinel (hercynite) are often associated with chlorite:

$$Grd + Phl + fluid\ 3 = Spl + Chl + fluid\ 4 \quad (2)$$

(Fig. 7f). During partial melting grandidierite can break down to spinel, chlorite and corundum. This corundum is often hydrated to give rise to diaspore, as observed in thin section:

$$Grd + melt = Spl + Chl + Crn/Dsp + fluid\ 5 \quad (3)$$

(Fig. 7g).

Grew (1996) found that sillimanite, andalusite, grandidierite and werdingite crystallized with K-feldspar and quartz in these rocks. Semroud et al.

(1976) suggested that silica deficiency is not an essential condition for the formation of grandidierite. The low stability limit of grandiderite appears to be close to the minimum melting curve of granite, and grandidierite is often associated with evidence of partial melting. The removal of such a melt could lead to silica deficiency favouring quartz-free grandiderite-bearing assemblages (Haslam 1980).

We conclude that hydration of early anhydrous assemblages was important in the Cape Andrahomana area, where grandidierite occurs sporadically. This hydration was probably associated with partial melting that gives rise to granite and pegmatite formation, and it took place during introduction of granitic veins derived from nearby crustal-melt granitic bodies.

Discussion and interpretations

We focus our discussion here on the source and movement of the fluids responsible for formation of the borosilicate phases. This leads inevitably to the problem of the origin of the diopsidites, and the role of fluid transport in the shear zones.

For assessment of the $P-T$ conditions of coexisting grandidierite–serendibite–sinhalite at Vohibola, no phases suitable for geothermobarometers are available; useful theoretical studies of MgAl-borates have been provided by Maghsudania (1991), Daniels et al. (1997) and Pöter & Schreyer (1998). We have used the compositions of mineral phases within the segregations, assuming that they represent equilibrium assemblages, but it is still difficult to estimate precisely the $P-T$ conditions of grandidierite, serendibite and sinhalite because of the lack of thermodynamic data for these minerals.

Grandidierite from Vohibola is remarkably unpleochroic, is enriched in Mg ($X_{Mg} = 0.94-0.97$), and has a high Al and B content in comparison with grandidierites from other localities in the world. Boron-enriched calc-silicate rocks and marbles like those at Ihosy and Ianapera in the Tranomaro belt (Nicollet 1985, 1990a, b) are also observed in other high-grade regions of the world, such as Mchinji, Malawi (Haslam 1980) and the Melville Peninsula in the Franklin district of Canada (Hutcheon et al. 1977). Dravite tourmaline containing small grains of serendibite from calc-silicate gneisses at Ihosy and Ianapera is interpreted as a breakdown product of serendibite (Nicollet 1990b).

The presence in grandidierite of phlogopite and diopside at Andrahomana suggests that the grandidierite could have originated during a melting process that transported boron in granitic veins that intruded calc-silicate rocks. However, in the outer margins of the phlogopite-rich pegmatitic lenses (Fig. 3), spinel and diopside frequently react to give serendibite and sinhalite. This reaction confirms that grandidierite is a prograde high-temperature mineral (Schau et al. 1986; Davidson et al. 1990; Carlson et al. 1995). It also shows that melting affected the diffusion process in the marginal zones of the lenses.

The widespread presence of diopside, scapolite, calcite and spinel in neighbouring rocks suggests an evaporitic metasedimentary origin (Moine et al. 1981). Marly metasediments were intruded and metasomatized by Proterozoic granites during granitization at Anosyan in the Fort Dauphin belt (Paquette et al. 1994), and diopside, phlogopite and spinel occur in Mg,Al-rich metasomatic skarns. The grandidierite and sapphirine in the Tranomaro belt formed in high-grade gneisses under conditions of $P = 4-6$ kbar and $T = 700-800$ °C.

Worldwide, serendibite is characteristic of skarn deposits in contact aureoles and in granulite–amphibolite-facies terranes. It may result from metamorphism of boron-rich sediments, as in calc-silicate marbles in several localities in the Grenville Province in the USA (Grew et al. 1990), and in phlogopite–plagioclase skarns at Tayozhnoye in the Aldan Shield, Russia (Grew et al. 1991). Alternatively, it may be the result of boron metasomatism along the contacts between carbonate sediments and other rocks. For example, it occurs in a calc-silicate contact zone between marble and granulite-facies granite at the type locality in Sri Lanka (Prior & Coomaráraswámy 1903; see also Mathavan & Fernando 2001), and in skarns together with grandidierite, kornerupine and tourmaline at the contact between magnesian marbles and desilicified aluminosilicate rocks in New York and California (USA), in the Taezhnyi deposit of southern Yakutia (Russia) and in the Pamirs (Tajikistan) (Aleksandrov & Troneva 2006).

We conclude that the serendibite and sinhalite at Vohibola in the Tranomaro belt formed under high-temperature conditions that enhanced the transport of boron along the contacts between carbonate rocks and granites and pegmatites. This process was probably terminated by the development of grandidierite (with grains up to 3 cm long), which filled open zones and fractures along the borders of phlogopite-rich lenses within diopsidites. Most textural reactions indicate that the borosilicates were product phases during the growth of the fractures. This corresponds to the last crystallization stage probably under granulite-facies metamorphic conditions. Later retrogressive metamorphism affected the rocks when B_2O_3-, H_2O- and CO_2-bearing fluids infiltrated the shear zones. These reactions occurred during isothermal decompression from c. 10 kbar to c. 3 kbar, as documented by Ackermand et al. (1989).

Comparable metasedimentary gneisses and granulites, usually enriched in Mg and Al as indicated by the presence of sapphirine, and in places containing rare borosilicates, occur in other separated fragments of Eastern Gondwana (Grew & Manton 1986). For example, in the Madurai block of southern India quartzites, biotite gneisses and calc-silicate rocks are interbedded with silica-poor pelitic granulites containing kornerupine with sapphirine, garnet, orthopyroxene and spinel that formed under peak conditions of 950–1050 °C and 10–11 kbar, the kornerupine forming during late retrogressive decompression (Sajeev et al. 2004). In the Central Highlands of Sri Lanka Mg,Al-rich kornerupine-bearing granulites form blocks within marbles and migmatites, the kornupine occurring in late breakdown assemblages with sapphirine, orthopyroxene and cordierite (Sajeev & Osanai 2004). In NE India in the Eastern Ghats belt boron-bearing grandidierite and kornerupine occur with sapphirine, sillimanite and corundum in an association that possibly formed at 800–850 °C and 7–8 kbar (Grew 1983).

Origin of the diopsidites

Diopsidites, which form a distinctive lithology in southern Madagascar including the Tranomaro belt, played an important role in the formation of the borosilicates. The diopsidites contain more than 90% diopside, which locally increases in amount to produce almost a monomineralic rock. A few per cent of phlogopite is commonly present, together with calcite, scapolite or anorthite, and in places with anhydrite, apatite and fluorite (Noizet 1969).

Particularly because of their conformable intercalation with marbles, calc-silicate rocks and metapelitic gneisses, they have been interpreted to have a sedimentary origin from carbonate–shale (Lacroix 1941; Moreau 1963), from isochemically recrystallized magnesian marls (Noizet 1969), and from carbonate–evaporitic rocks (Pierdzig 1992). Carbon and oxygen isotopic data of Boulvais et al. (1998) indicate that Tranomaro marbles were derived by decarbonation of impure siliceous limestones and dolostones coupled with fluid release from granitic intrusions. From a detailed chemical study Moine et al. (1985) concluded that the diopsidites in the Tranomaro belt were originally limestones and marls that had been metasomatized with loss of K, Na, Fe and in some cases Ca, and with introduction of Mg and Si. The very low contents of Ni and Cr were said to preclude an igneous origin. Moine et al. (1985) calculated the following metamorphic conditions. The parageneses in relevant lithologies and the Fe–Mg distribution in the pairs cordierite–garnet and cordierite–spinel constrained the main phase of regional metamorphism at $T = 700$–750 °C and fluid $P = 5$ kbar. Fluid inclusions in quartz in gneisses suggest a high concentration of CO_2 in the fluid phase. The presence of orthopyroxene in rocks of granitic composition indicates local attainment of granulite-facies conditions and a relatively low P_{H_2O}.

Because the pegmatitic lenses or pockets typically grade into enclosing diopsidite (Fig. 2), and because with their common diopside they are in metamorphic equilibrium with the host diopsidites, the infilling of the fractures with trace element-enriched fluids most probably took place during the regional metamorphism. Analyses of phlogopites from 14 deposits by De la Roche (1963) and Joo' (1971) showed that the fluorine contents ranged from 0.56 to 0.75 wt% with an average of 2.73 ± 1.44 wt%, and barium ranged from 46 to >3000 ppm with an average BaO content of 0.05 ± 0.02 wt%. Majmundar (1962) reported that phlogopites in diopsidites contain exceptionally high fluorine contents up to 3.00–3.75 wt%, and that the diopsidites have a high concentration of lithophile elements such as Be, Y, B, Mn and Sr, and low contents of Cr and Ni; in consequence, he concluded that the diopsidites were derived from sedimentary protoliths.

Joo' (1969) studied the chemical changes on passing from the diopsidites towards the pockets in the Benato deposit, and concluded that all the major elements necessary for the formation of the phlogopite and the associated diopside, calcite, scapolite, anhydrite and apatite in the pockets are present in sufficient quantity in the diopsidites. In particular, he emphasized that Al and Mg contents are high enough, but the low K content poses a problem, which he explained by an inhomogeneous distribution of K in the original sediments. Joo' (1969) also pointed out that the major mica concentrations are only rarely developed at the contact of the pockets with the adjacent gneisses, and that only exceptionally are they present next to marbles. Therefore he concluded that a hypothesis for formation of the pockets by metasomatic reaction between the marbles and diopsidites was not acceptable. In this respect, Joo' (1969) was in agreement with most French and Malagasy geologists, who believed that the diopsidites and their phlogopite pockets were derived by recrystallization of sediments. For example, Lacroix (1941) suggested an origin from carbonated shales or dolomitic marls, and from his geochemical studies Noizet (1969) proposed isochemical recrystallization of magnesian marls, the phlogopite-rich diopsidites being derived from marls that contained chlorite and illite, and those without phlogopite from sepiolite-bearing marls.

Because diopsidites are relatively rare worldwide in high-grade orogenic belts, although common in southern Madagascar, it is worth while here to discuss some occurrences that are most similar in occurrence and origin.

(1) In the Caraiba area of eastern Brazil Precambrian hypersthene gneisses and granitic orthogneisses contain layers of diopsidite (which contain spinel, apatite or titanite), diopside–forsterite marble and anhydrite–diopside rocks. Leake et al. (1979) considered that the diopsidites were derived from evaporites that were closely associated with dolomites, a common association in many young evaporite deposits. From their low iron levels and their oxygen and carbon isotope values Sighinolfi et al. (1980) concluded that the anhydrite-bearing diopsidites had an evaporitic origin. Sighinolfi & Fujimori (1974) described diopsidites that occur as lenses, layers and intercalations up to 10 m thick in granulite-facies hypersthene gneisses; they used chemical data and field relations to conclude that the diopsidites were formed by metamorphism of siliceous dolomites.

(2) In the Grenville orogen of eastern Canada diopsidites are interlayered with magnesian marble, quartzite, biotite–hornblende gneiss, diopside–hypersthene gneiss and granitic orthogneiss (Hogarth 1988). The diopsidites contain aluminous diopside, fluorapatite ± scapolite, actinolite, calcite and phlogopite. Within the diopsidites are layers and lenses of pegmatitic phlogopite that is, mined; a situation very similar to that at Vohibola. Hogarth (1988) suggested that the diopsidites were produced by metasomatism through reaction of the aluminous, magnesian marbles with the siliceous orthogneisses.

(3) In the Aldan Shield of Russia aluminous paragneisses are interlayered with scapolite diopsidites with phlogopite- and anhydrite-rich rocks, quartzites and scapolite marbles (Serdyuchenko 1975). Phlogopite lenses occur within the diopsidites, and thus are very similar to the Madagascar rocks (Noizet 1963). The diopsidic rocks enriched in diopside, phlogopite, scapolite and anhydrite owe their origin to the progressive metamorphism of evaporates (Serdyuchenko 1975; Rosen 1981).

Finally, a new type of diopsidite was described by Python et al. (2007). Diopsidite dykes (strongly depleted in Cr, Al, Ti and Na) up to a few tens of centimetres wide occur in the Oman ophiolite within mantle harzburgites and serpentinites, and particularly near asthenospheric diapirs. Python et al. (2007) interpreted the diopsidites as the result of very high-temperature circulation of seawater and carbonated fluids, which penetrated the mantle. Strangely, Santosh et al. (2010) suggested that the Oman dykes are comparable with Cr-rich diopside-rich dykes (no host rocks given) characterized by diopside, chromite and carbonates, which occur in the high-grade gneisses of the Palghat–Cauvery shear zone (PCSZ) in southern India, and they concluded that the diopside-rich dykes there accordingly provide a fingerprint of ocean plate stratigraphy within the PCSZ. We disagree with the comparison of Cr-depleted diopsidite dykes in the mantle peridotite of an ophiolite formed at an oceanic spreading centre with chromiferous diopside-rich dykes in a high-grade gneissic terrane, and we cannot understand the implications either of these have for ocean plate stratigraphy. To us, there seem to be two different compositional types of diopsidites in different geological settings and tectonic environments.

Regional structural controls on boron fluid compositions and transport

The prominent north–south-trending shear zones in southern Madagascar have been the conduits for introduction of H_2O and CO_2 (Pilli et al. 1997a, b), and for granitic fluids (Pili et al. 1999). We demonstrate here that the crustal shear zones also acted as channelways for the movement of boron. Boron can circulate along shear zones that act as fluid channels for partial melt liquids, and the maximum depletion of boron occurs in permeable zones that concentrate fluid flow. Also, the control of fluid flow by plastic deformation during metamorphism is critical to our understanding of the metamorphic processes (Ord & Oliver 1997). In many crustal shear zones volume loss is a result of thermo-hydromechanical critical behaviour, because pore collapse is regulated by a tendency for fluid pressure to remain close to lithostatic (O'Hara 1994). Boron fluids may be differentially introduced into host rocks depending on their rheology and their structural and tectonic setting. However, boron is a soluble element during hydrothermal alteration of sediments and igneous rocks; accordingly, it is a powerful flux for silicate melting (London et al. 1996), and it reduces the viscosity of the melts (e.g. Chorloton & Martin 1978; Pichavant 1987; Dingwell et al. 1992).

At Vohibola the host rocks of the grandidierite-bearing rocks are diopsidites derived from evaporitic sediments that were metasomatized by granitic activity (Moine et al. 1985). Most occurrences of grandidierite in southern Madagascar occur in minor shear zones (Ackermand et al. 1991a; Pili et al. 1999; our observations), such as those along the boundaries of the Tranomaro and Betroka belts, where granites were emplaced and isothermal uplift took place.

Most grandidierite occurrences, worldwide, are close to schörl tourmaline-bearing rocks associated with pegmatites and granites in a single magmatic hydrothermal province (London & Manning 1995). Granitic magmas of crustal melt origin play an important role in the distribution of boron from the lower to upper crust, because boron easily migrates from granites into reactive wall rocks and from the hydrosystems that surround the granites (London et al. 1996).

Noticeably, the borosilicates in the Tranomaro belt always occur in phlogopite-rich deposits associated with shear zones. According to Pierdzig (1992) the isotopic signature ($\delta^{18}O = 9‰$, $\delta^{13}C = -7‰$, $\delta D = -31‰$) of the mineralizing fluids in the shear zone responsible for the phlogopite deposits at Ampandandrava in the western Tranomaro belt of Nicollet (1990a) favours fluid derivation from an I-type syenite. Herd et al. (1984) demonstrated that boron for grandidierite formation in West Greenland was derived by metasomatism from granitic pegmatites, and that grandidierite grew contemporaneously with phlogopite, cordierite and plagioclase during high-temperature metamorphism. In ductile shear zones fluids such as H_2O and CO_2 circulate easily and favour anatectic melting during fluid–rock interaction and retrograde metamorphism (Pili et al. 1997a, b, 1999). Thus, the composition of grandidierite may be expected to vary as B_2O_3 and H_2O contents fluctuate towards and within the shear zones. Most peraluminous pegmatites and granites enriched in boron are typically generated as late- or post-tectonic products of partial melting in orogenic belts formed by continental collision (London et al. 1996). Boron and chlorine are incompatible, but their concentrations generally increase with water temperature. Considering that boron and chlorine are fugitive elements at magmatic temperatures, they tend to be concentrated in residual melts that infill fractures and veins, as we observe in the Tranomaro belt.

According to Santosh et al. (2007), fluorine-rich phlogopites or biotites are an indicator of UHT metamorphism. Phlogopites in metapelites (Al-orthopyroxene, sapphirine, osumilite, garnet, quartz) from the Napier complex, Antarctica contain up to c. 8 wt% F and textural relations suggest that they formed under UHT conditions (Motoyoshi & Hensen 2001); however, it should be noted that their phlogopite is a pre-peak metamorphic phase occurring as inclusions in the first four UHT phases listed above. At Ampandandrava phlogopites in diopsidites contain up to 3.75 wt% F (Majmundar 1962) or up to 1.7 wt% F (Stern & Klein 1983; Pili et al. 1999), at Benato to the north of Ampandandrava in the same shear zone phlogopites in diopsidites contain up to 4.9 wt% F (Pili et al. 1999), and at Sakamasy farther south in the same shear zone phlogopites contain up to 2.5 wt% F (Stern & Klein 1983; Pili et al. 1999). At Esira to the east of the Tranomaro belt phlogoites contain from 5.75 to 6.70 wt% F (de la Roche 1963). However, these phlogopites at Vohibola and elsewhere in southern Madagascar formed as a result of fluid infiltration and metasomatism of diopsidites, which took place after the peak granulite-facies metamorphism (at 750–700 °C according to Raith et al. 2008), and thus are unlikely to have formed under UHT conditions. At Andrahomana a granitic fluid source is suggested by the occurrence of borosilicates in granitic pegmatites related to tourmaline-bearing crustal melt granites, which in turn were most probably derived by partial melting of sediments (as at Vohibola) that are commonly characterized by elevated boron and fluorine contents (Moine et al. 1996; Pili et al. 1999; Raith et al. 2008), and the shear zones played an important role in the fluid transport.

Conclusions

We suggest that the distribution, assemblages and paragenetic history of the borosilicates grandidierite, serendibite and sinhalite that occur within shear zones in the high-grade terrane of SE Madagascar indicate two sources of boron; either directly from evaporitic or calc-silicate protolith sediments, or from granitic rocks that were originally derived by partial melting of the boron-bearing sediments. The distribution of B_2O_3 and H_2O in the Tranomaro belt depended on the rheology and permeability of the host rocks during deformation associated with amphibolite-facies metamorphism. The shear zones acted as conduits for transport of granitic partial melt fluids and boron-enriched fluids.

T.R. thanks the DAAD, Stiftung Volkswagenwerk and the Royal Society for supporting his research in Kiel (Germany) and Leicester (UK). B.F.W. and D.A. acknowledge respectively the Royal Society and the Deutsche Forschungsgemeinschaft for financial support to undertake research in the high-grade terranes of Madagascar. B. Mader (Kiel) and R. Wilson (Leicester) are thanked for their help in the microprobe laboratories. Finally, we are grateful for helpful and encouraging reviews from M. Santosh and T. Raharimahefa.

References

ACKERMAND, D., WINDLEY, B. F. & RAZAFINIPARANY, A. H. 1989. The Precambrian mobile belt of Southern Madagascar. In: DALY, J. S., CLIFF, R. A. & YARDLEY, B. W. D. (eds) Evolution of Metamorphic Belts. Geological Society, London, Special Publications, 43, 293–296.

ACKERMAND, D., WINDLEY, B. F. & RAZAFINIPARANY, A. H. 1991a. Abhängigkeit der Grandidieritbildungen von Gefügemilieu in hochgradig metamorphen Gesteinsserien in Süd-Madagaskar. Beiheft zur European Journal of Mineralogy, 3, 2.

ACKERMAND, D., WINDLEY, B. F. & RAZAFINIPARANY, A. H. 1991b. Kornerupine breakdown reactions in paragneisses from southern Madagascar. *Mineralogical Magazine*, **55**, 71–80.

ALEKSANDROV, S. M. & TRONEVA, M. A. 2006. Composition, mineral assemblages, and genesis of serendibite-bearing magnesian skarns. *Geochemistry International*, **44**, 665–680.

ANDRIAMAROFAHATRA, J., DE LA BOISSE, H. & NICOLLET, C. 1990. Datation U–Pb sur monazites et zircons du dernier épisode tectono-métamorphique granulitique majeur dans le Sud-Est de Madagascar. *Comptes Rendus de l'Académie des Sciences*, **310**, 1643–1648.

BEHIER, J. 1960. Contribution à la Minéralogie de Madagascar. *Annales Géologiques de Madagascar*, **29**, 1–78.

BEHIER, J. 1961. Travaux mineralogiques. *Rapport Annuel du Service Géologique de Madagascar pour 1961*, 183–184.

BERGER, A., GNOS, E., SCHREURS, G., FERNANDEZ, A. & RAKOTONDRAZAFY, M. 2006. Late Neoproterozoic, Ordovician and Carboniferous events recorded in monazites from southern–central Madagascar. *Precambrian Research*, **144**, 278–296.

BESAIRIE, H. 1973. *Madagascar: 1:2 000 000 Geological Map*. Service Géologique de Madagascar, Antananarivo.

BLACK, P. M. 1970. Grandidierite from Cuvier Island, New Zealand. *Mineralogical Magazine*, **37**, 615–617.

BOULVAIS, P., FOURCADE, S., GRUAU, G., MOINE, B. & CUNEY, M. 1998. Persistence of pre-metamorphic C and O isotopic signatures in marbles subject to Pan-African granulite-facies metamorphism and U–Th mineralization (Tranomaro, Southeast Madagascar). *Chemical Geology*, **150**, 247–262.

CARSON, C. J., DIRKS, P. G., HAND, M., SIMS, J. P. & WILSON, C. J. L. 1995. Compressional and extensional tectonics in low–medium pressure granulites from the Larsemann Hills, East Antarctica. *Geological Magazine*, **132**, 151–170.

CARTWRIGHT, I. & BUICK, I. S. 1994. Channeled fluid infiltration and variation in permeability in Reynolds Range marbles, Australia. *Journal of the Geological Society, London*, **151**, 583–586.

CHORLOTON, L. B. & MARTIN, R. F. 1978. The effect of boron on the granite solidus. *Canadian Mineralogist*, **16**, 239–244.

COLLINS, A. S. 2000. The tectonic evolution of Madagascar: its place in the East African Orogen. *Gondwana Research*, **3**, 549–552.

COLLINS, A. S. 2006. Madagascar and the amalgamation of central Gondwana. *Gondwana Research*, **9**, 3–16.

COLLINS, A. S. & WINDLEY, B. F. 2002. The tectonic evolution of central and northern Madagascar and its place in the final assembly of Gondwana. *Journal of Geology*, **110**, 325–339.

COLLINS, A. S., WINDLEY, B. F. & RAZAKAMANANA, T. 2000. Neoproterozoic extensional detachment in central Madagascar: implications for the collapse of the East African orogen. *Geological Magazine*, **137**, 39–51.

COLLINS, A. S., CLARK, C., SAJEEV, K., SANTOSH, M., KELSEY, D. E. & HAND, M. 2007. Passage through India: the Mozambique ocean suture, high-pressure granulites and the Palghat–Cauvery shear zone system. *Terra Nova*, **19**, 141–147.

DANIELS, P., KROSSE, S., WERDING, G. & SCHREYER, W. 1997. 'Pseudosinhalite' a new hydrous MgAl-borate: synthesis, phase characterization, crystal structure and *PT*-stability. *Contributions to Mineralogy and Petrology*, **128**, 261–271.

DAVIDSON, A., CARMICHAEL, D. M. & PATTINSON, D. R. M. 1990. Field guide to the metamorphism and geodynamics of the southwestern Grenville Province, Ontario. *International Geological Correlation Program Projects*, 235 and 304, 1–123.

DE GRAVE, J., DE PAEPE, P., DE GRAVE, E., VOCHTEN, R. & EECKHOUT, S. G. 2002. Mineralogical and Mössbauer spectroscopic study of a diopside occurring in the marbles of Andranondambo, southern Madagascar. *American Mineralogist*, **87**, 132–141.

DE LA ROCHE, H. 1963. Sur la composition chimique des phlogopites de Madagascar et sur la présence de variétés riches en fluor. *Comptes Rendus de la Semaine Géologique de Madagascar*, 175–178.

DE WIT, M. J., BOWRING, S. A., ASHWAL, L. D., RANDRIANASOLO, L. G., MOREL, V. P. I. & RAMBELOSON, R. A. 2001. Age and tectonic evolution of Neoproterozoic ductile shear zones in southwestern Madagascar, with implications for Gondwana studies. *Tectonics*, **20**, 1–45.

DINGWELL, D. B., KNOCHE, R. & WEBB, S. L. 1992. The effect of B_2O_3 on the viscosity of haplogranite liquids. *American Mineralogist*, **77**, 457–461.

GREW, E. S. 1983. A grandidierite–sapphirine association from India. *Mineralogical Magazine*, **47**, 401–403.

GREW, E. S. 1996. Borosilicates (exclusive of tourmaline) and boron in rock-forming minerals in metamorphic environments. *In*: GREW, E. S. & ANOVITZ, L. M. (eds) *Boron: Mineralogy, Petrology and Geochemistry*. Mineralogical Society of America, Reviews in Mineralogy, **33**, 771–788.

GREW, E. S. & MANTON, W. I. 1986. A new correlation of sapphirine granulites in the Indo-Antarctic metamorphic terrain: late Proterozoic dates from the Eastern Ghats province of India. *Precambrian Research*, **33**, 123–137.

GREW, E. S., YATES, M. G. & DE LORRAINE, W. 1990. Serendibite from the northwest Adirondack Lowlands, in Russell, New York, USA. *Mineralogical Magazine*, **54**, 133–136.

GREW, E. S., PERTSEV, N. N., BORONIKHIN, V. A., BORISOVSKIY, S. Y., YATES, M. G. & MARQUES, N. 1991. Serendibite in the Tayozhnoya deposit of the Aldan Shield, eastern Siberia, U.S.S.R. *American Mineralogist*, **76**, 1061–1080.

HASLAM, H. W. 1980. Grandidierite from a metamorphic aureole near Mchinji, Malawi. *Mineralogical Magazine*, **43**, 822–823.

HAYWARD, C. L., ANGEL, R. J. & ROSS, N. L. 1994. The structural redetermination and crystal-chemistry of sinhalite, MgAl BO_4. *European Journal of Mineralogy*, **6**, 313–321.

HEIDE, M. 1992. *Synthese und Stabilität von Grandidierit, $MgAl_3BSO_9$*. Diploma thesis, Ruhr Universität, Bochum.

HERD, R. K., WINDLEY, B. F. & ACKERMAND, D. 1984. Grandidierite from a pelitic xenolith in the Haddo House complex, N.E. Scotland. *Mineralogical Magazine*, **48**, 401–406.

HOGARTH, D. D. 1988. Chemical composition of fluorapatite and associated minerals from skarn near Gatineau, Quebec. *Mineralogical Magazine*, **52**, 347–358.

HUIJSMANS, P. P., BARTON, M. & VAN BERGEN, M. J. 1982. A pegmatite containing Fe-rich grandidierite, Ti-rich dumortierite and tourmaline from the Precambrian high-grade metamorphic complex of Rogaland, S.W. Norway. *Neues Jahrbuch für Mineralogie, Abhandlungen*, **143**, 249–261.

HUTCHEON, I., GUNTER, A. E. & LECHEMINANT, A. N. 1977. Serendibite from Penrhyn group marble, Melville Peninsula, District of Franklin, Canada. *Canadian Mineralogist*, **15**, 108–112.

ITO, M., SUZUKI, K. & YOGO, S. 1997. Cambrian granulite to upper amphibolite facies metamorphism of post-797 Ma sediments in Madagascar. *Journal of Earth and Planetary Sciences, Nagoya University*, **44**, 89–102.

JÖNS, N. 2006. *Metamorphic events during the formation of the East African Orogen: case studies from Madagascar and Tanzania*. PhD thesis, University of Kiel.

JOO', J. 1969. Essai d'interpretation génétique à propos de la répartition des éléments majeurs au voisinage de poche de mica. *Comptes Rendus de la Semaine Géologique de Madagascar*, 161–165.

JOO', J. 1971. Caractères cristallochimiques et géochimiques de la série phlogopite–eastonite de Madagascar. *Comptes Rendus de la Semaine Géologique de Madagascar*, 75–86.

KRÖNER, A., BRAUN, O. & JÄCKEL, P. 1996. Zircon geochronology of anatectic melts and residues from a high-grade pelitic assemblage at Ihosy, southern Madagascar. *Geological Magazine*, **133**, 311–323.

KRÖNER, A., HEGNER, E., COLLINS, A. S., WINDLEY, B. F., BREWER, T. S., RAZAKAMANANA, T. & PIDGEON, R. T. 2000. Age and magmatic history of the Antananarivo block, Central Madagascar, as derived from zircon geochronology and Nd isotopic systematics. *American Journal of Science*, **300**, 251–288.

KUSKY, T. M., ADELSALAM, M., STERN, J. S. & TUCKER, R. D. 2003. Evolution of the East African and related orogens, and the assembly of Gondwana. *Precambrian Research*, **123**, 81–85.

LACROIX, A. 1902. Note préliminaire sur une nouvelle espèce minérale (grandidiérite). *Bulletin Société de Français Minéralogie*, **25**, 85–86.

LACROIX, A. 1904. Sur la grandidérite. *Bulletin Société de Français Minéralogie*, **27**, 259–265.

LACROIX, A. 1922–1923. *Minéralogie de Madagascar*. Challamel, Paris.

LACROIX, A. 1941. Les gisements de phlogopite de Madagascar et les pyroxénites qui les renferment. *Annales Géologiques de Madagascar*, **11**, 1–119.

LAL, R. K. 1991. Empirical calibrations of the Ti-contents of biotite and muscovite for geothermobarometry and their application to low to high-grade metamorphic rocks. *Indo-Soviet Symposium on Experimental Mineralogy and Petrology, New Delhi, 1991, Abstracts*, **3**, 18.

LARDEAUX, J. M., MARTELAT, J. E., NICOLLET, C., PILI, E., RAKOTONDRAZAFY, R. & CARDON, H. 1999. Metamorphism and tectonics in southern Madagascar: an overview. *Gondwana Research*, **2**, 355–362.

LEAKE, B. E., FARROW, C. M. & TOWNEND, R. 1979. A pre-2,000 Myr old granulite facies metamorphosed evaporite from Caraiba, Brazil? *Nature*, **277**, 49–50.

LINDSLEY, D. H. 1983. Pyroxene thermometry. *American Mineralogist*, **68**, 477–493.

LONDON, D. & MANNING, D. A. C. 1995. Chemical variation and significance of tourmaline in southwest England. *Economic Geology*, **90**, 495–519.

LONDON, D., MORGAN, V. I. G. B. & WOLF, M. B. 1996. Boron in granitic rocks and their contact aureoles. *In*: GREW, E. S. & ANOVITZ, L. M. (eds) *Boron: Mineralogy, Petrology and Geochemistry*. Mineralogical Society of America, Reviews in Mineralogy, **33**, 299–330.

MAGHSUDANIA, H. 1991. *Synthesen sulfatiger Doppelschichtverbindungen vom Hydrotalkit-Typ*. PhD thesis, University of Marburg.

MAJMUNDAR, M. H. 1962. *Contribution à l'étude mineralogique et geochimique des pyroxenes et des micas dans les pyroxenites à phlogopite et dans les charnockites du Sud-est de Madagascar*. PhD Thesis, University of Nancy.

MARKL, G., BÄUERLE, J. & GRUJIC, D. 2000. Metamorphic evolution of Pan-African granulite facies metapelites from southern Madagascar. *Precambrian Research*, **102**, 47–68.

MARTELAT, J.-E., NICOLLET, C., LARDEAU, J.-M., VIDAL, G. & RAKOTONDRAZAFY, R. 1997. Lithospheric tectonic structures developed under high-grade metamorphism in the southern part of Madagascar. *Geodinamica Acta*, **10**, 94–114.

MARTELAT, J. E., LARDEAUX, J. M., NICOLLET, C. & RAKOTONDRAZAFY, R. 1999. Exhumation of granulites within a transpressive regime: an example from southern Madagascar. *Gondwana Research*, **2**, 363–367.

MARTELAT, J.-E., LARDEAU, J.-M., NICOLLET, C. & RAKOTONDRAZAFY, R. 2000. Strain pattern and late Precambrian deformation history in southern Madagascar. *Precambrian Research*, **102**, 1–20.

MATHAVAN, V. & FERNANDO, G. W. A. R. 2001. Reactions and textures in grossular–wollastonite–scapolite calc-silicate granulites from Maligawila, Sri Lanka: evidence for high-temperature isobaric cooling in the meta-sediments of the Highland Complex. *Lithos*, **59**, 217–232.

MCKIE, D. 1965. The magnesium aluminium borosilicates: kornerupine and grandidierite. *Mineralogical Magazine*, **34**, 346.

MEGERLIN, N. 1968. Sur une roche à kornérupine du Sud de Ianakafy (Centre Sud de Madagascar). *Comptes Rendus des Semaines Géologiques de Madagascar*, 67–70.

MOINE, B., SAUVAN, P. & JAROUSSE, J. 1981. Geochemistry of evaporite-bearing series: a tentative guide for the identification of meta-evaporites. *Contributions to Mineralogy and Petrology*, **76**, 401–412.

MOINE, B., RAKOTONDRATSIMA, C. & CUNEY, M. 1985. Les pyroxénites à urano-thorianite du Sud-Est de Madagascar: conditions physico-chimiques de la métasomatose. *Bulletin de Minéralogie*, **108**, 325–340.

MOINE, B., RAMANBAZAFY, A., CUNEY, M. & DE PARSEVAL, P. 1996. Signification des concentrations en F et en Cl dans les micas des granulites de Madagascar. *16th RST, Orléans, Société Géologique de France, Abstracts*, 100.

MOINE, B., RAKOTONDRAZAFY, A. F. M., RAMAMBAZAFY, A., RAKOTONDRATSIMA, C. & CUNEY, M. 1997. Controls on the urano-thorianite deposits in S-E Madagascar. *In*: COX, R. & ASHWAL, L. D. (eds) *Proterozoic Geology of Madagascar*. Gondwana Research Group. Miscellaneous Publications, **5**, 55–56.

MOREAU, M. 1963. Le gisement des minéralisations en thorianite à Madagascar. *Annales Géologiques de Madagascar*, **33**, 197–202.

MOTOYOSHI, Y. & HENSEN, B. J. 2001. F-rich phlogopite stability in ultra-high-temperature metapelites from the Napier Complex, East Antarctica. *American Mineralogist*, **86**, 1401–1414.

NICOLLET, C. 1985. Les gneiss rubanés à cordiérite et grenat d'Ihosy: un marqueur thermobarométrique dans le Sud de Madagascar. *Precambrian Research*, **28**, 175–185.

NICOLLET, C. 1990a. Crustal evolution of the granulites of Madagascar. *In*: VIELZEUF, D. & VIDAL, P. (eds) *Granulites and Crustal Evolution*. Kluwer, Dordrecht, 291–310.

NICOLLET, C. 1990b. Occurrences of grandidierite, serendibite and tourmaline near Ihosy, southern Madagascar. *Mineralogical Magazine*, **54**, 131–133.

NOIZET, G. 1963. Recherches récentes de phlogopite en U.R.S.S. *Comptes Rendus de la Semaine Géologique de Madagascar*, 239–240.

NOIZET, G. 1969. Sur l'origine et al classification des pyroxénites Androyennes du Sud de Madagascar. *Comptes Rendus de la Semaine Géologique de Madagascar*, 155–159.

O'HARA, K. D. 1994. Fluid–rock interaction in crustal shear zones: a directed percolation approach. *Geology*, **22**, 834–846.

OLESCH, M. & SEIFERT, F. 1976. Synthesis, powder data, and lattice constants of grandidierite (Mg,Fe)Al$_3$BSiO$_9$. *Neues Jahrbuch für Mineralogie, Monatshefte*, **111**, 513–518.

ORD, A. & OLIVER, N. H. S. 1997. Mechanical controls on fluid flow during regional metamorphism: some numerical models. *Journal of Metamorphic Geology*, **15**, 345–359.

PAQUETTE, J.-L., NÉDÉLEC, A., MOINE, B. & RAKOTONDRAZAFY, M. 1994. U–Pb, single zircon Pb-evaporation, and Sm–Nd isotopic study of a granulite domain in SE Madagascar. *Journal of Geology*, **102**, 523–538.

PICHAVANT, M. 1987. Effects of B and H$_2$O on liquidus phase relations in the haplogranite system at 1 kbar. *American Mineralogist*, **72**, 1056–1070.

PIERDZIG, S. 1992. *Granulitfazielle Gesteinsserien der Ampandrandava-Formation Südmadagaskars und die Entstehung ihrer Phlogopit-Mineralisationen*. PhD thesis, University of Bonn.

PILI, É., RICARD, Y., LARDEAUX, J.-M. & SHEPPARD, S. M. F. 1997a. Lithospheric shear zones and mantle–crust connections. *Tectonophysics*, **280**, 15–29.

PILI, É., SHEPPARD, S. M. F., LARDEAUX, J.-M., MARTELAT, J.-E. & NICOLLET, C. 1997b. Fluid flow v. scale of shear zones in the lower continental crust and the granulite paradox. *Geology*, **25**, 15–18.

PILI, É., SHEPPARD, S. M. F. & LARDEAUX, J.-M. 1999. Fluid–rock interaction in the granulites of Madagascar and lithospheric-scale transfer of fluids. *Gondwana Research*, **2**, 341–350.

PÖTER, B. & SCHREYER, W. 1998. Synthesis experiments in the ternary system MgO–B$_2$O$_3$–H$_2$O and resulting phase relations. *Abstract Supplement 1, Terra Nova*, **10**, 50.

PRADEEPKUMAR, A. P. & KRISHNANATH, R. 2000. A Pan-African 'humite epoch' in East Gondwana: implications for Neoproterozoic Gondwana geometry. *Journal of Geodynamics*, **29**, 43–62.

PRIOR, G. T. & COOMARÁRASWÁMY, A. K. 1903. Serendibite, a new borosilicate from Ceylon. *Mineralogical Magazine*, **13**, 224–227.

PYTHON, M., CEULENEER, G., ISHIDA, Y., BARRAT, J.-A. & ARAI, S. 2007. Oman diopsidites: a new lithology diagnostic of very high temperature hydrothermal circulation in mantle peridotite below oceanic spreading centres. *Earth and Planetary Science Letters*, **255**, 289–305.

RAHARIMAHEFA, T. & KUSKY, T. M. 2006. Structural and remote sensing studies of the southern Betsimisaraka suture, Madagascar. *Gondwana Research*, **10**, 186–197.

RAHARIMAHEFA, T. & KUSKY, T. M. 2009. Structural and remote sensing analysis of the Betsimisaraka suture in northeastern Madagascar. *Gondwana Research*, **15**, 14–27.

RAITH, M. M., RAKOTONDRAZAFY, R. & SENGUPTA, P. 2008. Petrology of corundum–spinel–sapphirine–anorthite rocks (sakenites) from the type locality in southern Madagascar. *Journal of Metamorphic Geology*, **26**, 647–667.

RAKOTONDRAZAFY, A. F. M., MOINE, B. & CUNEY, M. 1996. Mode of formation of hibonite (CaAl$_{12}$O$_{19}$) within the U–Th skarns from the granulites of SE Madagascar. *Contributions to Mineralogy and Petrology*, **123**, 190–201.

RAKOTONDRAZAFY, A. F. M., GIULIANI, G. ET AL. 2008. Gem corundum deposits of Madagascar: a review. *Ore Geology Reviews*, **34**, 134–154.

RAKOTONDRAZAFY, R. 1992. *Étude pétrologique de la série granulitique panafricaine de la région d'Ampandrandava: Sud de Madagascar*. PhD thesis, University of Antananarivo.

RAZAFINIPARANY, A. H., ACKERMAND, D. & WINDLEY, B. F. 1989. Grandidierite, serendibite and sinhalite: Products of boron-bearing fluid/crystal reactions in diopsidites and phlogopite in paragneiss, southeast Madagascar. *Terra Abstracts*, **1**, 302.

RAZAKAMANANA, T. 1999. *High-grade granulites of the Precambrian of southern Madagascar: petrology, tectonic and metallogenetic implications with regard to plate tectonic setting*. Thése d'etat (docteur-és-sciences), University of Toliara.

RAZAKAMANANA, T., WINDLEY, B. F. & ACKERMAND, D. 2001. Sapphirine–kornerupine granulites from the Precambrian of southern Madagascar: implications for the evolution of the deep crust in East Gondwana. *Gondwana Research*, **4**, 752–753.

RAZAKAMANANA, T., ACKERMAND, D. & WINDLEY, B. F. 2006. Metamorphic evolution of sapphirine-bearing rocks from Vohidava area, southern Madagascar: structural and tectonic implications. *In*: SCHWITZER, C., BRANDT, S., RAMILIJAONA, O., RAZANAHOERA, M. R., ACKERMAND, D., RAZAKAMANANA, T. & GANZHORN, J. U. (eds) *Proceedings of German–Malagasy*

Research Cooperation in Life and Earth Sciences. Concept Verlag, Berlin, 41–60.

ROZEN, O. M. 1981. Scapolite–plagioclase schists and the problem of Precambrian sulfate as illustrated by geochemical comparison of deposits of evaporite basins and metamorphic rocks of calcareous series. *Doklady Akademii Nauk SSSR*, **244**, 138–141.

SAJEEV, K. & OSANAI, Y. 2004. Ultrahigh-temperature metamorphism (1150 °C, 12 kbar) and multistage evolution of Mg,Al-rich granulites from the Central Highland complex, Sri Lanka. *Journal of Petrology*, **45**, 1821–1844.

SAJEEV, K., OSANAI, Y. & SANTOSH, M. 2004. Ultrahigh-temperature metamorphism followed by two-stage decompression of garnet–orthopyroxene–sillimanite granulites from Ganguvarpatti, Madurai block, southern India. *Contributions to Mineralogy and Petrology*, **148**, 29–46.

SANTOSH, M. & SAJEEV, K. 2006. Anticlockwise evolution of ultrahigh-temperature granulites within continental collision zone in southern India. *Lithos*, **92**, 447–464.

SANTOSH, M., TSUNOGAE, T., LI, J. H. & LIU, S. J. 2007. Discovery of sapphirine-bearing Mg–Al granulites in the North China craton: implications for Paleoproterozoic ultrahigh-temperature metamorphism. *Gondwana Research*, **11**, 263–285.

SANTOSH, M., MARUYAMA, S. & SATO, K. 2009. Anatomy of a Cambrian suture in Gondwana: Pacific-type orogeny in southern India? *Gondwana Research*, **16**, 321–341.

SCHAU, M., DAVIDSON, A. & CARMICHAEL, D. M. 1986. *Granulites and granulites*. Geological Association of Canada, Mineralogical Association of Canada, Canadian Geophysical Union, Joint Annual Meeting, Ottawa 1986. Field Trip Guidebook **6**.

SCHREYER, W. & WERDING, G. 1997. High-pressure behaviour of selected boron minerals and the question of boron distribution between fluids and rocks. *Lithos*, **41**, 251–266.

SEMROUD, B., FABRIÈS, J. & CONQUÉRÉ, F. 1976. La grandidiérite de Tizi-Ouchen (Algérie). *Bulletin de la Société Française de Minéralogie et Cristallographie*, **99**, 58–60.

SERDYUCHENKO, D. P. 1975. Some Precambrian scapolite-bearing rocks evolved from evaporites. *Lithos*, **8**, 1–7.

SHACKLETON, R. M. 1996. The final collision zone between East and West Gondwana: where is it? *Journal of African Earth Sciences*, **23**, 271–287.

SIGHINOLFI, G. P. & FUJIMORI, S. 1974. Petrology and chemistry of diopsidic rocks in granulite terrains from the Brazilian basement. *Atti della Societa Toscana di Scienze Naturali, Memorie Serie A*, **81**, 103–120.

SIGHINOLFI, G. P., KRONBERG, B. I., GORGONI, C. & FYFE, W. S. 1980. Geochemistry and genesis of sulphide–anhydrite-bearing Archean carbonate rocks from Bahia (Brazil). *Chemical Geology*, **29**, 323–331.

STERN, R. J. 1994. Arc assembly and continental collision in the Neoproterozoic East African Orogen: implications for the consolidation of Gondwanaland. *Annual Review of Earth and Planetary Sciences*, **22**, 319–351.

STERN, W. B. & KLEIN, H. H. 1983. Inconsistent chemical data on phlogopite: analysis and genesis of phlogopite from Madagascar. *Schweizerische Mineralogische und Petrographische Mitteilungen*, **63**, 187–202.

VAN DER VEER, D. G., SWIHART, G. H., SENGUPTA, P. K. & GREW, E. S. 1993. Cation occupancies in serendibite – a crystal-structure study. *American Mineralogist*, **78**, 195–203.

VON KNORRING, O. V., SAHAMA, TH. G. & LETHINEN, M. 1969. A note on grandidierite from Fort-Dauphin, Madagascar. *Bulletin of the Geological Society of Finland*, **41**, 79–84.

VRY, J. K. & CARTWRIGHT, I. 1994. Sapphirine–kornerupine rocks from the Reynolds Range, Central Australia. Constraints on the uplift history of a Proterozoic low-pressure terrain. *Contributions to Mineralogy and Petrology*, **116**, 78–91.

WERDING, G. & SCHREYER, W. 1978. Synthesis and crystal chemistry of kornerupine in the system $MgO-Al_2O_3-SiO_2-B_2O_3-H_2O$. *Contributions to Mineralogy and Petrology*, **67**, 247–259.

WERDING, G. & SCHREYER, W. 1984. Alkali-free tourmaline in the $MgO-Al_2O_3-B_2O_3-SiO_2-H_2O$. *Geochimica et Cosmochimica Acta*, **48**, 1331–1344.

WERDING, G. & SCHREYER, W. 1996. Experimental studies on borosilicates and selected borates. *In*: ANOVITZ, L. M. & GREW, E. S. (eds) *Boron: Mineralogy, Petrology and Geochemistry*. Mineralogical Society of America, Reviews in Mineralogy, **33**, 117–163.

WERDING, G., PERTSEV, W. N. & SCHREYER, W. P. 1990. Synthesis and phase relations of serendibite. *Transactions Doklady of USSR Academy of Sciences: Earth Science Sections*, **315**, 243.

WINDLEY, B. F., RAZAFINIPARANY, A. H., RAZAKAMANANA, T. & ACKERMAND, D. 1994. The tectonic framework of the Precambrian of Madagascar and its Gondwana connections: a review and reappraisal. *Geologische Rundschau*, **83**, 642–659.

Reconstruction and interpretation of giant mafic dyke swarms: a case study of 1.78 Ga magmatism in the North China craton

PENG PENG

State Key Laboratory of Lithospheric Evolution, Institute of Geology and Geophysics, Chinese Academy of Sciences, Beijing 100029, China (e-mail: pengpengwj@mail.iggcas.ac.cn)

Abstract: Short-lived giant mafic dyke swarms are keys to the interpretation of continental evolution and tectonics, reconstruction of continental palaeogeographical regimes, and petrogenesis of volcanism. The 1.78 Ga Taihang–Lvliang dyke swarm, one of the most significant and best-preserved Precambrian swarms in the central part of the North China craton (NCC), is reviewed and discussed. It is interpreted to have a radiating geometry that is compatible with the Xiong'er triple-junction rift, in which the Xiong'er volcanic province is proposed to be the extrusive counterpart of this swarm. It resulted in significant extension, uplift and magmatic accretion of the NCC, and it is comparable with the Phanerozoic large igneous provinces (LIPs) in areal extent (c. 0.3 Mkm^2) and estimated volume (c. 0.3 Mkm^3), short lifespan (<20 Ma), and intraplate setting. This North China LIP is unique in that it comprises large volumes of both mafic and intermediate components. It could have resulted from extensive mantle–crust interaction, probably driven by a large-scale mantle upwelling. A plume tectonic model is favoured by several lines of supporting evidence (i.e. massive volcanic flows correlated over large areas and a giant fanning dyke swarm with plume-affinitive chemistry). It could responsible for massive sulphide (Pb–Zn) and gold (Au–Ag) ore deposits in the Xiong'er volcanic province. Dismembered remnants of this magmatism in other block(s), with potential candidates in South America, Australia and India, could identify other cratonic blocks that were formerly connected to the North China craton.

There are more than 100 giant mafic dyke swarms that extend for more than 300 km on planet Earth; thus far, the largest recognized is the Mackenzie swarm of the northern Canadian Shield, which extends for >2000 km (Ernst & Buchan 2001). Giant dyke swarms provide information on large-scale extension occurring in the continental lithosphere, and thus they are important in interpreting continental evolution and plate or plume tectonics (e.g. Ernst *et al.* 1995; Hanski *et al.* 2006). They serve as conduits transporting large volumes of magma from the mantle and thus contribute to growth of the continental crust. They could be the key to the petrogenesis of an overall magmatic event as they may preserve different and sometimes more primitive magma compositions, less affected by assimilation. They also provide a powerful tool in reconstructing ancient continental palaeogeography because of their large areal extent, well-defined and often short duration, palaeomagnetic record and inherent geometry (e.g. Bleeker & Ernst 2006). However, after their emplacement, dyke swarms may have been overprinted by later tectonothermal events resulting in deformation, metamorphism, displacement and dismemberment, and thus reconstruction is necessary prior to interpretation. Here, a case study on a 1.78 Ga giant dyke swarm in the North China craton is reviewed and discussed.

Geological background

The North China craton (NCC, also known as the Sino-Korean craton; Fig. 1) formed as a result of amalgamation of Archaean blocks either in the late Palaeoproterozoic (c. 1.85 Ga; e.g. Zhao, G.-C. *et al.* 2001, 2005; Wilde *et al.* 2002; Guo *et al.* 2005; Kröner *et al.* 2005), or alternatively in the latest Archaean (c. 2.5 Ga) followed by Palaeoproterozoic remobilization and recratonization (e.g. rifting, collision and/or uplift) (Li *et al.* 2000, 2002; Zhai *et al.* 2000; Kusky & Li 2003; Zhai & Liu 2003; Kusky *et al.* 2007a, b; Zhai & Peng 2007). After 1.8 Ga, the NCC stabilized, followed by episodes of rifting (1.8–1.6 Ga and c. 0.9 Ga; e.g. Zhu *et al.* 2005; Peng *et al.* 2008b) and platform deposition (1.6–1.4 Ga; e.g. Zhao, Z.-P. *et al.* 1993). Whether the NCC was involved in a palaeo-supercontinent (e.g. Nuna) or not, and where it may have been positioned in a global configuration have been widely discussed (e.g. Wilde *et al.* 2002; Zhao, G.-C. *et al.* 2002, 2004, 2010; Peng *et al.* 2005; Hou *et al.* 2008a). Many of these major events first shaping and then modifying the NCC were accompanied by mafic dyke swarms; that is, the 2.5 Ga Taipingzhai–Naoyumen dykes, the 2.15 Ga Hengling dykes and sills, the 1.97 Ga Xiwangshan dykes, the 1.96 Ga Xuwujia dykes, the 1.78 Ga Taihang–Lvliang dykes, the 1.76 Ga Beitai dykes, the

From: KUSKY, T. M., ZHAI, M.-G. & XIAO, W. (eds) *The Evolving Continents: Understanding Processes of Continental Growth.* Geological Society, London, Special Publications, **338**, 163–178.
DOI: 10.1144/SP338.8 0305-8719/10/$15.00 © The Geological Society of London 2010.

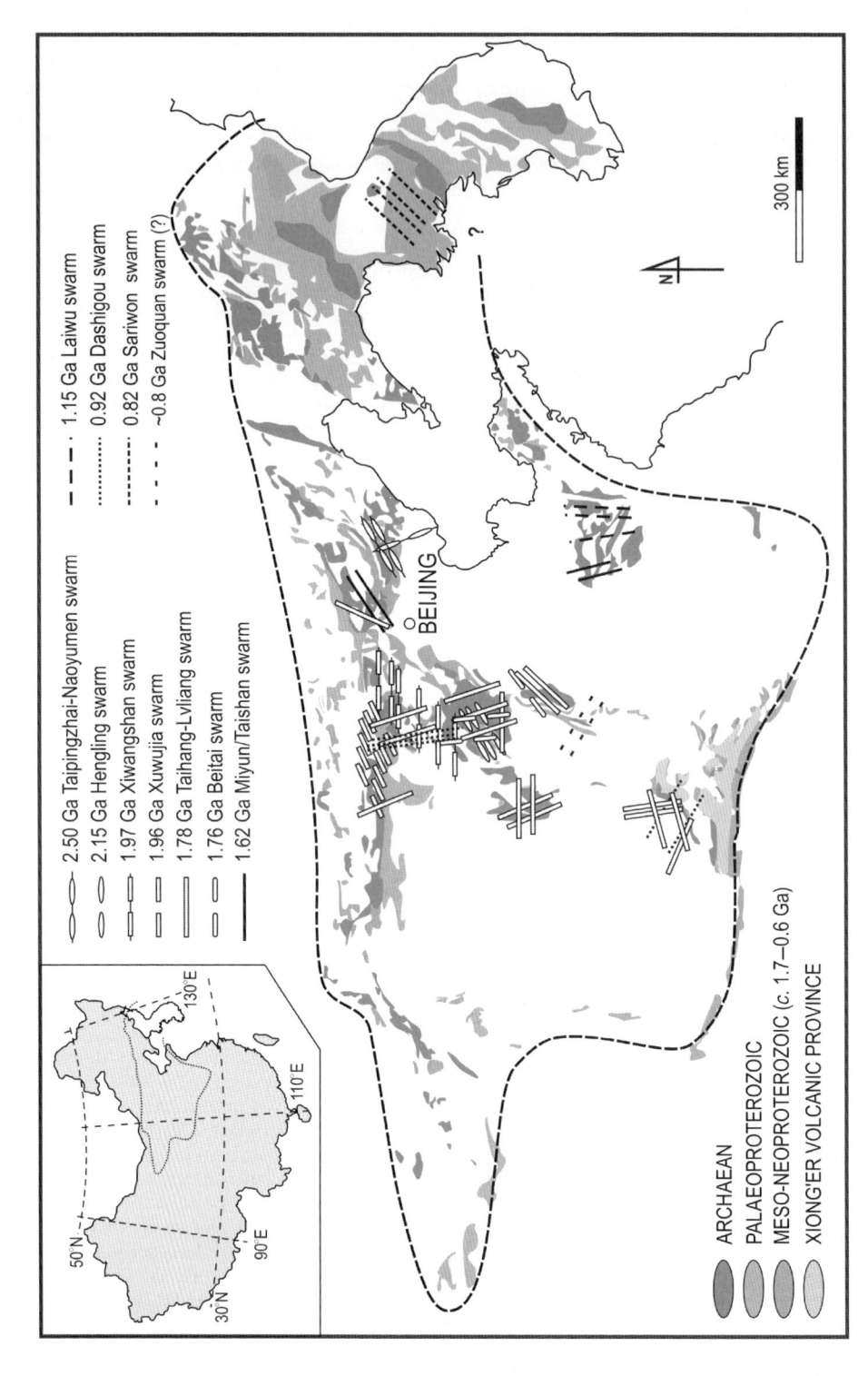

Fig. 1. Map showing the distribution of Precambrian mafic dykes and sills in the North China craton.

1.62 Ga Taishan–Miyun dykes, the 1.15 Ga Laiwu dykes, the 0.9 Ga Dashigou dykes, and the 0.9–0.8 Ga Sariwon and Zuoquan dykes (Fig. 1; Table 1). Each of these dyke swarms may provide important insights into the tectonothermal evolution of the Precambrian lithosphere of the NCC, and the possible palaeo-linkage(s) between the NCC and other craton(s). The 1.78 Ga Taihang–Lvliang swarm has a extent of c. 1000 km, and is the largest, most prominent and best-preserved Precambrian swarm in the NCC (e.g. Qian & Chen 1987; Halls et al. 2000; Hou et al. 2000, 2005, 2008a; Peng et al. 2004, 2007; Wang et al. 2004, 2008).

Brief introduction to the 1.78 Ga Taihang–Lvliang swarm

The Taihang–Lvliang dyke swarm (TLS) consists of NNW–SSE-trending (315°–345°) dykes evenly distributed throughout the central NCC, as well as a few NE–SW-(20–40°) and east–west-trending (250–290°) dykes (Fig. 2). It was followed by a younger NNW–SSE-trending swarm with distinct compositions (the Beitai swarm, 1765 Ma: Wang et al. 2004; Peng et al. 2006; Fig. 3a). The NE–SW-trending dykes occur mainly in the South Taihang Mountains. The east–west dykes are restricted to the Lvliang, southern Taihang, Huoshan and Zhongtiao Mountains, and they locally cut or branch off the NNW–SSE dykes. The east–west dykes can be further distinguished into two groups, one trending between 250° and 270° (mainly in the Lvliang and Taihang Mts) and another group trending 270–290° (mainly in the Zhongtiao and Huoshan Mts) (Fig. 2).

The TLS dykes are up to 60 km long and up to 100 m wide, with a typical width being c. 15 m. The dykes are vertical to subvertical and show sharp, chilled contacts with the country rocks. The systematic northward branching of the dykes indicates a magma flow direction from south to north (e.g. see Rickwood 1990). Mean ^{206}Pb–^{207}Pb ages reported for the TLS dykes are 1769 ± 3 Ma (zircon thermal ionization mass spectrometry (TIMS), Halls et al. 2000), 1778 ± 3 Ma [zircon sensitive high-resolution ion microprobe (SHRIMP), Peng et al. 2005], 1777 ± 3 Ma (zircon plus baddeleyite TIMS), and 1789 ± 28 Ma (baddeleyite TIMS) (Peng et al. 2006). An Ar–Ar age of about 1780 Ma is also available (Wang et al. 2004). The TLS dykes are composed of gabbro and dolerite, with a mineralogy dominated by plagioclase and clinopyroxene. They are tholeiitic in composition, varying from basalt to andesite, with minor occurrences of dacite and rhyolite. Peng et al. (2004, 2007) chemically divided the TLS dykes into three groups, followed by a fourth group, now identified as the Beitai swarm (Fig. 3a, b). It needs to be clarified that Wang et al. (2004) have also divided the dykes in South Taihang Mountains into three groups, on the basis of their chemistry, with their group 1 being compositionally similar to the NW group of the TLS dykes, their group 2 being similar to the Beitai swarm, and their group 3 being distinct, possibly another swarm.

Geometrical reconstruction of the Taihang–Lvliang swarm

Because the orientations of the TLS dykes have been modified after their intrusion [for instance, the Taihangshan Block has recorded a c. 15° anticlockwise rotation relative to the Ordos Block in the Mesozoic–Cenozoic (Fig. 2; e.g. Zhang, Y.-Q. et al. 2003; Huang et al. 2005)] the geometry of the dykes in these blocks needs to be reconstructed. In Figure 4a, the original orientations of dykes in the Taihang, South Taihang, Lvliang, Huoshan, Zhongtiao and Xiong'er Mountains are shown together with reconstructed orientations suggested by palaeomagnetic data (Zhang, Y.-Q. et al. 2003; Huang et al. 2005). Figure 4b shows the presumed dyke tracks after restoration of the rotations. The dykes constitute a radiating pattern, which could fit the geometry of the Xiong'er triple-rift (the Xiong'er volcanic province); that is, the majority are consistent with the rift arm extending into the central NCC. The other two groups of east–west-trending dykes are parallel to other two arms of this rift. Qian & Chen (1987) and Hou et al. (2000) suggested that the east–west dykes are late intrusions that were emplaced in a different stress field. However, these east–west dykes are distribute only in the areas with lower exposure depths. Peng et al. (2008a) argued for two groups of dykes intruding into coeval fissures, either in a changing stress regime such as from plume-generated uplift to the onset of rifting and breakup, or in a single stress field with two groups of conjugate fissures at the uppermost crustal level, based on the observation of NNW–SSE- and east–west-trending dykes in a reticular fissure system. Also, the local crosscutting relationships (east–west dykes cutting NNW–SSE dykes) would have arisen during continuous intrusion and uplift.

Although it is difficult to reconstruct the possible coeval dykes in other parts of the NCC (e.g. the Taishan Mts, Hou et al. 2008b), this fanning geometry clearly indicates a stress field radiating from a magma centre as a result of uplift; that is, not simply compression or extension (e.g. a north–south compression; Hou et al. 2006). It is revealed that the dykes were uplifted and exhumed from crustal levels up to 20 km, mainly deep in the

Table 1. *Precambrian mafic dyke swarms of the North China craton*

Swarm	Present orientation(s)	Rocks	Series	Distribution	Scale (km)	Ages
Taipingzhai–Naoyumen	NW–SE and ENE–WSW	Gabbro	Both tholeiitic and alkaline	Eastern Hebei	>100	2504 ± 11 Ma, 2516 ± 26 Ma (zircon U–Pb), Li et al. (2010)
Hengling	Deformed	Metagabbro (amphibolite schist)	Tholeiitic	Wutai Mts.	c. 100	2147 ± 5 Ma (zircon U–Pb), Peng et al. (2005)
Xiwangshan	Deformed (ENE–WSW to east–west)	Metagabbro (high-pressure granulite)	Tholeiitic	Sanggan River	c. 200–300	1973 ± 4 Ma (zircon U–Pb), Peng et al. (2005)
Xuwujia	Deformed (ENE–WSW)	Metagabbro (high-temperature granulite)	Tholeiitic	Liangcheng–Tuguiwula (Yinshan Mts.)	c. 200	1960 ± 4 Ma (zircon U–Pb), author's own unpublished data
Hengshan	Deformed (ENE–WSW)	Metagabbro (high-pressure granulite)	Tholeiitic	Hengshan Mts.	c. 100?	1914 ± 2 Ma, 1915 ± 4 Ma (zircon U–Pb), Kröner et al. (2006)
Taihang–Lvliang	NNW–SSE and east–west	Gabbro, dolerite	Tholeiitic	Central NCC and possibly other parts	c. 1000	1780–1770 Ma (present study)
Beitai	NNW–SSE to north–south	Gabbro	Tholeiitic	Hengshan–Taihang–South Taihang	>200	1765 ± 1 Ma (Ar–Ar whole rock), Wang et al. (2004)
Miyun	NE–SW	Gabbro, dolerite	Tholeiitic	Miyun–Chengde	c. 100	1620 Ma (zircon U–Pb), author's own unpublished data
Taishan	NNW–SSE to NE–SW	Gabbro, dolerite	Tholeiitic	Taishan Mts.	c. 200?	1619 ± 16 Ma (baddeleyite Pb–Pb), Li, H.-M., et al. pers. comm.
Dashigou	NNW–SSE	Gabbro	Alkaline	Hengshan–Wutaishan Mts.	c. 300?	917 ± 7 Ma (baddeleyite Pb–Pb), author's unpublished data
Sariwon	ENE–WSW to east–west	Gabbro, dolerite	Tholeiitic	Pyongnam Basin, North Korea	c. 150	816 ± 34 Ma (zircon U–Pb); 884 ± 15 Ma (baddeleyite Pb–Pb), Peng et al. (2008b) and author's own unpublished data
Zuoquan	NW–SE	Gabbro, dolerite	Tholeiitic	Zuoquan and adjacent area	Unknown	Neoproterozoic, according to geological relationship

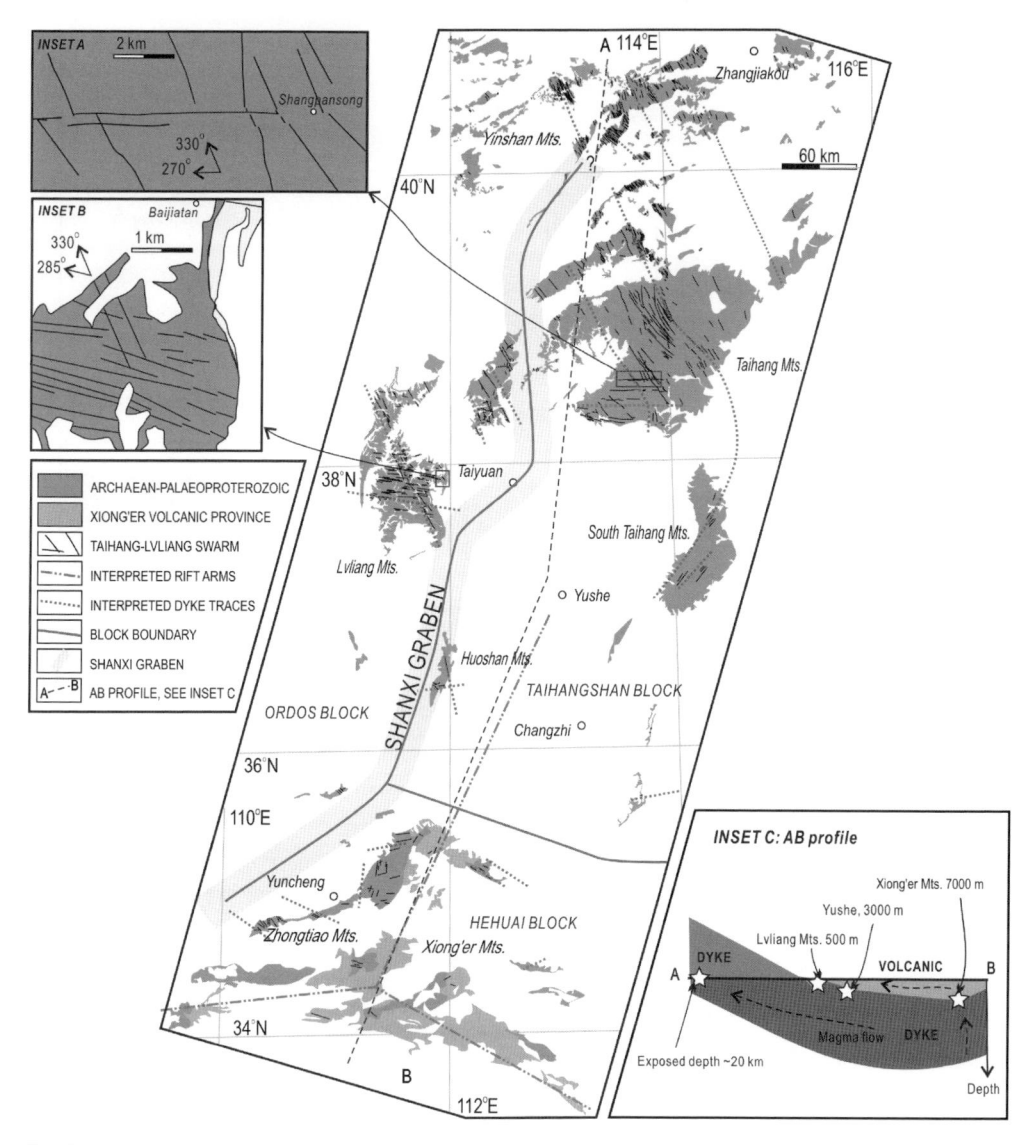

Fig. 2. Map showing the distribution of the Taihang–Lvliang swarm and Xiong'er volcanic province in the central NCC (after Peng et al. 2008a, with some younger dyke swarms, now dated and known to be unrelated, removed). Insets A and B show enlarged maps of local areas, and inset C is a profile of the study area based on the available geophysical (Wang 1995), geological (Peng et al. 2007) and palaeomagnetic data (Hou et al. 2000).

northern but shallow in the central study area according to the palaeomagnetic data (Hou et al. 2000) and a $P-T-t$ path (Peng et al. 2007). A north to south profile of the study area (Line AB in Fig. 2, inset c) can be constructed after incorporating seismic data in the south (e.g. Wang 1995). The increased uplift in the north of the study area could be partly responsible for the regional orientation changes of the dykes in the northern part of the central NCC (Fig. 4b); that is, the events resulting in this regional tilt could have distorted the orientations of the dykes in the northern part. Figure 4c shows an idealized image of the possible geometry of the TLS dykes and Xiong'er rift system at 1.78 Ga.

The Xiong'er volcanic province: extrusive counterpart of the Taihang–Lvliang swarm?

The Xiong'er volcanic province (XVP) has been thought to have no genetic relationship with the

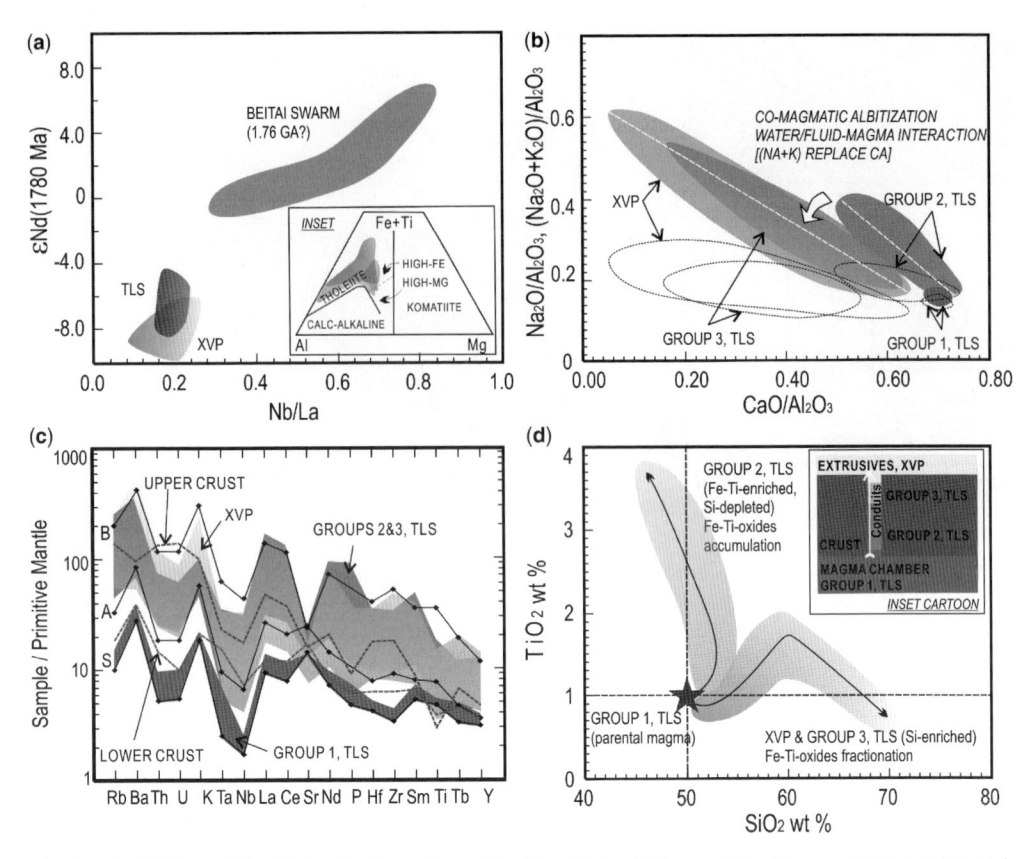

Fig. 3. (**a**) εNd (1780 Ma) v. Nb/La plot. Inset: Fe + Ti v. Al v. Mg (mol) diagram (after Peng *et al.* 2007). (**b**) Na$_2$O/Al$_2$O$_3$ (open fields) and (Na$_2$O + K$_2$O)/Al$_2$O$_3$ (coloured fields) v. CaO/Al$_2$O$_3$ plot. (**c**) Primitive mantle-normalized trace element spidergram of the Taihang–Lvliang dykes (TLS) and Xiong'er volcanic rocks (XVP): labelled curves refer to calculated liquid compositions after *in situ* crystallization (A) and fractional crystallization (B) from the starting composition (S) (after Peng *et al.* 2007). Primitive mantle-normalized values are after Sun & McDonough (1989). Data for the upper and lower crusts are after Rudnick & Gao (2001). (**d**) TiO$_2$ (wt%) v. SiO$_2$ (wt%) plot: the arrowed dashed lines show the differentiation trend, and the inset schematic illustration shows the relative differentiation depths of the TLS and XVP in the crust. Here Groups 1, 2 and 3 of TLS refer to the LT, NW and EW groups of Peng *et al.* (2007), respectively. Database is after Peng *et al.* (2008a).

TLS as it is compositionally dominated by intermediate rather than mafic volcanic rocks (e.g. Pirajno & Chen 2005; He *et al.* 2008, 2009, and references therein). The XVP is located in the south of the NCC, and has three branches: two branches along the southern margin of the NCC and a third one extending northward into the interior of the NCC (Fig. 2). The XVP has a thickness of 3–7 km (Zhao, T.-P. *et al.* 2002) and is dominated by thick and continuous lava flows, with rare, thin, sedimentary and volcaniclastic interlayers. There are also minor pillow lavas. The XVP is composed of diabasic and porphyritic rock. It is chemically tholeiitic and varies from basalt to andesite, dacite and rhyolite, with andesitic compositions being dominant; thus the XVP does not resemble a bimodal association. It consists of two volcanic cycles, both varying from mafic–intermediate to more silicic compositions; the age gap between the two cycles is not known. There are also some ultramafic bodies in the province, with associated Ni–Cu– PGE (platinum group elements) deposits, which are probably related to the XVP (Zhou *et al.* 2002). Ages of c. 1780–1770 Ma have been reported for the XVP (Zhao, T.-P. *et al.* 2004, 2005; Xu *et al.* 2007; He *et al.* 2009); however, some significantly younger ages have also been reported (He *et al.* 2009) but it remains unclear how these relate to the main XVP sequence.

The XVP is a host of important massive sulphide (Pb–Zn) and gold (Au–Ag) ore deposits in China, and there is evidence that the mineralization

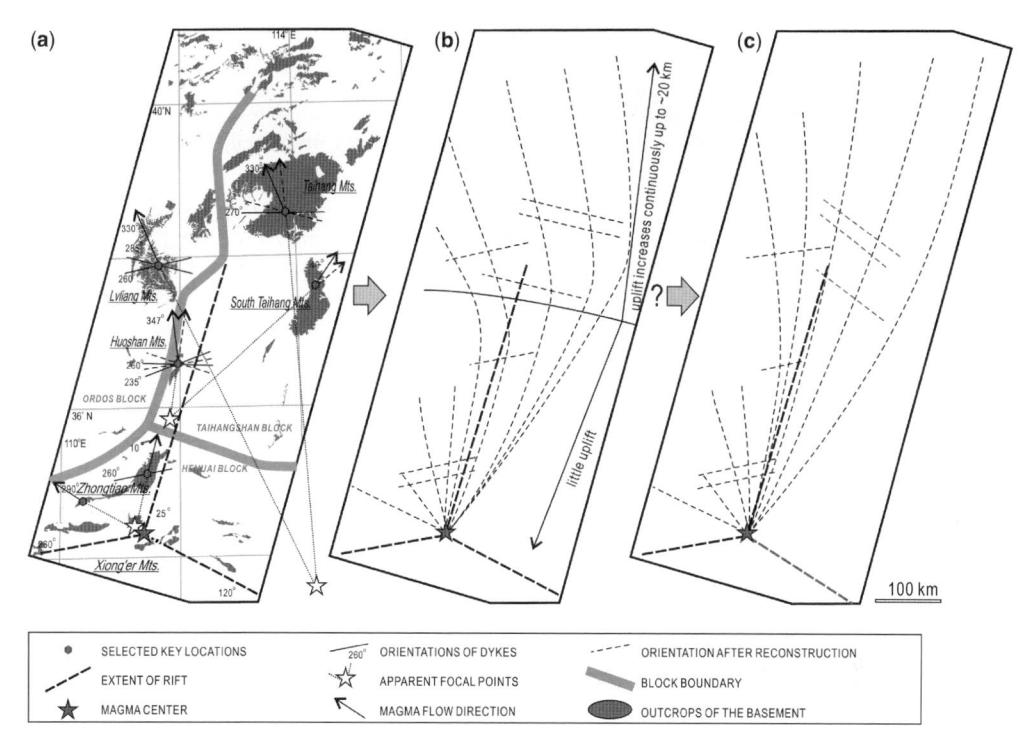

Fig. 4. Geometry of the TLS and XVP: (**a**) orientations of the dykes and apparent magmatic focal points in the selected locations, and the extent of the rifts; (**b**) presumed geometry of the dykes after reconstruction in relation to the configuration of the rift; (**c**) idealized geometry of the dykes and the rifts based on model discussed in text.

could be associated with the volcanism. Some may have formed by (magmatic) fluid–rock interaction during the late stages of volcanism (e.g. Zhao, J.-N. et al. 2002; Hou et al. 2003; Ren et al. 2003; Zhang, H.-C. et al. 2003; Weng et al. 2006; Cao et al. 2008), especially in some fissures and vent complexes (Pei et al. 2007). This water (fluid)–rock interaction also caused albitization of feldspar, as well as alteration of the whole-rock chemistry, resulting in a spilite–keratophyre-type affinity in some of the XVP rocks (Peng et al. 2008a).

A cogenetic relationship between the XVP and TLS is favoured (i.e. the XVP is the extrusive counterpart of the TLS), because of the following observations: (1) the feeder dykes of the XVP have similar ages and compositions to the dykes of the TLS; (2) the geometries of the TLS (radiating fan) and XVP (triple-junction) are compatible with each other, and they share the same magmatic centre (Fig. 4b); (3) the exposure depths of the TLS and XVP are spatially correlated with the exhumation of the central NCC (Fig. 2, inset c); (4) they share similar petrographic characteristics, chemical variations (e.g. SiO_2 contents of the TLS and XVP vary from 45 to 68 wt% and from 45 to 78 wt%, respectively), trace element patterns and isotopic compositions (Peng et al. 2008a). Further support for a cogenetic relationship comes from the observation that both the XVP rocks and a few of the TLS dykes have experienced co-magmatic albitization, resulted in the variations of certain major and trace elements (e.g. Na, Ca, K, Sr, Rb, etc., Peng et al. 2008a).

Thus the same parental magma, with varied degrees of differentiation and/or assimilation, can explain the petrogenesis of both the TLS and XVP. Figure 3c shows one possible model, which began with *in situ* crystallization, followed by assimilation and fractional crystallization, with an assemblage composed of plagioclase, clinopyroxene and olivine (Peng et al. 2008a). This model successfully interprets the variations of most trace elements and partly those of some major elements, excluding those altered by albitization. However, to explain the variations of some other elements (e.g. Fe and Ti) fractionation of Fe–Ti-oxides should be considered (Fig. 3d).

How then can we explain the apparent difference in dominant chemistry; that is, mafic (TLS) v.

intermediate (XVP)? The TLS is chemically divided into three groups: group 1 is minor and proposed as the parental magma composition of this magmatism; group 3 is intermediate-dominated and similar to the XVP; whereas group 2 forms most of the TLS and has very similar trace element patterns but partly distinguishable concentrations as compared with the XVP, especially for major elements (Table 2; Fig. 3c, d). The group 2 rocks show an Fe–Ti-enriched trend (Si-depleted), whereas the XVP rocks (and group 3) present a Si-enriched trend (Fig. 3d). It has been suggested that the liquid would have Fe–Ti enrichment when it differentiated in a closed system at a relatively low oxidation state, whereas it would be Si-enriched when it interacted with the oxidized and hydrated surroundings (Brooks et al. 1991). As the group 2 dykes are commonly exposed from greater depth than the group 3 dykes and the XVP (Peng et al. 2007), it is reasonable to propose that the corresponding liquid has evolved from being more mafic in the deeper crust to being more silicic at shallower depth (Fig. 3d). Also, as the iron-rich liquid is more dense and more difficult to erupt (e.g. Brooks et al. 1991), this more mafic liquid would crystallize along the margins of the dyke conduits, forming the mafic rocks (e.g. group 2). In the mean time, the remaining relatively more Si-rich liquid and also the fractionated liquid would interact and be incorporated with more oxidized and hydrated crust, fractionating more Fe–Ti-oxides, and produce more Si-rich liquid, forming the intermediate-dominated rocks (e.g. XVP and group 3) (Fig. 3d). Thus the differentiation of Fe–Ti-enriched liquid at great depth, as well as the continuous assimilation and fractionation (of plagioclase, clinopyroxene and olivine) during ascent, could be responsible for the average composition gap between the TLS and XVP (Table 2).

Another point is that some would argue that the XVP belongs to a calc-alkaline series, which makes it different from the tholeiitic TLS rocks (e.g. He et al. 2008, 2009, and references therein). However, this point of debate could be largely explained by the recognition of widespread albitization in the XVP and some of the TLS dykes (Han et al. 2006; Peng et al. 2008a; Fig. 3b). In this case, discrimination based on elements influenced by albitization (e.g. Na, K, Rb, Sr and Ca) should be avoided. In an Fe + Ti–Al–Mg diagram (Fig. 3a inset), both XVP and TLS samples plot in the tholeiitic field instead of the calc-alkaline field. It should be noted that most of the XVP samples, as well as many of the TLS samples, plot in the calc-alkaline field in a Th–Co diagram suitable for altered rocks (Zhao et al. 2009). However, this diagram is based on island arc rocks (Hastie et al. 2007), and has not yet been tested for continental volcanic rocks, especially those with high Th content.

A large igneous province: lines of evidence and particularities

Large igneous provinces (LIPs) are considered to be massive crustal and intraplate emplacements of predominantly mafic extrusive and intrusive rocks that originated via processes other than 'normal' sea-floor spreading (Coffin & Eldholm 1994, 2001). This definition has been extended to silicic provinces (Bryan et al. 2002; Sheth 2007; Bryan & Ernst 2008). Bryan & Ernst (2008) renewed this definition as those 'magmatic provinces with areal extents >0.1 Mkm^2, igneous volumes >0.1 Mkm^3 and maximum lifespans of c. 50 Ma that have intraplate tectonic settings or geochemical affinities, and are characterized by igneous pulse(s) of short duration (c. 1–5 Ma), during which a large proportion ($>75\%$) of the total igneous volume is emplaced'. It should be noted that a minimal areal extent of 0.05 Mkm^2 (Sheth 2007) or 1 Mkm^2 (Courtillot & Renne 2003), and a lifespan of c. 1 Ma (e.g. Courtillot & Renne 2003) or ≥ 40 Ma (e.g. Birkhold et al. 1999; Revillon et al. 2000) have been proposed.

Here, the area encompassed by Figure 2 is considered the estimated areal extent of the TLS, at c. 0.3 Mkm^2. The XVP is continuous over a north–south extent of 500 km and an east–west extent of 360 km (Fig. 2; Wang 1995; Zhao, T.-P. et al. 2002; Xu et al. 2007); thus its areal extent can be calculated as $\frac{1}{2} \times 500$ km $\times 360$ km = 0.09 Mkm^2, considering its triangular distribution. For the exposed areas, the areal extents are approximately 0.1 Mkm^2 and 0.02 Mkm^2 for the TLS and XVP, respectively (Fig. 2). The estimated magmatic volume of the TLS is calculated as: V (volume) = a (areal extent) $\times h$ (average height of the dykes) $\times \lambda$ (extension ratio). A height of 20 km is estimated based on an exposed depth (Peng et al. 2007) and the palaeomagnetic data (Hou et al. 2000). Although extension ratios ranging from 0.28 to 0.48% are available (Hou et al. 2006), three well-exposed profiles are further checked. These profiles, including the 25 km long Jvlebu–Zhangxiaocun (Datong), the 20 km long Hongqicun–Jiulongwan (Fengzhen) and the 50 km Doucun–Shengtangbu (Wutai) profiles, give extension ratios at 1.0%, 1.27% and 0.72%, respectively. The variation could be a result of uneven distribution and/or miscount. Here 1.0% is taken as the average extension ratio. Thus the estimated volume would be $V = 0.3$ (a, Mkm^2) $\times 20$ (h, km) $\times 1.0\%$ (λ) = 0.06 Mkm^3, and the estimated exposed volume would be $V = 0.1$ (exposed area, Mkm^2) $\times 20$ (h, km) $\times 1.0\%$ (λ) = 0.02 Mkm^3. The volcanic volume of

Table 2. *Average compositions of the Taihang–Lvliang swarm (TLS) and the Xiong'er volcanic province (XVP), and the upper and lower crust*

Contents (wt%)	Parental magma (Group 1, TLS)*	Group 2 (TLS) average*	Group 3 (TLS) TLS average*	TLS average*	XVP average*	TLS & XVP total average[†]	Lower crust[‡]	Upper crust[‡]
SiO_2	50.09	50.29	56.66	51.55	57.69	56.27	53.40	66.60
TiO_2	1.02	2.57	1.40	2.20	1.20	1.43	0.82	0.64
Al_2O_3	14.85	13.00	14.23	13.41	14.11	13.95	16.90	15.40
Fe_2O_3(total iron)	13.31	15.47	10.18	14.22	9.39	10.50	8.57	5.04
MnO	0.19	0.20	0.14	0.19	0.14	0.15	0.10	0.10
MgO	6.99	4.02	3.62	4.20	3.49	3.65	7.24	2.48
CaO	10.31	7.45	5.69	7.35	4.58	5.22	9.59	3.59
Na_2O	2.27	2.65	2.45	2.58	2.92	2.84	2.65	3.27
K_2O	0.61	2.28	3.06	2.29	3.30	3.07	0.61	2.80
P_2O_5	0.14	1.08	0.52	0.89	0.36	0.48	0.10	0.15
Total	99.76	99.01	97.95	98.86	97.17	97.56	99.98	100.07

*The TLS and XVP averages are based on a database from Peng *et al.* (2008*a*) (Groups 1, 2 and 3 correspond to the LT, NW and EW groups therein, respectively), and Group 1 (TLS) is proposed to represent the parental magma compositions.
[†]Total TLS and XVP average is calculated taking the estimated magma volumes as their weight.
[‡]The lower and upper crustal compositions are after Rudnick & Gao (2001).

the XVP is calculated using V (volume) $= a$ (areal extent) $\times h$ (average thickness of the volcanic rocks). Assuming the original XVP extrusive extent as a triangular pyramid and taking the maximum exposed thickness of 7 km as an overall maximum, the volume would be $V = \frac{1}{3} \times 0.09$ (a, Mkm^2) $\times 7$ (h, km) ≈ 0.2 Mkm^3, and the exposed volume would be $V = 0.02$ (exposed area, Mkm^2) $\times 5$ (average thickness, km) $= 0.1$ Mkm^3. Collectively, the TLS and XVP have an estimated total areal extent of c. 0.3 Mkm^2, a volume estimation of about 0.3 Mkm^3, an exposed area of c. 0.1 Mkm^2, and an exposed volume of c. 0.1 Mkm^3. All these estimates could be doubled or even tripled if remnants in other parts of the NCC are confirmed, especially for the TLS.

The TLS and XVP postdate the regional granulite-facies metamorphism and amphibolite-facies retrograde metamorphism at 1790–1780 Ma (e.g. Wang et al. 1995; Zhang et al. 2006). They are followed by 1.75–1.73 Ga post-magmatic syenitic intrusions (Ren et al. 2003) and the 1.76 Ga Beitai swarm (Peng et al. 2008a). Thus the lifespan and duration of TLS and XVP magmatism is roughly bracketed to 1780–1760 Ma (i.e. about 20 Ma), even if the Beitai swarm is included. However, the duration of the major pulse is unknown. Undoubtedly, the magmatism has an intraplate tectonic affinity as it occurs largely within the NCC undergoing extension, and has within-plate compositional characteristics (see Bryan & Ernst 2008).

In summary, the 1.78 Ga magmatism including the TLS and XVP fits the LIP definition except for not knowing the duration of the major pulse (e.g. Bryan & Ernst 2008). There are several schemes for classification of LIPs; for instance, oceanic v. continental (Coffin & Eldholm 2001), mafic v. silicic (Bryan & Ernst 2008), and volcanic v. plutonic (Sheth 2007). It may be noted that most LIPs (both silicic and mafic dominated) are compositionally bimodal, and may also show a spectrum of compositions from basalt to high-silica rhyolite (e.g. Sheth 2007; Bryan & Ernst 2008). However, this North China LIP is characterized by more intermediate rocks than mafic rocks (with very few silicic components). The mafic portions are about 85% and 20% for the TLS and XVP, respectively (based on a database given by Peng et al. 2008a). The total mafic portion (p) could be $p_{total} = V_{mafic}/V_{total} = (p_{TLS} \times V_{TLS} + p_{XVP} \times V_{XVP})/(V_{TLS} + V_{XVP}) \approx 35$ vol.%. In contrast, the intermediate portion could be c. 65 vol.%, as there are few other components. It should be mentioned that this mafic portion may be a minimum estimation, as the volume of TLS could be substantially underestimated.

Table 2 shows the average compositions of the TLS and XVP, their estimated total average, and their comparison with the average crust. Both the TLS and XVP averages, as well as their total average, show distinct characteristics from a potential melt from the lower or upper crust; for example, it would be the difficult for the low Al$_2$O$_3$ content but high TiO$_2$ content to originate from the crust (Table 2, Fig. 3c). However, the total average of TLS and XVP is intermediate in composition (Table 2), which makes them in somewhat different from mantle-derived associations. In this case, incorporation of crustal melts in the parental magma (chamber) could be possible. However, it is still hard to evaluate this, as there is a possibility of underestimating the volume of TLS and thus the mafic weight in the total average. Nevertheless, some characteristics of the XVP, for example, slightly lower Nb and Ta but higher Th, U and Nd contents, as well as distinct higher Si concentration, compared with the TLS counterparts, could be partly inherited from the upper crust (e.g. Table 2; Fig. 3c). Thus it is reasonable to suggest that there are volumes of crustal melts incorporated into the magma during its ascent, especially for the late-stage differentiates (XVP).

Constraints on regional evolution and geodynamics

Reconstruction of the 1.78 Ga magmatism in the NCC (the TLS and XVP) increases its importance for the regional evolution and geodynamics. It indicates a broad rigid block undergoing significant extension and uplift at 1.78 Ga. It suggests that this rigid NCC block had basically ceased basement evolution from 1.78 Ga until the Mesozoic–Cenozoic, when the geometry of the TLS was distorted (Fig. 4). This implies a major magmatic accretion event at 1.78 Ga in the NCC, consistent with results from some c. 1.8 Ga mantle xenoliths (e.g. Gao et al. 2002).

Such reconstruction can also constrain the tectonic settings of the NCC at 1.78 Ga, as various alternatives have been proposed; that is, syn-orogenic (post-collisional uplift in the central NCC and Andean-style collision along its southern margin) (Fig. 5a; e.g. Zhao, G.-C. et al. 2005, 2010; Wang et al. 2004, 2008; He et al. 2008, 2009) or non-orogenic (Fig. 5b, c; e.g. Li et al. 2000; Zhai et al. 2000; Kusky & Li 2003; Hou et al. 2006, 2008a; Kusky et al. 2007b; Peng et al. 2004, 2008a). The synorogenic hypothesis is mainly based on the subduction-influenced geochemistry (e.g. depletion in high field strength elements) of both the TLS and XVP rocks, and Andean-style calc-alkaline volcanism. Nevertheless, the chemistry could alternatively be interpreted as affected by assimilation of the continental crust,

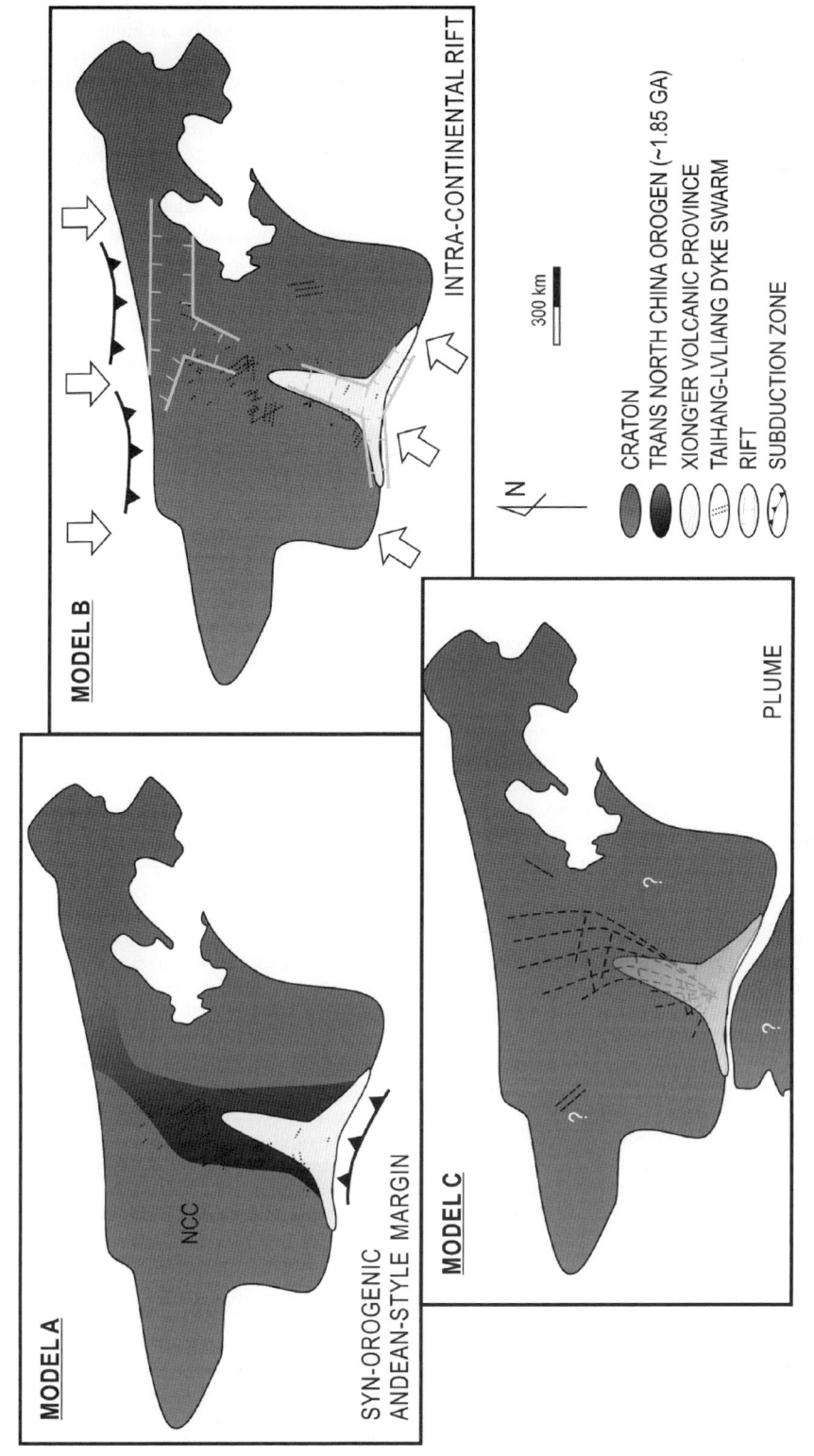

Fig. 5. Schematic illustrations showing tectonic models for the NCC at 1.78 Ga.

or inherited from a fertilized mantle region. Also, alteration and widespread albitization could explain a resemblance of the XVP to an Andean-style calc-alkaline association (in fact, it is tholeiitic, see details given by Peng et al. 2008a). As the TLS dykes have a radiating geometry (e.g. Fahrig 1987) and are very large scales (e.g. Ernst et al. 1995), they differ from synorogenic dykes; also, as the XVP developed in a rift with triple-junction geometry, with river–lake-facies sedimentary interlayers, it is less compatible with a continental margin environment.

However, instead of an intra-continental rift model (Fig. 5b), a plume model (Fig. 5c) with magma originating from a rifting centre is preferred here because it meets four out of five criteria suggested by Campbell (2001) to distinguish plume-associated volcanism: (1) uplift prior to volcanism (recorded by the 1.80–1.78 Ga regional $P-T-t$ paths and extensional deformation; e.g. Zhao, G.-C. et al. 2005; Guo et al. 2005; Zhang et al. 2006, 2007); (2) a radiating dyke swarm geometry; (3) massive volcanic flows correlated over a large area (>0.09 Mkm^2) and a long distance (>500 km); (4) plume-associated chemistry (an enriched magma source followed shortly by a depleted source with OIB affinity: Peng et al. 2007). This plume possibly could have responded to a massive mantle–crust interaction to produce volumes of both mafic and intermediate igneous rocks of the North China LIP, as well as widespread polymetallic mineralization in this area.

It should be addressed that this 1.78 Ga magmatism is centred in the southern most part of the NCC, and thus it is reasonable to predict missing parts of the TLS and/or XVP outside the NCC (Fig. 5c). According to a database of Ernst & Buchan (2001), roughly coeval (c. 1780 Ma) large mafic magmatic events are reported in South America (e.g. Uruguayan dykes in Rio de Plata craton, Halls et al. 2001; Avanavero dykes in Guyana shield, Norcross et al. 2000; Crepori gabbro–dolerite sills and dykes, Santos et al. 2002), Australia (e.g. Harts Range volcanic rocks and sills and Eastern Creek volcanic series, Sun 1997; Tewinga volcanic series, Page 1988; Mount Isa dykes, Parker et al. 1987; Hart doleritic sills, Page & Hoatson 2000), and possibly others (e.g. India: Dharwar dykes, Srivastava & Singh 2004). All the above units are potential candidates for the missing part(s) of the TLS and/or XVP, providing clues to which blocks may have been connected with the NCC.

Conclusions

Reconstruction and interpretation of ancient giant mafic dyke swarms could be a potential way to constrain ancient continental evolution and geodynamics. As a case study, a radiating geometry is reconstructed for the 1.78 Ga giant Taihang–Lvliang swarm of the NCC. It can match the geometry of the Xiong'er triple-junction rift, in which the Xiong'er volcanic province is specified as the extrusive counterpart of this swarm. This giant radiating dyke swarm clearly indicates significant extension, uplift, and magmatic accretion in the NCC. Also, it could provide clues to potential linkages between the NCC and other ancient block(s). This dyke swarm, as well as the volcanic counterpart, show affinities to Phanerozoic LIPs, and could be the remnants of an ancient North China LIP. However, this LIP is unique in that it is characterized by large volumes of both mafic (c. 35 vol.%) and intermediate (c. 65 vol.%) components, which suggests extensive mantle–crust interaction and notable differentiation in this ancient plume setting.

I thank my many colleagues who have worked on this subject, and have contributed much to it. I especially acknowledge M. G. Zhai, J. H. Guo, T. P. Zhao, J. H. Li, G. C. Zhao, G. T. Hou, S. W. Liu, Y. S. Wan, S. N. Lu, H. M. Li, Y. H. He, Y. J. Wang, T. Kusky, H. Halls, R. Ernst, W. Bleeker and S. Wilde. The paper has benefited from criticism by two anonymous reviewers. I thank B. Windley for his illuminating discussion and warm-hearted encouragement. This study is financially supported by China NSFC grant 40602024 and two previous grants awarded to M. G. Zhai and T. P. Zhao.

References

BIRKHOLD, A. B., NEAL, C. R., MAHONEY, J. J. & DUNCAN, R. A. 1999. The Ontong Java plateau: episode growth along the SE margin. AGU Fall Meeting, San Francisco, CA, EOS 80, F1103.

BLEEKER, W. & ERNST, R. 2006. Short-lived mantle generated magmatic events and their dyke swarms: the key unlocking Earth's paleogeographic record back to 2.6 Ga. In: HANSKI, E., MERTANEN, S., RAMÖ, T. & VUOLLO, J. (eds) Dyke Swarms – Time Markers of Crustal Evolution. Taylor & Francis, London, 3–26.

BROOKS, K. C., LARSEN, L. M. & NIELSEN, T. F. D. 1991. Important of iron–rich tholeiitic magmas at divergent plate margins: a reappraisal. Geology, 19, 269–272.

BRYAN, S. & ERNST, R. 2008. Revised definition of large igneous provinces (LIPs). Earth-Science Reviews, 86, 175–202.

BRYAN, S. E., RILEY, T. R., JERRAM, D. A., LEAT, P. T. & STEPHENS, C. J. 2002. Silicic volcanism: an undervalued component of large igneous provinces and volcanic rifted margins. In: MENZIES, M. A., KLEMPERER, S. L., EBINGER, C. J. & BAKER, J. (eds) Magmatic Rifted Margins. Geological Society of America, Special Papers, 362, 1–36.

CAMPBELL, I. H. 2001. Identification of ancient mantle plume. In: ERNST, R. E. & BUCHAN, K. L. (eds)

Mantle Plumes: Their Identification through Time. Geological Society of America, Special Papers, **352**, 5–21.

CAO, Y., LI, S.-R., SHEN, J.-F., YAO, M.-J., LI, Q.-K. & MAO, F.-L. 2008. Fluid–rock interaction in ore-forming process of Qianhe structure-controlled alteration-type gold deposit in western Henan Province. *Mineral Deposits*, **27**, 714–726 [in Chinese with English abstract].

COFFIN, M. F. & ELDHOLM, O. 1994. Large igneous provinces: crustal structure, dimensions and external consequences. *Reviews of Geophysics*, **32**, 1–36.

COFFIN, M. F. & ELDHOLM, O. 2001. Large Igneous Provinces: progenitors of some ophiolites? *In*: ERNST, R. E. & BUCHAN, K. L. (eds) *Mantle Plumes: Their Identification through Time.* Geological Society of America, Special Papers, **352**, 59–70.

COURTILLOT, V. E. & RENNE, P. R. 2003. On the ages of flood basalt events. *Comptes Rendus Geoscience*, **335**, 113–140.

ERNST, R. E. & BUCHAN, K. L. 2001. Large mafic magmatic events through time and links to mantle plume heads. *In*: ERNST, R. E. & BUCHAN, K. L. (eds) *Mantle Plumes: Their Identification through Time.* Geological Society of America, Special Papers, **352**, 483–575.

ERNST, R. E., HEAD, J. W., PARFITT, E., GROSFILS, E. & WILSON, L. 1995. Giant radiating dyke swarms on Earth and Venus. *Earth-Science Reviews*, **39**, 1–58.

FAHRIG, W. F. 1987. The tectonic settings of continental mafic dyke swarms: failed arm and early passive margin. *In*: HALLS, H. C. & FAHRIG, W. F. (eds) *Mafic Dyke Swarms.* Geological Association of Canada, Special Papers, **34**, 331–348.

GAO, S., RUDNICK, R. L., CARLSON, R. W., MCDONOUGH, W. F. & LIU, Y.-S. 2002. Re–Os evidence for replacement of ancient mantle lithosphere beneath the North China Craton. *Earth and Planetary Science Letters*, **198**, 307–322.

GUO, J.-H., SUN, M., CHEN, F.-K. & ZHAI, M.-G. 2005. Sm–Nd and SHRIMP U–Pb zircon geochronology of high-pressure granulites in the Sanggan area, North China Craton: timing of Palaeoproterozoic continental collision. *Journal of Asian Earth Sciences*, **24**, 629–642.

HALLS, H. C., LI, J.-H., DAVIS, D., HOU, G. T., ZHANG, B.-X. & QIAN, X.-L. 2000. A precisely dated Proterozoic paleomagnetic pole from the North China craton, and its relevance to paleo-continental construction. *Geophysical Journal International*, **143**, 185–203.

HALLS, H. C., CAMPAL, N., DAVIS, D. W. & BOSSI, J. 2001. Magnetic studies and U–Pb geochronology of the Uruguayan dyke swarm, Rio de la Plata craton, Uruguay: paleomagnetic and economic implications. *Journal of South American Earth Sciences*, **14**, 349–361.

HAN, Y.-G., ZHANG, S.-H., BAI, Z.-D. & DONG, J. 2006. Albitization of the volcanic rocks of the Xiong'er Group, Western Henan, and its implications. *Journal of Mineralogy and Petrology*, **26**, 35–42 [in Chinese with English abstract].

HANSKI, E., MERTANEN, S., RAMÖ, T. & VUOLLO, J. (eds) 2006. *Dyke Swarms – Time Markers of Crustal Evolution.* Taylor & Francis, London.

HASTIE, A. R., KERR, A. C., PEARCE, J. A. & MITCHELL, S. F. 2007. Classification of altered volcanic island arc rocks using immobile trace elements: development of the Th–Co discrimination Diagram. *Journal of Petrology*, **48**, 2341–2357.

HE, Y.-H., ZHAO, G.-C., SUN, M. & WILDE, S. 2008. Geochemistry, isotope systematics and petrogenesis of the volcanic rocks in the Zhongtiao Mountain: an alternative interpretation for the evolution of the southern margin of the North China Craton. *Lithos*, **102**, 158–178.

HE, Y.-H., ZHAO, G.-C., SUN, M. & XIA, X.-P. 2009. SHRIMP and LA-ICP-MS zircon geochronology of the Xiong'er volcanic rocks: implications for the Paleo-Mesoproterozoic evolution of the southern margin of the North China Craton. *Precambrian Research*, **168**, 213–222.

HOU, G.-T., LI, J.-H., QIAN, X.-L., ZHANG, B.-X. & HALLS, H. C. 2000. The paleomagnetism and geological significance of Mesoproterozoic dyke swarms in the central North China craton. *Science in China (D)*, **44**, 185–193.

HOU, G.-T., LIU, Y.-L., LI, J.-H. & JIN, A.-W. 2005. The SHRIMP U–Pb chronology of mafic dyke swarms: a case study of Laiwu diabase dykes in western Shandong. *Acta Petrologica et Mineralogica*, **24**, 179–185.

HOU, G.-T., WANG, C.-C., LI, J.-H. & QIAN, X.-L. 2006. Late Palaeoproterozoic extension and a paleo-stress field reconstruction of the North China Craton. *Tectonophysics*, **422**, 89–98.

HOU, G.-T., SANTOSH, M., QIAN, X.-L., LISTER, G. S. & LI, J.-H. 2008a. Configuration of the Late Palaeoproterozoic supercontinent Columbia: insights from radiating mafic dyke swarms. *Gondwana Research*, **14**, 395–409.

HOU, G.-T., LI, J.-H., YANG, M.-H., YAO, W.-H., WANG, C.-C. & WANG, Y.-X. 2008b. Geochemical constraints on the tectonic environment of the late Palaeoproterozoic mafic dyke swarms in the North China Craton. *Gondwana Research*, **13**, 103–116.

HOU, W.-R., XIAO, X.-G., ZHANG, H.-C., GAO, L. & GAO, D.-H. 2003. The genesis model of the gold–polymetallic deposit from the volcanic rocks in the Xiong'er rift. *Gold Geology*, **9**, 22–27 [in Chinese with English abstract].

HUANG, B.-C., SHI, R.-P., WANG, Y.-C. & ZHU, R.-X. 2005. Paleomagnetic investigation on Early–Middle Triassic sediments of the North China block: a new Early Triassic paleo-pole and its tectonic implications. *Geophysical Journal International*, **160**, 101–113.

KRÖNER, A., WILDE, S. A., LI, J.-H. & WANG, K.-Y. 2005. Age and evolution of a late Archaean to Palaeoproterozoic upper to lower crustal section in the Wutaishan/Hengshan/Fuping terrain of north China. *Journal of Asian Earth Sciences*, **24**, 577–576.

KRÖNER, A., WILDE, S. A. ET AL. 2006. Zircon geochronology and metamorphic evolution of mafic dykes in the Hengshan Complex of northern China: Evidence for late Palaeoproterozoic extension and subsequent high-pressure metamorphism in the North China Craton. *Precambrian Research*, **146**, 45–67.

KUSKY, T. M. & LI, J.-H. 2003. Palaeoproterozoic tectonic evolution of the North China craton. *Journal of Asian Earth Sciences*, **22**, 383–397.

KUSKY, T., LI, J.-H. & SANTOSH, M. 2007a. The Palaeoproterozoic North Hebei Orogen: North China craton's collisional suture with the Columbia supercontinent. *Gondwana Research*, **12**, 4–28.

KUSKY, T. M., WINDLEY, B. F. & ZHAI, M.-G. 2007b. Tectonic evolution of the North China block: from orogen to craton to orogen. *In*: ZHAI, M.-G., WINDLEY, B. F., KUSKY, T. M. & MENG, Q.-R. (eds) *Mesozoic Sub-Continental Lithospheric Thinning Under Eastern Asia*. Geological Society, London, Special Publications, **208**, 1–34.

LI, J.-H., QIAN, X.-L., HUANG, X.-N. & LIU, S.-W. 2000. The tectonic framework of the basement of north China craton and its implication for the early Precambrian cratonization. *Acta Geologica Sinica*, **16**, 1–10.

LI, J.-H., KUSKY, T. M. & HUANG, X.-N. 2002. Neoarchean podiform chromitites and mantle tectonites in ophiolitic mélange, North China Craton: a record of early oceanic mantle oceanic mantle processes. *GSA Today*, **12**, 4–11.

LI, T.-S., ZHAI, M.-G., PENG, P., CHEN, L. & GUO, J.-H. 2010. Ca. 2.5 billion years old coeval ultramafic–mafic and syenitic dykes in Eastern Hebei Region: implications for cratonization of the North China Craton. *Precambrian Research*, doi: 10.1016/j.precamres.2010.04.001.

NORCROSS, C., DAVIS, D. W., SPOONER, E. T. C. & RUST, A. 2000. U–Pb and Pb–Pb age constraints on Palaeoproterozoic magmatism, deformation and gold mineralization in the Omai area, Guyana Shield. *Precambrian Research*, **10**, 69–86.

PAGE, R. W. 1988. Geochronology of Early to Middle Proterozoic fold belts in northern Australia: a review. *Precambrian Research*, **40–41**, 1–19.

PAGE, R. W. & HOATSON, D. M. 2000. Geochronology of mafic–ultramafic intrusions. *In*: HOATSON, D. M. & BLAKE, D. H. (eds) *Geology and Economic Potential of the Palaeoproterozoic Layered Mafic–Ultramafic Layered Intrusions in the East Kimberley, Western Australia*. Australian Geological Survey Organisation Bulletin, **246**, 163–172.

PARKER, A. J., RICKWOOD, P. C. ET AL. 1987. Mafic dyke swarms of Australia. *In*: HALLS, H. C. & FAHRIG, W. F. (eds) *Mafic Dyke Swarms*. Geological Association of Canada, Special Papers, **34**, 401–417.

PEI, Y.-H., YAN, H.-Q. & MA, Y.-F. 2007. The relationship between paleo-volcanic apparatus and mineral resources of Xiong'er Group along Songxian–Ruzhou Zone in Henan Province. *Geology and Mineral Resources of South China*, **1**, 51–58 [in Chinese with English abstract].

PENG, P., ZHAI, M.-G., ZHANG, H.-F., ZHAO, T.-P. & NI, Z.-Y. 2004. Geochemistry and geological significance of the 1.8 Ga mafic dyke swarms in the North China Craton: an example from the juncture of Shanxi, Hebei and Inner Mongolia. *Acta Petrologica Sinica*, **20**, 439–456 [in Chinese with English abstract].

PENG, P., ZHAI, M.-G., ZHANG, H.-F. & GUO, J.-H. 2005. Geochronological constraints on the Palaeoproterozoic evolution of the North China Craton: SHRIMP zircon ages of different types of mafic dikes. *International Geology Review*, **47**, 492–508.

PENG, P., ZHAI, M.-G. & GUO, J.-H. 2006. 1.80–1.75 Ga mafic dyke swarms in the central North China craton: implications for a plume-related break-up event. *In*: HANSKI, E., MERTANEN, S., RAMÖ, T. & VUOLLO, J. (eds) *Dyke Swarms – Time Markers of Crustal Evolution*. Taylor & Francis, London, 99–112.

PENG, P., ZHAI, M.-G., GUO, J.-H., KUSKY, T. & ZHAO, T.-P. 2007. Nature of mantle source contributions and crystal differentiation in the petrogenesis of the 1.78 Ga mafic dykes in the central North China craton. *Gondwana Research*, **12**, 29–46.

PENG, P., ZHAI, M.-G., ERNST, R., GUO, J.-H., LIU, F. & HU, B. 2008a. A 1.78 Ga Large Igneous Province in the North China craton: The Xiong'er Volcanic Province and the North China dyke swarm. *Lithos*, **101**, 260–280.

PENG, P., ZHAI, M.-G., LI, Z., WU, F.-Y. & HOU, Q.-L. 2008b. Neoproterozoic (~820 Ma) mafic dyke swarms in the North China craton: implication for a conjoint to the Rodinia supercontinent? *Abstracts for the 13rd Gondwana Conference, Dali, China*, 160–161.

PIRAJNO, F. & CHEN, Y.-J. 2005. The Xiong'er Group: a 1.76 Ga Large Igneous Province in East–Central China? Available online at: www.largeigneousprovinces.org.

QIAN, X.-L. & CHEN, Y.-P. 1987. Late Precambrian mafic dyke swarms of the North China craton. *In*: HALLS, H. C. & FAHRIG, W. F. (eds) *Mafic Dyke Swarms*. Geology Association of Canada, Special Papers, **34**, 385–391.

REN, F.-G., LI, S.-B., ZHAO, J.-N., DING, S.-X. & CHEN, Z.-H. 2003. Te (Se) geochemical ore-hunting information from the gold deposits in the volcanic rocks of Xiong'er Group. *Geological Survey and Research*, **26**, 45–51 [in Chinese with English abstract].

REVILLON, S., ARNDT, N. T., CHAUVEL, C. & HALLOT, E. 2000. Geochemical study of ultramafic volcanic and plutonic rocks from Gorgona Island, Colombia: plumbing system of an oceanic plateau. *Journal of Petrology*, **41**, 1127–1154.

RICKWOOD, P. C. 1990. The anatomy of a dyke and the determination of propagation and magma flow directions. *In*: PARKER, A. J., RICKWOOD, P. C. & TUCKER, D. H. (eds) *Mafic Dykes and Emplacement Mechanisms*. Balkema, Rotterdam, 81–100.

RUDNICK, R.-L. & GAO, S. 2001. Composition of the continental crust. *In*: RUDNICK, R. L. (ed.) *Treatise on Geochemistry, Volume 3, The Crust*. Elsevier, Amsterdam, 1–64.

SANTOS, J. O. S., HARTMANN, L. A., MCNAUGHTON, N. J. & FLETCHER, I. R. 2002. Timing of mafic magmatism in the Tapajos Province (Brazil) and implications for the evolution of the Amazon craton: evidence from baddeleyite and zircon U–Pb SHRIMP geochronology. *Journal of South American Earth Sciences*, **15**, 409–429.

SHETH, H. C. 2007. 'Large Igneous Provinces (LIPs)': definition, recommended terminology, and a hierarchical classification. *Earth-Science Reviews*, **85**, 117–124.

SRIVASTAVA, K. R. & SINGH, R. K. 2004. Trace element geochemistry and genesis of Precambrian sub-alkaline mafic dikes from the central Indian craton: evidence for mantle metasomatism. *Journal of Asian Earth Sciences*, **23**, 373–389.

SUN, S.-S. 1997. Chemical and isotopic features of Palaeoproterozoic mafic igneous rocks of Australia: implications for tectonic processes. *In*: RUTLAND, R. W. R. & DRUMMOND, B. J. (eds) *Palaeoproterozoic Tectonics and Metallogenesis: Comparative Analysis of Parts of the Australian and Fennoscandian Shields.* Australian Geological Survey Organization Record, **44**, 119–122.

SUN, S.-S. & MCDONOUGH, W. F. 1989. Chemical and isotopic systematics of oceanic basalts: implications for mantle composition and processes. *In*: SAUNDERS, A. D. & NORRY, M. J. (eds) *Magmatism in the Ocean Basins.* Geological Society, London, Special Publications, **42**, 313–354.

WANG, S.-S., SANG, H.-Q., QIU, J., CHEN, M.-E. & LI, M.-R. 1995. The metamorphic age of pre-Changcheng system in Beijing–Tianjin area and a discussion about the lower limit age of Changcheng system. *Scientia Geologica Sinica*, **30**, 348–354.

WANG, T.-H. 1995. Evolutionary characteristics of geological structure and oil–gas accumulation in Shanxi–Shaanxi area. *Journal of Geology and Mineral Resource of North China*, **10**, 283–398 [in Chinese with English abstract].

WANG, Y.-J., FAN, W.-M., ZHANG, Y.-H., GUO, F., ZHANG, H.-F. & PENG, T.-P. 2004. Geochemical, $^{40}Ar/^{39}Ar$ geochronological and Sr–Nd isotopic constraints on the origin of Palaeoproterozoic mafic dikes from the southern Taihang Mountains and implications for the ca. 1800 Ma event of the North China Craton. *Precambrian Research*, **135**, 55–77.

WANG, Y.-J., ZHAO, G.-C., CAWOOD, P. A., FAN, W.-M., PENG, T.-P. & SUN, L.-H. 2008. Geochemistry of Palaeoproterozoic (~1770 Ma) mafic dikes from the Trans-North China Orogen and tectonic implications. *Journal of Asian Earth Sciences*, **33**, 61–77.

WENG, J.-C., LI, Z.-M., YANG, Z.-Q. & LI, W.-Z. 2006. Hydrothermally modified Pb–Zn deposit: a new deposit type in volcanic rocks of the Xiong'er Group, Henan, China. *Geological Bulletin of China*, **25**, 502–505 [in Chinese with English abstract].

WILDE, S. A., ZHAO, G.-C. & SUN, M. 2002. Development of the North China craton during the Late Archaean and its final amalgamation at 1.8 Ga: some speculation on its position within a global Palaeoproterozoic Supercontinent. *Gondwana Research*, **5**, 85–94.

XU, Y.-G., CHUNG, S.-L., JAHN, B.-M. & WU, G.-Y. 2001. Petrologic and geochemical constraints on the petrogenesis of Permian–Triassic Emeishan flood basalts in southwestern China. *Lithos*, **58**, 145–168.

XU, Y.-H., ZHAO, T.-P., PENG, P., ZHAI, M.-G., QI, L. & LUO, Y. 2007. Geochemical characteristics and geological significance of the Paleoproterozoic volcanic rocks from the Xiaoliangling Formation in the Lvliang area, Shanxi Province. *Acta Petrologica Sinica*, **23**, 1123–1132 [in Chinese with English abstract].

ZHAI, M.-G. & LIU, W.-J. 2003. Palaeoproterozoic tectonic history of the North China craton: a review. *Precambrian Research*, **122**, 183–199.

ZHAI, M.-G. & PENG, P. 2007. Palaeoproterozoic events in the North China Craton. *Acta Petrologica Sinica*, **23**, 2665–2682.

ZHAI, M.-G., BIAN, A.-G. & ZHAO, T.-P. 2000. Amalgamation of the supercontinental of the North China craton and its break up during late–middle Proterozoic. *Science in China (Series D)*, **43**, 219–232.

ZHANG, H.-C., XIAO, R.-G., AN, G.-Y., ZHANG, L., HOU, W.-R. & FEI, H.-C. 2003. Hydrothermal mineralization of Au(Ag)-polymetallic ore deposit in the volcanic rock series of the Xiong'er Group. *Geology in China*, **34**, 400–405 [in Chinese with English abstract].

ZHANG, H.-F., ZHAI, M.-G. & PENG, P. 2006. Zircon SHRIMP U–Pb age of the Palaeoproterozoic high-pressure granulites from the Sanggan area, the North China craton and its geologic implications. *Earth Science Frontiers*, **13**, 190–199 [in Chinese with English abstract].

ZHANG, J., ZHAO, G.-C. *ET AL.* 2007. Structural, geochronological and aeromagnetic studies of the Hengshan–Wutai–Fuping mountain belt: implications for the tectonic evolution of the trans-North China Orogen. *Abstracts of National Conference on Petrology and Geodynamics 2007*, 200–201.

ZHANG, Y.-Q., MA, Y.-S., YANG, N., SHI, W. & DONG, S.-W. 2003. Cenozoic extensional stress evolution in North China. *Journal of Geodynamics*, **36**, 591–613.

ZHAO, G.-C., WILDE, S. A., CAWOOD, P. A. & SUN, M. 2001. Archaean blocks and their boundaries in the North China craton: lithological, geochemical, structural and $P–T$ path constraints and tectonic evolution. *Precambrian Research*, **107**, 45–73.

ZHAO, G. C., SUN, M. & WILDE, S. A. 2002. Review of global 2.1–1.8 Ga orogens: implications for a Pre-Rodinia supercontinent. *Earth-Science Reviews*, **59**, 125–162.

ZHAO, G. C., SUN, M., WILDE, S. A. & LI, S. Z. 2004. A Paleo-Mesoproterozoic supercontinent: assembly, growth and breakup. *Earth-Science Reviews*, **67**, 91–123.

ZHAO, G.-C., SUN, M., WILDE, S. A. & LI, S.-Z. 2005. Late Archaean to Palaeoproterozoic evolution of the North China craton: key issues revisited. *Precambrian Research*, **136**, 177–202.

ZHAO, G.-C., HE, Y.-H. & SUN, M. 2009. The Xiong'er volcanic belt at the southern margin of the North China Craton: petrographic and geochemical evidence for its outboard position in the Paleo-Mesoproterozoic Columbia Supercontinent. *Gondwana Research*, **16**, 170–180.

ZHAO, J.-N., REN, F.-G. & LI, S.-B. 2002. Characters and significance of amygdaloidal fabric copper ore in Dasheping copper mine, Ruyang, Henan Province. *Progress in Precambrian Research*, **25**, 97–104 [in Chinese with English abstract].

ZHAO, T.-P., ZHOU, M.-F., ZHAI, M.-G. & XIA, B. 2002. Palaeoproterozoic rift-related volcanism of the Xiong'er group, North China craton: implications for the breakup of Columbia. *International Geology Review*, **44**, 336–351.

ZHAO, T.-P., ZHAI, M.-G., XIA, B., LI, H.-M., ZHANG, Y.-X. & WANG, Y.-S. 2004. Zircon U–Pb SHRIMP dating for the volcanic rocks of the Xiong'er Group:

constraints on the initial formation age of the cover of the North China craton. *Chinese Science Bulletin*, **49**, 2495–2502.

ZHAO, T.-P., WANG, J.-P. & ZHANG, Z.-H. (eds) 2005. *Proterozoic Geology of Mt. Wangwushan and Adjacent Areas, China*. China Dadi Publishing House, Beijing.

ZHAO, Z.-P. ET AL. 1993. *Precambrian Crustal Evolution of the Sino-Korean Para-platform*. Science Press, Beijing.

ZHOU, M.-F., YANG, Z.-X., SONG, X.-Y., KEAYS, R. R. & LESHER, C. M. 2002. Magmatic Ni–Cu–(PGE) sulphide deposits in China. *CIM Special Volume*, **54**, 619–636.

ZHU, S.-X., HUANG, X.-G. & SUN, S.-F. 2005. New progress in the research of the Mesoproterozoic Changcheng system (1800–1400 Ma) in the Yanshan range, North China. *Journal of Stratigraphy*, **19**, 437–449 [in Chinese with English abstract].

The role of geochronology in understanding continental evolution

ALFRED KRÖNER[1,2]

[1]*Institut für Geowissenschaften, Universität Mainz, 55099 Mainz, Germany*
(e-mail: kroener@uni-mainz.de)

[2]*Beijing SHRIMP Centre, Chinese Academy of Geological Sciences, 26 Baiwanzhuang Road, 100037 Beijing, China*

Abstract: Geochronology has become one of the most essential tools in reconstructing processes of continental growth and evolution, and *in situ* dating of minerals has become common practice through the development of high-resolution ion microprobes and laser ablation inductively coupled plasma mass spectrometry techniques. Zircon has established itself as the most robust and reliable mineral to record magmatic and metamorphic processes. The combination of mineral ages with Sm–Nd, Lu–Hf and O isotopic systematics constrains magma sources and their evolution, and a picture is emerging that supports the beginning of modern-style plate tectonics in the early Archaean. Major fields for future research in geochronology include the search for very old crustal remnants, the establishment of Precambrian supercontinents, reconstruction of magmatic and tectonic processes in accretionary orogens, verification of ancient high-pressure rocks, and the reconstruction of detailed metamorphic histories by dating minerals in their original textural settings.

Geochronology is an increasingly important tool in tectonics to understand the timing of crust formation, stabilization and destruction as well as orogenic processes and other features related to the chronological evolution of the Earth's rigid lithosphere. Dating techniques and methods have developed dramatically over the last 30 years, particularly since precise single mineral dating became possible, and our understanding of the behaviour of isotopic systems in minerals at different temperatures and pressures has improved dramatically. When first concepts of continental growth and evolution were linked to geochronology (e.g. Hurley *et al.* 1967) the number of radiometric ages was scarce, and scientists used whole-rock and multi-grain analytical techniques, thus generating data whose significance often remained obscure. Yet, in those days important discoveries were made, such as delineation of cratons and mobile belts (Hurley & Rand 1969), recognition of the antiquity of many tonalite–trondhjemite–granodiorite (TTG) assemblages (Moorbath 1977), and recognition that continental evolution was not continuous but episodic (Condie 1980; Cahen *et al.* 1984; Condie *et al.* 2009*a*, *b*). With the improvement in dating methods and the availability of more precise ages, refinement in our understanding of plate tectonics, regional geology and continental development led to global models for crustal evolution such as those of Windley (1995) and Goodwin (1991).

Further improvements in dating and isotopic techniques such as single-zircon thermal ionization mass spectrometry (TIMS) and whole-rock Sm–Nd analysis provided more detailed information and revealed the complexity of many magmatic and metamorphic processes, and made it possible to distinguish between juvenile and reworked crustal segments. The design and development of the high-resolution ion microprobe (Clement *et al.* 1977) and its application to *in situ* dating of zircon and other minerals (Compston *et al.* 1982; Kinny *et al.* 1994; Williams *et al.* 1996) was a major breakthrough in geochronology that, in combination with cathodoluminescence (CL) and back-scattered electron (BSE) imaging, helped significantly in understanding magmatic and metamorphic processes and made it possible to recognize and date complex zircon domains such as several growth generations, inherited cores, and metamorphic overgrowths (Figs 1, 2) and relate this to specific geological events. Useful summaries on zircon and its role in dating crustal rocks can be found in a recent volume of *Elements*, edited by Harley & Kelly (2007).

New techniques in laser-ablation inductively coupled plasma mass spectrometry (LA-ICP-MS; Hirata & Nesbitt 1995) also permit *in situ* dating of single minerals and may become more widely used in the near future when multi-collector instruments are available. The latest development is in the use of nanotechnology in mineral dating, and first results in using a high-resolution ion microprobe with good lateral resolution of up to 0.5 μm (Cameca-Nanosims) for zircon geochronology are

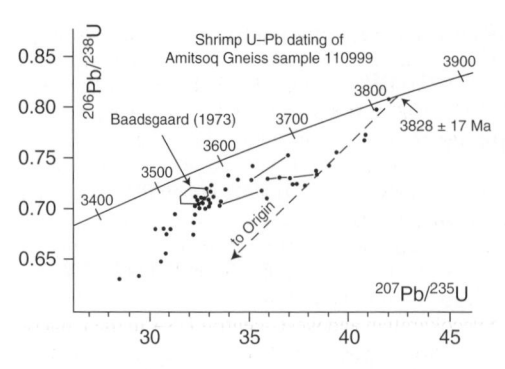

Fig. 1. Concordia diagram for zircon dating results from a polyphase, migmatitic Amitsoq gneiss sample, Godthaab District, West Greenland. The conventional multigrain TIMS analysis is from Baadsgaard (1973), whereas the single-grain SHRIMP analyses are from Kinny (1986). The >3800 Ma age is interpreted to reflect crystallization of the tonalitic protolith, with overgrowth–recrystallization during a granulite-facies event at c. 3650 Ma. [For details of interpretation see Kinny (1986).]

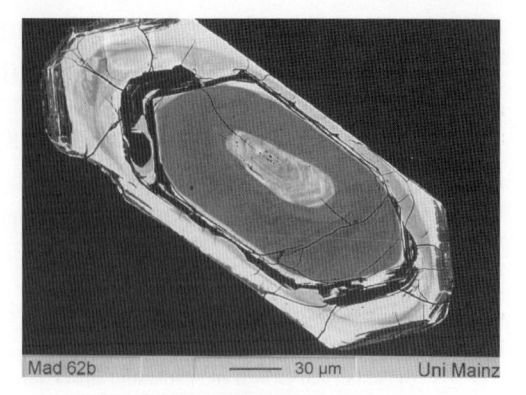

Fig. 2. Cathodoluminescence photograph of zircon from granite NW of Carion, central Madagascar. Dark (U-rich) magmatic core (c. 1007 Ma) with remnant of inherited xenocryst (light grey) is surrounded by broad, light grey (U-poor) magmatic overgrowth rim formed at time of granite intrusion some 789 Ma ago (A. Kröner, unpubl. data).

now available (Stern et al. 2005; Nakahata et al. 2008). This technology makes it possible to date very small parts and narrow rims of zircon, and may be particularly useful in deciphering complex polyphase magmatic and metamorphic processes.

Although minerals can be dated directly in polished thin sections using secondary ionization mass spectrometry (SIMS), another new development not yet widely used in geochronology and crustal studies is the application of a microdrill to isolate *in situ* minerals from a thin section where they occur in their original magmatic or metamorphic environment and can then be dated by SIMS after transfer onto a sample mount (e.g. Kröner et al. 2000; Möller et al. 2002, 2003).

Recognition of episodic juvenile crustal growth

The most widely available current techniques to recognize whether a rock formed through magmatic processes in the mantle or in the continental crust are whole-rock Sm–Nd isotopic analysis (DePaolo 1981) and Lu–Hf isotopic analyses of zircon separates (Patchett et al. 1981) or *in situ* analysis of dated single zircons (Kinny et al. 1991). Because Nd is less compatible than Sm, the ratio of these two elements changes significantly during mantle melting to produce felsic crustal rocks, and the resulting different isotopic evolution paths for depleted mantle and felsic crust can be used to estimate when crustal material separated from its mantle source, provided the Sm/Nd ratio has not been modified during subsequent intracrustal processes. This so-called Nd model age is commonly used to estimate the age of juvenile crust-forming events and has been extremely useful not only to estimate rates of crustal growth but also to recognize crustal age provinces, particularly in Precambrian terrane. The Lu–Hf isotopic system works in a similar way (e.g. Kinny et al. 1991).

In the ideal case of a mantle melt differentiating into gabbro from which more differentiated granitoid rocks may evolve, as is common during subduction processes, the zircon age of a given juvenile igneous rock and its whole-rock Nd model age should be almost identical, and the Nd isotopic composition should reflect the mantle source ($\varepsilon_{Nd(t)}$ value is positive). However, in cases where the mantle melt becomes underplated below continental lithosphere and interacts with older crustal material before differentiating further, significant differences in the zircon and Nd model ages may result, depending on the age of the crustal material and the proportion of crust–mantle interaction. In this case the Nd model age no longer reflects the time of separation of a melt from the mantle and is better referred to as a mean crustal residence age (Arndt & Goldstein 1987), and the $\varepsilon_{Nd(t)}$ value is generally low and commonly below zero. In cases of intracrustal melting the Nd model age does not reflect the time of magmatism but provides an averaged value for the crustal source and, depending on the age of this source, the $\varepsilon_{Nd(t)}$ value may be strongly negative. The melting event can then be determined only by direct mineral or whole-rock dating.

Mixing of mantle and crustal sources is a very common process during continental growth and evolution, and Nd mean crustal residence ages are therefore not always reliable indicators of the timing of crust-forming events. However, in combination with zircon ages the Sm–Nd isotopic system is very useful in recognizing distinct crustal provinces (e.g. Milisenda *et al.* 1988; Hegner & Kröner 2000; Dickin & McNutt 2003).

The integration of U–Pb age and Hf isotope tracer information contained within zircon has proved to be a powerful approach for studies of magmatic processes and crustal evolution (e.g. Stevenson & Patchett 1990; Corfu & Noble 1992; Davis *et al.* 2005; Scherer *et al.* 2007). High Hf contents in zircon, its relative immobility during alteration and metamorphism, and thus its capacity to retain original chemical and isotopic information through sedimentary recycling and high-grade metamorphism make it an ideal tracer for petrogenesis. Moreover, through low Lu/Hf ratios zircon preserves close to the initial Hf isotopic composition inherited at the time of crystallization (Kemp *et al.* 2005, 2009). The earlier-used single-zircon digestion method (e.g. Davis *et al.* 2005, and references therein) has now been largely replaced by *in situ* measurement of Hf isotopes by laser ablation (Wu *et al.* 2006, and references therein), allowing polyphase zircon crystallization histories to be deciphered and to be linked with specific tectonic settings (e.g. Gerdes & Zeh 2009; Kemp *et al.* 2009; Mišković & Schaltegger 2009). Such studies also help in sedimentary provenance studies (e.g. Howard *et al.* 2009), to portray the evolution of high-pressure metamorphic rocks (e.g. Xia *et al.* 2009), and are also able to decipher crust–mantle evolution through the study of zircons from kimberlites and mantle xenoliths (e.g. Griffin *et al.* 2000; Zheng *et al.* 2009). The *in situ* analysis of oxygen isotopes in zircon adds a further dimension to crustal evolution studies (e.g. Valley *et al.* 2005; Liu *et al.* 2009).

The Re–Os isotopic system is particularly useful in dating mafic and ultramafic rocks, sulphide ores, black shales and molybdenite (e.g. Walker *et al.* 1988; Luck & Allègre 1991; Stein *et al.* 2001) and to characterize mantle melting events (Walker *et al.* 1989). Its use in reconstructing crustal evolution scenarios is, however, limited, whereas it has great potential to portray mantle evolution. Meisel *et al.* (2001) have determined the Re–Os isotopic systematics of spinel- and garnet-bearing mantle xenoliths from North and Central America, Europe, southern Africa, Asia, and the Pacific region to define the isotopic composition of a hypothetical primitive upper mantle (PUM).

Major episodic periods of continental crustal growth have long been identified by groupings in magmatic emplacement ages (e.g. Percival *et al.* 2001), from zircons in well-mixed sediments (e.g. Rino *et al.* 2004), and from a combination of magmatic ages and Nd mean crustal residence ages (e.g. Reymer & Schubert 1986). Reymer & Schubert (1986, 1987) estimated crustal growth rates by combining Nd model ages with volumetric calculations of juvenile arc-crust generation and arrived at anomalously high crust-production rates for certain periods of Earth history. They estimated the average arc accretion rate in the Mesozoic–Cenozoic to have been 20–40 km^3 km^{-1} Ma^{-1}, whereas the Canadian shield in the Archaean, the Svecokarelian and west–central USA in the Palaeoproterozoic, and the Arabian–Nubian Shield in the Neoproterozoic yielded abnormally high crust-production rates of 160–310 km^3 km^{-1} Ma^{-1}. Similarly high crust production in the Palaeozoic was proposed for the Central Asian Orogenic Belt (Sengör *et al.* 1993).

However, these models did not consider crustal loss through tectonic erosion in subduction zones, which is estimated at about 90 km^3 km^{-1} Ma^{-1} globally (Clift & Vannucchi 2004). Many of these estimates are now also in doubt either because the duration of crust formation was underestimated, as in the case of the Central Asian Orogenic Belt (c. 1100–250 Ma, Kröner *et al.* 2007), or because detailed geochronology has shown that many of the assumed juvenile terranes contained remnants of older crust that were revealed as zircon xenocrysts in arc magmatic rocks. Conventional multi-grain zircon geochronology usually does not easily detect inherited components, but the increasing use of *in situ* single-zircon dating in combination with CL has produced remarkable results in detecting old xenocrysts in seemingly juvenile rocks that question the sole use of whole-rock Sm–Nd isotopic systematics in assessing rates of crustal growth. For example, Kennedy *et al.* (2004, 2005) undertook sensitive high-resolution ion microprobe (SHRIMP) zircon dating as part of a regional survey in the northern Arabian shield and detected a large number of xenocrystic zircons up to Palaeoproterozoic in age in seemingly juvenile Neoproterozoic volcanic and granitoid rocks. Hargrove *et al.* (2006) SHRIMP-dated zircons from igneous rocks in the Bi'r Umq suture zone in western Saudi Arabia and found xenocrysts as old as Archaean, yet the whole-rock $\varepsilon_{Nd(t)}$ value for these rocks was only slightly lower than for comparable rocks without inheritance. Similarly, Ali *et al.* (2009) found a large number of xenocrystic zircons in Neoproterozoic mafic volcanic rocks in the Eastern Desert of Egypt, previously considered to have formed in an intraoceanic arc setting. Our work in the Palaeozoic arc terranes of the Central Asian Orogenic Belt in Kazakhstan and Mongolia produced similar results. Kröner *et al.* (2008) reported an early Archaean zircon xenocryst from a Silurian diorite of the Stepnyak arc terrain in northern Kazakhstan, and numerous Precambrian xenocrysts were found in arc-related rocks of central and southern Mongolia (Kröner et al. 2007).

Increasing recognition of zircon inheritance in seemingly juvenile rocks poses a problem in the tectonic interpretation of arc terranes. If the zircons are derived from older crustal rocks underlying the arc and were incorporated into arc magmatic rocks during melt generation and/or ascent, such as in the Andes (Schmitt et al. 2002; Schilling et al. 2006) or in Japan (Sano et al. 2000; Fujii et al. 2008), the tectonic interpretation usually follows the Andean or Japan model, and the rock in question is interpreted as part of a continental margin magmatic arc. In this case a significant part of the (concealed) lower part of the arc may consist of older continental crust, and calculations of arc production rates are erroneous.

However, it is not certain that all inherited zircons are derived from continental material in the root zone of such arc complexes. Detrital zircons from pelitic sediments on the ocean floor or clastic sediments in slope or trench environments may be transported deep into the mantle during subduction–accretion processes (Shibata et al. 2008; Scholl & von Huene 2009) as has been shown by the discovery of zircon xenocrysts in eclogitic mélanges (e.g. Rubatto et al. 1998). In rare cases such zircons even survive diamond-zone metamorphism (e.g. Clauoé-Long et al. 1991). Tange & Takahashi (2004) have shown that zircon may survive in the mantle up to 1500 °C and 20 GPa, equivalent to about 600 km depth in the Earth. These zircons may become incorporated into arc magmas during slab melting (Shimoda et al. 1998) and then appear in calc-alkaline magmatic arc rocks that would erroneously be interpreted to have formed in a continental margin setting. Survival of ancient crustal zircons and diamonds in mantle melts, presumably from deeply subducted crust, is also documented by the presence of xenocrysts in kimberlites derived from an eclogitic source (Jacob & Foley 1999; Spetsius et al. 2008) and from gabbroic rocks of the ocean crust (Pilot et al. 1998; Belyatsky et al. 2008; Bortnikov et al. 2008; Kostitsyn et al. 2009) and in ophiolites (Wright & Wyld 1986; Whattam et al. 2006; P. Jian, pers. comm.).

Mineral ages and metamorphic history

Much progress has been made in linking mineral ages in metamorphic rocks with textures and metamorphic evolution (e.g. Vance et al. 2003). However, questions remain as to whether such minerals as metamorphic zircon begin to form during prograde metamorphism in the amphibolite facies, during peak metamorphic conditions near the amphibolite–granulite-facies transition, or during the retrograde evolution (Fraser et al. 1997; Roberts & Finger 1997; Parrish 2001). Premetamorphic igneous or detrital zircons may be overgrown by metamorphic rims during prograde metamorphism in a hydrous environment (Fig. 2), provided Zr-rich solutions are present, perhaps as a result of the breakdown of phases such as garnet and hornblende (Fraser et al. 1997; Rubatto & Hermann 2007). Also, older zircons may show partial or complete dissolution (Hoskin & Black 2000), as can be seen by significant rounding and recrystallization of the pyramidal ends in amphibolite- to granulite-facies rocks (Kröner et al. 1999; see Fig. 3b, c). The advantage of zircon as a chronometer is that it may preserve its original U–Th–Pb isotopic composition even under extreme crustal conditions because of its high closure temperature of well above 900 °C (Lee et al. 1997; Gardés & Montel 2009). Möller et al. (2002) investigated zircons from an ultrahigh-temperature domain (>950 °C) in southern Norway and obtained perfectly concordant magmatic ages from which they concluded that thermally driven diffusion does not appear to be a viable and important mechanism for the mobility of Pb in natural, non-metamict zircon under even the most extreme crustal conditions. Thus, non-metamict zircons are still the most reliable chronometer to record rock formation and high-grade metamorphism, and are an important tool in monitoring crustal growth and orogeny.

Zircon may also grow during low and very low-grade metamorphism at around 250 °C, provided suitable fluids are available. This has been demonstrated from zircon rims formed under greenschist-facies conditions around older detrital grains (Rasmussen 2005) and from altered detrital zircons where nano-crystalline material replaces U-rich, metamict zircon via modification of whole grains or selective alteration of parts of grains (Hay & Dempster 2009). Such growth appears to occur preferentially in pelitic rocks and may be linked to halogen concentrations, facilitating zirconium mobility in hydrothermal systems (Rasmussen 2005).

Parrish (2001) has summarized the effects of prograde and retrograde metamorphism on the U–Pb isotopic systematics in zircon, allanite, monazite, titanite, and rutile, but the thermochemistry of reactions involving many of these mineral phases is still poorly understood (Watson 1996; see Kelsey et al. 2008, for discussion). Therefore, the interpretation of many mineral ages is still ambiguous (Foster & Parrish 2003); in particular, those where significant Pb loss is apparent (Mezger & Krogstad

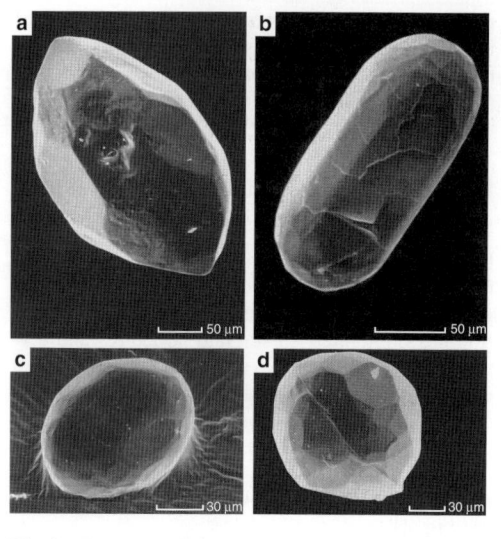

Fig. 3. Zircon morphologies in rocks of different metamorphic grade under the electron microscope. (**a**) Near-euhedral detrital grain from granulite-facies metaquartzite of early Archaean Beitbridge Group, Limpopo Belt, South Africa. The well-preserved morphology suggests short sedimentary transport, and the dry rock inhibited recrystallization of the zircon, on which the pitmarks of sedimentary transport can still be seen. (**b**) Recrystallized magmatic grain from upper amphibolite-facies Archaean Alldays granitic gneiss in Limpopo belt, South Africa. The original sharp pyramidal terminations were dissolved through interaction with metamorphic fluids, and the new rounded metamorphic surfaces are characterized by many small facets. This grain may be misinterpreted as detrital. (**c**) Severely recrystallized magmatic zircon from granulite-facies charnockitic gneiss of the Highland Group, Sri Lanka, showing complete disappearance of the original prismatic habit and strong rounding and multifaceted new growth as a result of reaction with metamorphic fluid. This may easily be mistaken for a detrital grain. (**d**) New metamorphic growth of ball-shaped, multifaceted grain with large facets in pelitic metasediment from granulite-facies Highland Group, Sri Lanka.

1997) or where the closure temperatures of the dated minerals are uncertain. Gardés & Montel (2009) have recently provided a new theoretical model for diffusive isotope loss that permits assessment of the opening and resetting temperatures of radiochronometers during metamorphism. The temperature at which the daughter isotope loss begins is the opening temperature, and the temperature at which the daughter isotope is completely lost is the resetting temperature. As can be expected, zircon and monazite have the highest opening temperatures, varying between $c.$ 680 and $>900\ °C$ depending on grain size, and also have the highest resetting temperatures, between $c.$ 850 and 1400 °C (Gardés & Montel 2009), making them the most reliable chronometers for high-grade rocks.

Monazite occurs in many magmatic and metamorphic rocks, particularly in gneisses and pelitic metasediments, and in view of its low diffusion rate for Pb (Cherniak et al. 2004; Gardés et al. 2006) is thus extensively used to date crustal processes. As with zircon, the development of in situ analytical techniques [SIMS and electron microprobe analysis (EPMA)] has led to a better understanding of the monazite Th–Pb and U–Pb chronometers and their use to reconstruct geological histories (for reviews see Goncalves et al. 2005; Williams et al. 2007). However, monazite often displays distinct compositional and age domains (Fig. 3) together with complex chemical zoning patterns (see Rasmussen & Muhling 2007, for review) and if these are not properly understood, age dating may lead to geologically meaningless results (Seydoux-Guillaume et al. 2003). Combinations of SHRIMP and EPMA dating of monazite domains have been successfully employed to unravel complex magmatic and metamorphic growth processes (e.g. Dahl et al. 2005; Fujii et al. 2008). Allanite in situ dating by high-resolution ion microprobe has also been successfully employed in understanding metamorphic rates and orogenic processes (Janots et al. 2009).

Relatively little is known about monazite growth during low-temperature processes (Rasmussen & Muhling 2007), whereas many successful age studies were performed on monazite from high-grade metamorphic rocks (e.g. Braun & Bröcker 2004; Santosh et al. 2009). Kirkland et al. (2009) have presented an impressive study of zircon and monazite growth in shear zones and concluded that diversity in zircon rim ages compared with monazite reflects dissimilar growth behaviour.

Rutile occurs as an accessory phase in high-grade metamorphic rocks, igneous rocks and siliciclastic sediments, and has been shown to yield precise U–Pb ages (Schärer et al. 1986; Mezger et al. 1989). In view of its relatively high closure temperature for Pb diffusion it is particularly suitable for dating high-grade metamorphism (Vry & Baker 2006) and assessing metamorphic and magmatic temperatures (Zack et al. 2004).

Titanite is a common accessory phase in many felsic rocks and a U–Pb geochronometer widely used for studying metamorphic processes (Frost et al. 2000). In view of its widespread occurrence in Archaean granitoid gneisses it has been successfully used to date early Precambrian magmatic and tectonic events (e.g. Crowley et al. 2002). Amelin (2009) has demonstrated the usefulness of titanite in crustal evolution studies through the combined use of the U–Pb and Sm–Nd systems in Archaean rocks. He showed that the Nd isotope composition at the time of titanite formation can be used to infer initial Nd isotopic values of the rocks at the time of zircon crystallization. Very importantly, this approach does not involve the assumption that the Sm–Nd system in the rocks remained closed since their formation, whereas this assumption underlies all initial Nd determinations based on whole-rock analyses.

Further minerals suitable for U–Pb dating and occurring as accessory minerals in many rocks, particularly mafic dykes, are baddeleyite, xenotime and zirconolite. Baddeleyite (ZrO_2) is common accessory phase in gabbroic rocks, mafic dykes and alkaline intrusive rocks, and is high in U (200–1000 ppm), contains little or no common lead, has a high closure temperature like zircon, and is very resistant to lead loss, thus making it an ideal U/Pb chronometer for mafic rocks. Also, it is unlikely to occur as xenocrysts and is progressively replaced by secondary zircon during prograde metamorphism so that relict (magmatic) baddeleyite cores and zircon rims (metamorphic) can be dated (e.g. Wingate et al. 1998). The mineral is commonly dated by both TIMS and SIMS (e.g. Krogh et al. 1984; Wingate et al. 1998). Xenotime (YPO_4) may form during magmatic and diagenetic processes, and occurs as a minor accessory phase in pegmatites, pelitic metasediments and hydrothermal ore deposits (Schärer et al. 1999). Like zircon and monazite, it is considered to be extremely resistant to diffusive Pb loss in most geological environments (Cherniak 2006), and this is is supported by the preservation of concordant and precise $^{207}Pb/^{206}Pb$ ages in detrital and diagenetic xenotime with complex post-depositional history (Rasmussen et al. 2007). Zirconolite ($CaZrTi_2O_7$), although occurring in a variety of rock types, has been neglected as a chronometer, but Rasmussen & Fletcher (2004) and Rasmussen et al. (2009) have recently presented precise in situ SHRIMP U–Pb ages for mafic dykes in Western Australia and consider it an ideal mineral for dating as it contains negligible initial common Pb, has high U concentrations, and does not occur as xenocrysts.

The $^{40}Ar/^{39}Ar$ isotopic system is mostly applied to reconstruct cooling and uplift histories in orogenic terranes and generally reflects the post-peak metamorphic evolution of a given rock (McDougall & Harrison 1988). The combination of Ar–Ar mineral dating with whole-rock Sm–Nd and/or Rb–Sr and single zircon U–Pb geochronology has led to the discovery of extremely fast uplift and cooling histories in several high- and ultrahigh-pressure metamorphic terranes such as in subducted mélanges and collisional belts (e.g. Dong et al. 2004; Lin et al. 2007; Agard et al. 2009). Ar–Ar mineral dating is also an important tool in palaeomagnetism, as the closure temperature of hornblende and biotite is relatively close to the blocking temperature of the magnetic carrier (McDougall & Harrison 1988).

Protolith recognition and interpretation

A very difficult task in high-grade metamorphic terranes is the correct identification of a gneiss protolith (Werner 1987; Passchier et al. 1990). This is particularly difficult where intense ductile deformation has obliterated all original structures and textures, and a pervasive layering has resulted from a combination of structural and metamorphic processes (Myers 1978; Fig. 5). Regularly banded quartzo-feldspathic gneisses, for example, have often been interpreted to be derived from a sedimentary source (e.g. Voll & Kleinschrodt 1991; see Fig. 4).

It is indeed difficult to recognize in the field whether a grey, layered gneiss was derived from a greywacke-type sedimentary rock or a granitoid intrusion, and to distinguish between a clastic sedimentary rock and tuff or lava in a supracrustal succession. Chemical data, in combination with petrology, may provide patterns that allow distinction between sedimentary and igneous rocks (Nesbitt & Young 1984; Werner 1987). However, a faster method is to look at the heavy mineral fractions and examine the morphology of zircon and other phases. Although detrital zircons may become oval to spherical and thus resemble zircons from high-grade metamorphic terranes (Fig. 3c), they usually have a pitted surface, whereas metamorphic recrystallization leads to multifaceted grains (e.g. Corfu et al. 2003; see Fig. 3d). Single zircon dating of such rocks, in combination with CL, will show whether the ages reflect a single, magmatic population of texturally homogeneous grains and the host rock is therefore of igneous origin, or whether the ages and zircon internal textures are highly variable and reflect input from heterogeneous sources and are thus likely to be detrital (e.g. Nutman et al. 2004). Not surprisingly, zircon dating has shown in many high-grade terranes such as Sri Lanka, southern India, southern Africa and West Greenland that there are many more rocks of granitoid than of sedimentary origin (e.g. Myers 1970; Kröner et al. 1994; Nutman et al. 2000; Collins et al. 2007). This has significant implications for models of orogenic evolution and crustal growth.

A similar problem concerns the correct interpretation of banded gneisses, which are predominantly of igneous origin and are particularly abundant in Archaean terranes but also occur in younger high-grade domains (e.g. Sri Lanka, Kröner et al. 1994). Such rocks result from intense ductile deformation

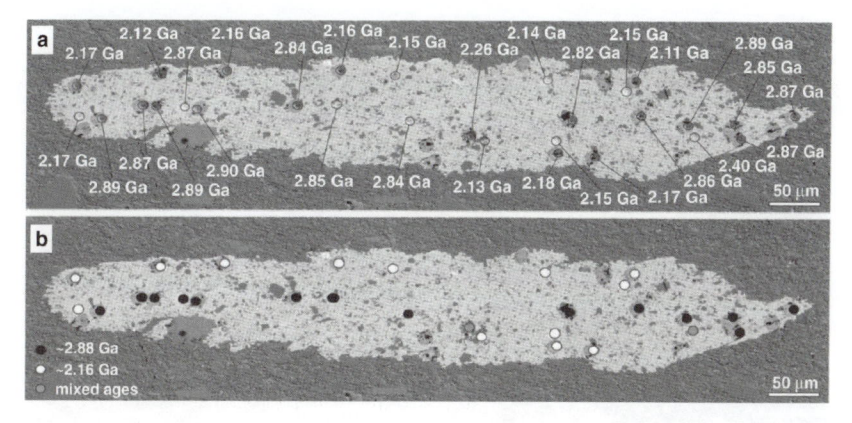

Fig. 4. Elongate, inclusion-rich monazite crystal with variable Pb/Pb ages (from Rasmussen *et al.* 2007). (**a**) BSE image showing location of SHRIMP analytical pits and corresponding ^{207}Pb/^{206}Pb ages. (**b**) Summary of ion microprobe age data showing an older population (2.88 Ga; black circles) in the core corresponding to La-poor and Sm- and Y-rich monazite, and a second generation (2.16 Ga; white circles) in the La-rich rim.

where previous intrusive relationships are now obliterated (Myers 1978; Passchier *et al.* 1990) and where the age range in a given rock can be determined only by detailed single-grain zircon geochronology. Whereas the age range in such gneisses found in root zones of Proterozoic and Phanerozoic arc complexes is relatively narrow, reflecting arc magmatism during continuing subduction (Schulmann *et al.* 2005; Corrigan *et al.* 2006; Greene *et al.* 2006), there are very significant age differences in many Archaean TTG banded gneisses, commonly extending over hundreds of millions of years (e.g. Nutman *et al.* 1996, 2000; Iizuka *et al.* 2007), even in a single rock specimen (e.g. Compston & Kröner 1988; see Fig. 6). These ages record long periods of granitoid magmatism that

Fig. 5. Finely banded and strongly retrograded mylonitic granitoid gneiss from the Aktyuz Complex, northern Tianshan, Kyrgyzstan. This rock was previously interpreted as an Archaean metasediment on account of its intense deformation and metamorphism but is, in fact, 840 Ma old (Kröner *et al.* 2009).

Fig. 6. (a) Banded early Archaean Ngwane gneiss from Ancient Gneiss Complex, Swaziland (sample AGC 150 of Compston & Kröner 1988). (b) Concordia diagram showing SHRIMP analyses of single zircons from 2 kg sample collected from gneiss shown in (a). [For details of age interpretation see Compston & Kröner (1988).]

are difficult to reconcile with models of their formation in modern-style arc complexes (Hamilton 1998) and seem more compatible with formation through repeated melting events above long-lived plumes (see discussion on micro- and mesoplume tectonics by Van Kranendonk 2007). The established genetic relationships between such long-lived Archaean TTG magmatism and felsic rocks in adjacent greenstone belts also show that modern relatively short-lived plate margin magmatism related to subduction and accretion may not be the best analogue for long-lived greenstone belts such as Barberton in southern Africa, and the Pilbara in West Australia (Van Kranendonk *et al.* 2007).

A major problem in the interpretation of many orogenic domains is the question of whether medium- to high-grade gneisses, often migmatitic or showing complicated deformational features, represent a pre-orogenic basement or were formed during accretion and/or collision and may represent deep parts of the orogenic system. Such cases are particularly frequent in the vast region of the Neoproterozopic to Palaeozoic Central Asian Orogenic Belt (Kröner *et al.* 2007; Windley *et al.* 2007; see also Li *et al.* 2008; Zubkov *et al.* 2008) and in NW China (Charvet *et al.* 2007; Wang *et al.* 2008). This largely stems from the fact that it was customary in some parts of the world to classify medium-grade rocks as Proterozoic and high-grade rocks as Archaean. Also, misinterpretation of bulk-zircon fraction analyses occasionally resulted in old ages, interpreted as protolith formation, whereas, in fact, these are inherited or detrital ages and do not reflect the age of formation of the rocks in which they occur. A classic example of such misinterpretations is the high-grade terrane bordering the Siberian craton in the south, which had long been considered to be Palaeoproterozoic (Petrova & Levizkii 1984) and has since been shown by modern single-grain zircon geochronology to be early Palaeozoic in age (Salnikova *et al.* 1998; Gladkochub *et al.* 2008). A more extreme case is represented by the Buteel gneiss in northern Mongolia, originally mapped as Archaean and now dated at 211–240 Ma as part of a Triassic metamorphic core complex (Donskaya *et al.* 2008). Similar examples have been desribed from southern Mongolia (Demoux *et al.* 2009) and the Tianshan of NW China (e.g. Zhu & Song 2006). Our recent work in the northern Tianshan of Kyrgyzstan has shown that most rocks previously interpreted as Archaean and Palaeoproterozoic on the basis of metamorphic grade and bulk-zircon analyses (e.g. Bakirov & Maksumova 2001) are, in fact, Neoproterozoic or Palaeozoic in age (Kröner *et al.* 2009).

Terrane accretion and geochronology

One of the most common modern applications of geochronology and isotope geochemistry is in terrane identification and amalgamation to reconstruct the history of orogenic belts, crustal growth processes and the configuration of ancient supercontinents. Apart from differences in lithology and tectonometamorphic evolution, isotopic data provide a fundamental tool to recognize and define crustal segments with different ages, origins, and histories. This is particularly important in ancient and strongly deformed terranes where palaeomagnetism and palaeontology cannot be used. Terrane identification based on precise zircon ages, often in combination with Sm–Nd whole-rock and/or Lu–Hf zircon isotopic systematics, has been successfully applied in several Archaean cratons such as the Kaapvaal (Schoene *et al.* 2009), West Greenland (Nutman & Friend 2007), Superior Province (Percival *et al.* 2001) and North China (Jahn *et al.* 2008; Liu *et al.* 2008). Condie (2007) discussed terrane lifespans in accretionary orogens from the Archaean to the Phanerozoic, mainly based on zircon ages, and estimated highly variable accretion rates depending on the tectonic setting. These estimates may be misleading because tectonic erosion on active plate margins (Clift & Vanucchi 2004; Scholl & von Huene 2009) was not considered. Nevertheless, Condie's (2007) compilation of whole-rock Sm–Nd isotopic systematics reiterates the episodic nature of juvenile crust production on Earth.

Accretionary orogens are particularly heterogeneous geochronologically in terms of the crustal components added through their lifespans. One of the difficult tasks in analysing such orogens is to reconstruct the original palaeogeography during their evolution (e.g. Hall 2009). Geochronology and isotope geochemistry play an important role here in being able to distinguish between juvenile and inherited crustal components, which often become intimately mixed during the accretion process. It is very difficult in such orogens, for example, to distinguish in the field between a basement terrane (microcontinent) and the strongly deformed and metamorphosed root zone of an arc complex (e.g. Salnikova *et al.* 2001; Demoux *et al.* 2009). Particularly good examples are the Arabian–Nubian shield and the Central Asian Orogenic Belt, where the literature is full of erroneous identifications of metamorphic terranes as Precambrian basement blocks, whereas geochronology shows many of these to have formed during the Neoproterozoic or Palaeozoic subduction–accretion history of these orogens (Kröner *et al.* 1987, 2007; Hu *et al.* 2000; Salnikova *et al.* 2001; Johnson & Woldehaimanot 2003). Geochronology may also help to test the terrane hypothesis in accretionary belts. For example, Badarch *et al.* (2002) subdivided the accretionary collage of Mongolia into a large number of distinct terranes, whereas an increasing number of precise age data are incompatible with many of these, particularly in southern Mongolia (Kröner *et al.* submitted; Lehmann *et al.* submitted).

Geochronology has also played a major role in identifying ancient supercontinents and in reconstructing their amalgamation histories. The basis for initial Rodinia reconstructions was to consider most Grenville-age mobile belts as collisional orogens resulting from amalgamation of older crustal blocks (Hoffman 1991; Condie 2001). Subsequent refinement in the geochronological record showed that many terranes or belts assigned to the Grenville age bracket were, in fact, older or younger and were unlikely to have been part of the supercontinent (e.g. De Waele et al. 2003; Kröner & Cordani 2003), and this, together with improved palaeogeographical considerations, led to a significantly modified Rodinia model (Li et al. 2008). The amalgamation history of Gondwana is similarly based on geochronological similarities in the Pan-African and Brasiliano orogens of South America, Africa, Madagascar, southern India, Sri Lanka, East Antarctica and Australia (Fitzsimons 2000; Kröner 2001; Meert & Liebermann 2008). The still largely speculative late Palaeoproterozoic supercontinent Columbia (Rogers & Santosh 2002; Zhao et al. 2002; Mints 2007) is almost entirely based on comparisons of 2.1–1.6 Ga age provinces and is still vaguely defined. It is obvious that age and isotopic data alone cannot identify crustal segments that once constituted part of a supercontinent and are now dispersed over the globe but, in combination with structural data, specific rock associations and palaeomagnetism, geochronology may play an important role in reconstructing continental configurations in the Precambrian.

Recognition of continental reactivation through geochronology

Reworking of continental crust predominantly occurs at the margins of orogenic belts, in wide intracontinental shear zones and during intraplate orogeny, and may be so severe as to completely obliterate the older depositional, magmatic, deformational and metamorphic history (Holdsworth et al. 2001). However, because of its high closure temperature for the U–Pb isotopic system, well-preserved zircon is an ideal chronometer to monitor such reworking. For instance, Jacobs et al. (2003) have shown that metamorphic rocks in Dronning Maud Land, Antarctica, formed during a high-grade Grenville-age event and were structurally rejuvenated and metamorphically overprinted during the late Neoproterozoic to early Palaeozoic Pan-African episode, probably during amalgamation of Gondwana. The old zircons retain their isotopic memory but are often surrounded by Pan-African overgrowth, whereas new zircon growth is also recorded. Similar features are known from many other belts, such as the Limpopo in southern Africa, where pre-orogenic zircon ages are preserved yet the host rocks are strongly modified and do not permit reconstruction of the older history (Kröner et al. 1999). Well-documented examples occur in the Pan-African Mozambique belt of Tanzania, where refoliated Archaean granitoid gneisses are preserved as a large enclave NNW of Dar-es-Salaam within high-grade Neoproterozoic rocks (Muhongo et al. 2001). Elsewhere in the belt extensive structural reworking and metamorphic reconstitution of Archaean to Palaeoproterozoic rocks during the Pan-African event has led to a series of banded gneisses where all structures are parallel, only Pan-African metamorphism is preserved, and the only way to recognize age differences is by detailed zircon geochronology (e.g. Sommer et al. 2005; Vogt et al. 2006). These data imply that the Tanzania craton originally extended eastwards almost entirely across the present Mozambique belt and was extensively reworked, often beyond recognition, during collision of East and West Gondwana. There is much speculation on how cratonic crust can be reworked, and one possibility is subcrustal mantle delamination, providing the heat to sufficiently 'soften' the overlying craton to become susceptible again to ductile deformation (Houseman & Molnar 2001).

The ultimate cause for destabilization and reworking of cratonic continental lithosphere is still a matter of significant speculation, but here again geochronology and isotope geochemistry have provided information on possible mechanisms. In central Asia and northern China the model of large-scale mantle lithosphere delamination receives support from the primitive isotopic systematics of many anorogenic granitoids (Jahn et al. 2004; Kovalenko et al. 2004) suggesting an ultimate mantle source for these melts.

In NE Africa a large region between the Neoproterozoic East African orogen east of the River Nile and the Tuareg shield in the western Hoggar Massif of Algeria is occupied by a gneiss–migmatite terrane of predominantly Pan-African age in which remnants of much older crust have been identified, such as Jabal Ouweinat at the Egypt–Libya border. This vast region has been interpreted as the 'Saharan Metacraton' and is considered to represent a destabilized cratonic block partly reworked in the late Neoproterozoic but still recognized as a semi-rigid crustal entity in terms of rheology as well as isotopic and geochronological characteristics (Abdelsalam et al. 2002). Liégeois et al. (2003) linked many of the Pan-African granitoid intrusions and migmatites to linear lithospheric delamination along mega-shear zones, leading to lower crustal melting as a result of asthenospheric rise. It is likely that cratonic reworking has occurred repeatedly through Earth history, and geochronology in combination with isotope geochemistry, petrology, rheology and seismology may help to further understand this process.

Does geochronology and isotope geochemistry tell us when plate tectonics began?

There is currently much debate on when modern-style plate tectonics began (Stern 2005; Cawood et al. 2006; Condie & Pease 2008), and isotopic evidence has contributed considerably to this discussion. Stern (2005, 2008) argued that ophiolites, blueschists and UHP rocks first appeared abundantly in the Neoproterozoic, thus documenting the emergence of plate tectonics as it operates today, and Hamilton (1998) proposed a non-plate-tectonic Earth in the Archaean. In contrast, Cawood et al. (2006) and Condie & Kröner (2008) listed criteria, including age and other isotopic data, that argue for subduction tectonics since at least the Mesoarchaean, if not earlier. Important arguments were the recycling of ancient zircons in subducted sediments and ocean crust and oxygen isotope data supporting that ancient eclogites found in kimberlite must thus have gone through a subduction cycle (Jacob & Foley 1999; Spetsius et al. 2008). Volodichev et al. (2004) reported the first SHRIMP-dated Neoarchaean eclogites, and although these rocks were not derived from subducted oceanic crust they nevertheless provide evidence for HP metamorphism early in Earth history. Shirey et al. (2008) summarized isotopic and trace element evidence for the operation of subduction and recycling since 3.9 Ga, and Harrison et al. (2005) speculated from mineral inclusions in the oldest known zircons that subduction already operated some 4 Ga ago.

Condie et al. (2009a) speculated from a compilation of global geochronological data that a major period of magmatic and tectonic quiescence occurred on Earth between 2.45 and 2.20 Ga ago and that plate tectonics was effectively shut down during this period for about 250 Ma, leading to what those workers named 'stagnant lid tectonics'. However, no viable mechanism was proposed for the shutdown and revival of subduction.

Conclusions

Geochronology has become one of the most essential tools in reconstructing processes of continental growth and evolution, and there has been a shift away from dating whole-rock samples to investigating the time and temperature of mineral growth and relating this to tectonic processes. Zircon, because of its high closure temperature for the U–Th–Pb isotopic system, has established itself as the most robust and reliable mineral to record magmatic and metamorphic processes although there is still debate on when precisely, within a specific tectonic environment, this growth occurred. In situ dating of minerals has become common practice through the development of high-resolution ion microprobes and LA-ICP-MS techniques, and this has led to a wealth of reliable age data resulting in much improved models for continental evolution and orogenesis. The combination of mineral ages with Sm–Nd, Lu–Hf and O isotopic systematics provides important constraints on magma sources and their evolution, and a picture is emerging that supports the beginning of modern-style plate tectonics in the early Archaean. Major fields for future research in geochronology include the search for very old crustal remnants, the establishment of Precambrian supercontinents, reconstruction of magmatic and tectonic processes in accretionary orogens, verification of ancient high-pressure rocks, and the reconstruction of detailed metamorphic histories by dating minerals in their original textural settings.

This is a contribution to honour Brian Windley, with whom I had the privilege and pleasure of doing joint field work in Madagascar and Mongolia. I thank B. Rasmussen, M. T. D. Wingate, G. Brügmann and A. Nutman for valuable discussions and references, and S. A. Wilde, M. Santosh and E. Hegner for constructive reviews improving the manuscript. This is Contribution 624 of the Mainz Research Centre on Earth System Science.

References

ABDELSALAM, M. G., LIÉGEOIS, J.-P. & STERN, R. J. 2002. The Saharan metacraton. *Journal of African Earth Sciences*, **34**, 119–136.

AGARD, P., YAMATO, P., JOLIVET, L. & BUROV, E. 2009. Exhumation of oceanic blueschists and eclogites in subduction zones: timing and mechanisms. *Earth-Science Reviews*, **92**, 53–79.

ALI, K. A., STERN, R. J., MANTON, W. I., KIMURA, J. & KHAMEES, H. A. 2009. Geochemistry, Nd isotopes and U–Pb SHRIMP zircon dating of Neoproterozoic volcanic rocks from the Central Eastern Desert of Egypt: new insights into the ~750 Ma crust-forming event. *Precambrian Research*, **171**, 1–22.

AMELIN, Y. 2009. Sm–Nd and U–Pb systematics of single titanite grains. *Chemical Geology*, **261**, 53–61.

ARNDT, N. T. & GOLDSTEIN, S. L. 1987. Use and abuse of crust-formation ages. *Geology*, **15**, 893–895.

BAADSGAARD, H. 1973. U–Th–Pb dates on zircons from the early Precambrian Amitsoq gneisses, Godthaab district, West Greenland. *Earth and Planetary Science Letters*, **19**, 22–28.

BADARCH, G., CUNNINGHAM, W. D. & WINDLEY, B. F. 2002. A new subdivision for Mongolia: implications for Phanerozoic crustal growth in Central Asia. *Journal of Asian Earth Sciences*, **21**, 87–110.

BAKIROV, A. B. & MAKSUMOVA, R. A. 2001. Geodynamic evolution of the Tian Shan lithosphere. *Russian Geology and Geophysics*, **42**, 1359–1366.

Belyatsky, B., Lepekhina, E., Antonov, A., Shuliatin, O. & Sergeev, S. 2008. Age and genesis of accessory zircon from MAR gabbroids. *Geophysical Research Abstracts*, **10**, EGU2008-A-01314.

Bortnikov, N. S., Zinger, T. F., Sharkov, E. V., Lepekhina, E. N., Antnov, A. V. & Sergeev, S. A. 2008. Cenozoic and Precambrian accessory zircons in gabbroids of the 3rd layer of oceanic crust in axial part of the Mid-Atlantic Ridge, 6°N: U–Pb SIMS SHRIMP data. *American Geophysical Union, Eos, Transactions, AGU*, **89**(53), Fall Meeting Suppl., F2824.

Braun, I. & Bröcker, M. 2004. Monazite dating of granitic gneisses and leucogranites from the Kerala Khondalite Belt, southern India: implications for late Proterozoic crustal evolution in East Gondwana. *International Journal of Earth Sciences*, **93**, 13–22.

Cahen, L., Snelling, N. J., Delhal, J. & Vail, J. R. 1984. *The Geochronology and Evolution of Africa*. Clarendon Press, Oxford.

Cawood, P. A., Kröner, A. & Pisarevsky, S. 2006. Precambrian plate tectonics: criteria and evidence. *GSA Today*, **16**, 4–11.

Charvet, J., Shu, L. & Laurent-Charvet, S. 2007. Paleozoic structural and geodynamic evolution of eastern Tianshan (NW China): welding of the Tarim and Junggar plates. *Episodes*, **30**, 162–186.

Cherniak, D. J. 2006. Pb and rare earth element diffusion in xenotime. *Lithos*, **88**, 1–14.

Cherniak, D. J., Watson, B. E., Grove, M. & Harrison, T. M. 2004. Pb diffusion in monazite: a combined RBS/SIMS study. *Geochimica et Cosmochimica Acta*, **68**, 829–840.

Clauoé-Long, J. C., Sobolev, N. V., Shatsky, V. S. & Sobolev, A. V. 1991. Zircon response to diamond-pressure metamorphism in the Kokchetav massif, USSR. *Geology*, **19**, 710–713.

Clement, S. W. J., Compston, W. & Newstead, G. 1977. Design of a large, high-resolution ion microprobe. In: Benninghoven, A. (ed.) *Proceedings of the International Secondary Ion Mass Spectroscopy Conference, Münster, 1977*. Springer, Heidelberg, 1–6.

Clift, P. & Vannucchi, P. 2004. Controls on tectonic accretion versus erosion in subduction zones: implications for the origin and recycling of the continental crust. *Reviews of Geophysics*, **42**, RG2001, doi:10.1029/2003RG000127.

Collins, A. S., Santosh, M., Braun, I. & Clark, C. 2007. Age and sedimentary provenance of the Southern Granulites, South India. U–Th–Pb SHRIMP secondary ion mass spectrometry. *Precambrian Research*, **155**, 125–138.

Compston, W. & Kröner, A. 1988. Multiple zircon growth within early Archaean tonalitic gneiss from the Ancient Gneiss Complex, Swaziland. *Earth and Planetary Science Letters*, **87**, 13–28.

Compston, W., Williams, I. S. & Clement, S. W. J. 1982. U–Pb ages within single zircons using a sensitive high mass-resolution ion microprobe. In: *30th Annual Conference, American Society of Mass Spectrometry*. American Society for Mass Spectrometry, Honolulu, 593–595.

Condie, K. C. 1980. Origin and early development of the Earth's crust. *Precambrian Research*, **11**, 183–197.

Condie, K. C. 2001. Continental growth during formation of Rodinia at 1.35–0.9 Ga. *Gondwana Research*, **4**, 5–16.

Condie, K. C. 2007. Accretionary orogens in space and time. In: Hatcher, R. D. Jr., Carlson, M. P., McBride, J. H. & Martínez Catalán, J. R. (eds) *4-D Framework of Continental Crust*. Geological Society of America, Memoirs, **200**, 145–158.

Condie, K. C. & Kröner, A. 2008. When did plate tectonics begin? Evidence from the geologic record. In: Condie, K. C. & Pease, V. (eds) *When Did Plate Tectonics Begin on Planet Earth?* Geological Society of America, Special Papers, **440**, 281–294.

Condie, K. C. & Pease, V. (eds) 2008. *When Did Plate Tectonics Begin on Planet Earth?* Geological Society of America, Special Papers, **440**, 1–294.

Condie, K. C., O'Neill, C. & Aster, R. C. 2009a. Evidence and implications for a widespread magmatic shutdown for 250 My on Earth. *Earth and Planetary Science Letters*, **282**, 294–298.

Condie, K. C., Belousova, E., Griffin, W. L. & Sircombe, K. N. 2009b. Granitoid events in space and time: constraints from igneous and detrital zircon age spectra. *Gondwana Research*, **15**, 222–284.

Corfu, F. & Noble, S. R. 1992. Genesis of the southern Abitibi Greenstone belt, Superior Province, Canada: evidence from zircon Hf isotope analysis using a single filament technique. *Geochimica et Cosmochimica Acta*, **56**, 2081–2097.

Corfu, F., Hanchar, J. M., Hoskin, P. W. O. & Kinny, P. 2003. Atlas of zircon tectures. In: Hanchar, J. M. & Hoskin, P. W. O. (eds) *Zircon*. Mineralogical Society of America, Reviews in Mineralogy and Geochemistry, **53**, 469–500.

Corrigan, D., Hajnal, Z., Németh, B. & Lucas, S. B. 2006. Tectonic framework of a Paleoproterozoic arc–continent to continent–continent collisional zone, Trans-Hudson Orogen, from geological and seismic reflection studies. *Canadian Journal of Earth Sciences*, **42**, 421–434.

Crowley, J. L., Myers, J. S. & Dunning, G. R. 2002. Timing and nature of multiple 3700–3600 Ma tectonic events in intrusive rocks north of the Isua greenstone belt, southern West Greenland. *Geological Society of America Bulletin*, **114**, 1311–1325.

Dahl, P. S., Hamilton, M. A., Jercinovic, M. J., Terry, M. P., Williams, M. L. & Frei, R. 2005. Comparative isotopic and chemical geochronometry of monazite, with implications for U–Th–Pb dating by electron microprobe: an example from metamorphic rocks of the eastern Wyoming Craton (U.S.A.). *American Mineralogist*, **90**, 619–638.

Davis, D. W., Amelin, Y., Nowell, G. M. & Parrish, R. R. 2005. Hf isotopes in zircon from the western Superior province, Canada: implications for Archaean crustal development and evolution of the depleted mantle reservoir. *Precambrian Research*, **140**, 132–156.

Demoux, A., Kröner, A., Liu, D. & Badarch, G. 2009. Precambrian crystalline basement in southern Mongolia as revealed by SHRIMP zircon dating. *International Journal of Earth Sciences*, **98**, 1365–1380.

DePaolo, D. J. 1981. Neodymium isotopes in the Colorado Front Range and crust–mantle evolution in the Proterozoic. *Nature*, **291**, 193–196.

DE WAELE, B., WINGATE, M. T. D., FITZSIMONS, I. C. W. & MAPANI, B. S. E. 2003. Untying the Kibaran knot: a re-assessment of Mesoproterozoic correlations in southern Africa based on SHRIMP U–Pb data from the Irumide belt. *Geology*, **31**, 509–512.

DICKIN, A. P. & MCNUTT, R. H. 2003. An application of Nd isotope mapping in structural geology: delineating an allochthonous Grenvillian terrane at North Bay, Ontario. *Geological Magazine*, **140**, 539–548.

DONG, S., CHEN, J. & HUANG, D. 2004. Differential exhumation of tectonic units and ultrahigh-pressure metamorphic rocks in the Dabie Mountains, China. *Island Arc*, **7**, 174–183.

DONSKAYA, T. V., WINDLEY, B. F. ET AL. 2008. Age and evolution of late Mesozoic metamorphic core complexes in southern Siberia and northern Mongolia. *Journal of the Geological Society, London*, **165**, 405–421.

FITZSIMONS, I. C. W. 2000. A review of tectonic events in the East Antarctic shield, and their implications for three separate collisional orogens. *Journal of African Earth Sciences*, **31**, 3–23.

FOSTER, G. & PARRISH, R. R. 2003. Metamorphic monazite and the generation of $P-T-t$ paths. *In*: VANCE, D., MÜLLER, W. & VILLA, I. M. (eds) *Geochronology; Linking the Isotopic Record with Petrology and Textures*. Geological Society, London, Special Publications, **220**, 25–47.

FRASER, G., ELLIS, D. & EGGINS, S. 1997. Zirconium abundance in granulite-facies minerals, with implications for zircon geochronology in high-grade rocks. *Geology*, **25**, 607–610.

FROST, B. R., CHAMBERLAIN, K. R. & SCHUMACHER, J. C. 2000. Sphene (titanite): phase relations and role as a geochronometer. *Chemical Geology*, **172**, 131–148.

FUJII, M., HAYASAKA, Y. & TERADA, K. 2008. SHRIMP zircon and EPMA monazite dating of granitic rocks from the Maizuru terrane, southwest Japan: correlation with East Asian Paleozoic terranes and geological implications. *Island Arc*, **17**, 322–341.

GARDÉS, E. & MONTEL, J.-M. 2009. Opening and resetting temperatures in heating geochronological systems. *Contributions to Mineralogy and Petrology*, **158**, 185–195.

GARDÉS, E., JAOUL, O., MONTEL, J. M., SEYDOUX-GULLIAUME, A. M. & WIRTH, R. 2006. Pb diffusion in monazite: an experimental study of $Pb^{2+}+Th^{4+} \leftrightarrow 2Nd^{3+}$ interdiffusion. *Geochimica et Cosmochimica Acta*, **70**, 2325–2336.

GERDES, A. & ZEH, A. 2009. Zircon formation versus zircon alteration – new insights from combined U–Pb and Lu–Hf *in-situ* LA-ICP-MS analyses, and consequences for the interpretation of Archean zircon from the Central Zone of the Limpopo Belt. *Chemical Geology*, **261**, 230–243.

GLADKOCHUB, D. P., DONSKAYA, T. V. ET AL. 2008. Petrology, geochronology, and tectonic implications of c. 500 Ma metamorphic and igneous rocks along the northern margin of the Central Asian Orogen (Olkhon terrane, Lake Baikal, Siberia). *Journal of the Geological Society, London*, **165**, 235–246.

GONCALVES, P., WILLIAMS, M. L. & JERCINOVIC, M. J. 2005. Electron-microprobe age mapping of monazite. *American Mineralogist*, **90**, 578–585.

GOODWIN, A. M. 1991. *Precambrian Geology*. Academic Press, London.

GREENE, A. R., DEBARI, S. M., KELEMEN, P. B., BLUSZTAJN, P. & CLIFT, P. 2006. A detailed geochemical study of island arc crust: the Talkeetna arc section, south–central Alaska. *Journal of Petrology*, **47**, 1051–1093.

GRIFFIN, W. L., PEARSON, N. J., BELOUSOVA, E., JACKSON, S. E., O'REILLY, S. Y., VAN ACHTERBERG, E. & SHEE, S. R. 2000. The Hf isotope composition of cratonic mantle: LAM-MC-ICPMS analysis of zircon megacrysts in kimberlites. *Geochimica et Cosmochimica Acta*, **64**, 133–147.

HALL, R. 2009. The Eurasian SE Asian margin as a modern example of an accretionary orogen. *In*: CAWOOD, P. A. & KRÖNER, A. (eds) *Earth Accretionary Systems in Space and Time*. Geological Society, London, Special Publications, **318**, 351–372.

HAMILTON, W. B. 1998. Archean tectonics and magmatism were not products of plate tectonics. *Precambrian Research*, **91**, 143–179.

HARGROVE, U. S., STERN, R. J., KIMURA, J. I., MANTON, W. I. & JOHNSON, P. 2006. How juvenile is the Arabian–Nubian Shield? Evidence from Nd isotopes and pre-Neoproterozoic inherited zircon in the Bi'r Umq suture zone, Saudi Arabia. *Earth and Planetary Science Letters*, **252**, 308–326.

HARLEY, S. L. & KELLY, N. M. (eds) 2007. Zircon – Tiny but timely. *Elements*, **3**, 13–54.

HARRISON, T. M., BLICHERT-TOFT, J., MUELLER, W., ALBARÈDE, F., HOLDEN, P. & MOJZSIS, S. J. 2005. Heterogeneous Hadean hafnium: evidence of continental crust at 4.4 to 4.5 Ga. *Science*, **310**, 1947–1950.

HAY, D. C. & DEMPSTER, T. J. 2009. Zircon behaviour during low-temperature metamorphism. *Journal of Petrology*, **50**, 571–589.

HEGNER, E. & KRÖNER, A. 2000. Review of Nd isotopic data and xenocryst and detrital zircon ages from the pre-Variscan basement in the eastern Bohemian Massif: speculations on palinspasic reconstructions. *In*: FRANKE, W., HAAK, V., ONCKEN, O. & TANNER, D. (eds) *Orogenic Processes: Quantification and Modelling in the Variscan Belt*. Geological Society, London, Special Publications, **179**, 113–129.

HIRATA, T. & NESBITT, R. W. 1995. U–Pb isotope geochronology of zircon – evaluation of the laser probe-inductively coupled plasma-mass spectrometry technique. *Geochimica et Cosmochimica Acta*, **59**, 2491–2500.

HOFFMAN, P. F. 1991. Did the breakout of Laurentia turn Gondwana inside-out? *Science*, **252**, 1409–1412.

HOLDSWORTH, R. E., HAND, M., MILLER, J. A. & BUICK, I. S. 2001. Continental reactivation and reworking: an introduction. *In*: MILLER, J. A., HOLDSWORTH, R. E., BUICK, I. S. & HAND, M. (eds) *Continental Reactivation and Reworking*. Geological Society, London, Special Publications, **184**, 1–12.

HOSKIN, P. W. O. & BLACK, L. P. 2000. Metamorphic zircon formation by solid-state recrystallization of protolith igneous zircon. *Journal of Metamorphic Geology*, **18**, 423–439.

HOUSEMAN, G. & MOLNAR, P. 2001. Mechanism of lithospheric rejuvenation associated with continental orogeny. *In*: MILLER, J. A., HOLDSWORTH, R. E., BUICK, I. S. & HAND, M. (eds) *Continental*

Reactivation and Reworking. Geological Society, London, Special Publications, **184**, 13–38.

HOWARD, K. E., HAND, M., BAROVICH, K. M., REID, A., WADE, B. P. & BELOUSOVA, E. A. 2009. Detrital zircon ages: improving interpretation via Nd and Hf isotopic data. *Chemical Geology*, **262**, 277–292.

HU, A., JAHN, B., ZHANG, G., CHEN, Y. & ZHANG, Q. 2000. Crustal evolution and Phanerozoic crustal growth in northern Xinjiang: Nd isotopic evidence. Part I. Isotopic characterization of basement rocks. *Tectonophysics*, **328**, 15–51.

HURLEY, P. M. & RAND, J. R. 1969. Pre-drift continental nuclei. *Science*, **164**, 1229–1242.

HURLEY, P. M., DE ALMEIDA, F. F. M. ET AL. 1967. Test of continental drift by comparison of radiometric ages. *Science*, **157**, 495–500.

IIZUKA, T., KOMIYA, T. ET AL. 2007. Geology and zircon geochronology of the Acasta Gneiss Complex, northwestern Canada: new constraints on its tectonothermal history. *Precambrian Research*, **153**, 179–208.

JACOB, D. E. & FOLEY, S. F. 1999. Evidence for Archean ocean crust with low high field strength element signature from diamondiferous eclogite xenoliths. *Lithos*, **48**, 317–336.

JACOBS, J., FANNING, C. M. & BAUER, W. 2003. Timing of Grenville-age vs. Pan-African medium- to high grade metamorphism in western Dronning Maud Land (East Antarctica), and implications for the palaeogeography of Kalahari in Rodinia. *International Journal of Earth Sciences*, **92**, 301–315.

JAHN, B.-M., CAPDEVILA, R., LIU, D., VERNON, A. & BADARCH, G. 2004. Sources of Phanerozoic granitoids in the transect Bayanhongor–Ulaan Baatar, Mongolia: geochemical and Nd isotopic evidence, and implications for Phanerozoic crustal growth. *Journal of Asian Earth Sciences*, **23**, 629–653.

JAHN, B. M., LIU, D., WAN, Y., SONG, B. & WU, J. 2008. Archean crustal evolution in the Jiadong Peninsula, China, as revealed by zircon SHRIMP geochronology, elemental and Nd-isotope geochemistry. *American Journal of Science*, **308**, 232–269.

JANOTS, E., ENGI, M., RUBATTO, D., BERGER, A., GREGORY, C. & RAHN, M. 2009. Metamorphic rates in collisional orogeny from *in situ* allanite and monazite dating. *Geology*, **37**, 11–14.

JOHNSON, P. R. & WOLDEHAIMANOT, B. 2003. Development of the Arabian–Nubian Shield: perspectives on accretion and deformation in the northern East African orogen and the assembly of Gondwana. *In*: YOSHIDA, M., WINDLEY, B. F. & DASGUPTA, S. (eds) *Proterozoic East Gondwana: Supercontinent Assembly and Breakup*. Geological Society, London, Special Publications, **206**, 289–325.

KELSEY, D. E., CLARK, C. & HAND, M. 2008. Thermobarometric modelling of zircon and monazite growth in melt-bearing systems: examples using model metapelitic and metapsammitic granulites. *Journal of Metamorphic Geology*, **26**, 199–212.

KEMP, A. I. S., WORMALD, R. J. & PRICE, R. C. 2005. Hf isotopes in zircon reveal contrasting sources and crystallisation histories for alkaline to peralkaline granites of Temora, southeastern Australia. *Geology*, **33**, 797–800.

KEMP, A. I. S., FOSTER, G. L., SCHERSTÉN, A., WHITEHOUSE, M. J., DARLING, J. & STOREY, C. 2009. Concurrent Pb–Hf isotope analysis of zircon by laser ablation multi-collector ICP-MS, with implications for the crustal evolution of Greenland and the Himalayas. *Chemical Geology*, **261**, 244–260.

KENNEDY, A., JOHNSON, P. R. & KATTAN, F. H. 2004. *SHRIMP geochronology in the northern Arabian Shield, Part I: Data acquisition*. Saudi Geological Survey, Open-File-Report **SGS-OF-2004-11**.

KENNEDY, A., JOHNSON, P. R. & KATTAN, F. H. 2005. *SHRIMP geochronology in the northern Arabian Shield, Part II: Data acquisition, 2004*. Saudi Geological Survey, Open-File-Report **SGS-OF-2005-10**.

KINNY, P. D. 1986. 3820 Ma zircons from a tonalitic Amitsoq gneiss in the Godthab district of southern West Greenland. *Earth and Planetary Science Letters*, **79**, 337–347.

KINNY, P. D., COMPSTON, W. & WILLIAMS, I. S. 1991. A reconnaissance ion-probe study of hafnium isotopes in zircons. *Geochimica et Cosmochimica Acta*, **55**, 849–859.

KINNY, P. D., MCNAUGHTON, N. J., FANNING, C. M. & MAAS, R. 1994. 518 Ma sphene (titanite) from the Khan pegmatite, Namibia, southwest Africa: a potential ion-microprobe standard. 8th International Conference on Geochronology, Cosmochronology and Isotope Geology, Berkeley. *US Geological Survey Circular*, **1107**, 171.

KIRKLAND, C. L., WHITEHOUSE, M. J. & SLAGSTAD, T. 2009. Fluid-assisted zircon and monazite growth within a shear zone: a case study from Finnmark, Arctic Norway. *Contributions to Mineralogy and Petrology*, **158**, 675–681, doi:10.1007/s00410-009-0401-x.

KOSTITSYN, YU. A., BELOUSOVA, E. A., BORTNIKOV, N. S. & SHARKOV, E. V. 2009. Zircons in gabbroid formations from the axial zone of the mid-Atlantic ridge: U–Pb age and ^{176}Hf/^{177}Hf ratio (Results of investigations by the laser ablation method). *Doklady Earth Sciences*, **429**, 1305–1309.

KOVALENKO, V. I., YARMOLYUK, V. V., KOVACH, V. P., KOTOV, A. B., KOZAKOV, I. K., SALNIKOVA, E. B. & LARIN, A. M. 2004. Isotope provinces, mechanisms of generation and sources of the continental crust in the Central Asian mobile belt: geological and isotopic evidence. *Journal of Asian Earth Sciences*, **23**, 605–627.

KROGH, T. E., DAVIS, D. W. & CORFU, F. 1984. Precise U–Pb zircon and baddeleyite ages from the Sudbury area. *In*: PYE, E. G., NALDRETT, A. J. & GIBLIN, P. E. (eds) *The Geology and Ore Deposits of the Sudbury Structure*. Ontario Geological Survey Special Publication, **1**, 431–447.

KRÖNER, A. 2001. The Mozambique belt of East Africa and Madagascar: significance of zircon and Nd model ages for Rodinia and Gondwana supercontinent formation and dispersal. *South African Journal of Geology*, **105**, 151–167.

KRÖNER, A. & CORDANI, U. 2003. African, southern Indian and South American cratons were not part of the Rodinia supercontinent: evidence from field relationships and geochronology. *Tectonophysics*, **375**, 325–352.

KRÖNER, A., GREILING, R. ET AL. 1987. Pan-African crustal evolution in the Nubian segment of northeast

Africa. *In*: KRÖNER, A. (ed.) *Proterozoic Lithospheric Evolution*. American Geophysical Union, Geodynamics Series, **17**, 235–257.

KRÖNER, A., KEHELPANNALA, K. V. W. & KRIEGSMAN, L. 1994. Origin of compositional banding and causes for crustal thickening in the high-grade terrain of Sri Lanka. *Precambrian Research*, **66**, 21–38.

KRÖNER, A., JAECKEL, P., BRANDL, G., NEMCHIN, A. A. & PIDGEON, R. T. 1999. Single zircon ages for granitoid gneisses in the Central Zone of the Limpopo Belt, southern Africa and geodynamic significance. *Precambrian Research*, **93**, 299–337.

KRÖNER, A., O'BRIEN, P. J., NEMCHIN, A. A. & PIDGEON, R. T. 2000. Zircon ages for high pressure granulites from South Bohemia, Czech Republic, and their connection to Carboniferous high temperature processes. *Contributions to Mineralogy and Petrology*, **138**, 127–142.

KRÖNER, A., WINDLEY, B. F. ET AL. 2007. Accretionary growth and crust formation in the Central Asian Orogenic Belt and comparison with the Arabian–Nubian Shield. *In*: HATCHER, R. D. JR., CARLSON, M. P., MCBRIDE, J. H. & MARTÍNEZ CATALÁN, J. R. (eds) *4-D Framework of Continental Crust*. Geological Society of America, Memoirs, **200**, 181–209.

KRÖNER, A., HEGNER, E., LEHMANN, B., HEINHORST, J., WINGATE, M. T. D., LIU, D. Y. & ERMELOV, P. 2008. Palaeozoic arc magmatism in the Central Asian Orogenic Belt of Kazakhstan: SHRIMP zircon ages and whole-rock Nd isotopic systematics. *Journal of Asian Earth Sciences*, **32**, 118–130.

KRÖNER, A., ALEXEIEV, D. V. ET AL. 2009. New single zircon ages of Precambrian and Palaeozoic rocks from the Northern, Middle and Southern Tianshan belts in Kyrgyzstan. *In*: *Abstracts Volume, International Excursion and Workshop Tectonic Evolution and Crustal Structure of the Tien Shan Belt and Related Terrains in the Central Asian Orogenic Belt*. Central Asian Institute of Applied Geosciences, Bishkek, Kyrgyzstan, 30–31.

KRÖNER, A., SCHULMANN, K. ET AL. submitted. Lithostratigraphic and geochronological constraints on the evolution of the Central Asian Orogenic Belt in SW Mongolia: early Paleozoic rifting followed by late Paleozoic accretion. *American Journal of Science*.

LEE, J. K. W., WILLIAMS, I. S. & ELLIS, D. J. 1997. Pb, U and Th diffusion in natural zircon. *Nature*, **390**, 159–161.

LEHMANN, J., SCHULMANN, K., LEXA, O., STIPSKÁ, P., KRÖNER, A., TOMURHUU, D. & DORJSUREN, O. submitted. Structural constraints on the accretionary Paleozoic history of the SW Mongolian Altai. *American Journal of Science*.

LI, T. ET AL. 2008. *Geological Map of Central Asia and adjacent areas 1:2 500 000, 9 sheets*. Geological Publishing House, Beijing.

LI, Z. X., BOGDANOVA, S. V. ET AL. 2008. Assembly, configuration, and break-up history of Rodinia: a synthesis. *Precambrian Research*, **160**, 179–210.

LIÉGEOIS, J.-P., LATOUCHE, L., BOUGHRARA, M., NAVEZ, J. & GUIRAUD, M. 2003. The LATEA metacraton (Central Hoggar, Tuareg shield, Algeria): behaviour of an old passive margin during the Pan-African orogeny. *Journal of African Earth Sciences*, **37**, 161–190.

LIN, L.-H., WANG, P.-L., LO, C.-H., TSAI, C.-H. & JAHN, B.-M. 2007. ^{40}Ar–^{39}Ar thermochronological constraints on the exhumation of ultrahigh-pressure metamorphic rocks in the Sulu terrane of eastern China. *International Geology Review*, **47**, 872–886.

LIU, D., WILDE, S. A., WAN, Y., WU, J., ZHOU, H., DONG, C. & YIN, X. 2008. New U–Pb and Hf isotopic data confirm Anshan as the oldest preserved segment of the North China craton. *American Journal of Science*, **308**, 200–231.

LIU, D., WILDE, S. A. ET AL. 2009. Combined U–Pb, hafnium and oxygen isotope analysis of zircons from meta-igneous rocks in the southern North China Craton reveal multiple events in the late Mesoarchean–early Neoarchean. *Chemical Geology*, **261**, 140–154.

LUCK, J. M. & ALLÈGRE, C. J. 1991. Osmium isotopes in ophiolites. *Earth and Planetary Science Letters*, **107**, 406–415.

MCDOUGALL, I. & HARRISON, T. M. 1988. *Geochronology and thermochronology by the ^{40}Ar/^{39}Ar method*. Oxford Monographs on Geology and Geophysics, **9**.

MEERT, J. G. & LIEBERMANN, B. S. 2008. The Neoproterozoic assembly of Gondwana and its relationship to the Ediacaran–Cambrian radiation. *Gondwana Research*, **14**, 5–21.

MEISEL, T., WALKER, R. J., IRVING, A. J. & LORAND, J.-P. 2001. Osmium isotopic compositions of mantle xenoliths: a global perspective. *Geochimica et Cosmochimica Acta*, **65**, 1311–1323.

MEZGER, K. & KROGSTAD, E. J. 1997. Interpretation of discordant U–Pb zircon ages: an evaluation. *Journal of Metamorphic Geology*, **15**, 127–140.

MEZGER, K., HANSON, G. N. & BOHLEN, S. R. 1989. High-precision U–Pb ages of metamorphic rutile: application to the cooling history of high-grade terranes. *Earth and Planetary Science Letters*, **96**, 106–118.

MILISENDA, C. C., LIEW, T. C., HOFMANN, A. W. & KRÖNER, A. 1988. Isotopic mapping of age provinces in Precambrian high-grade terrains: Sri Lanka. *Journal of Geology*, **96**, 608–615.

MINTS, M. 2007. Paleoproterozoic supercontinent: origin and evolution of accretionary and collisional orogens exemplified in Northern cratons. *Geotectonics*, **41**, 257–280.

MIŠKOVIĆ, A. & SCHALTEGGER, U. 2009. Crustal growth along a non-collisional cratonic margin: a Lu–Hf isotopic survey of the eastern Cordilleran granitoids of Peru. *Earth and Planetary Science Letters*, **279**, 303–315.

MÖLLER, A., O'BRIEN, P. J., KENNEDY, A. & KRÖNER, A. 2002. Polyphase zircon in ultrahigh-temperature granulites (Rogaland, SW Norway): constraints for Pb diffusion in zircon. *Journal of Metamorphic Geology*, **20**, 727–740.

MÖLLER, A., O'BRIEN, P. J., KENNEDY, A. & KRÖNER, A. 2003. Linking growth episodes of zircon and metamorphic textures to zircon chemistry: an example from the ultrahigh-temperature granulites of Rogaland (SW Norway). *In*: VANCE, D., MÜLLER, W. & VILLA, I. M. (eds) *Geochronology: Linking the Isotopic Record with Petrology and Textures*. Geological Society, London, Special Publications, **220**, 65–81.

MOORBATH, S. 1977. The oldest rocks and the growth of continents. *Scientific American*, **236**, 92–104.

MUHONGO, S., KRÖNER, A. & NEMCHIN, A. A. 2001. Single zircon evaporation and SHRIMP ages for granulite-facies rocks in the Mozambique belt of Tanzania. *Journal of Geology*, **109**, 171–189.

MYERS, J. S. 1970. Gneiss types and their significance in the repeatedly deformed and metamorphosed Lewisian Complex of western Harris, Outer Hebrides. *Scottish Journal of Geology*, **6**, 186–199.

MYERS, J. S. 1978. Formation of banded gneisses by deformation of igneous rocks. *Precambrian Research*, **6**, 43–64.

NAKAHATA, N., TSUTSUMI, Y. & SANO, Y. 2008. Ion microprobe U–Pb dating of zircon with a 15 micrometer spatial resolution using NanoSIMS. *Gondwana Research*, **14**, 587–596.

NESBITT, H. W. & YOUNG, G. M. 1984. Early Proterozoic climates and plate motions inferred from major element geochemistry of lutites. *Nature*, **299**, 715–717.

NUTMAN, A. P. & FRIEND, C. R. L. 2007. Adjacent terranes with ca. 2715 and 2650 Ma high-pressure metamorphic assemblages in the Nuuk region of the North Atlantic Craton, southern West Greenland: complexity of Neoarchaean collisional orogeny. *Precambrian Research*, **155**, 159–203.

NUTMAN, A. P., MCGREGOR, V. R., FRIEND, C. R. L., BENNET, V. C. & KINNY, P. D. 1996. The Itsaq gneiss complex of southern West Greenland; the world's most extensive record of early crustal evolution. *Precambrian Research*, **78**, 1–39.

NUTMANN, A. P., FRIEND, C. R. L., BENNETT, V. C. & MCGREGOR, V. R. 2000. The early Archacan Itsaq Gneiss Complex of southern West Greenland: the importance of field observations in interpreting dates and isotopic data constraining early terrestrial evolution. *Geochimica et Cosmochimica Acta*, **64**, 3035–3060.

NUTMAN, A. P., FRIEND, C. R. L., BARLER, S. L. L. & MCGREGOR, V. R. 2004. Inventory and assessment of Palaeoarchaean gneiss terrains and detrital zircons in southern West Greenland. *Precambrian Research*, **135**, 281–314.

PARRISH, R. R. 2001. The response of mineral chenonometers to metamorphism and deformation in orogenic belts. *In*: MILLER, J. A., HOLDSWORTH, R. E., BUICK, I. S. & HAND, M. (eds) *Continental Reactivatin and Reworking*. Geological Society, London, Special Publications, **184**, 289–301.

PASSCHIER, C. W., MYERS, J. S. & KRÖNER, A. 1990. *Field Geology of High-grade Gneiss Terrains*. Springer, Berlin.

PATCHETT, P., KOUVO, O., HEDGE, C. E. & TATSUMOTO, M. 1981. Evolution of continental crust and mantle heterogeneity: evidence from Hf isotopes. *Contributions to Mineralogy and Petrology*, **78**, 279–297.

PERCIVAL, J. A., STERN, R. A. & SKULSKI, T. 2001. Crustal growth through successive arc magmatism: reconnaissance U–Pb SHRIMP data from the northeastern Superior Province, Canada. *Precambrian Research*, **109**, 203–238.

PETROVA, Z. I. & LEVITZKII, V. I. 1984. *Petrology and Geochemistry of Baikal Granulite Complexes*. Nauka, Novosibirsk [in Russian].

PILOT, J., WERNER, C. D., HAUBRICH, F. & BAUMANN, N. 1998. Palaeozoic and Proterozoic zircons from the mid-Atlantic ridge. *Nature*, **393**, 676–679.

RASMUSSEN, B. 2005. Zircon growth in very low grade metasedimentary rocks: evidence for zirconium mobility at 250 °C. *Contributions to Mineralogy and Petrology*, **150**, 146–155.

RASMUSSEN, B. & FLETCHER, I. R. 2004. Zirconolite: a new U–Pb chronometer for mafic igneous rocks. *Geology*, **32**, 785–788.

RASMUSSEN, B. & MUHLING, J. R. 2007. Monazite begets monazite: evidence for dissolution of detrital monazite and reprecipitation of syntectonic monazite during low-grade regional metamorphism. *Contributions to Mineralogy and Petrology*, **154**, 675–689.

RASMUSSEN, B., FLETCHER, I. R., MUHLING, J. R., THORNE, W. S. & BROADBENT, G. C. 2007. Prolonged history of episodic fluid flow in giant hematite ore bodies: evidence from *in situ* U–Pb geochronology of hydrothermal xenotime. *Earth and Planetary Science Letters*, **258**, 249–259.

RASMUSSEN, B., MUELLER, A. G. & FLETCHER, I. R. 2009. Zirconolite and xenotime U–Pb age constraints on the emplacement of the Golden Mile Dolerite sill and gold mineralization at the Mt Charlotte mine, Eastern Goldfields Province, Yilgarn Craton, Western Australia. *Contributions to Mineralogy and Petrology*, **157**, 559–572.

REYMER, A. & SCHUBERT, G. 1986. Rapid growth of some major segments of continental crust. *Geology*, **14**, 299–302.

REYMER, A. & SCHUBERT, G. 1987. Phanerozoic and Precambrian crustal growth. *In*: KRÖNER, A. (ed.) *Proterozoic Lithospheric Evolution*. American Geophysical Union, Geodynamics Series, **17**, 1–9.

RINO, S., KOMIYA, T., WINDLEY, B. F., KATAYAMA, I., MOTOKI, A. & HIRATA, T. 2004. Major episodic increases of continental growth determined from zircon ages in river sands; implications for mantle overturns in the early Precambrian. *Physics of the Earth and Planetary Interiors*, **146**, 369–394.

ROBERTS, M. P. & FINGER, F. 1997. Do U–Pb ages from granulites reflect peak metamorphic conditions? *Geology*, **25**, 319–322.

ROGERS, J. J. W. & SANTOSH, M. 2002. Configuration of Columbia, a Mesoproterozoic supercontinent. *Gondwana Research*, **5**, 5–22.

RUBATTO, D. & HERMANN, J. 2007. Zircon behaviour in deeply subducted rocks. *Elements*, **3**, 31–35.

RUBATTO, D., GEBAUER, D. & FANNING, M. 1998. Jurassic formation and Eocene subduction of the Zermatt–Saas–Fee ophiolites: implications for the geodynamic evolution of the Central and Western Alps. *Contributions to Mineralogy and Petrology*, **132**, 269–287.

SALNIKOVA, E. B., SERGEEV, S. A., KOTOV, A. B., YAKOVLEVA, S. Z., REZNITSKII, L. Z. & VASIL'EV, E. P. 1998. U–Pb zircon dating of granulite metamorphism in the Slyudyanskiy complex, Eastern Siberia. *Gondwana Research*, **1**, 195–205.

SALNIKOVA, E. B., KOZAKOV, I. K. ET AL. 2001. Age of Paleozoic granites and metamorphism in the Tuvino–Mongolian Massif of the Central Asian mobile belt: loss of a Precambrian microcontinent. *Precambrian Research*, **110**, 143–164.

SANO, Y., HIDAKA, H., TERADA, K., SHIMIZU, H. & SUZUKI, M. 2000. Ion microprobe U–Pb zircon geochronology of the Hida gneiss: finding of the oldest minerals in Japan. *Geochemical Journal*, **34**, 135–153.

SANTOSH, M., TSUNOGAE, T., TSUTSUMI, Y. & IWAMURA I, M. 2009. Microstructurally controlled monazite chronology of ultrahigh-temperature granulites from southern India: implications for the timing of Gondwana assembly. *Island Arc*, **18**, 248–265.

SCHÄRER, U., XU, R. H. & ALLÈGRE, C. J. 1986. U–Th–Pb systematics and ages of Himalayan leucogranites, south Tibet. *Earth and Planetary Science Letters*, **77**, 35–48.

SCHÄRER, U., DE PARSEVEL, P., POLVÉ, M. & DE SAINT BLANQUAT, M. 1999. Formation of the Trimouns talc–chlorite deposit (Pyrenees) from persistent hydrothermal activity between 112 and 97 Ma. *Terra Nova*, **11**, 30–37.

SCHERER, E. E., WHITEHOUSE, M. J. & MÜNKER, C. 2007. Zircon as a monitor of crustal growth. *Elements*, **3**, 19–24.

SCHILLING, F. R., TRUMBULL, R. B. ET AL. 2006. Partial melting in the central Andean crust: a review of geophysical, petrophysical, and petrologic evidence. *In*: ONCKEN, O., CHONG, G. ET AL. (eds) *The Andes*. Springer, Berlin, 459–474.

SCHMITT, A. K., LINDSAY, J. M., DA SILVA, S. & TRUMBULL, R. B. 2002. U–Pb zircon chronostratigraphy of early-Pliocene ignimbrites from La Pacana, north Chile: implications for the formation of stratified magma chambers. *Journal of Volcanology and Geothermal Research*, **120**, 43–53.

SCHOENE, B., DE WIT, M. & BOWRING, S. A. 2009. Mesoarchean assembly and stabilization of the eastern Kaapvaal craton: a structural–thermochronological perspective. *Terra Nova*, **21**, 219–228.

SCHOLL, D. W. & VON HUENE, R. 2009. Subduction zone recycling and truncated or missing tectonic and depositional domains expected at continental suture zones. *In*: CAWOOD, P. A. & KRÖNER, A. (eds) *Earth Accretionary Systems in Space and Time*. Geological Society, London, Special Publications, **318**, 105–125.

SCHULMANN, K., KRÖNER, A., HEGNER, E., WERNDT, I., KONOPASEK, J., LEXA, O. & STIPSKÁ, P. 2005. Chronological constraints on the pre-orogenic history, burial and exhumation of deep-seated rocks along the eastern margin of the Variscan orogen, Bohemian Massif, Czech Republic. *American Journal of Science*, **305**, 407–448.

SENGÖR, A. M. C., NATAL'IN, B. A. & BURTMANN, V. S. 1993. Evolution of the Altaid tectonic collage and Palaeozoic crustal growth in Eurasia. *Nature*, **364**, 299–306.

SEYDOUX-GUILLAUME, A. M., GONCALVES, P., WIRTH, R. & DEUTSCH, A. 2003. Transmission electron microscope study of polyphase and discordant monazites; site-specific specimen preparation using the focused ion beam technique. *Geology*, **31**, 973–976.

SHIBATA, T., ORIHASHI, Y., KIMURA, G. & HASHIMOTO, Y. 2008. Underplating of mélange evidenced by the depositional ages: U–Pb dating of zircons from the Shimanto accretionary complex, southwest Japan. *Island Arc*, **17**, 376–393.

SHIMODA, G., TATSUMI, Y., NOHDA, S., ISHIZAKA, K. & JAHN, B. M. 1998. Setouchi high-Mg andesites revisited: geochemical evidence for melting of subducting sediments. *Earth and Planetary Science Letters*, **160**, 479–492.

SHIREY, S. B., KAMBER, B. S., WHITEHOUSE, M. J., MUELLER, P. A. & BASU, A. R. 2008. A review of the isotopic and trace element evidence for mantle and crustal processes in the Hadean and Archean: implications for the onset of plate tectonic subduction. *In*: CONDIE, K. C. & PEASE, V. (eds) *When Did Plate Tectonics Begin on Planet Earth?* Geological Society of America, Special Papers, **440**, 1–30.

SOMMER, H., KRÖNER, A., HAUZENBERGER, C. & MUHONGO, S. 2005. Reworking of Archaean and Palaeoproterozoic crust in the Mozambique belt of central Tanzania as documented by SHRIMP zircon geochronology. *Journal of African Earth Sciences*, **43**, 447–463.

SPETSIUS, Z. V., TAYLOR, L. A., VALLEY, J. W., DEANGELIS, M. T., SPICUZZA, M., IVANOV, A. S. & BANZERUK, V. I. 2008. Diamondiferous xenoliths from crustal subduction: garnet oxygen isotopes from the Nyurbinskaya pipe, Yakutia. *European Journal of Mineralogy*, **20**, 375–385.

STEIN, H. J., MARKEY, R. J., MORGAN, J. W., HANNAH, J. L. & SCHERSTÉN, A. 2001. The remarkable Re–Os chronometer in molybdenite: how and why it works. *Terra Nova*, **13**, 479–486.

STERN, R. A., FLETCHER, I. R., RASMUSSEN, B., MCNAUGHTON, N. J. & GRIFFIN, B. J. 2005. Ion microprobe (NanoSIMS 50) Pb-isotope geochronology at <5 μm scale. *International Journal of Mass Spectrometry*, **244**, 125–134.

STERN, R. J. 2005. Evidence from ophiolites, blueschists, and ultra-high pressure metamorphic terranes that the modern episode of subduction tectonics began in Neoproterozoic time. *Geology*, **33**, 557–560.

STERN, R. J. 2008. Modern-style plate tectonics began in Neoproterozoic time: An alternative interpretation of Earth's tectonic history. *In*: CONDIE, K. C. & PEASE, V. (eds) *When Did Plate Tectonics Begin on Planet Earth?* Geological Society of America, Special Papers, **440**, 265–280.

STEVENSON, R. K. & PATCHETT, P. J. 1990. Implications for the evolution of continental crust from Hf isotope systematics of Archaean detrital zircons. *Geochimica et Cosmochimica Acta*, **54**, 1683–1697.

TANGE, Y. & TAKAHASHI, E. 2004. Stability of the high-pressure polymorph of zircon (ZrSiO$_4$) in the deep mantle. *Physics of the Earth and Planetary Interiors*, **143–144**, 223–229.

VALLEY, J. W., LACKEY, J. S. ET AL. 2005. 4.4 billion years of crustal maturation: oxygen isotopes in magmatic zircon. *Contributions to Mineralogy and Petrology*, **150**, 561–580.

VANCE, D., MÜLLER, W. & VILLA, I. M. (eds) 2003. *Geochronology: Linking the Isotopic Record with Petrology and Textures*. Geological Society, London, Special Publications, **220**.

VAN KRANENDONK, M. J. 2007. Tectonics of early Earth. *In*: VAN KRANENDONK, M. J., SMITHIES, R. H. & BENNETT, V. C. (eds) *Earth's Oldest Rocks*. Elsevier, Amsterdam, 1105–1116.

VAN KRANENDONK, M. J., SMITHIES, R. H., HICKMAN, A. H. & CHAMPION, D. C. 2007. Review: secular tectonic evolution of Archean continental crust: interplay between horizontal and vertical processes in the formation of the Pilbara craton, Australia. *Terra Nova*, **19**, 1–38.

VOGT, M., KRÖNER, A., POLLER, U., SOMMER, H., MUHONGO, S. & WINGATE, M. T. D. 2006. Archaean and Palaeoproterozoic gneisses reworked during a Neoproterozoic (Pan-African) high-grade event in the Mozambique belt of East Africa: structural relationships and zircon ages from the Kidatu area, central Tanzania. *Journal of African Earth Sciences*, **45**, 139–155.

VOLODICHEV, O. I., SLABUNOV, A. I., BIBIKOVA, E. V., KONILOV, A. N. & KUZENKO, T. I. 2004. Archean eclogites in the Belomorian mobile belt, Baltic shield. *Petrology*, **12**, 540–560.

VOLL, G. & KLEINSCHRODT, R. 1991. Sri Lanka: structural, magmatic and metamorphic development of a Gondwana fragment. *In*: KRÖNER, A. (ed.) *The Crystalline Crust of Sri Lanka*. Geological Survey Department, Sri Lanka, Professional Papers, **5**, 22–51.

VRY, J. K. & BAKER, J. A. 2006. LA-MC-ICPMS Pb–Pb dating of rutile from slowly cooled granulites: confirmation of the high closure temperature for Pb diffusion in rutile. *Geochimica et Cosmochimica Acta*, **70**, 1807–1820.

WALKER, R. J., SHIREY, S. B. & STECHER, O. 1988. Comparative Re–Os, Sm–Nd and Rb–Sr isotope and trace element systematics for Archean komatiite flows from Munro Township, Abitibi belt, Ontario. *Earth and Planetary Science Letters*, **87**, 1–12.

WALKER, R. J., CARLSON, R. W., SHIREY, S. B. & BOYD, F. R. 1989. Os, Sr, Nd, and Pb isotope systematics of southern African peridotite xenoliths: implications for the chemical evolution of subcontinental mantle. *Geochimica et Cosmochimica Acta*, **53**, 1583–1595.

WANG, B., FAURE, M., SHU, L., CLUZEL, D., CHARVET, J., DE JONG, K. & CHEN, Y. 2008. Paleozoic tectonic evolution of the Yili Block, western Chinese Tianshan. *Bulletin Géologique de France*, **179**, 483–490.

WATSON, E. B. 1996. Dissolution, growth and sirvical of zircons during crustal fusion: kinetic principles, geological models and implications for isotopic inheritance. *Transactions of the Royal Society of Edinburgh, Earth Sciences*, **87**, 43–56.

WERNER, C. D. 1987. Saxonian granulites – igneous or lithogenous. A contribution to the geochemical diagnosis of the original rocks in high-metamorphic complexes. *In*: GERSTENBERGER, H. (ed.) *Contributions to the Geology of the Saxonian Granulite Massif (Sächsisches Granulitgebirge)*. Zentralinstitut für Isotopen- und Strahlenforschung, Leipzig, **133**, 221–250.

WHATTAM, S. A., MALPAS, J., SMITH, I. E. M. & ALI, J. R. 2006. Link between SSZ ophiolite formation, emplacement and arc inception, Northland, New Zealand: U–Pb SHRIMP constraints; Cenozoic SW Pacific tectonic implications. *Earth and Planetary Science Letters*, **250**, 606–632.

WILLIAMS, I. S., BUICK, I. S. & CARTWRIGHT, I. 1996. An extended episode of early Mesoproterozoic metamorphic fluid flow in the Reynolds Range, central Australia. *Journal of Metamorphic Geology*, **14**, 29–47.

WILLIAMS, M. L., JERCINOVIC, M. J. & HETHERINGTON, C. J. 2007. Microprobe monazite geochronology: understanding geological processes by integrating composition and chronology. *Annual Review of Earth and Planetary Science*, **35**, 137–175.

WINDLEY, B. F. 1995. *The Evolving Continents*. 3rd edn. Wiley, Chichester.

WINDLEY, B. F., ALEXEIEV, D., XIAO, W. J., KRÖNER, A. & BADARCH, G. 2007. Tectonic models for accretion of the Central Asian Orogenic Belt. *Journal of the Geological Society, London*, **164**, 31–47.

WINGATE, M. T. D., CAMPBELL, I. H., COMPSTON, W. & GIBSON, G. M. 1998. Ion microprobe U–Pb ages for Neoproterozoic basaltic magmatism in south–central Australia and implications for the breakup of Rodinia. *Precambrian Research*, **87**, 135–159.

WRIGHT, J. E. & WYLD, S. A. 1986. Significance of xenocrystic Precambrian zircon contained within the southern continuation of the Josephine ophiolite: Devlis Elbow ophiolite remnant, Klamath Mountains, northern California. *Geology*, **14**, 671–674.

WU, F. Y., YANG, Y. H., XIE, L. W., YANG, J. H. & XU, P. 2006. Hf isotopic compositions of the standard zircons and baddeleyites used in U–Pb geochronology. *Chemical Geology*, **234**, 105–126.

XIA, Q. X., ZHENG, Y. F., YUAN, H. & WU, F. Y. 2009. Contrasting Lu–Hf and U–Th–Pb isotope systematics between metamorphic growth and recrystallization of zircon from eclogite-facies metagranites in the Dabie orogen, China. *Lithos*, **112**, 477–496.

ZACK, T., MOARES, R. & KRONZ, A. 2004. Temperature dependence of Zr in rutile: empirical calibration of a rutile thermometer. *Contributions to Mineralogy and Petrology*, **148**, 471–488.

ZHAO, G., CAWOOD, P. A., WILDE, S. A. & SUN, M. 2002. Review of global 2.1–1.8 Ga orogens: implications for a pre-Rodinia supercontinent. *Earth-Science Reviews*, **59**, 125–162.

ZHENG, J. P., GRIFFIN, W. L. ET AL. 2009. Neoarchean (2.7–2.8 Ga) accretion beneath the North China Craton: U–Pb age, trace elements and Hf isotopes of zircons in diamondiferous kimberlites. *Lithos*, **112**, 188–202.

ZHU, Y. F. & SONG, B. 2006. Petrology and SHRIMP chronology of mylonitized Tianger granite, Xinjiang: also about the dating of hydrothermal zircon rim in granite. *Acta Petrologica Sinica*, **22**, 135–144 [in Chinese with English abstract].

ZUBKOV, V. P., KONJUCHOV, A. G., BISKE, G. S. & VLASOV, N. G. (eds) 2008. *Geological Map of the Kyrgyz Republic, 1:500 000*. State Agency for Geology and Mineral Resources, Bishkek; VSEGEI, St. Petersburg [in Russian].

Chromitites from the Fiskenæsset anorthositic complex, West Greenland: clues to late Archaean mantle processes

HUGH ROLLINSON[1]*, CLAIRE REID[2] & BRIAN WINDLEY[3]

[1]*School of Science, University of Derby, Kedleston Road, Derby DE22 1GB, UK*

[2]*University of Gloucestershire, Francis Close Hall, Cheltenham GL50 2RH, UK*

[3]*Department of Geology, University of Leicester, Leicester LE1 7RH, UK*

Corresponding author (e-mail: h.rollinson@derby.ac.uk)

Abstract: Chromitites in the late Archaean Fiskenæsset anorthosite complex are characterized by a most unusual mineral assemblage: highly calcic plagioclase, iron-rich aluminous chromites and primary amphibole. In particular, the chromite compositions are atypical of chromitites in layered igneous intrusions. However, rare occurrences of this mineral assemblage are found in modern arcs and it is proposed here that the late Archaean calcic anorthositic chromitites formed by the partial melting of unusually aluminous harzburgite in a mantle wedge above a subduction zone. This melting process produced a hydrous, aluminous basalt, which fractionated at depth in the crust to produce a variety of high-alumina basalt compositions, from which the anorthosite complex with its chromitite horizons formed as a cumulate within the continental crust. The principal trigger for the late precipitation of chromite is thought to have been the removal of Al from the basaltic melt through plagioclase crystallization, and the build-up of Cr through an absence of clinopyroxene. It is proposed that the aluminous mantle source of the parent magma was produced by the melting of a harzburgitic mantle refertilized by small-volume, aluminous slab melts. This process ceased at the end of the Archaean because the dominant mechanism of crust generation changed such that melt production shifted from the slab into the mantle wedge, thus explaining why highly calcic anorthosites are almost totally restricted to the Archaean.

Highly calcic megacrystic anorthosites containing plagioclase with a high anorthite content (An_{80-90}) are rare and are, almost entirely, restricted to the Archaean (de Wit & Ashwal 1997). Anorthosites that appear later in Earth history have a very different character and contain plagioclases that are much less calcic (An_{40-60}, Phinney *et al.* 1988; Ashwal 1993). Highly calcic anorthosite complexes are found in most Archaean cratons, where they tend to form part of layered igneous complexes that are typified by relatively thin ultramafic sequences and thick gabbroic–anorthositic units, commonly enclosed in tonalitic gneisses. These anorthosites host very large plagioclase megacrysts (average diameter 6–8 cm) that are equidimensional rather than the more typical lath-shaped crystals and commonly contain chromitite horizons containing unusually Fe-rich chromites (Rollinson 1995), which bear little compositional similarity to other igneous chromites.

The origin of Archaean megacrystic anorthosites is poorly understood. Similar highly calcic anorthosites have been reported from the lunar highlands, where they are the dominant lithology, although the consensus now is that they have formed in a different manner from terrestrial Archaean anorthosites (Steele *et al.* 1977; O'Hara 2000; Taylor 2009).

Thus in Archaean anorthosites, the conditions under which the highly calcic plagioclase crystallizes and the processes that control the composition of and trigger the precipitation of the Fe-rich chromites have not yet been adequately explained. However, these rocks are important, for they form an integral part of some late Archaean high-grade tonalite–trondhjemite–granodiorite (TTG) gneiss terranes and as such hold clues to the origin and tectonic setting of these segments of ancient, lower continental crust (Phinney *et al.* 1988).

In this paper we present new geochemical results for chromitites from the late Archaean Fiskenæsset anorthosite complex in West Greenland, with a view to using their chemistry to determine the composition of their parent melt. In addition, we seek to explain the origin and genesis of the Archaean anorthosite association and, in keeping with the overall theme of this volume, its bearing on the process of Archaean crustal evolution.

Geological setting

The late Archaean Fiskenæsset layered anorthosite complex is located close to the small coastal village of Fiskenæsset in SW Greenland. The complex has been mapped over an area of about

From: KUSKY, T. M., ZHAI, M.-G. & XIAO, W. (eds) *The Evolving Continents: Understanding Processes of Continental Growth*. Geological Society, London, Special Publications, 338, 197–212.
DOI: 10.1144/SP338.10 0305-8719/10/$15.00 © The Geological Society of London 2010.

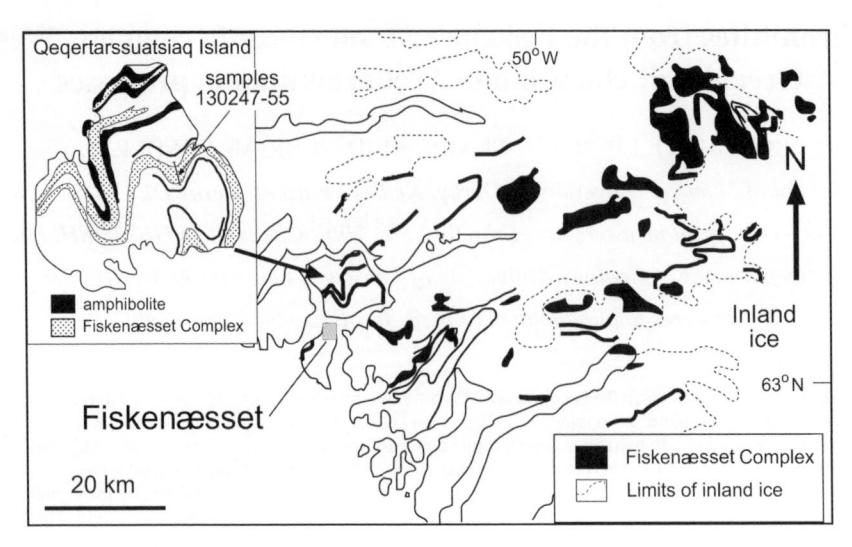

Fig. 1. Geological map of the Fiskenæsset area of West Greenland, showing (inset) the detailed geology of Qeqertarssuatsiaq Island and the sample localities for the material used in this study (after Myers 1975; Steele *et al.* 1977).

2500 km² from the coast to the inland ice (Fig. 1). The late Archaean rocks of the Fiskenæsset region form part of the lower crustal section of the Bjørnesund block, one of six Meso- to Neo-Archaean crustal blocks in West Greenland (Windley & Garde 2009). The Fiskenæsset Complex comprises a layered igneous complex that is bordered by metavolcanic amphibolites and engulfed by the protoliths of TTG gneisses, all of which have been multiply deformed and metamorphosed to upper amphibolite to granulite facies. The regional metamorphism has been dated using the Pb–Pb whole-rock isochron method at 2850 ± 100 Ma (Black *et al.* 1973).

Chromitites from the Fiskenæsset Complex were first described by Ghisler & Windley (1967) and subsequently by Ghisler (1976). Detailed descriptions of the Fiskenæsset complex as a whole were given by Windley *et al.* (1973), Windley & Smith (1974), Myers (1975, 1976a, b, 1985), Myers & Platt (1977) and Steele *et al.* (1977). The original igneous stratigraphy of the complex was determined by Windley *et al.* (1973) and Myers (1975), who showed that it comprises, from bottom to top, a lower gabbro, ultramafic rocks, gabbro, anorthosite and an upper gabbro (Fig. 2). The anorthosite unit near the top of the intrusion is the thickest unit. In the Majorqap qâva area, inland of Fiskenæsset, the total present-day stratigraphic thickness is thought to be about 540 m (Myers 1985), although before deformation it would have been greater. Elsewhere in the thickest, best-preserved fold hinges the complex is up to 2 km thick. The contact relationships show that the layered complex is intrusive into pillow-bearing amphibolites (Escher & Myers 1975) and may have had a sill-like form. Weaver *et al.* (1981, 1982) concluded on geochemical grounds that the pillowed amphibolites were originally mid-ocean ridge basalt (MORB)-type oceanic crust, although this was contested (Henderson 1984). Ashwal *et al.* (1989) calculated a Sm–Nd isochron age of 2860 ± 50 Ma for the complex, which they interpreted as the primary igneous crystallization age, and an initial ε_{Nd} of 2.9 ± 0.4, consistent with a late Archaean depleted mantle source.

The samples used in this study were collected from chromitite seams in the upper half of the anorthosite unit in the eastern part of Qeqertarssuatsiaq Island (Fig. 1). Single chromitite layers are between 0.5 and 10 cm thick alternating with anorthosite layers of similar thickness (Fig. 3a, b). The anorthosite unit has a total length of exposure of about 125 km and the chromitite horizons can be followed for several tens of kilometres within the anorthosite layer. The samples used here have the Geological Survey of Greenland numbers 132047–132055 (here abbreviated to 47–55) and were collected from the northern limb of an isoclinal fold (Fig. 1).

Petrography

The principal mineral phases in the Qeqertarssuatsiaq chromitites are plagioclase, amphibole and chromite (see Tables 1–3). In some samples the

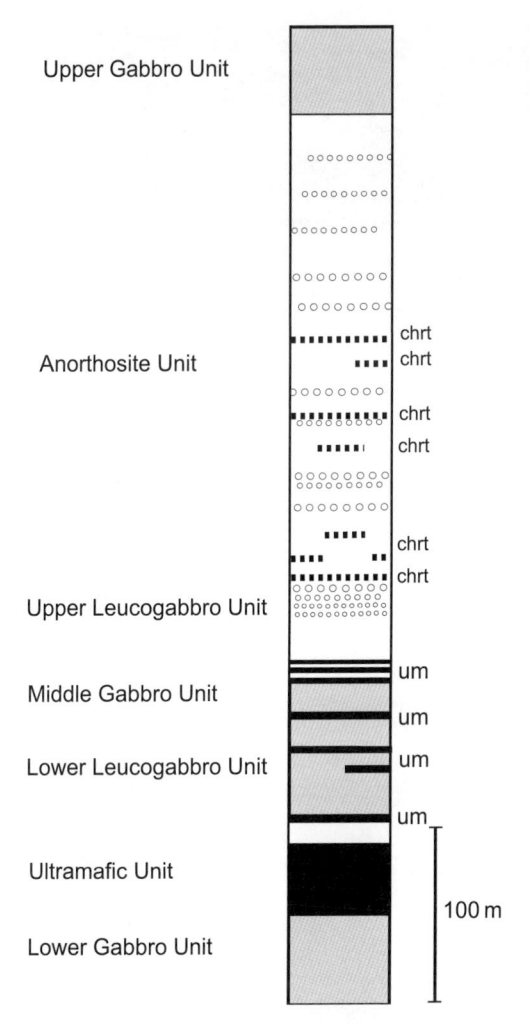

Fig. 2. The stratigraphy of the Fiskenæsset complex (after Ashwal & Myers 1994). um, ultramafic horizon; chrt, chromitite horizon.

plagioclase forms megacrysts, thought to be original igneous phenocrysts, and the chromite and amphibole form interstitial phases (Fig. 3c–f).

Chromite

Chromite makes up between 8 and 32% of the samples investigated in this study. The majority of chromite grains are subhedral to euhedral in form, often with straight grain boundaries, and they vary in size from about 100 μm to 2 mm in diameter (Fig. 3g, h). The grains are unaltered apart from a very small amount of chloritic alteration. Two types of chromite grain have been identified, inclusion-rich and inclusion-free grains (Ghisler 1970, 1976; Steele *et al.* 1977). Inclusion-rich chromites are zoned, sometimes with an inclusion-free core surrounded by an inclusion-rich zone and then an inclusion-free rim (Fig. 3g). The inclusion-free grains are smaller and are a few hundred micrometres across. In addition, very small chromite grains, about 5 μm across, are found as inclusions in plagioclase.

A variety of inclusion types were identified in the inclusion-bearing chromites using the scanning electron microscope. These include plagioclase, amphibole and an unidentified Mg–Ca-rich phase, which could be dolomite. There are also rounded inclusions of phlogopite 5–10 μm across, the shape of which suggests that they may have formed as melt inclusions during the crystallization of the chromite. In addition there are 'inclusions' that probably owe their origin to exsolution. These include rutile and magnesian spinel.

Plagioclase

Plagioclase is the dominant phase in these rocks, making up between 40 and 75% of the volume. In some samples it forms megacrysts up to 2.5 cm in diameter (Fig. 3f). Elsewhere it forms elliptical masses or thin bands from 0.5 to 0.3 mm in thickness. Some of these may be deformed megacrysts. Both the plagioclase bands and megacrysts are now made up of granular plagioclase grains up to 2 mm across, indicative of later recrystallization. Inclusions of chromite are present in some plagioclase grains and in sample 52 inclusions of biotite are also present. The granular plagioclase grains show some slight zoning and although no consistent pattern of zoning was observed, several grains had slightly more anorthite-rich rims.

Amphibole

Amphibole makes up between 12 and 30% of these samples. It tends to be associated with the chromite and is found in bands up to 2 cm thick. It commonly forms subhedral grains up to 1.5 mm in diameter. Amphibole grains may contain rounded inclusions of chromite, rutile and biotite and, more rarely, plagioclase. We found no evidence of clinopyroxene-rich cores to the amphiboles.

Analytical methods

The majority of mineral analyses reported here were made on a Cameca SX100 electron microprobe at the University of Manchester. Operating conditions were 15 kV and beam currents between 20 and 100 nA. The amphibole analyses reported here were made on a Cambridge Instruments Microscan V microprobe in the Department of Geology and

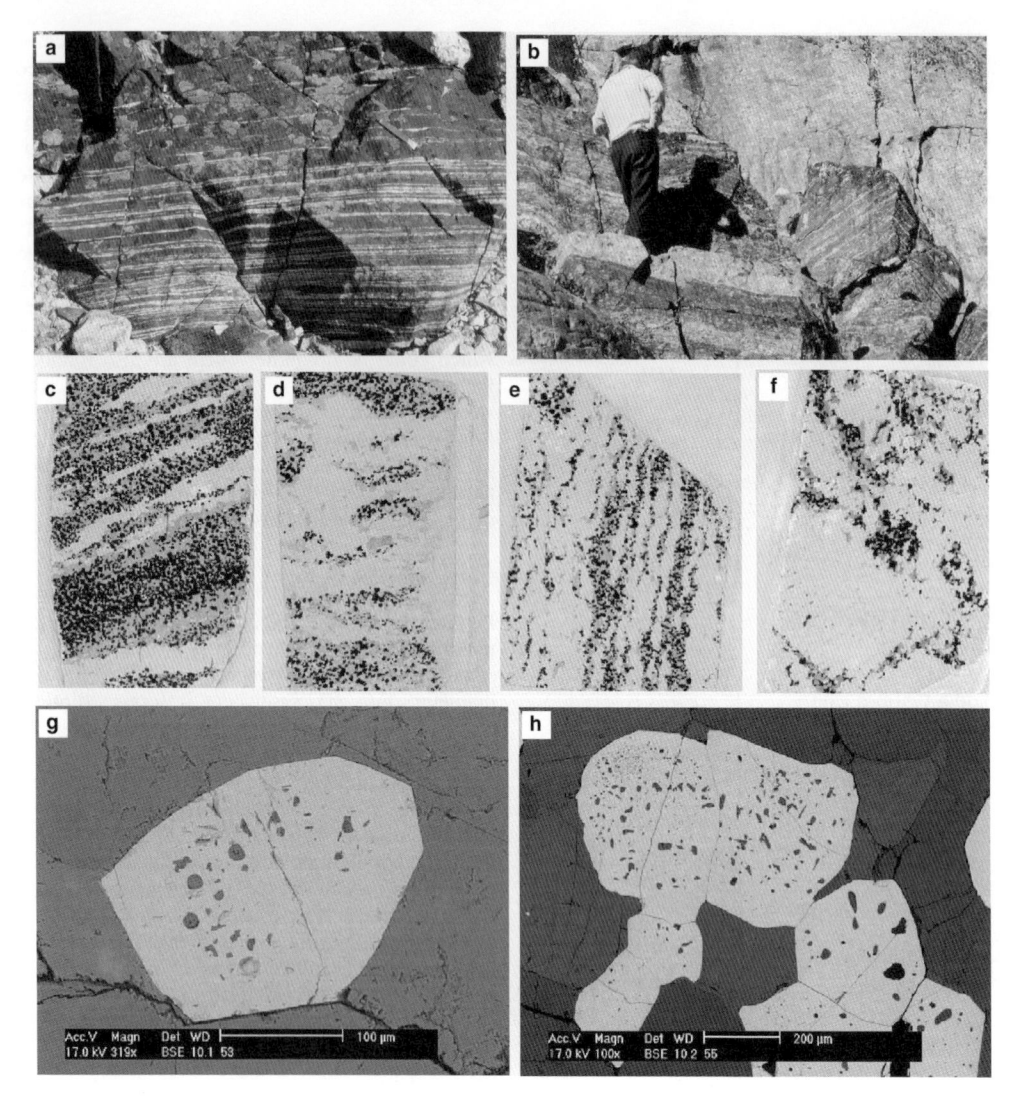

Fig. 3. (**a**) Banded chromitite from the Fiskenæsset anorthosite on Qeqertarssuatsiaq Island; pencil for scale. (**b**) Banded chromitite from the Fiskenæsset anorthosite on Qeqertarssuatsiaq Island. The person standing on the outcrop is the late Professor John Sutton. (**c–f**) Textures in the chromititic anorthosites as seen in thin section, showing plagioclase grains (colourless) and chromite–amphibole bands (grey and black grains). The width of each of these figures is 2.5 cm. From (c) to (f) these samples are 54, 51, 55 and 53. (**g**) Scanning electron microscope image of a zoned chromite grain from sample 53, showing an inclusion-free rim, an inclusion zone containing rounded phlogopite inclusions and an inclusion-free core. (**h**) Scanning electron microscope image of inclusion-rich chromite grains from sample 55. The larger rounded inclusions are amphibole.

Geophysics, University of Edinburgh. A variety of natural and synthetic standards were used. Mineral textures and inclusions were imaged and analysed using a Philips XL40 scanning electron microscope at the Geological Survey of Denmark and Greenland (GEUS), Copenhagen. All analyses were made at an acceleration voltage of 17 kV. Fe(III) in spinels was calculated using the charge balance equation of Droop (1987).

Mineral chemistry

Chromite

On a cr-number v. fe-number $[Cr/(Cr + Al)$ v. $Fe(II)/(FeII + Mg)]$ diagram chromite compositions plot at intermediate cr-number values (0.442–0.680) and high fe-number values (0.698–0.967) and define an fe-number–cr-number

Table 1. Representative chromite core compositions from the Fiskenæsset Complex

Sample:	47	47	49	49	50	50	51	51	52	52	53	53	54	54	55	55
SiO$_2$	0.01	0.01	0.02	0.02	0.00	0.00	0.00	0.10	0.01	0.02	0.01	0.00	0.03	0.03	0.03	0.01
TiO$_2$	0.14	0.06	0.13	0.09	0.07	0.24	0.41	0.08	0.30	0.20	0.79	0.67	0.10	0.30	0.27	0.13
Al$_2$O$_3$	22.50	22.37	24.93	24.37	22.13	21.77	24.70	25.70	12.98	12.39	16.08	17.00	26.86	25.64	22.05	21.72
Cr$_2$O$_3$	34.97	35.25	35.15	35.23	34.11	34.71	32.38	35.53	36.75	36.81	36.36	37.68	33.88	34.71	36.22	36.15
MgO	3.42	3.49	5.05	4.96	3.18	3.72	4.23	5.37	0.71	0.97	2.59	2.66	5.64	5.41	4.08	4.09
CaO	0.01	0.01	0.02	0.01	0.01	0.02	0.00	0.09	0.00	0.01	0.01	0.01	0.01	0.00	0.01	0.01
MnO	0.53	0.54	0.36	0.33	0.46	0.46	0.37	0.34	0.58	0.57	0.40	0.36	0.32	0.36	0.33	0.35
FeO	36.70	36.69	34.29	33.90	39.04	38.23	37.20	32.09	46.98	47.41	42.30	39.93	32.81	33.23	36.54	36.23
NiO	0.06	0.07	0.10	0.09	0.08	0.09	0.08	0.10	0.04	0.05	0.03	0.00	0.14	0.11	0.09	0.08
ZnO	0.46	0.44	0.12	0.10	0.18	0.18	0.20	0.11	0.16	0.18	0.14	0.13	0.12	0.10	0.11	0.13
Total	98.79	98.91	100.17	99.10	99.25	99.41	99.59	99.50	98.51	98.60	98.69	98.44	99.89	99.89	99.73	98.90

Cations to 24 oxygens

SiO$_2$	0.003	0.002	0.004	0.004	0.000	0.000	0.000	0.024	0.000	0.004	0.004	0.000	0.008	0.008	0.008	0.004
TiO$_2$	0.028	0.013	0.023	0.020	0.015	0.047	0.082	0.016	0.064	0.041	0.161	0.139	0.020	0.055	0.055	0.027
Al$_2$O$_3$	7.001	6.951	7.503	7.426	6.870	6.740	7.512	7.741	4.290	4.091	5.166	5.457	8.010	7.695	6.783	6.740
Cr$_2$O$_3$	7.301	7.349	7.095	7.198	7.106	7.205	6.609	7.178	8.149	8.156	7.838	8.112	6.777	6.988	7.472	7.527
MgO	1.344	1.371	1.923	1.911	1.248	1.454	1.631	2.048	0.295	0.407	1.054	1.082	2.124	2.053	1.590	1.605
CaO	0.002	0.001	0.008	0.004	0.004	0.004	0.000	0.024	0.000	0.004	0.004	0.004	0.000	0.000	0.004	0.000
MnO	0.119	0.121	0.078	0.070	0.105	0.101	0.082	0.071	0.136	0.136	0.092	0.085	0.067	0.079	0.074	0.078
Fe(II)	6.467	6.419	5.976	5.991	6.597	6.441	6.307	5.862	7.610	7.442	6.969	6.929	5.770	5.896	6.343	6.317
Fe(III)	1.634	1.672	1.347	1.337	2.005	1.954	1.723	0.998	3.413	3.670	2.678	2.164	1.173	1.183	1.632	1.662
NiO	0.012	0.015	0.020	0.020	0.015	0.019	0.016	0.020	0.008	0.011	0.008	0.000	0.027	0.024	0.019	0.016
ZnO	0.090	0.086	0.023	0.020	0.035	0.035	0.039	0.020	0.034	0.038	0.027	0.027	0.024	0.020	0.019	0.023
fe-no.	0.828	0.824	0.757	0.758	0.841	0.816	0.795	0.741	0.963	0.948	0.869	0.865	0.731	0.742	0.800	0.797
cr-no.	0.510	0.514	0.486	0.492	0.508	0.517	0.468	0.481	0.655	0.666	0.603	0.598	0.458	0.476	0.524	0.528

Table 2. *Representative plagioclase analyses from the Fiskenæsset Complex*

Sample:	47	47	49	49	50	50	52	52	54	54	55	55
SiO_2	46.45	46.19	45.19	45.53	46.09	46.18	56.20	55.28	45.61	45.46	45.00	46.51
TiO_2	0.00	0.00	0.00	0.01	0.00	0.01	nd	nd	0.01	0.00	0.00	0.01
Al_2O_3	33.81	33.61	34.41	34.64	33.67	33.68	27.24	27.59	34.88	34.70	34.54	34.31
Cr_2O_3	0.00	0.02	0.00	0.00	0.00	0.00	nd	nd	0.02	0.00	0.00	0.11
MgO	0.01	0.00	0.00	0.01	0.00	0.30	nd	nd	0.01	0.01	0.00	0.00
CaO	17.89	17.96	18.73	18.62	17.93	17.77	9.66	10.32	18.47	18.51	18.52	18.19
MnO	0.00	0.00	0.03	0.00	0.03	0.01	nd	nd	0.00	0.02	0.00	0.00
FeO	0.04	0.05	0.06	0.05	0.03	0.00	0.11	0.04	0.11	0.07	0.14	0.15
Na_2O	1.55	1.51	1.04	1.13	1.62	1.53	6.25	5.93	1.11	1.05	1.14	0.98
K_2O	0.00	0.00	0.01	0.00	0.01	0.02	nd	nd	0.02	0.00	0.00	0.00
Total	99.75	99.35	99.47	99.97	99.37	99.19	99.46	99.16	100.24	99.82	99.35	100.26

Cations to 32 oxygens

Si	8.572	8.564	8.388	8.404	8.548	8.568	10.156	10.040	8.392	8.400	8.364	8.536
Ti	0.000	0.000	0.000	0.000	0.000	0.000	0.000	0.000	0.000	0.000	0.000	0.000
Al	7.352	7.344	7.528	7.536	7.360	7.364	5.804	5.908	7.564	7.556	7.568	7.420
Cr	0.000	0.004	0.000	0.000	0.000	0.000	0.000	0.000	0.004	0.000	0.000	0.016
Mg	0.000	0.000	0.000	0.000	0.000	0.000	0.000	0.000	0.004	0.004	0.000	0.000
Ca	3.540	3.568	3.724	3.680	3.564	3.532	1.872	2.008	3.644	3.664	3.688	3.576
Mn	0.000	0.000	0.004	0.000	0.004	0.000	0.000	0.000	0.000	0.004	0.000	0.000
Fe	0.004	0.008	0.008	0.008	0.004	0.000	0.016	0.008	0.016	0.012	0.020	0.024
Na	0.552	0.544	0.376	0.404	0.580	0.548	2.188	2.088	0.396	0.376	0.412	0.348
K	0.000	0.000	0.000	0.000	0.004	0.004	0.000	0.000	0.004	0.000	0.000	0.000
An	0.865	0.868	0.908	0.901	0.859	0.865	0.461	0.490	0.902	0.907	0.900	0.911

Abbreviation: nd, not determined.

Table 3. *Representative amphibole analyses from the Fiskenæsset Complex*

Sample:	47	47	49	49	50	50	51	51	52	52	53	53	54	54	55	55
SiO_2	42.78	42.18	43.86	43.67	43.64	42.99	44.19	43.74	43.32	42.77	44.31	43.42	43.86	44.01	43.38	43.44
TiO_2	1.11	1.40	1.24	1.29	1.14	1.43	1.29	1.20	0.54	0.66	0.99	1.30	0.93	1.29	1.17	1.18
Al_2O_3	15.11	14.59	12.74	12.72	12.97	13.08	12.85	13.15	12.63	12.77	11.67	12.51	13.41	13.19	13.35	13.26
Cr_2O_3	1.38	1.45	1.52	1.56	1.43	1.49	1.33	1.49	1.42	1.77	2.23	1.66	2.23	1.61	1.98	2.26
FeO	7.48	7.97	7.07	7.31	8.21	8.85	8.34	8.94	10.94	11.36	8.13	9.13	6.45	6.91	7.12	7.33
MnO	0.13	0.14	0.14	0.16	0.20	0.18	0.18	0.21	0.17	0.20	0.12	0.13	0.12	0.14	0.12	0.11
MgO	14.25	14.22	15.53	15.44	15.03	14.63	15.00	14.69	13.32	12.95	15.77	14.24	16.46	15.78	15.53	15.53
NiO	0.11	0.08	0.15	0.12	0.11	0.09	0.10	0.10	0.09	0.09	0.07	0.06	0.15	0.13	0.13	0.12
CaO	11.94	11.86	11.48	11.40	11.46	11.20	10.94	10.74	11.81	11.49	11.86	11.70	11.74	11.58	11.92	11.88
ZnO	0.05	0.05	0.02	0.01	0.02	0.01	0.00	0.03	0.00	0.03	0.03	0.03	0.00	0.02	0.05	0.03
Na_2O	1.82	1.98	1.83	1.80	1.70	1.78	1.84	1.89	1.43	1.51	1.52	1.56	1.95	1.96	1.91	1.79
K_2O	0.47	0.49	0.21	0.20	0.38	0.48	0.30	0.30	0.57	0.77	0.51	0.55	0.33	0.22	0.63	0.74
Total	96.61	96.41	95.78	95.67	96.29	96.21	96.35	96.47	96.22	96.36	97.21	96.29	97.63	96.84	97.30	97.66

Cations to 23 oxygens

SiO_2	6.222	6.180	6.409	6.396	6.381	6.319	6.439	6.390	6.426	6.366	6.435	6.389	6.294	6.360	6.282	6.276
TiO_2	0.121	0.154	0.136	0.142	0.125	0.158	0.142	0.131	0.060	0.074	0.108	0.144	0.100	0.140	0.128	0.128
Al_2O_3	2.590	2.520	2.194	2.196	2.235	2.267	2.207	2.265	2.208	2.242	1.998	2.171	2.268	2.247	2.279	2.259
Cr_2O_3	0.159	0.168	0.176	0.181	0.165	0.173	0.153	0.172	0.166	0.208	0.256	0.193	0.253	0.184	0.226	0.259
FeO	0.910	0.976	0.864	0.895	1.005	1.088	1.016	1.092	1.357	1.414	0.987	1.123	0.774	0.835	0.862	0.886
MnO	0.013	0.014	0.015	0.016	0.020	0.019	0.018	0.021	0.018	0.021	0.012	0.013	0.012	0.014	0.012	0.011
MgO	3.088	3.104	3.383	3.371	3.275	3.206	3.258	3.198	2.946	2.873	3.413	3.122	3.521	3.399	3.352	3.343
NiO	0.013	0.010	0.017	0.014	0.013	0.010	0.011	0.012	0.011	0.011	0.008	0.007	0.018	0.015	0.016	0.014
CaO	1.861	1.862	1.797	1.789	1.796	1.763	1.708	1.681	1.877	1.833	1.846	1.845	1.805	1.793	1.850	1.839
ZnO	0.005	0.005	0.002	0.001	0.003	0.001	0.000	0.003	0.000	0.003	0.004	0.003	0.000	0.002	0.005	0.004
Na_2O	0.513	0.563	0.517	0.510	0.483	0.507	0.518	0.535	0.411	0.435	0.428	0.446	0.542	0.550	0.537	0.503
K_2O	0.087	0.092	0.039	0.037	0.071	0.091	0.055	0.056	0.107	0.145	0.095	0.102	0.060	0.041	0.117	0.136
Na + K	0.600	0.655	0.556	0.547	0.554	0.598	0.574	0.592	0.518	0.580	0.523	0.548	0.602	0.591	0.653	0.639
mg-no.	0.772	0.761	0.796	0.790	0.765	0.747	0.762	0.745	0.685	0.670	0.776	0.735	0.820	0.803	0.795	0.791

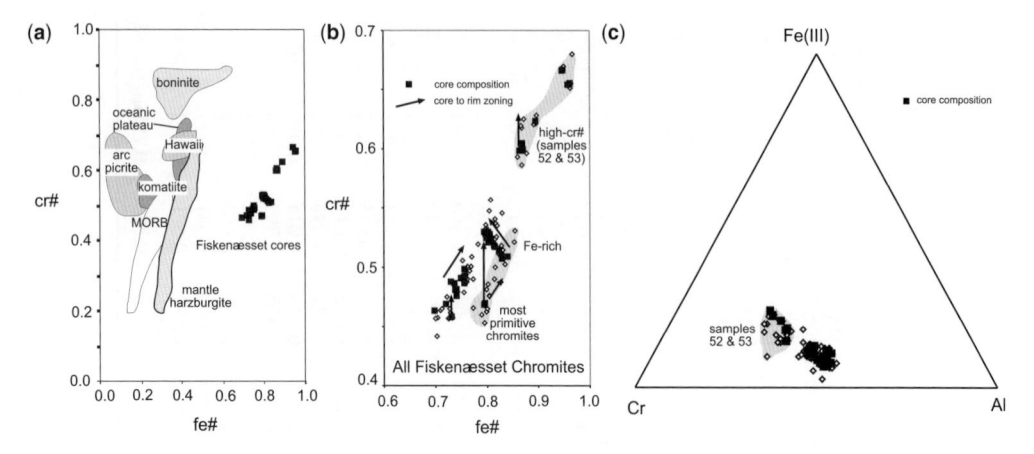

Fig. 4. (a) cr-number v. fe-number [Cr/(Cr + Al) v. Fe(II)/(FeII + Mg)] diagram for chromites from volcanic rocks from a variety of modern tectonic settings (after Rollinson 1995) and the Oman ophiolite mantle harzburgite (Le Mée *et al.* 2004) plotted relative to the core compositions of the Fiskenæsset chromites. (b) cr-number v. fe-number diagram for Fiskenæsset chromites showing the compositional change in single grains; the shaded fields identify the Fe-rich chromites and the high-cr-number chromites found in samples 52 and 53. (c) Triangular trivalent element plot for the Fiskenæsset chromites. The shaded field shows the compositions of the high-Fe(III) chromites in samples 52 and 53.

enrichment trend (Fig. 4b). When compared with chromites from modern volcanic rocks (Fig. 4a) the Fiskenæsset chromites are amongst the most Fe-rich of terrestrial mafic and ultramafic rocks at low to intermediate Cr ratios, and are much more Fe-rich than chromites from typical mantle-derived melts. For example, at cr-number 0.5 MORB chromites have an fe-number of about 0.3, whereas the Fiskenæsset samples have an fe-number of between 0.75 and 0.85 (Fig. 4a).

In detail we identify three compositional types of chromite, each of which follows a cr-number–fe-number enrichment trend (Fig. 4b): (1) chromites from samples 52 and 53 have high cr-numbers (0.586–0.698) and very high fe-numbers (0.860–0.967); these chromites are particularly low in Al_2O_3; (2) chromites with intermediate cr-numbers (0.453–0.531) and high fe-numbers (0.773–0.854); (3) chromites with intermediate cr-numbers (0.442–0.557) and lower fe-numbers (0.698–0.830); sample 49 lies on this trend. Samples 47, 50, 51, 54 and 55 contain chromite grains that plot on both trends (2) and (3).

On an Fe(III)–Cr–Al ternary plot (Fig. 4c) samples 52 and 53 are also distinctive and plot with high Fe(III) (9–23% Fe(III)/(Fe(III) + Cr + Al) compared with less than 14% in the other samples. They are also enriched in Ti (TiO_2 = 0.30–0.53 wt%) relative to the other samples (TiO_2 = 0.10–0.0.22 wt%). Sample 47 is distinctive inasmuch it is slightly enriched in ZnO (0.46 wt%) and MnO (0.52 wt%) relative to other chromites (average ZnO 0.13 wt%, average MnO 0.39 wt%).

A comparison of chromite core compositions with all chromite analyses (Fig. 4b) shows that the grains are not strongly zoned, but that chromite cores tend to be more Cr- and Fe-poor relative to the rims. The inclusion-rich group of chromites have cores that are within the low-Cr, high-Fe group (group (2) above). These observations suggest that the earliest chromites (i.e. the most geochemically primitive) were the inclusion-rich grains and that the inclusion-free grains were later. The earliest chromites to form therefore were relatively low in Cr, and belonged to the high-Fe group (group (2) above), and so had a cr-number of *c.* 0.46 and an fe-number of *c.* 0.8. The earliest grains are also slightly enriched in V. Later chromites were enriched in both Cr and Fe, reflecting the chemical evolution of their parent magmas during crystallization.

When plotted against stratigraphic height (Fig. 5) the chromites show a trend towards lower cr-numbers and lower fe-numbers, but there is a marked excursion towards high cr-numbers and fe-numbers in the middle of the section (samples 52 and 53). If it is assumed that each major chromitite band represents a separate batch of magma, then over time the parental magma evolved towards lower cr-numbers and fe-numbers, apart from samples 52 and 53 when a different batch of magma produced chromites with higher cr-numbers and fe-numbers. In detail, chromites within the single chromitite bands evolved slightly differently (Fig. 4b). In samples 47 and 51 (later grains) the chromites evolved to more magnesian and more Cr-rich compositions, whereas in

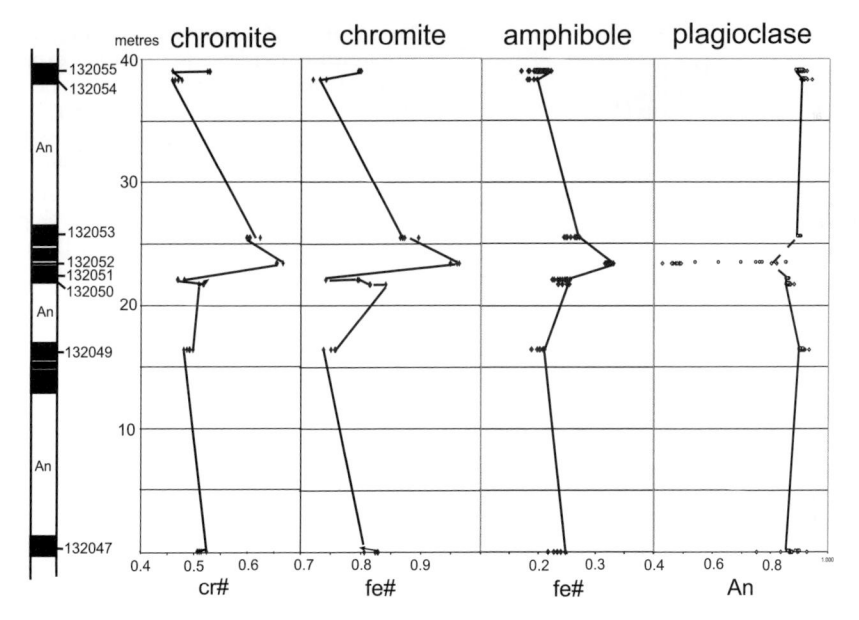

Fig. 5. Compositional change with stratigraphic height in the Fiskenæsset chromitites (cr-number and fe-number), amphibole (fe-number) and plagioclase (An). Stratigraphic sketch after Steele *et al.* (1977).

samples 49, 51 (early grains) and 54 chromites evolved to more Fe- and Cr-rich compositions. Other samples show no Fe-enrichment and show only Cr-enrichment (samples 50, 52, 53 and 55). The overall fe-number–cr-number trend in Figure 5 is thought to reflect compositional change in the parent magma, whereas the subtle details preserved in chromite core–rim compositions reflect the local environment of the chromite band and reflect both magmatic and metamorphic subsolidus exchange.

Plagioclase

Plagioclases from the Fiskenæsset anorthosite are highly calcic with An contents up to An_{91}. Some samples show a spread of plagioclase compositions, with as much as 17% An variation, although core compositions vary only from $An_{86.5}$ to $An_{90.8}$. The only exception is sample 52, in which plagioclase compositions vary from $An_{42.7}$ to $An_{81.7}$. This wide range of compositions probably reflects the metamorphic recrystallization of the originally igneous plagioclase megacrysts that characterize this sample. The presence of biotite in this sample might indicate the former presence of metasomatic fluids. Inclusions of plagioclase in chromite in samples 54 and 55 are slightly more calcic than measured core compositions. For example, sample 54 inclusions in chromite have a composition of $An_{91.4}$ whereas plagioclase cores are only $An_{90.8}$, and sample 55 contains inclusions in chromites with the composition $An_{90-91.6}$ and has plagioclase cores of An_{90}.

Within the sample suite examined in this study there is a small increase in the An content of the plagioclases with stratigraphic height from about $An_{86.5}$ to An_{91}, although the trend within the intrusion as a whole is from higher to lower An content with increasing stratigraphic height (Steele *et al.* 1977). There is a significant excursion to lower An values in sample 52, coinciding with changes in chromite and amphibole compositions, which, as noted above, may relate to a batch of magma with a different composition.

Amphibole

The Fiskenæsset amphiboles are calcic and magnesian and plot as pargasites on the Mg-number v. Si classification diagram of Leake *et al.* (1997) for calcic amphiboles (Fig. 6). Amphiboles contain between 1.1 and 1.8 wt% Cr_2O_3. Amphibole inclusions in chromite in samples 53, 54 and 55 also plot within the field of measured amphibole compositions (Fig. 6). A detailed study of zoning in sample 55 shows that the cores of the largest grains have the highest fe-number and that fe-numbers are lower at the rims of large grains and in the smaller grains. We interpret this to indicate that there was Fe–Mg exchange between amphibole and chromite, so that the rims of large amphibole grains and entire smaller grains have become more magnesian. Hence we use the highest fe-numbers as indicative

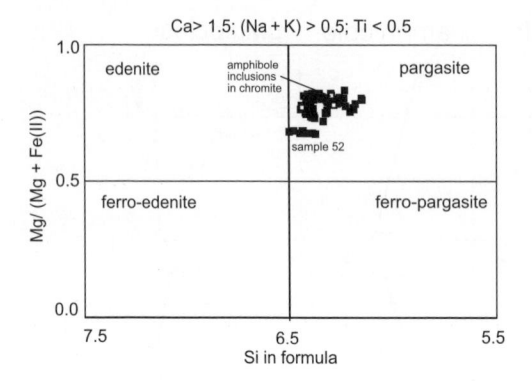

Fig. 6. Mg/(Mg + Fe(II)) v. Si in the formula unit classification diagram (after Leake et al. 1997) for the Fiskenæsset amphiboles. All samples plot as pargasites, although amphiboles in sample 52 plot with lower Mg/(Mg + Fe(II)).

of the earliest amphibole compositions. A plot of amphibole fe-number with stratigraphic height shows some variation. Values initially decrease up-section, then increase to a maximum value in sample 52 and then decrease again. Steele et al. (1977) show a progressive increase in fe-number in amphibole from the ultramafic to anorthositic units. There is little change in TiO_2 content of the amphiboles up-section apart from sample 52, in which concentrations are lower.

We calculated equilibration temperatures for adjacent amphibole–plagioclase rim compositions in three samples using the Holland & Blundy (1994) edenite–richterite thermometer (their thermometer B). At a pressure of 5 kbar our results show average temperatures for samples 49, 52 and 54 of 886 ± 30 °C, 758 ± 30 °C and 906 ± 31 °C, respectively (1σ uncertainties). The discrepancy between these results probably reflects the different An content of the plagioclases. In samples 49 and 54 the plagioclase composition exceeds An_{90}, the recommended limit of the thermometer. Thus the lower temperature estimate of c. 760 ± 30 °C for sample 52, based on plagioclase with compositions within the limits of the thermometer, is to be preferred as the closure temperature of plagioclase–amphibole Na–Si Ca–Al exchange. It is probable that these closure temperatures reflect the metamorphic cooling of the intrusion.

The crystallization history of the Fiskenæsset anorthositic chromitites

A central assumption of this paper is that the layering and many of the macro-scale textures observed in the Fiskenæsset anorthosite complex are igneous in origin and are the product of processes within a layered igneous intrusion (Myers 1985). However, given that the Fiskenæsset area has experienced a regional amphibolite- to granulite-grade metamorphism it is important to consider the extent to which the mineral compositions described here have been modified as a result of metamorphic processes.

Of particular concern is the origin of the amphibole, for some previous workers have proposed that it is of igneous origin (Windley et al. 1981), whereas others have argued for an origin through later metamorphic recrystallization (Myers 1985). Here we prefer a primary igneous origin for the amphibole in this section of the intrusion, although we recognize that it may have also experienced some recrystallization during metamorphism. We base our argument on three observations.

(1) The presence of amphibole inclusions within chromite suggests that they represent primary igneous inclusions.

(2) Calcic amphiboles are present in unmetamorphosed chromitites in the mantle section of the Oman ophiolite, both as a matrix phase within the chromitite and as inclusions within the chromite (Schiano et al. 1997; Rollinson 2008). This latter observation underlines the fact that chromite can form from a hydrous magma in equilibrium with amphibole.

(3) Primary, igneous, amphibole-bearing anorthosites have been described from south Sweden, and attributed to an origin from a hydrous basalt (Claeson & Meurer 2004).

For these reasons we propose a primary igneous mineralogy for the Fiskenæsset anorthositic chromitites, which comprised chromite, plagioclase and amphibole. This indicates that the parental melt was hydrous. From the textures of these rocks and the inclusion history of the various phases we infer the following crystallization sequence: plagioclase → plagioclase + chromite → plagioclase + chromite + amphibole → amphibole + chromite.

During the later deformation and amphibolite-facies metamorphism the plagioclase recrystallized, giving rise to a granular texture within former plagioclase megacrysts. However, in the pure anorthosites, in the absence of other phases the original An content would be retained. Elsewhere, grain boundary re-equilibration took place between the major phases at c. 760 °C, leading to a readjustment of Fe–Mg, Cr–Al and Ca–Na ratios. Thus mineral core compositions of the larger grains are thought to be more likely to preserve the original compositions of the igneous phases.

Discussion

The Fiskenæsset anorthositic chromitites are distinctive in two ways. First, they contain unusually calcic plagioclases (up to An$_{91}$) and, second, they contain unusually Fe-rich chromites (up to fe-number = 0.85 at cr-number c. 0.6). Any model for their origin must explain both of these mineral-chemical features. Previous studies have shown that there is no consensus on the composition of the melt from which they have crystallized and a variety of melt compositions have been proposed. These include aluminous tholeiite (Weaver et al. 1981), tholeiite (Ashwal et al. 1983) and komatiite (Ashwal 1993).

Here we distinguish between the melt from which the anorthosite crystallized and the 'primary' melt, from which the 'anorthosite melt' was produced. We shall argue that the primary melt was mafic and a product of mantle melting, whereas the melt parental to the anorthosite was derived from this primary melt by differentiation processes.

The composition of the melt from which the Fiskenæsset anorthositic chromitites crystallized

We constrain the composition of the melt from which the Fiskenæsset anorthositic chromitites formed from our own mineral-chemical observations, from previous geochemical investigations and from the results of experimental petrology as follows.

(1) Our mineral-chemical observations show that the melt parental to the anorthosites contained calcic plagioclase, amphibole and Fe-rich, Ti-poor chromite as liquidus phases, and on this basis we infer that it was aluminous, calcic, relatively iron and chrome rich, but relatively poor in silica, alkalis, Ti and Mg.

(2) Our most primitive chromite compositions are Fe-rich and aluminous (fe-number = 0.8, cr-number = 0.46) and they evolve to more Fe-rich and more Cr-rich compositions during crystal fractionation. We note that unlike chromites in mafic and ultramafic rocks these anorthositic chromites form late in the crystallization sequence, after the start of the plagioclase crystallization and simultaneous with amphibole crystallization. Thus the variation in cr-number of the chromite, recorded, for example, in Figure 4b, is related to the variation in the amount of plagioclase crystallizing along with the chromite, as plagioclase crystallization will deplete the melt in Al and so increase the cr-number in the chromite. Similarly, the Fe/Mg ratio of the chromite is governed by the degree of amphibole crystallization, such that the amphibole preference for Mg over Fe enhances the Fe-rich character of the chromites (Fig. 4b).

(3) The Ti and Al contents of chromite reflect those of the magmas from which they crystallized (Kamenetsky et al. 2001). In the case of the most primitive chromites (Al$_2$O$_3$ c. 25 wt%; TiO$_2$ c. 0.16 wt%) this would suggest a low-Ti, aluminous melt. Kamenetsky et al. (2001) also showed that the Ti and Al contents of chromites of mantle-derived melts can be used to indicate the former tectonic setting of the magma. However, these results apply only when chromite is either an early liquidus phase or in equilibrium with a Ti–Al-free phase such as olivine. In this study chromite crystallized together with other aluminous phases (plagioclase and amphibole) and another Ti-phase (amphibole), and has exsolved a Ti-rich phase (rutile), and so the Ti and Al content cannot be used to characterize the former tectonic setting with any certainty.

(4) In detail there were several batches of melt with slightly different compositions, and these melts gave rise to the different mineral compositions observed. These different batches of melt imply that the magma parental to the anorthosite is the product of a prior fractionation process.

(5) The melt was hydrous. A variety of experimental studies have shown that the An content of plagioclase in mafic magmas is controlled by variations in melt composition, melt water content, temperature and pressure (Beard & Borgia 1989; Sisson & Grove 1993). However, a recent study by Takagi et al. (2005) has shown that at constant composition, in a low-alkali, high-alumina, arc tholeiite (17 wt% Al$_2$O$_3$), there is a linear relationship between the An content of plagioclase and the water content of the melt. In addition, high-An plagioclase is the liquidus phase in these melts with up to 5% H$_2$O at low pressure, decreasing to about 3% at 5 kbar, suggesting that, in the study of Takagi et al. (2005), the optimal conditions for the crystallization of high-An plagioclase are 2–3 kbar. Explaining the high An content of the plagioclase through a hydrous mafic magma is also consistent with the presence of primary amphibole in these anorthosites. The highly calcic plagioclases reported by Takagi et al. (2005) (>An$_{90}$) are comparable with the compositions reported here for Fiskenæsset (An$_{86.5-90.8}$).

(6) The melt was enriched in light REE (LREE). Following Henderson et al. (1976), we inverted the REE concentrations of four highly calcic plagioclases (An$_{89}$), measured by Henderson et al. (1976) and Ashwal et al. (1989), to estimate the composition of the parental melt. We used the partition coefficients for Ca-rich plagioclases from basaltic melts from the experimental study of Phinney & Morrison (1990; table 2, column B). Our results show irregular REE patterns, enriched in the LREE relative to the middle (MREE) and heavy REE (HREE). There is no Eu anomaly (Fig. 7). The enrichment in LREE indicates that the parental melt was different from the Archaean tholeiites (now preserved as amphibolites) into

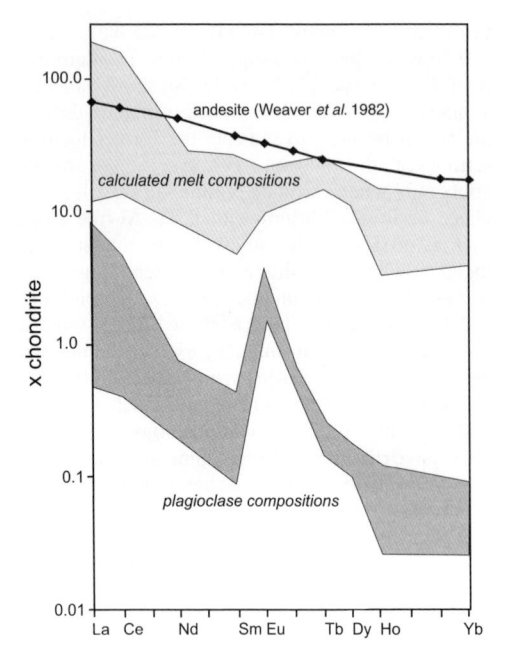

Fig. 7. Chondrite-normalized REE plot for the field of melt compositions inferred to be parental to the Fiskenæsset anorthite-rich plagioclases. Melt compositions were calculated from the plagioclase REE compositions measured by Henderson et al. (1976) and Ashwal et al. (1989), using the partition coefficient data of Phinney & Morrison (1990). An andesite from the upper amphibolites is shown for comparison (Weaver et al. 1982).

which the Fiskenæsset intrusion was emplaced (Weaver et al. 1982). However, there is some similarity between the calculated melt compositions and a single sample of andesite also reported by Weaver et al. (1982) from the Upper Fiskenæsset amphibolites (Fig. 7).

The composition of the melt parental to the 'anorthositic magma'

It is noted above that the rocks of the Fiskenæsset intrusion show subtle differences in mineral chemistry indicating that they formed from several batches of melt each with slightly different compositions. These different batches of melt imply that the magmas from which the anorthosite formed are themselves derived from a parental magma through fractional crystallization. The nature of this magma can be characterized as follows.

(1) It was mafic or ultramafic. This is indicated by the presence of olivine- and pyroxene-rich cumulates in the lower part of the Fiskenæsset succession.

(2) It was mantle-derived. The isotopic study of Ashwal et al. (1989) showed that the anorthositic melt has an ε_{Nd} typical of depleted mantle in the late Archaean.

(3) It was LREE enriched and probably came from an enriched mantle source. The LREE-enriched nature of the melt, inferred from inverting the plagioclase compositions, means that it is not a typical Archaean tholeiite.

(4) The parental melt must have been highly aluminous and highly calcic. This means that the parental melt was either a small melt fraction of mantle peridotite, for typical mantle has low concentrations of Al and Ca, or from an enriched mantle source.

Comparison with arc magmas

The hydrous nature of the magma parental to Fiskenæsset anorthositic chromitites suggests that it might have formed from a basaltic magma in an arc-type setting. There is some support for this view from both the amphibole–plagioclase chemistry and the chromite chemistry. Takagi et al. (2005) reported that extremely An-rich plagioclases ($>An_{90}$) are commonly found in arc basalts. Similarly, Arculus & Wills (1980) reported cumulate xenoliths from the Lesser Antilles Arc, which are rich in plagioclase, amphibole and ferrian chromite. The plagioclase is highly calcic (up to An_{100}) and the amphibole compositions include pargasite and tschermakite. These observations are consistent with the results of experimental petrology reported above, which show that highly calcic plagioclases form in hydrous melts. In addition, a plot of chromite compositions from a variety of primitive arc basalts shows that, in contrast to the main arc tholeiite field of Barnes & Roedder (2001) some primitive arc basalts have chromites with very high fe-numbers (>0.6) at relatively low cr-numbers (0.4–0.6, Fig. 8a). Although chromite core compositions from Fiskenæsset do not overlap with the main arc tholeiite field of Barnes & Roedder (2001) they do show some similarities to the Fe-rich arc-basalt chromites.

Given that chromite Cr/Al ratios tend to faithfully reflect that of their parent magma, it is possible to interpret the arc-basalt chromite data in Figure 8a in terms of two or more cr-number–fe-number trends, originating from the composition of mantle harzburgite. If the chromite chemistry of mantle harzburgite can be used to reflect the bulk harzburgite chemistry, then the more aluminous chromite trend, the trend on which the Fiskenæsset samples lie, could derive from a more aluminous (more fertile) mantle harzburgite (Fig. 8a, fertile mantle). The less aluminous chromites, more typical of arc basalts, appear to be derived from a less aluminous, more depleted, mantle (Fig. 8a, depleted mantle).

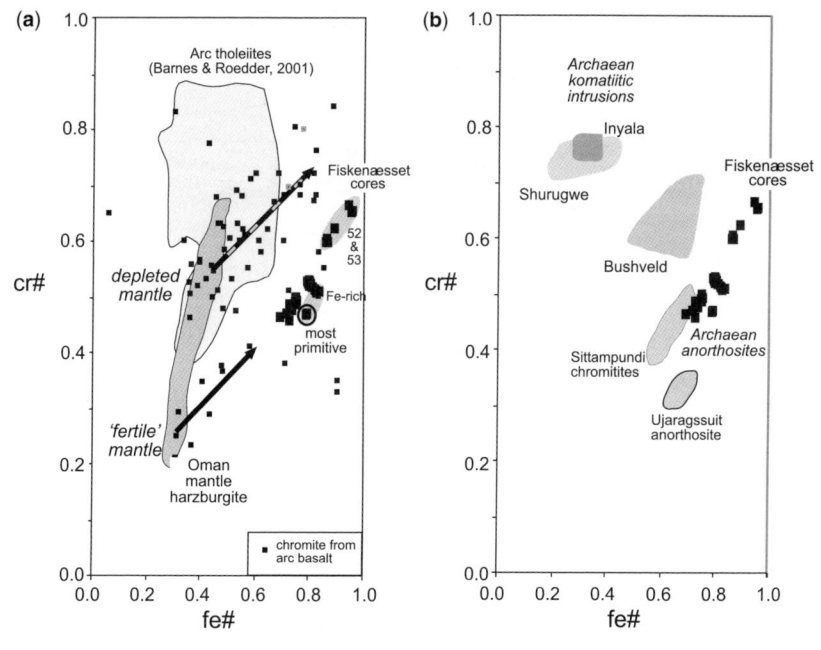

Fig. 8. (a) cr-number v. fe-number plot of Fiskenæsset chromites and chromites from arc basalts. The Fiskenæsset chromites are represented by the core compositions and the high-Fe and high-cr-number samples are indicated by shading. The large, light shaded field is the 90 percentile field for arc basalts from the compilation of Barnes & Roedder (2001). The darker shaded field is for harzburgites from the mantle in the Oman ophiolite (Le Mée et al. 2004). Data points for arc-basalt chromites are from Stewart et al. (1996), Heath et al. (1998), Groves et al. (2003) and Righter et al. (2008). The arrows show the possible evolution of melts from aluminous (fertile) mantle in the Archaean (lower arrow) and from less aluminous (depleted) mantle in recent arcs (upper arrow). (b) cr-number v. fe-number plot of Fiskenæsset chromites (core compositions), plotted relative to data for other Archaean chromititic anorthosites from Sittampundi (data from U. K. Bhui, pers. comm.) and Ujaragssuit, West Greenland (Rollinson et al. 2002), the Bushveld intrusion (data from Eales & Reynolds 1986) and from chromitites associated with Archaean komatiitic intrusions from Shurugwe, Zimbabwe (Stowe 1987) and Inyala, Zimbabwe (Rollinson 1997).

These observations lead to two conclusions: first, that the Fiskenæsset chromitites were cumulates derived from an arc basalt, which was derived from a relatively aluminous mantle source; second, that the 'fertile' mantle source of the Fiskenæsset magmas is not representative of the source of modern arc magmas. The origin of the fertile mantle source is not known and could be either relatively undepleted mantle or, as will be proposed below, mantle that has been refertilized by the migration of aluminous melts within the mantle wedge.

The similarity in composition to other Archaean anorthositic chromitites (Fig. 8b) such as Sittampundi (U. K. Bhui, pers. comm.) and Ujaragssuit Nunat (Rollinson et al. 2002) would indicate that aluminous mantle of this type was a particular type of Archaean mantle, for anorthositic chromitites are clearly different from komatiitic chromitites (Fig. 8b). However, the record of rare Palaeozoic and Mesozoic calcic anorthosites (see Windley & Garde 2009) means that processes of this type are not exclusive to the Archaean.

Summary of petrogenetic evolution

The arguments presented above suggest the following petrogenetic processes for the formation of calcic anothositic chromitites (see Fig. 9).

(1) The partial melting of unusually aluminous mantle harzburgite was initiated by the migration of hydrous fluids; the mantle was most probably a mantle wedge in a subduction setting and the water derived through the dehydration of subducted oceanic crust.

(2) The product of the partial melting was a hydrous, aluminous basalt.

(3) The hydrous, aluminous basalt was fractionated in magma chambers in the lower crust to produce a variety of high-alumina basalt compositions, each with slightly different compositions.

(4) The anorthosite complex with its chromitite horizons formed as a cumulate within the continental crust, probably in an arc setting.

(5) The trigger for the late precipitation of chromite in the anorthosite was the removal of Al from

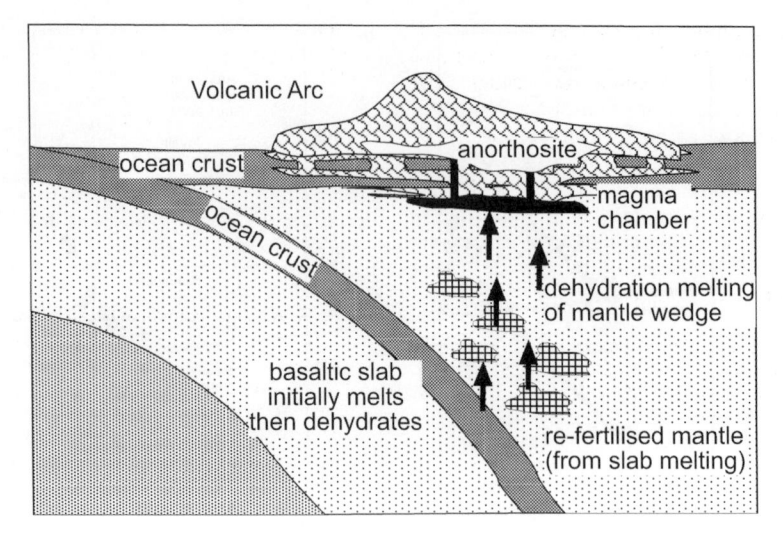

Fig. 9. Schematic illustration showing the principal processes involved in the formation of Archaean anorthosites.

the basaltic melt, via plagioclase crystallization, and the absence of clinopyroxene crystallization, so that Cr was retained in the melt. In addition, Matveev & Ballhaus (2002) have stressed the importance of the presence of water during the crystallization of chromite.

Mid-Archaean arcs in West Greenland

Recent studies of mid-Archaean rocks in West Greenland show the importance of arcs in the geological evolution of this region, consistent with the inferences of this study. This, however, represents something of a paradigm shift in West Greenland geology, as the Archaean gneisses of this region have been regarded as a 'classical' high-grade gneiss terrane, representing typical Archaean lower crust. Garde (2007) reported field and geochemical evidence for the presence of an oceanic arc embedded in younger TTG gneisses in the Akia terrane, southern West Greenland. Similarly, Polat et al. (2008) described mid-Archaean, supra-arc, oceanic crust from the Ivistaartoq greenstone belt of West Greenland. In both instances anorthositic rocks were reported in these associations. In the Akia terrane Garde (2007) mapped leucogabbros and anorthosites in association with the supracrustal rocks now interpreted as ocean arc. Polat et al. (2008) reported hornblende-anorthosite inclusions in gabbros from the Ivisaartoq greenstone belt.

Calcic anorthosites and crust generation in the Archaean

There remains the question of why calcic anorthosites appear to be relatively common in the Archaean and rare later in Earth history. It is significant in this regard that Archaean calcic anorthosites were consistently intruded by TTG felsic crust (Myers 1976b). Current models of Archaean crust generation suggest that TTG magmas were formed in a subduction environment, consistent with the model for calcic anorthosites presented here. In detail, one possible mechanism for Archaean TTG magma genesis invokes the melting of the subducting slab to produce a siliceous aluminous melt that migrated through the mantle wedge to form TTG magmas in the crust (Rollinson 2007). It is possible that aluminous, fertile mantle, proposed here as the source of the parent magma to the calcic anorthosites, was produced in this environment through small-volume, aluminous slab melts becoming frozen in the mantle wedge, only to be subsequently involved in a later mantle melting event. We propose that the anorthositic complexes formed in the magma chambers of island arcs, which were later invaded by TTG magmas (see Garde 2007).

Such a scenario requires a cycle of melting and dehydration of a subducting slab. Initially, arc basalts are formed through slab dehydration and the melting of the mantle wedge. This is followed by a small amount of slab melting leading to the refertilization of the mantle wedge with aluminous melts. Further dehydration melting would lead to the production of aluminous tholeiites, parental to the anorthosites. Finally, further, extensive slab melting led to the formation of large volumes of TTG magma.

This process ceased at the end of the Archaean because the principal mechanism of crust generation changed and slab melting largely ceased (Rollinson

2007). Only rarely now are arc magmas so aluminous as to give rise to 'Fiskenæsset-type' chromites (Fig. 8a), thereby explaining the cessation of anorthosite cumulates at the end of the Archaean. The rare instances of geologically recent calcic anorthosites are reported from Andean-type continental margins: the Peninsular Ranges anorthosites and the Peruvian anorthosites (see references given by Windley & Garde 2009), and these may provide a suitable analogue for mid-Archaean crustal growth in West Greenland.

The work reported in this study was the subject of a Masters thesis at the University of Gloucestershire, by C.R. Microprobe analyses at Manchester and Edinburgh were undertaken through NERC funding; C. Hayward, D. Plant (Manchester) and P. Hill (Edinburgh) are thanked for their assistance. P. and C. Appel (GEUS) are thanked for their hospitality to C.R. during her stay in Copenhagen and for assistance with the SEM analysis. We thank A. Polat and L. Ashwal for their helpful comments on the manuscript.

References

ARCULUS, R. J. & WILLS, K. J. A. 1980. The petrology of plutonic blocks and inclusions from the Lesser Antilles Island arc. *Journal of Petrology*, **21**, 743–799.

ASHWAL, L. D. 1993. *Anorthosites*. Springer, Berlin.

ASHWAL, L. D. & MYERS, J. S. 1994. Archaean anorthosites. In: CONDIE, K. C. (ed.) *Archaean Crustal Evolution*. Developments in Precambrian Geology, **11**, 315–356.

ASHWAL, L. D., MORRISON, D. A., PHINNEY, W. C. & WOOD, J. 1983. Origin of Archaean anorthosites: evidence from the Bad Vermilion Lake complex, Ontario. *Contributions to Mineralogy and Petrology*, **82**, 259–273.

ASHWAL, L. D., JACOBSEN, S. B., MYERS, J. S., KALSBEEK, F. & GOLDSTEIN, S. J. 1989. Sm–Nd age of the Fiskenæsset Anorthosite Complex, West Greenland. *Earth and Planetary Science Letters*, **91**, 261–270.

BARNES, S. J. & ROEDER, P. L. 2001. The range of spinel compositions in terrestrial mafic and ultramafic rocks. *Journal of Petrology*, **42**, 2279–2302.

BEARD, J. S. & BORGIA, A. 1989. Temporal variations of mineralogy and petrology in cognate gabbroic enclaves at Arenal volcano, Costa Rica. *Contributions to Mineralogy and Petrology*, **103**, 110–122.

BLACK, L. P., MOORBATH, S., PANKHURST, R. J. & WINDLEY, B. F. 1973. $^{207}Pb/^{206}Pb$ whole rock age of the Archaean granulite facies metamorphic event in west Greenland. *Nature*, **244**, 50–53.

CLAESON, D. T. & MEURER, W. P. 2004. Fractional crystallisation of hydrous basaltic 'arc-type' magmas and the formation of amphibole-bearing gabbroic cumulates. *Contributions to Mineralogy and Petrology*, **147**, 288–304.

DE WIT, M. J. & ASHWAL, L. D. 1997. *Greenstone Belts*. Oxford University Press, Oxford.

DROOP, G. T. R. 1987. A general equation for estimating Fe^{3+} concentrations in ferromagnesian silicates and oxides from microprobe analyses, using stoichiometric criteria. *Mineralogical Magazine*, **51**, 431–435.

EALES, V. & REYNOLDS, I. M. 1986. Cryptic variations within chromitites of the upper critical zone, northwestern Bushveld complex. *Economic Geology*, **81**, 1056–1066.

ESCHER, J. C. & MYERS, J. S. 1975. New evidence concerning the original relationships of early Precambrian volcanics and anorthosites in the Fiskenæsset region, southern west Greenland. *Rapport Grønlands Geologiske Undersøgelse*, **75**, 72–76.

GARDE, A. A. 2007. A mid-Archaean island arc complex in the eastern Akia terrain, Godthåbsfjord, southern West Greenland. *Journal of the Geological Society, London*, **164**, 565–579.

GHISLER, M. 1970. Pre-metamorphic folded chromite deposits of stratiform type in the early Precambrian of West Greenland. *Mineralium Deposita*, **5**, 223–236.

GHISLER, M. 1976. *The Geology, Mineralogy and Geochemistry of the Pre-orogenic Archaean Stratiform Chromite Deposits at Fiskenæsset, West Greenland*. Monograph Series on Mineral Deposits **14**.

GHISLER, M. & WINDLEY, B. F. 1967. The chromite deposits of the Fiskenæsset region, West Greenland. Grønlands Geologiske Undersøgelse Rapport, **12**.

GROVES, T. L., ELKINS-TANTON, L. T., PARMAN, S. W., CHATTERJEE, N., MUNTENER, O. & GAETANI, G. A. 2003. Fractional crystallization and mantle-melting controls on calcalkaline differentiation trends. *Contributions to Mineralogy and Petrology*, **145**, 515–533.

HEATH, E., MACDONALD, R., BELKIN, H., HAWKESWORTH, C. & SIGURDSSON, H. 1998. Magmagenesis at Soufrière Volcano, St Vincent, Lesser Antilles Arc. *Journal of Petrology*, **39**, 1721–1764.

HENDERSON, P. 1984. Comment on 'Geochemistry and petrogenesis of the Fiskenæsset anorthosite complex, southern West Greenland: Nature of the parent magma' by B. L. Weaver, J. Tarney and B. Windley. *Geochimica et Cosmochimica Acta*, **48**, 415–416.

HENDERSON, P., FISHLOCK, S., LAUL, J. C., COOPER, T. D., CONARD, R. L., BOYNTON, W. V. & SCHMITT, R. A. 1976. Rare earth element abundances in rocks and minerals from the Fiskenæsset complex, West Greenland. *Earth and Planetary Science Letters*, **30**, 37–49.

HOLLAND, T. & BLUNDY, J. 1994. Non-ideal interactions in calcic amphiboles and their bearing on amphibole–plagioclase thermometry. *Contributions to Mineralogy and Petrology*, **116**, 433–447.

KAMENETSKY, V. S., CRAWFORD, A. J. & MEFFRE, S. 2001. Factors controlling chemistry of magmatic spinel: an empirical study of associated olivine, Cr-spinel and melt inclusions from primitive rocks. *Journal of Petrology*, **42**, 655–671.

LEAKE, B., WOOLLEY, A. R. ET AL. 1997. Nomenclature of amphiboles; Report of the Subcommittee on Amphiboles of the International Mineralogical Association Commission on New Minerals and Mineral Names. *Mineralogical Magazine*, **61**, 295–321.

LE MÉE, L., GIRARDEAU, J. & MONNIER, C. 2004. Mantle segmentation along the Oman ophiolite fossil mid-ocean ridge. *Nature*, **432**, 167–172.

MATVEEV, S. & BALLHAUS, C. 2002. Role of water in the origin of podiform chromitite deposits. *Earth and Planetary Science Letters*, **203**, 235–243.

MYERS, J. S. 1975. *Igneous stratigraphy of Archaean anorthosite at Majorqap qava, near Fiskenæsset, South-West Greenland*. Grønlands Geologiske Undersøgelse Rapport, **74**.

MYERS, J. S. 1976a. Channel deposits of peridotite, gabbro and chromitite from turbidity currents in the stratiform Fiskenæsset anorthosite complex, southwest Greenland. *Lithos*, **9**, 281–291.

MYERS, J. S. 1976b. Stratigraphy of the Fiskenæsset anorthosite complex, southern West Greenland, and comparison with the Bushveld and Stillwater complexes. *Rapport Grønlands Geologiske Undersøgelse*, **80**, 87–92.

MYERS, J. S. 1985. *Stratigraphy and structure of the Fiskenæsset Complex, West Greenland*. Grønland Geologiske Undersøgelse Bulletin, **150**.

MYERS, J. S. & PLATT, R. G. 1977. Mineral chemistry of layered Archaean anorthosite at Majorqap qava, near Fiskenæsset, southwest Greenland. *Lithos*, **10**, 59–72.

O'HARA, M. J. 2000. Flood basalts, basalt floods or topless Bushvelds? Lunar petrogenesis revisited. *Journal of Petrology*, **41**, 1545–1651.

PHINNEY, W. C. & MORRISON, D. A. 1990. Partition coefficients for calcic plagioclase: implications for Archaean anorthosites. *Geochimica et Cosmochimica Acta*, **54**, 1639–1654.

PHINNEY, W. C., MORRISON, D. A. & MACZUGA, D. E. 1988. Anorthosites and related megacrystic units in the evolution of Archaean crust. *Journal of Petrology*, **29**, 1283–1323.

POLAT, A., FREI, R., APPEL, P. W. U., DILEK, Y., FRYER, B., ORDONEZ-CALDERON, J. C. & YANG, Z. 2008. The origin and compositions of Meso-Archaean oceanic crust: evidence from the 3075 Ma Ivisaartoq greenstone belt, S.W. Greenland. *Lithos*, **100**, 293–321.

RIGHTER, K., CHESLEY, J. T., CAIAZZA, C. M., GIBSON, E. K., JR & RUIZ, J. 2008. Re and Os concentrations in arc basalts: the roles of volatility and source region fO_2 variations. *Geochimica et Cosmochimica Acta*, **72**, 926–947.

ROLLINSON, H. R. 1995. The relationship between chromite chemistry and the tectonic setting of Archaean ultramafic rocks. *In*: BLENKINSOP, T. G. & TROMP, P. L. (eds) *Sub-Saharan Economic Geology*. Geological Society of Zimbabwe, Special Publication, **3**, 7–23.

ROLLINSON, H. R. 1997. The Archaean komatiite-related Inyala chromitite, southern Zimbabwe. *Economic Geology*, **92**, 98–107.

ROLLINSON, H. R. 2007. *Early Earth Systems: a Geochemical Approach*. Blackwell, Oxford.

ROLLINSON, H. R. 2008. The geochemistry of mantle chromitites from the northern part of the Oman ophiolite – inferred parental melt compositions. *Contributions to Mineralogy and Petrology*, **156**, 273–288. doi:10.1007/s00410-008-0284-2.

ROLLINSON, H. R., APPEL, P. W. U. & FREI, R. 2002. A metamorphosed, early Archaean chromitite from west Greenland: implications for the genesis of Archaean anorthositic chromites. *Journal of Petrology*, **43**, 2143–2170.

SCHIANO, P., CLOCCHIATTI, R., LORAND, J.-P., MASSARE, D., DELOULE, E. & CHAUSSIDON, M. 1997. Primitive basaltic melts included in podiform chromites from the Oman ophiolite. *Earth and Planetary Science Letters*, **146**, 489–497.

SISSON, T. W. & GROVE, T. L. 1993. Experimental investigations of the role of H_2O in calc-alkaline differentiation and subduction zone magmatism. *Contributions to Mineralogy and Petrology*, **113**, 143–166.

STEELE, I. M., BISHOP, F. C., SMITH, J. V. & WINDLEY, B. F. 1977. The Fiskenæsset complex, West Greenland, Part III. Chemistry of silicates and oxide minerals from oxide bearing rocks, mostly from Qeqertarssuatsiaq. Bulletin, Grønlands Geologiske Undersøgelse, **124**.

STEWART, R. B., PRICE, R. C. & SMITH, I. E. M. 1996. Evolution of high-K arc magma, Egmont volcano, Taranaki, New Zealand: evidence from mineral chemistry. *Journal of Volcanology and Geothermal Research*, **74**, 275–295.

STOWE, C. W. 1987. Chromite deposits of the Shurugwe greenstone belt, Zimbabwe. *In*: STOWE, C. W. (ed.) *Evolution of Chromium Ore Fields*. Van Nostrand Reinhold, New York, 71–88.

TAKAGI, D., SATO, H. & NAKAGAWA, M. 2005. Experimental study of low alkali tholeiite at 1–5 kbar: optimal condition for the crystallisation of high-An plagioclase in hydrous arc tholeiite. *Contributions to Mineralogy and Petrology*, **149**, 527–540.

TAYLOR, G. J. 2009. Ancient lunar crust: origin, composition and implications. *Elements*, **5**, 17–22.

WEAVER, B. L., TARNEY, J. & WINDLEY, B. 1981. Geochemistry and petrogenesis of the Fiskenæsset anorthosite complex, southern West Greenland: nature of the parent magma. *Geochimica et Cosmochimica Acta*, **45**, 771–725.

WEAVER, B. L., TARNEY, J., WINDLEY, B. F. & LEAKE, B. E. 1982. Geochemistry and petrogenesis of Archaean metavolcanic amphibolites from Fiskenæsset, S. W. Greenland. *Geochimica et Cosmochimica Acta*, **46**, 2203–2215.

WINDLEY, B. F. & GARDE, A. A. 2009. Arc generated blocks with crustal sections in the North Atlantic Craton of west Greenland: crustal growth in the Archaean with modern analogues. *Earth-Science Reviews*, **93**, 1–30.

WINDLEY, B. F. & SMITH, J. V. 1974. The Fiskenæsset complex, west Greenland, Part II. General mineral chemistry from Qeqertarssuatsiaq. Bulletin, Grønlands Geologiske Undersøgelse, **108**.

WINDLEY, B. F., HERD, R. K. & BOWDEN, A. A. 1973. *The Fiskenæsset Complex, West Greenland Part 1. A Preliminary Study of the Stratigraphy, Petrology and Whole Rock Chemistry from Qeqertarssuatsiaq*. Geological Survey of Greenland, Bulletin, **106**.

WINDLEY, B. F., BISHOP, F. C. & SMITH, J. S. 1981. Metamorphosed layered igneous complexes in Archean granulite–gneiss belts. *Annual Review of Earth and Planetary Sciences*, **9**, 175–198.

A buried Palaeoproterozoic spreading ridge in the northern Nagssugtoqidian orogen, West Greenland

ADAM A. GARDE[1]* & JULIE A. HOLLIS[1,2]

[1]*Geological Survey of Denmark and Greenland, Øster Voldgade 10, Copenhagen-K 1350, Denmark*

[2]*Present address: Northern Territory Geological Survey, PO Box 3000, Darwin, NT 0801, Australia*

Corresponding author (e-mail: aag@geus.dk)

Abstract: Ophiolitic rocks on two small island groups in the northern Nagssugtoqidian orogen, West Greenland, shed new light on the eastern Laurentian Palaeoproterozoic plate-tectonic collage. The islands expose amphibolite-facies (520–550 °C, 2.6–3.0 kbar) tholeiitic pillow lava, black chloritic shale, manganiferous banded iron formation (BIF), podded chert, jasper, graded andalusite–staurolite schist with numerous sills, and terrigenous sandstone, an association that occurs today in oceanic spreading zones undergoing burial in forearc trenches. Deformation is weak, with upright folds preserving original younging. U–Pb detrital zircon ages indicate derivation from Archaean, 1890 Ma, and 1850 Ma sources. Occurring north of the Archaean Aasiaat domain that partly escaped Nagssugtoqidian deformation, the sequence pinpoints a new SSE-dipping Palaeoproterozoic subduction zone that links up in the NE with the Disko Bugt suture at the Nagssugtoqidian–Rinkian boundary. This solves an enigma of the previous plate-tectonic model, in which the Arfersiorfik magmatic arc in the central Nagssugtoqidian orogen was located within its own suture zone, and helps to explain why the Nagssugtoqidian–Rinkian system is so broad. It also explains previously recognized geochemical and Pb-isotopic compositional breaks in orthogneisses in the Kangaatsiaq region, and greatly simplifies correlation with adjacent eastern Canadian orogens in Baffin Island and Labrador.

It has long been recognized that plate tectonics have governed the geological evolution of the Earth at least as far back as the Palaeoproterozoic, and probably also during most or all of the Archaean era. Remnants of Palaeoproterozoic and Archaean ophiolites in variable states of preservation (but nevertheless indicative of plate boundaries) are known from many parts of the Earth (e.g. Kusky 2004), and there is also growing evidence that the bulk of the orthogneisses of Precambrian cratons was accreted by arcs, which were oceanic at first and subsequently turned into the convergent margins of growing microcontinents (e.g. Windley & Garde 2009). However, it is an entirely different matter to understand the original crustal architecture of these ancient orogens, which commonly expose only their deeply eroded, intensely deformed and metamorphosed roots. The rarity of well-preserved upper-crustal sections of these ancient orogens hampers efforts to reliably relate processes at work in the deep, old crust with upper-crustal arc and collisional plate-tectonic processes such as those operating today.

This paper addresses the plate-tectonic architecture in the northern part of the Palaeoproterozoic Nagssugtoqidian orogen in West Greenland and its importance for correlation with contemporaneous orogens in eastern Canada. Together with the adjacent Rinkian fold belt the Nagssugtoqidian orogen forms the middle of three southward-younging Palaeoproterozoic orogenic systems, which were successively developed in North-West, Central West and South Greenland, respectively, from about 1.95 to 1.8 Ga (Fig. 1a; Lahtinen *et al.* 2008). Following rifting of the Rae and North Atlantic cratons in western Greenland and eastern Canada in the early part of the Palaeoproterozoic, the oldest Inglefield mobile belt (c. 1.95 Ga) developed on the northern margin of the Rae craton. At around 1.85 Ga, collision of the Greenland parts of the Rae and North Atlantic cratons resulted in the Rinkian fold belt and Nagssugtoqidian orogen, which are located adjacent to each other in West Greenland and preserve only little juvenile Palaeoproterozoic crust. Finally, rapid northeastward subduction between 1.85 and 1.8 Ga beneath the southern margin of the North Atlantic craton produced the convergent Ketilidian orogen with a major juvenile, continental arc (the Julianehåb batholith) as well as a slightly younger forearc system and eventually crustal melt granites aged c. 1.75–1.72 Ga on the outboard side of the orogen.

From: KUSKY, T. M., ZHAI, M.-G. & XIAO, W. (eds) *The Evolving Continents: Understanding Processes of Continental Growth.* Geological Society, London, Special Publications, **338**, 213–234.
DOI: 10.1144/SP338.11   0305-8719/10/$15.00 © The Geological Society of London 2010.

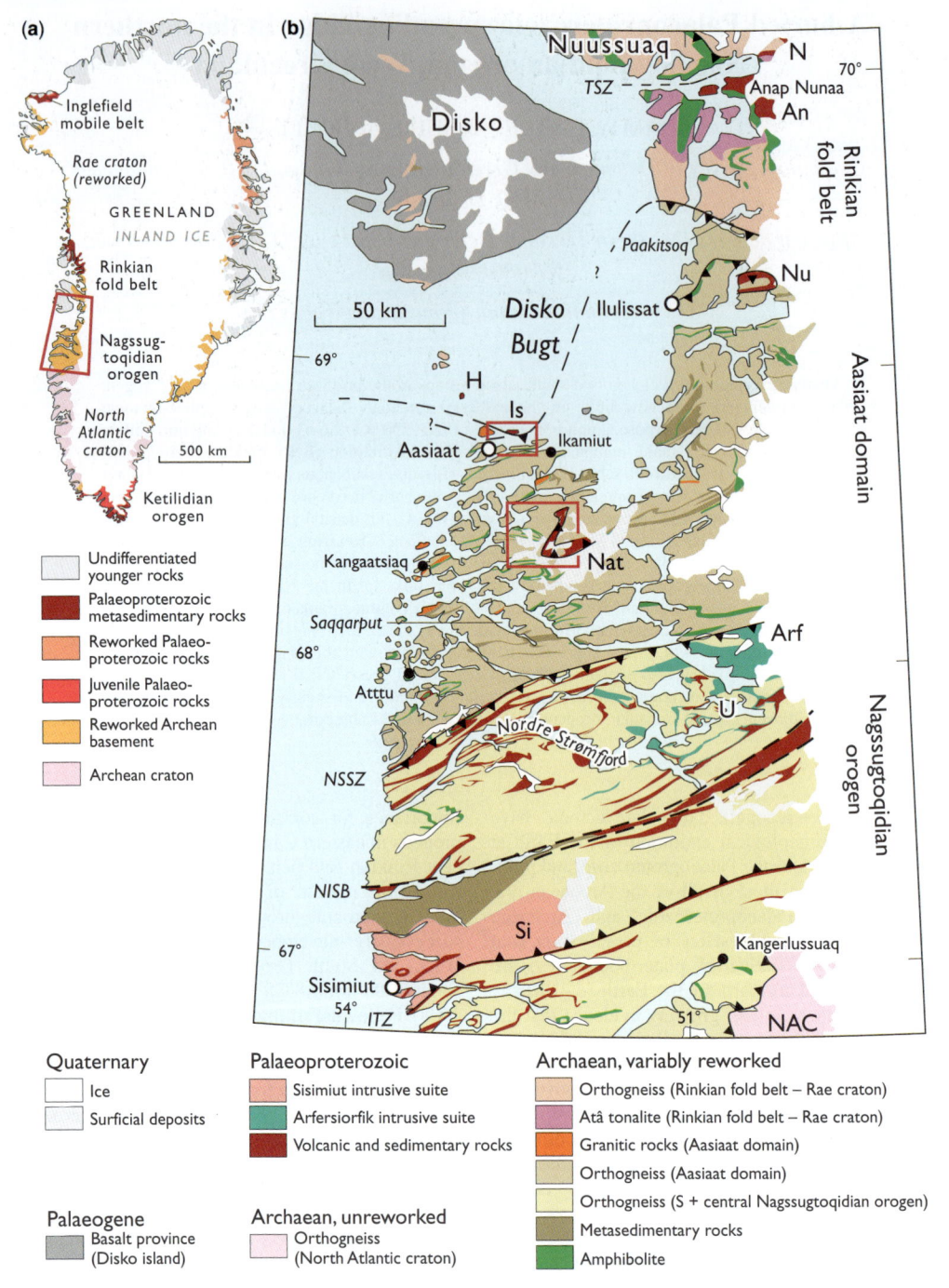

Fig. 1. (**a**) Index map showing the main Archaean and Palaeoproterozoic components of Greenland with the three Palaeoproterozoic belts in the west and south mentioned in the introduction; the red frame indicates the location of (b). The relationship between Palaeoproterozoic rocks in West Greenland and those in northern East Greenland (reworked by the Caledonian orogeny) is unknown. (**b**) Simplified geological and tectonic map of the Nagssugtoqidian orogen and the southernmost part of the Rinkian fold belt. Red frames show the positions of local maps in Figures 3 and 4. NAC, North Atlantic craton. Tectonic units: ITZ, Ikertôq thrust zone; NISB, Nordre Isortoq steep belt; NSSZ, Nordre Strømfjord shear zone; TSZ, Torsukattak shear zone. Palaeoproterozoic lithological units: An, Anap nunâ Group; Arf,

In this paper we describe a rare suite of well-preserved upper-crustal rocks in the northern part of the deeply eroded Nagssugtoqidian orogen, which we interpret as an accretionary wedge of volcanic, exhalative and sedimentary rocks from an ocean-floor–arc-trench setting. The recognition of this important association sheds new light on the complex Rinkian–Nagssugtoqidien transition zone in central West Greenland and not least on the correlation of this region with several Palaeoproterozoic orogens in eastern Canada. This correlation has also recently been addressed by Connelly et al. (2006) and St.-Onge et al. (2009).

The Nagssugtoqidian orogen and Rinkian fold belt in West Greenland

Both the Nagssugtoqidian orogen, which is exposed at upper amphibolite and granulite facies, and the adjacent Rinkian fold belt in central–northern West Greenland are major Palaeoproterozoic collisional orogens with counterparts in eastern Canada. They both comprise extensive reworking of Mesoarchaean continental crust, and they overlap significantly in time (Connelly et al. 2006). The southern front of the Nagssugtoqidian orogen was originally described by Ramberg (1949), who recognized the northward-increasing deformation of a major dyke swarm exposed along the NE–SW-trending, 150 km long Kangerlussuaq fjord at the northern margin of the North Atlantic craton (Fig. 1b). The orogen was originally described as a Palaeoproterozoic deformation zone within Archaean rocks, based on mapping and research projects in the 1960s and 1970s, mostly in its southern and central parts (e.g. Henderson 1969; Escher et al. 1976; Watterson 1978; Korstgaard 1979). Subsequent work around Nordre Strømfjord in the centre of the orogen showed that it also comprises clastic and metavolcanic supracrustal rocks of pre-Nagssugtoqidian Proterozoic age and remnants of one or more juvenile Palaeoproterozoic magmatic arcs, and Kalsbeek et al. (1987) proposed the existence of a central hidden suture.

The first plate-tectonic model of the Nagssugtoqidian orogen

Following additional studies by the Danish Lithosphere Centre in the 1990s Connelly et al. (2000) and van Gool et al. (2002) divided the orogen into a northern, central and southern part, separated from each other by steep, linear high-strain zones (Fig. 1). Remnants of two Palaeoproterozoic magmatic arcs were recognized in the NE and SW of the central, most high-grade part, namely the Arfersiorfik and Sisimiut intrusive suites (Fig. 1b; intrusive ages between 1921 and 1885 Ma, and 1921 and 1873 Ma, respectively; see e.g. Kalsbeek et al. 1987; Connelly et al. 2000; data of J. Hollis cited by van Gool & Marker 2007). Contemporaneous supracrustal rocks were also described, mainly mafic amphibolites and mica schists with minor slivers of marble and olivine-rich ultrabasic rocks. The plate-tectonic architecture of the orogen was interpreted in terms of a northern and a southern continent that were initially separated by a small ocean in the central part of the orogen, and it was proposed that a composite south-dipping suture was established during the closure of the ocean (Fig. 2). According to van Gool et al. (2002) and van Gool & Marker (2007) the south-dipping subduction zone gave rise to both of the magmatic arcs. Those workers also proposed that the suture obtained its present, complex outcrop pattern during the final, collisional stage, as outlined by an imbricated and folded stack of Palaeoproterozoic supracrustal rocks and the Arfersiorfik intrusive suite collectively termed the Ussuit unit. However, this plate-tectonic model has two major problems.

The first problem is rooted within the central Nagssugtoqidian orogen itself, and is a geometrical misfit between the proposed subduction zone and the Arfersiorfik intrusive suite. The Sisimiut arc in the SW of the central part of the orogen is located some 50 km to the south of the suture zone; that is, with a position consistent with a south-dipping subduction zone in the middle of the central part, assuming significant collisional crustal shortening after the magmatic accretion of the arc (Fig. 2). However, the Arfersiorfik arc to the NE is located directly above the relict subduction zone supposed to have fed it. This is unsatisfactory from a simple geometric consideration, even after allowing for crustal shortening during the ensuing continent–continent collision. A typical modern magmatic arc is located up to about 300 km behind its correlated trench, corresponding to a subduction angle of around 25° and melt generation at c. 150 km depth.

A second problem of the model by van Gool et al. (2002) is its inability to provide a realistic

---

Fig. 1. (Continued) Arfersiorfik intrusive suite; H, Hunde Ejland; Is, volcanic–sedimentary rock association on Isuamiut–Qaqqarsuatsiaq and Equutiit Killiat islands; Nat, Naternaq supracrustal rocks; Nu, Nunatarsuaq supracrustal rocks; Si, Sisimiut intrusive suite; U, Ussuit unit. In the division of the Nagssugtoqidian orogen according to van Gool et al. (2002) and van Gool & Marker (2007), its northern, central and southern parts are separated by the NSSZ and ITZ, respectively.

1. Rifting   Kangâmiut dykes (2040 Ma)

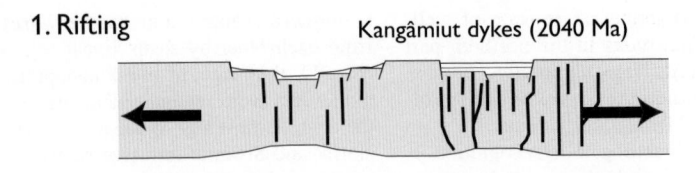

2. Drifting/sediment deposition (c. 2000–1920 Ma)

3. Subduction/calc-alkaline magmatism (c. 1920–1870 Ma)

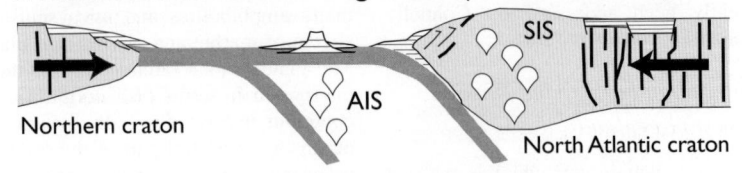

4. Peak metamorphism during collision (1860–1840 Ma)

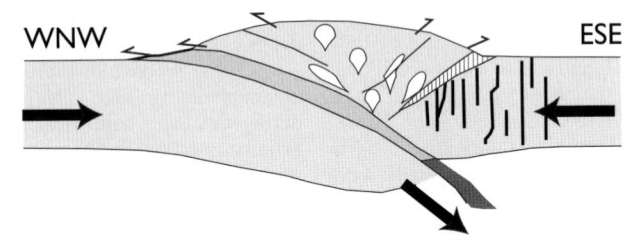

5. N–S contraction – large-scale folding (c. 1825 Ma)

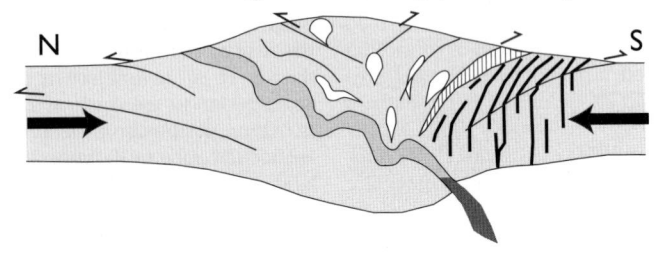

6. Sinistral movement in steep belts (c. 1775 Ma)

Fig. 2. The previous plate-tectonic model for the Nagssugtoqidian orogen with one composite south-dipping suture in its central part. Simplified from van Gool *et al.* (2002) and van Gool & Marker (2007).

match between the single south-dipping suture in central West Greenland with the two established palaeosutures in eastern Canada (namely, those in southern Baffin Island and northeastern Labrador; see St-Onge et al. 2009).

New observations in the northern Nagssugtoqidian orogen and their plate-tectonic significance

An undeformed Archaean domain within the northern Nagssugtoqidian orogen

Systematic mapping in the northern Nagssugtoqidian orogen in 2001–2003 and published as 1:100 000 scale geological map sheets (Garde 2004, 2006; van Gool 2005) uncovered some additional problems with the plate-tectonic model described in the previous section and shown in Figure 2. First of all, it was discovered that a large crustal block about 50 km wide immediately to the north of the Nordre Strømfjord shear zone, which consists of Archaean orthogneiss and metavolcanic belts, has altogether escaped Nagssugtoqidian deformation (the Aasiaat domain, Fig. 1b; St-Onge et al. 2009). The Aasiaat domain did record Palaeoproterozoic heating under amphibolite-facies conditions, and c. 1800–1750 Ma pegmatites were formed locally, but the granulite-facies metamorphism occurring in its southern part is Archaean in age. The unexpected absence of Nagssugtoqidian deformation and only moderate Palaeoproterozoic heating is based on (1) structural studies (Piazolo et al. 2004; Mazur et al. 2006), (2) the presence of metamorphosed but completely undeformed, east–west-trending mafic dykes of presumed Palaeoproterozoic age NE of the settlement Attu (see Garde 2004), and (3) U–Pb zircon geochronology of a well-preserved, synkinematic granite vein on the limb of a major fold belonging to the latest established fold phase at Saqqarput c. 30 km south of Kangaatsiaq. This yielded an Archaean age (2748 ± 19 Ma, Thrane & Connelly 2006), showing that the deformation took place in the Archaean.

Intensely deformed, sillimanite-grade Palaeoproterozoic metavolcanic and metasedimentary rocks at Naternaq in the northern Nagssugtoqidian orogen

The investigations in 2001–2003 included partial remapping of the Naternaq (Lersletten) supracrustal belt about 40 km SE of Aasiaat (Figs 1 & 3), although not all parts of the belt could be visited or studied in detail. The belt was previously considered to be of Archaean age, but detrital zircons from a fine-grained muscovite–sillimanite schist have yielded an ion microprobe U–Pb maximum depositional age of 1904 ± 8 Ma (Thrane & Connelly 2006). This age is compatible with derivation from a volcanic arc above the 1921–1885 Ma Arfersiorfik intrusive suite in the central Nagssugtoqidian orogen. Apart from local pegmatites and dykes, the Naternaq supracrustal belt is the only Palaeoproterozoic component known between the central Nagssugtoqidian orogen and Aasiaat. Its plate-tectonic significance is addressed in a later section.

The Naternaq supracrustal belt (Fig. 3) is up to c. 4 km wide and comprises about equal proportions of mafic, fine-grained, generally intensely deformed amphibolite and fine-grained mica ± garnet ± staurolite ± sillimanite schist locally with pseudomorphed andalusite, and small lenses of fine-grained siliceous and carbonate-rich rocks of possible chemical sedimentary origin. The southeastern part of the belt also comprises a sediment-hosted volcanogenic–exhalative pyrite–pyrrhotite deposit with associated fine-grained siliceous and sulphide-rich rocks, as well as a small occurrence of carbonate-facies banded iron formation that is likewise presumed to be of volcanogenic–exhalative origin (Østergaard et al. 2002).

The overall tectonic structure of the Naternaq belt can be described as an overturned syncline that has been refolded into a large V-shaped body. Its NNE–SSW-trending western limb forms a conspicuous tectonic discordance against the underlying Archaean orthogneiss and amphibolite, which trend east–west in this area (Fig. 3), although the contact itself is concealed by moraine. Where the contact is exposed in the hinge zone and on the southern limb, it is tectonic and displays intense deformation. Whereas the Archaean tonalitic orthogneisses and narrow metavolcanic belts structurally below the Naternaq belt are dominated by east–west-trending structures and generally flat-lying lineations and fold axes, the southern part of the Naternaq belt itself comprises a series of 100 m scale, upright to overturned SE-plunging folds that locally refold the belt and are accompanied by common SE-plunging mineral and extension lineations. Irregular bodies of tourmaline-bearing pegmatite are commonly associated with these late hinge zones and locally grade into coarse, leucocratic granite.

Amphibolite-facies (hemi)pelagic volcanic, exhalative and sedimentary rocks with Palaeoproterozoic terrigenous input in the northern Nagssugtoqidian orogen

Two groups of small islands about 10 km NE of Aasiaat, Isuamiut–Qaqqarsuatsiaq and Equutiit

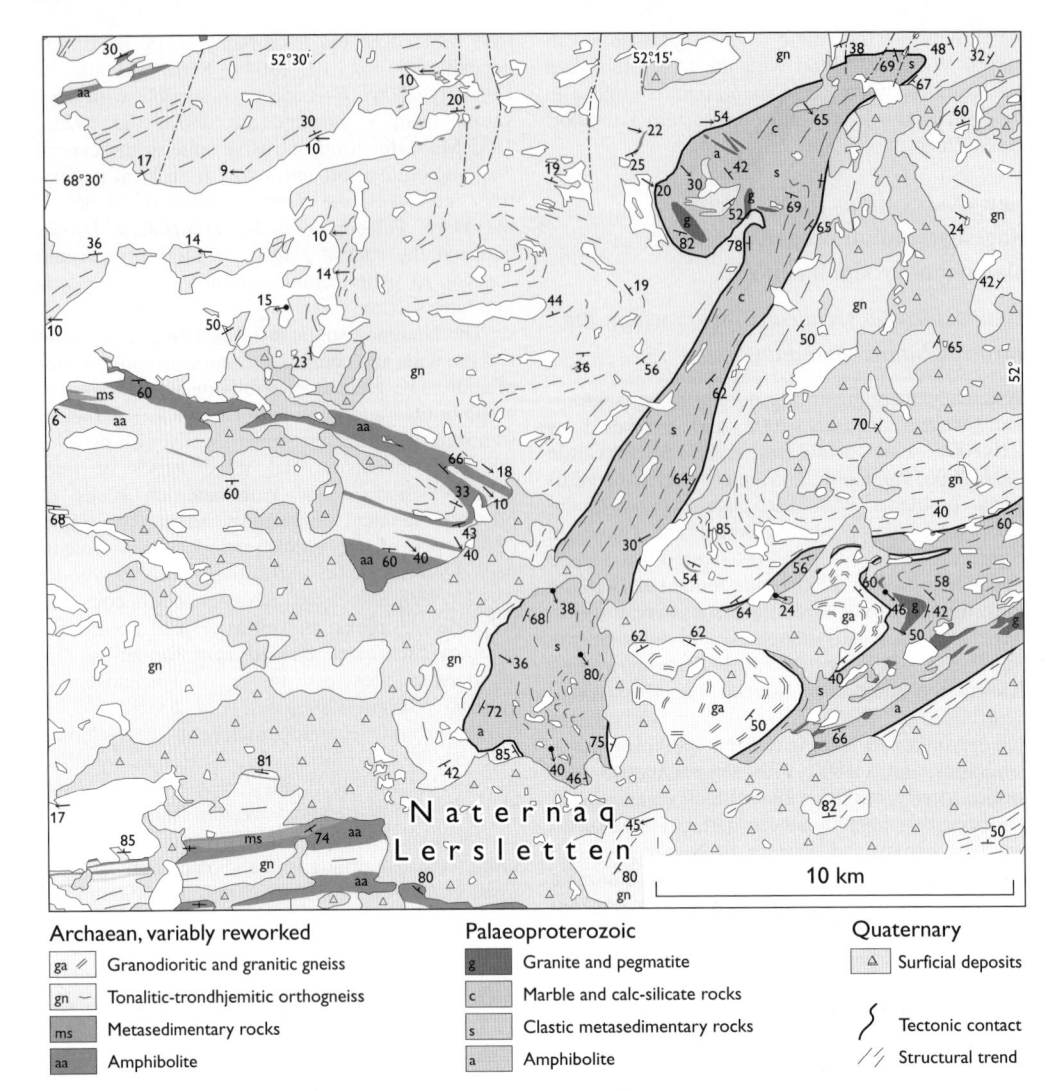

Fig. 3. Geological map of the Naternaq supracrustal belt and surroundings (from Garde 2004, 2006; position shown in Fig. 1b), displaying a tectonic discordance between the NNE–SSW-striking western part of the Palaeoproterozoic Naternaq supracrustal belt and the underlying, east–west-striking Archaean basement rocks. The refolded syncline of the Naternaq supracrustal belt contains NE-plunging, upright to overturned fold structures and common SE-plunging mineral and extension lineations, which are absent from most of the surrounding Archaean basement.

Killiat, expose exceptionally well-preserved and in places virtually undeformed metavolcanic and metasedimentary rocks, metamorphosed at middle amphibolite facies (Fig. 4). The islands were mapped and studied in 1948–1949 by Ellitsgaard-Rasmussen (1954) and briefly visited again in 2003, when a few samples of representative rocks types were collected (Garde et al. 2004). Similar rocks occur on Hunde Ejland, an isolated group of small islands c. 20 km north of Aasiaat (Fig. 1b).

The original map by Ellitsgaard-Rasmussen (Fig. 5) displays a regular pattern of alternating, folded and metamorphosed volcanic and sedimentary rocks, and shows that the two groups of islands comprise different rock associations, although their origin was not addressed. The Isuamiut–Qaqqarsuatsiaq islands are dominated by fine-grained mafic pillow lava and sills (Fig. 6a) that typically comprise massive actinolite–albite–epidote ± garnet. The sills preserve chilled margin contacts with the chloritic schists, grading into more massive medium-grained actinolite–albite rocks that dominate the outcrops of mafic rocks. The other main component is dark,

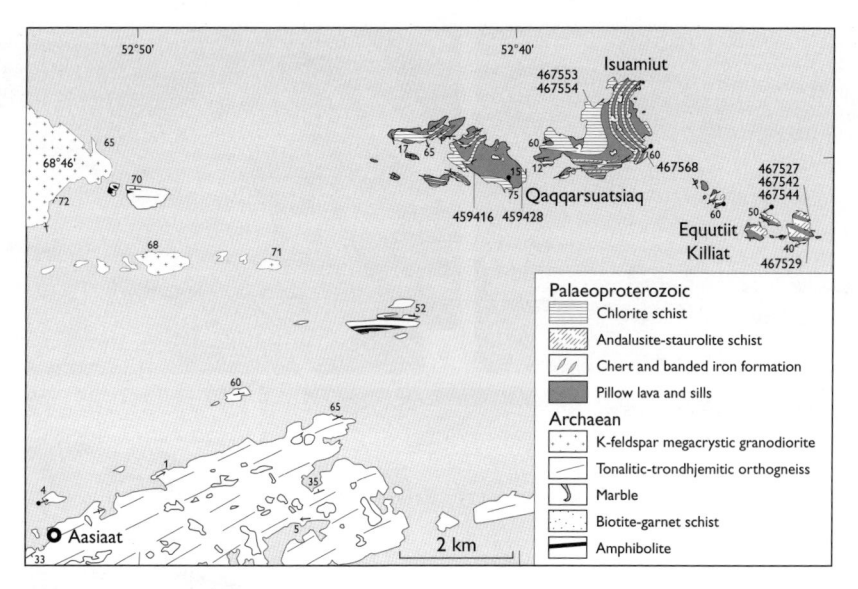

Fig. 4. Geological map of the Isuamiut–Qaqqarsuatsiaq–Equutiit Killiat islands exposing Palaeoproterozoic (meta)volcanic and sedimentary rocks NE of Aasiaat and adjacent Archaean basement rocks on other islands (see Fig. 1b for location). Modified from Garde (2006). The Palaeoproterozoic rocks were originally mapped by Ellitsgaard-Rasmussen (1954).

fine-grained, relatively siliceous garnet–biotite–chlorite ± muscovite schist with very fine-grained disseminated graphite (Fig. 6b). Local beds of quartz-rich sandstone occur within the dark chloritic schist and typically range from a few centimetres to a few decimetres in thickness (Fig. 6b). A sandstone unit up to 10 m thick was observed at the western end of the island NW of Qaqqarsuatsiaq. Some of the coarsest chloritic schists locally preserve centimetre-scale graded bedding and ripple marks (Fig. 6c). In some cases the schists are interlayered on a scale of less than 1 m with finely laminated, fine- to medium-grained actinolite–albite ± garnet siliceous rocks that we interpret as metamorphosed tuffs. Other important lithological components are repeated, elongate lenses of well-preserved (meta-)chert and jasper (Fig. 6d), manganese-rich banded iron formation (Fig. 6e), and iron-rich carbonate rocks with actinolite and locally diopside.

The Equutiit Killiat islands predominantly consist of aluminous andalusite–staurolite ± garnet schists with local ripple marks (Fig. 6f) and more or less ubiquitous, well-preserved graded bedding (Fig. 6g) or locally finely laminated beds (Fig. 6j), intruded by numerous fine-grained mafic sills up to a few metres in thickness (Fig. 6h, k).

Primary sedimentary and volcanic structures on both groups of islands, such as graded bedding at

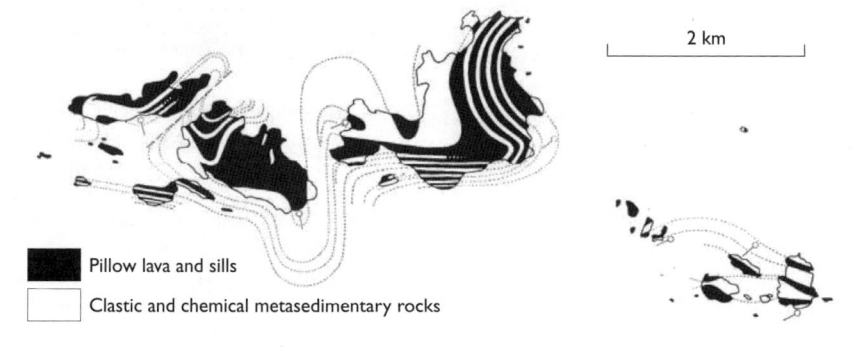

Fig. 5. Ellitsgaard-Rasmussen's (1954) simplified structural map of the Isuamiut–Qaqqarsuatsiaq–Equutiit Killiat islands NE of Aasiaat, displaying alternating igneous and sedimentary units and simple folds. Scale and labelling inserted by the present authors.

Fig. 6. Distinctive ocean-floor to arc-trench assemblages at middle amphibolite facies NE of Aasiaat. Ocean-floor pillow lavas and (hemi)pelagic sediments, chert and banded iron formation on the Isuamiut–Qaqqarsuatsiaq islands (**a–e**), and terrigenous graded and laminated clastic deposits with numerous sills on the Equutiit Killiat islands, interpreted as representing a buried spreading ridge (**f–h, i, j**). (a) Pillow lava, northern Isuamiut; (b) fine-grained

Equutiit Killiat, truncated ripple laminations on Qaqqarsuatsiaq and volcanic pillows on Isuamiut, consistently display structurally upward younging directions. The rocks generally possess a weak to moderate cleavage axial planar to one set of upright to overturned, south-plunging folds on scales ranging from metres to hundreds of metres. South of the Isuamiut–Qaqqarsuatsiaq–Equutiit Killiat islands there is a strong southward increase of the Nagssugtoqidian strain, which can be observed on small orthogneiss islands. The orthogneisses at Aasiaat still farther south display an intense, vertical, roughly east–west-trending foliation and rootless isoclinal folds (Fig. 4; Garde et al. 2004).

Mafic igneous and sedimentary rocks with probable VMS-style mineralization on Hunde Ejland–Kitsissuarsuit, c. 15 km NNW of Aasiaat

The island group of Hunde Ejland c. 15 km NNW of Aasiaat (Fig. 4) is underlain by amphibolite-facies, low-strain, very fine-grained volcanic and hypabyssal mafic rocks as well as aluminous and siliceous clastic rocks that resemble the rock associations found on the Isuamiut–Qaqqarsuatsiaq–Equutiit Killiat islands (J. A. M. van Gool, pers. comm.). These rocks have not been investigated in detail and remain undated, but it appears plausible that they may be of Palaeoproterozoic age and are related to the former rocks. In this context it is worth noticing that several sulphidic, iron-rich quartz vein samples with probable VMS-style Cu–Zn–Au–Ag–Hg–Se mineralization have been found by a local inhabitant in 2004 during the 'Ujarassiorit' mineral hunt programme organized by the Greenlandic Home Rule government. A chemical analysis by induced neutron activation of one of these samples from Activation Laboratories Inc., Canada, is included in Table 1. This type of mineralization may be found today in 'black smokers' on the ocean floor and has also been reported from the Mesoproterozoic of North China (Li & Kusky 2007). In particular, spreading ridges undergoing burial form favourable sites of mineralization, because the sedimentary overburden makes the entrapment of mineralizing fluids more effective (Mottl et al. 1994; M. Hannington, pers. comm.). More work on Hunde Ejland is obviously required to substantiate the age, setting and significance of the mineralization and its host rocks.

Metamorphism

In contrast to the garnet–chlorite–biotite–muscovite schists on Isuamiut and Qaqqarsuatsiaq, the detrital rocks on Equutiit Killiat c. 3 km to the ESE are staurolite–muscovite–biotite $\pm$ chlorite schists, locally with garnet or andalusite and rarely both. In all cases, garnet, staurolite and andalusite porphyroblasts, where present, overgrow the primary bedding structures (e.g. graded bedding) and are unoriented in relation to fold structures, although they are wrapped by and intergrown with muscovite and minor biotite in an S_0-parallel fabric. Locally, staurolite is wrapped by thin rims of chlorite in andalusite-bearing schists, consistent with limited post-peak retrogression below the temperature stability field of staurolite.

The undeformed or only weakly deformed texture of both mafic and metasedimentary rocks, with metamorphic mineral assemblages overgrowing both bedding and local ductile structures, indicates that these rocks were contact metamorphosed (see also Ellitsgaard-Rasmussen 1954). Based on the observed intrusive relationships the likely heat source was the emplacement of the mafic sills themselves and perhaps associated deeper-level intrusive rocks.

A staurolite–garnet–andalusite–muscovite–chlorite–biotite–plagioclase–quartz schist (sample 467542) from Equutiit Killiat was studied to constrain the conditions of metamorphism. Across the island the same lithology shows variations in this mineral assemblage including garnet-absent, andalusite-absent and chlorite-absent varieties. Andalusite porphyroblasts 2–5 mm in size are wrapped by the muscovite-rich foliated matrix, which also comprises plagioclase, quartz and minor chlorite. Staurolite and garnet porphyroblasts 3–10 mm in size are intergrown with, but not wrapped by the fabric, and thus interpreted as formed subsequently to andalusite growth but still forming a stable peak assemblage with the latter. The whole-rock composition (see caption to Fig. 7) was modelled in the system MnNCKFMASH (MnO–NaO–CaO–K_2O–FeO–MgO–Al_2O_3–SiO_2–H_2O), on the basis of excess fluid (H_2O), using the THERMOCALC internally

---

Fig. 6. (*Continued*) chloritic schist with graded sandstone bed, south Qaqqarsuatsiaq; (c) climbing ripple marks in chloritic schist, SW Qaqqarsuatsiaq; (d, e) chert and manganiferous banded iron formation, south Qaqqarsuatsiaq; (f) climbing ripple marks in staurolite schist, Equutiit Killiat; (g) andalusite–staurolite schist with repeated graded bedding and an upright fold, western Equutiit Killiat; arrows mark andalusite porphyroblasts up to 5 cm large in bedtops; (h) deformed mafic sill in andalusite–staurolite schist; (i) finely laminated staurolite schist; (j) intrusive contact between underside of sill and finely laminated staurolite schist, all NE Equutiit Killiat. Coin for scale is 2.8 cm across.

Table 1. Chemical compositions of Palaeoproterozoic volcanic and sedimentary rocks from the Isuamiut–Qaqqarsiatsiaq–Equtiit Killiat islands, and a mineralized vein from Hunde Ejland

Location:	Isuamiut–Qaqqarsiatsiaq					Equtiit Killiat			Hunde Ejland
Northing:	68°45′.968	68°46′.531	68°45′.287	68°46′.531	68°45′.806	68°45′.275	68°45′.275	68°45′.464	68°45′.464	68°52′
Easting:	−52°41′.548	−52°37′729	−52°33′444	−52°37′729	−52°39′.847	−52°32′285	−52°32′285	−52°32′396	−52°32′396	−53°05′
Sample:	459416	467553	467568	467554	459428	467529	467527	467542	467544	2004-442
Rock:	Chloritic schist	Chert	BIF	Pillow lava	Sill	Sill	Staurolite schist	Staurolite schist	Staurolite schist	Quartz-rich vein
SiO$_2$	60.894	91.216	52.026	49.463	49.387	49.547	55.131	53.744	53.227	63.66
TiO$_2$	0.579	0.025	0.484	0.873	1.124	1.125	1.121	1.199	1.326	0.15
Al$_2$O$_3$	17.141	0.217	10.032	14.613	14.110	13.746	24.191	24.218	23.980	6.24
Fe$_2$O$_3$*	7.443				12.163			10.952	10.322	16.80
Fe$_2$O$_3$		2.075	12.226	1.513		1.260	2.686			0.01
FeO		0.690	7.370	8.290		10.590	4.980			0.07
MnO	0.947	0.086	3.243	0.126	0.280	0.196	0.082	0.091	0.034	0.80
MgO	2.212	0.159	4.360	8.597	7.869	7.514	2.085	2.609	1.293	3.05
CaO	1.450	2.481	4.853	11.968	10.166	10.698	0.874	0.315	0.340	0.17
Na$_2$O	1.270	0.010	0.380	2.070	2.820	2.800	2.090	0.840	2.750	0.09
K$_2$O	4.303	0.000	0.633	0.081	0.675	0.023	3.759	3.221	2.526	5.99
P$_2$O$_5$	0.154	0.003	0.238	0.066	0.084	0.082	0.079	0.093	0.105	
Volatiles	2.15	2.19	2.78	1.49	0.53	1.59	2.40	1.45	2.36	
Sum	98.54	99.15	98.63	99.15	99.21	99.17	99.48	98.73	98.26	97.07
Sc	17.7	0.6	10.8	48.5	28.6	43.1		28.1	26.4	1.9
V	78	8	83	313	297	311		157	103	10
Cr	44	2	63	385	89	51		98	133	102
Co	20.0	18.1	28.1	51.5	53.6	57.6		29.7	34.3	21.8
Ni	35	8	26	148	83	88		70	82	13
Cu	24.3	40.6	49.8	128.3	46.4	180.6		26.0	3.8	17 800
Zn	64.5	8.8	76.8	62.7	86.8	78.3		76.4	59.2	17 400

Ga	22.3	0.5	13.2	15.8	18.3	16.9	29.4	28.1
Rb	118.0	0.1	30.0	0.5	3.1	0.6	125.9	92.4
Sr	62	4	12	291	122	171	52	198
Y	12.9	4.5	33.7	18.8	20.7	19.8	14.2	11.1
Zr	188.6	6.6	147.7	16.0	26.2	34.5	133.3	119.8
Nb	11.0	0.3	14.2	2.2	4.9	4.0	12.8	13.0
Cs	4.4	0.0	2.3	0.0	0.2	0.0	3.1	1.9
Ba	776	1	99	32	23	35	833	511
La	10.3	5.2	26.2	2.0	5.8	4.5	46.0	35.1
Ce	29.4	11.8	54.5	6.0	14.3	12.0	92.0	70.1
Pr	2.70	1.40	6.86	1.11	2.02	1.93	11.54	8.58
Nd	9.5	5.8	25.3	5.8	9.6	9.3	39.2	29.8
Sm	2.02	1.09	5.41	1.99	2.87	2.81	7.23	5.65
Eu	0.39	0.26	1.26	0.81	0.96	0.94	1.28	1.42
Gd	2.47	1.14	6.24	2.43	3.23	3.01	6.76	5.28
Tb	0.38	0.15	0.98	0.45	0.55	0.55	0.80	0.63
Dy	2.32	0.78	6.05	3.18	3.24	3.53	3.63	2.86
Ho	0.47	0.14	1.16	0.70	0.72	0.72	0.61	0.44
Er	1.30	0.33	3.06	1.88	1.84	1.88	1.49	1.08
Tm	0.21	0.04	0.46	0.30	0.28	0.28	0.20	0.16
Yb	1.39	0.21	2.84	1.83	1.78	1.76	1.25	0.91
Lu	0.24	0.03	0.41	0.25	0.25	0.24	0.21	0.14
Hf	5.18	0.08	4.11	0.62	0.79	1.25	3.67	3.23
Ta	2.1	0.1	1.3	0.5	0.5	0.7	1.1	1.0
Pb	2.9	0.8	7.9	0.8	6.9	1.1	8.7	22.1
Th	10.1	0.2	5.6	0.2	0.8	0.7	9.9	7.2
U	2.1	0.2	2.0	0.1	0.2	0.2	2.0	1.1

Ga		3.9
Rb		0.2
Sr		2
Y		2.5
Zr		n.a.
Nb		0.1
La		4.1
Ce		9.0
Hf		0.08
Pb		370.0
Th		0.6
U		0.1
S %		3.9
Au ppb		128
Ag		22.6
As		4.4
Cd		30.4
Ge		0.43
Hg		7.29
In		1.79
Mo		8.30
Sb		0.52
Se		85.1
Sn		25.6
Te		1.68
W		2.31

Trace elements in ppm except S and Au. Position of sample 2004-442, Hunde Ejland is approximate. n.a., not analysed.

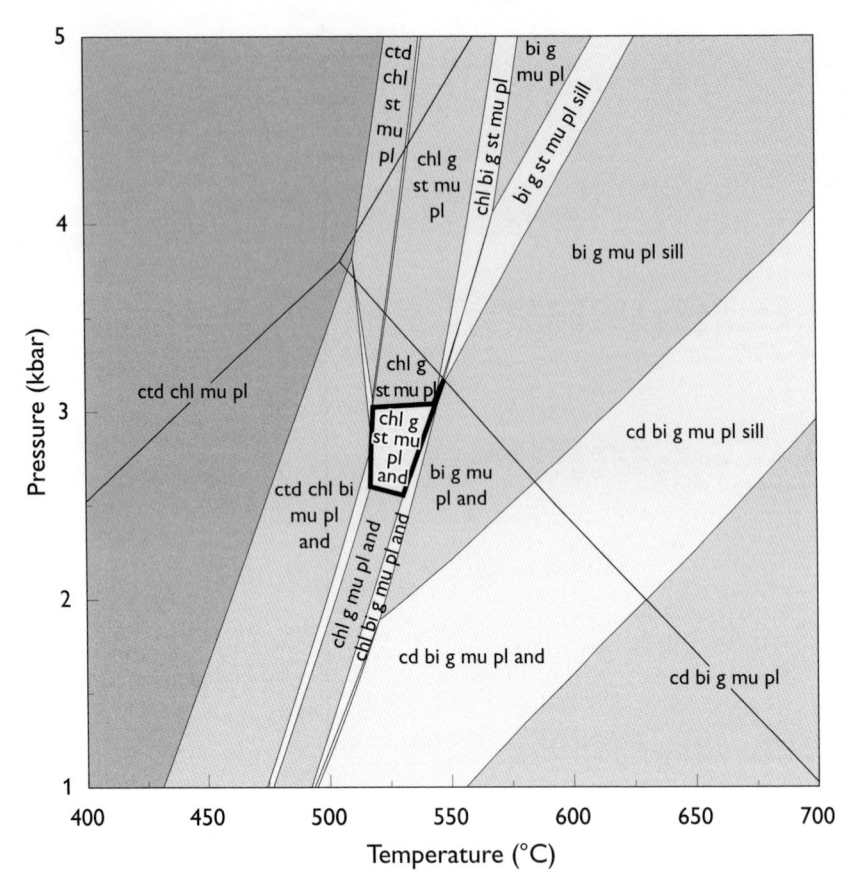

Fig. 7. Quantitative pseudosection constructed for a garnet–staurolite–andalusite–muscovite–chlorite–plagioclase–quartz schist from Equutiit Killiat (sample 467542; see Table 1 and Fig. 4) assuming excess H$_2$O. The sample has the modal composition (in wt%) MnO 0.09, Na$_2$O 1.01, CaO 0.41, K$_2$O 2.53, FeO 7.34, MgO 4.79, Al$_2$O$_3$ 17.59, SiO$_2$ 66.23. Quartz is present in all phase fields. The depth of greyscale indicates the variance of the phase fields, with white representing divariant fields and the darkest grey representing quintivariant fields. The *PT* stability field of the peak metamorphic assemblage is outlined in bold.

consistent thermodynamic dataset and computer program (Powell & Holland 1988; Holland & Powell 1998; and subsequent upgrades). Although Mn is only a minor constituent of the bulk-rock composition, it was found important to include this component as it significantly expands the stability fields of garnet and staurolite to lower pressures, which is particularly relevant in this case (see Wei *et al.* 2004). Given a peak metamorphic assemblage involving garnet–staurolite–andalusite–chlorite–muscovite–plagioclase–quartz, Figure 7 illustrates that this sample experienced peak metamorphic conditions in a narrow window in the range 2.6–3.0 kbar and 520–550 °C.

The apparent difference in metamorphic grade between Isuamiut–Qaqqarsuatsiaq (garnet–chlorite schists) and Equutiit Killiat (staurolite schists) can in fact be attributed to compositional differences. Quantitative modelling in the system MnNCKFMASH reveals that the lower Al contents of chloritic schists on Isuamiut and Qaqqarsuatsiaq push aluminosilicate stability up-*T* by *c.* 100 °C and staurolite stability up-*T* by *c.* 60 °C. Thus chloritic mineral assemblages, as observed, would be expected at similar *PT* conditions to those experienced by the more aluminous staurolite schists on Equutiit Killiat.

Geochemistry

Major and trace element compositions of amphibolite-facies pillow lava, sills, chert, banded iron formation and clastic sedimentary rocks are shown in Table 1, and rare earth element (REE) and multi-element diagrams in Figure 8. Major elements were determined by X-ray fluorescence (XRF; except Na, (which was determined by atomic absorption spectrometry), and trace elements by inductively coupled plasma mass spectrometry (ICP-MS) on fused glass discs (for analytical details see Kystol & Larsen 1999; Hollis et al. 2006). The pillow lava and sills are tholeiitic; their major element composition is very similar to the two wet chemical analyses of 'greenstones' by Ellitsgaard-Rasmussen (1954). Relative to normal mid-ocean ridge basalt (N-MORB) they have slightly lower concentrations of high field strength elements (HFSE), but they do not possess negative Nb–Ta anomalies. The pillow lava (sample 467554) has a slightly arched REE pattern at c. 5–8 times chondrite, whereas the two sills display minor enrichment in light rare earth elements (LREE; samples 459428 and 467529). The large ion lithophile elements (LILE) display a scatter, such as is very commonly observed in metamorphosed rocks. The absence of a negative Nb–Ta anomaly and the flat REE pattern of the pillow lava are considered diagnostic and suggest a rift environment on the ocean floor (i.e. MORB), although more analyses would be desirable to substantiate this. The small LREE enrichment in the two sills might, in isolation, indicate an arc tholeiitic origin, but may equally well be interpreted as being due to minor addition of these elements to a primitive tholeiite by sea-floor hydrothermal or metamorphic fluids.

The banded iron formation from Isuamiut (Fig. 4) is unmagnetic and rich in hematite. The analysed sample (467568) contains 12.23wt% Fe_2O_3, 7.37wt% FeO and as much as 3.24wt% MnO (Table 1), pointing to a volcanogenic–exhalative origin. Its REE composition is close to that of post-Archaean average shale (Fig. 8c; McLennan 1989), suggesting that it contains some detrital material. The chert from the same island has a very low content of major elements (except SiO_2) and trace elements, and REE contents of about 0.1 times post-Archaean average shale with a small positive Eu anomaly (sample 467553, Fig. 8c). Its REE composition resembles that of younger Palaeoproterozoic banded iron formation in the Black Hills, Dakota, USA (Frei et al. 2008), compatible with chemical precipitation on the sea-floor and adsorption of REE from seawater, as is also suggested by its texture and close association with the manganiferous banded iron formation.

The chloritic schist (sample 459416) from Qaqqarsuatsiaq is relatively siliceous and not nearly as aluminous as the turbidites on the Equutiit Killiat islands (see below). Its high content of MnO (0.95wt%), almost flat REE pattern and negative Eu anomaly are compatible with a mixture of

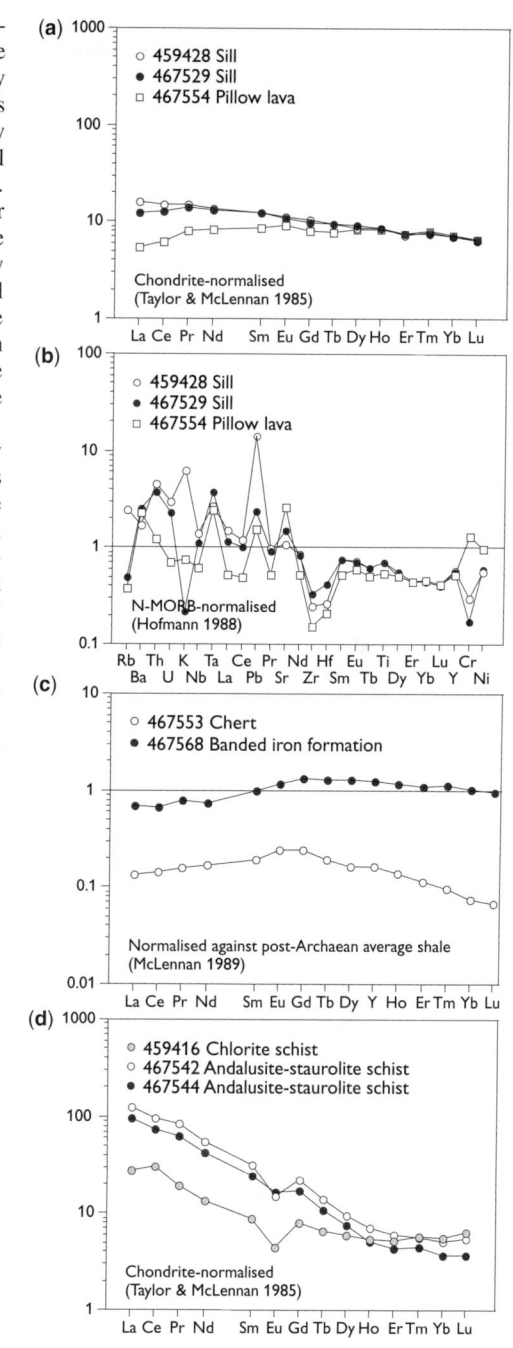

Fig. 8. Multi-element diagrams normalized against chondrite, MORB and post-Archaean average shale for various rocks from the Isuamiut–Qaqqarsuatsiaq–Equutiit Killiat islands. (Note the absence of Nb and Ta anomalies in the tholeiitic pillow lava and sills relative to MORB. Note also the low REE content and almost flat chondrite-normalized REE pattern of chloritic schist 459416, interpreted as a hemipelagic mudstone.)

sea-floor weathering products from the adjacent mafic igneous rocks, siliceous chemical sediment, and a small component of terrigenous detritus (as shown by the ages of its few zircon grains; see below). The small apparent positive Ce anomaly may be related to oxidation–reduction processes on the sea-floor (see, e.g. Frei et al. 2008).

Major element analyses of three samples of staurolite-bearing schists from Equutiit Killiat (samples 467527, 467542 and 467544, Table 1, Fig. 8b) document their aluminous character. In a chondrite-normalized REE diagram the last two rocks display about 100 times LREE enrichment, a clear negative Eu anomaly, and low heavy REE (HREE) contents (about 5 times chondrite) with an upward-concave pattern. Their REE composition is, thus, compatible with a derivation from weathered intermediate to felsic volcanic (and/or intrusive) rocks with relative loss of plagioclase, and derived from an arc environment. The samples contain around 125 ppm Zr, which may suggest a relatively small clastic sedimentary input to the sedimentary precursors. In summary, their aluminous, relatively low-silica composition coupled with their low HREE and low Zr contents are consistent with an interpretation as distal turbidites from intermediate volcanic and/or continental sources. These might either be tonalite–trondhjemite–granodiorite (TTG)-type Archaean continental crust (as suggested by the upward-concave HREE pattern) or a Palaeoproterozoic magmatic arc, or both.

Geochronology

Heavy mineral separations were performed on an andalusite–staurolite schist from Equutiit Killiat and a garnet–chlorite–biotite schist (sample 459416) from Qaqqarsuatsiaq. A very small number of variably rounded detrital zircon grains typically about 100 μm in length were extracted from the latter, whereas none were recovered from the andalusite–staurolite schist. The heavy mineral separates were obtained by crushing, sieving, washing and panning, followed by heavy liquid and magnetic separation. Zircons were hand-picked, mounted in epoxy resin, and polished through their mid-sections. The grains were mounted together with reference zircon 91500 (Ontario, Canada, weighted average $^{207}Pb/^{206}Pb$ age of 1065 Ma, Wiedenbeck et al. 1995). Zircon morphologies (Table 2) were identified using back-scattered electron imaging using a Philips XL 40 scanning electron microscope at GEUS, operating at 20 kV and a working distance of 10 mm. U–Pb zircon data were collected using a Cameca IMS 1270 secondary ion mass spectrometer at the NORDSIM laboratory, Swedish Museum of Natural History, Stockholm. Analytical procedures and common lead corrections are similar to those described by Whitehouse et al. (1997, 1999). U–Pb concordia ages were calculated using IsoPlot (Ludwig 2000).

Ion microprobe U–Pb age determinations of single zircon grains yielded a wide range of ages (Table 2, Fig. 9a). A small Archaean population with ages in the range c. 2750–2850 Ma closely matches the ages of orthogneisses and Archaean detrital metasedimentary rocks from the immediately adjacent Ikamiut area (Kalsbeek et al. 1987; Hollis et al. 2006). Two Palaeoproterozoic age populations are so distinct that a common age was calculated for each group although, being detrital, they need not represent magmatic ages. The older group yielded a concordia age (2σ) of 1890 ± 7 Ma (seven grains, MSWD of concordance and equivalence = 0.73, probability of concordance = 0.73), and the younger group 1847 ± 10 Ma (five grains, MSWD of concordance and equivalence = 1.5, probability of concordance = 0.13). Three grains that yielded still younger ages and large analytical errors were discarded because of suspected early, late magmatic or metamorphic lead loss. The age of the older population (group 1 in Fig. 9b) is identical within error to the age of the Naternaq supracrustal belt as determined from sample 464435 by ion microprobe (1904 ± 8 Ma, MSWD = 1.9, 13 analyses published by Thrane & Connelly 2006). The younger age population (group 2 in Fig. 9b) matches the previously recorded peak of Nagssugtoqidian metamorphism at c. 1850 Ma. However, these zircon grains are morphologically and chemically similar to those in group 1, and the relatively low metamorphic grade (subsolidus middle amphibolite facies) and the low concentration of Zr in this sample are likely to have prevented new zircon growth during metamorphism. Therefore, we interpret both of the Proterozoic age groups in sample 459416 as representing detrital grains. The c. 1890 Ma age population is readily interpreted as representing originally igneous grains, which may have been derived from the Arfersiorfik arc or a contemporaneous Canadian arc. The c. 1847 Ma population (group 2) may represent igneous zircon from a hitherto unrecognized or completely eroded arc of this age. Alternatively, it might consist of transported metamorphic zircon grains that were formed in its source area during the Nagssugtoqidian thermal maximum, but this interpretation is not supported by the relatively high Th/U ratios also in group 2. In conclusion, sample 459416 contains material from both Archaean and Palaeoproterozoic sources, and the youngest age population of 1847 ± 10 Ma is interpreted as the maximum depositional age, and shows that Garde et al. (2004) incorrectly inferred an Archaean age

Table 2. *Ion microprobe (secondary ion mass spectrometry) U–Pb zircon age data for chloritic schist 459416, Qaqqarsuatsiaq island*

Analysis No.	Morphology	U (ppm)	Pb (ppm)	$^{206}Pb/^{204}Pb$	Th/U	$^{207}Pb/^{235}U$	1σ (%)	$^{206}Pb/^{238}U$	1σ (%)	±	$^{207}Pb/^{206}Pb$	1σ (%)	$^{207}Pb/^{235}U$	2σ (Ma)	$^{206}Pb/^{238}U$	2σ (Ma)	$^{207}Pb/^{206}Pb$	2σ (Ma)	Disc. (%)	f^{206} (%)
13	pr, m, osc	107	35	65 573	0.58	3.388	1.9	0.2605	1.8	0.93	0.0943	0.70	1502	15	1492	24	1515	13	0.6	{0.03}
17	pr, m, h	101	41	15 168	0.80	4.272	1.9	0.3026	1.8	0.94	0.1024	0.66	1688	16	1704	27	1668	12	−1.0	{0.12}
11	pr, e, osc	61	25	5414	0.68	4.617	2.0	0.3090	1.8	0.89	0.1084	0.90	1752	17	1736	27	1772	16	0.9	0.35
21	pr, e, osc	373	157	5387	0.58	5.090	1.8	0.3275	1.8	0.98	0.1127	0.38	1835	16	1826	28	1844	7	0.4	0.35
3	m, osc	392	203	9504	1.57	5.141	1.8	0.3307	1.8	0.98	0.1127	0.34	1843	15	1842	28	1844	6	0.1	0.2
23	pr, e, osc	565	225	8497	0.41	5.005	1.8	0.3216	1.8	0.98	0.1129	0.33	1820	15	1798	28	1846	6	1.2	0.22
25	pr, e, c, rext	1660	735	3464	0.82	5.114	1.8	0.3270	1.8	0.99	0.1134	0.22	1838	15	1824	28	1855	4	0.8	0.54
19	pr, frac, m, osc	268	102	2325	0.24	4.999	1.8	0.3195	1.8	0.96	0.1135	0.50	1819	16	1787	28	1856	9	1.8	0.8
1	pr, m, osc	347	145	12 271	0.48	5.289	1.8	0.3339	1.8	0.98	0.1149	0.34	1867	16	1857	29	1878	6	0.5	0.15
15	pr, e, osc	350	146	14 317	0.40	5.384	1.8	0.3385	1.8	0.98	0.1154	0.32	1882	16	1879	29	1885	6	0.2	0.13
16	pr, m, osc	585	257	623 243	0.58	5.464	1.8	0.3430	1.8	0.99	0.1155	0.24	1895	15	1901	29	1888	4	−0.3	{0.00}
24	pr, e, osc	628	251	56 392	0.19	5.460	1.8	0.3423	1.8	0.99	0.1157	0.23	1894	15	1898	29	1891	4	−0.2	0.03
7	pr, e, osc	520	228	33 104	0.53	5.532	1.8	0.3462	1.8	0.99	0.1159	0.26	1906	16	1917	29	1894	5	−0.6	0.06
9	pr, frag, m, osc	457	195	63 882	0.48	5.449	1.8	0.3408	1.8	0.98	0.1159	0.35	1893	16	1891	29	1895	6	0.1	0.03
22	pr, frag, m, osc	216	94	68 807	0.51	5.489	1.8	0.3427	1.8	0.98	0.1162	0.39	1899	16	1900	29	1898	7	0.0	{0.03}
20	pr, m, osc	2068	1607	41 710	1.58	11.40	1.9	0.4798	1.8	0.91	0.1723	0.78	2556	18	2526	37	2580	13	1.2	0.04
6	pr, m, osc	1023	575	37 375	0.39	10.42	1.8	0.4340	1.8	1.00	0.1742	0.16	2473	17	2324	35	2598	3	6.0	0.05
14	rmd, eq, m, osc	247	158	9830	0.00	14.13	1.8	0.5368	1.8	0.98	0.1909	0.40	2758	17	2770	40	2750	6	−0.4	0.19
26	pr, e, osc	161	110	32 919	0.55	13.39	1.8	0.5078	1.8	0.98	0.1913	0.32	2708	17	2647	39	2753	5	2.2	0.06
4	m, h	83	51	6158	0.49	12.35	1.8	0.4581	1.8	0.96	0.1955	0.49	2631	17	2431	36	2789	8	7.6	0.3
8	frag, m, h	50	37	22 854	0.48	15.23	1.9	0.5553	1.8	0.96	0.1990	0.54	2830	18	2847	41	2818	9	−0.6	{0.08}
2	pr, eq, m, osc	203	165	42 422	1.08	14.85	1.8	0.5400	1.8	0.99	0.1994	0.28	2806	17	2784	40	2821	5	0.8	0.04
5	pr, m, osc	92	70	234 040	0.61	15.62	1.8	0.5579	1.8	0.97	0.2030	0.46	2854	18	2858	41	2851	7	−0.2	{0.01}

f^{206}(%), the fraction of common ^{206}Pb, estimated from the measured ^{204}Pb (no correction applied from numbers in curly brackets, where ^{204}Pb is very low and significant). Disc. (%), conventional degree of discordance of the zircon analysis (at the centre of the error ellipse). Zircon morphology: c, core; e, end; eq, equant; frac, fractured; frag, fragmented; h, homogeneous; m, middle; osc, oscillatory zoned; pr, prismatic; rext, recrystallized; md, rounded.

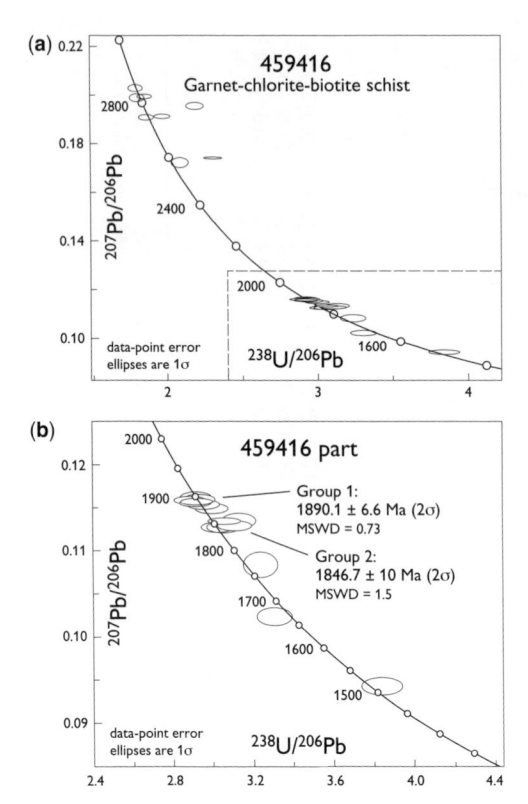

Fig. 9. U–Pb Tera–Wasserburg diagrams for detrital zircon from chloritic schist 459416, Qaqqarsuatsiaq island. Location shown in Figure 4. (**a**) Large spread of Archaean and Palaeoproterozoic ages; (**b**) two Palaeoproterozoic age groups, both interpreted as detrital. The c. 1850 Ma age of group 2 is the maximum age of deposition on the north side of the Aasiaat domain, which is contemporaneous with the collision between its southern margin and the North Atlantic craton.

of the islands in the absence of geochronological data. Assuming an arc-trench setting, the sedimentary rocks containing c. 1850 Ma detritus seem to have been deposited simultaneously with continuing arc magmatism and subduction.

Origin of the volcanic and sedimentary association at the Isuamiut–Qaqqarsuatsiaq and Equutiit Killiat islands

The two rock associations: (1) tholeiitic pillow lava and sills, very fine-grained black mudstone, manganiferous banded iron formation, podded chert and jasper, iron-rich carbonate rocks and terrigenous sandstone on the Isuamiut–Qaqqarsuatsiaq islands, and (2) contact metamorphosed aluminous turbidites with numerous tholeiitic sills on the Equutiit Killiat islands, are considered highly diagnostic for their depositional environments and plate-tectonic setting. These rocks occur within a small geographical area just 3 by 7 km in size, in a zone of low strain and relatively low metamorphic grade, where critical sedimentary and igneous features such as graded bedding, climbing ripples, pillow structures, chilled contacts, and delicate chert fabrics are well preserved and can be identified with certainty. The first rock association matches that found in modern arc-trench settings, where ocean-floor lavas and pelagic, fine-grained mud, chert, chemical sediments and locally exhalative rocks are transported to the edge of the accretionary wedge and mixed with detrital material, which is delivered into the trench from the adjacent arc by turbidity currents.

A prime example of a modern arc-trench setting that comprises virtually all the different lithological units of the first association on the Isuamiut–Qaqqarsuatsiaq islands occurs in the Inyama area of southwestern Japan. Here, the Eurassic plate was subducted under the Philippine Sea plate in Middle–Late Jurassic time (Isozaki et al. 1990; Matsuda & Isozaki 1991; B. F. Windley, pers. comm.). At Inyama a detailed investigation of the biostratigraphy and precise age determinations of the pelagic radiolaria area played a key role in the identification of the trench component and its implications. The age range of the chert constrains the time required for the ocean floor to travel from the volcanic ridge to the trench. In addition, the radiolarian age data also showed that the regular alternation between different rock types at the margin of the arc (in the Japanese case an alteration between chert and clastic sedimentary rocks) is due to tectonic repetition of strata with the same age range, as the top of the ocean floor was stepwise scraped off the subducting slab and added to the bottom or margin of the arc (Fig. 10; see also Matsuda & Isozaki 1991, Fig. 9). In the Japanese case, it has been shown that most of the pillow lava underlying the pelagic and hemipelagic sediments is subducted, and the sheeted dyke complex to be expected beneath the pillow lava is never preserved. Another well-documented example of ocean-floor deposits mixed with arc detritus in the accretionary complex between the arc core and the subducting slab occurs on the Neoproterozoic island of Anglesey, Wales (Kawai et al. 2006, 2007). Here, MORB-like pillow lava is also preserved. However, the tectonic structure is much more complex than in the previous example and includes a tongue of blueschists that was extruded up-dip from beneath the arc into the accretionary complex. Other similar examples from around the world in variable states of preservation have been reported (e.g. by Kusky 2004).

The second rock association on the Equutiit Killiat islands matches a special case, where the

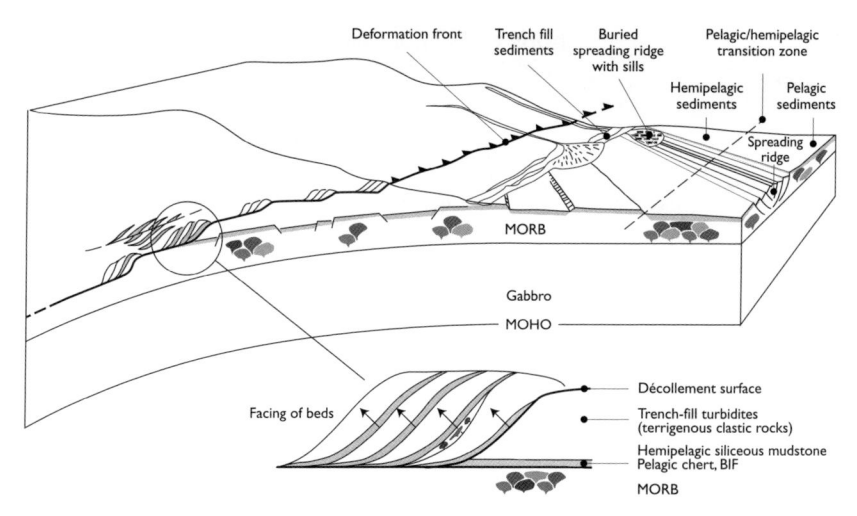

Fig. 10. Schematic illustration of Plate tectonics of an accretionary wedge, trench, ocean floor and spreading ridge, illustrating possible configurations of the two related depositional environments on the islands of Isuamiut–Qaqqarsuatsiaq and Equutiit Killiat. The former islands preserve a typical arc-trench environment, where ocean-floor pillow lava of tholeiitic composition and hemipelagic clastic and chemical sedimentary rocks are mixed with sandy terrigenous turbidites. The latter islands preserve a mixture of distal arc-derived aluminous turbidites with numerous tholeiitic sills, which together represent a buried oceanic spreading ridge. The figure is based on the study of a Jurassic ocean-floor–trench assemblage in southwestern Japan by Matsuda & Isozaki (1991, fig. 9), and also shows how the top of the ocean-floor–trench assemblages are scraped off the subducting plate and preserved as tectonic slices in the accretionary wedge.

volcanic spreading ridge itself is being buried by distal turbidites from an adjacent arc or continent (Fig. 10). A modern example of this ocean-floor scenario has been documented from Ocean Drilling Program (ODP) Leg 139 in the northeastern Pacific Ocean. That study was carried out where the Middle Valley of the Juan de Fuca ridge is currently undergoing burial by detritus that is, transported as density currents from the coast of North America several hundred kilometres away (Mottl *et al.* 1994). The rapid sediment accumulation in the Middle Valley has forced a change in the magmatic activity of the ridge from extrusive (pillow lavas) to intrusive (numerous sills in the turbidites). A Palaeogene example of the same setting was described by Kusky & Young (1999) from the Resurrection Peninsula ophiolite of southern Alaska's Chugach terrane, which likewise comprises distal turbidites. In the cases of the latter and the Equutiit Killiat islands, the spreading ridge was not only undergoing burial but eventually also was subducted, or else it would not have been preserved. In Figure 10 the burial of the spreading ridge is shown as older than the development of the trench. However, depending on the local sea-floor topography and the sediment load, it is possible that major turbidite fans may be able to travel across an active trench and onto the ocean floor beyond it.

A new plate-tectonic model for the Nagssugtoqidian orogen

The recognition of the stable Aasiaat domain in the northern Nagssugtoqidian orogen and the Palaeoproterozoic volcanic and sedimentary rocks at Naternaq and on the Isuamiut–Qaqqarsuatsiaq–Equutiit Killiat islands, as well as the identification of the latter as (hemi)pelagic rocks and a buried spreading ridge in an accretionary wedge, have important implications for the plate-tectonic interpretation of the Nagssugtoqidian orogen. The new information requires a suture to be present between the Isuamiut–Qaqqarsuatsiaq–Equutiit Killiat islands and the reworked Archaean rocks of the Aasiaat domain, and thus prompts a revision of the plate-tectonic model for the Nagssugtoqidian orogen that was proposed by van Gool *et al.* (2002). The recent recognition that the Nagssugtoqidian orogen and the Rinkian fold belt represent the southern and northern parts of one major collisional orogenic system between the Rae and North Atlantic cratons in western Greenland and eastern Canada, and the proposal of a suture zone at Paakitsoq in the southern Rinkian fold belt by Connelly *et al.* (2006) were major first steps in this revision.

A revision of the plate-tectonic scenario for the northern Nagssugtoqidian orogen must honour the currently known distribution of Palaeoproterozoic

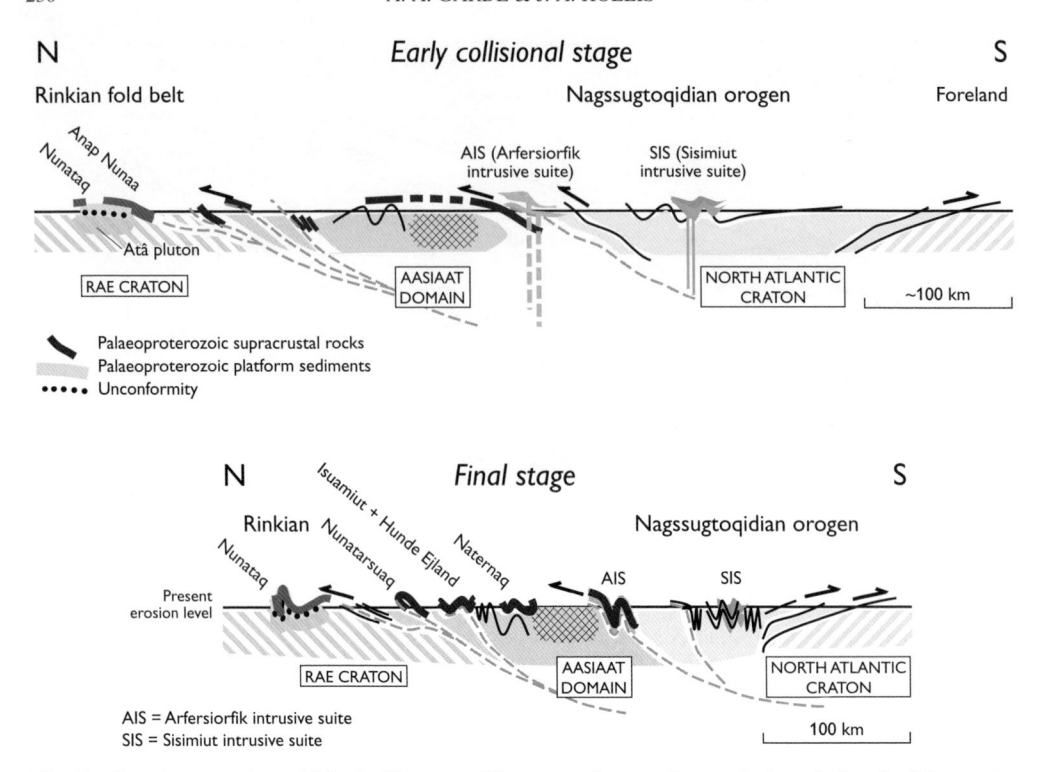

Fig. 11. New plate-tectonic model for the Nagssugtoqidian orogen from north to south along the length of the map in Figure 1b. The new model has two SSE-dipping sutures (namely, subduction zones), each of which can be correlated with a Palaeoproterozoic magmatic arc. The northern subduction zone (Disko Bugt suture) fed the oceanic Arfersiorfik plutonic suite south of the Aasiaat domain, whereas the previously established southern subduction zone (central Nagssugtoqidian suture) fed the continental Sisimiut intrusive suite that was emplaced into the North Atlantic craton.

ocean-floor–trench–arc supracrustal associations within the Nagssugtoqidian orogen and in the poorly defined transition zone to the Rinkian fold belt to the north. It must also account for the presence and locations of the two Palaeoproterozoic plutonic arcs in the central Nagssugtoqidian orogen. Finally, it must resolve the correlation with known suture zones in eastern Canada much more readily than previously published models.

A new plate-tectonic model for the Nagssugtoqidian orogen and the southern part of the Rinkian fold belt, which is simple and meets the above requirements, is shown in Figure 11. A preliminary outline was presented by Garde et al. (2007) and St.-Onge et al. (2009), where the Aasiaat domain and the new Disko Bugt suture were first labelled. In the new model, there are two south- to SSE-dipping subduction zones in the central and northern Nagssugtoqidian orogen, which are separated by the previously unidentified Aasiaat domain or microcontinent (see also Connelly & Thrane 2005).

As shown in Figures 1 and 11 the northern, newly identified Disko Bugt suture on the north side of the Aasiaat domain can be linked with the suture at Paakitsoq proposed by Connelly et al. (2006). It forms a network of thrust faults in the Ilulissat area, which is presumed to have been initiated by thrusting during early collision between the Rae craton and the Aasiaat domain, and subsequently modified by further thrusting and folding. The presence of well-preserved but undated Palaeoproterozoic supracrustal rocks at Hunde Ejland similar to those occurring at the Isuamiut–Equutiit Killiat islands may suggest that there was a similar splay system north of Aasiaat, as indicated in Figure 1b. The 1895 Ma Arfersiorfik intrusive suite is interpreted to have been fed by the former subduction zone along the Disko Bugt suture. The Arfersiorfik intrusive suite would represent the plutonic root system of an arc located in the ocean that separated the Aasiaat domain from the North Atlantic craton prior to the Nagssugtoqidian collision.

The southern, central Nagssugtoqidian suture between the Aasiaat domain and the North Atlantic craton is left unchanged from the previous model, except that its subduction zone would have fed only the Sisimiut intrusive suite. The location of this suite within Archaean rocks means that it is a

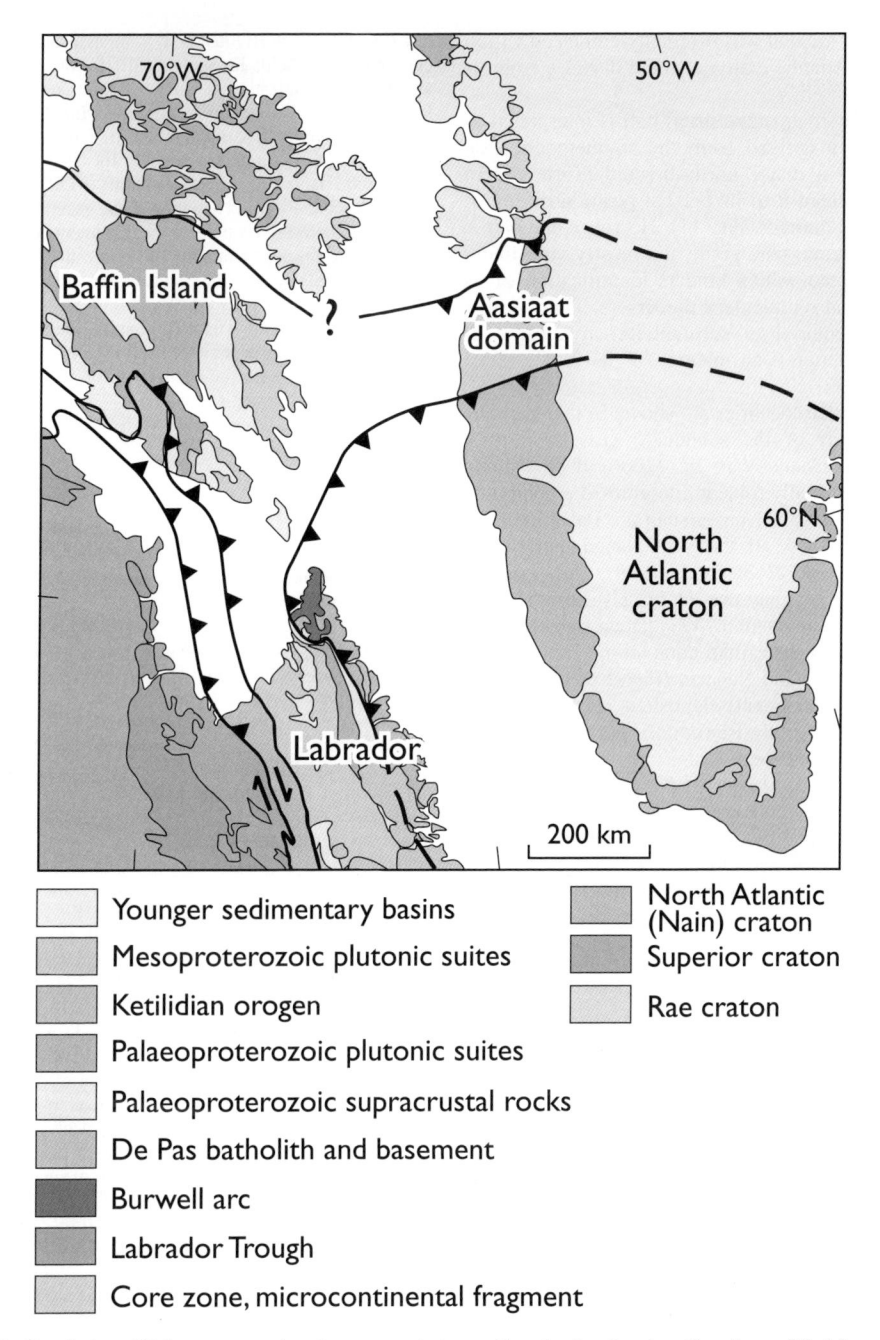

Fig. 12. Correlation of Palaeoproterozoic palaeosutures between Greenland and eastern Canada, modified from St.-Onge *et al.* (2009). The proposed Disko Bugt suture north of the Aasiaat domain may be correlated with the east–west-trending suture on southern Baffin Island, and the central Nagssugtoqidian suture with the north–south-striking suture of the Torngat orogen in Labrador. It should be noted that the North Atlantic craton in southern Greenland is the overriding plate, whereas the Superior craton in eastern Canada is the lower plate.

continental arc that was originally emplaced within the North Atlantic craton at some distance from its northern margin.

The Naternaq supracrustal belt is interpreted as having been derived from the southern collision zone along a major north-directed thrust system that transported it to its present position on top of the Aasiaat domain (Fig. 11). However, the belt is at high metamorphic grade and mostly very intensely deformed, which hinders identification of its volcanic and sedimentary precursors. Thus, further work is required to substantiate an oceanic or suprasubduction-zone nature of its fine-grained amphibolites.

Three other localities are shown in the northern, Rinkian part of the schematic cross-section of Figure 11. Connelly et al. (2006) demonstrated that the isoclinally folded, fine-grained amphibolite and mica schist at Nunatarsuaq are Palaeoproterozoic in age (these are the Nunatarsuaq supracrustal rocks of Garde & Steenfelt 1999). These poorly preserved rocks may thus be broadly equivalent to those at the Isuamiut–Qaqqarsuatsiaq–Equutiit Killiat islands. The Anap nunâ Group farther north at Anap Nunaa and Nunataq (Figs 1 & 11) consists of platform sediments deposited at the southern passive margin of the Rae craton, and a basal unconformity is preserved (Garde & Steenfelt 1999; Higgins & Soper 1999). The Anap nunâ Group is correlated with the Karrat Group in the central and northern parts of the Rinkian fold belt (Henderson & Pulvertaft 1987; St-Onge et al. 2009), although the unconformity itself between the Karrat Group and its Archaean basement has been identified with certainty in only one place at Maarmorilik, north of Nuussuaq (Garde 1978).

Correlation with eastern Canada

The new plate-tectonic model for the Nagssugtoqidian orogen and southern Rinkian fold belt offers a straightforward correlation with the two established Palaeoproterozoic sutures in SE Baffin Island and northern Labrador as shown in Figure 12 (see also St.-Onge et al. 2009). The northern suture matches the Baffin Island suture, and the southern one may be linked with the suture of the Torngat orogen in northern Labrador. However, as pointed out by St.-Onge et al. (2009), the southern craton (i.e. the North Atlantic craton) is the overriding plate in West Greenland, whereas the northern craton (Rae) is the overriding plate in eastern Canada.

This work was conducted while J. Hollis was an employee at the Geological Survey of Denmark and Greenland. We thank M. Sylvest Eegholm, Mærsk Olie og Gas A/S, S. Mazur, University of Wroclaw, and J. A. M. van Gool, Scandinavian Highlands A/S, for assistance and discussions in the field; K. Secher, GEUS, for information about the mineralization at Hunde Ejland; and, not least, B. F. Windley, University of Leicester, for drawing our attention to equivalent modern arc-trench settings. We also thank M. Whitehouse and L. Ilievsky at the ion microprobe laboratory in Stockholm (NORDSIM) for help with acquisition and handling of the geochronological data. The laboratory is operated by agreement between the joint Nordic research councils (NOS-N), the Geological Survey of Finland, and the Swedish Museum of Natural History. T. Kusky, M. St.-Onge and an anonymous reviewer are thanked for their comments, which improved the manuscript. This paper is NORDSIM Contribution 222 and is published with the permission of the Geological Survey of Denmark and Greenland.

References

CONNELLY, J. N. & THRANE, K. 2005. Rapid determination of Pb isotopes to define Precambrian allochthonous domains: an example from West Greenland. *Geology*, **33**, 953–956.

CONNELLY, J. N., VAN GOOL, J. A. M. & MENGEL, F. C. 2000. Temporal evolution of a deeply eroded orogen: The Nagssugtoqidian orogen, West Greenland. *Canadian Journal of Earth Sciences*, **37**, 1121–1142.

CONNELLY, J. N., THRANE, K., KRAWIEC, A. W. & GARDE, A. A. 2006. Linking the Palaeoproterozoic Nagssugtoqidian and Rinkian orogens through the Disko Bugt region of West Greenland. *Journal of the Geological Society, London*, **163**, 319–335.

ELLITSGAARD-RASMUSSEN, K. 1954. *On the geology of a metamorphic complex in West Greenland. The islands of Anarssuit, Isuamiut, and Eqûtit*. Bulletin Grønlands Geologiske Undersøgelse, **5**.

ESCHER, A., SØRENSEN, K. & ZECK, H. P. 1976. Nagssugtoqidian mobile belt in West Greenland. *In*: ESCHER, A. & WATT, W. W. (eds) *Geology of Greenland*. Geological Survey of Greenland, Copenhagen, 76–95.

FREI, R., DAHL, P. S., DUKE, E. F., FREI, K. M., HANSEN, T. R., FRANDSSON, M. M. & JENSEN, L. A. 2008. Trace element and isotopic characterization of Neoarchean and Paleoproterozoic iron formations in the Black Hills (South Dakota, USA): assessment of chemical change during 2.9–1.9 Ga deposition bracketing the 2.4–2.2 Ga first rise of atmospheric oxygen. *Precambrian Research*, **162**, 441–474.

GARDE, A. A. 1978. *The lower Proterozoic Mârmorilik formation, east of Mârmorilik, West Greenland*. Meddelelser om Grønland, **200(3)**.

GARDE, A. A. 2004. *Geological map of Greenland, 1:100 000, Kangaatsiaq 68 V.1 Syd*. Geological Survey of Denmark and Greenland, Copenhagen.

GARDE, A. A. 2006. *Geological map of Greenland, 1:100 000, Ikamiut 68 V.1 Nord*. Geological Survey of Denmark and Greenland, Copenhagen.

GARDE, A. A. & STEENFELT, A. 1999. Precambrian geology of Nuussuaq and the area north-east of Disko Bugt, West Greenland. *In*: KALSBEEK, F. (ed.) *Precambrian Geology of the Disko Bugt Region, West Greenland*. Geology of Greenland Survey Bulletin, **181**, 7–40.

GARDE, A. A., CHRISTIANSEN, M. J., HOLLIS, J. A., MAZUR, S. & VAN GOOL, J. A. M. 2004. Low-pressure metamorphism during Archaean crustal growth: a low-strain zone in the northern Nagssugtoqidian orogen, West Greenland. *In*: SØNDERHOLM, M. & HIGGINS, A. K. (eds) Geological Survey of Denmark and Greenland Bulletin, **4**, 73–76.

GARDE, A. A., HOLLIS, J. A. & MAZUR, S. 2007. Palaeoproterozoic greenstones and pelitic schists in the northern Nagssugtoqidian orogen, West Greenland: evidence for a second subduction zone? *Geological Association of Canada–Mineralogical Association of Canada (GAC–MAC) Annual Meeting, Yellowknife 2007, Program with Abstracts*.

HENDERSON, G. 1969. *The Precambrian rocks of the Egedesminde–Christianshåb area, West Greenland*. Rapport Grønlands Geologiske Undersøgelse, **23**.

HENDERSON, G. & PULVERTAFT, T. C. R. 1987. *The lithostratigraphy and structure of a Lower Proterozoic dome and nappe complex. Descriptive text to 1:100 000 sheets Mârmorilik 71 V. 2 Syd, Nûgâtsiaq 71 V. 2 Nord and Pangnertôq 72 V. 2 Syd*. Geological Survey of Greenland, Copenhagen.

HIGGINS, A. K. & SOPER, N. J. 1999. The Precambrian supracrustal rocks of Nunataq, north-east Disko Bugt, West Greenland. *In*: KALSBEEK, F. (ed.) *Precambrian Geology of the Disko Bugt Region, West Greenland*. Geology of Greenland Survey Bulletin, **181**, 79–86.

HOFMANN, A. W. 1988. Chemical differentiation of the Earth: the relationships between mantle, continental crust, and oceanic crust. *Earth and Planetary Science Letters*, **90**, 297–314.

HOLLAND, T. J. B. & POWELL, R. 1998. An internally consistent thermodynamic dataset for phases of petrological interest. *Journal of Metamorphic Geology*, **16**, 309–343.

HOLLIS, J. A., KEIDING, M., STENSGAARD, B. M., VAN GOOL, J. A. M. & GARDE, A. A. 2006. Evolution of Neoarchaean supracrustal belts at the northern margin of the North Atlantic craton, West Greenland. *In*: GARDE, A. A. & KALSBEEK, F. (eds) *Precambrian Crustal Evolution and Cretaceous–Palaeogene faulting in West Greenland*. Geological Survey of Denmark and Greenland Bulletin, **11**, 9–31.

ISOZAKI, Y., MARUAYAMA, S. & FURUOKA, F. 1990. Accreted oceanic materials in Japan. *Tectonophysics*, **181**, 179–205.

KALSBEEK, F., PIDGEON, R. T. & TAYLOR, P. N. 1987. Nagssugtoqidian mobile belt of West Greenland; a cryptic 1850 Ma suture between two Archaean continents; chemical and isotopic evidence. *Earth and Planetary Science Letters*, **85**, 365–385.

KAWAI, T., WINDLEY, B. F., TERABAYASHI, M., YAMAMOTO, H., MARUYAMA, S. & ISOZAKI, Y. 2006. Mineral isograds and metamorphic zones of the Anglesey blueschist belt, UK: implications for the metamorphic development of a Neoproterozoic subduction–accretion complex. *Journal of Metamorphic Geology*, **24**, 591–602.

KAWAI, T., WINDLEY, B. F. *ET AL.* 2007. Geotectonic framework of the Blueschist Unit on Anglesey–Lleyn, UK, and its role in the development of a Neoproterozoic accretionary orogen. *Precambrian Research*, **153**, 11–28.

KORSTGAARD, J. (ed.) 1979. *Nagssugtoqidian geology*. Rapport Grønlands Geologiske Undersøgelse, **89**.

KUSKY, T. M. (ed.) 2004. *Precambrian Ophiolites and Related Rocks*. Developments in Precambrian Geology, **13**.

KUSKY, T. & YOUNG, C. 1999. Emplacement of the Resurrection Peninsula ophiolite in the southern Alaska forearc during a ridge–trench encounter. *Journal of Geophysical Research*, **104**, 29 025–29 054.

KYSTOL, J. & LARSEN, L. L. 1999. Analytical procedures in the Rock Geochemical Laboratory of the Geological Survey of Denmark and Greenland. *Geological Survey of Greenland Bulletin*, **184**, 59–62.

LAHTINEN, R., GARDE, A. A. & MELEZHIK, V. A. 2008. Palaeoproterozoic evolution of Fennoscandia and Greenland. *Episodes*, **31**, 20–28.

LI, J. & KUSKY, T. M. 2007. World's largest known Precambrian fossil black smoker chimneys and associated microbial vent communities, North China: Implications for early life. *Gondwana Research*, **12**, 84–100.

MATSUDA, T. & ISOZAKI, Y. 1991. Well-documented travel history of Mesozoic pelagic chert in Japan: from remote ocean to subduction zone. *Tectonics*, **10**, 475–499.

LUDWIG, K. R. 2000. *Isoplot/Ex version 2.2: A Geochronological Toolkit for Microsoft Excel*. Berkeley Geochronological Center Special Publication.

MAZUR, S., PIAZOLO, S. & ALSOP, G. I. 2006. Structural analysis of the northern Nagssugtoqidian orogen, West Greenland: an example of complex tectonic patterns in reworked high-grade metamorphic terrains. *In*: GARDE, A. A. & KALSBEEK, F. (eds) *Precambrian Crustal Evolution and Cretaceous–Palaeogene Faulting in West Greenland*. Geological Survey of Denmark and Greenland Bulletin, **11**, 163–178.

MCLENNAN, S. M. 1989. Rare earth elements in sedimentary rocks: Influence of provenance and sedimentary processes. *In*: LIPIN, B. R. & MCKAY, T. A. (eds) *Geochemistry and Mineralogy of Rare Earth Elements*. Mineralogical Society of America, Reviews in Mineralogy, **21**, 169–200.

MOTTL, M. J., DAVIS, E. E., FISHER, A. T. & SLACK, J. F. (eds) 1994. *Proceedings of the Ocean Drilling Program, Scientific Results, 139*. Ocean Drilling Program, College Station, TX.

ØSTERGAARD, C., GARDE, A. A., NYGAARD, J., BLOMSTERBERG, J., NIELSEN, B. M., STENDAL, H. & THOMAS, C. W. 2002. The Precambrian supracrustal rocks in the Naternaq (Lersletten) and Ikamiut areas, central West Greenland. *In*: HIGGINS, A. K., SECHER, K. & SØNDERHOLM, M. (eds) *Geology of Greenland Survey Bulletin*, **191**, 24–32.

PIAZOLO, S., ALSOP, G. I., VAN GOOL, J. & NIELSEN, B. M. 2004. Using GIS to unravel high strain patterns in high grade terranes: a case study of indentor tectonics from West Greenland. *In*: ALSOP, G. I., HOLDSWORTH, R. E., MCCAFFREY, K. J. W. & HAND, M. (eds) *Flow Processes in Faults and Shear Zones*. Geological Society, London, Special Publications, **224**, 63–78.

POWELL, R. & HOLLAND, T. J. B. 1988. An internally consistent thermodynamic dataset with uncertainties and correlations: 3. Application, methods, worked examples and a computer program. *Journal of Metamorphic Geology*, **6**, 173–204.

RAMBERG, H. 1949. On the petrogenesis of the gneiss complexes between Sukkertoppen and Christianshaab, West Greenland. *Meddelelser fra Dansk Geologisk Forening*, **11**, 312–327.

ST.-ONGE, M., VAN GOOL, J. A. M., GARDE, A. A. & SCOTT, D. J. 2009. Correlation of Archaean and Palaeoproterozoic units between northeastern Canada and western Greenland: constraining the pre-collisional upper plate accretionary history of the Trans-Hudson orogen. *In*: CAWOOD, P. A. & KRONER, A. (eds) *Accretionary Orogens in Space and Time*. Geological Society, London, Special Publications, **318**, 193–235.

TAYLOR, S. R. & MCLENNAN, S. M. 1985. *The Continental Crust: its Composition and Evolution*. Blackwell, Oxford.

THRANE, K. & CONNELLY, J. N. 2006. Zircon geochronology from the Kangaatsiaq–Qasigiannguit region, the northern part of the 1.9–1.8 Ga Nagssugtoqidian orogen, West Greenland. *In*: GARDE, A. A. & KALSBEEK, F. (eds) *Precambrian Crustal Evolution and Cretaceous–Palaeogene Faulting in West Greenland*. Geological Survey of Denmark and Greenland Bulletin, **11**, 7–40.

VAN GOOL, J. A. M. 2005. *Geological map of Greenland, 1:100 000, Kangersuneq 68 V.2 Nord*. Geological Survey of Denmark and Greenland, Copenhagen.

VAN GOOL, J. A. M. & MARKER, M. 2007. *Explanatory notes to the geological map of Greenland, 1:100 000, Ussuit 67 V.2 Nord*. Geological Survey of Denmark and Greenland Map Series, **3**.

VAN GOOL, J. A. M., CONNELLY, J. N., MARKER, M. & MENGEL, F. C. 2002. The Nagssugtoqidian Orogen of West Greenland: tectonic evolution and regional correlations from a West Greenland perspective. *Canadian Journal of Earth Sciences*, **39**, 665–686.

WATTERSON, J. 1978. Proterozoic intraplate deformation in the light of South-East Asian neotectonics. *Nature*, **273**, 636–640.

WEI, C. J., POWELL, R. & CLARKE, G. L. 2004. Calculated phase equilibria for low- and medium-pressure metapelites in the KFMASH and KMnFMASH systems. *Journal of Metamorphic Geology*, **22**, 495–508.

WHITEHOUSE, M. J., CLAESSON, S., SUNDE, T. & VESTIN, J. 1997. Ion microprobe U–Pb zircon geochronology and correlation of Archaean gneisses from the Lewisian Complex of Gruinard Bay, northwestern Scotland. *Geochimica et Cosmochimica Acta*, **61**, 4429–4438.

WHITEHOUSE, M. J., KAMBER, B. S. & MOORBATH, S. 1999. Age significance of U–Th–Pb zircon data from early Archaean rocks of west Greenland—a reassessment based on combined ion-microprobe and imaging studies. *Chemical Geology*, **160**, 201–224.

WIEDENBECK, M., ALLE, P. ET AL. 1995. Three natural zircon standards for U–Th–Pb, Lu–Hf, trace element and REE analyses. *Geostandards Newsletter*, **19**, 1–23.

WINDLEY, B. F. & GARDE, A. A. 2009. Arc-generated blocks with crustal sections in the North Atlantic craton of West Greenland: new mechanism of crustal growth in the Archean with modern analogues. *Earth-Science Reviews*, **93**, 1–30.

Precambrian key tectonic events and evolution of the North China craton

MINGGUO ZHAI[1,2], TIE-SHENG LI[1,2], PENG PENG[1,2], BO HU[1,2], FU LIU[1,2] & YANBIN ZHANG[1,2]

[1]*State Key Laboratory of Lithospheric Evolution, Institute of Geology and Geophysics, Chinese Academy of Sciences, PO Box 9825, Beijing, 100029, China*

[2]*Key Laboratory of Mineral Resources, Institute of Geology and Geophysics, Chinese Academy of Sciences, PO Box 9825, Beijing, 100029, China*

*Corresponding author (e-mail: mgzhai@mail.igcas.ac.cn)

Abstract: The North China craton (NCC) is one of oldest cratons in the world, with crust up to *c.* 3.8 Ga old, and has a complicated evolution. The main Early Precambrian geological events and key tectonic issues are as follows. (1) Old continental nuclei have been recognized in the NCC, and the oldest remnants of granitic gneiss and supracrustal rocks are 3.8 Ga old. The main crustal growth in the NCC took place at 2.9–2.7 Ga. The NCC can be divided into several microblocks, which are separated by Archaean greenstone belts that represent continental accretion surrounding the old continental nuclei. (2) By 2.5 Ga, the microblocks amalgamated to form a coherent craton by continent–continent, arc–continent or arc–arc collisions. The tectonic processes in Neoarchaean and modern times appear to differ more in degree than in principle. Extensive intrusion of K-granite sills and mafic dykes and regional upper amphibolite- to granulite-facies metamorphism occurred, and marked the beginning of cratonization in the NCC. Coeval ultramafic–mafic and syenitic dykes of *c.* 2500 Ma in Eastern Hebei indicate that the NCC became a stable, thick and huge continent at the end of the Archaean, and probably was a part of the Neoarchaean supercontinent that has been suggested by previous studies. (3) In the period between 2500 and 2350 Ma, the NCC was tectonically inactive, but the development of a Palaeoproterozoic volcanic and granitic rocks occurred between 2300 and 1950 Ma. The volcanic–sedimentary rocks are termed Palaeoproterozoic mobile belts; these have a linear distribution, and were affected by strong folding and metamorphism at 1900–1850 Ma, and intruded by granites and pegmatites at 1850–1800 Ma. The Palaeoproterozoic mobile belts formed and evolved within the craton or continental margin (epicontinental geosyncline). Some 2.30–1.95 Ga rift-margin, passive continental margin deposits, analogous arc or back-arc assemblages, as well as HP and HT–UHT metamorphic complexes seem to be comparable with many in the late Phanerozoic orogenic belts. Regarding Palaeoproterozoic orogeny in other cratons, it is possible that a global Palaeoproterozoic orogenic event occurred, existed and resulted in the formation of a pre-Rodinian supercontinent at *c.* 2.0–1.85 Ga. (4) In contrast, the *c.* 1800 Ma event is an extension–migmatization event, which includes uplift of the lower crust of the NCC as a whole, the emplacement of mafic dyke swarms, continental rifting, and intrusion of an orogenic magmatic association. This event has been considered to be related to the break-up of the pre-Rodinian supercontinent at 1.8 Ga, attributed to a Palaeoproterozoic plume. (5) As HP and HT–UHT metamorphic rocks occur widely in the NCC, their high pressure of 10–14 kbar has attracted attention from researchers, and several continental collisional models have been proposed. However, it is argued that these rocks have much higher geothermal gradient and much slower uplift rate than those in Phanerozoic orogenic belts. Moreover, HP and HT–UHT rocks commonly occur together and are not distributed in linear zones, suggesting that the geological and tectonic implications of these data should be reassessed.

The North China craton (NCC) covers over 300 000 km², and is one of oldest cratons in the world with crust up to *c.* 3.8 Ga old, and has a complicated evolutionary history (Liu *et al.* 1992; Zhao, Z. P. 1993; Windley 1995; Wan *et al.* 2001; Zhai 2004; Kusky *et al.* 2007*a*). In recent years, its early tectonic evolution has attracted increasing attention from researchers. The key issues studied include early Precambrian high-pressure granulites, ultrahigh-temperature metamorphic rocks, Archaean ophiolite, Archaean coeval ultramafic–syenitic dykes, Palaeoproterozoic mafic dyke swarms, and Precambrian plate tectonics and plume tectonics (Zhai *et al.* 1992, 2002; Li, J. H. *et al.* 1997; Zhai 1997; Zhao, G. C. *et al.* 1999, 2007; Kusky *et al.* 2001, 2007*a*;

From: KUSKY, T. M., ZHAI, M.-G. & XIAO, W. (eds) *The Evolving Continents: Understanding Processes of Continental Growth.* Geological Society, London, Special Publications, **338**, 235–262.
DOI: 10.1144/SP338.12   0305-8719/10/$15.00 © The Geological Society of London 2010.

Kröner et al. 2005; Peng et al. 2005; Wilde et al. 2005; Santosh et al. 2007; Lim, T. S. et al. 2010).

Pre-3.8 Ga rocks occur in the centre of the Archaean craton within Proterozoic belts, and it has been thought that these were Archaean cratons surrounded by Proterozoic belts. The Archaean cratons of the world are often quoted as containing two types of terrane: (1) gneiss-dominated belts metamorphosed largely to a high metamorphic grade; (2) well-preserved, low-grade, volcanic-dominated greenstone belts (Windley 1995). It is, in general, accepted that modern-style plate-tectonic processes were active by the Palaeo-Mesoproterozoic when large stable cratons had formed, against which trailing margins (Dann 1991; Loukola-Ruskeeniemi et al. 1991), Andean-type margins and collisional belts could develop (Windley 1995). Summarizing different models of Archaean continental evolution, Windley (1986, 1995) suggested that the most viable model basically combines a back-arc marginal basin setting for greenstone belt formation and the main arc (plutonic batholith) interpretation of the high-grade gneissic complex. Recently, Kusky (2004a, b) and Condie & Kröner (2008) summarized some geological indicators and suggested that modern plate tectonics were operational, at least in some places on Earth, by 3.0 Ga or even earlier, and that they became widespread by 2.7 Ga, not as a single global 'event' at a distinct time. The North China craton provides considerable positive and negative geological evidence for the above-mentioned models.

Some previous studies (Ouyang & Zhang 1998; Wu et al. 1998; Geng 1998; Zhai et al. 2005) reported several continental nuclei in the NCC greater than c. 3.0–3.8 Ga old, which are surrounded by late Archaean greenstone belts, although in these belts nearly all the rocks underwent amphibolite- to granulite-facies metamorphism. Nd, Pb and Hf isotopic geochemical and geochronological studies have shown that there are two discernible peaks at c. 2.7 and 2.9 Ga, representing main crustal growth epochs (Zhang et al. 1998). Zhai (2004) and Kusky et al. (2007a) also emphasized that c. 2.5–2.55 Ga is another important growth epoch.

Almost all Archaean rocks in high-grade regions or greenstone belts in the NCC underwent high-grade metamorphism at amphibolite or granulite facies, some at high-pressure granulite facies and ultrahigh-temperature granulite facies. They commonly have a multistage metamorphic history. For example, upper amphibolite- and granulite-facies rocks are commonly overprinted by low- to mid-amphibolite-facies assemblages. Two important metamorphic events in the NCC took place at 2600–2450 Ma and 1950–1800 Ma (Zhai et al. 2000a; Zhai & Liu 2001a, b; Zhao 2001), which indicate, respectively, that the united North China craton fundamentally formed by amalgamation of several microblocks at the end of the Archaean and the cratonization of a present-scale NCC was finally accomplished in the Palaeoproterozoic (Cheng 1994; Zhai 2008).

Palaeoproterozoic mobile belts with ages of 2300–1950 Ma, such as those of Jinyü and Jiaoliao, represent intracratonic orogenesis with some Phanerozoic orogenic characteristics. The evolution of the mobile belts includes basin rifting, subduction and final collision at c. 1900 Ma. After this, extensive migmatization and metamorphism occurred, termed the 1800 Ma event, followed by continental rift formation and the intrusion of mafic dyke swarms. This event led to the metamorphosed basement of the NCC being uplifted as a whole and becoming exhumed to the surface by a series of complicated detachment structures.

Oldest rocks and continental nucleus

Oldest rocks

Liu et al. (1992), Song et al. (1996) and Wan et al. (2001) reported zircon U–Pb isotopic data of c. 3.8 Ga from the NCC, indicating the presence of old continental crust at two localities, near Tiejiashan, NE China and near Caozhuang, eastern Hebei (Fig. 1).

The Neoarchaean Anshan complex is divided into three parts: the Tiejiashan gneiss, the Anshan supracrustal series and the Anshan gneiss (Zhai et al. 1990). The Anshan supracrustal series contain abundant banded iron formation rocks with Sm–Nd isochron ages of 2.7–2.66 Ga. The granitoid protoliths of the Anshan gneiss were intruded extensively into the Anshan supracrustal series at 2.55–2.47 Ga, based on their zircon U–Pb ages. The Tiejiashan gneiss constitutes the basement of the Anshan supracrustal series and includes a series of granitic and trondhjemitic orthogneisses and deformed felsic veins (Fig. 2b). Their zircon U–Pb ages range from c. 2960 Ma in the central part to c. 3306 Ma in the northern part (Liu et al. 1992). Two samples of sheared granitic gneiss were collected from Chentaigou village in the northwestern margin of the Tiejiashan gneiss. Liu et al. (1992) reported that ion microprobe U–Pb analyses show that the samples contain two generations of zircons, with ages of 3805 ± 5 Ma and 3300 Ma. The zircon in a layered siliceous supracrustal rock near Chengtaigou yielded a sensitive high-mass resolution ion microprobe (SHRIMP) age of 3362 ± 5 Ma that was interpreted as an age of detrital zircon (Song et al. 1996). Wan et al. (2001) further reported that the oldest Tiejiashan gneiss occurs as a large lens in the trondhjemitic country

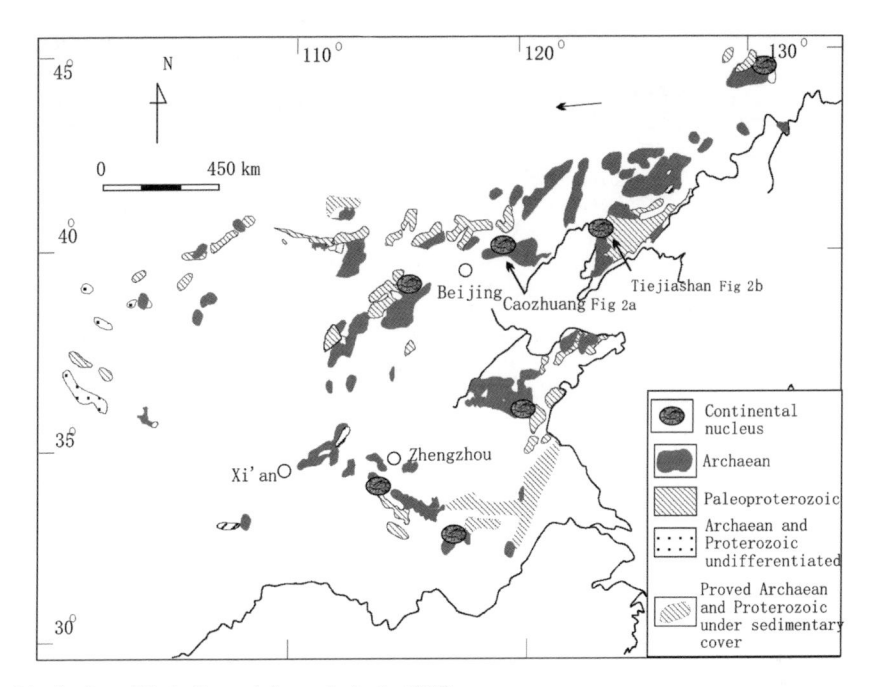

Fig. 1. Distribution of Early Precambrian rocks in the NCC.

gneiss, and the latter yielded a 3.1 Ga SHRIMP zircon U–Pb age. *In situ* zircon Hf isotopic analyses indicate that these old granitoid rocks were derived from juvenile crust with age peaks of crustal growth at c. 3.4, 3.6 and 3.9 Ga (Wu *et al.* 2008). These data show that the Anshan complex has a very complicated geological history.

The oldest supracrustal remnant in the NCC is the Caozhuang complex in Eastern Hebei Province (Fig. 2a). The Caozhuang complex can be divided into two parts; the Caozhuang Group and the Huangboyü banded grey gneiss. The Caozhuang Group is a slab 1.9 km long and 400–500 m wide that is a complicated synformal fold structure. The Huangboyü grey gneiss was intruded into the supracrustal rocks of the Caozhuang Group prior to intense multi-stage deformations. Zircons from the Huangboyü gneiss have zircon U–Pb ages of 3.0–3.3 Ga (Zhao, Z. P. 1993). The main supracrustal rocks include amphibolite, serpentinized marble, fuchsitic quartzite, metamorphosed calc-silicate rock, banded iron formation, biotite gneiss, and sillimanite–plagioclase gneiss. These rocks were metamorphosed to upper amphibolite facies to granulite facies (Yan *et al.* 1991). The amphibolites yielded Sm–Nd isochron ages of c. 3.5 Ga (Huang *et al.* 1983; Jahn & Zhang 1984). Zircons from the fuchsitic quartzite are colourless to lilac and the rounded shape of the grains is attributed to abrasion during sedimentary transport. Eighty-two analyses on 61 zircons using the SHRIMP technique yielded four age populations of 3.83–3.82, 3.8–3.78, 3.72–3.7 and 3.68–3.6 Ga. Higher U zircons rims were distinguished, which imply high-grade metamorphic events at 2.5 Ga and 1.89 Ga (Liu *et al.* 1992). Wu *et al.* (2005) reported *in situ* zircon Hf analyses from the fuchsitic quartzite. The data show that the Lu–Hf system remained closed during later thermal disturbances. Zircons with concordant ages have Hf isotopic model ages of about 3.8 Ga, suggesting a recycling of this ancient crust.

Continental nucleus and crustal growth

Besides these two oldest crust remnants described above, several additional old continental nuclei in the NCC have also been proposed; for example, the Huai'an complex in the western–central NCC, the Yishui complex in the eastern NCC, the Xinyang in the southwestern NCC and the Longgang complex in the northeastern NCC (Fig. 1a). They have isotopic ages 3.5–3.3 Ga (Kröner *et al.* 1987; Guo *et al.* 1991), 3.1–2.97 Ga (Wu *et al.* 1998) and 3.1–3.0 Ga (Zhai & Windley 1990), respectively.

The old continental nuclei, in general, are surrounded by late Archaean and Proterozoic supracrustal rocks as in other cratons (Bai *et al.* 1993). Zhai & Liu (2001*a*) reported that estimated volumetric crustal growth of the NCC was about 90% by 2.5–2.45 Ga, based on new geological maps,

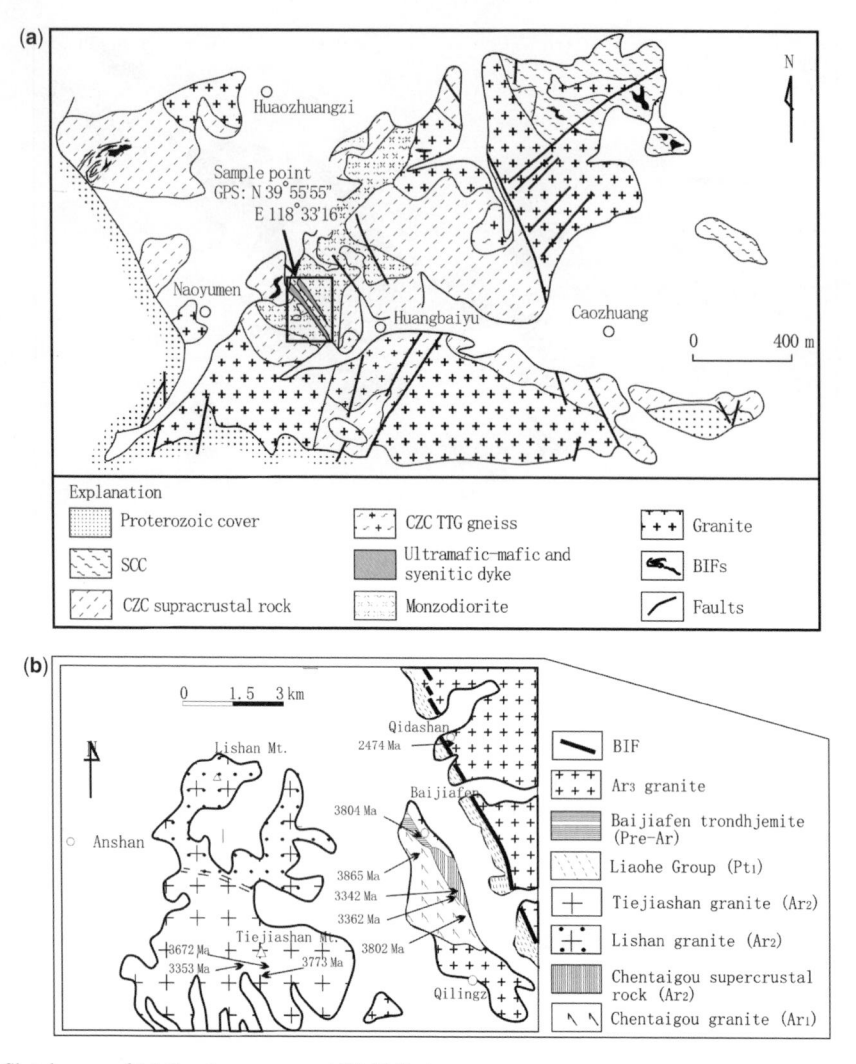

Fig. 2. Sketch maps of (**a**) Caozhuang area and (**b**) Tiejiashan area.

geophysical data and geochronological data. The Nd isotope, Hf isotope and trace element characteristics of early Precambrian rocks in the NCC and their implication for crustal growth have been discussed by various workers (Jahn 1990; Geng & Liu 1997; Zhang 1998; Liu et al. 2004; Wan et al. 2005; Wu et al. 2005; Zhai et al. 2007). The Sm–Nd T_{DM} ages can roughly indicate the formation age of the crust. Figure 3b is a histogram of Nd T_{DM} ages from mafic igneous rocks. The samples with ages >3.0 Ga account for c. 15% of the total, whereas those with ages <2.5 Ga account for only c. 7%, and samples with T_{DM} ages between 3.0 and 2.5 Ga account for 78% of the total. There are two discernible peaks, at c. 2.7 and 2.9 Ga (Fig. 3b). The diagram of $\varepsilon Nd_{(t)}$ v. t/Ga (Fig. 3a) shows two characteristics: all values of $\varepsilon Nd_{(t)}$ are positive, and there is an obvious change of $\varepsilon Nd_{(t)}$ with the change of t. The values of $\varepsilon Nd_{(t)}$ deviate from the depleted mantle evolution curve at about 3.0 Ga, which is attributable to contamination of crustal materials, and indicates that a thick continental crust existed in the NCC during the Neoarchaean. Rare earth elements (REE) also demonstrate the same tendency; for example, the higher La/Nb ratios of pre-3.0 Ga mafic rocks indicate the presence of a considerable continent crust by that time (Jahn 1990). The most mafic granulites and amphibolites from the NCC display REE patterns similar to those of basalts from island arc, continental margin and within-continent settings, indicating that these rocks formed in different tectonic settings.

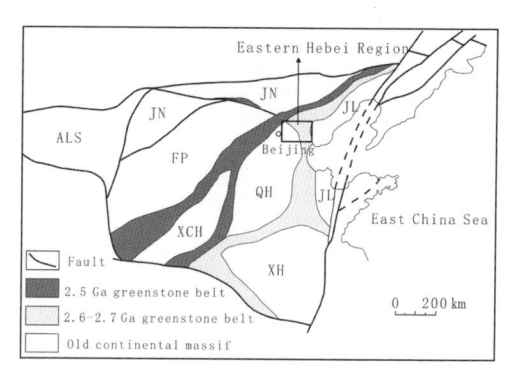

Fig. 4. Distribution of the greenstone belts and microblocks in the NCC.

greenstone belts of amphibolite facies include the Yanlingguan in the central–eastern NCC, the Dengfeng in the southwestern NCC, the Dongwufenzi in the northwestern NCC, the Wutaishan in the central–western NCC and the Qingyuan (Hongtoushan) in the northeastern NCC. The Yanlingguan greenstone belt formed at $c.$ 2700–2900 Ma, and is composed of bimodal volcanic rocks and sedimentary rocks metamorphosed to amphibolite facies. Mafic rocks are mainly amphibolites, which are of tholeiitic type. Most rocks exhibit slight light REE (LREE) enrichment and others have flat or LREE-depleted patterns (Wan et al. 1998; Zhai 2004). Some hornblendites and pyroxenites are similar to komatiites in chemistry, but typical spinifex texture is rare, possibly because of metamorphic recrystallization (Cheng & Xu 1991; Cheng & Kusky 2006). The metamorphosed sedimentary rocks include banded iron formations (BIFs), metapelites, and/or marble. The other four greenstone belts formed at the end of the Neoarchaean, although they contain some 2700–2900 Ma volcanic–sedimentary rocks and experienced a long evolution history (Zhai 1997; Geng 1998; Zhao et al. 1999; Geng et al. 2002). Their volcanic rocks have compositions ranging from basalt to andesite and to rhyolite, and show geochemical characteristics of an island arc association. The BIFs in the Wutaishan greenstone belt form industrial deposits of iron ore, and the Qingyuan greenstone belt contains Cu–Zn massive sulphide deposits with a few BIFs.

High-grade granulite–gneiss complexes are extensively developed in the NCC. The high-grade regions, in general, represent ancient crust, and are surrounded by greenstone belts. The main high-grade regions include the Baishanzhen and the Anshan in NE China, the Qian'an in Eastern Hebei, the Taishan in Shandong, the Huai'an at the junction of Hebei–Shanxi–Inner Mongolia, and the Lushan in central Henan. The greenstone belts and high-grade regions were generally intruded by

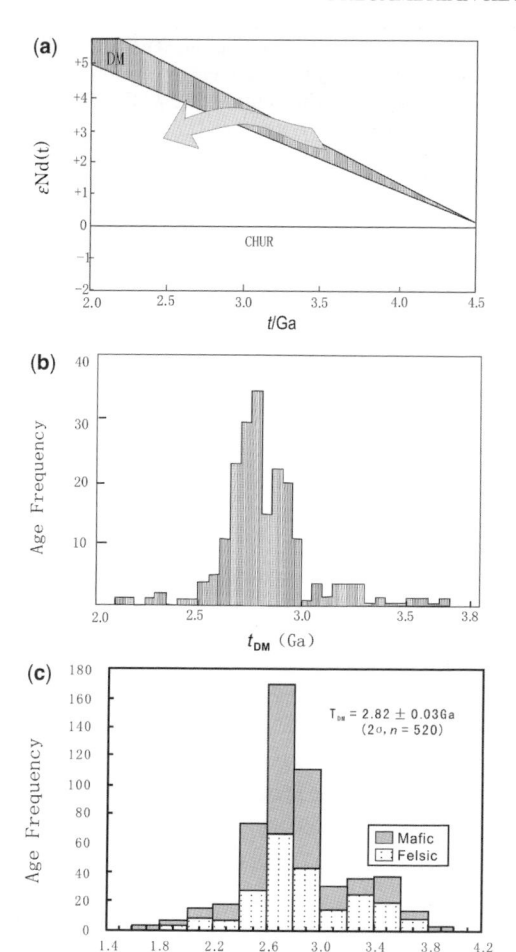

Fig. 3. $\varepsilon Nd_{(t)}$–t/Ga diagram (a), Sm–Nd t_{DM} age histogram (b), and Hf isotopic model ages (c) of mafic igneous rocks from the NCC.

However, few samples have mid-ocean ridge basalt (MORB) characteristics. The Hf isotopic model ages range from $c.$ 1950 to 3800 Ma; the main range is 2600–3000 Ma with a peak at 2820 Ma (Fig. 3c). However, zircon U–Pb ages from magmatic rocks, mainly orthogneisss, demonstrate several ranges of 3600–3800, 3000–3200 and 2700–2900 Ma, with a peak value at 2500–2600 Ma, the latest of which indicates a extensively crust partial melting event in the NCC.

Archaean regions and cratonization

Archaean regions in the NCC

Figure 4 shows the distribution of greenstone belts in the NCC. According to rock associations, granitic gneisses and charnockites, and underwent deformation together. The high-grade regions are composed of orthogneisses (80–90%), rootless metagabbros (5–10%) and slabs of supracrustal rocks (10–15%), associated with strong and complicated deformation. The supracrustal rocks comprise mafic granulites or amphibolites, BIFs, and medium-grained biotite gneisses with chemistries of slate–greywacke and intermediate–acid volcanic rocks (Sills *et al.* 1987). The Qian'an complex of granulite facies and the Anshan complex of amphibolite facies are characterized by abundant BIFs, although BIFs in high-grade regions are uncommon in the world. Field relations and geochemical characteristics show that these BIF-bearing supracrustal rocks and related intrusions were produced by subduction-related processes. Early arcs and back-arc basins that were engulfed by later plutonic rocks developed in the active continental margin (Zhai & Windley 1990; Windley 1995; Kröner *et al.* 2001).

Some previous studies (Deng *et al.* 1996; Ouyang & Zhang 1998; Wu *et al.* 1998; Zhai *et al.* 2005) proposed that the NCC can be divided into several microblocks with continental nuclei greater than *c.* 3.0–3.8 Ga old, which are surrounded by late Archaean and Proterozoic rocks. The greenstone belts are interpreted as arc–back-basin associations whereas the 3.0–2.5 Ga high-grade regions are thought to represent a continental margin or island arc association. The suggested microblocks are outlined by the greenstone belts (Fig. 4) and are, from east to west, the Jiaoliao (JL), Qianhuai (QH), Fuping (FP), Ji'ning (JN) and Alashan (ALS) blocks, with another two blocks, Xuhuai (XH) and Xuchang (XCH), in the south. Recent studies have led to a consensus that the basement of the NCC was composed of several blocks or terranes that were finally amalgamated to form a coherent craton (Zhao, Z. P. 1993; Wu *et al.* 1998; Zhai *et al.* 2000*a*; Kusky & Li 2003; Zhao, G. C. *et al.* 2005; Zhai & Peng. 2007).

The Archaean Eastern Hebei terrane: remnants of ancient oceanic crust or island arc

The Archaean Eastern Hebei region is a key area for understanding the early Precambrian evolution of the NCC (Fig. 5). Archaean rocks are well developed in the Eastern Hebei region, and include TTG and granitic gneisses, gabbroic intrusive rocks, and BIF-bearing supracrustal rocks metamorphosed to granulite and high-grade amphibolite facies (Geng 1998; Wu *et al.* 1998; Li, J. H. *et al.* 1999; Zhai & Liu 2003). These rocks constitute three lithological–tectonic units: the Paleoarchaean Caozhuang complex (CZC), the Meso-Neoarchaean Shuichang complex (SCC) and the Neoarchaean Zunhua

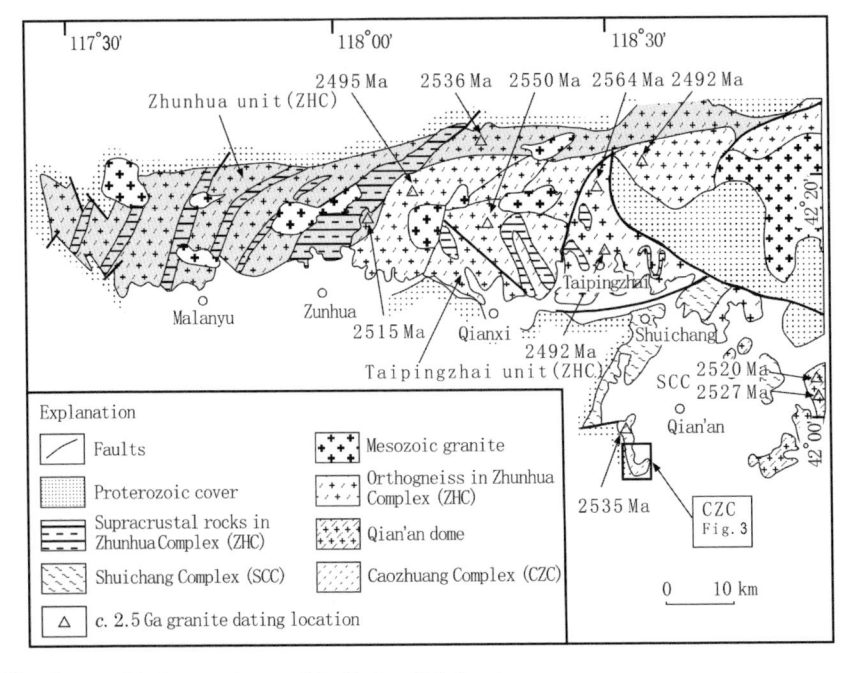

Fig. 5. Sketch map of Archaean outcrops of the Eastern Hebei region.

complex (ZHC) (Fig. 5). The CZC represents the oldest sial crust of 3.8–3.5 Ga in the NCC, occupying only c. 4 km². A Quaternary sequence covers the complex in the east, and the Palaeoproterozoic Changcheng System unconformably overlies the complex to the west and south. To the north, the CZC is in contact with the SCC by a ductile shear zone that is multiply deformed, and the CZC was finally thrust up on the SCC (Zhao, Z. P. 1993). The CZC includes metamorphosed supracrustal rocks and tonalitic gneisses named the Caozhuang Group and Huangboyü gneiss, respectively. The Caozhuang Group is composed of thin interlayered metavolcanic (30–53 vol.%) and sedimentary (65–70 vol.%) rocks, which include amphibolites, BIFs, serpentine marbles, fuchsite quartz schists and a few tremolitites and diopsidites. The protoliths of the amphibolites and tremolitites are tholeiitic or komatiitic. The amphibolites contain slightly high REE abundance with slightly enriched LREE patterns, and their contents of Cr (c. 226 ppm), Ni (c. 144 ppm) and MgO (up to 9.44%) are also slightly high. These basalts were proposed to be derived from larger-proportion mantle melting (Zhai et al. 2005). The Huangboyü tonalitic gneisses underwent very complicated deformation. Their zircon U–Pb ages are 3.4–3.2 Ga (Li, Z. Z. et al. 1980; Zhao, Z. P. 1993). The CZC is similar in its rock association and chemistry to other Meso-Neoarchaean high-grade complexes in the NCC (Li, Z. Z. et al. 1980; Wang, R. M. et al. 1985; Zhao, Z. P. 1993). Zhai & Liu (2003) suggested that this primary sial crust possibly was an old island arc related to subduction of intraoceanic crust. The SCC occurs as a dome with area of c. 700 km², and is, therefore, traditionally termed the Qian'an migmatite–granitoid uplift terrane (Qian et al. 1985) or the Qian'an gneiss dome (Geng et al. 2006a). The U–Pb zircon ages of diorite and granodiorite of the Qian'an dome are 2499 Ma and 2494 Ma respectively (Liu et al. 1990). It is unconformably covered by the Palaeoproterozoic Changcheng System and Mesozoic sediments on its western and eastern margins respectively. Its northern boundary is a fault that separates it from the Neoarchaean ZHC. The meta-supracrustal rocks and some banded orthogneisses commonly occur as complicated folded slabs along the margins of the terrane, especially on the western and northern margins. The main supracrustal rocks are mafic granulites, pyroxene amphibolites, pyroxene-bearing biotite–plagioclase gneisses, and a few garnet–sillimanite gneisses and garnet-bearing fayalite peridotites (eulysite). Abundant BIFs constitute important iron deposits in China. The supracrustal rocks underwent metamorphism of granulite facies with moderate pressure. The supracrustal rocks and intrusive granitic sills have Sm–Nd and zircon U–Pb ages of 3280–3049 Ma (Geng 1998). The metamorphosed supracrustal rocks and banded gneisses of the SCC are considered to be a typical example of an old island arc complex, based on the rock association, deformation and geochemistry (Sills et al. 1987; Wang, R. M. et al. 1985; Windley 1995). However, granitic and charnockitic bodies occurring along the northern margin yield much younger zircon U–Pb ages (2647–2495 Ma) than those of supracrustal rocks (Liu et al. 1990; Zhao, Z. P. 1993).

The ZHC (Li, T. S. 1999) can be divided into two parts: the linear Zunhua unit of high-grade amphibolite facies (partially granulite facies) and the dome-shaped occurrences of the Taipingzhai unit of granulite facies. The Zunhua unit occurs as strata trending NE–SW, and it mainly comprises supracrustal sequences with some granitoid sills. The Taipingzhai unit occurs as domes with complicated folds and underwent granulite-facies metamorphism; it mainly comprises TTG orthogneisses, gabbroic rocks and some small lenses of supracrustal rocks. The supracrustal rocks in the two units are almost the same in geochemistry, metamorphic history and geochronology. The main rock types include amphibolites or two-pyroxene granulites, biotite felsic gneisses or intermediate–acid granulites, and BIFs (He et al. 1991). The granitic gneisses of the Zunhua unit have zircon SHRIMP U–Pb ages of 2495–2536 Ma (Geng et al. 2006a). Metavolcanic rocks demonstrate an evolving trend from basaltic to andesitic and to rhyolitic, and have similar geochemistry to modern island arc volcanic rocks (Zhai et al. 1990; Wan et al. 1998; Li, J. H. et al. 1999). The Taipingzhai unit comprises tonalititc granulites and some metagabbros (mafic granulites). Their Rb–Sr and Sm–Nd isochron ages are 2470 ± 70 Ma and 2480 ± 125 Ma (Jahn & Zhang 1984; Jahn 1990), and they have single-grain zircon U–Pb ages of 2480–2530 Ma (Pidgeon 1980), and a zircon SHRIMP U–Pb age of 2564 Ma (Geng et al. 2006a). The Sm–Nd mineral–rock isochron ages for mafic rocks from Zunhua and Taipingzhai are 2591 ± 142 Ma and 2644 ± 112 Ma respectively (Li, T. S. 1999). The metamorphic history of mafic granulites is characterized by a mineral reaction texture in which clinopyroxene is surrounded by small garnets, indicating an anticlockwise PT path (Li, T. S. 1999). Therefore, Li, T. S. (1999) and Zhai & Liu (2001a) suggested that the Zunhua unit and the Taipingzhai unit jointly constitute a Neoarchaean island arc terrane, respectively representing the upper part and the root part. From the above, it can be seen that the eastern Hebei high-grade region is mainly composed of three ancient island arc terranes that formed during the Palaeoarchaean, Mesoarchaean and Neoarchaean periods respectively (Zhai & Liu 2003). This seems to indicate a tectonic process of island arc terrane accretion achieved by arc–arc or arc–micro continent collision (Zhai et al. 2005).

However, Wang, R. M. et al. (1985) suggested that the Zunhua unit is a high-grade metamorphosed greenstone belt. The geochemical characteristics of the metamorphosed mafic rocks have been modified by later metamorphism, and possibly they were oceanic tholeiitic. Kusky et al. (2001, 2004a, b), Li, J. H. et al. (2002), Huang et al. (2004), Kusky (2004a, b) and Huson et al. (2004) reported that ultramafic rocks occurring as lenses in Dongwanzi, in the northeastern part of the ZHC, were Archaean oceanic mantle. After opposition from Zhai et al. (2002), Zhang, Q. et al. (2003) and Zhao, G. C. et al. (2007) on the occurrence, geochemistry and rock association of the Dongwanzi ultramafic body, Kusky et al. (2007b) emphasized that the ophiolitic rocks had been intruded by five generations of magmatic rocks, and combined a series of ultramafic lenses from Dongwanzi, via Zunhua to Wutaishan and other basic metamorphic rocks, suggesting a Late Archaean ophiolite mélange zone. Li, J. H. & Kusky (2007) further suggested that the Wutaishan complex represents an island arc assemblage, and the Hutuo Complex represents a foreland complex. These two complexes and the Zunhua ophiolites together constitute an Archaean collisional zone in the central NCC, where finally the eastern Hebei terrane and western block amalgamated to form a coherent Archaean NCC. This model is different from those of Archaean arc–arc or arc–continent amalgamation (Zhai 1997, 2004) or Palaeoproterozoic continental collision (Zhao, G. C. et al. 1999, 2005).

Condie & Kröner (2008) proposed some tracking petrological assemblages as geological indicators in Early Precambrian geotectonic process for modern plate tectonics; these are ophiolites, arc or back-arc assemblages, accretionary prisms and ocean plate stratigraphy, a foreland basin, blueschists, a passive continental margin, and ultrahigh-pressure metamorphic rocks and paired metamorphic belts. Some of these petrological assemblages have been thought to occur in the Eastern Hebei region. For example, the BIFs and associated supracrustal rocks to some extent are similar to ocean plate stratigraphy. The TTG orthogneisses are similar to the Andean-type continental margin rock association, and the orthogneisses are converted by later deformation and associated high-grade metamorphism from the tonalities and granodiorites, although these magmatic and tectonic processes were probably more intense in the Archaean than in late Phanerozoic time (Windley 1986). However, ophiolites, blueschist and ultrahigh-pressure metamorphic rocks are difficult to find in the NCC. Voluminous granulites, TTG gneisses, komatiites, and the absence of UHP rocks and blueschists probably suggest that plume tectonics is important in the Archaean. In any case, it is reasonable that the processes of accretion and crustal growth in ancient and modern times may be different more in degree than in principle.

Cratonization and amalgamation event

Mainly based on early Precambrian geology of the Eastern Hebei region and combining other Archaean terranes, Cheng (1994) proposed a two-stage cratonization model of the NCC. The first stage took place in the Neoarchaean, when several microblocks were amalgamated and underwent granulite-facies metamorphism to form a present-scale NCC. The second stage took place at c. 2.5 Ga (the time of the boundary between the Archaean and Proterozoic), when the small ocean basins and continental rifts within the craton closed, with amphibolite-facies metamorphism and migmatization. Zhao, Z. P. (1993) suggested that the NCC fundamentally formed since c. 3.0 Ga and evolved to a platform-style craton at the end of the Neoarchaean, when it experienced sodium (TTG intrusive event), sodium–potassium (TTG and granitic intrusive event) and potassium (granitic intrusive event) cratonizations. The above-mentioned models are established on, respectively, the horizontal and vertical crustal growth.

Coeval ultramafic–syenitic dykes. Extensive granitic intrusive bodies or sills and mafic dyke swarms are marks of cratonization (Brown 1979; Zhao, Z. P. 1993; Windley 1995). Emplacement of ultramafic–mafic and alkaline dykes has long been considered to mark a post-orogenic or anorogenic event, and thus can be used to constrain the time of cratonization. The Neoarchaean granitoids and mafic dykes in the eastern Hebei region have been described in detail (Qian et al. 2005; Geng et al. 2006a).

Extremely rare Archaean coeval ultramafic–mafic and syenitic dykes have been found in the CZC, Eastern Hebei region (Li, T. S. et al. 2010). Figure 2a is a geological sketch map of the CZC. The > 3.0–3.3 Ga rocks occupy half of the area of outcrop, and include BIF-bearing supracrustal rocks and TTG–granitic gneisses. The other half consists of Neoarchaean granitic rocks, including hypersthene granite with an age of 2.52–2.56 Ga, granite of 2.53 Ga and monzodiorite of 2.60 Ga (Zhao, Z. P. 1993). The country rocks of the ultramafic–mafic and syenitic dykes are gneissic monzodiorite and BIF-bearing supracrustal rocks. The intrusive boundaries are clear. The dykes cut the gneissosity of the country rocks and show typical spherical weathering in the field. They occur as subvertical intrusions with a NW310° strike, and are up to 300 m long and about 10 m wide.

The ultramafic–mafic dykes are dark grey coloured. They consist mainly of clinopyroxene, olivine, and minor orthopyroxene, hornblende, biotite, Fe–Ti oxides and spinel, and may be with or without plagioclase. Depending on the existence of the plagioclase, the dykes can be subdivided into olivine pyroxenite and olivine gabbro, which have medium granular texture and gabbroic texture, respectively. Reaction rims of orthopyroxene and hornblende replacing olivine are common. Also observed in the dykes is exsolution of orthopyroxene along cleavages of clinopyroxene. The syenitic dykes (syenitite) consist predominantly of orthoclase and biotite, with minor plagioclase (<7%), quartz (<5%) and Fe–Ti oxides, without pyroxene. The olivine gabbro sample 04QA09 and syenitite sample 04QA08 for zircon SHRIMP U–Pb dating were collected in the Caozhuang–Naoyumen area at 39°55′55″N, 118°33′16″E. The Th/U ratios are 0.83–1.90 and 0.91–1.78 for zircons from the olivine gabbro and syenitite dykes, respectively. Because of low zircon amounts, only six zircon grains were determined from the olivine gabbro, and they yielded a weighted mean $^{207}Pb/^{206}Pb$ age of 2516 ± 26 Ma (Fig. 6a), interpreted as the crystallization age. Fifteen zircon grains from the syenitite dyke yielded a weighted mean $^{207}Pb/^{206}Pb$ age of 2504 ± 11 Ma (Fig. 6b), also interpreted as the crystallization age of the syenitite. Zircons from the olivine gabbro have $^{176}Hf/^{177}Hf$ ratios varying from 0.281286 to 0.281368, yielding single-stage Hf model ages between 2668 and 2740 Ma, with a mean age of 2705 Ma, and $\varepsilon_{Hf}(t)$ values between 1.9 and 4.2, with a weighted mean value of 3.13 ± 0.40. Zircons from the syenitite have $^{176}Hf/^{177}Hf$ ratios varying from 0.281280 to 0.281382, yielding single-stage Hf model ages between 2646 and 2705 Ma, with a mean value of 2677 Ma, and $\varepsilon_{Hf}(t)$ values between 2.7 and 4.4, with a weighted mean value of 3.56 ± 0.23.

The olivine gabbros with Mg-numbers values of 59–63 are similar to high-magnesian tholeiitic basalt. They demonstrate relatively LREE-enriched patterns without Eu anomalies ($La/Yb_N = 9.28$–9.78, $Gd/Yb_N = 2.8$, $\delta Eu = 0.96$–1.00), with enrichment in large ion lithophile elements (LILE) and depletion in high field strength elements (HFSE). They also have high Cr and Ni contents, and have Zr/Hf and Nb/Ta ratios similar to those of the primitive mantle. The syenitites are alkaline in composition, with 8.67–8.88 wt% $Na_2O + K_2O$, and show high total REE contents (543–854 ppm) and strongly LREE-enriched patterns with minor Eu negative anomalies ($La/Yb_N = 50$–101, $Gd/Yb_N = 5.2$–5.3, $\delta Eu = 0.80$–0.82). Spider diagrams of trace elements show a strong LILE enrichment relative to HFSE. Zr/Hf and Nb/Ta ratios are 47–49 and 26–76, respectively, and are much higher than those of

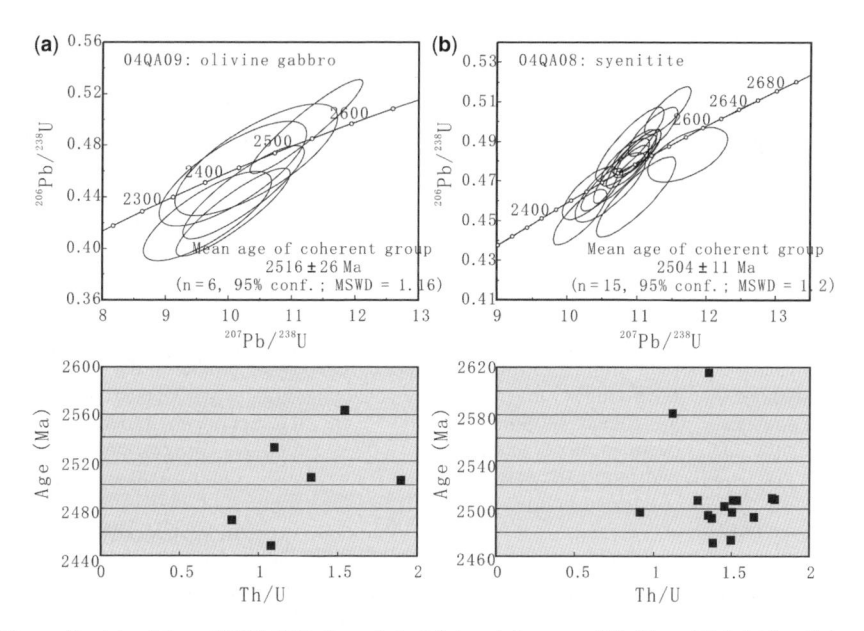

Fig. 6. Concordia plots of zircon SHRIMP U–Pb analytical data and diagrams of Th/U age (Ma): (**a**) for olivine gabbro (04QA09; (**b**) for syenitite (04QA08).

primitive mantle and higher than the average Nb/Ta ratio of post-Archaean continental crust. The petrological and geochemical features indicate that these dykes were derived from a deep subcontinental lithospheric mantle source, which implies that the NCC probably had a large-scale continental crust with a considerable thickness. Although it is difficult to estimate the depth of the magma source, we may confirm that the coeval ultramafic–mafic and syenitic dykes should be derived from a deep mantle based on their petrology and geochemistry.

Archaean–Proterozoic boundary and cratonization. The coeval olivine gabbro and syenitite dykes yield c. 2.5 Ga zircon SHRIMP U–Pb ages. It is note worthy that 2.5 Ga is the boundary age between the Archaean and Proterozoic. In the period between c. 2500 and 2350 Ma, the NCC was tectonically inactive, but the development of Palaeoproterozoic volcanic and granitic rocks occurred between 2350 and 1950 Ma. The Archaean rocks and Proterozoic rocks demonstrate an enormous difference in rock association and geochemistry. The connotation of the Archaean–Proterozoic boundary is global cratonization and basically formation of continents at their present scale. Windley (1995), Condie et al. (2001) and Rogers & Santosh (2004) suggested that there was a c. 2.5 Ga Neoarchaean supercontinent, followed by a continental break-up in the Palaeoproterozoic.

Extensive granitic intrusive bodies or sills and mafic dyke swarms are marks of cratonization (Zhao, Z. P. 1993; Windley 1995). Emplacement of ultramafic–mafic and alkaline dykes has long been considered to mark a post-orogenic or anorogenic event, and thus can be used to constrain the time of cratonization. The Neoarchaean granitoids and mafic dykes in the eastern Hebei region have been described in detail (Qian et al. 2005; Geng et al. 2006a). Figure 5 shows the locations of the Late Neoarchaean granites with SHRIMP zircon U–Pb ages, which are widely distributed in the Eastern Hebei region and intruded into various metamorphic supracrustal rocks and magmatic rocks (Geng et al. 2006a; J. H. Yang, pers. comm.). The granites occur as sills and small bodies, accompanied by strong migmatization and pegmatite intrusion. As the youngest magmatic intrusion in Archaean, the coeval ultramafic–mafic and syenitic dykes were intruded into the c. 2.5–2.6 Ga granite sills and migmatite and pegmatite, indicating that the Archaean cratonization was finally accomplished.

The other microcontinental blocks in the NCC are the JL, QH, FP, JN, XCH, XH and ALS microblocks (Fig. 4). The rock types and distribution in these microblocks display distinct differences. For example, old rocks with ages up to 3.8 Ga and abundant Mesoarchaean BIFs are present only in the QH block. Rocks older than Neoarchaean are not exposed in the JN and FP blocks, although they may exist in the deep crust, based on geophysical data. Neoarchaean volcanism and magmatism in these blocks took place at 2.9–2.7 Ga and 2.6–2.45 Ga, but their occurrence in different blocks varies greatly. Volcanic activity from 2.9 to 2.7 Ga was, in general, strong in all blocks, especially in the JL, QH and XCH blocks, associated with abundant BIFs. However, BIFs are not common in the ALS block (Geng et al. 2006b). Volcanic activity at 2.5 Ga was weak in the JN block; however, it was rather intense in the JL, FP and XCH blocks. Basic–intermediate–acid volcanic rocks in the FP and XCH areas are closely associated with the BIFs. In the JL block, however, volcanic rocks contain massive sulphide Cu–Zn ores. All these differences indicate that these microblocks possibly developed in different tectonic settings; that is, they had not been amalgamated into a coherent craton until at least c. 2.5 Ga. It is noteworthy that nearly all Archaean rocks underwent metamorphism at c. 2.5–2.55 Ga, and voluminous granitic sills with an age of c. 2.5 Ga were intruded into neighbouring blocks, suggesting amalgamation prior to intrusion. For example, a series of granite bodies are located along the contact zone between the JL and QH blocks (Wu et al. 1998), indicating that the microblocks were assembled and constituted a combined NCC by the end of the Neoarchaean (Li, J. H. et al. 1999). Recent published data reveal that the granite sills and pegmatites with 2457–2570 Ma zircon U–Pb ages are extensively developed in all six microblocks (Zhang, W. J. et al. 2000; Guan et al. 2002; Luo et al. 2004; Shen et al. 2004; Wang, Z. J. et al. 2004; Yang et al. 2004; Gao et al. 2005; Jian et al. 2005; Kröner et al. 2005; Li, J.-H. et al. 2005; Wan et al. 2005; Wilde et al. 2005; Lu, X. P. et al. 2006; Zhang, H. F. et al. 2006 Liu, S. W. et al. 2007a, b). For example, the Molihong high-potassium granite and Caiyu hypersthene-bearing granite in the JL block, the Angou granite and Shipaihe granodiorite in the XCH block, the Wanzi granite and Pingshan granite in the FP block, the Guyang granite in the JN block, and the Shanhaiguan granite and Qianan granite in the QH block are representative intrusive bodies with c. 2.5 Ga zircon U–Pb ages, and clearly marked the Neoarchaean cratonization of the NCC (Zhao, Z. P. 1993; Wang, Y. J. et al. 2004; Li, J. H. et al. 2006; Zhai & Peng 2007). Therefore, it is reasonable that the NCC at the present scale formed at the end of the Archaean. Qian et al. (1985) proposed an important tectonic unconformity event between the Archaean basement and Palaeoproterozoic metamorphic sedimentary rocks (khondalite series). Although both of them underwent high-grade metamorphism and complicated deformation, they show substantial differences in rock association, geochemistry, structural style and other features. The khondalite series are, as a common Precambrian rock association, distributed throughout the NCC, and are called the Fengzhen Formation in the JN block, the Lüliang Formation in the FP block, the Fengzishan Formation in the JL block, the Huoshan Formation in the XH block, the Louzishan Formation in the QH block, and the Lushan Formation in the XCH block. Detailed geological and geochemical studies support that the khondalite series were mainly formed at $c.$ 2.30 Ga; this series is mostly composed of argillo-arenaceous rocks, and therefore requires a relatively large-scale Archaean cratonic basement (Wan *et al.* 2000*a, b*).

Palaeoproterozoic mobile belts and high-temperature–high-pressure metamorphism

Palaeoproterozoic mobile belts

The NCC behaved as a stable continent block without tectonic–thermal action during the period $c.$ 2500–2350 Ma, similar to other cratons in the world (Condie *et al.* 2001). A Palaeoproterozoic volcanic and granitic event with an age of 2300–1970 Ma is widespread in the NCC. These Palaeoproterozoic volcanic–sedimentary rocks typically occur as linear fold belts, commonly unconformable over Archaean basement, and underwent low-grade metamorphism and intrusion by granites. Their occurrences are different from the Archaean greenstone–granite belts and demonstrate Phanerozoic orogenic characteristics in some places; therefore Zhai & Peng (2007) termed them Palaeoproterozoic mobile belts. Three Palaeoproterozoic mobile belts in the NCC are shown in Figure 7, which are located, respectively, in northeastern China, central–western China and northwestern China. The representative rock sequences are the Liaohe Group and the Fengzishan Group in the Liaoji Mobile Belt, the Lüliang Group, Hutuo Group and Zhongtiao Group in the Jinyü Mobile Belt, and the Fengzhen Group in the Fengzhen Mobile Belt.

The volcanic–sedimentary rocks in the Liaoji Mobile Belt are called the Liaohe Group in NE China and the Fenzishan Group in Eastern Shandong Peninsula, and are, respectively, metamorphosed to greenschist–amphibolite facies and amphibolite–granulite facies. The rocks are divided into two formations. The lower formation is composed of basic–acid volcanic rocks and sedimentary rocks. The volcanic rocks have bimodal

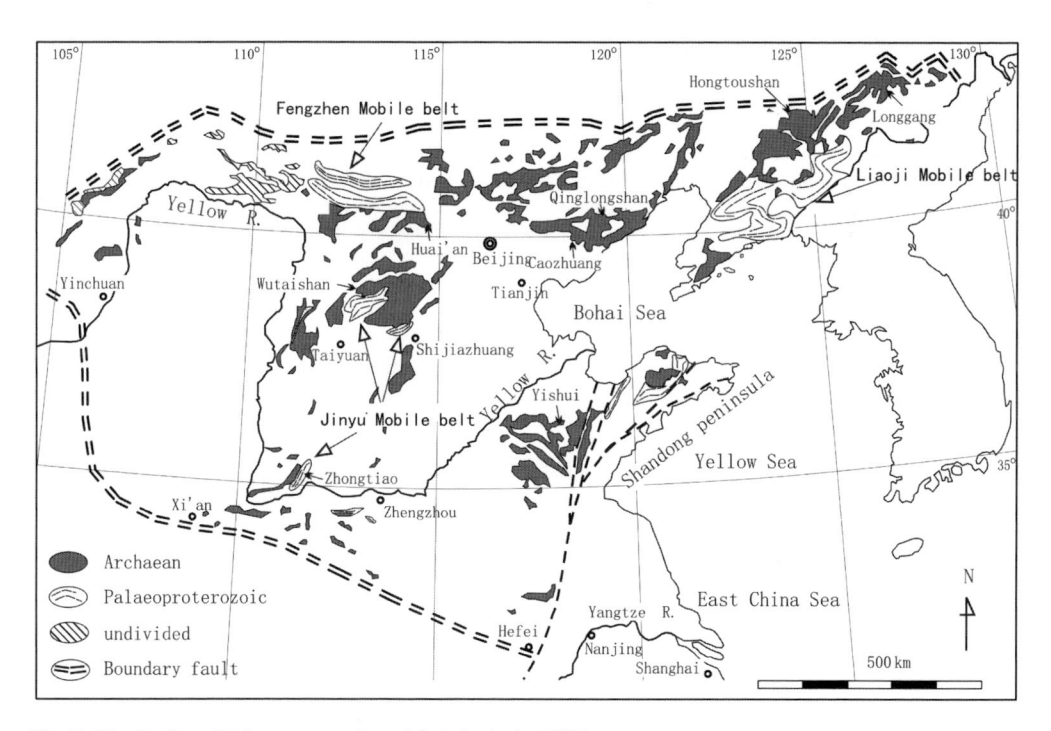

Fig. 7. Distribution of Palaeoproterozoic mobile belts in the NCC.

petrological and geochemical characteristics. The basic rocks have the chemistry of continental tholeiitic basalt and the felsic volcanic rocks show Si oversaturation and are enriched in Na and K. The sodium-felsic rocks contain thick boron formations with tourmalinization. The sedimentary rocks include argillaceous schists, siltstones and marbles, in which graphites commonly occur. The upper formation is layered detrital–carbonate sedimentary rocks containing a major magnesite mineral deposit. Zhang, Q. S. (1984) suggested that these rocks formed in a rift environment. However, Bai et al. (1993) emphasized that basic volcanic rocks in the Liaohe Group have the characteristics of marine basalts and sedimentary rocks are turbiditic. Three metamorphism stages have been described by Li, S. Z. et al. (1998): vertically progressive, laterally progressive and extensional. Deformed Palaeoproterozoic granite and undeformed granites yield zircon U–Pb ages of 2160 Ma and 1850 Ma, respectively, and the later age shows that the latest deformation was earlier than 1850 Ma. The Liaoji Mobile Belt has been suggested to be a continental rift basin or a small old ocean basin located between the Longgang massif in NE China and the Nangrim massif in the northern Korea peninsula. With a change in tectonic regime from extension to compression, the formation of the Liaohe Mobile Belt is related to a change from subduction to collision (Bai et al. 1993; Chen et al. 2003; Lu, X. P. et al. 2006).

The Jinyü Mobile Belt includes several Palaeoproterozoic formations, which are developed in the region 34°15′–39°N and 110°45′–114°50′E. The rocks have similar protoliths, tectonic style and metamorphic history, and are named the Hutuo Group, Lüliang Group and Zhongtiao Group in central–northern Shanxi, western Shanxi, and southwestern Shanxi–northwestern Henan Provinces. The Hutuo Group is composed of conglomerates, sandstones, pelites, basic volcanic rocks, and basic and felsic tuffs. Some basic volcanic rocks or basic dykes have three groups of SHRIMP zircon U–Pb ages of c. 2450–2390 c. 2200–2000 and c. 1900–1800 Ma (J. S. Wu et al. pers. comm., 2008; Z. J. Wang, pers. comm., 2008). We prefer to interpret them, respectively, as residual, magmatic and metamorphic ages (Zhai & Peng 2007), although other researchers have different opinions. On the basis of the rock associations and geochemistry, these rocks have been interpreted to have been deposited in a continental rift (Zhao, Z. P. 1993; Miao et al. 1999) or foreland basin (Li, J. H. et al. 2000, 2006). The sedimentary rocks in the Zhongtiao Group are mainly coarse- to medium-grained clastic rocks, pelites and carbonates, and the volcanic rocks are basalts, basaltic andesites and minor dacites. Sun & Hu (1993) reported that the geochemical characteristics of basic and intermediate–acid volcanic rocks are similar to those of island arc volcanic rocks. Their metamorphic grade is greenschist or amphibolite facies. The geochronological data for the metamorphosed volcanic rocks, granites and related Cu-ores are consistent, and can be divided into three rock-forming periods at 2.36–2.32 Ga, 2.16–2.01 Ga and 2.09–2.06 Ga, and a metamorphic period at 1.9–1.83 Ga. The geochronological data are similar to those of the Hutuo Group. Basaltic rocks in the Lüliang Group show geochemical characteristics of continental basalt with $\varepsilon Nd(t_{DM})$ of +3.0. The felsic volcanic rocks are rich in LILE and ΣREE have high La/Yb values (Geng et al. 2003). Zircon U–Pb ages are 2360–2031 Ma for basic volcanic rocks, and 2050–2031 Ma for granites (Yu et al. 1996, 1997; Wan et al. 2000a, b). The rocks of the Lüliang Group also record a 1866–1850 Ma metamorphic event, associated with charnockite intrusion at 1800 Ma (Yu et al. 1997; Geng et al. 2003).

The Fengzhen Group in the Fengzhen Mobile Belt is a thick sequence of metasedimentary rocks rich in aluminium, containing little volcanic material, which underwent granulite-facies metamorphism (Qian et al. 1985; Shen et al. 1992; Wu et al. 2000). The two formations of the Fengzhen Group are a lower metamorphosed detrital–pelite formation and an upper carbonate formation. The lower formation includes mainly quartzite with or without garnet, graphite gneiss, mica–quartz schist and garnet–sillimanite gneiss, which are called khondalite sequences. The upper formation is mainly composed of marbles with a few mica schists and calc-silicate rocks. Although some geologists believe that the Fengzhen Group is Archaean in age based on geological features (Zhao, Z. P. 1993; Lu, L. Z. et al. 1995; Qian 1996), zircon U–Pb ages of the khondalites support an interpretation of their being Palaeoproterozoic (Wan et al. 2000a; Wu et al. 2000; Zhao, G. C. et al. 2005). The protoliths of these rocks were deposited either in a cratonic basin (Condie et al. 1992; Qian 1996; Wan et al. 2000a; Xu et al. 2005) or in a marginal sea (Lu, L. Z. et al. 1995). Another opinion worth noting is that the clockwise PT path of the khondalites from moderate-pressure granulite facies to amphibolite facies may be considered to represent a continental collision (Wu et al. 2000). Recently, high-pressure and ultrahigh-temperature granulites have also been reported in the Fengzhen Group (Lu, L. Z. et al. 1995; Guo et al. 2006; Santosh et al. 2007). The ages of ultrahigh-temperature metamorphism and retrograde high-temperature metamorphism of sapphirine-bearing Mg–Al granulites are 1910–1900 Ma and 1850–1870 Ma. The garnet granites are commonly associated with the khondalites, showing intrusive or transitional contact. The garnet granites and khondalitic rocks have similar geochemical characteristics. The zircon U–Pb ages for garnet granites are 1912–1892 Ma (Guo et al. 1999a, 2005). Zhai et al. (2003a) suggested that the garnet granites are crustal partial melting granites related to an ultrahigh- or high-temperature metamorphic event, the protolith rocks are khondalitic rocks, and the estimated depth of anatexis is 35–40 km.

In short, the Palaeoproterozoic mobile belts in the NCC formed and evolved within a craton or continental margin at c. 2350–1950 Ma. Some 2350–1900 Ma rift-margin and passive continental margin deposits (St-Onge & Lucas 1990; Windley 1995), ophiolites (Kontinen 1987; Helmstaedt & Scott 1992; Kusky 2004a; Furnes et al. 2007), orogens (Hoffman 1988; Condie 2007) and BIF-bearing foreland basins (Hoffman 1987; Giles et al. 2002) in other cratons seem to be comparable with many in the late Phanerozoic (Windley 1995). Compared with other Palaeoproterozoic basins and belts in the world, we agree that these rocks herald a new major stage in Earth history, and their geotectonic environment is different from comparable Archaean examples (Windley 1986). The Palaeoproterozoic mobile belts in the NCC have Phanerozoic orogenic characteristics, indicating that plate tectonics was operative in the NCC at a much smaller scale.

High-pressure granulites and retrogressed eclogites

High-pressure mafic granulites and retrogressed eclogites were discovered in 1992 and 1995, respectively (Zhai et al. 1992, 1995). The first locations are situated at the junction of the QH, JN and FP blocks, in the western–central NCC. The high-pressure rocks occur within granitoid gneisses that commonly underwent strong deformation and mylonitization, associated with granite sills and pegmatite dykes. Therefore, the high-pressure granulites, retrogressed eclogites and their country rocks were named the Sanggan–Chengde structural belt or central zone, which has been proposed to be an Early Precambrian continent–continent collision belt between the QH block and FP–JN blocks, or a western block and an eastern block, similar to a modern high-pressure orogenic belt (Zhai et al. 1992; Guo et al. 1993; Zhai 1997; Zhao et al. 1999). The high-pressure rocks have two types of occurrence, as follows. (1) Long flat tectonic slabs occur at the northern boundary of the Sanggan–Chengde structural belt (Guo et al. 1993; Li et al. 1997). The high-pressure granulite layers developed together with two-pyroxene mafic granulites and intermediate to acid (tonalitic) granulites; for example, at Manjinggou, NW Hebei. They are locally sheared and mylonitized. (2) Lenses within tonalitic gneisses, migmatites and granites are distributed throughout the Sanggan–Chengde structural belt. The lenses are normally several metres to more than 10 m long, and commonly occur in groups. Locally, tens of lenses different sizes occur in narrow belts, for example at Baimashi in northern Shanxi, or as deformed or broken dykes in garnet-bearing felsic orthogneiss, for example at Mashikou in eastern Shandong (Fig. 8).

Early Precambrian high-pressure granulites in the NCC, which were first defined by Zhai et al. (1992), are based on garnet coexisting with plagioclase and pyroxene in the mafic granulites (Carney et al. 1991; Harley 1988), indicating a metamorphic mineral reaction of garnet in quartz tholeiite. Carswell & O'Brien (1993) defined these metamorphic mineral assemblages as garnet granulite facies. The typical mineral assemblages of the high-pressure mafic rocks in the NCC are clinopyroxene (Cpx) + Grt ± Plg + Qtz + rutile, Cpx + Grt + Hb + Plg ± Qtz and Cpx ± orthopyroxene (Opx) + Grt + Plg ± Qtz. The garnets are surrounded by a symplectite of fine-grained Opx + Cpx + Plg, indicating decompression. The high-pressure rocks also underwent a strong metamorphic overprint in the amphibolite facies and, as a result, some mafic granulites have been completely or partially retrogressed to amphibolites. Liu et al. (1996) suggested two metamorphic episodes of high-pressure granulite facies in the NCC with and without orthopyroxene, whereas O'Brien et al. (2005) proposed that these garnet granulites may have formed in the same metamorphic dynamic process but were situated at different crustal depths. A minor amount of eclogites were retrogressed to garnet granulites, which are found in Baimashi, Hengshan, Shanxi Province only. Mineral assemblages and reaction textures reveal three metamorphic stages that are successively of eclogite, high-pressure granulite and amphibolite facies. The mineral assemblage of the first eclogite stage is composed of Ca-rich garnet + pseudomorphed omphacite + rutile + quartz, which are preserved as inclusion minerals in garnets. Pseudomorph omphacite is composed of a very fine-grained aggregation of hypersthene, diopside and albite, which retains the crystal form and whole chemical composition of omphacite. Some matrix clinopyroxenes have been broken down to a vermiform albite and Na-poor clinopyroxene symplectite, indicating previous omphacite. The estimated P–T conditions of the three metamorphic stages for the high-pressure granulites are 1.2–1.4 GPa and c. 800 °C, 0.7–0.9 GPa and c. 820 °C

Fig. 8. Photographs showing the gabbroic dykes of high-pressure granulite facies and their country rocks. (**a–c**) HP granulite dyke in orthogneiss; (**d**) garnet porphyroblast with symplectite of HP granulite; (**e**) country gneiss with folded dykes; (**f**) country augen gneiss; (**g**) folded country gneiss; (**h**) HP granulite lens in deformed country gneiss.

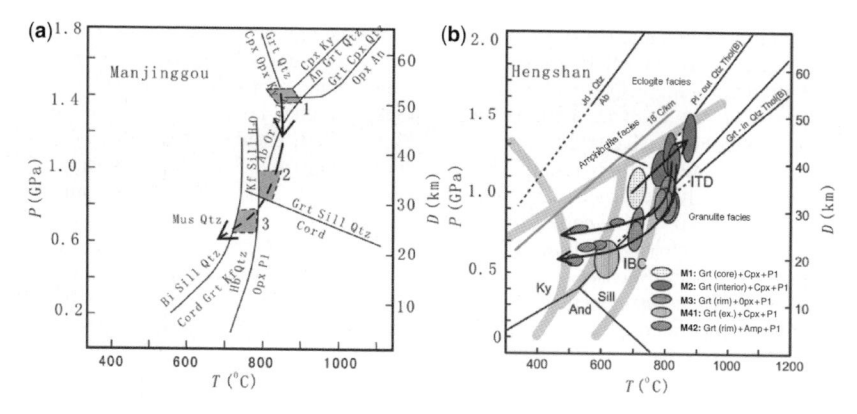

Fig. 9. P–T path of HP granulites from Manjinggou (a) and Hengshan (b).

and 0.5–0.7 GPa and c. 600 °C, with a decompressional PT path (Fig. 9), revealing that the granulites were uplifted from depths of 45–50 km to <20 km (Zhai et al. 1992; Guo et al. 1999b, 2005; Zhai & Liu 2001b). Protoliths of the mafic high-pressure granulites and retrogressed eclogites are gabbroic dykes and small bodies, as found by studies on petrology and geochemistry (Zhai & Liu 2001b; Kröner et al. 2005; Zhao, G. C. et al. 2005). The retrogressed eclogites have similar metamorphic P–T path; the conditions are c. 1.6–1.7 GPa and c. 700–800 °C, 0.8–1.1 GPa and c. 820 °C, 0.5–0.7 GPa and c. 600 °C (Zhai et al. 1995, 2000b).

Several geochronological methods have been used to determine the age of these events. The SHRIMP zircon U–Pb data concentrate in three ranges: 2000–1900 Ma, 1860–1830 Ma and 1810–1760 Ma (Guo et al. 2005; Kröner et al. 2005; Peng et al. 2005; Zhang, H. F. et al. 2006), which, respectively, represent high-pressure granulite-facies, moderate-pressure granulite-facies and amphibolite-facies metamorphic stages. The whole-rock Nd t_{DM} ages are 2640–2510 Ma, and the mineral Sm–Nd isochron ages are 1860–1820 Ma, which have been interpreted by Zhai & Liu (2003) and Guo et al. (2005) to be ages of magma source and granulite-facies metamorphism. However, the timing of the high-pressure metamorphic stage is still controversial. For example, one interpretation is that the age of high-pressure metamorphism is c. 2.5 Ga and the age of moderate-pressure granulite-facies metamorphism is c. 1.83 Ga, as indicated by symplectite minerals (Zhang, G. H. et al. 1998; Hu et al. 1999; Mao et al. 1999; Li, J. H. et al. 2000), and another is that the age of high-pressure metamorphism is c. 1.82–1.86 Ga and that the protolith age is c. 2.5 Ga (Zhao, G. C. et al. 1999, 2000; Guo et al. 2001, 2005; Kröner et al. 2001).

Recently, more locations of high-pressure granulites have been reported (Ma & Wang 1994; Li, J. H. et al. 1998; Liu et al. 1998; Zhai et al. 2000b; Tang et al. 2003), in Sifangdong, Miyun, Chengde and Jianping in the northern NCC from west to east; in Yangyuan, Fuping, Zanhuang and Xingtai in the central NCC; in Lushan, Xinyang in the southern NCC; and in Qixia, Laixi, Laiyang and Pingdu in the eastern NCC. Therefore, high-pressure granulites are developed almost throughout the area where ancient metamorphic basement rocks crop out in the NCC, not only along the Sanggan structural belt or in the central zone (Fig. 10).

High- and ultrahigh-temperature granulite-facies rocks

High-temperature (HT) granulite-facies rocks are extensively distributed in the NCC (Fig. 8). The typical rocks are khondalites, garnet–pyroxene granulites and hypersthene granites. The mineral

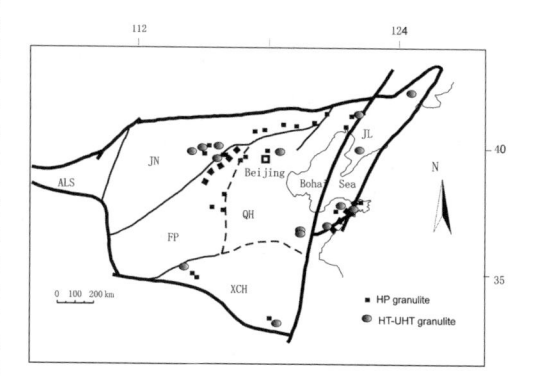

Fig. 10. Distribution of high-pressure granulites and high–ultrahigh-temperature granulites in the NCC.

assemblages of Grt + sillimanite (Sill) + cordierite (Cord) + Plg + Qtz ± rutile ± spinel (Spl), Grt + Cpx + Opx ± graphite + Plg + Qtz, biotite (Bi) + Plg + K-feldspar + Opx ± Cord are marks of high-temperature metamorphism. Hypersthene granites are commonly associated with granulite-facies rocks, and used to be called charnockite or enderbite. They were derived from partial melting from old crust and closely followed high-temperature metamorphism (Shen et al. 1992; Bai et al. 1993; Lu, L. Z. et al. 1995; Zhai et al. 2003a). Ultrahigh-temperature (UHT) granulites have been reported by Guo et al. (2006) and Santosh et al. (2007), which are sapphirine (Spr)bearing khondalites. Two localities are Tuguiwula and Wuchuan in the Fengzhen Mobile Belt, Inner Mongolia. The representative UHT mineral assemblages are Spr + Grt + Spl, Grt + Spl + Sill, Spr + Qtz + Sill + Spl, and the estimated PT conditions for this assemblage are c. 1.0 GPa and >1050 °C. Secondary mineral assemblages are Bi (high-Ti) + Spr + Sill + Spl, Sil + Cord + Qtz + Grt and Spl + Opx + Qtz, and the estimated PT conditions for these are c. 1.0 GPa and 900–1000 °C. The third mineral assemblage is Spl + Plg ± Sill + magnetite (c. 0.7–0.8 GPa, 850–900 °C), and the occurrence of chlorite and high-Fe biotite may represent a later mineral assemblage. HT and UHT rocks have a complicated metamorphic history in three or four stages, and commonly underwent initial isobaric cooling followed by isothermal decompression (Fig. 11; Guo et al. 2006; Santosh et al. 2007). However, the metamorphic process from LT–LP middle crustal rocks to hot lower crustal granulites was not been recorded because of overprinting by a late HT metamorphism. Lu, L. Z. et al (1995) suggested that khondalites in the Fengzhen Mobile Belt also underwent a high-pressure metamorphic stage before the HT–UHT stage with remnants of kyanite (Ky), but the Ky remnants are very difficult to find. Zhou et al. (2004) reported HP granulite-facies metamorphosed pelite with Grt + Ky + perthite in khondalitic rocks of the Fengzishan Group in the Liaoji Mobile Belt, eastern NCC. Wan et al. (2000a, b, 2003) summarized that the khondalitic rocks were deposited at 2400–2200 Ma and metamorphosed at 1880–1820 Ma, and Zhang, H. F. et al. (2006) and Santosh et al. (2007) further suggested that metamorphic ages can be subdivided into a HT–UHT metamorphic stage at 1930–1900 Ma and a secondary HT–MT metamorphic stage at 1850–1820 Ma.

Relationship of HP and HT–UHT metamorphic rocks and basement uplift

The times of metamorphism of the HP granulites and HT–UHT rocks are similar, as mentioned above, and these two kinds of granulite-facies rocks commonly occur in the same area. For example, in Laiyang and Laixi in Shandong, Huangtuyao in Inner Mongolia, Manjinggou in northern Hebei, and Gushan in northeastern Shanxi, the mafic garnet granulites are next to the khondalites. Liu & Li (2008) emphasized this distribution of HP granulites and HT–UHT metamorphic rocks, and suggested a Palaeoproterozoic paired zone in the southern part of Central Inner Mongolia. Moreover, the possible transitional rocks from HP to HT–UHT may be found; for example, garnet–two-pyroxene granulites with or without graphite are located between HP granulites and khondalites in Songtan and Haojiatan, eastern Shandong, and their PT condition is 1.0–1.1 GPa and c. 750-850 °C, between those of the HP and HT–UHT rocks (Zhai et al. 2000b). The HT–HP khondalites with Ky + Grt + K-feldspar in the HT khondalite sequence (Lu, L. Z. 1995; Zhou et al. 2004) indicate that the HP

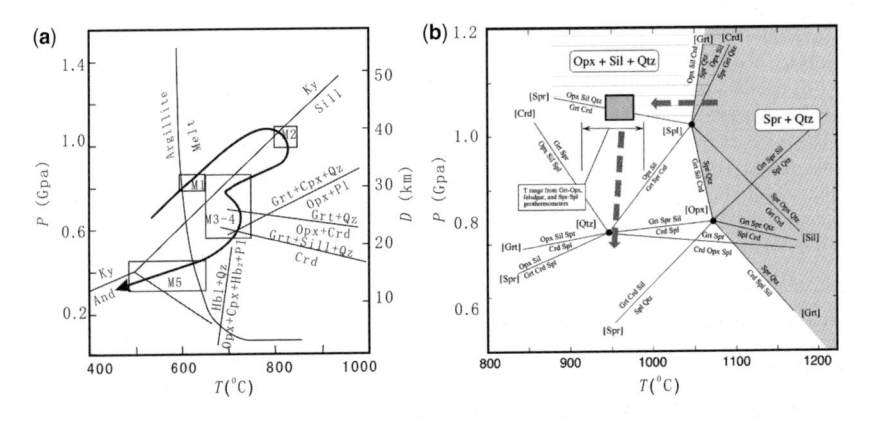

Fig. 11. P–T path of UHT rocks from Wuchan (a) and Tuguiwula (b).

and HT–UHT rocks underwent a continuously changing range of $P-T$ conditions. Both rocks have a peak metamorphic age at c. 1900 Ma followed by a c. 1850 Ma retrograde metamorphism, and underwent deformation with a foliation striking ENE–WSW and dipping NNW at c. 70°.

The absence of blueschists and ultrahigh-pressure metamorphic terranes make it difficult to support the operation of modern plate tectonics in the Early Precambrian, although various researchers have suggested a continental collision model based on HP granulites (Zhai et al. 1992, 1995; Zhao, G. C. et al. 1999; Li, J. H. et al. 2000; Kusky et al. 2001). The high geothermal gradient seems to indicate a shallow subduction. The estimated geothermal gradients are c. 16 °C km^{-1} for UP rocks and c. 22–25 °C km^{-1} for HT–UHT rocks, but c. 6 °C km^{-1} for UHP rocks in the Dabieshan orogenic belt. The estimated uplift rates are 0.33–0.5 mm a^{-1} for UP and UT–UHT rocks in the NCC, but 3–5 mm a^{-1} for rocks in the Dabieshan orogenic belt. Therefore, the uplift rates of early Precambrian high-grade metamorphic rocks in the NCC are much slower than and very different from the rates in Phanerozoic orogenic belts. The uplift rates of early Precambrian high-grade metamorphic rocks are estimated from three main metamorphic epochs of HP granulite, MP granulite and amphibolite facies, representing uplift from lower, middle–lower and middle crustal levels. The time for uplift to the surface is ascertained from 1780 Ma unmetamorphosed mafic dykes and 1780–1760 Ma unmetamorphosed volcanic rocks in the Palaeoproterozoic Xiong'er–Yanshan aulacogens that unconformably covered the high-grade metamorphic basements. The fact that HT–UHT rocks are associated with HP granulites indicates a more complicated tectonic process than modern plate tectonics. Various workers (Zhai & Liu 2003; Zhai et al. 2005; Zhai & Peng 2007) gave attention to the occurrence and metamorphic ages of HP and HT–UHT rocks in the NCC, and indicated the further possibility that these rocks represent the lowermost–lower crustal rocks with a high geothermal gradient, and were probably distributed in a broad area rather than in a narrow zone. These lower crustal rocks were rather slowly uplifted to the surface relative to the rate for Phanerozoic orogenic belts by an unknown tectonic mechanism.

Mafic dyke swarms and Palaeo-Mesoproterozoic continental rifting

Mafic dyke swarms

The Palaeoproterozoic mafic dyke swarm, extending over 1000 km in the NCC, plays an important

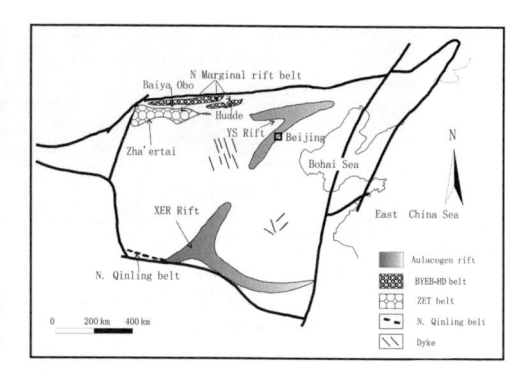

Fig. 12. Sketch map showing distribution of Palaeoproterozoic mafic dykes and rifts.

role in understanding Palaeoproterozoic evolution of the NCC (Fig. 12; Chen & Shi 1983; Qian & Chen 1987; Zhang, J. S. et al. 1994; Hou et al. 1998, 2001; Halls et al. 2000; Li, J. H. et al. 2001; Peng et al. 2005). The dykes are vertical to subvertical with chilled margins. Single dykes are up to 60 km long, and 0.5–100 m wide, and the density of the dykes is several to tens of dykes per kilometre. The dykes have NW–SE and east–west orientations in the western–central part and NE–SW and NW–SE orientations in the eastern part. They consist of clinopyroxene and plagioclase, with Fe–Ti oxides, biotite, alkali-feldspar, apatite and quartz as accessory phases. Olivine phenocrysts can be also observed in the dykes. The dykes vary from alkaline to tholeiitic and a few from basalt to dacite in composition. All of the dykes have a high content of total rare earth elements (ΣREE) and are characterized by slight to moderate LREE enrichment. They are relatively enriched in LILE (except Sr) and depleted in HFSE. Three groups of dykes have been identified by Peng et al. (2005) on the basis of their chemistry. (1) high-Mg to high-Fe tholeiite basalt, with a low FeO(total)–TiO$_2$–P$_2$O$_5$ content; (2) high-Fe basalt, with a high FeO(total)–TiO$_2$–P$_2$O$_5$ content; (3) high-Fe tholeiite basalt to andesite. These three groups derived from enriched mantle, showing different differentiation trends with different degrees of crustal assimilation.

Published ages for dykes include a single-zircon dilution U–Pb age of 1769 ± 3 Ma (Halls et al. 2000; Li et al. 2001), and a SHRIMP zircon U–Pb age of 1778 ± 3 Ma for a sample (SX020) from the central part (Peng et al. 2005). The zircon grains of sample SX020 are brownish and translucent, and some are prismatic, mostly shorter than 150 μm. BSE images of the zircon grains clearly show oscillatory zoning (Fig. 13). The sample dyke yields a baddeleyite isotopic dilution U–Pb age of 1777.6 ± 3.4 Ma (Fig. 14; Peng et al.

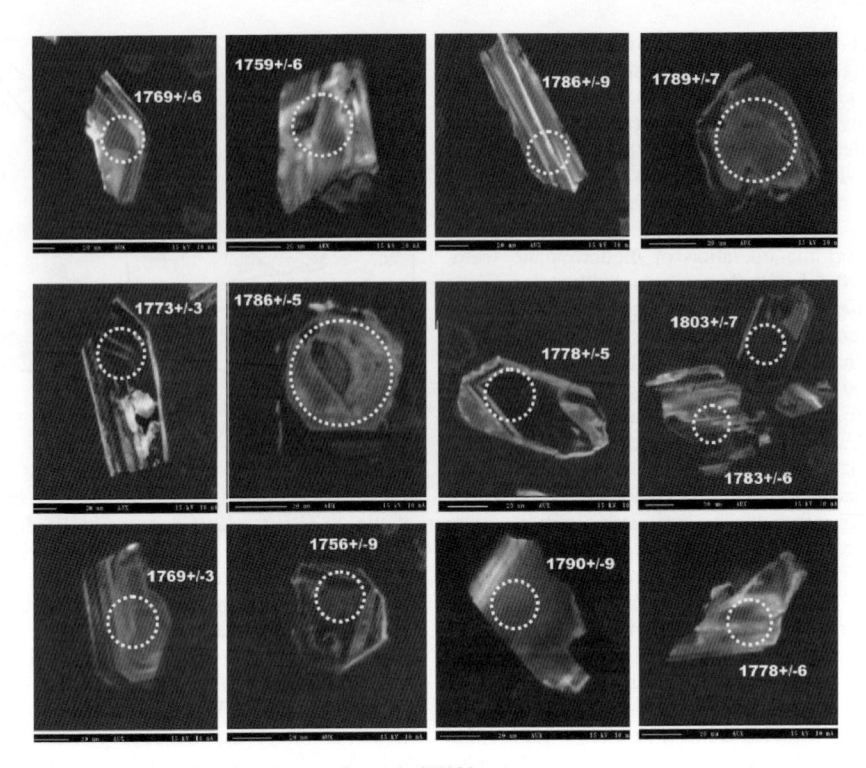

Fig. 13. Zircon CL images and analysed spots of sample SX020.

2006). Another dyke sample yields a baddeleyite isotopic dilution U–Pb age of 1789 ± 28 Ma. K–Ar, Rb–Sr and Sm–Nd ages are centralized on the Late Palaeoproterozoic (Qian & Chen 1987; Hou et al. 1998, 2001), and ^{40}Ar–^{39}Ar ages at about 1780 Ma have also been reported (Wang, Y. J. et al. 2004). These dates suggest that the mafic dykes formed no later than 1.78 Ga.

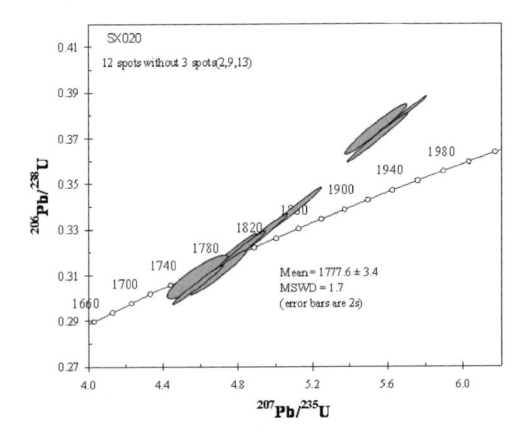

Fig. 14. Baddeleyite isotopic dilution U–Pb age of 1777.6 ± 3.4 Ma for dyke (SX020) (after Peng et al. 2006).

Rifting event

An important rifting event occurred within the NCC at c. 1780–1640 Ma closely followed basement uplift and mafic dyke swarm intrusion. The Xiong'er (XER) and Yanshan (YS) aulacogen rifts are distributed in the northern and southern NCC (Fig. 12), and the northern marginal rifts and northern Qinling marginal rift are distributed along the northern and southern margins of the NCC, respectively (Kusky & Li 2003; Zhai et al. 2003b). The Palaeoproterozoic continental rifting probably represents a limited break-up event of the rigid continent.

Aulacogen rifts. The volcanic rocks in the XER aulacogen are widely developed in the southern part of the NCC, overlying the early Precambrian metamorphic basement, and are overlain by Meso-Neoproterozoic and Palaeozoic terrigenous detrital rocks, carbonate and tillite. The volcanic rocks crop out over an area of >6000 km^2 with a thickness ranging from 3000 to 7000 m, and are composed of basalt, andesite, trachyandesite, dacite and volcanic clastic rocks. The volcanic rocks define two magma cycles: (1) from mafic to intermediate, and intermediate to felsic; (2) from mafic–intermediate to intermediate. The volcanic rocks typically exhibit thick and continuous flow layers with thin sedimentary layers in some places, and are overlain by a conglomerate–sandstone–mudstone sequence containing volcanic clastic rocks. Geochemically, most of the volcanic rocks are high in LILE but exhibit negative anomalies in HFSE. The $\varepsilon Nd_{(t)}$ values calculated at $t = 1780$ Ma range from -5.4 to -9.7. Nd T_{DM} ages range from 2.4 to 3.2 Ga, whereas the $^{87}Sr/^{86}Srt$ values show large variations (Peng et al. 2008). Zircons from five volcanic rocks were dated by the U–Pb SHRIMP method (Zhao, T. P. et al. 2004a). The results indicate that the Xiong'er Group formed at 1.80–1.75 Ga. Although geochemical characteristics of the Xiong'er volcanic rocks are partly similar to island arc affinity (e.g. Chen et al. 1992), more geologists prefer to consider that these rocks formed in an intracontinental setting, based on the sedimentary analyses and rock associations (Sun et al. 1985; Bai et al. 1993; Zhao, Z. P. 1993; Pirajno & Chen 2005). Zhao, T. P. et al. (2001) suggested that the volcanic eruption centre of XER was in western Henan Province, and the aulacogen rift extended to the west, east and north, forming a triple point. Finally, dioritic intrusions indicate magmatic activity at the end of rifting. Recently, Zhao, T. P. et al. (2002), Peng et al. (2006, 2008) and Kusky & Santosh (2010) proposed that the Xiong'er volcanic rocks could be linked to a break-up event of the NCC, associated with the disassembly of the supercontinent Columbia.

The YS aulacogen mainly trends NE–SW to east–west, and branches into the Taihang Mountains to the south. The sedimentary–volcanic sequence is termed the Changcheng System, and is unconformably covered by the Mesoproterozoic Jixian System, Neoproterozoic Qingbaikou System and Palaeozoic sediments of platform-type sedimentary basins. The Changcheng System is subdivided into five groups, which are, from bottom to top, the Changchenggou, Chuanlinggou, Tuanshanzi, Dahongyu and Gaoyuzhuang Groups. The dominant rocks are thick layered conglomerates, quartz sandstones, greywackes, fine-grained sandstones, mudstones, shales and carbonates. The Tuanshan Group and the Dahongyu Group consist of volcanic lavas and clastic rocks. The volcanic rocks are alkali basalts with a small amount of acid rocks, showing the characteristics of a continental bimodal volcanic series (Yu et al. 1994, 1996). SHRIMP U–Pb ages of detrital zircons for feldspathic quartzites from the lower part of the Changzhougou Group and the Chuanlinggou Group are 2580–2360 Ma with a peak of 2500 Ma, and 2600–2350 Ma and 1900–1800 Ma, respectively (Wan et al. 2003). Therefore Wan et al. suggested that the depositional time of the Changcheng System was younger than c. 1800 Ma. Volcanic rocks of the Tuanshanzi Group and Dahongyu Group yielded zircon U–Pb ages of 1720–1620 Ma (Li, H. K. et al. 1995; Lu, S. N. et al. 1995, 2002), which are younger than the Xiong'er volcanic rocks.

Continental margin rift basins. Several continental rift basins are developed along the northern margin of the NCC: the Bayan Obo, Langshan–Zhaertai and Huade basins (Fig. 12). Their sedimentary successions are similar, with multi-depositional cycles of greywacke, sandstone, mudstone, shale, carbonate and evaporite, termed the Bayan Obo Formation, Langshan–Zhaertai Formation and Huade Formation. The Bayan Obo Formation and the Langshan–Zhaertai Formation contain a minor amount of alkali volcanic rocks (Li, Q. L. et al. 2007; Peng & Zhai 2004), whereas the Huade Formation contains no volcanic rock (Hu et al. 2009). Most of the rocks underwent low-grade metamorphism and deformation, as a result of which some of the mudstones and shales changed to garnet- or andalusite-bearing slates and mica–quartz schists.

The Bayan Obo Formation is well known for its abundant REE–Nb–Fe mineral deposits (Ren & Wang 2000). The Bayan Obo rift basin is located north of the Zhaertai Basin and NW of the Huade Basin; its length is c. 800 km and thickness c. 10 000 m. Clastic rocks and argillaceous rocks are dominant, occupying over 90 vol.% of all sediment sequences. Other rocks are carbonatites and volcanic rocks. Volcanic activity was concentrated in two intervals in the lower part and upper middle part of the formation. The volcanic rocks are characterized by high abundances of K and Na, and their contents of LiO_2, Nb_2O_5 and BaO are much higher than in normal igneous rocks. The abnormal REE enrichment is related to carbonatites. The carbonatites commonly contain schohartite, aegirite–augite, zircon, monazite, bastnasite, apatite and spinel, and their contents of FeO, P_2O_5 and K_2O are high. They are also enriched in Nb, Ta, Ce, Ti, Th, Ba and Zr. The $\delta_{13}C$ values range from $-6.57‰$ to $+0.36‰$, and $\delta_{18}O$ values range from $+8.28‰$ to $+19.36‰$ (Chen and Shao 1987; Wang & Li 1987). Their LREE contents are high, similar to those of alkaline continental basalts. These characteristics indicate that the carbonatites came from a deep source.

LA-ICP-MS U–Pb ages of detrital zircons for sandstones in the lower part of the Bayan Obo Group are concentrated at 1850–2000 Ma and c. 2500 Ma, and basalts yield a zircon U–Pb age of 1852 ± 4 Ma (Wang, Y. X. et al. 2002; Fan, H. R. pers. comm.). Recently, Li, Q. L. et al. (2007) and Hu et al. (2009) reported LA-ICP-MS U–Pb ages of detrital zircons from sandstones in the Zhaertai Formation and the Huade Formation.

The detrital zircon ages of the Zhaertai Formation are concentrated at 2550–2400 Ma, and a basalt age is 1743 Ma. The ages of the Huade Formation show two main peaks at 1800 ± 50 Ma and 1850 ± 50 Ma, as well as two minor age peaks at c. 2500 Ma and c. 2000 Ma. The youngest concordant age of the detrital zircons from meta-pebbly arkose at the bottom of the Huade Group is 1758 ± 7 Ma, which constrains the oldest depositional age of this sequence. All of the above-mentioned ages are similar to those of the Yanshan and Xiong'er aulacogen rifts, and correlated to main crustal growth events of the NCC. Therefore, we suggest that the Bayan Obo, Langshan–Zhaertai and Huade formations represent Palaeo-Mesoproterozoic stable shallow–hypabyssal sedimentary basins in the NCC. These three basins commonly constitute a passive continental margin rift system at the northern margin of the NCC. This rift initiated at about the same time as the Yanshan and Xiong'er aulacogen rifts.

Anorogenic magmatic intrusion

In the North China craton, a rapakivi–anorthosite anorogenic magma association is exposed in the Archaean–Palaeoproterozoic metamorphic basement rocks in the northern NCC. Rapakivi granites mainly occur in Miyun, Chicheng, Luanping and Kuandian and through the Beijing, northern Hebei and Liaoning provinces. Their country rocks are Archaean metamorphic rocks of 2521 Ma and Palaeoproterozoic granitic and migmatitic gneisses (Liu et al. 2007; Yang et al. 2008). The representative rapakivi granite body is the Miyun pluton, with an exposed area of 25 km². It was intruded into the Archaean Miyun complex, which includes orthogneiss, metabasites of granulite facies and BIFs. The pluton has characteristically porphyritic and mega-porphyritic textures with ovoid alkali feldspars distributed homogeneously throughout the granite. These alkali feldspars range from 10–40 mm to 600 mm in diameter and most of them are mantled by plagioclase, with sharp irregular contacts. The margin of the pluton is changed to fine-grained granite. The main mafic minerals are hornblende and biotite. Yang et al. (2005) and Liu et al. (2007) reported zircon LA-ICPMS U–Pb ages of 1681 ± 10 Ma and 1679 ± 10 Ma, respectively, and a zircon SHRIMP U–Pb age of 1685 ± 15 Ma for the Miyun pluton. The zircons have an $\varepsilon_{Hf}(t)$ value of −5.0, indicating that the rapakivi pluton was derived from a crustal source. The two-stage model ages (T_{DM2}) of rapakivi granite are about 2.6–2.8 Ga, similar those of the host Archaean gneiss, indicating that the rapakivi granite was derived from partial melting of the crust during the Neoarchaean.

The Damiao anorthosite complex Hebei Province, is the only massif-type intrusion in the NCC. This complex is composed of dominant anorthosite, and leucogabbro, gabbro, norite and mangerite. The host rocks are the Archaean–early Palaeoproterozoic Dantazi complex, consisting of supracrustal rocks and orthogneisses metamorphosed to granulite–amphibolite facies. The mafic granulites have a 'white eye-socket' feature, as a result of the decompressional corona texture of fine-grained Plg + Cpx + Opx ± Amp surrounding garnet, and a 'red eye-socket' feature, as a result of the corona growth of garnet, indicating high–moderate metamorphic pressure. Zhao, T. P. et al. (2004b) reported single-zircon U–Pb ages of 1693 ± 7 Ma and 1715 ± 6 Ma; the zircons were separated from norite and mangerite, respectively.

The crystallization ages of the rapakivi granites and the anorthosite complex are very consistent with each other, showing that the emplacement of the anorogenic magma association occurred at 1715–1685 Ma in the NCC. Some researchers have paid attention to the relationship of Palaeoproterozoic events in the NCC: the c. 1800 Ma basement uplift, the 1780 Ma mafic dyke swarm, c. 1800–1650 Ma XER and YS (Changcheng System) rifting with alkali volcanic rocks at 1720–1620 Ma, and the 1715–1685 Ma anorogenic magma association. The anorogenic magma is genetically related to those of the mafic dyke swarm and the volcanic rocks in the Palaeoproterozoic YS and XER aulacogen rifts (Xie & Wang 1988; Zhao, Z. P. 1993; Yu et al. 1994; Rämö et al. 1995; Zhai & Liu, 2003; Zhai et al. 2003b; Peng et al. 2008). Therefore, it is possible that the Palaeoproterozoic events were caused by an upwelling mantle and are an indicator of the break-up of a Palaeoproterozoic continent.

Discussion and conclusion

(1) Old continental nuclei are recognized in the NCC, and the oldest remnants of granitic gneiss and supracrustal rocks are 3.8 Ga old. The main crustal growth in the NCC took place at 2.9–2.7 Ga. The NCC can be divided into seven microblocks, which are separated by Archaean greenstone belts that represent continental accretion surrounding the old continental nuclei.

(2) By 2.5 Ga, the microblocks amalgamated to form a coherent craton by continent–continent, arc–continent or arc–arc collisions. The tectonic processes in Neoarchaean and modern times appear to be different in degree rather than in principle. Extensive intrusion of K-granite sills and mafic dykes and regional metamorphism at upper amphibolite–granulite facies occurred, and

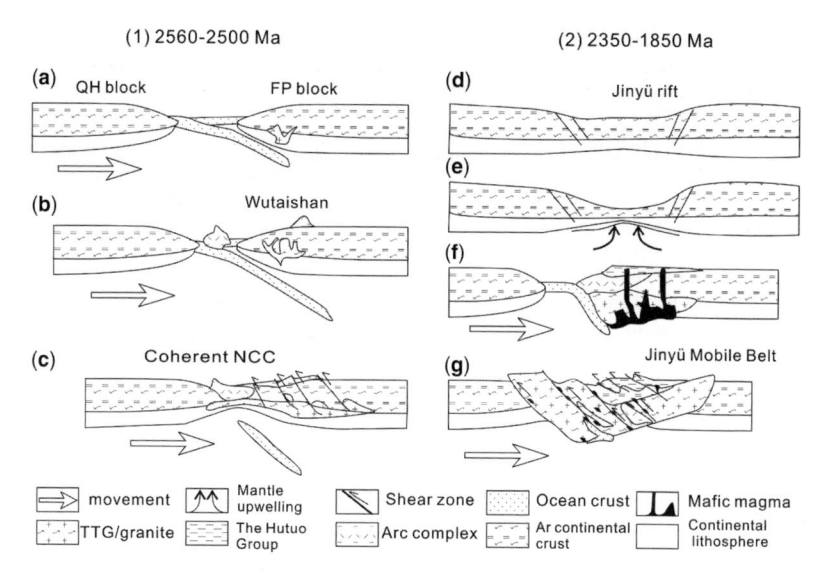

Fig. 15. Tectonic model showing Neoarchaean–Palaeoproterozoic evolution of the North China craton.

marked the realization of cratonization in the NCC. Coeval ultramafic–mafic and syenitic dykes of c. 2500 Ma in eastern Hebei indicate that the NCC was a stable, thick and huge continent at the end of the Archaean, and probably was a part of the Neoarchaean supercontinent that has been suggested by previous studies (Condie et al. 2001; Rogers & Santosh 2004; Kusky & Santosh 2010).

(3) In the period between c. 2500 Ma and 2350 Ma, the NCC was tectonically inactive, but the development of Palaeoproterozoic volcanic and granitic rocks occurred between 2300 and 1950 Ma. Volcanic–sedimentary rocks have been termed Palaeozoic mobile belts by Zhai & Liu (2003), and occur in a linear distribution with strong folding and metamorphism at 1900–1850 Ma, and intrusion by granites and pegmatites at 1850–1800 Ma. The Palaeoproterozoic mobile belts formed and evolved within a craton or continental margin (epicontinental geosyncline). Some 2.30–1.95 Ga rift-margin, passive continental margin deposits, analogous to arc or back-arc assemblages, as well as HP and HT–UHT metamorphic complexes, seem to be comparable with many in the late Phanerozoic orogenic belts (Fig. 15). Regarding Palaeoproterozoic orogeny in other cratons (Zhao, G. C. et al. 2002), a global Palaeoproterozoic orogenic event possibly existed and resulted in the formation of a pre-Rodinian supercontinent at c. 2.0–1.85 Ga.

(4) In contrast, the c. 1800 Ma event is an extension–migmatization event, which includes uplift of the lower crust of the NCC as a whole, mafic dyke swarms, continental rifting and intrusion of an orogenic magmatic association. This event has been considered to be related to the break-up of a pre-Rodinian supercontinent at 1.8 Ga, attributed to a Palaeoproterozoic plume.

(5) HP and HT–UHT metamorphic rocks occur widely in the NCC. Their high pressure of 10–14 kbar has attracted the attention of researchers, and several continental collisional models have been proposed. However, it is still argued that these rocks have a much higher geothermal gradient c. $16\,°C\,km^{-1}$ and much slower uplift rate (0.33–0.5 mm a^{-1} than those in Phanerozoic orogenic belts. Moreover, HP and HT–UHT rocks commonly occur together and their distributions are not in linear zone c. All these observations suggest that we should rethink geological and tectonic models for the evolution of the North China craton.

This study represents the research results of a project (Grant 2006CB403504) supported by the Ministry of Science & Technology, and projects (Grants c. 40672128, 90714003 and 40721062) supported by the National Nature Science Foundation of China. We thank J. Guo, N. Jiang, G. Zhao, J. Li, S. Wilde, T. Kusky and A. Kröner for their co-operation, help and discussion, and especially T. Kusky for polishing the English. I expressly thank my old friend and teacher Brian Windley; we started our co-operation in 1981 and I have learnt a lot from him. I offer this paper for his outstanding contribution to Earth Science.

References

BAI, J., HUANG, X. G., DAI, F. Y. & WU, C. H. 1993. *The Precambrian evolution of China*. Geological Publishing House, Beijing, 199–203.

BROWN, G. C. 1979. The changing pattern of batholith emplacement during earth history. *In*: ATHERTON, M. P. & TARNEY, J. (eds) *Origin of Granite Batholiths*. Shiva, Nantwich, 106–115.

CARNEY, J. N., TRELOAR, P. J., BARTON, C. M., CROW, M. J. & EVAN, J. A. 1991. Deep-crustal granulites with migmatic and mylonitic fabric from the Zambezi Belt, northeastern Zimbabwe. *Journal of Metamorphic Geology*, **9**, 461–479.

CARSWELL, D. A. & O'BRIEN, P. J. 1993. High-pressure quartz feldspathic garnet granulites in the Moldanubian zone, Bohemian Massif, in lower Austria: Their $P-T$ conditions for formation, uplift history and geotectonic significance. *Journal of Petrology*, **3**, 13–25.

CHEN, H. & SHAO, J. A. 1987. Petrogenesis and geotectonic background of carbonatites in Bayan Obo region. *In*: SHENYANG INSTITUTE OF GEOLOGY AND MINERAL RESOURCES (ed.) *Collected Works of Plate Tectonics in Northern China (Part 2)*. Geological Publishing House, Beijing, 73–79 [in Chinese with English abstract].

CHEN, R. D., LI, X. S. & ZHANG, F. S. 2003. Several problems about geology in East Liaoning, China. *Geology in China*, **30**, 209–213 [in Chinese with English abstract].

CHEN, Y. J., FU, S. G. & QIANG, L. Z. 1992. The tectonic environment for the formation of the Xiong'er Group and the Xiyanghe Group. *Geological Review*, **38**, 325–333.

CHEN, X.-D. & SHI, L.-B. 1983. Primary study on diabase dyke swarms in Wutai–Taihang area. *Chinese Science Bulletin*, **16**, 1002–1005.

CHENG, S. H. & KUSKY, T. 2006. Komatiites from west Shandong, North China craton: implications for plume tectonics. *Gondwana Research*, **12**, 77–83.

CHENG, Y. Q. (ed.) 1994. *Geology of China*. Geological Publishing House, Beijing [in Chinese].

CHENG, Y. Q. & XU, H. F. 1991. Some new ideas on the komatiite in Archean Yanlingguan group in Xintai, Shandong. *Regional Geology of China*, **135**, 31–32 [in Chinese with English abstract].

CONDIE, K. C. 2007. Accretionary orogens in space and time. *In*: HATCHER, R. D. JR, CARLSON, M. P., MCBRIDE, J. H. & MARTINÉZ CATALÁN, J. R. (eds) *4-D Framework of Continental Crust*. Geological Society of America, Memoirs, **200**, 145–158.

CONDIE, K. C. & KRÖNER, A. 2008. When did plate tectonics begin? Evidence from the geologic record. *In*: CONDIE, K. C. & PEASE, V. (eds) *When Did Plate Tectonics Begin an Planet Earth?* Geological Society of America, Special papers, **440**, 281–294.

CONDIE, K. C., BORYTA, M. D., LIU, J. Z. & QIAN, X. L. 1992. The origin of khondalites: geochemical evidence from the Archean to Early Proterozoic granulite belt in the North China craton. *Precambrian Research*, **59**, 207–223.

CONDIE, K. C., DES MARAIS, D. J. & ABBOT, D. 2001. Precambrian superplumes and supercontinents: a record in black shales, carbon isotopes and paleoclimates. *Precambrian Research*, **106**, 239–260.

DANN, J. C. 1991. Early Proterozoic ophiolite, Central Arizona. *Geology*, **19**, 594–597.

DENG, J. F., ZHAO, H. L. & MO, X. X. 1996. *Continental roots–plume tectonics of China*. Geological Publication House, Beijing, 1–97 [in Chinese].

FURNES, H., DE WIT, M., STAUDIGEL, H., ROSING, M. & MUEHLENBACH, K. 2007. A vestige of Earth's oldest ophiolite. *Science*, **15**, 1704–1707, doi: 10.1126/science.1139170.

GENG, Y. S. 1998. Archean granite pluton events of Qianan area, East Hebei province and its evolution. *In*: CHEN, Y.Q. (ed.) *Corpus on Early Precambrian Research of the North China Craton*. Geological Publishing House, Beijing, 105–121 [in Chinese].

GENG, Y. S. & LIU, D. Y. 1997. Precambrian geochronology. *In*: YU, J. S. (ed.) *Isotopic Geochemistry in China*. Scientific Press, Beijing, 1–35 [in Chinese with English abstract].

GENG, Y. S., WAN, Y. S. & YANG, C. H. 2003. The Paleoproterozoic rift-type volcanism in Luliangshan area, Shanxi province, and its geological significance. *Acta Geoscientia Sinica*, **24**, 97–104 [in Chinese with English abstract].

GAO, S., ZHOU, L., LING, W. L., LIU, Y. S. & ZHOU, D. W. 2005. Age and geochemistry of volcanic rocks of Angou Group at the Archean–Proterozoic boundary. *Earth Sciences—Journal of China University of Geoscience*, **30**, 259–263 [in Chinese with English abstract].

GENG, Y. S., WAN, Y. S. & SHEN, Q. H. 2002. Early Precambrian basic volcanism and crustal growth in the North China Craton. *Acta Geologica Sinica*, **76**, 199–206.

GENG, Y. S., LIU, F. L. & YANG, C. H. 2006a. Magmatic event at the end of the Archean in eastern Hebei Province and its geological implication. *Acta Geologica Sinica*, **80**, 819–833.

GENG, Y. S., WANG, X. S., SHEN, Q. H. & WU, C. M. 2006b. Redefinition of the Alxa Group-complex (Precambrian metamorphic basement) in the Alxa area, Inner Mongolia. *Geology in China*, **33**, 138–145 [in Chinese with English abstract].

GILES, D., BETTS, P. & LISTER, G. 2002. Far-field continental back-arc setting for the 1.80–1.67 Ga basins of NE Australia. *Geology*, **30**, 823–826.

GUAN, H., SUN, M., WILDE, S. A., ZHOU, X. H. & ZHAI, M. G. 2002. SHRIMP U–Pb zircon geochronology of the Fuping Complex: implications for formation and assembly of the North China Craton. *Precambrian Research*, **113**, 1–18.

GUO, J. H., LIU, Y. G. & XIA, Y. L. 1991. Old continental nuclei—Early Archean Huai'an complex in the NCC: their Pb isotopic characteristics. *In*: Institute of Geology, Academia Sinica (ed.) *Annual Bulletin (89–90) of Open Laboratory of Lithosphere Tectonic Evolution, Academia Sinica*. Chinese Scientific and Technological Press, Beijing, 115–118.

GUO, J. H., ZHAI, M. G., LI, Y. G. & YAN, Y. H. 1993. The early Precambrian Manjinggou melange zone of high-pressure granulite facies in Huai'an: its geological characteristics, petrology and isotopic chronology. *Acta Petrologica Sinica*, **9**, 329–341 [in Chinese with English abstract].

GUO, J. H., SHI, X., BIAN, A. G., XU, R. H., ZHAI, M. G. & LI, Y. G. 1999a. Pb isotopic composition of feldspar and U-Pb age of zircon from early Proterozoic granite in Sanggan area, North China Craton: metamorphism, crustal melting and tectono-thermal event. *Acta Petrologica Sinica*, **15**, 199-207 [in Chinese with English abstract].

GUO, J. H., ZHAI, M. G. & LI, Y. G. 1999b. Metamorphism, P-T path and geotectonic implication of garnet amphibolite and granulite from Hengshan, Shanxi Province. *Acta Geologica Sinica*, **34**, 311-325.

GUO, J. H., WANG, S. S., SANG, H. Q. & ZHAI, M. G. 2001. ^{40}Ar-^{39}Ar age spectra of garnet porphyroblast: Implications for metamorphic age of high-pressure granulite in the North China craton. *Acta Petrologica Sinica*, **17**, 436-442 [in Chinese with English abstract].

GUO, J. H., SUN, M., CHEN, F. K. & ZHAI, M. G. 2005. Sm-Nd and SHRIMP zircon U-Pb geochronology of high-pressure granulites in the Sanggan area, North China Craton: timing of Paleoproterozoic continental collision. *Journal of Asian Earth Sciences*, **24**, 629-642.

GUO, J. H., CHEN, Y., PENG, P., LIU, F., CHEN, L. & ZHANG, L. Q. 2006. Sapphirine-bearing granulite in Daqingshan, Inner Mongolia: 1.8 Ga UHT metamorphic event. *In*: QIU, J. S., XIAO, E., HU, J. & LI, Z. (eds) *Abstract Volume of 2006 Petrology and Earth Dynamics in China*. Nanjing University, Nanjing, 215-218 [in Chinese].

HALLS, H. C., LI, J.-H., DAVIS, D., HOU, G. T., ZHANG, B.-X. & QIAN, X.-L. 2000. A precisely dated Proterozoic paleomagnetic pole from the North China Craton, and its relevance to paleocontinental construction. *Geophyical Journal International*, **143**, 185-203.

HARLEY, S. 1988. Proterozoic granulites from the Rauer Group, East Antarctica. I. Decompressional pressure-temperature paths deduced from mafic and felsic gneisses. *Journal Petrology*, **5**, 1059-1095.

HE, G. P., LU, L. Z., YE, H. W., JIN, S. Q. & YE, T. S. 1991. *Early Precambrian Metamorphic Evolution of Eastern Hebei and East-southern Inner Mongolia*. Jilin University Press, Changchun, 25-218.

HELMSTAEDT, H. H. & SCOTT, D. J. 1992. The Proterozoic ophiolite problem. *In*: CONDIE, K. C. (ed.) *Proterozoic Crustal Evolution*. Elsevier, Amsterdam, 55-95.

HOFFMAN, P. F. 1987. Early Proterozoic foredeeps, foredeep magmatism, and Superior-type iron-formations on the Canadian Shield. *In*: KRÖNER, A. (ed.) *Proterozoic Lithospheric Evolution*. American Geophysical Union, Geodynamics Series, **17**, 85-98.

HOFFMAN, P. F. 1988. United plates of America, the birth of a craton: early Proterozoic assembly and growth of Laurentia. *Annual Review of Earth and Planetary Science*, **16**, 543-603.

HOU, G.-T., ZHANG, C. & QIAN, X. L. 1998. The formation mechanism and tectonic stress field of the Mesoproterozoic mafic dyke swarms in the north China craton. *Geological Review*, **44**, 309-314 [in Chinese with English abstract].

HOU, G.-T., LI, J.-H. & QIAN, X.-L. 2001. Geochemical characteristics and tectonic setting of Mesoproterozoic dyke swarms in northern Shanxi. *Acta Petrologica Sinica*, **17**, 352-357 [in Chinese with English abstract].

HU, B., ZHAI, M. G., GUO, J. H., PENG, P., LIU, F. & LIU, S. 2009. LA-ICP-MS U-Pb geochronology of detrital zircons from the Huade Group at the northern margin of the North China Craton and its tectonic significance. *Acta Petrologica Sinica*, **25**, 193-211 [in Chinese with English abstract].

HU, S. L., GUO, J. H., DAI, T. M. & PU, Z. P. 1999. Continuous laser-probe ^{40}Ar/^{39}Ar age dating on garnet and plagioclase: constraint to metamorphism of high-pressure basic granulite from Sanggan area, the North China Craton. *Acta Petrologica Sinica*, **15**, 518-523 [in Chinese with English abstract].

HUANG, X., BAI, Y. L. & DEPAOLO, D. J. 1983. Sm-Nd isotope study of early Archaean rocks, Qian'an, Hebei province, China. *Geochimica et Cosmochimica Acta*, **50**, 625-631.

HUANG, X. N., LI, Z. H., KUSKY, T. M. & CHEN, Z. 2004. Microstructures of the Zunhua 2.50 Ga podiform chromite, North China Craton and implications for the deformation and rheology of the Archean oceanic lithospheric mantle. *In*: KUSKY, T. M. (ed.) *Precambrian Ophiolites and Related Rocks. Developments in Precambrian Geology*, **13**, 321-337.

HUSON, R., KUSKY, T. M. & LI, Z. H. 2004. Geochemical and petrographic characteristics of the Central Belt of the Archean Dongwanzi ophiolite complex. *In*: KUSKY, T. M. (ed.) *Precambrian Ophiolites and Related Rocks. Developments in Precambrian Geology*, **13**, 283-320.

JAHN, B. M. 1990. Early Precambrian basic rocks of China. *In*: HALL, R. P. & HUGHES, D. J. (eds) *Early Precambrian Basic Magmatism*. Blackie, Glasgow, 294-316.

JAHN, B. M. & ZHANG, Z. Q. 1984. Radiometric ages (Rb-Sr, Sm-Nd, U-Pb) and REE geochemistry of Archaean granulite gneisses from eastern Hebei province, China. *In*: KRÖNER, A., HANSON, G. N. & GOODWIN, A. M. (eds) *Archaean Geochemistry*. Springer, Berlin, 183-204.

JIAN, P., ZHANG, Q., LIU, D. Y., JIN, W. J., JIA, X. Q. & QIAN, Q. 2005. SHRIMP dating and geological significance of late Archean High-Mg diorite (sanukite) and hornblende-granite at Guyang of Inner Mongolia. *Acta Petrologica Sinica*, **21**, 151-157 [in Chinese with English abstract].

KONTINEN, A. 1987. An early Proterozoic ophiolite - The Jormua mafic-ultramafic complex, northeastern Finland. *Precambrian Research*, **35**, 313-341.

KRÖNER, A., COMPSTON, W., ZHANG, G. W., GUO, A. L. & CUI, W. Y. 1987. Single zircon ages for Archaean rocks from Henan, Hebei and Inner Mongolia, China and tectonic implication. *In*: *Proceedings of the International Symposium on Tectonic Evolution and Dynamics of Continental Lithosphere*. Peking University, Beijing, 43-44.

KRÖNER, A., WILDE, A. S., O'BRIEN, P. J. O. & LI, J. H. 2001. The late Archaean to Palaeoproterozoic Hengshan and Wutai complexes of northern China. *In*: CASSIDY, K. F., DUNPHY, J. M. & KRANENDONK, M. J. V. (eds) *4th International Archaean Symposium (Extended Abstracts)*. SGSO-Geoscience Australia, Perth, 327.

KRÖNER, A., WILDE, S. A., LI, J. H. & WANG, K. Y. 2005. Age and evolution of a late Archean to Paleoproterozoic upper to lower crustal section in the Wutaishan/Hengshan/Fuping terrane of northern China. *Journal Asian Earth Sciences*, **24**, 577-595.

KUSKY, T. M. 2004a. Precambrian ophiolites and related rocks, introduction. In: KUSKY, T. M. (ed.) *Precambrian Ophiolites and Related Rocks*. Developments in Precambrian Geology, **13**, 1–35.

KUSKY, T. M. 2004b. What, if anything, have we learned about Precambrian ophiolites and early Earth processes? In: KUSKY, T. M. (ed.) *Precambrian Ophiolites and Related Rocks*. Developments in Precambrian Geology, **13**, 727–737.

KUSKY, T. M. & LI, J. H. 2003. Paleoproterozoic tectonic evolution of the North China Craton. *Journal of Asian Earth Science*, **22**, 383–397.

KUSKY, T. M. & SANTOSH, M. 2010. The Columbia connection in North China. *Journal of the Geological Society, London*, (in press).

KUSKY, T. M., LI, J. H. & TUCKER, R. D. 2001. The Archaean Dongwanzi ophiolite complex, North China craton: 2.505-billion-year-old oceanic crust and mantle. *Science*, **292**, 1142–1145.

KUSKY, T. M., LI, J. H., RAHARIMAHEFA, T. & CARLSON, R. W. 2004a. Re–Os isotope chemistry and geochronology of chromite from mantle podiform chromites from the Zunhua ophiolitic mélange belt, N. China: correlation with the Dongwanzi ophiolite. In: KUSKY, T. M. (ed.) *Precambrian Ophiolites and Related Rocks*. Developments in Precambrian Geology, **13**, 275–282.

KUSKY, T. M., LI, Z. H., GLASS, A. & HUANG, H. A. 2004b. Archean ophiolites and ophiolite fragments of the North China Craton. In: KUSKY, T. M. (ed.) *Precambrian Ophiolites and Related Rocks*. Developments in Precambrian Geology, **13**, 223–274.

KUSKY, T. M., WINDLEY, B. F. & ZHAI, M. G. 2007a. Tectonic evolution of the North China Block: from orogen to craton to orogen. In: ZHAI, M. G., WINDLEY, B. F., KUSKY, T. & MENG, Q. R. (eds) *Mesozoic Sub-Continental Thinning Beneath Eastern North China*. Geological Society, London, Special Publications, **280**, 1–34.

KUSKY, T. M., LI, J. H. & SANTOSH, M. 2007b. The Paleoproterozoic North Hebei Orogen: North China Craton's collisional suture with Columbia supercontinent. *Gondwana Research*, **12**, 4–28.

LI, J. H. & KUSKY, T. M. 2007. A late Archean foreland and thrust in the North China craton: implications for early collisional tectonics. *Gondwana Research*, **12**, 47–66.

LI, H. K., LI, H. M. & LU, S. N. 1995. Grain zircon U–Pb ages for volcanic rocks from Tunashanzi formation of Changcheng system and their geological implications. *Geochemica*, **24**, 43–48 [in Chinese with English abstract].

LI, J. H., HE, W. Y. & QIAN, X. L. 1997. Genetic mechanism and tectonic setting of Proterozoic mafic dyke swarm: its implication to palaeo-plate reconstruction. *Geological Journal of China University*, **3**, 2–8 [in Chinese with English abstract].

LI, J. H., ZHAI, M. G. & LI, Y. G. 1998. Discovery of Archaean high-pressure granulite in Luanping-Chengde, northern Hebei Province: geological implication. *Acta Petrologica Sinica*, **14**, 34–41.

LI, J. H., QIAN, X. L. & LIU, S. W. 1999. Geochemical characteristics of khondalite series in middle part of the North China Craton and its significance to continental cratonization. *Science in China (D)*, **29**, 193–203.

LI, J. H., QIAN, X. L., HUANG, C. N. & LIU, S. W. 2000. Tectonic framework of North China Block and its cratonization in the Early Precambrian. *Acta Petrologica Sinica*, **16**, 1–10 [in Chinese with English abstract].

LI, J. H., HOU, G.-T., QIAN, X. L., HALLS, H. C. & DON, D. 2001. Single-zircon U–Pb age of the initial Mesoproterozoic basic dyke swarms in Hengshan mountain and its implication for the tectonic evolution of the North China Craton. *Geological Review*, **47**, 234–238 [in Chinese with English abstract].

LI, J. H., KUSKY, T. M. & HUANG, X. 2002. Neoarchaean podiform chromites and harzburgite tectonite in ophiolitic mélange, North China Craton remnants of Archean oceanic mantle. *GSA Today*, **12**, 4–11.

LI, J.-H., YANG, C. G., DU, L. L., WAN, Y. S. & LIU, Z. X. 2005. SHRIMP U–Pb geochronology evidence for the formation time of the Wanzi Group at Pingshan County, Hebei Province. *Geological Review*, **51**, 201–207 [in Chinese with English abstract].

LI, J. H., NIU, X. L., QIAN, X. L. & TIAN, Y. Q. 2006. Division of Archean/Proterozoic boundary and its implication for geological evolution in Wutai Mountain area, North China. *Geotectonica et Metallogena*, **30**, 419–412 [in Chinese with English abstract].

LI, Q. L., CHEN, F. K., GUO, J. H., LI, X. L., YANG, Y. H. & SIEBEL, W. 2007. Zircon ages and Nd–Hf isotopic composition of the Zhaertai Group (Inner Mongolia): evidence for early Proterozoic evolution of the northern North China Craton. *Journal of Asian Earth Sciences*, **30**, 573–590.

LI, S. Z., LIU, Y. J., YANG, Z. S. & MA, R. 1998. Relations between deformation and metamorphic recrystallization in metapelite of Liaohe Group. *Acta Petrologica Sinica*, **14**, 351–365 [in Chinese with English abstract].

LI, T. S. 1999. *Taipingzhai–Zunhua Neoarchaean island arc terrain and continental growth in eastern Hebei, North China*. PhD thesis, Institute of Geology, Chinese Academy of Sciences, Beijing, 10–129 [in Chinese with English abstract].

LI, T. S., ZHAI, M. G., PENG, P., CHEN, L. & GUO, J. H. 2010. Ca. 2.5 billion years old coeval ultramafic–mafic and syenitic dykes in Eastern Hebei Region: implications for cratonization of the North China Craton. *Precambrian Research*. [submitted].

LI, Z. Z., BAI, Y. L. & GU, D. L. 1980. Some aspects of Pre-Sinian strata in eastern Hebei. *Acta Geologica Sinica*, **3**, 11–25.

LIU, D. Y., SHEN, Q. H., ZHANG, Z. Q., JAHN, B. M. & AURVAY, B. 1990. Archaean crustal evolution in China: U–Pb geochronology of the Qian'an complex. *Precambrian Research*, **48**, 223–244.

LIU, D. Y., NUTMAN, A. P. W., COMPSTON, W., WU, J. S. & SHEN, Q. H. 1992. Remnants of ≥3800 Ma crust in the Chinese part of the Sino-Korean Craton. *Geology*, **20**, 339–342.

LIU, D. Y., WAN, Y. S., WU, J. S., WILDE, S. A., ZHOU, H. Y., DONG, C. Y. & YIN, X. Y. 2007. Eoarchean rocks and zircons in the North China Craton. In: VAN KRANENDONK, M. J., SMITHIES, H. R. H. & BENNETT, V. (eds) *Earth's Oldest Rocks*. Developments in Precambrian Geology, **15**, 251–273.

LIU, S. J. & LI, J. H. 2008. Paleoproterozoic paired zone in southern part of Central Inner Mongolia and its tectonic implication. *In*: XIAO, W. J., ZHAI, M. G. & LI, X. H. (eds) *Program and Abstracts of the Gondwana 13*. IGG CAS, Dali, China, 132.

LIU, S. W., SHEN, Q. H. & GENG, Y. S. 1996. Metamorphic evolution of two types of garnet-granulites in northwestern Hebei Province and analyses by Gibbs method. *Acta Petrologica Sinica*, **12**, 261–275.

LIU, S. W., LU, Y. J., FENG, Y. G., ZHANG, C., TIAN, W., YAN, Q. R. & LIU, X. M. 2007a. Geology and zircon U–Pb isotopic chronology of Dantazi Complex, Northern Hebei Province. *Geological Journal of China University*, **13**, 484–497 [in Chinese with English abstract].

LIU, S. W., LU, Y. J., FENG, Y. G., LIU, X. M., YAN, Q. R., ZHANG, C. & TIAN, W. 2007b. Zircon and monazite geochronology of the Hongqiyingzi complex, northern Hebei, China. *Geological Bulletin of China*, **26**, 1086–1100 [In Chinese with English Abstract].

LIU, W. J., ZHAI, M. G. & LI, Y. G. 1998. Metamorphism of the high-pressure basic granulites in Laixi, eastern Shandong, China. *Acta Petrologica Sinica*, **14**, 449–459.

LIU, Y. S., GAO, S., YUAN, X. L., ZHOU, L., LIU, X. M., WANG, X. C. & WANG, L. S. 2004. U–Pb zircon ages and Nd, Sr and Pb isotopes of lower crustal xenoliths from North China: insights on evolution of the lower continental crust. *Chemical Geology*, **211**, 87–109.

LOUKOLA-RUSKEENIEMI, K., HEINO, T., TALVITIE, J. & VANNE, J. 1991. Base metal-rich metamorphosed black shales, associated with Proterozoic ophiolites in the Kainuu, Finland: a genetic link with the Outokumpu rock assemblage. *Mineralium Deposita*, **26**, 143–151.

LU, L. Z., XU, X. C. & LIU, F. L. 1995. *The Precambrian khondalite series in northern China*. Changchun Publishing House, Changchun, 1–99.

LU, S. N., YANG, C. L. & LI, H. K. 1995. Sm–Nd isotopic information on the metamorphosed Precambrian basement of the North China platform. *Journal of Geology and Mineral Resources of North China*, **10**, 143–153 [in Chinese with English abstract].

LU, S. N., YANG, C. L., LI, H. K. & LI, H. M. 2002. A group of rifting events in the terminal Paleoproterozoic in the North China Craton. *Gondwana Research*, **5**, 123–132.

LU, X. P., WU, F. Y., GUO, J. H., WILDE, S. A., YANG, J. H., LIU, X. M. & ZHANG, X. O. 2006. Zircon U–Pb geochronological constraints on the Paleoproterozoic crustal evolution of the Eastern Block in the North China Craton. *Precambrian Research*, **146**, 138–164.

LUO, Y., SUN, M., ZHAO, G. C., LI, S. Z., XU, P., YE, K. & XIA, X. M. 2004. LA-ICP-MS U–Pb ages of the Liaohe Group in the Eastern Block of the North China Craton: constraints on the evolution of the Jiao-Liao-Ji Belt. *Precambrian Research*, **134**, 349–371.

MA, J. & WANG, R. M. 1994. Reviews in garnet–clinopyroxene geothermometers and geobarometers with their application to granulite: the composition of Miyun (Zunhua) and Xuanhua granulite forming condition. *In*: QIAN, X. L. & WANG, R. M. (eds) *Geological Evolution of Granulite Facies Zone in Northern North China*. Seismological Press, Dali, Beijing, 71–88 [in Chinese with English abstract].

MAO, D. B., ZHONG, C. T., CHEN, Z. H., LIN, Y. X., LI, H. M. & HU, X. D. 1999. The isotope ages and their geological implications of high-pressure basic granulites in north region to Chengde, Hebei Province, China. *Acta Petrologica Sinica*, **15**, 524–531 [in Chinese with English abstract].

MIAO, P. S., ZHANG, Z. F., ZHANG, J. Z., ZHAO, Z. X. & XU, S. C. 1999. Discussion of Paleoproterozoic sedimentary sequences. *Geology in China*, **18**, 405–413 [in Chinese with English abstract].

O'BRIEN, P. J., WALTE, N. & LI, J. H. 2005. The petrology of two distinct granulite types in the Hengshan Mts, China, and tectonic implications. *Journal of Asian Earth Sciences*, **24**, 615–627.

OUYANG, Z. Y. & ZHANG, F. Q. 1998. Upper mantle inhomogeneity and origin of North China craton. *Chinese Scientific Bulletin*, **43**, 20 [in Chinese with English abstract].

PENG, P., ZHAI, M. G., ZHANG, H. F. & GUO, J. H. 2005. Geochronological constraints on Paleoproterozoic evolution of the North China craton: SHRIMP zircon ages of different types of mafic dikes. *International Geological Review*, **47**, 492–508.

PENG, P., ZHAI, M. G. & GUO, J. H. 2006. 1.80–1.75 Ga mafic dyke swarms in the central North China craton: implications for a plume-related break-up event. *In*: HANSKI, E., MERTANEN, S., RÄMÖ, T. & VUOLLO, J. (eds) *Dyke Swarms – Time Markers of Crustal Evolution*. Taylor & Francis, London, 99–112.

PENG, P., ZHAI, M. G., ERNST, R. E., GUO, J. H., LIU, F. & HU, B. 2008. A 1.78 Ga large igneous province in the North China craton: The Xiong'er volcanic province and North China dyke swarm. *Lithos*, **101**, 260–280.

PENG, R. M. & ZHAI, Y. S. 2004. The characteristics of hydrothermal exhalative mineralization of the Langshan-Zhaertai belt, Inner Mongolia, China. *Earth Science Frontiers*, **11**, 257–268 [in Chinese with English abstract].

PIDGEON, R. T. 1980. 2480 Ma old zircons from granulite facies rocks from east Hebei province, North China. *Geological Review*, **26**, 198–207.

PIRAJNO, F. & CHEN, Y. J. 2005. The Xiong'er Group: a 1.76 Ga large igneous province in East–central China? Available online at: http://www.largeigneousprovinces.org.

QIAN, X. L. 1996. The nature of the early Precambrian continental crust and its tectonic evolution model. *Acta Petrologica Sinica*, **12**, 169–178.

QIAN, X. L. & CHEN, Y.-P. 1987. Late Precambrian mafic dyke swarms of the North China craton. *In*: HALLS, H. C. & FAHRIG, W. F. (eds) *Mafic Dykes Swarms*, Geological Association of Canada, Special Papers, **34**, 385–391.

QIAN, X. L., CUI, W. Y. & WANG, S. Q. 1985. Evolution of the Inner Mongolia–eastern Hebei Archaean granulite belt in the North China craton. *In*: Department of Geology (ed.) *The Records of Geological Research*. Beijing University Press, Beijing, 20–29.

QIAN, X. L., LI, J. H. & CHENG, S. H. 2005. A review on Precambrian tectonic evolution of continental crust.

Geological Journal of China Universities, **11**, 145–153 [in Chinese with English abstract].

RÄMÖ, O. T., HAAPALA, I., VAASJOKI, M., YU, J. H. & FU, H. Q. 1995. 1700 Ma Shachang complex, northeast China: Proterozoic rapakivi granite not associated with Palaeoproterozoic orogenic crust. *Geology*, **23**, 815–818.

REN, Y. C. & WANG, K. Y. 2000. Study of super REE–Fe–Nb deposits in Banyan Obo. *In*: TU, G. C. (ed.) *Super Mineral Deposits in China (1)*. Scientific Press, Beijing, 10–26 [in Chinese with English abstract].

ROGERS, J. J. W. & SANTOSH, M. 2004. *Continents and Supercontinents*. Oxford University Press, New York, 289.

SANTOSH, M., TSUNOGAE, T., LI, J. H. & LIU, S. J. 2007. Discovery of sapphirine-bearing Mg–Al granulites in the North China Craton: implications for Paleoproterozoic ultrahigh temperature metamorphism. *Gondwana Research*, **11**, 263–285.

SHEN, Q. H., XU, H. F., ZHANG, Z. Q., GAO, J. F., WU, J. S. & JI, C. L. 1992. *Precambrian Granulites in China*. Geological Publishing House, Beijing, 16–31 and 214–223.

SHEN, Q. H., SONG, B., XU, H. F., GENG, Y. S. & SHEN, K. 2004. Emplacement and metamorphism ages of the Caiyu and Dashan igneous bodies, Yishui County, Shandong Province: Zircon SHRIMP chronology. *Geological Review*, **50**, 275–284 [In Chinese with English abstract].

SILLS, J. D., WANG, K. Y., YAN, Y. H. & WINDLEY, B. F. 1987. The Archaean high-grade gneiss terrane in E. Hebei Province, NE China: geological framework and conditions of metamorphism. *In*: PARK, R. G. & TARNEY, J. (eds) *The Evolution of the Lewisian and Comparable Precambrian High-grade Terrain*. Geological Society, London, Special Publications, **27**, 297–305.

SONG, B., ALLEN, P. N., LIU, D. Y. & WU, J. S. 1996. 3800 to 2500 Ma crustal evolution in the Anshan area of Liaoning Province, northeastern China. *Precambrian Research*, **78**, 79–94.

ST-ONGE, M. R. & LUCAS, S. B. 1990. Evolution of the Cape Smith Belt: early Proterozoic continental underthrusting, ophiolite obduction, and thick-skinned folding. *In*: LEWRY, J. F. & STAUFFER, M. B. (eds) *The Early Proterozoic Trans-Hudson Orogen of North America*. Geological Association of Canada, Waterloo, Ont., 313–351.

SUN, D. Z. & HU, W. X. 1993. *The tectonic framework of Precambrian in Zhongtiaoshan*. Geological Publishing House, Beijing, 108–117 [in Chinese with English abstract].

SUN, S., ZHANG, G. W. & CHEN, Z. M. 1985. *Evolution of Precambrian Crust in the Southern North China Fault Block*. Metallurgical Industry Press, Beijing, 258 [in Chinese].

TANG, J., ZHENG, Y. F., WU, Y. B. & ZHOU, J. B. 2003. SHRIMP zircon U–Pb dating and O-isotope study of metamorphosed rocks in western East Shandong. *In*: ZHENG, Y. F. (ed.) *Proceeding for Geodynamics in Subduction and Collision Symposium (Abstracts)*, Chinese University of Science and Technology, Hefei, 115–116 [in Chinese].

WAN, Y. S., GENG, Y. S. & WU, J. S. 1998. The geochemical character of the early Precambrian basalts in the North China craton. *In*: CHENG, Y. Q. (ed.) *Corpus on early Precambrian research of the North China craton*. Geological Publishing House, Beijing, 39–59 [in Chinese].

WAN, Y. S., GENG, Y. S., LIU, F. L. & SHEN, Q. H. 2000a. Age and composition of the khondalite series of the North China Craton and its adjacent Area. *Progress in Precambrian Research*, **23**, 221–237 [in Chinese with English abstract].

WAN, Y. S., GENG, Y. S. & SHEN, Q. H. 2000b. Khondalite series-geochronology and geochemistry of the Jiehekou Group in Luliang area, Shanxi province. *Acta Petrologica Sinica*, **16**, 49–58 [in Chinese with English abstract].

WAN, Y. S., SONG, B. ET AL. 2001. Geochronology and geochemistry of 3.8–2.5 Ga Archean rock belt in Dongshan scenic park, Ahshan area. *Acta Geologica Sinica*, **75**, 363–370 [in Chinese with English abstract].

WAN, Y. S., ZHANG, D. Q. & SONG, T. R. 2003. Detrital zircon SHRIMP dating for the Changzhougou Formation in the Changcheng System in Shisanling, Beijing: constraints to material source of cover of the North China Craton and sedimentation time. *Chinese Science Bulletin*, **18**, 1970–1975.

WAN, Y. S., LIU, D. Y. & SONG, B. 2005. Geochemical and Nd isotopic composition of 3.8 Ga meta-quartz dioritic and trondhjemitic rocks from the Archean area and their geological significance. *Journal of Asian Earth Science*, **24**, 563–575.

WANG, J. & LI, S. Q. 1987. Langshan–Bayan Obo rift system and characteristics of mineralization. *In*: *Collected Works of Plate Tectonics in Northern China (Part 2)*, Geological Publishing House, Beijing, 59–72 [in Chinese with English abstract].

WANG, R. M., HE, S. Y., CHEN, Z. Z., LI, P. F. & DAI, F. X. 1985. Geochemical evolution and metamorphic development of the Early Precambrian in Eastern Hebei, China. *Precambrian Research*, **27**, 111–129.

WANG, Y. J., FAN, W. M., ZHANG, Y. H., GUO, F., ZHANG, H. F. & PENG, T. P. 2004. Geochemical. $^{40}Ar/^{39}Ar$ geochronological and Sr–Nd isotopic constraints on the origin of Paleoproterozoic mafic dykes from the southern Taihang Mountains and implications for the ca. 1800 Ma event of the North China Craton. *Precambrian Research*, **135**, 55–77.

WANG, Y. X., QIU, Y. Z., GAO, J. Y. & ZHANG, Q. 2002. Proterozoic anorogenic magmatic rocks in Bayan Obo mine district, Inner Mongolia and constraints to mineralization. *Science in China, Supplement*, 21–32.

WANG, Z. J., SHEN, Q. H. & WAN, Y. S. 2004. SHRIMP U–Pb zircon geochronology of the Shipaihe 'Metadiorite Mass' from Deng Feng County, Henan Province. *Acta Geoscientica Sinica*, **25**, 295–298 [in Chinese with English abstract].

WILDE, S. A., CAWOOD, P. A., WANG, K. Y. & NEMCHIN, A. A. 2005. Granitoid evolution in the Late Archean Wutai Complex, North China Craton. *Journal of Asian Earth Sciences*, **24**, 597–613.

WINDLEY, B. F. 1986. *The Evolving Continents*. 2nd edn. Wiley, Chichester, 1–65.

WINDLEY, B. F. 1995. *The Evolving Continents*. 3rd edn. Wiley, Chichester, 377–385 and 459–462.

WU, C. H., LI, H. M., ZHONG, C. T. & ZOU, Y. C. 2000. Single zircon U–Pb ages of Fuping gneiss and Wanzi gneiss. *Progress in Precambrian Research*, **23**, 129–139 [in Chinese with English abstract].

WU, F. Y., YANG, J. H., LIU, X. M., LI, T. S., XIE, L. W. & YANG, Y. H. 2005. Hf isotopic characteristics of 3.8 Ga zircon and time of early crust of North China Craton. *Chinese Science Bulletin*, **50**, 1996–2003.

WU, F. Y., ZHANG, Y. B., YANG, J. H., XIE, L. W. & YANG, Y. H. 2008. Zircon U–Pb and Hf isotopic constraints on the Early Archean crustal evolution in Anshan of the North China Craton. *Precambrian Research*, **167**, 339–362.

WU, J. S., GEN, Y. S., SHEN, Q. H., WAN, Y. S., LIU, D. Y. & SONG, B. 1998. *Archaean geology characteristics and tectonic evolution of the China–Korea Paleocontinent*. Geological Publishing House, Beijing, 192–211 [in Chinese].

XIE, G. H. & WANG, J. W. 1988. Primitive isotopic age study for Damiao anorthosite. *Geochemistry*, **1**, 13–17 [in Chinese with English abatract].

XU, Z. Y., LIU, Z. H., HU, F. X. & YANG, Z. S. 2005. Geochemical characteristics of the calc-silicate rocks in khondalite series in Daqingshan Area, Inner Mongolia. *Journal of Jilin University (Earth Science Edition)*, **35**, 681–689 [in Chinese with English abstract].

YAN, Y. H., ZHAI, M. G. & GUO, J. H. 1991. Cordierite–sillimanite mineral assemblage of granulite facies in North China Craton: a indicator to low-pressure granulite facies metamorphism. *Acta Petrologica Sinica*, **2**, 19–27.

YANG, C. H., DU, L. L., WAN, Y. S. & LIU, Z. X. 2004. SHRIMP zircon U–Pb chronology of Tonglitic Gneiss in Banqiaogou area, Pingshan County, Hebei Province. *Geological Journal of China Universities*, **10**, 514–522 [in Chinese with English abstract].

YANG, J.-H., WU, F.-Y., LIU, X.-M. & XIE, L.-W. 2005. Zircon U–Pb ages and Hf isotopes and their geological significance of the Miyun rapakivi granites from Beijing, China. *Acta Petrologica Sinica*, **21**, 1633–1644 [in Chinese with English abstract].

YANG, J.-H., WU, F. Y., WILDE, S. A. & ZHAO, G. C. 2008. Petrogenesis and geodynamics of Late Archean magmatism in eastern Hebei, eastern North China Craton: geochronological, geochemical and Nd–Hf isotopic evidence. *Precambrian Research*, **167**, 125–149.

YU, J. H., FU, H. Q., ZHANG, F. L. & WANG, F. X. 1994. Petrogenesis of potassium alkaline volcanic associated with rapakivi granites in the Proterozoic rift of Beijing, China. *Mineralogy and Petrology*, **50**, 83–96.

YU, J. H., FU, H., ZHANG, F., WAN, F., HAAPALA, I., RAMO, T. O. & VAASJOKI, M. 1996. *Anorogenic rapakivi granites and related rocks in northern North China Craton*. China Science and Technology Press, Beijing, 29–31.

YU, J. H., WANG, D. Z. & WANG, S. Y. 1997. Zircon U–Pb ages of the Luliang Group and its main matamorphism. *Geological Review*, **21**, 154–160 [in Chinese with English abstract].

ZHAI, M. G. 1997. Recent advances in the study of granulites from the North China Craton. *International Geological Review*, **39**, 325–341.

ZHAI, M. G. 2004. Precambrian geological events in the North China Craton. *In*: MALPAS, J., FLETCHER, C. J. N., ALI, J. R. & AITCHISON, C. (eds) *Aspects on Tectonic Evolution of China*. Geological Society, London, Special Publications, **226**, 57–72.

ZHAI, M. G. 2008. Lower crust and lithosphere beneath the North China Craton before the Mesozoic lithospheric disruption. *Acta Petrologica Sinica*, **24**, 2185–2204.

ZHAI, M. G. & LIU, W. J. 2001a. The formation of granulite and its contribution to evolution of the continental crust. *Acta Petrologica Sinica*, **17**, 28–38 [in Chinese with English abstract].

ZHAI, M. G. & LIU, W. J. 2001b. An oblique cross-section of Precambrian crust in the North China Craton. *Physics and Chemistry of the Earth (A)*, **26**, 781–792.

ZHAI, M. G. & LIU, W. J. 2003. Palaeoproterozoic tectonic history of the North China Craton: a review. *Precambrian Research*, **122**, 183–99.

ZHAI, M. G. & PENG, P. 2007. Paleoproterozoic events in North China Craton. *Acta Petrologica Sinica*, **23**, 2665–2687 [in Chinese with English abstract].

ZHAI, M. G. & WINDLEY, B. F. 1990. The Archaean and Early Proterozoic banded iron formations of North China: their characteristics, geotectonic relations, chemistry and implications for crustal growth. *Precambrian Research*, **48**, 267–286.

ZHAI, M. G., WINDLEY, B. F. & SILLS, J. D. 1990. Archaean gneisses, amphibolites, banded iron-formation from Anshan area of Liaoning, NE China: their geochemistry, metamorphism and petrogenesis. *Precambrian Research*, **46**, 195–216.

ZHAI, M. G., GUO, J. H., YAN, Y. H., LI, Y. G. & ZHANG, W. H. 1992. Discovery and preliminary study of Archaean high-pressure basic granulites in North China. *Science in China (B)*, **12**, 1325–1330 [in Chinese].

ZHAI, M. G., GUO, J. H., LI, J. H., YAN, Y. H., LI, Y. G. & ZHANG, W. H. 1995. The discoveries of retrograde eclogites in North China craton in Archaean. *Chinese Science Bulletin*, **40**, 1590–1594.

ZHAI, M. G., BIAN, A. G. & ZHAO, T. P. 2000a. The amalgamation of the supercontinent of North China craton at the end of the Neoarchaean, and its break-up during the late Palaeoproterozoic and Mesoproterozoic. *Science in China (D)*, **43**, Supplement, 219–232.

ZHAI, M. G., CONG, B. L., GUO, J. H., LI, Y. G. & WANG, Q. C. 2000b. Sm–Nd geochronology and petrography of garnet pyroxene granulites in the northern Sulu region of China and their geotectonic implication. *Lithos*, **52**, 23–33.

ZHAI, M. G., ZHAO, G. C. & ZHANG, Q. 2002. Does the Dongwanzi 2505 Ma ophiolite exist? Comment on 'The Archean Dongwanzi ophiolite complex, North China Craton: 2.505-billion-year-old oceanic crust and mantle' by T. M. Kusky *et al.* (2001). *Science*, **295**, 923a.

ZHAI, M. G., GUO, J. H., LI, Y. G., LIU, W. J., PENG, P. & SHI, X. 2003a. Two linear granite belts in the central–western North China Craton and their implication for Late Neoarchaean–Palaeoproterozoic continental evolution. *Precambrian Research*, **127**, 267–283.

ZHAI, M. G., SHAO, J. A., HAO, J. & PENG, P. 2003b. Geological signature and possible position of the North

China block in the Supercontinent Rodinia. *Gondwana Research*, **6**, 171–183.

ZHAI, M. G., GUO, J. H. & LIU, W. J. 2005. Neoarchaean to Paleoproterozoic continental evolution and tectonic history of the North China craton. *Journal of Asian Earth Sciences*, **24**, 547–561.

ZHAI, M. G., GUO, J. H., LI, Z., HOU, Q. L., PENG, P., FAN, Q. C & LI, T. S. 2007. Linking Sulu orogenic belt to Korean Peninsula: evidences of metamorphism, Precambrian basement and Paleozoic basins. *Gondwana Research*, **12**, 388–403.

ZHANG, G. H., ZHOU, X. H., SUN, M., CHEN, S. H. & FENG, J. L. 1998. Sr–Nd–Pb isotopic characteristics of granulite and pyroxenite xenoliths in Hannuoba basalt, Hebei and geological implication. *Acta Petrologica Sinica*, **15**, 190–197 [in Chinese with English abstract].

ZHANG, H. F., ZHAI, M. G. & PENG, P. 2006. Zircon SHRIMP U–Pb age of the Paleoproterozoic high-pressure granulites from the Sanggan area, the North China Craton and its geologic implications. *Earth Science Frontiers*, **13**, 190–199.

ZHANG, J. S., DRIKS, P. H. G. M. & PASSCHIER, C. W. 1994. Extensional collapse and uplift of a polymetamorphic granulite terrane in the Archaean of North China. *Precambrian Research*, **67**, 37–57.

ZHANG, Q., NI, Z. Y. & ZHAI, M. G. 2003. Comments on the ophiolites in Eastern Hebei. *Earth Science Frontiers*, **10**, 429–437.

ZHANG, Q. S. 1984. *Geology and Metallogeny of the Early Precambrian in China*. Jilin People Press, Changchun, 196–230.

ZHANG, W. J., LI, L. & GENG, M. S. 2000. Petrology and dating of Neo-Archean intrusive rocks from Guyang area, Inner Mongolia. *Earth Science Journal of China University Geoscience*, **25**, 221–226 [in Chinese with English abstract].

ZHANG, Z. Q. 1998. On main growth epoch of early Precambrian crust of the North China craton based on the Sm–Nd isotopic characteristics. *In*: CHENG, Y. Q. (ed.) *Corpus on early Precambrian research of the North China craton*. Geological Publishing House, Beijing, 133–136.

ZHAO, G. C. 2001. Palaeoproterozoic assembly of the North China craton. *Geological Magazine*, **138**, 89–91.

ZHAO, G. C., CAWOOD, P. A., WILDE, S. A., SUN, M. & LU, L. Z. 1999. Thermal evolution of two textural types of mafic granulites in the North China craton: evidence for both mantle plume and collisional tectonics. *Geological Magazine*, **136**, 223–240.

ZHAO, G. C., CAWOOD, P. A., WILDE, S. A., SUN, M. & LU, L. 2000. Metamorphism of basement rocks in the Central zone of the North China craton: implications for Paleoproterozoic tectonic evolution. *Precambrian Research*, **103**, 55–88.

ZHAO, G. C., CAWOOD, P. A., WILDE, S. A. & SUN, M. 2002. Review of 2.1–1.8 Ga orogens: implication for pre-Rodinia supercontinent. *Earth-Science Review*, **59**, 125–162.

ZHAO, G. C., SUN, M., WILDE, S. A. & LI, S. Z. 2005. Late Archean to Paleoproterozoic evolution of the North China Craton: key issues revisited. *Precambrian Research*, **136**, 177–202.

ZHAO, G. C., WILDE, S. A., LI, S. Z., SUN, M., GRANT, M. L. & LI, X. P. 2007. U–Pb zircon age constraints on the Dongwanzi ultramafic–mafic body. North China, confirm it is not an Archean ophiolite. *Earth and Planetary Science Letters*, **255**, 85–93.

ZHAO, T. P., ZHOU, M. F., JIN, C. W., GUAN, H. & LI, H. M. 2001. Discussion on the age of the Xiong'er Group in the southern margin of the North China Craton. *Scientia Geologica Sinica*, **36**, 326–334 [in Chinese with English abstract].

ZHAO, T. P., ZHOU, M.-F., ZHAI, M. G. & XIA, B. 2002. Paleoproterozoic rift-related volcanism of the Xiong'er group, North China Craton: implications for the breakup of Columbia. *International Geology Review*, **44**, 336–351.

ZHAO, T. P., CHEN, F., ZHAI, M. G. & NI, Z. Y. 2004a. Zircon U–Pb ages of Damiao Proterozoic anorthosite and geological implication. *Acta Petrologica Sinica*, **20**, 685–690 [in Chinese with English abstract].

ZHAO, T. P., ZHAI, M. G., XIA, B., LI, H. M., ZHANG, Y. X. & WAN, Y. S. 2004b. Study on the zircon SHRIMP ages of the Xiong'er Group volcanic rocks: constraint on the starting time of covering strata in the North China Craton. *Chinese Science Bulletin*, **9**, 2495–2502.

ZHAO, Z. P. 1993. *Evolution of Precambrian Crust of Sino-Korean Platform*. Scientific Press, Beijing, 366–368.

ZHOU, X. W., WEI, C. J., GENG, Y. S. & ZHANG, L. F. 2004. Discovery and geological significance of meta-pelite of high-pressure granulite facies in Qixia, northern Shandong. *Scientific Bulletin*, **49**, 1424–1430.

Mesoarchaean to Palaeoproterozoic growth of the northern segment of the Itabuna–Salvador–Curaçá orogen, São Francisco craton, Brazil

E. P. OLIVEIRA[1]*, N. J. McNAUGHTON[2] & R. ARMSTRONG[3]

[1]*Institute of Geosciences, PO Box 6152, University of Campinas—UNICAMP, 13083-970, Campinas, SP, Brazil*

[2]*John de Laeter Centre of Mass Spectrometry, School of Applied Physics, Curtin University of Technology, Perth, WA 6845, Australia*

[3]*Research School of Earth Sciences, ANU, Canberra, ACT 0200, Australia*

*Correspondig author (e-mail: elson@ige.unicamp.br)

Abstract: The geology of the northern segment of the Itabuna–Salvador–Curaçá orogen, São Francisco craton, is reviewed, and new U–Pb ages, and Nd isotope and major and trace element data are combined to improve understanding of its tectonic evolution. The results indicate that oceanic crust and island arc sequences accreted at about 3.30 Ga to form the Mundo Novo greenstone belt, and between 2.15 and 2.12 Ga to form the Rio Itapicuru and Rio Capim greenstone belts. At about 3.08–2.98 Ga, mafic crust underwent partial melting to form the Retirolândia and Jacurici tonalite–trondhjemite–granodiorite belts of the Serrinha block. From 2.69 to 2.58 Ga an Andean-type arc with ocean crust remnants formed the Caraiba complex possibly at the Gavião block margin. Between 2.11 and 2.105 Ga, the Rio Itapicuru arc collided with the Retirolândia–Jacurici microcontinent, possibly involving slab breakoff. Oblique convergence between 2.09 and 2.07 Ga led to collision of the Serrinha microcontinent with the Caraíba–Gavião superblock and reworked the Caraíba arc to granulites, locally at ultrahigh-temperature conditions. At the same time, arc dacites spread over the Rio Itapicuru greenstone belt, and the 3.12–3.0 Ga Uauá terrane, crosscut by 2.58 Ga mafic dykes, extruded from south to north, possibly together with the 2.15 Ga Rio Capim greenstone belt.

When and how the first continental crust formed, and how crustal fragments assemble to form continents and supercontinents are subjects of increasing recent debate (e.g. Rogers & Santosh 2004; Rollinson 2007; Van Kranendonk *et al.* 2007*a*; Condie & Pease 2008; O'Neil *et al.* 2008).

Phanerozoic orogenic belts worldwide are the outcome of contraction between converging plates, either oceanic–oceanic, oceanic–continental, or continental–continental, and several orogenic belts in the continents contain geological records of all of these possibilities (e.g. the Himalayas and the Caledonides; see Windley 1995), implying that continents grow mainly by lateral accretion of landmasses. Archaean terranes, on the other hand, are well known for their abundant high-grade gneisses and greenstone–TTG (tonalite–tondhjemite–granodiorite) association, often without any clear connection with plate collision (e.g. East Pilbara craton; see Van Kranendonk *et al.* 2007*b*), suggesting that vertical tectonics and other processes may have contributed to continental growth during Earth's evolution.

The time of onset of modern-type plate tectonics and associated crust formation processes is uncertain, but the geological record indicates that in some places it was already in operation by the end of the Palaeoarchaean (Kusky & Polat 1999; Smithies *et al.* 2007; Pease *et al.* 2008), and that the best sites for understanding processes of continental growth are the orogens.

Brazil does not have any active orogen, but the country is underlain by Precambrian basement with ages spanning at least 3.5 Ga of Earth's history. The oldest rocks so far recognized are 3.4 Ga grey gneisses in the Gavião block of the São Francisco craton (Martin *et al.* 1991, 1997; Nutman & Cordani 1993) and 3.5 Ga migmatite inliers within Neoproterozoic gneisses of northeastern Brazil (Dantas *et al.* 2004). Indications of much older crust have come recently from 3.8 Ga detrital zircons in quartzites of the Iron Quadrangle in Minas Gerais state (Hartmann *et al.* 2006) and as 3.6 Ga zircon xenocrysts in granites of the São Francisco craton (Rios *et al.* 2008). However, extensive belts of Palaeoarchaean to Palaeoproterozoic crust are well exposed in the northeastern part of the São Francisco craton, and they may provide insights into how continental crust evolved during a time of profound changes in planet Earth.

From: Kusky, T. M., Zhai, M.-G. & Xiao, W. (eds) *The Evolving Continents: Understanding Processes of Continental Growth*. Geological Society, London, Special Publications, 338, 263–286.
DOI: 10.1144/SP338.13 0305-8719/10/$15.00 © The Geological Society of London 2010.

The Itabuna–Salvador–Curaçá orogen is one of these belts (Fig. 1), for which we present here a review, new sensitive high-resolution ion microprobe (SHRIMP) U–Pb ages, and, when relevant, whole-rock Nd isotope and geochemical data. We focus on the northern part of the orogen because Late Palaeoarchaean to Palaeoproterozoic units are represented here, and more detailed geological maps and abundant geochronological data are available owing to the area's high mineral exploration interest.

The São Francisco craton

The main geological features of the São Francisco craton have been outlined by Teixeira & Figueiredo (1991) and Teixeira et al. (2000). In general, the São Francisco craton consists of Archaean to Palaeoproterozoic high-grade (migmatite, granulite) gneisses and granite–greenstone supracrustal terranes overlain by Meso- to Neoproterozoic platform-type cover (Fig. 1). Typical greenstone belts with spinifex-textured komatiites occur in the iron quadrangle of Minas Gerais state in the southernmost part of the craton (e.g. Lobato et al. 2001), and in the Umburanas region in the north (Cunha & Fróes 1994). In the northern part of the São Francisco craton greenstone belts such as the Rio Itapicuru, Mundo Novo, Rio Capim and Contendas-Mirante (Mascarenhas 1979; Mascarenhas & Sá 1982; Jardim de Sá et al. 1984; Davison et al. 1988) contain no komatiites, and geochronological data have demonstrated that they range in age from c. 3.3 Ga (Mundo Novo greenstone belt, Peucat et al. 2002) to c. 2.1 Ga (e.g. Rio Itapicuru, Silva et al. 2001). The youngest greenstone belts may represent former island arcs and accretionary wedges accreted onto older continental blocks.

The high-grade terranes, mostly exposed in Bahia state (Fig. 1), are traditionally separated into the Neoarchaean Jequié migmatite–granulite complex (Alibert & Barbosa 1992; Barbosa & Sabaté 2004), the Mesoarchaean to Palaeoproterozoic Serrinha block (Mascarenhas 1979; Oliveira et al. 2004a; Mello et al. 2006; Rios et al. 2009), and the Neoarchaean Itabuna–Salvador–Curaçá belt (Barbosa & Sabaté 2004). The last comprises a northern segment known as the Salvador–Curaçá belt (Santos & Souza 1985; 1 in Fig. 1) and a southern segment named the Atlantic coast granulite belt (Mascarenhas 1979) or Itabuna complex (Figueiredo 1989; Teixeira & Figueiredo 1991; 2 in Fig. 1). The two segments are currently referred to as the Itabuna–Salvador–Curaçá orogen (Barbosa et al. 2001, 2008; Delgado et al. 2003; Oliveira et al. 2004a), as shown in Figure 1.

The Itabuna–Salvador–Curaçá orogen

The Itabuna–Salvador–Curaçá orogen is made up of a continuous belt of high-grade metamorphic rocks that crop out for over 800 km in the eastern part of the São Francisco craton, in Bahia state (Fig. 1). At the latitude of Salvador city, the belt apparently splits into two arms; one trends NE along the coastline and the other trends north inland. The southern segment of the Itabuna–Salvador–Curaçá belt is located between the Archaean Jequié block and the Atlantic Ocean, whereas in its northern segment, the inland arm of the belt is sandwiched between the Archaean Gavião block to the west and the Serrinha block to the east. The orogen has been the subject of several studies motivated mostly by its high metallogenetic potential (Cu, Cr and Au; e.g. Silva et al. 2001; Oliveira et al. 2004b; Mello et al. 2006) and geotectonic correlation.

Tectonic models for evolution of the Itabuna–Salvador–Curaçá orogen

All tectonic models proposed for the orogen during the Palaeoproterozoic involve a final stage of continent–continent collision similar to that for Phanerozoic orogens (Figueiredo 1989; Barbosa 1990; Padilha & Melo 1991; Teixeira & Figueiredo 1991; Silva 1992; Ledru et al. 1994, 1997; Teixeira et al. 2000; Silva et al. 2001; Oliveira et al. 2002, 2004b; Barbosa & Sabaté 2004).

According to Barbosa (1990), Ledru et al. (1994) and Barbosa & Sabaté (2004) the southern segment of the orogen is undergoing final collision between the Gabon massif, in West Africa, and the Jequié

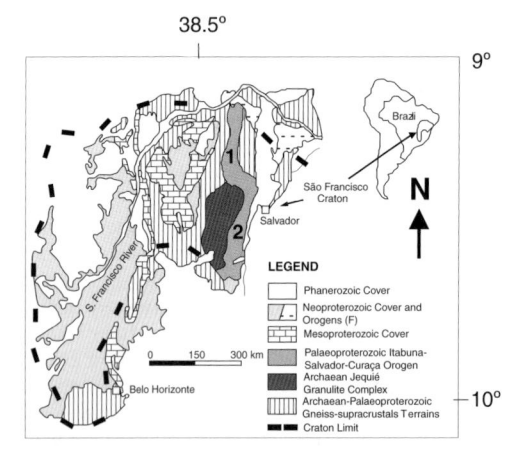

Fig. 1. The São Francisco craton with location of the Itabuna–Salvador–Curaçá orogen, showing its northern (1) and southern (2) segments. Adapted from Oliveira & Tarney (1995).

microcontinent, which occurred in the following sequence: 2.6–2.4 Ga, accretion of a continental arc; 2.4–2.2 Ga, intrusion of shoshonitic to alkaline monzonites and monzodiorites; 2.07–2.08 Ga, continent–continent collision and granulite-facies metamorphism.

For the orogen's northern segment the geochronological and geological data point to a long cycle of basement reworking (Sabaté et al. 1994; Silva et al. 1997; Oliveira et al. 2000, 2004a, b), accretion of continental and oceanic arcs, and plutonic complexes to Archaean blocks, and final continental collision (Teixeira & Figueiredo 1991; Melo et al. 1991; Ledru et al. 1997; Silva et al. 2001; Oliveira et al. 2004a, b).

Barbosa & Sabaté (2004) proposed that four Archaean tectonic blocks, namely Gavião, Serrinha, Jequié and the Itabuna–Salvador–Curaçá orogen, collided during the Palaeoproterozoic–Transamazonian orogeny.

Below we present descriptions of units that build up the northern segment of the orogen, and we present a review of age dating and combine new age data with Nd isotope and relevant major and trace element whole-rock geochemistry to discuss Archaean to Proterozoic regional evolution.

Rock units of the northern segment of the Itabuna–Salvador–Curaçá orogen

The northern segment of the Itabuna–Salvador–Curaçá orogen (Fig. 2) consists of a central domain made up of granulite-facies igneous and sedimentary rocks of the Caraíba complex, bordered to the west by gneisses, migmatites and supracrustal rocks of the Gavião block, and to the east by gneisses, migmatites and greenstone belts of the Serrinha block.

Granodiorite to tonalite granulites of the Caraíba complex in the central domain are intrusive into, or are in tectonic contact with aluminous gneisses, amphibolites, graphite-rich gneisses, banded iron formations, marble and calc-silicate rocks of the Tanque Novo complex to the north, and the Ipirá complex to the south (e.g. Delgado et al. 2003; Kosin et al. 2003). Sapphirine-bearing aluminous gneisses of the Ipirá complex have achieved metamorphic equilibrium at 900–950 °C at 7.0–8.0 kbar pressures (Leite et al. 2009), and they are representatives of ultrahigh-temperature (UHT) metamorphism in the belt. These units, along with peridotites, pyroxenites, ferrogabbros, gabbronorites and leucogabbros of the São José do Jacuípe suite, in the south, were later intruded by syn-collisional north–south-trending granite to syenite plutons, of which the Itiuba Syenite is the main representative (S in Fig 2). The São José do Jacuípe mafic–ultramafic suite has been interpreted as an ocean-floor sequence intruded by calc-alkaline plutons of the Caraíba complex (Teixeira 1997), and according to Delgado et al. (2003) and Oliveira et al. (2004a) the Caraíba complex may be a Neoarchaean Andean-type continental arc. New U–Pb SHRIMP ages will be presented in the next section for these two units.

Migmatites, banded gneisses, orthogneisses, mafic dykes and mafic–ultramafic complexes of the Serrinha block, to the east of the central domain, are overlain by, or are in tectonic contact with supracrustal sequences of the Rio Itapicuru and Rio Capim greenstone belts, and of the Caldeirão shear belt (Fig. 2). Granites intrude all units. The Uauá, Jacurici and Retirolândia gneiss–migmatite complexes (Fig. 2) are Mesoarchaean subdomains in the Serrinha block. New U–Pb SHRIMP ages for the two latter subdomains will be given in the geochronology section. The Uauá subdomain was intruded by at least two mafic dyke swarms; the younger one is bimodal (i.e. it is made up of tholeiite and norite dykes). These dykes change their NE–SW trend to NW–SE as they approach the 10 km thick Caldeirão shear belt (CB in Fig. 2), where they have undergone boudinage, disruption and metamorphosis to amphibolites. To the east, the Uauá subdomain is separated from the Rio Capim greenstone belt by 100–300 m thick shear zones that show syn-mylonitization asymmetric folds indicating a right-lateral sense of shearing. These structural characteristics support the interpretation that the Uauá domain is an exotic block displaced, or extruded, from south to north during Palaeoproterozoic oblique collision (Oliveira et al. 2001, 2004b). The other two gneiss–migmatite subdomains referred to above (Jacurici and Retirolândia) also have mafic dykes but correlations with the Uauá dykes are uncertain owing to lack of high-precision U–Pb ages. In the Retirolândia subdomain mafic dykes crop out mostly along the NNW–SSE-trending contact with the Rio Itapicuru greenstone belt, where both dykes and host migmatites were deformed and converted into banded gneisses. This feature will be recalled below when discussing the regional evolution.

The Rio Itapicuru greenstone belt in the Serrinha block (1 in Fig. 2) is a low-grade metamorphic supracrustal sequence c. 180 km long and 30 km wide, divided by Kishida & Riccio (1980) into three lithostratigraphic units: (1) the basal mafic volcanic unit composed of massive and pillowed basaltic flows interlayered with chert, banded iron formation, and carbonaceous shale; (2) the intermediate to felsic volcanic unit with metadacites, metandesites and metapyroclastic rocks; (3) a metasedimentary pelitic–psammitic unit composed mainly of metapelites and minor chemical

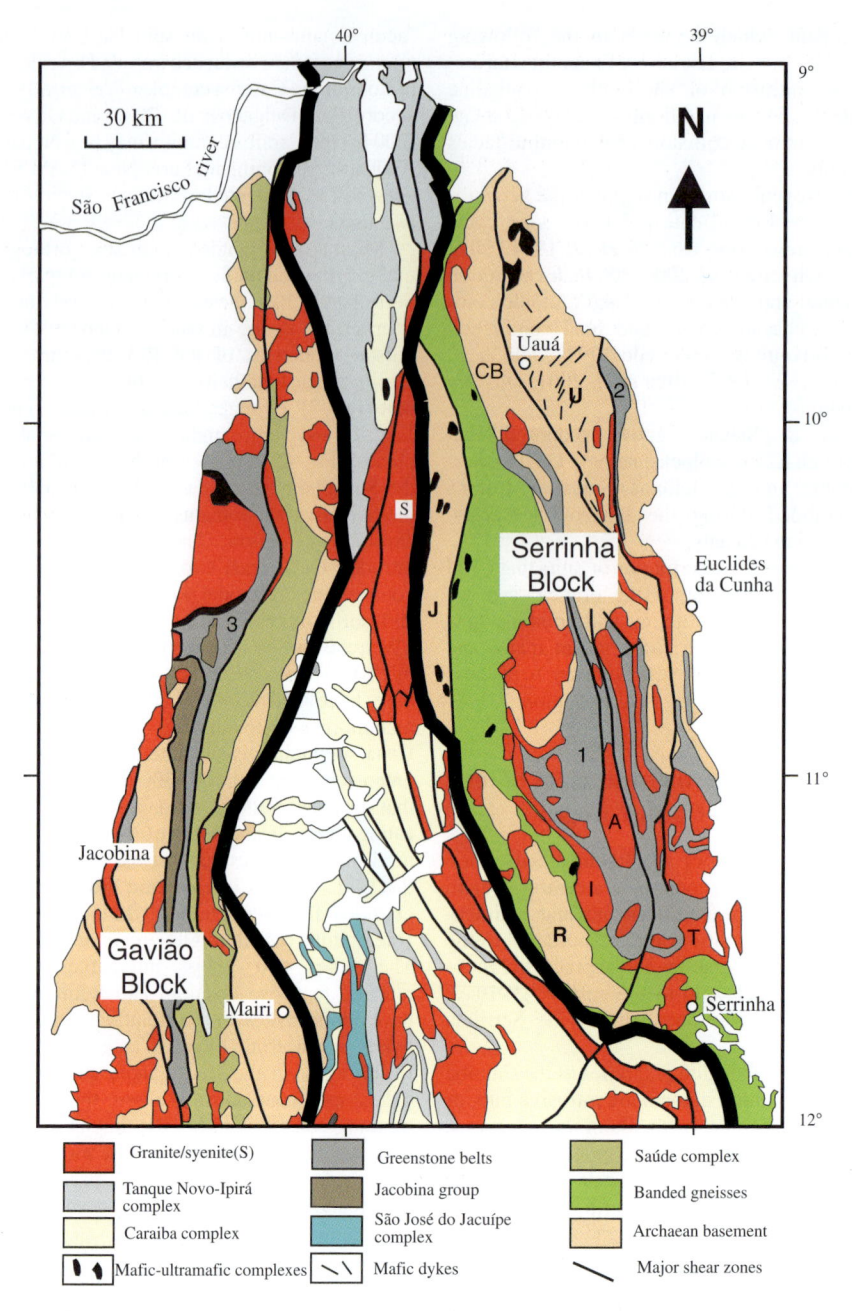

Fig. 2. The northern segment of the Itabuna–Salvador–Curaçá orogen (modified after Kosin *et al.* 2003) with its major tectonic divisions. Greenstone belts: 1, Rio Itapicuru; 2, Rio Capim; 3, Mundo Novo. basement subdomains in the Serrinha nucleus: R, Retirolândia; J, Jacurici; U, Uauá. Granitic bodies: S, Itiúba syenite; I, Itareru tonalite; A, Ambrósio dome; T, Teofilândia tonalite; CB, Caldeirão belt. Bold black lines indicate the approximate limit of the granulite-facies core of the orogen.

sedimentary rocks. The Rio Capim greenstone belt (2 in Fig. 2), on the other hand, is a relatively small, 4 km wide, *c.* 20 km long, north–south- to NW–SE-trending belt of deformed and metamorphosed mafic to felsic volcanic rocks and associated pelitic rocks, intruded by a few plutons ranging in composition from gabbro–diorite to granite (Winge 1981; Jardim de Sá *et al.* 1984).

From NW to SE the belt shows mineral assemblages indicative of increasing metamorphic grade from low amphibolite to granulite facies (Jardim de Sá et al. 1984). The Rio Capim greenstone belt lies in contact with Archaean rocks of the Uauá subdomain, to the west, along hundred metre thick upright shear zones; to the east it is overlain by Neoproterozoic sedimentary rocks of the Sergipano belt.

The central domain of the orogen's northern segment is bounded to the west by supracrustal rocks of the Saúde complex (paragneisses, aluminous schists, quartzites, metaconglomerates, calc-silicate rocks, banded iron formations and gondite), and gneisses and migmatites of the Mairi complex (Kosin et al. 2003). Far to the west (Fig. 2) there are three other units: the Mundo Novo greenstone belt [quartzites, metacherts, phyllites, metapelites, iron formations, metaconglomerates, metabasalts and metadacites (3 in Fig. 2)], the Jacobina Group (metaconglomerates, quartzites, graphite schist and mafic–ultramafic complexes), and migmatites and gneisses of the Archaean Gavião block and Mairi complex.

Several mafic–ultramafic bodies occur in the northern segment of the orogen, a number of which host mineral deposits such as the Cu-rich norites and hypersthenites of Caraíba (Oliveira & Tarney 1995), chromite peridotites of Campo Formoso and Medrado (Marques & Ferreira Filho 2003), and chromite serpentinites of Santa Luz (Oliveira et al. 2007).

Geochronological and geochemical data

Here we examine only the high-precision geochronological data on zircon grains available in the literature (U–Pb SHRIMP, isotope dilution, and less often Pb evaporation) and present new results for key units. For some rocks, such as mafic dyke swarms and a few mafic–ultramafic complexes, for which there are few conclusive zircon ages, we use Sm–Nd or Pb–Pb isochrons. Whenever possible, zircon ages are combined with Sm–Nd isotope data and whole-rock major and trace element data to establish reliable constraints for tectonic interpretations.

The new zircon U–Pb data were collected mostly at the Perth SHRIMP consortium of Curtin University of Technology, the University of Western Australia, and the Geological Survey of Western Australia, and less often at the Canberra SHRIMP laboratory of the Australian National University. To prevent cross contamination, zircon concentrates were obtained using jaw crushers and disc mill previously polished with automatic steel brushers, followed by manual panning. The Sm–Nd isotope data were acquired at the isotope laboratories of Kansas University (USA) and University of Brasilia (Brazil) by thermal ionization mass spectrometry following the general procedures of Patchett & Ruiz (1987). Depleted mantle Nd model ages (T_{DM}) and ε_{Nd} values were calculated according to DePaolo (1988). Whole-rock major and trace element analyses were obtained at the geochemistry laboratory of University of Campinas—UNICAMP by X-ray fluorescence spectrometry, and sometimes also by inductively coupled plasma mass spectrometry (ICP-MS) for the REE. Representative Nd isotope and major and trace element analyses are given in Tables 1 and 2. The full dataset including SHRIMP spot analyses is available upon request.

Gavião block

This block and the Serrinha block contain the oldest gneisses and migmatites of the São Francisco craton (e.g. Martin et al. 1991, 1997; Nutman & Cordani 1993; Paixão & Oliveira 1998; Cordani et al. 1999; Oliveira et al. 1999, 2004a; Rios et al. 2009). They are considered representatives of the microcontinents that collided to form the Itabuna–Salvador–Curaçá orogen.

In the extreme NE part of the Gavião block, which is the relevant area for this review, Mougeot et al. (1996) reported zircon U–Pb ages in the range of 3450–3402 Ma for a migmatite, whereas Peucat et al. (2002) obtained zircon Pb evaporation minimum ages of 3025 and 3040 Ma for gneisses of the Mairi Complex. For the latter, Nd model ages fall in the range 3.20–3.46 Ga, with $\varepsilon_{Nd(3040\ Ma)}$ between −0.71 and −4.21 (Table 1).

Mundo Novo greenstone belt

Mascarenhas & Silva (1994) described this greenstone belt in detail but only recently has precise age dating become available. Peucat et al. (2002) obtained a zircon U–Pb SHRIMP age of 3305 ± 9 Ma for a metadacite, showing that this is one of most ancient supracrustal sequences in the São Francisco craton. The metadacite Nd model ages vary from 3.58 to 3.37 Ga, with $\varepsilon_{Nd(3305\ Ma)}$ in the range 0.49 to −1.94 (Table 1). Although no whole-rock major and trace element data are available for the volcanic rocks, the positive to slightly negative $\varepsilon_{Nd(T)}$ values suggest that the unit may belong to an arc assemblage accreted onto the Gavião block.

Serrinha block basement gneisses and mafic dykes

Age data for the Serrinha block are relatively abundant. In the Uauá subdomain, in the north (Fig. 2), the oldest rocks so far recognized are orthogneisses

Table 1. Nd isotope data for units of the northern segment of the Itabuna–Salvador–Curaçá orogen

Sample	Rock unit	U–Pb age (Ma)	$^{147}Sm/^{144}Nd$	$^{143}Nd/^{144}Nd$ ($\pm 1\sigma$)	T_{DM} (Ga)	$\varepsilon_{Nd}(0)$	$\varepsilon_{Nd}(T)$
Mairi Complex							
PO-91	Mairi migmatite–gneiss	>3040	0.1018	0.510518 (21)	3.46	−41.35	−4.21
PO-103	Mairi orthogneiss	>3040	0.0974	0.510608 (39)	3.20	−39.60	−0.71
Peucat-526	Mairi gneiss	>3040	0.0869	0.510357 (11)	3.24	−44.50	−1.49
Mundo Novo greenstone belt							
PO-119	Mundo Novo metadacite	3305	0.1193	0.510848 (27)	3.58	−34.92	−1.94
Peucat-600	Mundo Novo metadacite	3305	0.1231	0.511044 (7)	3.38	−31.09	0.49
Serrinha nucleus							
PO-45.2	Jacurici G1 tonalite	2983	0.0831	0.510540 (20)	2.92	−40.92	2.75
PO-47.1	Jacurici G1 tonalite	2983	0.0753	0.510203 (18)	3.13	−47.50	−0.85
PO-9	Retirolândia TTG	3085	0.0945	0.510614 (14)	3.11	−39.48	1.15
PO-64	Retirolândia TTG	3085	0.1163	0.510997 (14)	3.22	−32.01	−0.04
UAEO-44	Uauá enderbitic granulite	3000	0.1255	0.511164 (22)	3.28	−28.75	−1.25
UAEO-137.3	Uauá enderbitic granulite	2933	0.1052	0.510828 (13)	3.12	−35.32	−0.75
PO-37	Uauá enderbitic granulite	3000	0.0876	0.510504 (20)	3.07	−41.63	0.54
PO-82	Uauá banded gneiss	–	0.0772	0.509999 (36)	3.41	−51.48	
JUB 21	Capim tonalite	3120	0.1224	0.510949 (10)	3.54	−32.94	−3.09
JM-180B	Capim tonalite	3120	0.0921	0.510214 (15)	3.57	−47.28	−5.27
Caraíba complex							
PO-102	S. J. Jacuipe enderbite	2695	0.1080	0.511086 (20)	2.82	−30.27	0.49
OOM-1D	G_2 granodiorite	2574	0.0989	0.511198 (21)	2.44	−28.09	4.32
SAF-08	G_2 granodiorite	2574	0.0715	0.510499 (11)	2.72	−41.73	−0.26
PPIN-3	Curaçá G_2 granodiorite	2574	0.0796	0.510616 (13)	2.75	−39.44	1.18
AERO-C	Airport G_2 granodiorite	2574	0.0893	0.510826 (13)	2.71	−35.35	1.89
Rio Capim greenstone belt							
Craitu-2	Capim dacite	2148	0.0947	0.511245 (11)	2.29	−27.17	0.97
Craitu-3	Capim dacite	2148	0.0951	0.511258 (11)	2.28	−26.92	1.11
CAZ-15.B	High-grade metapelite		0.1616	0.511998 (13)	3.11	−12.48	

Sample	Description	Age (Ma)	$^{147}Sm/^{144}Nd$	$^{143}Nd/^{144}Nd$	T_{DM} (Ga)	$\varepsilon_{Nd}(0)$	$\varepsilon_{Nd}(T)$
UA96-8.4	Low-grade metapelite		0.1133	0.511380 (14)	2.52	−24.54	
UA96-5.1	Basic granulite		0.1016	0.511230 (15)	2.46	−27.46	
Rio Itapicuru greenstone belt							
GBRI-6.3	Massive basalt	2145	0.1906	0.512727 (14)	2.06	1.73	3.42
GBRI-12	Pillowed basalt	2145	0.1976	0.512851 (13)		4.15	3.92
GBRI-10	Dacite	2081	0.1215	0.511747 (15)	2.13	−17.38	2.70
GBRI-11	Dacite	2081	0.1240	0.511736 (11)	2.21	−17.60	1.82
THB-1.6	Teofilândia tonalite	2130	0.1017	0.511427 (11)	2.19	−23.62	2.38
THB-3.3	Teofilândia tonalite	2130	0.1057	0.511474 (10)	2.20	−22.70	2.20
THB-4.2A	Barrocas tonalite	2127	0.1511	0.512109 (15)	2.27	−10.31	2.13
FB-135.60A	Barrocas tonalite	2127	0.1189	0.511678 (15)	2.18	−18.73	2.53
JCI-136b	Itareru tonalite	2109	0.0977	0.511263 (27)	2.33	−26.82	0
SLM-16	Itareru tonalite	2109	0.1010	0.511305 (13)	2.34	−25.92	−0.08
CLAB-247	Ambrósio granodiorite	2080	0.0849	0.510811 (10)	2.64	−35.64	−5.82
CLAB-255	Ambrósio granodiorite	2080	0.0941	0.511011 (13)	2.58	−31.73	−4.35
PO-18.2	Morrodo Lopes granite	2072	0.1152	0.511201 (26)	2.85	−28.03	−6.37
JCI-133	Morrodo Lopes granite	2072	0.1272	0.511255 (33)	3.17	−26.98	−8.52
Saúde complex, Tanque Novo and Ipirá groups							
PO-97	Saúde Al-gneiss		0.1143	0.511339 (19)	2.61	−25.34	
AW 08	Tanque Novo Al-gneiss		0.1235	0.511434 (26)	2.72	−29.16	
SAF-04	Tanque Novo Al-gneiss		0.0725	0.510507 (12)	2.73	−41.57	
Caldeirão belt							
MS-1.2	Monte Santo Al-gneiss		0.1011	0.510701 (26)	3.18	−37.78	
UAEO-57.2	Caldeirao Al-gneiss		0.1166	0.511003 (36)	3.22	−31.89	
Granitoids in the Caraíba complex							
AERO-B	Airport G3 granite	2027	0.0830	0.510719 (13)	2.71	−37.44	−7.88
PO-43	Itiúba syenite	2084	0.0919	0.510766 (18)	2.85	−36.52	−8.51
ITIUBA 4	Itiúba syenite	2084	0.0982	0.510992 (13)	2.70	−32.11	−5.77
PO-110	Capela qtz-monzonite	2078	0.1142	0.510993 (11)	3.15	−32.09	−10.12

Sample numbers preceded by 'Peucat' are from Peucat *et al.* (2002).

Table 2. *Representative major and trace element analyses for selected gneisses, amphibolites and volcanic rocks of the northern segment of the Itabuna–Salvador–Curaçá orogen*

	Uauá orthogranulites			Uauá thrust orthogranulites			Retirolândia TTG			Jacurici TTG			Caraíba orthogneisses			Caraíba A-14 amphibolites		
Sample:	EO-115	EO-44	EO-44.1	EO-137.3	EO-137.4	EO-78	RETI 2B	RETI 3	PO-64	PO-39	PO-45.1	PO-46	PO-102	PO-74	AERO-C	A14-04	A14-05	A14-07
SiO$_2$	55.90	61.13	67.78	54.79	59.19	71.72	68.44	72.21	66.92	69.13	67.52	68.10	69.14	60.67	69.95	46.40	41.60	44.50
TiO$_2$	0.94	0.92	0.49	1.05	0.82	0.34	0.31	0.25	0.38	0.35	0.54	0.46	0.32	0.79	0.55	1.12	1.42	1.19
Al$_2$O$_3$	17.66	17.34	16.66	17.92	19.12	14.46	16.71	14.52	16.72	15.72	15.46	15.54	16.07	15.02	14.19	13.10	15.40	14.80
Fe$_2$O$_3^t$	8.77	6.98	3.82	9.40	5.07	2.96	2.65	2.45	3.36	2.83	4.42	3.37	2.63	5.04	4.27	14.70	17.30	14.90
MnO	0.11	0.09	0.06	0.13	0.07	0.04	0.04	0.03	0.06	0.03	0.06	0.03	0.03	0.09	0.04	0.22	0.38	0.26
MgO	2.84	2.37	1.46	3.23	1.54	0.72	0.95	1.57	1.24	1.26	1.68	0.97	1.12	1.34	1.08	7.70	9.10	9.80
CaO	7.24	6.61	4.92	6.04	3.55	2.83	3.31	1.98	4.10	2.77	3.05	3.29	3.25	3.20	2.77	12.30	12.50	13.20
Na$_2$O	4.38	4.31	4.17	4.19	5.04	3.78	5.49	4.65	4.86	5.09	4.83	4.96	5.28	4.16	3.68	1.30	0.30	1.10
K$_2$O	0.40	0.04	0.80	3.15	4.98	2.84	1.24	2.03	1.22	1.59	1.73	1.38	1.34	7.37	3.26	0.10	0.06	0.65
P$_2$O$_5$	0.47	0.03	0.15	0.55	0.34	0.09	0.14	0.03	0.12	0.07	0.16	0.18	0.11	0.48	0.15	0.12	0.04	0.09
LOI	1.20						0.32	0.41	0.35	0.37	0.22	0.60	0.17	0.02				
Total	99.91	99.82	100.30	100.45	99.73	99.79	99.60	100.10	99.35	99.21	99.68	98.87	99.45	98.18	99.94	95.92	96.54	99.18

Trace elements (ppm) XRF

Ba	302	421	497	2511	5264	1271	168	377	396	597	588	861	467	5807	1374	68	21	78
Ce							38	11	25.4				23	365	114.1	8.1	5.5	5.2
Cr	62	38	60	40	b.d.l	12	12.7	23.8	22.3	18.6	18.8	22.9	37	26	3.7	187	247	488
Ga	22						20.8	20.1	19.9				21	25	18.3	16.3	16	16.8
La							23	11	20.1				19	290	59.9	2.9	1.5	0.2
Nb	5	8	b.d.l	9	8	b.d.l	7	10.8	5.3	3	9.1	3.5	2	8	4.1	2	4.8	2.9
Nd							14	b.d.l	8	16.2	18.6	17.4	11	180	42.4	6.5	5.3	7.6
Ni	31	27	13	b.d.l	b.d.l	b.d.l	8.2	10.8	12.3	7.8	12.1	9.6	10	28	7.7	134.6	140.5	190.3
Pb							8	14.1	8.5	10.3	13.8	10.8	19	12				

	1	2	5	68	83	44	46	65	47	35.6	78.4	28.4	24	172	70.4	2.8	1.9	20.8
Rb	1	2	5	68	83	44	46	65	47	35.6	78.4	28.4	24	172	70.4	2.8	1.9	20.8
Sr	550	467	319	862	739	294	428	348	401.8	417	251	509	624	2072	378	30	24.9	54.3
Th							7.1	6.4	3.6	6.9	9.8	4		9	10.4	b.d.l	0.5	b.d.l
V	137	93	41	206	62	26	25.4	25.8	40.7	34.2	62.9	40	34	67	49.6	326	405	396
Y	14	12	5	28	26	8	9.7	31	10.6	3.8	17	5.6	4	27	15.1	26.9	39.2	31.4
Zn	102	96	63	89	77	46	51	49	64.8	57.2	77.9	65.7	54	78	46.8	104.4	126.4	114.4
Zr	197	199	166	195	421	132	131	114	108.4	162.6	186.5	173.8	74	331	272.4	63		56.2
Sr/Y	39.3	38.9	63.8	30.8	28.4	36.8	44.1	11.2	37.9	109.7	14.8	90.9	141.9	77	25		50	

Rare earth elements, ICP-MS

La																2.83	1.44	2.19
Ce																7.31	4.35	6.60
Pr																1.07	0.74	1.21
Nd																6.25	5.10	6.77
Sm																2.21	2.58	2.52
Eu																0.83	1.09	0.95
Gd																3.19	4.85	3.65
Tb																0.59	0.89	0.64
Dy																3.95	5.69	4.28
Ho																0.93	1.29	0.98
Er																2.59	3.56	2.73
Tm																0.40	0.55	0.43
Yb																2.56	3.56	2.79
Lu																0.39	0.55	0.42

b.d.l., below detection limit.

from the Caldeirão belt with a SHRIMP age of 3152 ± 5 Ma (Oliveira et al. 2002). However, banded gneiss from the central area of the Uauá area has given the oldest depleted-mantle Nd model age of all the gneisses, of 3.4 Ga (Table 1), suggesting the existence of much older crust in the area than ever reported. Indeed, farther south, Rios et al. (2008) have recently demonstrated the presence of 3.6 Ga zircon xenocrysts in the 2155 ± 3 Ma Quijingue trondhjemite. Mesoarchaean rocks are widespread in the Uauá subdomain. Paixão & Oliveira (1998) obtained a 3161 ± 65 Ma whole-rock Pb isochron for anorthosites of the Lagoa da Vaca layered complex and a zircon Pb-evaporation age of 3072 ± 20 Ma for orthogranulites, and Cordani et al. (1999) presented zircon U–Pb SHRIMP ages between 3.12 and 3.13 Ga for the Capim tonalite. Several other Archaean felsic igneous bodies occur in the Uauá subdomain, of which the Uauá quarry enderbitic granulite and a gneissic granodiorite to the SE of Uauá town were respectively dated at 2933 ± 3 Ma and 2991 ± 22 Ma (Oliveira et al. 2002); the latter entrains xenoliths of the country-rock banded gneisses. Two mafic dyke swarms intrude the basement; samples from the young one yielded a reference Sm–Nd isochron of 2586 ± 66 Ma (Oliveira et al. 1999). The granulite-facies felsic igneous bodies with intercalated garnet-bearing mafic granulites are strongly foliated, often with horizontal, or south-dipping low-angle foliation; they show stretching lineations and asymmetric folds indicative of having been thrust northward. Additionally, these granulites show slightly negative to positive $\varepsilon_{Nd(3000\,Ma)}$ values (Table 1) and subduction zone geochemical characteristics (Fig. 3). This rock package has thus the potential to be exhumed tectonic slices of the root of a Mesoarchaean continental arc.

Farther to the south, tonalite to granodiorite orthogranulite of the Jacurici subdomain and grey gneiss palaeosome of the Retirolândia subdomain have given zircon U–Pb SHRIMP ages of 2983 ± 6 Ma and 3085 ± 6 Ma respectively (Fig. 4). $\varepsilon_{Nd(T)}$ values are mostly positive for Retirolândia and Jacurici rocks (Table 1). Unlike granulites of the Uauá subdomain, which show a calc-alkaline trend in the K–Na–Ca diagram, the Retirolândia and Jacurici samples cluster in the trondhjemite field (Fig. 5), suggesting that they belong to the TTG suite.

Caraíba complex

Owing to its large exposure area in the orogen (Fig. 2), the Caraiba high-grade complex is of prime importance to understanding regional evolution. Silva et al. (1997) were the first to recognize that zircon grains from granulite-facies orthogneisses of the São José de Jacuípe area (southern half of the central domain in Fig. 2) have old cores with average ages of 2695 ± 12 and 2634 ± 19 Ma rimmed by 2072 Ma metamorphic overgrowths. This finding was a major breakthrough because previous Rb–Sr age determinations for the Caraíba complex were ambiguous, producing either Archaean or Palaeoproterozoic ages (see summary by Figueiredo 1989).

To improve our understanding of the Caraíba complex, we present new zircon U–Pb SHRIMP ages for tonalite orthogneiss from the airport outcrop of the Caraíba copper mine, a type locality of the Caraíba complex, and for a leucogabbro of the São José do Jacuipe mafic–ultramafic complex, respectively in the northernmost and southern area of the central domain in Figure 2. In the airport outcrop granulite-facies G$_2$ tonalite gneiss entrains

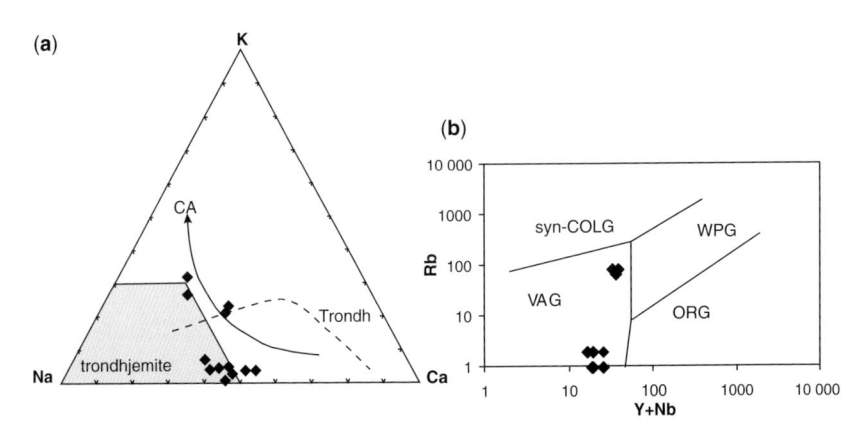

Fig. 3. Geochemical characteristics of the Uauá orthogranulites. (a) K–Na–Ca diagram with the trondhjemite and calc-alkaline trends after Martin (1994). (b) tectonic setting discrimination diagram of Pearce et al. (1984); syn-COLG, syn-collisional granites; VAG, volcanic arc granites; WPG, within-plate granites; ORG, ocean-ridge granites.

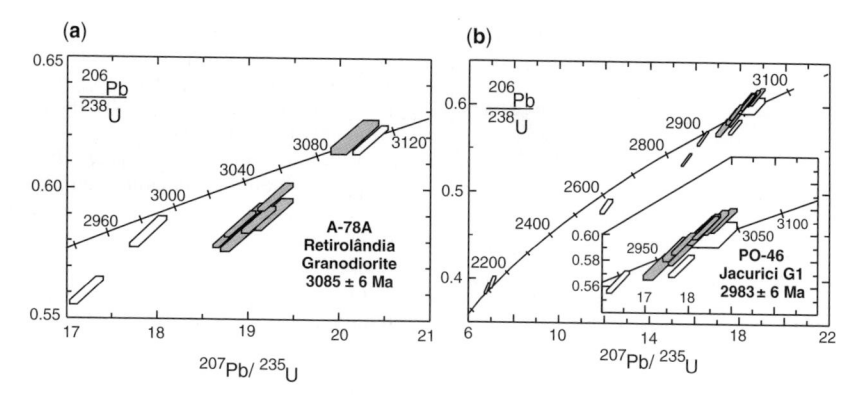

Fig. 4. SHRIMP U–Pb zircon ages for TTG orthogneisses from Retirolândia and Jacurici.

disrupted amphibolite enclaves and is crosscut by pink granite sheets related to the third regional deformation phase, as described by D'el-Rey Silva et al. (2007). On the basis of a Sm–Nd isochron those workers suggested that the amphibolite protolith crystallized at about 2.6 Ga. Furthermore, according to D'el-Rey Silva et al. (2007), zircon U–Pb isotope dilution analyses have given a less precise age of 2248 ± 36 Ma for the host tonalite. Our SHRIMP results for the same tonalite indicate igneous zircon cores with an average age of 2574 ± 6 Ma and metamorphic rims with an age of 2074 ± 14 Ma (Fig. 6). The $\varepsilon_{Nd(T)}$ value for the same sample is 0.20 and for comparable tonalites in the Caraíba area the $\varepsilon_{Nd(T)}$ values fall in the range −0.27 to 4.33 (Table 1). Zircons from the São José do Jacuípe leucogabbro yielded two main age groups (Fig. 7). Most zircon cores cluster around an age of 2583 ± 8 Ma, whereas four metamorphic zircon rims gave an age of 2082 ± 17 Ma; the 2583 Ma age is interpreted as the age of igneous crystallization.

Figure 8 shows that the Caraíba tonalites and similar rocks from the São José do Jacuípe area, in the south, all follow the calc-alkaline trend and are akin to volcanic arc granites. More interesting, close to the Caraíba mine, Oliveira (1990) recognized light rare earth element (LREE)-depleted amphibolites associated with metaperidotites (A14 body). We reanalysed some of these amphibolites by ICP-MS and confirmed their similarity to Phanerozoic incompatible element-depleted ocean-floor basalts (normal mid-ocean ridge basalt; N-MORB) (Fig. 9); this amphibolite group is the only one that shows such a geochemical signature amongst at least four other regional groups, and for this reason it is unlikely that depletion of LREE was due to high-grade metamorphism. In summary,

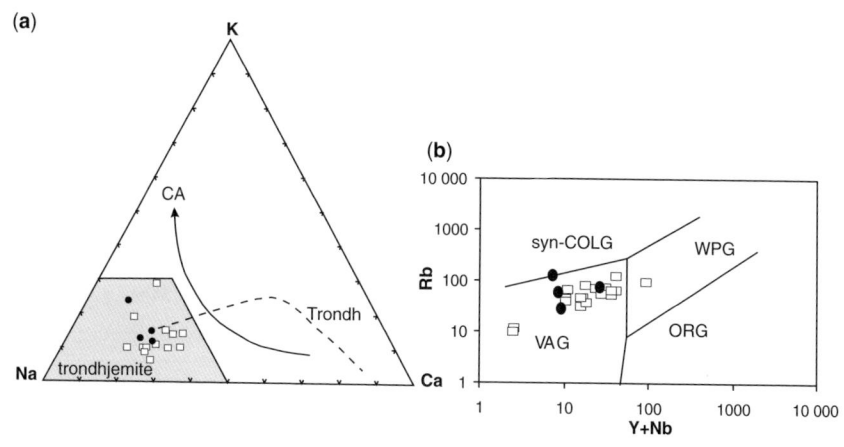

Fig. 5. Geochemical characteristics of the Retirolândia (□) and Jacurici (●) orthogneisses. (**a**) K–Na–Ca diagram with the trondhjemite and calc-alkaline trends after Martin (1994). (**b**) Tectonic setting discrimination diagram of Pearce et al. (1984). Abbreviations as in Figure 3.

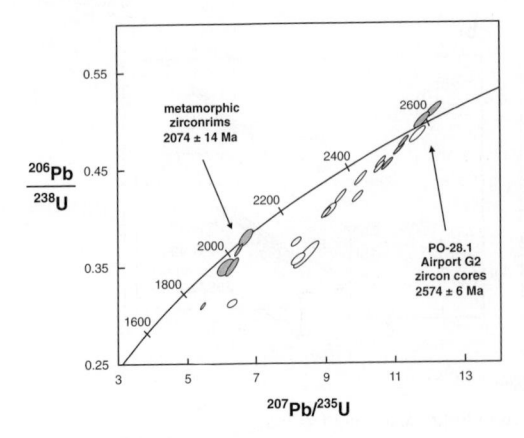

Fig. 6. SHRIMP U–Pb zircon ages for the Caraíba airport G₂ granulitic granodiorite, showing ages for the igneous protolith and granulite-facies metamorphic rims.

the juvenile Nd isotope signature and arc-like major and trace element characteristics are compelling evidence that granite batholiths of the Caraíba complex may have originally been formed in arc settings. Whether the arc was intra-oceanic or of the continental type is uncertain owing to tectonic imbrication of the rock units, but in both alternatives oceanic plates like those of the Phanerozoic record had come into existence in this part of South America by the end of the Archaean, and the evidence for this is probably preserved as accreted metamorphosed ocean-floor basalts in the Caraiba area.

Rio Capim greenstone belt

This greenstone belt and the Rio Itapicuru greenstone belt occur in the Serrinha block. The Rio Capim belt is composed mostly of amphibolites and tuffaceous metadacites structurally interleaved with garnet–biotite–sillimanite–cordierite schists, graphite schist, biotite gneiss, banded iron formation, ferrugineous and carbonate metacherts, and a few intrusions of gabbro, diorite and granite. The metadacites may occasionally contain gneiss xenoliths. Oliveira *et al.* (1998) presented a Pb–Pb isochron of 2153 Ma for metadacites, and isotope dilution U–Pb zircon ages of 2138 Ma and 2126 Ma for leucogabbro and diorite respectively. New zircon U–Pb SHRIMP data for a metadacite yield an age of 2148 ± 9 Ma (Fig. 10), confirming a Palaeoproterozoic age for the greenstone belt and temporal correlation with the Rio Itapicuru greenstone belt that will be considered below. The $\varepsilon_{Nd(2148)}$ values for two metadacites from continuous outcrops are 0.97 and 1.11 (Table 1); their trace element geochemistry is indistinguishable from that of volcanic arc rocks (Fig. 11). Furthermore, depleted-mantle Nd model ages for garnet–sillimanite schists and other igneous rock of the belt are in the range 3.1–2.2 Ga (Table 1), suggesting clast contributions from sources much older than solely from the belt. Overall, field relationships, geochronology and geochemical data indicate that at least a significant portion of the Rio Capim greenstone belt, especially the felsic volcanic rocks, appears to have originated in a

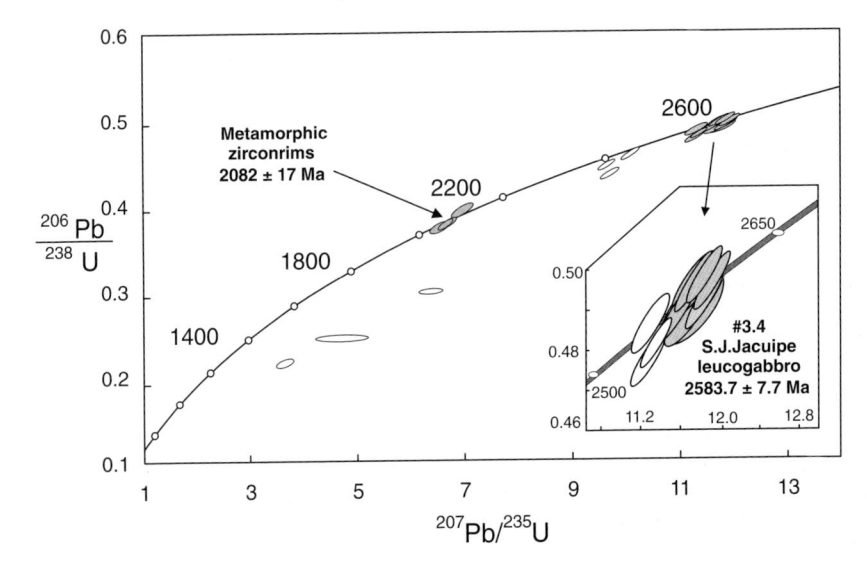

Fig. 7. SHRIMP U–Pb zircon ages for the São José do Jacuípe leucogabbro, showing ages for the igneous protolith and granulite-facies metamorphic rims.

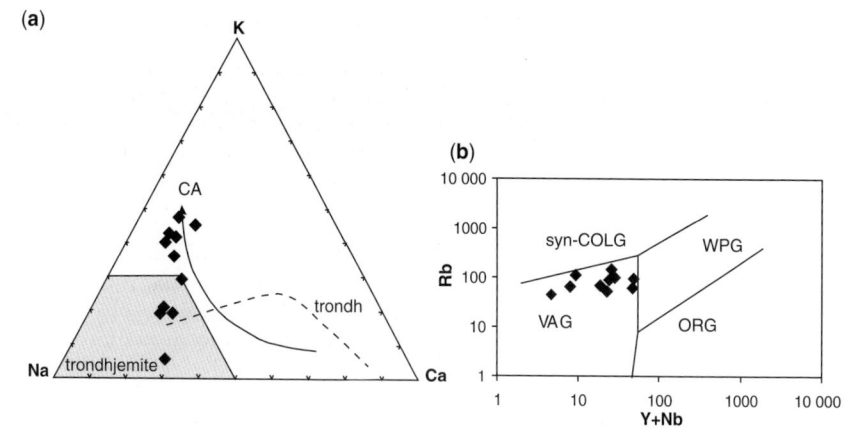

Fig. 8. Geochemical characteristics of the Caraíba complex. (**a**) K–Na–Ca diagram with the trondhjemite and calc-alkaline trends after Martin (1994). (**b**) Tectonic setting discrimination diagram of Pearce *et al.* (1984). Abbreviations as in Figure 3.

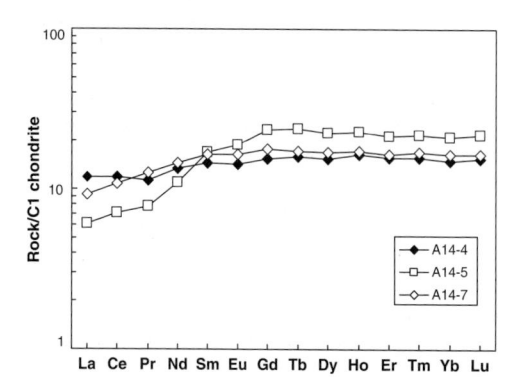

Fig. 9. Rare earth element patterns for the A-14 amphibolites at the Caraíba mine. C1 chondrite after Sun & McDonough (1989).

continental volcanic arc. The sharp contact of the Rio Capim belt against Mesoarchaean rocks of the Uauá subdomain suggests that the former was accreted to the latter and that the shear zones are potential suture zones.

Rio Itapicuru greenstone belt

This volcanic–sedimentary sequence is currently interpreted as a back-arc basin developed upon continental crust (Silva 1992; Silva *et al.* 2001), although the location of the coeval arc has not yet been determined. Field relationships along the Itapicuru River show that the greenstone belt is made up of a basal unit of massive and pillowed basalts, succeeded by andesites, dacites and pyroclastic rocks of the felsic unit, and by clastic and chemical sedimentary rocks of the upper unit, all of them showing greenschist-facies metamorphism (Kishida & Riccio 1980; Davison *et al.* 1988). In other localities, basalts and chemical sediments are interlayered (Silva *et al.* 2001). Several granitoids intrude the belt, especially the basalts and the sedimentary rocks, and together they form a regional pattern of domes and keels, like Archaean granite–greenstone terranes. The previously available ages indicate the following event sequence:

2209 ± 60 Ma, ocean-floor metabasalts (Pb isochron, Silva *et al.* 2001);

2170 ± 60 Ma, arc andesites and dacites (Pb isochron, Silva *et al.* 2001);

2163 ± 5 to 2152 ± 6 Ma, intrusion of TTG and calc-alkaline plutons into basalts (Pb evaporation and U–Pb, Eficeas granodiorite, Rios 2002; Nordestina trondhjemite, Cruz Filho *et al.* 2005; Trilhado granodiorite, Mello *et al.* 2006);

2130 ± 7 to 2127 ± 5 Ma, intrusion of calc-alkaline plutons into basalts (Pb evaporation and U–Pb, Barrocas granodiorite, Chauvet *et al.* 1997a;

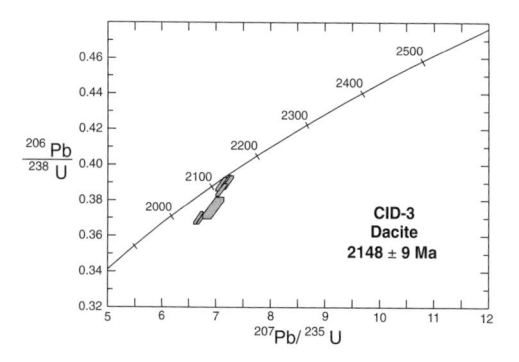

Fig. 10. SHRIMP U–Pb zircon age for the Rio Capim greenstone belt dacite.

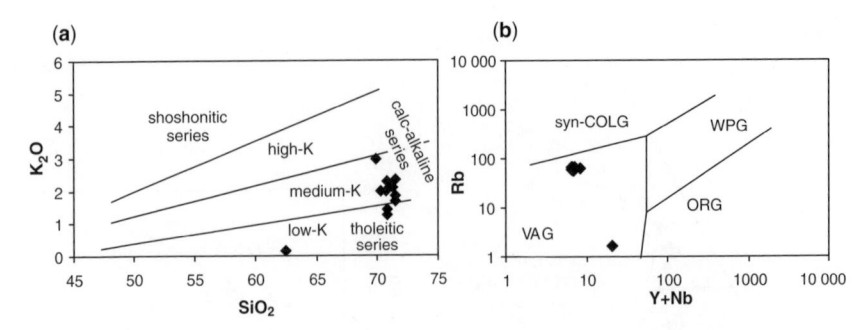

Fig. 11. Geochemical characteristics of the Capim dacites. (**a**) K$_2$O v. SiO$_2$ diagram for volcanic rocks with fields after Peccerillo & Taylor (1976). (**b**) Tectonic setting discrimination diagram of Pearce *et al.* (1984). Abbreviations as in Figure 3.

Barrueto 2002; Teofilândia granodiorite, Mello *et al.* 2006);

2111 ± 10 to 2106 ± 6 Ma, intrusion of high-K to ultrapotassic plutons along the basement–greenstone boundary (U–Pb, Morro do Afonso syenite, Rios *et al.* 2007; Itareru tonalite, Carvalho & Oliveira 2003; Fazenda Gavião granodiorite, Costa 2008; Cansanção monzonite, Rios 2002);

2098 ± 2 to 2086 ± 2 Ma, intrusion of syenite plutons into the basement (Pb evaporation and U–Pb, Serra do Pintado, Agulhas, Bananas, Conceição *et al.* 2002);

2080 ± 2 Ma, syntectonic emplacement of granodiorite–basement domes (U–Pb, Ambrósio, Mello *et al.* 2006);

2072 ± 1 Ma, late tectonic intrusion of potassic plutons (U–Pb, Morro do Lopes, Rios *et al.* 2000);

2054 ± 2 to 2049 ± 4 Ma, gold mineralization in shear zones (Ar–Ar, Vasconcelos & Becker 1992; Mello *et al.* 2006).

The Nd isotope data of Silva *et al.* (2001) indicate T$_{DM}$ model ages of *c.* 2.20 Ga for basalts and 2.12 Ga for andesites, with $\varepsilon_{Nd(T)}$ values mostly positive (+4 for basalts and +2 for andesites). Barrueto *et al.* (1998) and Barrueto (2002) have also found positive $\varepsilon_{Nd(T)}$ values for the Teofilândia granodiorite (+1.84 to +2.38) and Barrocas granodiorite (+2.13 to +2.58), and T$_{DM}$ model ages in the range 2.28–2.19 Ga; on the basis of geochemical signatures and lack of basement xenoliths those workers suggested that the plutons may have formed in an intra-oceanic arc. The 2109 Ma Itareru tonalite contains amphibolite and paragneiss xenoliths and shows Nd model ages between 2.40 and 2.33 Ga; its $\varepsilon_{Nd(2109)}$ values are in the range −0.79 to zero (Table 1), suggesting interaction of the juvenile parental magma with pre-existing crust. On the other hand, the younger plutons such as the Ambrosio basement–granodiorite dome and the Morro do Lopes granite, which are intrusive respectively into supracrustal rocks and basement gneisses, have much older Nd model ages (2.58–3.21 Ga) and negative $\varepsilon_{Nd(t)}$ values (−8.53 to −6.37) (Table 1). These plutons are interpreted as basement partial melts or as having significant contributions from the basement (Oliveira *et al.* 2004*a*).

Because the ages given above for basalts and andesite–dacites are whole-rock Pb isochrons (Silva *et al.* 2001), we present here new zircon U–Pb SHRIMP data for these rocks. The ages found for a massive metabasalt and a porphyritic metabasalt close to the Itapicuru river are 2145 ± 8 Ma and 2142 ± 6 Ma respectively (Fig. 12). One dacite from the Itapicuru river gave an age of 2081 ± 9 Ma (Fig. 13). This dacite sample contains three older zircon grains with ages of 3364 ± 10 Ma (107% of discordance), 3064 ± 6 Ma (98% of discordance) and 3017 ± 8 Ma (92% of discordance), suggesting magma interaction with much older crust. The new ages have several implications for models of regional evolution, and a more comprehensive account will be published elsewhere. For instance, the basalt unit has been intruded by a few granitic plutons with zircon ages (2163–2152 Ma) older than that of the basalts (*c.* 2145 Ma), suggesting that either the ages of these plutons were calculated on populations of inherited zircon grains, or that there might exist basalts with distinct ages. If the first alternative is valid, then where are the *c.* 2160–2150 Ma crust remnants? One plausible possibility is represented by igneous rocks such as the Quijingue pluton (2155 ± 8 Ma), which are intrusive into the basement gneisses (Rios *et al.* 2008). The second hypothesis may be realistic, except for some plutons such as the Nordestina trondhjemite and Trilhado granodiorite, which crop out very close to the dated basalts, and no shear zones separate the granites from the basalts. Another interesting issue refers to plausible tectonic settings of the basalts and andesite–dacites. If the basalts have originated in a *c.* 2145 Ma back-arc basin upon continental crust as postulated

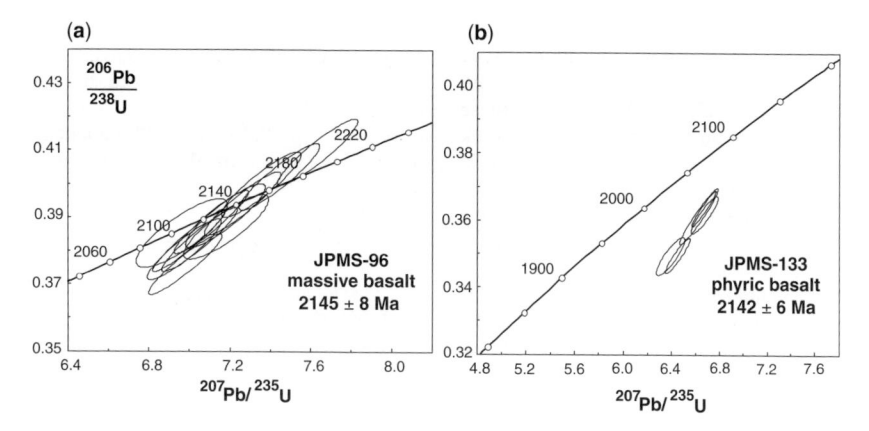

Fig. 12. SHRIMP U–Pb zircon ages for metabasalts from the Rio Itapicuru greenstone belt.

by Silva (1992) and Silva *et al.* (2001) then the coeval arc may potentially be represented by rocks of the *c.* 2150 Ma Rio Capim greenstone belt, which lies to the ENE of the Rio Itapicuru greenstone belt; this implies a subduction zone dipping west. Moreover, the new age for dacites of the Rio Itapicuru greenstone belt (*c.* 2080 Ma) and their bimodal distribution, with adakite to the west and calc-alkaline rocks to the east (Ruggiero 2008; Ruggiero & Oliveira 2010), suggest another younger subduction zone dipping west.

Jacobina group

The geochronological information on sedimentary rocks of this group is relevant for the evolution of the orogen. The group crops out along the contact with the Gavião block, in the west, and comprises metaconglomerates, quartzites and graphite schist intruded by mafic–ultramafic complexes. Mougeot *et al.* (1996) have dated detrital zircons from quartzites and found two main zircon populations: an old (3.45–3.35 Ga) and a young (*c.* 2080 Ma) population. Although the Gavião block may apparently be the main source of clasts, palaeocurrent studies point to transport from east to west. This feature led Mougeot *et al.* (1996) to suggest that at least part of the Jacobina basin was filled with clasts sourced in the orogenic belt to the east. According to Ledru *et al.* (1997), the Jacobina group is a foreland basin associated with the Salvador–Curaçá belt, here named the northern segment of the Itabuna–Salvador–Curaçá orogen.

Saúde complex, Tanque Novo and Ipirá groups

These units are composed mostly of high-grade metamorphosed sedimentary rocks. According to Leite (2002) and Kosin *et al.* (2003) the three units are coeval, although no U–Pb geochronological data are available. However, Nd model ages for aluminous paragneisses of the Tanque Novo group in the north and the Saúde complex in the south indicate maximum ages between 2.6 and 2.7 Ga for protolith deposition (Table 1). Electron microprobe dating of monazite in sapphirine-bearing granulite paragneisses indicates UHT metamorphism at *c.* 2.08–2.05 Ga (Leite *et al.* 2009), which sets a minimum age for these units.

Caldeirão shear belt

This belt is the western limit of the Uauá subdomain in the Serrinha block. It is made up of upright mylonitized sequence of quartzites, sillimanite–garnet–cordierite–micaschist or gneisses, and amphibolites structurally interleaved with basement gneisses and migmatites. Zircons of a basement gneiss have yielded a SHRIMP ^{207}Pb/^{206}Pb age of 3152 ± 5 Ma (Oliveira *et al.* 2002), whereas detrital zircon

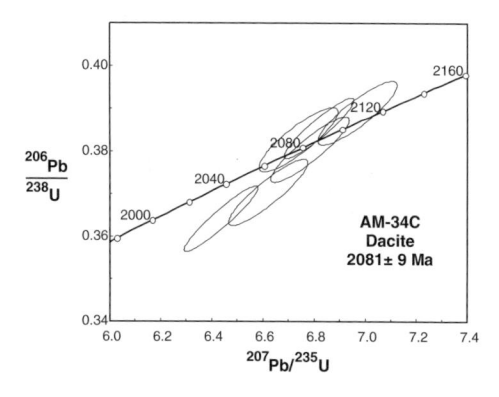

Fig. 13. SHRIMP U–Pb zircon age for dacite from the Rio Itapicuru greenstone belt.

grains of a quartzite yielded core age populations of 3204, 3097 and 3051 Ma, and metamorphic rims of 2076 ± 10 Ma, as described by Oliveira et al. (2002) and Mello et al. (2006). According to these workers deposition of the original sediments took place between 2687 Ma (i.e. the age of the youngest detrital zircon grain) and the metamorphism age of 2076 Ma. The Nd model ages of paragneisses are always older than 3.1 Ga (Table 1). Deformation in the belt appears to have lasted until at least 2039 ± 2 Ma as inferred from U–Pb dating of synmetamorphic titanite grains of a mafic dyke that has undergone boudinage (Oliveira et al. 2000).

Plutonic felsic bodies intruded in the Caraíba complex

Several granitoid plutons were emplaced into high-grade orthogneisses of the Caraíba complex, some of which are north–south elongated bodies. In the Caraíba mine area in the north they are generally designated as G_3 intrusions (e.g. Jardim de Sá et al. 1982), and the 2084 Ma Itiúba syenite (Oliveira et al. 2004b) is the type intrusion (S in Fig. 2). Farther south the plutons were mapped as syn- to post-transcurrent faulting granitoids (e.g. Melo et al. 1991); the 2060 Ma Bravo granite (Barbosa et al. 2008) is a representative body. The SHRIMP ages for two additional plutons are presented here: these are for the syn-transcurrent faulting Capela do Alto Alegre quartz-monzonite in the south and the Caraíba airport G_3 granite in the north. The results indicate ages of 2078 ± 6 Ma for the Capela do Alto Alegre quartz-monzonite and 2027 ± 16 Ma for the Caraíba airport granite (Fig. 14). The former contains 2429–2670 Ma zircon grains probably inherited from orthogneisses of the Caraíba complex.

The Itiúba syenite presents Nd model ages in the interval 2.70–2.81 Ga and negative $\varepsilon_{Nd(T)}$ values (−5.77 and −8.51), whereas the Capela do Alto Alegre quartz-monzonite has T_{DM} = 3.16 Ga and $\varepsilon_{Nd(T)}$ = −10.12 (Table 1), and the Airport G_3 T_{DM} = 2.71 Ga and $\varepsilon_{Nd(T)}$ = −7.88, suggesting an origin by partial melting of ancient lithosphere and/or interaction of the parental magma with continental crust.

The high-grade metamorphism

The best age estimates for the high-grade amphibolite- to granulite-facies metamorphism in the northern segment of the Itabuna–Salvador–Curaçá orogen come from the studies by Silva et al. (1997), Oliveira et al. (2002, 2003), Mello et al. (2006) and the present paper. The ages presented on metamorphic zircon rims range from 2082 ± 17 to 2074 ± 14 Ma. No metamorphic ages older than 2100 Ma have been reported so far in the orogen, except for the c. 3.0 Ga Uauá granulites in the Serrinha block as commented on above. This implies that most if not all of the high-grade metamorphism and associated deformation in the belt took place in the Palaeoproterozoic.

Discussion

Field relationships combined with U–Pb zircon ages, whole-rock Nd isotope, and major and trace element data unravel a complex history of continental crust growth in the northern segment of the Itabuna–Salvador–Curaçá orogen from the late Palaeoarchaean to the Palaeoproterozoic. A synthesis of this is presented here.

Accretion of oceanic crust and/or island arcs took place about 3300 Ma in the Mundo Novo greenstone belt (Mascarenhas & Silva 1994; Peucat et al. 2002; present paper), between c. 2150 and 2127 Ma in the Rio Itapicuru greenstone belt (Barrueto et al. 1998; Silva et al. 2001; Barrueto

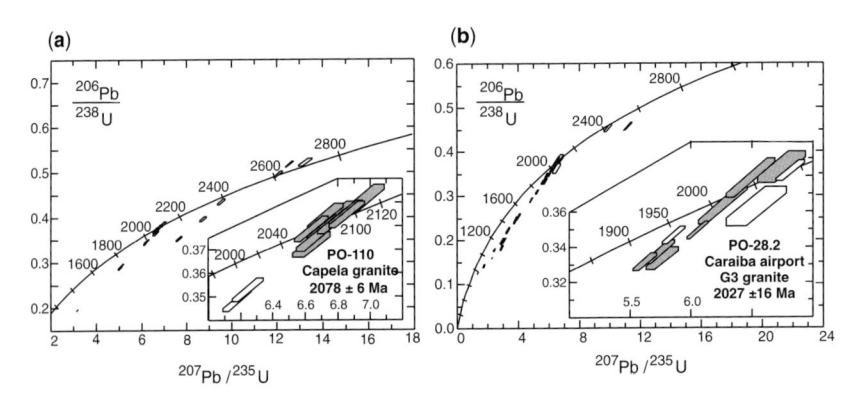

Fig. 14. SHRIMP U–Pb zircon ages for the Capela do Alto Alegre quartz-monzonite (**a**) and Caraíba airport G_3 granite (**b**).

2002; Mello et al. 2006; present paper), and about 2148 Ma in the Rio Capim greenstone belt (Oliveira et al. 1998, 2004a; present paper).

The igneous protoliths of the Caraíba granulite complex that occupies the orogen's central domain may have originated in an Andean-type arc between c. 2690 and 2580 Ma based on zircon U–Pb SHRIMP ages presented by Silva et al. (1997) and in this paper, as well on the calc-alkaline signature and positive to slightly negative $\varepsilon_{Nd(T)}$ values (+4.33 to −0.67) of the rocks. Remnants of incompatible element-depleted ocean-floor basalts are also likely to have been obducted onto the continental arc in some parts of the Caraíba complex, such as the A-14 amphibolites of the Caraíba mine area, and the 2583 Ma São José do Jacuípe mafic–ultramafic complex. However, we do not yet know whether the Neoarchaean Caraíba arc originated by oceanic crust subduction beneath the Serrinha block to the east or the Gavião block to the west, or neither. The ages of the Caraíba complex (2.69–2.58 Ga) are similar to ages (2.6–2.4 Ga) reported by Barbosa & Sabaté (2004) for arc accretion in the southern Itabuna–Salvador–Curaçá belt, and the two granulite-facies belts may be part of a single and continuous belt, on the basis of geological mapping (Fig. 1). According to Barbosa (1990) and Barbosa & Sabaté (2004) the areal distribution of shoshonite and calc-alkaline plutons, tholeiites and Fe–Ti-rich basalts respectively from west to east, in the southern belt, suggests westward subduction of oceanic crust. On these grounds, the Caraiba arc might have formed at the margin of the Gavião block (or palaeocontinent), and later between c. 2.08 and 2.07 Ga the two already amalgamated tectonic entities collided with the Serrinha block or microcontinent.

In the Serrinha block, protoliths of basement gneisses of Retirolândia and Jacurici subdomains have Mesoarchaean ages between 3085 and 2983 Ma. The rocks have major and trace element characteristics (i.e. high silica content, low MgO at SiO_2 >63%, high Sr/Y ratios at low Y contents, and data that cluster in the trondhjemite field of the Ca–Na–K diagram) very similar to those of other Archaean TTG series worldwide, especially the high-SiO_2 adakite type (Martin et al. 2005). The gneisses' geochemical signature coupled with their dominant positive $\varepsilon_{Nd(T)}$ values (+0.04 to +2.75) suggest a model in which young basaltic crust, either subducted (Martin 1999; Smithies 2000) or underplated (Hou et al. 2004; Wang et al. 2005), underwent partial melting to produce the TTG suite.

However, the same petrogenetic model cannot be entirely applicable to the Mesoarchaean granitoids and granulite-facies orthogneisses of the Uauá subdomain in the Serrinha block. The calc-alkaline trend shown by the Uauá granulites and Capim tonalite on the Ca–Na–K diagram, their low MgO contents at SiO_2 <62%, moderate to high Y, low Sr/Y, Ni and Cr, and positive $\varepsilon_{Nd(3000\ Ma)}$ values for the Uauá granulites (+1.23 to +1.80) or slightly negative $\varepsilon_{Nd(3120\ Ma)}$ for the Capim tonalite (−3.08 to −5.27) are comparable with Phanerozoic continental margin calc-alkaline batholiths. These features distinguish the Uauá basement gneisses from Archaean TTG and sanukitoids according to the geochemical criteria suggested by Martin et al. (2005). The Uauá subdomain can be interpreted as an exotic block or terrane with a more complex evolution than the other two basement subdomains. This conclusion is based on its distinct structural characteristics (e.g. it is bounded by upright shear zones, and has inner domains of thrust gneisses and granulites) and igneous rock associations (e.g. the presence of a layered anorthosite complex and the vast occurrence of mafic dykes). Moreover, the kinematic indicators in shear zones (sinistral to the west, dextral to the east) and the geochronological record (Uauá rocks are at least 900 Ma older than the Capim greenstone belt to the east) support the conclusion that the Uauá block was displaced from south to north during collision in the Palaeoproterozoic.

The timing of collision between the Serrinha block and the Caraiba complex may be bracketed between c. 2085 and 2060 Ma on the basis of SHRIMP U–Pb dating of metamorphic overgrowths in zircons (e.g. Silva et al. 1997; present paper) and emplacement of syn-collisional granites (Itiúba syenite, Capela do Alto Alegre monzonite and Bravo granite, according to Oliveira et al. 2004b, the present paper and Barbosa et al. 2008, respectively). However, no ophiolite remnant or ultrahigh-pressure metamorphic rocks have yet been found to allow us to define a suture zone between the Caraíba complex and the Serrinha block. The Caraíba complex and associated units (Tanque Novo–Ipirá supracrustal rocks, São José do Jacuípe mafic–ultramafic suite) have been intensively reworked during collision when they underwent deformation, high-grade metamorphism (sometimes at ultrahigh temperature, Leite et al. 2009) and intrusion of sheet-like granitic bodies.

Collision in the Serrinha block was more complicated and probably involved an early arc–continent collision before the final collision that affected the entire Itabuna–Salvador–Curaçá belt. This assumption is based on various types of evidence found in the Rio Itapicuru greenstone belt.

First, the greenstone belt comprises metamorphosed basalts, andesites, dacites and associated clastic and chemical metasedimentary rocks intruded by TTG and calc-alkaline plutons. The basalt geochemistry is indistinguishable from that of basalts of the ocean–continent transition

(Donatti Filho & Oliveira 2007), whereas the andesites and dacites are chemically similar to arc-related adakites and calc-alkaline volcanic rocks (Ruggiero 2008; Ruggiero & Oliveira 2010). Although the ages of some plutons intrusive into the basalts must be revised [i.e. they are older ($c.$ 2155 Ma) than the host basalts ($c.$ 2145 Ma)], the rock association, geochemistry and positive $\varepsilon_{Nd(T)}$ values are characteristic of arc assemblages. On these grounds the belt can be interpreted as a Palaeoproterozoic arc, possibly oceanic, accreted onto the Archaean Serrinha basement or microcontinent.

Second, the arc must have collided with the Serrinha microcontinent before the final continent–continent collision at about 2.08–2.06 Ga. In this regard, the potassic plutons with ages in the range 2111–2106 Ma may be an important piece of evidence of arc–continent collision. Indeed, as summarized by Oliveira et al. (2008) and Oliveira (2009), the basement–greenstone transition in the west is marked by mylonitized banded gneisses (mafic dykes and migmatite) and retrograde garnet-amphibolites (mafic dykes) in the basement, and intrusion of high-K to ultrapotassic plutons (e.g. Itareru high-K to shoshonitic tonalite, Carvalho & Oliveira 2003; Morro do Afonso ultrapotassic syenite, Rios et al. 2007; Fazenda Gavião high-K granodiorite, Costa 2008) along the basement–greenstone contact. Of prime significance is the 90 km long sheet-like Itareru tonalite (I in Fig. 2). Elongated potassic plutons in other orogens are typically intruded along terrane boundaries (Musumeci 1999) and some of them have been interpreted as mantle–crust partial melts triggered by heat coming from the upwelling asthenosphere during slab breakoff, or slab-tearing. Possible examples are the Bergell pluton and coeval igneous bodies in the Alps (e.g. Von Blanckenburg & Davies 1995; Oberli et al. 2004), syenites in China (Peng et al. 2008), volcanic rocks in Turkey (Keskin et al. 2008), and high Ba–Sr granitoids in the British Caledonides (Fowler et al. 2008). Slab breakoff (Davies & von Blanckenburg 1995) is the natural fate of the oceanic part of a passive continental margin as it subducts underneath an oceanic arc or a continental plate. If slab breakoff takes place at shallow depth then K-rich magmas (lamprophyre, shoshonite, high-K calc-alkaline granitoids to ultrapotassic syenites) may be common and emplaced along a relatively narrow belt of the resulting orogen. The model predicts also that the subducted continental crust will be exhumed. A similar tectonic scenario can be envisaged for the collision of the Rio Itapicuru arc with the Serrinha microcontinent, wherein the banded gneisses and the retrograde garnet amphibolite dykes in the basement may be the exhumed continental margin and the K-rich plutons may be slab breakoff-related melts.

However, a new arc may have been accreted onto the old one, or alternatively new arc lavas may have been formed to account for the occurrence of 2081 Ma adakites and calc-alkaline dacites in the north–central sector of the greenstone belt.

Third, deformation in the north–central sector of the Rio Itapicuru greenstone belt is generally accepted as having begun with east–west- to NW–SE-directed compression (D_1) of bedding-parallel shear zones followed by north–south-trending, left-lateral strike-slip motion (D_2) along the belt (e.g. Davison et al. 1988; Alves da Silva et al. 1993; Chauvet et al. 1997b). Although further geochronological and structural studies are needed, the bedding-parallel shear zones might be related to the arc–continent collision referred to above, whereas D_1 and D_2 relate to the final continent–continent collision. The 2080 Ma Ambrosio basement–granodiorite composite dome (Lacerda 2000; Mello et al. 2006) and other coeval bodies are the type intrusions during D_1. Older 2.15–2.10 Ga plutons were highly to partially deformed during D_1 and D_2 depending upon deformation partitioning and the rheology contrast between granitoids and supracrustal country rocks. Indeed, the strike-slip-related, north–south-trending regional upright shear zones are mostly localized in the supracrustal rocks.

During continental collision, foreland basins may have been deposited onto the Gavião block margin and Caraíba complex, such as the upper part of the Jacobina basin with 2.08 Ga detrital zircon grains transported from east to west (e.g. Mougeot et al. 1996; Ledru et al. 1997).

In the late stage of collision, the orogen's northern segment was entirely affected by sinistral wrench-tectonics as the result of oblique collision, probably between two major continental blocks represented by the Gavião block in the west and the West Africa proto-craton, or Gabon craton, in the east (e.g. Figueiredo 1989; Barbosa & Sabaté 2004). During this stage, some blocks or terranes may have been displaced for hundreds of kilometres from their original position in a manner similar to modern-type extrusion tectonics. The Uauá block (subdomain) in the Serrinha block was one of these extruded blocks (Oliveira et al. 2001). A minimum age of 2039 ± 2 Ma was suggested for the extrusion tectonics on the basis of U–Pb dating of synmetamorphic titanite from the boudinaged Uauá mafic dyke in the Caldeirão shear belt (Oliveira et al. 2000). However, the escape of blocks appears to have begun much earlier as inferred from intrusion of elongated granitic bodies in the orogen (e.g. Itiúba syenite 2084 Ma) and by 2074 Ma metamorphic overgrowth rims in detrital zircons of the Caldeirão belt quartzites (Oliveira et al. 2002; Mello et al. 2006).

Figure 15 illustrates our model for the evolution of the northern segment of the Itabuna–Salvador–Curaçá orogen from the Mesoarchaean to the Palaeoproterozoic. The suggested geological scenario involves the accretion of the Neoarchaean Caraíba complex onto the Late Palaeoarchaean Gavião block, the likely fragmentation of the Mesoarchaean Serrinha microcontinent to form the Rio Itapicuru

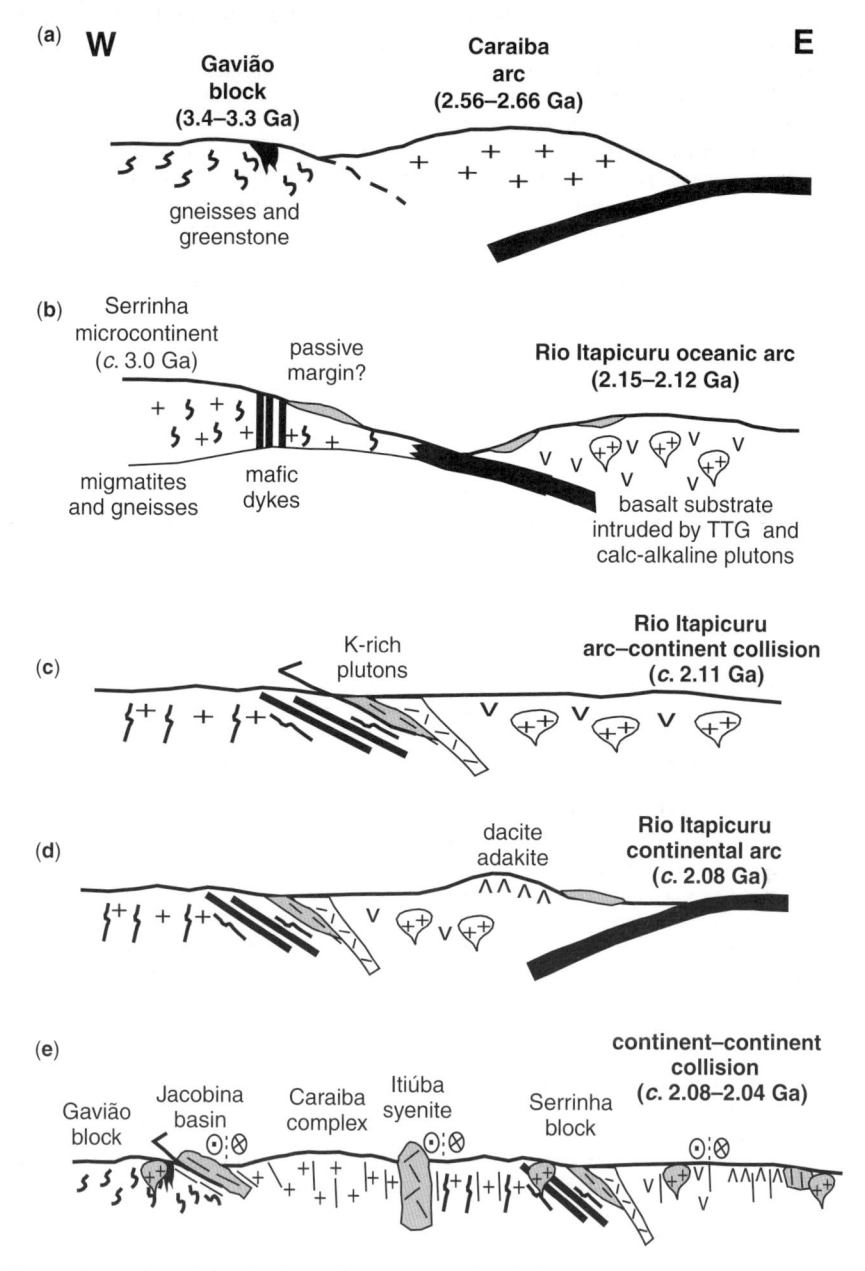

Fig. 15. Proposed tectonic evolution for the northern segment of the Itabuna–Salvador–Curaçá orogen. (**a**) Neoarchaean accretion of the Caraíba arc onto the Palaeoarchaean Gavião block; (**b**) fragmentation of the Mesoarchaean Serrinha microcontinent to open the Palaeoproterozoic Rio Itapicuru ocean basin, followed by basin closure and island-arc formation; (**c**) arc–continent collision with emplacement of K-rich plutons; (**d**) continental arc with adakite and calc-alkaline volcanic rocks; (**e**) final continent–continent oblique collision, orogen-parallel block displacement, and syn- to post-collisional granite–syenite emplacement.

oceanic basin and its subsequent closure, and the final Palaeoproterozoic continent–continent oblique collision.

In summary, orogenic belts worldwide are probably the best places to understand continental growth processes and how they change with time (e.g. Windley 1995; Condie 2007, 2008; Chardon et al. 2009; Windley & Garde 2009). The northern segment of the Itabuna–Salvador–Curaçá orogen in the São Francisco craton of Brazil is an example of such a region. It has the rock record of at least 1 Ga of continental growth from c. 3.3 to 2.0 Ga, a time when planet Earth underwent major global changes and the first supercontinents probably started to form (e.g. Windley 1995; Aspler & Chiarenzelli 1998; Rogers & Santosh 2004; Melezhik et al. 2005). Although further studies are needed to improve our knowledge on the evolution of this ancient orogen, the geological record presented here demonstrates that continental growth processes that are generally proposed for Phanerozoic orogens, such as subduction of oceanic crust to form island arcs and continental arcs, accretion of several arcs and exotic terranes to form accretionary belts, and collision of two or more continents to form supercontinents, can be recognized as far back as the Mesoarchaean.

Differences from modern processes do exist. A key example is the Palaeoproterozoic Rio Itapicuru greenstone belt in the Serrinha block. This belt comprises dome and keel structures (i.e. typical of Archaean terranes) juxtaposed with Archaean gneissic basement by overthrusting followed by lateral displacement along major shear zones, as in modern-type orogens. No ophiolite has been reported in the orogen, although minor relicts of metamorphosed N-MORB type and a probable 'transitional-type ophiolite' have been suggested, respectively in the Caraíba complex (this paper) and in the Serrinha block (Oliveira et al. 2007). Also, no high-pressure metamorphic assemblages have been found so far. Instead, intermediate-pressure, high- and ultrahigh-temperature assemblages are common in the reworked central part of the orogen (e.g. Leite et al. 2009). Explanations for these differences from modern-type orogens can probably be found in models that emphasize the space and time variability of uppermost mantle temperature in controlling plate interactions and continental growth (e.g. Chardon et al. 2009), and in the assumption that the tectonic transition from Archaean- to modern-type orogens has been gradual (e.g. Cagnard et al. 2008).

R. Van Schmus (Kansas University) and M. Pimentel (Universidade de Brasília) have given E.P.O. permission to use their isotope laboratory facilities, where all Nd isotope data were collected, and they are gratefully acknowledged. The Brazilian National Research Council (CNPq) and São Paulo State Research Foundation (FAPESP) provided research grants. The comments of T. Kusky and an anonymous referee greatly improved the original manuscript. Several papers of B. Windley and coworkers from Leicester University, published in the late 1970s and early 1980s, have profoundly influenced the senior author in the study of Precambrian orogenic belts and Earth evolution.

References

ALIBERT, C. & BARBOSA, J. 1992. Ages U–Pb déterminés à la SHRIMP sur des zircons du complexe de Jequié, craton du São Francisco, Bahia, Brésil. In: Societé Géologique de France, Réunion des Sciences de la Terre, 14. Toulouse, France, p. 4.

ALVES DA SILVA, F. C., CHAUVET, A. & FAURE, M. 1993. Early Proterozoic orogeny (Transamazonian) and syntectonic granite emplacement in the Rio Itapicuru greenstone belt, Bahia, Brazil. Comptes Rendus de l'Académie des Sciences, Série II, 316, 1139–1146.

ASPLER, L. B. & CHIARENZELLI, J. R. 1998. Two Neoarchean supercontinents? Evidence from the Paleoproterozoic. Sedimentary Geology, 120, 75–104.

BARBOSA, J. S. F. 1990. The granulites of the Jequié Complex and Atlantic Mobile Belt, southern Bahia, Brazil – An expression of Archean–Proterozoic plate convergence. In: VIELZEUF, D. & VIDAL, P. (eds) Granulites and Crustal Evolution. Springer, Berlin, 195–221.

BARBOSA, J. S. F. & SABATÉ, P. 2004. Archaean and Paleoproterozoic crust of the São Francisco Craton, Bahia, Brazil: geodynamic features. Precambrian Research, 133, 1–27.

BARBOSA, J. S. F., OLIVEIRA, E. P., CORRÊA GOMES, L. C., MARINHO, M. M. & MELO, R. C. 2001. I Workshop sobre o Orógeno Itabuna–Salvador–Curaçá. 10–16 de setembro 2001. CBPM, Companhia Baiana de Pesquisa Mineral, Salvador.

BARBOSA, J. S. F., PEUCAT, J.-J. ET AL. 2008. Petrogenesis of the late-orogenic Bravo granite and surrounding high-grade counry rocks in the Palaeoproterozoic orogen of Itabuna–Salvador–Curaçá block, Bahia, Brazil. Precambrian Research, 167, 35–52.

BARRUETO, H. R. 2002. Petrogênese das intrusões compostas de Teofilândia e Barrocas, Greenstone Belt do Rio Itapicuru, Bahia, Brasil. PhD thesis, Instituto de Geociências, UNICAMP, Campinas.

BARRUETO, H. R., OLIVEIRA, E. P. & DALLAGNOL, R. 1998. Trace element and Nd isotope evidence for juvenile, arc-related granitoids in the southern portion of the Paleoproterozoic Rio Itapicuru Greenstone Belt (RIGB), Bahia, Brazil. Proceedings XL Congresso Brasileiro de Geologia, Sociedade Brasileira de Geologia, Belo Horizonte, 520.

CAGNARD, F., GAPAIS, D. & BARBEY, P. 2008. Transition between 'Archaean-type' to 'Modern-type' orogenic belts: insights from Palaeoproterozoic times. Geophysical Research Abstracts, 10, EGU2008-A-0734-1.

CARVALHO, M. J. & OLIVEIRA, E. P. 2003. Geologia do Tonalito Itareru, Bloco Serrinha, Bahia: uma intrusão sin-tectônica do início da colisão continental no

Segmento Norte do Orógeno Itabuna–Salvador–Curaçá. *Revista Brasileira de Geociências, Supplement 1*, **33**, 55–68.

CHARDON, D., GAPAIS, D. & CAGNARD, F. 2009. Flow of ultra-hot orogens: a view from the Precambrian, clues for the Phanerozoic. *Tectonophysics*, doi: 10.1016/j.tecto.2009.03.008.

CHAUVET, A., GUERROT, C., ALVES DA SILVA, F. C. & FAURE, M. 1997a. Géochronologie ^{207}Pb/^{206}Pb et ^{40}Ar/^{39}Ar des granites paléoprotérozoiques de la ceinture de roches vertes du Rio Iapicuru (Bahia, Brésil). *Comptes Rendus de l'Académie des Sciences, Série II*, **324**, 293–300.

CHAUVET, A., SILVA, F. C. A., FAURE, M. & GUERROT, C. 1997b. Structural evolution of the Paleoproterozoic Rio Itapicuru granite–greenstone belt (Bahia, Brazil): the role of synkinematic plutons in the regional tectonics. *Precambrian Research*, **84**, 139–162.

CONCEIÇÃO, H., RIOS, D. C. ET AL. 2002. Zircon geochronology and petrology of alkaline–potassic syenites, southwestern Serrinha block, East São Francisco Craton, Brazil. *International Geology Review*, **44**, 117–136.

CONDIE, K. C. 2007. Accretionary orogens in space and time. *In*: HATCHER, R. D. JR., CARLSON, M. P., MCBRIDE, J. H. & MARTÍNEZ CATALÁN, J. R. (eds) *4-D Framework of Continental Crust*. Geological Society of America, Special Papers, **200**, 145–158.

CONDIE, K. C. 2008. Did the character of subduction change at the end of the Archean? Constraints from convergent-margin granitoids. *Geology*, **36**, 611–614.

CONDIE, K. C. & PEASE, V. 2008. *When did plate tectonics begin on planet Earth?* Geological Society of America, Special Papers, **440**.

CORDANI, U. G., SATO, K. & NUTMAN, A. 1999. Single zircon SHRIMP determination from Archean tonalitic rocks near Uauá, Bahia, Brazil. *Proceedings II South American Symposium on Isotope Geology*, SEGEMAR Servicio Geológico Minero Argentino, Córdoba, 27–30.

COSTA, F. G. 2008. *Petrogênese do granodiorito Fazenda Gavião: registro de uma colisão arco-continente no greenstone belt do Rio Itapicuru, Craton do São Francisco, Bahia*. Masters dissertation, Instituto de Geociências, UNICAMP, Campinas.

CRUZ FILHO, B. E., CONCEIÇÃO, H., ROSA, M. L. S., RIOS, D. C., MACAMBIRA, M. J. B. & MARINHO, M. M. 2005. Geocronologia e assinatura isotópica (Rb–Sr e Sm–Nd) do batólito trondhjemítico Nordestina, núcleo Serrinha, nordeste do estado da Bahia. *Revista Brasileira de Geociências*, **35**, 1–8.

CUNHA, J. C. & FRÓES, R. J. B. 1994. *Komatiítos com textura spinifex do Greenstone Belt de Umburanas, Bahia*. Companhia Baiana de Pesquisa Mineral, Série Arquivos Abertos, **7**.

DANTAS, E. L., VAN SCHMUS, W. R. ET AL. 2004. The 3.4–3.5 Ga São José do Campestre massif, NE Brazil: remnants of the oldest crust in South America. *Precambrian Research*, **130**, 113–137.

DAVIES, J. H. & VON BLANCKENBURG, F. 1995. Slab breakoff: a model of lithosphere detachment and its test in the magmatism and deformation of collisional orogens. *Earth and Planetary Science Letters*, **129**, 85–102.

DAVISON, I., TEIXEIRA, J. B. G., SILVA, M. G., ROCHA NETO, M. B. & MATOS, F. M. V. 1988. The Rio Itapicuru Greenstone Belt, Bahia, Brazil: structure and stratigraphical outline. *Precambrian Research*, **42**, 1–17.

DELGADO, I. M., SOUZA, J. D. ET AL. 2003. Geotectônica do Escudo Atlântico. *In*: BIZZI, L. A., SCHOBBENHAUS, C., VIDOTTI, R. M. & GONÇALVES, J. H. (eds) *Geologia, Tectônica e Recursos Minerais do Brasil*. CPRM, Brasilia, 227–334.

D'EL-REY SILVA, L. J. H., DANTAS, E. L., TEIXEIRA, J. B. G., LAUX, J. H. & SILVA, M. G. 2007. U–Pb and Sm–Nd geochronology of amphibolites from the Curaçá Belt, São Francisco Craton, Brazil: tectonic implications. *Gondwana Research*, **12**, 454–467.

DEPAOLO, D. J. 1988. *Neodymium Isotope Geochemistry. An Introduction*. Springer, Berlin.

DONATTI FILHO, J. P. & OLIVEIRA, E. P. 2007. Trace-element geochemistry of basalts from the Rio Itapicuru Greenstone Belt, Bahia, and the tectonic setting revisited. *In*: *Proceedings XI Simpósio Nacional Estudos Tectônicos, Sociedade Brasileira de Geologia*, Natal, 296–299.

FIGUEIREDO, M. C. H. 1989. Geochemical evolution of eastern Bahia, Brazil: a probable Early Proterozoic subduction-related magmatic arc. *Journal of South American Earth Sciences*, **2**, 131–145.

FOWLER, M. B., KOCKS, H., DARBYSHIRE, D. P. F. & GREENWOOD, P. B. 2008. Petrogenesis of high Ba–Sr plutons from the Northern Highlands Terrane of the British Caledonian Province. *Lithos*, **105**, 129–148.

HARTMANN, L. A., ENDO, I. ET AL. 2006. Provenance and age delimitation of Quadrilátero Ferrífero sandstones based on zircon U–Pb isotopes. *Journal of South American Earth Sciences*, **20**, 273–285.

HOU, Z. Q., GAO, Y. F., QU, X. M., RUI, Z. Y. & MO, X. X. 2004. Origin of adakitic intrusives generated during mid-Miocene east–west extension in southern Tibet. *Earth and Planetary Science Letters*, **220**, 139–155.

JARDIM DE SÁ, E. F., ARCHANJO, C. J. & LEGRAND, J.-M. 1982. Structural and metamorphic history of part of the high-grade terrain in the Curaçá Valley, Bahia, Brazil. *Revista Brasileira de Geociências*, **12**, 251–262.

JARDIM DE SÁ, E. F. J., SOUZA, Z. S., FONSECA, V. P. & LEGRAND, J. M. 1984. Relações entre 'greenstone belts' e terrenos de alto grau: o caso da faixa Rio Capim, NE da Bahia. *In*: *Proceedings XXXIII Congresso Brasileiro de Geologia, Sociedade Brasileira de Geologia*, Rio de Janeiro, 2615–2629.

KESKIN, M., GENÇ, S. C. & TÜYSÜZ, O. 2008. Petrology and geochemistry of post-collisional Middle Eocene volcanic units in North–Central Turkey: evidence for magma generation by slab breakoff following the closure of the Northern Neotethys Ocean. *Lithos*, **104**, 267–305.

KISHIDA, A. & RICCIO, L. 1980. Chemostratigraphy of lava sequences from the Rio Itapicuru Greenstone Belt, Bahia, Brazil. *Precambrian Research*, **11**, 161–178.

KOSIN, M., MELO, R. C., SOUZA, J. D., OLIVEIRA, E. P., CARVALHO, M. J. & LEITE, C. M. M. 2003. Geologia do Bloco Serrinha e do segmento norte do Orógeno

Itabuna–Salvador–Curaçá. *Revista Brasileira de Geociências, Supplement 1*, **33**, 15–26.

KUSKY, T. M. & POLAT, A. 1999. Growth of granite–greenstone terranes at convergent margins and stabilization of Archean cratons. *Tectonophysics*, **305**, 43–73.

LACERDA, C. M. M. 2000. *Evolução Estrutural e Petrogenética do Domo Granodiorítico de Ambrósio, Greenstone Belt do Rio Itapicuru, Bahia*. PhD thesis, Instituto de Geociências, UNICAMP, Campinas.

LEDRU, P., COCHERIE, A., BARBOSA, J., JOHAN, V. & ONSTOTT, T. 1994. Ages du métamorphisme granulitique dans le craton du São Francisco (Brésil). Implications sur la nature de l'orogène transamazonien. *Comptes Rendus de l'Académie des Sciences, Série II*, **318**, 251–257.

LEDRU, P., MILÉSI, J. P., JOHAN, V., SABATÉ, P. & MALUSKI, H. 1997. Foreland basins and gold-bearing conglomerates: a new model for the Jacobina Basin (São Francisco province, Brazil). *Precambrian Research*, **86**, 155–176.

LEITE, C. M. M. 2002. *A evolução geodinâmica da orogenese paleoproterozóica nas regiões de Capim Grosso–Jacobina e Pintadas–Mundo Novo (Bahia, Brasil): Metamorfismo, anatexia crustal e tectônica*. PhD thesis, Instituto de Geociências, Universidade Federal da Bahia, Salvador.

LEITE, C. M. M., BARBOSA, J. S. F., GONCALVES, P., NICOLLET, C. & SABATÉ, P. 2009. Petrological evolution of silica-undersaturated sapphirine-bearing granulite in the Paleoproterozoic Salvador–Curaçá Belt, Bahia, Brazil. *Gondwana Research*, **15**, 49–70.

LOBATO, L. M., RODRIGUES, L. C. R., ZUCCHETTI, M., NOCE, C. M., BALTAZAR, O. F., SILVA, L. C. & PINTO, C. P. 2001. Brazil's premier gold province. Part I: the tectonic, magmatic and structural setting of the Archaean Rio das Velhas greenstone belt, Quadrilátero Ferrífero. *Mineralium Deposita*, **36**, 228–248.

MARQUES, J. C. & FERREIRA FILHO, C. F. 2003. The chromite deposit of the Ipueira–Medrado sill, São Francisco craton, Bahia State, Brazil. *Economic Geology*, **98**, 87–108.

MARTIN, H. 1994. The Archaean grey gneisses and the genesis of the continental crust. *In*: CONDIE, K. C. (ed.) *The Archean Crustal Evolution*. Developments in Precambrian Geology, **11**, 205–259.

MARTIN, H. 1999. Adakitic magmas: modern analogs of Archaean granitoids. *Lithos*, **46**, 411–429.

MARTIN, H., SABATÉ, P., PEUCAT, J. J. & CUNHA, J. C. 1991. Un Segment de Croute Continentale d'Âge Archée an ancien (3, 4 milliards d'années): le Massif de Sete Voltas (Bahia-Brésil). *Comptes Rendus de l'Académie des Sciences, Série II*, **313**, 531–538.

MARTIN, H., PEUCAT, J. J., SABATÉ, P. & CUNHA, J. C. 1997. Crustal evolution in the early Archaean of South America: example of the Sete Voltas Massif, Bahia State, Brazil. *Precambrian Research*, **82**, 35–62.

MARTIN, H., SMITHIES, R. H., RAPP, R., MOYEN, J.-F. & CHAMPION, D. 2005. An overview of adakite, TTG and sanukitoid: relationships and some implications for crustal evolution. *Lithos*, **79**, 1–24.

MASCARENHAS, J. F. 1979. Estruturas do tipo greenstone belt no leste da Bahia. *In*: *Geologia e Recursos Minerais do Estado da Bahia. Textos Básicos*, **2**. Salvador, SME/COM, 25–53.

MASCARENHAS, J. H. & SÁ, J. H. S. 1982. Geological and metallogenic patterns in the Archean and Early Proterozoic of Bahia State, Eastern Brazil. *Revista Brasileira de Geociências*, **12**, 193–214.

MASCARENHAS, J. F. & SILVA, E. F. A. 1994. *Greenstone Belt de Mundo Novo: caracterização e implicações metalogenéticas e geotectônicas no Cráton do São Francisco*. Companhia Baiana de Pesquisa Mineral. Série Arquivos Abertos, **5**.

MELEZHIK, V. A., FALLICK, A. E., HANSKI, E. J., KUMP, L. R., LEPLAND, A., PRAVE, A. R. & STRAUSS, H. 2005. Emergence of the aerobic biosphere during the Archean–Proterozoic transition: challenges of future research. *GSA Today*, **15**, 4–11.

MELLO, E. F., XAVIER, R. P., MCNAUGHTON, N. J., HAGEMANN, S. G., FLETCHER, I. & SNEE, L. 2006. Age constraints on felsic intrusions, metamorphism and gold mineralization in the Paleoproterozoic Rio Itapicuru greenstone belt, NE Bahia State, Brazil. *Mineralium Deposita*, **40**, 849–866.

MELO, R. C., SILVA, L. C. & FERNANDES, P. C. 1991. Estratigrafia. *In*: MELO, R. C. (org.) *Pintadas – Folha SC.24-Y-D-V*. Programa de Levantamentos Geológicos Básicos do Brasil, CPRM, Brasilia, 23–47.

MOUGEOT, R., RESPAUT, J. P., LEDRU, P., MILESI, J. P. & JOHAN, V. 1996. U–Pb geochronological constraints for the evolution of the Paleoproterozoic Jacobina auriferous basin (São Francisco Province, Bahia, Brazil). *In*: *Proceedings XXXIX Congresso Brasileiro de Geologia, Sociedade Brasileira de Geologia*, Salvador, Brazil, Vol. 6, 582–584.

MUSUMECI, G. 1999. Magmatic belts in accretionary magins, a key for tectonic evolution: the Tonalite Belt of North Victoria Land (East Antarctica). *Journal of the Geological Society, London*, **156**, 177–189.

NUTMAN, A. P. & CORDANI, U. C. 1993. Shrimp U–Pb zircon geochronology of Archean granitoids from the Contendas–Mirante area of the São Francisco Craton, Bahia, Brazil. *Precambrian Research*, **163**, 179–188.

OBERLI, F., MEIER, M., BERGER, A., ROSENBERG, C. L. & GIERÉ, R. 2004. U–Th–Pb and ^{230}Th/^{238}U disequilibrium isotope systematics: precise accessory mineral chronology and melt evolution tracing in the Alpine Bergell intrusion. *Geochimica et Cosmochimica Acta*, **68**, 2543–2560.

OLIVEIRA, E. P. 1990. *Petrogenesis of mafic–ultramafic rocks from the Precambrian Curaçá terrane, Brazil*. PhD thesis, University of Leicester.

OLIVEIRA, E. P. 2009. Arc-continent collision in the Palaeoproterozoic Rio Itapicuru greenstone belt, São Francisco Craton, Brazil. *In*: GLEN, R. A. & MARTIN, C. (Compilers) *International Conference on Island-Arc Continent Collisions: The Macquarie Arc Conference, April 2009*, Geological Society of Australia Abstracts, **92**, 104–105.

OLIVEIRA, E. P. & TARNEY, J. 1995. Genesis of the copper-rich Caraiba norite–hypersthenite complex, Brazil. *Mineralium Deposita*, **30**, 351–373.

OLIVEIRA, E. P., LAFON, J.-M. & SOUZA, Z. S. 1998. A Paleoproterozoic age for the Rio Capim volcano-plutonic sequence, Bahia, Brazil: whole-rock Pb–Pb, Pb-evaporation and U–Pb constraints. *In*: *Proceedings*

XL *Congresso Brasileiro de Geologia, Sociedade Brasileira de Geologia*, Belo Horizonte, 14.

OLIVEIRA, E. P., LAFON, J.-M. & SOUZA, Z. S. 1999. Archaean–Proterozoic transition in the Uauá Block, NE São Francisco Craton, Brazil: U–Pb, Pb–Pb and Nd isotope constraints. *In: VII Simpósio Nacional de Estudos Tectônicos*. Sociedade Brasileira de Geologia, Lençóis, 38–40.

OLIVEIRA, E. P., SOUZA, Z. S. & CORRÊA-GOMES, L. C. 2000. U–Pb dating of deformed mafic dyke and host gneiss: implications for understanding reworking processes on the western margin of the Archaean Uauá Block, NE São Francisco Craton, Brazil. *Revista Brasileira de Geociências*, **30**, 149–152.

OLIVEIRA, E. P., CARVALHO, M. J. & DUARTE, M. I. D. 2001. Extrusion of the Archaean Uauá Block in the northern segment of the Itabuna–Salvador–Curaçá orogen, Bahia, and implications for diamond prospecting. *Revista Brasileira de Geociências*, **31**, 643–644.

OLIVEIRA, E. P., MELLO, E. F. & MCNAUGHTON, N. 2002. Reconnaissance U–Pb geochronology of early Precambrian quartzites from the Caldeirão belt and their basement, NE São Francisco Craton, Bahia, Brazil: implications for the early evolution of the Palaeoproterozoic Salvador–Curaçá Orogen. *Journal of South American Earth Sciences*, **15**, 284–298.

OLIVEIRA, E. P., MCNAUGHTON, N., ARMSTRONG, R. & FLETCHER, I. 2003. U–Pb SHRIMP age of the Caraiba, Medrado and S. José do Jacuipe mafic–ultramafic complexes, Paleoproterozoic Itabuna–Salvador–Curaçá orogen, São Francisco craton, Brazil. *In: Proceedings IV Symposium on South American Isotope Geology, Sociedade Brasileira de Geologia, Salvador, Brazil, Volume II*, 752–754.

OLIVEIRA, E. P., CARVALHO, M. J. & MCNAUGHTON, N. 2004a. Evolução do segmento norte do Orógeno Itabuna–Salvador–Curaçá: cronologia da acresção de arcos, colisão continental e escape de terrenos. *Revista Geologia USP – Série Científica*, **4**, 41–53.

OLIVEIRA, E. P., WINDLEY, B. F., MCNAUGHTON, N., PIMENTEL, M. & FLETCHER, I. R. 2004b. Contrasting copper and chromium metallogenic evolution of terranes in the Palaeoproterozoic Itabuna–Salvador–Curaçá Orogen, São Francisco Craton, Brazil: new zircon (SHRIMP) and Sm–Nd (model) ages and their significance for orogen-parallel escape tectonics. *Precambrian Research*, **128**, 143–165.

OLIVEIRA, E. P., ESCAYOLA, M., SOUZA, Z. S., BUENO, J. F., ARAÚJO, M. G. S. & MCNAUGHTON, N. 2007. The Santa Luz chromite-peridotite and associated mafic dykes, Bahia-Brazil: remnants of a transitional-type ophiolite related to the Palaeoproterozoic (>2.1 Ga) Rio Itapicuru greenstone belt? *Revista Brasileira de Geociências, Supplement 4*, **37**, 28–39.

OLIVEIRA, E. P., COSTA, F. G. ET AL. 2008. Banded gneisses, retrograde amphibolite and K-rich plutons as evidences of arc–continent collision with slab break-off in the Palaeoproterozoic Rio Itapicuru Greenstone Belt, Bahia. *In: Proceedings 44 Congresso Brasileiro de Geologia, Sociedade Brasileira de Geologia*, Curitiba, 50.

O'NEIL, J., CARLSON, R. W., FRANCIS, D. & STEVENSON, R. K. 2008. Neodymium-142 evidence for Hadean mafic crust. *Science*, **321**, 1828–1831.

PADILHA, A. V. & MELO, R. C. 1991. Estruturas e Tectônica. *In*: MELO, R. C. (ed.) *Pintadas, folha SC.24-Y-D-V, Estado da Bahia*. CPRM, Programa Levantamentos Geológicos Básicos do Brasil, Escala 1:250.000. Projeto Mapas Metalogenéticos e de Previsão de Recursos Minerais.

PAIXÃO, M. A. P. & OLIVEIRA, E. P. 1998. The Lagoa da Vaca complex: an Archaean layered anorthosite body on the western edge of the Uauá Block, Bahia, Brazil. *Revista Brasileira de Geociencias*, **28**, 201–208.

PATCHETT, P. J. & RUIZ, J. 1987. Nd isotopic ages of crust formation and metamorphism in the Precambrian of eastern and southern Mexico. *Contributions to Mineralogy and Petrology*, **96**, 523–528.

PEARCE, J. A., HARRIS, N. B. W. & TINDLE, A. G. 1984. Trace element discrimination diagrams for the tectonic interpretation of granitic rocks. *Journal of Petrology*, **25**, 956–983.

PEASE, V., PERCIVAL, J., SMITHIES, H., STEVENS, G. & VAN KRANENDONK, M. 2008. When did plate tectonics begin? Evidence from the orogenic record. *In*: CONDIE, K. C. & PEASE, V. (eds) *When did Plate Tectonics Begin on Planet Earth?* Geological Society of America, Special Papers, **440**, 199–228.

PECCERILLO, R. & TAYLOR, S. R. 1976. Geochemistry of Eocene calc-alkaline volcanic rocks from the Kastamonu area, northern Turkey. *Contributions to Mineralogy and Petrology*, **58**, 63–81.

PEUCAT, J. J., MASCARENHAS, J. F., BARBOSA, J. S., SOUZA, F. S., MARINHO, M. M., FANNING, C. M. & LEITE, C. M. M. 2002. 3.3 Ga SHRIMP U–Pb zircon age of a felsic metavolcanic rock from the Mundo Novo greenstone belt in the São Francisco craton, Bahia (NE Brazil). *Journal of South American Earth Sciences*, **15**, 363–373.

PENG, P., ZHAI, M., GUO, J., ZHANG, H. & ZHANG, Y. 2008. Petrogenesis of Triassic post-collisional syenite plutons in the Sino-Korean craton: an example from North Korea. *Geological Magazine*, **145**, 637–647.

RIOS, D. C. 2002. Granitogênese no Núcleo Serrinha, Bahia, Brasil: Geocronologia e Litogeoquímica. PhD thesis, Instituto de Geociências, Universidade Federal da Bahia, Salvador.

RIOS, D. C., DAVIS, D. W., CONCEIÇÃO, H., MACAMBIRA, M. J. B., PEIXOTO, A. A., CRUZ FILHO, B. E. & OLIVEIRA, L. L. 2000. Ages of granites of the Serrinha Nucleus, Bahia (Brazil): an overview. *Revista Brasileira de Geociências*, **30**, 74–77.

RIOS, D. C., CONCEIÇÃO, H. ET AL. 2007. Paleoproterozoic potassic–ultrapotassic magmatism: Morro do Afonso syenite pluton, Bahia, Brazil. *Precambrian Research*, **154**, 1–30.

RIOS, D. C., DAVIS, D. W. ET AL. 2008. 3.65–2.10 Ga history of crust formation from zircon geochronology and isotope geochemistry of the Quijingue and Euclides plutons, Serrinha nucleus, Brazil. *Precambrian Research*, **167**, 53–70.

RIOS, D. C., DAVIS, D. W., CONCEIÇÃO, H., ROSA, M. L. S., DAVIS, W. J. & DICKIN, A. P. 2009. Geologic evolution of the Serrinha nucleus granite–greenstone terrane (NE Bahia, Brazil) constrained by U–Pb single zircon geochronology. *Precambrian Research*, **170**, 175–201.

Rogers, J. J. W. & Santosh, M. 2004. *Continents and Supercontinents*. Oxford University Press, Oxford.

Rollinson, H. 2007. *Early Earth Systems – A Geochemical Approach*. Blackwell, Oxford.

Ruggiero, A. 2008. *A unidade Maria Preta: Geologia, geoquímica e petrogênese de rochas vulcânicas e sub-vulcânicas intermediárias a félsicas no greenstone belt do Rio Itapicuru, Bahia*. Masters dissertation, Instituto de Geociências, UNICAMP, Campinas.

Ruggiero, A. & Oliveira, E. P. 2010. Caracterização de vulcânicas adakíticas e cálcio-alcalinas no Greenstone belt do Rio Itapicuru, Bahia: petrogênese e implicações geodinâmicas. *Revista Brasileira de Geociências*, **40**, in press.

Sabaté, P., Peucat, J.-J., Melo, R. C. & Pereira, L. H. M. 1994. Datação por Pb-evaporação de monozircão em ortognaisse do Complexo Caraíba: expressão do acrescimento crustal transamazônico do Cinturão Salvador–Curaçá (Cráton do São Francisco, Bahia, Brasil). *In*: *38° Congresso Brasileiro de Geologia, Sociedade Brasileira de Geologia, Camboriú*, Vol. *1*, 219–220.

Santos, R. A. & Souza, J. D. 1985. *Projeto Mapas Metalogenéticos e de Previsão de Recursos Minerais. Serrinha, Folha SC.24-Y-D*. Convênio DNPM/CPRM, Brasília.

Silva, L. C., McNaughton, N. J., Melo, R. C. & Fletcher, I. R. 1997. U–Pb SHRIMP ages in the Itabuna–Caraíba TTG high-grade complex: the first window beyond the Paleoproterozoic overprinting of the eastern Jequié craton, NE Brazil. *In*: *Proceedings 2nd International Symposium on Granites and Associated Mineralizations*. Sociedade Brasileira de Geologia, Salvador-Bahia, 282–283.

Silva, M. G. 1992. O greenstone belt do Rio Itapicuru: uma bacia do tipo back-arc fóssil. *Revista Brasileira de Geociências*, **22**, 157–166.

Silva, M. G., Coelho, C. E. S., Teixeira, J. B. G., Alves da Silva, F. C., Silva, R. A. & Souza, J. A. B. 2001. The Rio Itapicuru greenstone belt, Bahia, Brazil: geologic evolution and review of gold mineralization. *Mineralium Deposita*, **36**, 345–357.

Smithies, R. H. 2000. The Archaean tonalite–trondhjemite–granodiorite (TTG) series is not an analogue of Cenozoic adakite. *Earth and Planetary Science Letters*, **182**, 115–125.

Smithies, R. H., van Kranendonk, M. J. & Champion, D. C. 2007. The Mesoarchaean emergence of modern style subduction. *Gondwana Research*, **11**, 50–68.

Sun, S.-S. & McDonough, W. F. 1989. Chemical and isotope systematics of oceanic basalts: implications for mantle composition and processes. *In*: Saunders, A. D. & Norry, M. J. (eds) *Magmatism in the Ocean Basins*. Geological Society, London, Special Publications, **42**, 313–345.

Teixeira, L. R. 1997. *O Complexo Caraíba e a Suíte São José do Jacuípe no cinturão Salvador–Curaçá (Bahia-Brasil): Petrologia, geoquímica e potencial metalogenético*. PhD thesis, Instituto de Geociências, Universidade Federal da Bahia, Salvador.

Teixeira, W. & Figueiredo, M. C. H. 1991. An outline of Early Proterozoic crustal evolution in the São Francisco Craton, Brazil: a review. *Precambrian Research*, **53**, 1–22.

Teixeira, W., Sabaté, P., Barbosa, J., Noce, C. M. & Carneiro, M. A. 2000. Archean and Paleoproterozoic tectonic evolution of the São Francisco Craton. *In*: Cordani, U. G., Milani, E. J., Thomaz Filho, A. & Campos, D. A. (eds) *Tectonic Evolution of South America. 31st International Geological Congress, Rio de Janeiro*, 101–137.

Van Kranendonk, M. J., Smithies, R. H. & Bennett, V. C. 2007a. *Earth's Oldest Rocks*. Developments in Precambrian Geology, **15**.

Van Kranendonk, M. J., Smithies, R. H., Hickman, A. H. & Champion, D. C. 2007b. Paleoarchean development of a continental nucleus: the East Pilbara terrane of the Pilbara Craton, Western Australia. *In*: Van Kranendonk, M. J., Smithies, R. H. & Bennett, V. C. (eds) *Earth's Oldest Rocks*. Developments in Precambrian Geology, **15**, 307–337.

Vasconcelos, P. & Becker, T. 1992. A idade da mineralização aurífera no depósito da Fazenda Brasileiro, Bahia, Brasil. *In*: *Workshop em Metalogênese: Pesquisas atuais e novas tendências*. Universidade Estadual de Campinas, Campinas, Boletim de Resumos, 29.

Von Blanckenburg, F. & Davies, J. H. 1995. Slab breakoff: a model for syncollisional magmatism and tectonics in the Alps. *Tectonics*, **14**, 120–131.

Wang, Q., McDermott, F., Xu, J. F., Bellon, H. & Zhu, Y. T. 2005. Cenozoic K-rich adakitic volcanic rocks in the Hohxil area, northern Tibet: lower-crust melting in an intracontinental setting. *Geology*, **33**, 465–468.

Windley, B. F. 1995. *The Evolving Continents*. 3rd edn. Wiley, New York.

Windley, B. F. & Garde, A. A. 2009. Arc-generated blocks with crustal sections in the North Atlantic craton of West Greenland: crustal growth in the Archean with modern analogues. *Earth-Science Reviews*, **93**, 1–30.

Winge, M. 1981. *A Sequência Vulcano-sedimentar do Grupo Capim – Bahia: Caracterização geológica e modelo metalogenético*. Masters dissertation, Universidade de Brasília.

A Review of the geology and tectonics of the Kohistan island arc, north Pakistan

MICHAEL G. PETTERSON

Department of Geology, Leicester University, University Road, Leicester LE1 7RH, UK
(e-mail: mp329@le.ac.uk)

Abstract: This paper summarizes some 30 years of more intense recent work and almost 100 years of geological observations in Kohistan. The paper is divided into two section: an earlier factual-based section with minimal interpretation, and a later section summarizing a range of ideas based on the data as well as presenting new thoughts and interpretations. Kohistan is a $c.$ 30 000 km^2 terrane situated in northern Pakistan. The great bulk of Kohistan represents growth and crustal accretion during the Cretaceous at an intra-oceanic island arc dating from $c.$ 134 Ma to $c.$ 90 Ma (Early to Late Cretaceous). This period saw the extrusion of $c.$ 15–20 km of arc volcanic and related sedimentary rocks as well as the intrusion of the oldest parts of the Kohistan batholith, lower crustal pluton intrusion, crustal melting and the accretion of an ultramafic mantle–lower crust sequence. The crust had thickened sufficiently by $c.$ 95 Ma to allow widespread granulite-facies metamorphism to take place within the lower arc. At around 90 Ma Kohistan underwent a $c.$ 5 Ma high-intensity deformation caused by the collision with Eurasia. The collision created crustal-scale folds and shears in the ductile zone and large-scale faults and thrusts in the brittle zone. The whole terrane acquired a strong penetrative foliation fabric. Kohistan, now an Andean margin, was extended and intruded by a diapiric-generated crustal-scale mafic–ultramafic intrusion (the Chilas Complex) with a volume of 0.2×10^6 km^3 that now occupies much of the mid–lower crust of Kohistan and had a profound impact on its thermal structure. The Andean–post-collisonal ($c.$ 90–26 Ma) period also saw the intrusion of the stage 2 and 3 components of the batholith and the extrusion of the Dir Group and Shamran/Teru volcanic rocks. Collision with India at $c.$ 55–45 Ma saw the rotation, upturning, underplating and whole-scale preservation of the terrane. The seismic structure of Kohistan has some similarities to that of mature arcs such as the Lesser and Greater Antilles and Japan, although Kohistan has a higher proportion of high-velocity granulites in the lower crust. The chemical composition of Kohistan is very different from that of average continental crust, although it is similar to an analogue obducted arc within Alaska (Talkeetna), suggesting that 'mature' continental crust undergoes a series of geochemical processes and reworking to transform an initial stage 1 'primitive arc crust'. Most of Kohistan is gabbroic in composition, particularly within the lower and middle crust. A high proportion of the 'basement' volcanic units is also basaltic to basaltic andesite with smaller proportions of boninite, andesite to rhyolite, ignimbrite and volcaniclastic material. Post Eurasian-collision 'cover' volcanic rocks are highly evolved, comprising predominant rhyolites, ignimbrites and related volcaniclastic rocks. Most lithological units throughout the crustal section have an arc-like geochemical composition (e.g. high LREE/HREE and LFSE/HFSE ratios) although some have oceanic (main ocean and back-arc) characteristics. Isotopic compositions indicate that the great bulk of igneous rocks have an ultimate sub-arc mantle source. In broad terms the Kohistan terrane represents a juvenile mantle extract addition to the Phanerozoic continental crust with a total volume of $c.$ 1.2×10^6 km^3 (equivalent to $c.$ 1/50 the volume of the Ontong–Java Plateau or Alaska).

The Kohistan terrane, north Pakistan (Fig. 1) is one of the world's best examples of a complete section through island arc crust from the uppermost supracrustal rocks to the lowermost crust and adjoining mantle. As such it is a remarkable natural laboratory that has been studied by tens if not hundreds of researchers. This contribution is a summary of research that focuses on: (1) the basic nature, in terms of seismic velocity, geochemistry and general character, of the lithological units that make up the arc; (2) the evolution and accretion of the crust over time; (3) age constraints; (4) major-scale crustal processes affecting the arc as it transformed from an intra-oceanic arc to the hinterland of a major continent–continent collision. Estimates are made regarding thicknesses (of single units and the whole crust) and volumes, to assess the relative contribution of Kohistan to the growth of the Earth's continental crust. The evolution and changing character of arc accretion, thermal structures, metamorphism, deformation, volcanism and magmatism, geochemistry and magma source are discussed, and new opportunities for further research are suggested. The paper is divided into two sections in the hope that this approach will be attractive to both experienced researchers and

From: KUSKY, T. M., ZHAI, M.-G. & XIAO, W. (eds) *The Evolving Continents: Understanding Processes of Continental Growth.* Geological Society, London, Special Publications, **338**, 287–327.
DOI: 10.1144/SP338.14   0305-8719/10/$15.00 © The Geological Society of London 2010.

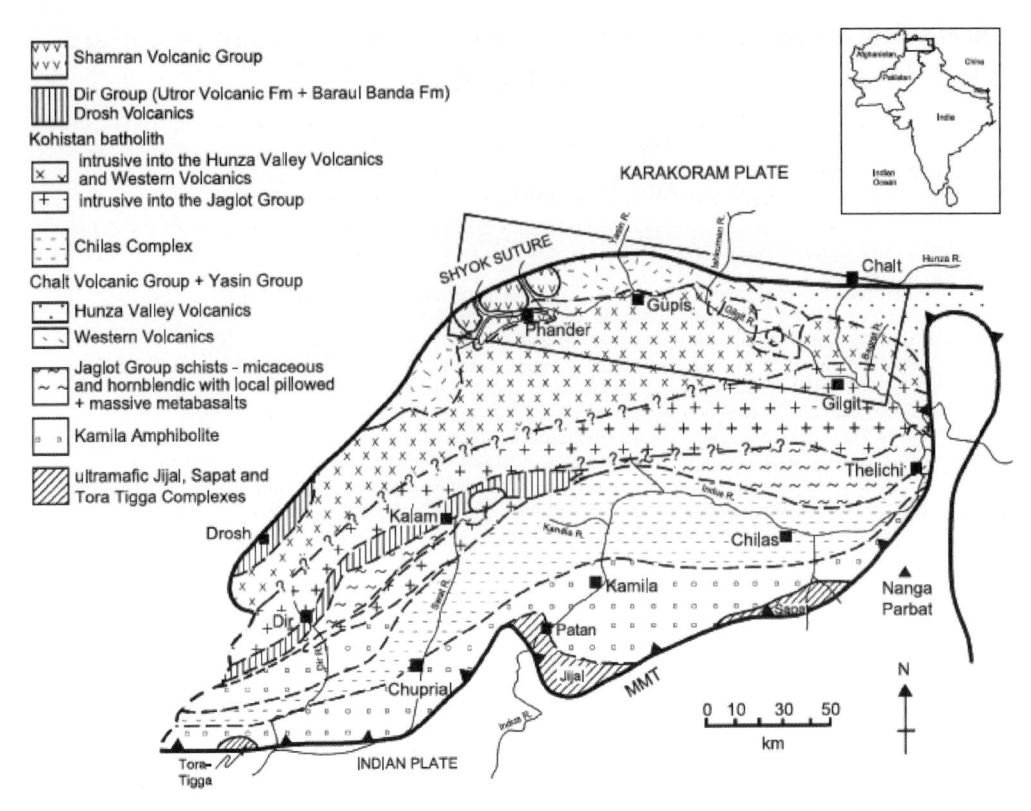

Fig. 1. Geology of Kohistan (from Petterson & Treloar 2004, acknowledgements Elsevier). General geology of Kohistan ranging from ultramafic–mafic bodies in the south (Jijal–Chilas units) to more felsic calc-alkaline volcanic and plutonic rocks in the north (Kohistan batholith, Chalt, Western Shamran and Jaglot Volcanic Groups). Box shows position of more detailed map reproduced in Fig. 13 for the Shamran–Teru Volcanic Group. Inset shows position of Kohistan in north Pakistan, and adjacent countries of Afghanistan and India (Ladakh).

workers who may be new to Kohistan. The paper results from an exhaustive and up-to-date literature review together with new insights from the author, who has been associated with Kohistan since 1981. Section 1 attempts to document published data with minimal interpretation, whereas section 2 focuses on documenting ideas, old and new, that interpret these data in terms of crustal genesis and evolution. Readers are directed to Sengor & Natalin (1996) and Kazmi & Jan (1997) for general overviews of the regional geotectonic setting of Pakistan, the NW part of the Greater Himalayan orogen, and crustal accretion with time throughout Asia.

History of research

This paper draws on *c.* 35 years of recent intensive research and almost 100 years of geological observations. The earliest observations (that are recorded) date back to the work of Hayden (1914), Ivanac *et al.* (1956), Wadia (in the 1920–1930s) Gansser (in the 1960s–1970s; Desio 1966; Desio *et al.* 1964, 1977; Desio & Zanettin 1970; Desio & Martina 1972) and Desio (from *c.* 1960 to the 1980s; Gansser 1964, 1979, 1981), who recorded fundamental aspects of field geology and structure and drafted early maps. The second stage of research took place mainly in the 1960s and 1970s, involving Pakistani scientists such as Tahirkheli and Jan and published in volumes such as those by Tahirkheli & Jan (1979) and Tahirkheli (1982). These workers began to understand the significance of Kohistan as an obducted island arc, and continued to work through the third and final stage of research starting with Bard (1983) and Bard *et al.* (1980) and continuing to the present day. It was during the earlier parts of stage 3 from the late 1970s into the 1980s that Brian Windley, together with his associates (such as Mike Coward) researched Kohistan intensively, attracting research grants, and with them teams of PhD students and Post-Doctorate researchers who have left a lasting legacy and very important contribution to our present-day understanding of Kohistan.

Section 1: general characteristics of the Kohistan terrane

As with any topic with a significant history of research it can be difficult to extract from the literature 'fact' from interpretation. This section aims to summarize the key facts relating to the general character of the Kohistan terrane as the foundation of the paper on which to assess the relative merits of interpretations and models.

Geographical and tectonic setting

Kohistan (Fig. 1) is located within the Northern Area district of Pakistan, surrounded by Afghanistan to the west, India (Ladakh region) to the immediate east and SE, and China (Xinjiang and Tibet) to the north and far east. The geographical centre of Kohistan is at approximately 35°30′N, 73°E. Kohistan occupies an area of $c.$ 30 000 km^2, or about the same size as Belgium Albania, Burundi or Taiwan, contains the worlds ninth highest mountain (Nanga Parbat, 8125 m high), and is predominantly a deeply dissected mountainous area with steep, gorge-like mountain slopes, valley floors at an elevation of 1000–2500 m and a dry mountain climate (precipitation around 7–15 cm a^{-1}).

In geotectonic terms Kohistan is bound by two suture-scale faults: the Shyok Suture to the north and the Indus–Tsangpo Suture to the south. These structures separate Kohistan from distinctively different, much larger and older geological entities (i.e. the southern part of Eurasia to the north and the northern part of the Indo-Pakistan plate to the south). Kohistan is part of a terrane that extends westwards into Afghanistan and eastwards into Ladakh; the strike length is some 700 km long (strike varying between NE–SW, east–west and NW–SE, with a maximum width between sutures (perpendicular to strike) of almost 200 km. By far the most complete part of the Kohistan–Ladakh terrane, in terms of geological units exposed, is situated in Pakistan, with Afghanistan hosting a very small area and Ladakh hosting predominantly upper crustal units (Treloar et al. 1990; Rolland et al. 2002). Kohistan is unique because it hosts one of the most (if not the most) complete crustal sections in the world, from the mantle to the uppermost crust. The only other comparable crustal section in the world is the Jurassic Talkeetna terrane in Alaska (Barker & Grantz 1982; DeBari & Coleman 1989; Greene et al. 2006; Hacker et al. 2008). The batholithic units within Kohistan are part of a $c.$ 2700 km long batholith extending from Afghanistan to northern Burma, and present in Kohistan, Ladakh and Tibet.

Kohistan–Ladakh is described as a terrane *sensu stricto* because it is bounded on all sides by sutures separating it from other terranes that have experienced a significantly different geological history. The Nanga Parbat–Haramosh syntaxis almost splits the Kohistan–Ladakh terrane in two along the eastern border of Kohistan with Baltistan. In regional terms Kohistan is one piece of a tectonic jigsaw that saw the amalgamation of many terranes to form the present-day Asian supercontinent (e.g. Sengor & Natalin 1996). Suturing of Gondwanan and intra-oceanic terranes to Asia occurred largely from the Jurassic and Cretaceous, affecting areas right across the proto-Asian continent (e.g. Mongolia and Afghanistan, Sengor & Natalin 1996).

Lithostratigraphy of Kohistan

Many papers have described in detail the various units of Kohistan (e.g. Tahirkheli et al. 1979; Bard 1983; Coward et al. 1986; Treloar et al. 1996; Searle et al. 1999; Bignold et al. 2006; Jagoutz et al. 2006; Dhuime et al. 2007, 2009; Garrido et al. 2007; Takahashi et al. 2007; Khan et al. 2009) and it is beyond the scope of this paper to duplicate these descriptions. This section focuses on the key lithostratigraphical characteristics of each unit from oldest to youngest and/or south to north in as systematic a fashion as possible. Readers are referred to Figure 1 (map of Kohistan) for the location of the various defined units, and to Figure 2 to understand the structural context of Kohistan.

Lowermost ultramafic–mafic units as typified by the Jijal Complex

The lowermost part of Kohistan, immediately to the north of the Indus Suture, comprises a number of ultramafic–mafic metamorphosed igneous complexes, the largest of which is the Jijal Complex (others include the Tora Tiga, Sapt and Babusar Complexes) bounded to the south by the Indus Suture and to the north by the Kamilla Amphibolites. The Jijal Complex is some 30 km NW–SE by 20 km NE–SW and is well exposed adjacent to the Karakoram Highway between Jijal and Pattan. The Jijal Complex is composed predominantly of lower dunites, harzburgites, websterites and pyroxenites, and upper garnet granulites (Jan & Howie 1980, 1981; Tahirkheli 1982; Jan & Windley 1990; Garrido et al. 2006; Dhuime et al. 2007, 2009). Metamorphic textures are most common although original igneous textures such as mineral layering and cumulus textures are recognized.

The Kamila Amphibolites

The most comprehensive lithostratigraphical accounts of the Kamila Amphibolites have been

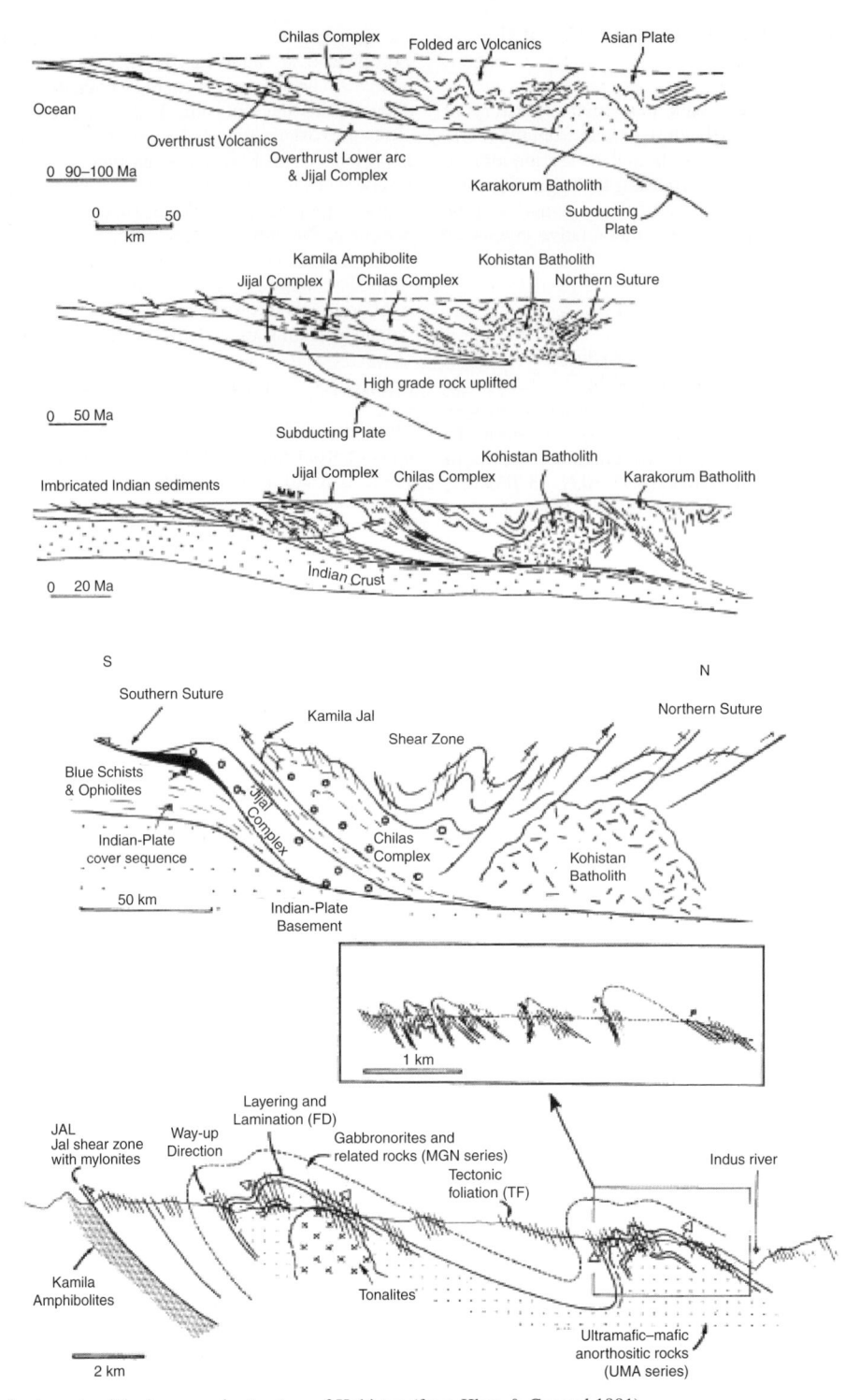

Fig. 2. Aspects of the larger-scale structure of Kohistan (from Khan & Coward 1991).

given by Tahirkheli (1982), Jan (1988, 1991), Khan *et al.* (1989), Treloar *et al.* (1990, 1996) and Dhuime *et al.* (2009). This unit extends some 250 km east–west by up to 10–45 km wide and is often the unit immediately north of the Indus Suture (unless Jijal-like ultramafic rocks are present). The unit is bounded to the south by either the Indus Suture or the Jijal Complex (probable intrusive relationship, Jijal into Kamila) and to the north by the Chilas Complex (probable intrusive relationship, Chilas into Kamila). The unit is characterized by metavolcanic (predominantly basalts and basaltic andesites) and metaplutonic rocks (gabbros, norites, and diorites). The unit is affected by a crustal-scale shear zone and can be highly deformed. Less deformed areas (such as around Chuprial) exhibit original igneous and volcanic features such as layering, lava–lava contacts, pillow structures, volcanic breccias and hyaloclastites.

The Chilas Complex

Chilas is a large-volume, basic–ultrabasic plutonic body measuring more than 300 km long by up to 40 km wide (e.g. Treloar *et al.* 1996), in a similar class to other large global plutonic bodies (e.g. Skaergaard, Stillwater, Bushveld, Border Ranges Complex). It most probably has intrusive relationships with the Kamila Amphibolites to the south and the Jaglot Group to the north. More than 85% of the unit comprises relatively monotonous gabbro-norite (the gabbro-norite association) consisting of plagioclase, orthopyroxene, clinopyroxene, magnetite, ilmenite with or without magnetite, scapolite, biotite, quartz, K-feldspar and hornblende. Around 15% of the Chilas Complex comprises ultramafic rocks (dunite, troctolite, peridotite, pyroxenite, anorthosite and gabbro-norite) within a unit termed 'the ultramafic association' or UMA. The UMA exhibits a range of original igneous textures including mineral layering, slumping, graded bedding, and syndepositional faults. The UMA is particularly well exposed around Chilas close to Nanga Parbat. Most workers have observed the gabbronorites being intruded by the UMA, although examples of the converse situation exist. Readers are directed to Tahirkheli (1982), Jan (1988), Khan *et al.* (1989, 1993), Treloar *et al.* (1996), and Garrido *et al.* (2006), Jagoutz *et al.* (2006, 2007) and Takahashi *et al.* (2007) for further details.

The Jaglot Group

The Jaglot Group extends in a semi-contiguous fashion for some 250 km east–west and up to 20–30 km north–south. It has a probable intrusive relationship with the Chilas Complex to the south and, where not in intrusive contact with the Kohistan batholith is conformable with the Chalt Volcanic Group to the north. The Jaglot Group is a mixed, largely metasedimentary unit comprising metasandstones, carbonates, siltstones, mudstones and turbidites with local metabasalt, andesite and rhyolite volcanic rocks. Khan *et al.* (2007) described a 1–4 km wide, 15 km long basalt and dolerite dyke swarm, oriented NW–SE, at the base of a volcanic-rich unit (Thelichi Formation) within the Jaglot Group. The Jaglot Group has only relatively recently been recognized as an important cross-terrane unit. Readers are directed to Tahirkheli (1982), Treloar *et al.* (1996), Bignold *et al.* (2006) and Khan *et al.* (2007) for further details.

Kohistan batholith

The Kohistan batholith extends over 270 km east–west (as mentioned above, the batholith in Kohistan is only part of a *c.* 2700 km long Trans-Himalayan batholith) and up to 50–60 km north–south. The batholith comprises a wide range of lithologies from hornblendite to leucogranite, but predominantly is gabbroic, gabbroic diorite and granite or trondhjemite in composition. Batholithic intrusions are largely small–medium- to medium–large-volume plutons with significant vertical and horizontal dimensions. Intrusive bodies also take the form of sills, dykes, sheets, lopoliths and other smaller geometries. Many intrusions are complex and composite multi-intrusive bodies with up to four or five phases of intrusion. The most common major minerals are plagioclase, alkali feldspar, hornblende, biotite and quartz. The batholith is largely intrusive into the Chalt Volcanic Group (see below) and the Jaglot Group metasedimentary-dominated unit, although Khan *et al.* (1998) also reported batholithic units intruding Kamila Amphibolites around Babusar. The batholith has been particularly uplifted and eroded close to Nanga Parbat and also west of Gilgit where volcano-sedimentary sequences unconformably overlie the batholith (e.g. Sullivan *et al.* 1993). Even the earliest pioneers recognized that there were probably three distinct phases of batholithic intrusion (e.g. Wadia 1932; Ivanac *et al.* 1956) that could be differentiated on the basis of the presence or absence of well-developed penetrative fabrics with earlier gneissic-like gabbros and granitoids intruded by undeformed plutons that in turn were intruded by late granite sheets. For more complete field descriptions of the batholith, readers are directed to studies by Coward *et al.* (1982*a*, *b*), Tahirkheli (1982), Petterson & Windley (1986, 1991, 1992), Petterson *et al.* (1991*a*) and Searle *et al.* (1999). More recent papers that include new data on the batholith include those by Schaltegger et al. (2002), Heuberger et al. (2007) and Khan et al. (2009).

Chalt Volcanic Group

The Chalt Volcanic Group (CVG) is the northernmost Kohistan-wide conformable geological unit, occupying an area some 330 km east–west by up to 30 km north–south. The northern contact is conformable with the overlying Yasin volcano-sedimentary unit or the Shyok Suture and the southern contact is predominantly obscured by the Kohistan batholith, although the CVG should conformably rest upon the Jaglot Group. Mapping data in this area indicate a general younging direction northwards, suggesting that the CVG is younger than volcanic rocks cropping out around Jaglot (Petterson & Treloar 2004). The CVG is divisible into an eastern Hunza Formation and western Ghizar Formation. The Hunza Formation is dominated by pillowed and unpillowed boninites, basalt to andesite lavas with minor rhyolites, thin ignimbrites, primary and reworked volcanoclastic units and minor intrusive rocks. The Ghizar Formation is situated west of Gilgit and is a much more heterogeneous sequence of volcanic rocks than the Hunza Formation, comprising basaltic to andesitic (and minor rhyodacitic) tuffs and primary volcaniclastic rocks, and reworked volcaniclastic rocks with locally important lava-dominated sequence. A proximal Strombolian–Vulcanian centre has been identified in Ishkuman valley, west of Gilgit. Readers are directed in particular to (studies by Petterson et al. (1991b) and Petterson & Treloar (2004) for detailed descriptions of the volcanic sequences and to those by Tahirkheli (1982), Pudsey et al. (1985a), Coward et al. (1986), Pudsey (1986) and Searle et al. (1999) for more general or localized descriptions of these units.

Yasin volcano-sedimentary formation

The Yasin Formation, where exposed, rests conformably upon the CVG. This unit is exposed rather episodically adjacent to the Shyok Suture, being absent in some areas. The Yasin Formation comprises a mixed assemblage of volcaniclastic and non-volcanic turbidites, sandstones, siltstones and mudstones with minor carbonate units, some of which contain Albian–Aptian *Orbitolina* fossils. This clastic-dominated sequence also contains minor basalt and andesite lavas (Hayden 1914; Desio 1959, 1963a, b, 1964, 1974, 1975, 1977, 1980; Tahirkheli 1982; Pudsey et al. 1985b; Pudsey 1986; Robertson & Collins 2002).

More localized volcano-sedimentary units: the Shamran/Teru Volcanic Formation and Dir Group

Within Kohistan there exist a number of stratigraphical units that (1) are localized (relative to the Kohistan-wide units described above), (2) have unconformable bases, largely upon uplifted and eroded Kohistan batholith plutons, (3) are undeformed and (4) have evolved, highly silicic magmatic components. The Dir Group is the best described and exposed, being situated within a c. 150 km long (ENE–SSW) by up to 10 km wide belt stretching within west–central Kohistan from the Afghanistan border through Dir and east of Kalam. The field aspects of the Dir Group have been most comprehensively described by Tahirkheli (1982) and Sullivan et al. (1993) and comprise a lower, rather monotonous sequence of turbiditic sandstones and siltstones, with minor Thanetian (60–55 Ma) carbonate horizons bearing *Miscellanea miscella* and *Actinosiphon tibeticus* (the Baraul Band Slates) and an upper heterogeneous sequence of volcanic breccias, sandstones and mudstones, ignimbrites and rhyolites intruded by younger parts of the Kohistan batholith (the Utror Volcanics). Within the higher reaches of the Ghizar valley there exists a sequence of ignimbrites, rhyolites and silicic volcanoclastic rocks that map out as a series of high-altitude outliers resting unconformably upon an eroded volcano-batholithic basement; the Shamran or Teru volcanic rocks have been described by Sullivan et al. (1993), Treloar et al. (1996), Danishwar et al. (2001), Khan (2001), Bignold & Treloar (2003), Khan et al. (2004, 2009) and Petterson & Treloar (2004)

Kohistan lithostratigraphic summary

The brief descriptions above demonstrate that the Kohistan terrane is a discrete geological entity. Simple field lithological descriptions show that the terrane is predominantly of igneous origin (although it has experienced high degrees of metamorphism and deformation) and predominantly of basic (gabbroic, noritic, amphibolitic–gabbroic or basaltic, granulitic–gabbroic or basaltic; basaltic and basaltic andesitic) composition. Silicic andesitic to rhyodacitic igneous rocks are most prolific in northern Kohistan, although granitic intrusions are present within the Kamila Amphibolite unit (e.g. Dhuime et al. 2009). The youngest, more localized rocks that rest unconformably on the Kohistan-wide units are dominated by rhyolites and ignimbrites. All except the youngest, more localized units, are located within discrete lithostratigraphical packages that have a long axis oriented approximately east-west and a short axis oriented approximately north-south. The bulk of Kohistan is intrusive-plutonic in character although total thicknesses of volcanic units represented by the CVG are significant There is a relatively small volume of sedimentary or metasedimentary material present and much of this has a volcanic component to it.

Age of Kohistan

Table 1 summarizes key ages from Kohistan, both absolute and palaeontological ages. The table focuses on ages that define the original age of the rock or a major subsequent thermal event at granulite-facies (>800 °C) metamorphism. References for this discussion are presented in Table 1 and involve a wide range of workers and over three decades of research. The data, when viewed for the whole of Kohistan, are most instructive. Palaeontological data demonstrate that the northernmost Yasin sediment unit is Albian-Aptian (99–125 Ma) in age and one of the youngest units, the Dir Group, which rests unconformably upon more regional Kohistan units, is Thanetian in age (c. 60–55 Ma). These are key (relatively) fixed chronological pins that are indisputable, suggesting that much of the basement of Kohistan is Albian-Aptian or older and that the cover resting on basement (Dir Group) is early Tertiary in age. The oldest absolute ages are c. 117–118 Ma (Sm-Nd) for the Jijal Complex, which fits within the lower range of Albian-Aptian Cretaceous times and may give clues about the earlier stages of Kohistan. The longest lasting unit in Kohistan is the batholith, which has an oldest age of c. 112–102 Ma and youngest age of c. 26 Ma (e.g. Casnedi et al. 1978; Petterson & Windley 1985; George et al. 1993; Heuberger et al. 2007; Khan et al. 2009), representing c. 86 Ma of magmatism. This can be further divided (from workers such as Petterson & Windley 1985; Treloar et al.1989; George et al. 1993) into four sub-groupings within the batholiths. The deformed plutonic rocks are dated by the older age of c. 102 Ma, the youngest granite and leucogranite sheets are dated predominantly at around 50–26 Ma, undeformed gabbros and gabbroic diorites at c. 80–60 Ma and undeformed granites and granodiorites at c. 60–40 Ma. The Kohistan batholith is described as having three main growth stages comprising deformed plutons (stage 1), undeformed gabbros, granites and diorites (stage 2), and the latest stage granite sheets (stage 3) (Petterson & Windley 1985). The Jijal Complex exhibits two age modes: one at c. 117–118 Ma and the other at c. 99–91 Ma (average c. 95 Ma), reflecting an earlier primary magmatic event and a later granulite-facies metamorphic event. The c. 95 Ma age is also reflected in the highest-grade metamorphism within the Kamila Amphibolite unit, indicating that the Kamila volcanic units were erupted earlier than 95 Ma. The Chilas Complex has been repeatedly dated at 80–c. 86 Ma with a modal age of c. 85 Ma. There are no ages for the CVG but we do know that this unit is constrained to some degree by the Jijal Complex ages and the Albian-Aptian Yasin sedimentary unit. The youngest dated units are the cover Dir Group and Shamran/Teru volcanic and sedimentary units, which are constrained between c. 63 and 32 Ma with a mean age of c. 50 Ma, possibly suggestive of episodic volcanism between the early-mid-Paleocene and the late Eocene. In simple terms it appears that Kohistan began to form during the Early Cretaceous and that magmatic, crustal genesis activity ended around 26 Ma (Oligocene period).

Structure, deformation, metamorphism, cooling ages, sutures and lithostratigraphic thickness

It is almost inevitable that a terrane like Kohistan trapped between two sutures within a collisional tectonic environment will have undergone very significant deformation, and this is indeed the case. This section focuses on the cross-Kohistan structural, rather than localized structure within a particular unit, and abstracts heavily from data and models of, for example, Coward et al. (1982a, b, 1986, 1987), Tahirkheli (1982), Bard (1983), Pudsey et al. (1985a, b), Pudsey (1986), Treloar et al. (1990), Khan & Coward (1991), Khan et al. (1997, 2009), Searle et al. (1999), Robertson & Collins (2002), Schaltegger et al. (2002), Bignold & Treloar (2003), Petterson & Treloar (2004), Dhuime et al. (2007, 2009) and Heuberger et al. (2007), particularly the work of Mike Coward and his colleagues summarized in Figure 2. Kohistan is dominated by upright, large-scale structures with folds that have half-wavelengths of several tens of kilometres, crustal-scale shear systems (e.g. the Kamila shear zone) and large-scale thrusts and thrust stacks. In general the northern part of Kohistan (including the eastern part of the Shyok Suture zone) is dominated by steep, southerly dipping structures whereas southern Kohistan exhibits moderate to steep northerly dipping structures. The basement of Kohistan (i.e. minus the Dir Group and Shamran/Teru Volcanics) with the exception of the younger stages 2 and 3 of the Kohistan batholith, exhibits a penetrative fabric with strong tectonic foliation and lineation structures, particularly when close to major shear and fault-thrust zones. Recent work by Garrido et al. (2006) and Dhuime et al. (2007, 2009), as presented in Figures 3 and 4a-c, has elucidated further the highly complex

Table 1. *Summary of absolute and palaeontological age constraints for Kohistan*

Geological unit	Age	Method	References
Teru/Shamran Volcanics	58 ± 1 Ma	Ar/Ar	Treloar et al. 1989
	43.8 ± 0.5 Ma	Ar/Ar	Khan et al. 2004
	32.5 ± 0.4 Ma	Ar/Ar	Khan et al. 2004
	64.9–63.1 ± 0.9–2.5 Ma	U–Pb	Khan et al. 2009
Dir Group	Thanetian (60–55 Ma)	*Miscellanea miscella* fossils	Sullivan et al. 1993
	55 ± 2 Ma	Ar/Ar	Treloar et al. 1989
Yasin sediments	Albian–Aptian (99–125 Ma)	*Orbitolina* fossils	Pudsey et al. 1986
			Pudsey et al. 1985a, b, 1986
Chalt Volcanic Groups			
Kohistan batholith	111.52 ± 0.40 Ma	U–Pb	Heuberger et al. 2007
	49.80 ± 0.15 Ma	U–Pb	Heuberger et al. 2007
	47.4 ± 0.5 Ma	U–Pb	Heuberger et al. 2007
	64.5 ± 0.5 Ma	U–Pb	Khan et al. 2009
	41.0 ± 0.5 Ma	U–Pb	Khan et al. 2009
	102 +/12 Ma	Rb–Sr WR	Petterson & Windley 1985
	54 ± 4 Ma	Rb–Sr WR	Petterson & Windley 1985
	40 ± 6 Ma	Rb–Sr WR	Petterson & Windley 1985
	34 ± 14 Ma	Rb–Sr WR	Petterson & Windley 1985
	29 ± 6 Ma	Rb–Sr WR	Petterson & Windley 1985
	75 Ma	Ar/Ar	Treloar et al. 1989
	54.1 ± 1 Ma	Ar/Ar	Treloar et al. 1989
	63 ± 1 Ma	Ar/Ar	Treloar et al. 1989
	65 ± 1 Ma	Ar/Ar	Treloar et al. 1989
	26.2 ± 1.2 Ma	Rb–Sr WR	George et al. 1993
	49.1 ± 11 Ma	Rb–Sr WR	George et al. 1993
Chilas Complex	85.73 ± 0.15 Ma	U–Pb zircon	Schaltagger et al. 2002
	82.8 ± 1.1 Ma	U–Pb zircon	Schaltagger et al. 2002
	85–80 Ma (cooling ages)	Ar/Ar	Treloar et al. 1989
	69.5 ± 9.3 Ma	Sm–Nd	Yamamoto & Nakamura 1996
	84 ± 0.5 Ma	U–Pb zircon	Zeitler & Chamberlain 1991
	84 Ma	U–Pb zircon	Zeitler 1985
Kamila Amphibolites	95–100 Ma	Sm–Nd and Rb–Sr garnet	Anczkiewicz & Vance 2000
Jijal Complex	98.9 ± 0.4 Ma	U–Pb zircon	Schaltagger et al. 2002
	97.1 +/0.2 Ma	U–Pb zircon	Schaltagger et al. 2002
	91.8 ± 1.4 Ma	U–Pb zircon	Schaltagger et al. 2002
	117 ± 7 Ma	Sm–Nd	Dhuime et al. 2007
	118 ± 12 Ma	Sm–Nd	Yamamoto & Nakamura 2000
	94 ± 4.7 Ma	Sm–Nd	Yamamoto & Nakamura 2000
	91 ± 6.3 Ma	Sm–Nd	Yamamoto & Nakamura 1996
	95–100 Ma	Sm–Nd and Rb–Sr garnet	Anczkiewicz & Vance 2000

folded and thrust–sheared nature of the lower part of the Kohistan crust, with dunites, websterites, garnet granulites, hornblendites and amphibolites forming discrete lithostratigraphical units, cut by granite dykes and intrusions and affected by polyphase deformation. The 85 Ma Chilas Complex has less well-developed penetrative fabrics when compared with those of the Jijal, Kamila, Jaglot, CVG and Yasin lithostratigraphic units (Treloar et al. 1996). Petterson & Windley (1992) showed clearly that the undeformed 75 Ma Jutal–Nomal dykes cross-cut penetrative fabrics within the 102 Ma Matum Das stage 1 gneissic trondhjemite (Fig. 5) as well as the upright, southerly dipping folds and shears of the Chalt–Ghizar–Yasin lithostratigraphic units. These data suggest that the earliest deformation stage within Kohistan had largely finished by 85 Ma and was definitely over by 75 Ma.

The basement of Kohistan (as defined above) is metamorphosed from granulite to greenschist grade. There is a general metamorphic gradient from south to north with the Jijal Complex exhibiting the highest granulite-grade metamorphism, with some localized amphibolite-grade retrogression.

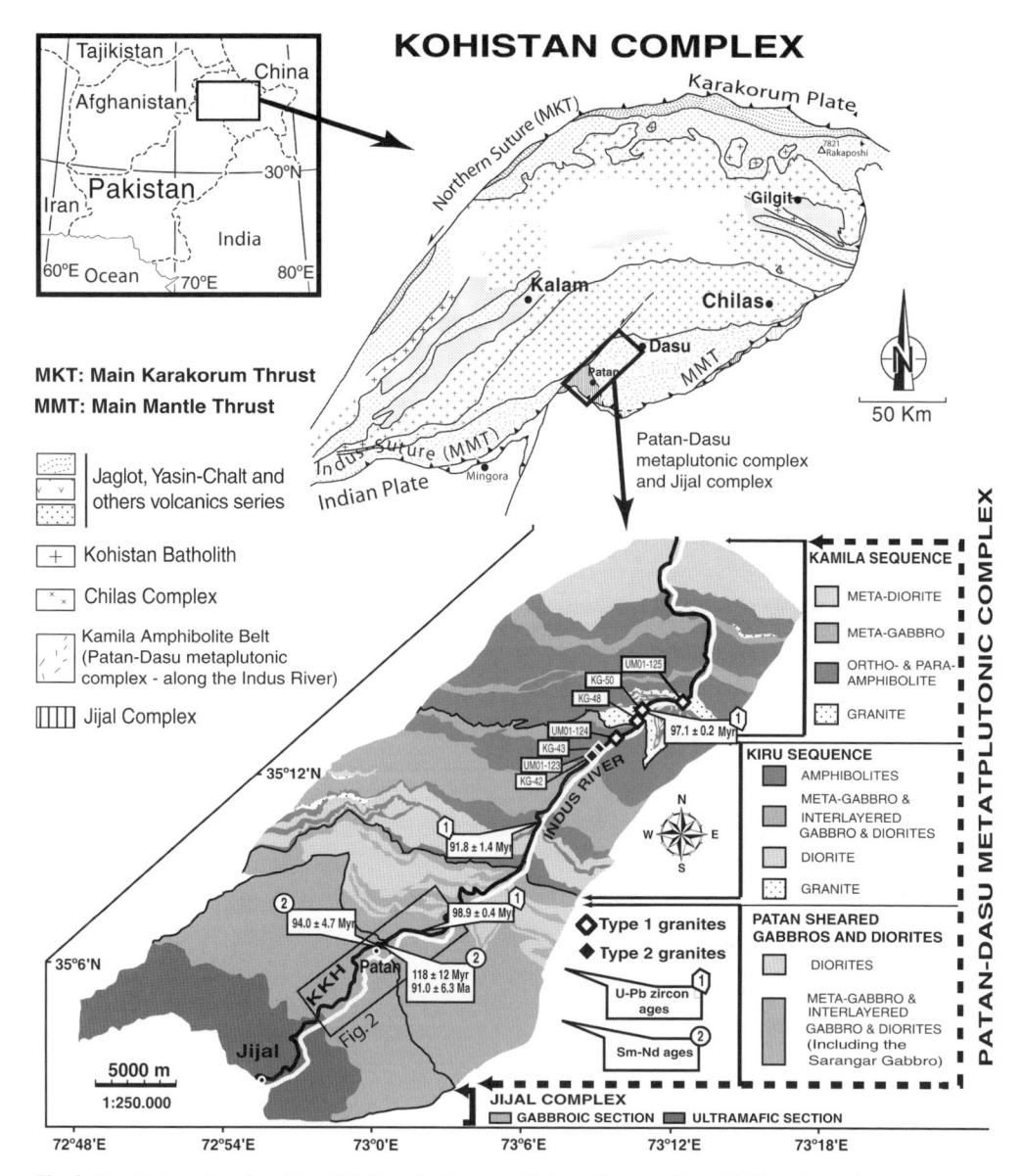

Fig. 3. Detailed stratigraphy of the Jijal Complex between Jijal and Dasu, southern Kohistan (note the range in lithologies from ultramafic and mafic plutonic rocks to diorite and granite) (Garrido *et al.* 2006, acknowledgements Oxford University Press).

Peak $P-T$ estimates are: 810–825 °C and 1–1.2 GPa (10–12 kbar, 30–40 km depth; Padron-Navarta *et al.* 2008); 700–800 °C and 0.6–1.2 GPa (6–12 kbar, 20–36 km depth; Yoshino & Okudaira 2004); 700–950 °C, >1 GPa (>10 kbar, *c.* 30 km depth; Yamamoto & Nakamura 2000); 949 °C and 17 kbar (1.7 GPa, *c.* 50 km depth; Yamamoto 1993). Metamorphic grade within the Kamila Amphibolite belt is lowest in the central part and increases towards both the Jijal and Chilas Complexes. Peak metamorphic $P-T$ is estimated at 550–680 °C and 4.5–6.5 kbar (0.45–0.65 GPa, 12–20 km depth; Jan 1988). The Chilas Complex is predominantly granulite grade, although estimates of peak $P-T$ conditions are lower than those for Jijal; for example, 700 °C, 0.6–0.7 GPa (6–7 kbar, *c.* 20 km depth; Yoshino & Okudaira 2004); 750–850 °C and 5–6.5 kbar (0.5–0.6 GPa,

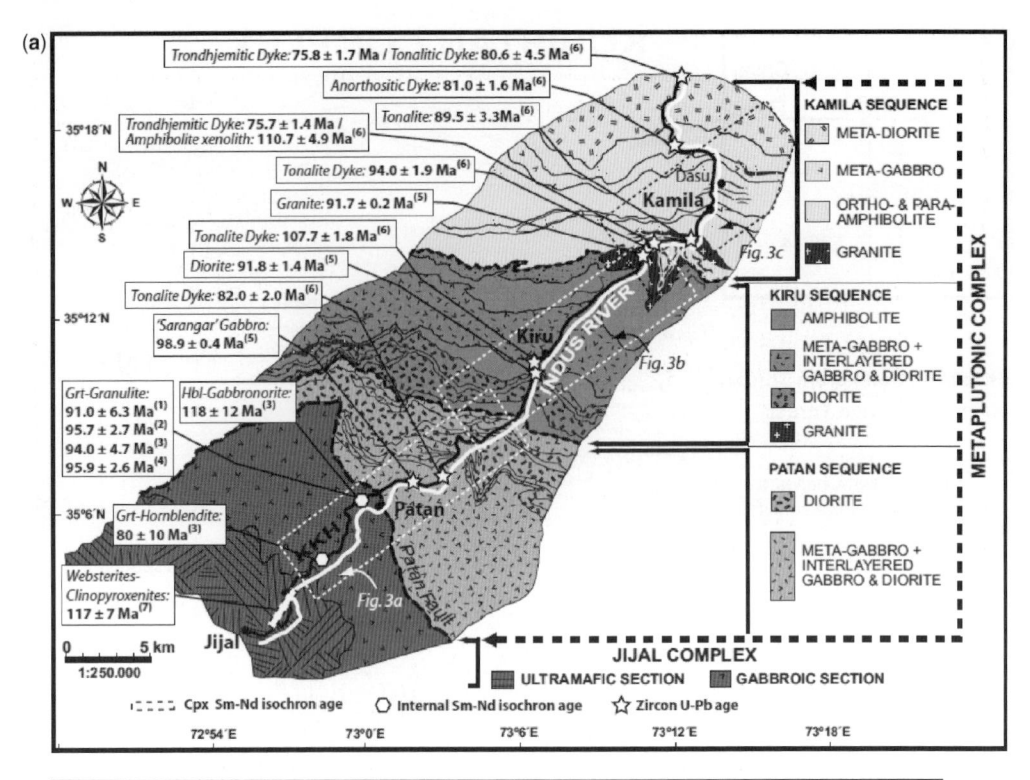

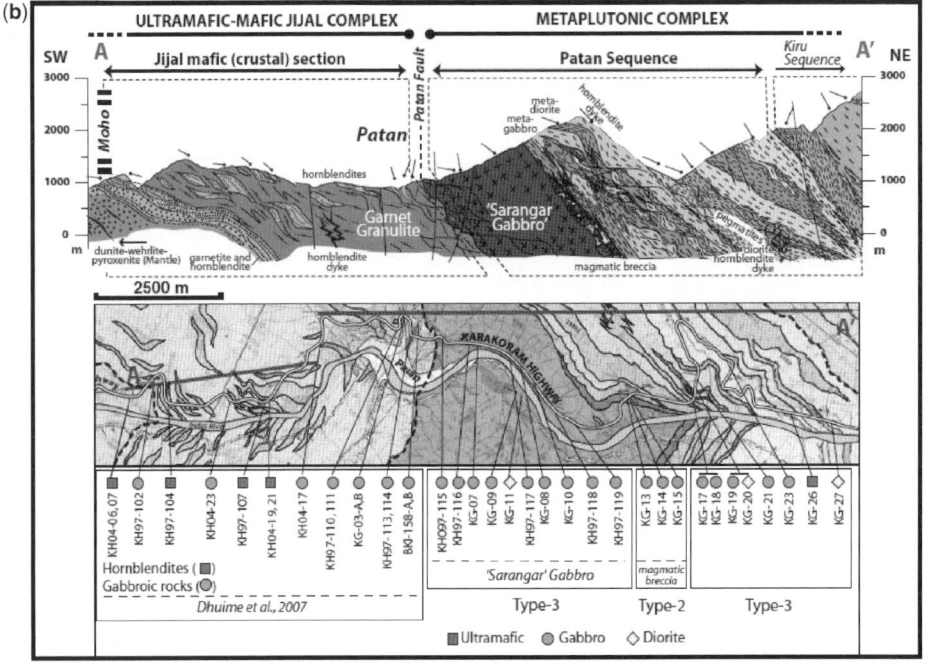

Fig. 4. (a–c) Details of mapping and structure of the ultramafic and mafic components of the Jijal Complex and southern part of the Kamila Amphibolite Belt (from Dhuime *et al.* 2009, acknowledgements Oxford University Press).

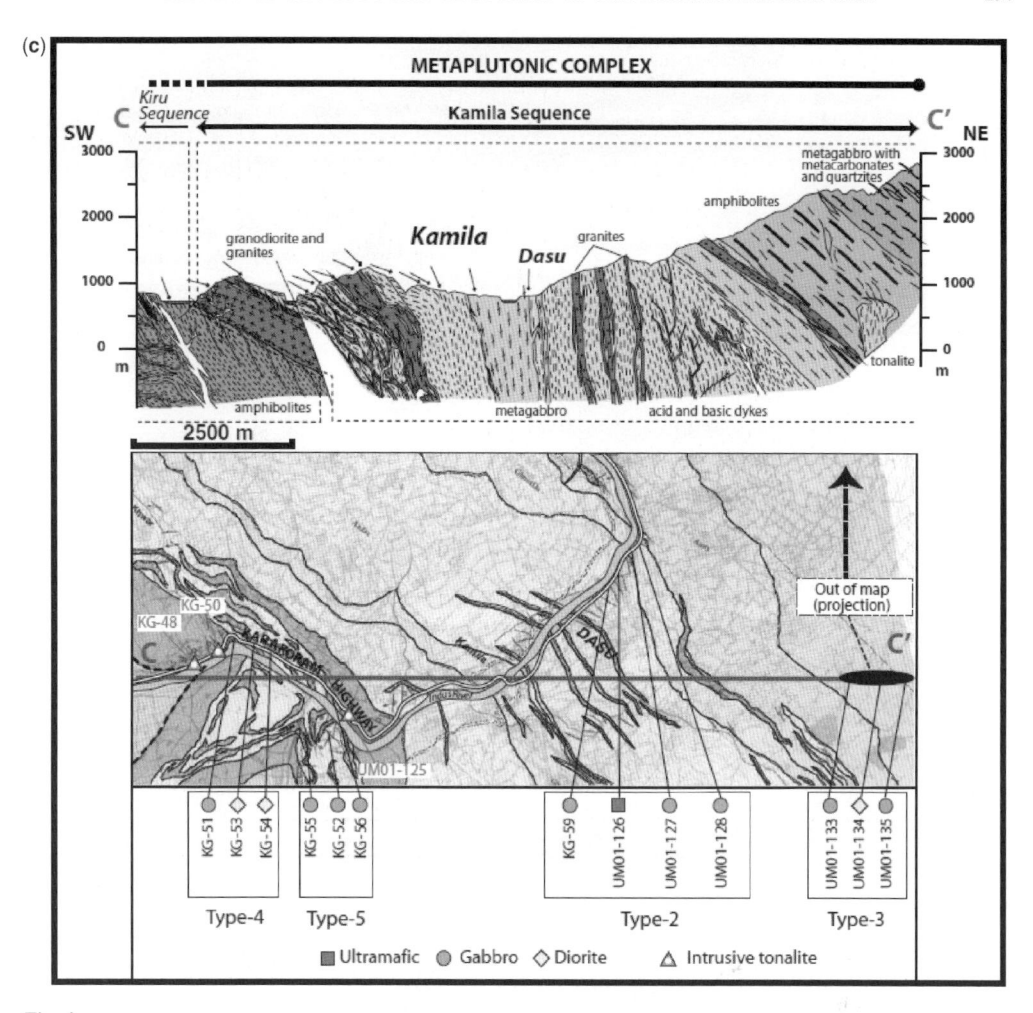

Fig. 4.

c. 18 km depth; Jan 1979; Jan & Howie 1980; Bard 1983). North of the Chilas Complex amphibolite- to greenschist-grade metamorphism predominates, although parts of the Jaglot Group are at garnet grade and contain magmatic anatectic melts. Cooling ages for Kohistan (e.g. Treloar et al. 1989; Yamamoto 1993; Yamamoto & Nakamura 1996, 2000; Yamamoto et al. 2005) suggest that southern Kohistan and possibly the bulk of the terrane cooled through the 500 °C isotherm around 80 Ma, and that the eastern part of northern Kohistan cooled through 300 °C at c. 40 Ma with the western part of north Kohistan cooling through 300 °C much earlier than this [Sullivan et al. (1993) argued that northwestern Kohistan underwent significant cooling and uplift prior to the extrusion of the c. 58 Ma Dir Group, which rests unconformably on eroded basement]. Table 1 illustrates dates for peak metamorphism within the Jijal Complex that have an average value of c. 95 Ma; at this time the Jijal ultramafic rocks were at a depth of c. 20–50 km, most probably 30–40 km. Cooling ages around Nanga Parbat (e.g. Zeitler 1985; Treloar et al. 1989) demonstrate the rapid, recent rise of this mountain and the Indian crust, which have also rapidly uplifted adjacent areas of Kohistan.

Combined structural, metamorphic and cooling age data indicate that the key cause of deformation and metamorphism had to happen prior to: (1) 80 Ma, when the bulk of Kohistan cooled through 500 °C; (2) 75 Ma, when the Jutal dykes clearly cross-cut pervasive penetrative fabrics and the major structures; (3) 85 Ma, when the Chilas Complex was intruded, as this unit is significantly less deformed than the rest of the Kohistan basement. Treloar et al. (1996) argued that the Chilas

Fig. 5. Dyke swarm around Jutal village with undeformed 75 Ma dykes cutting a *c.* 102 Ma deformed trondhjemite (from Petterson & Windley 1993, acknowledgements Geological Society, London).

Complex intruded during the final stages of the first deformation stage. Many workers have agreed that the collision between India and Kohistan–Asia did not occur prior to 55 Ma (Beck *et al.* 1995; Searle *et al.* 1999), although some have stated that this occurred as early as 70 Ma (e.g. Yin 2006) or as late as 34 Ma (e.g. Aitchison *et al.* 2007). This evidence would rule out the main Himalayan collision as being the causal factor for generating the bulk of the penetrative structures, fabrics and metamorphism present in Kohistan, although Khan *et al.* (2009) argued that Kohistan collided first with India at 61 Ma. The most popular tectonic model for the geotectonic evolution of Kohistan argues that important structures and deformation fabrics were formed when Kohistan collided with Eurasia to form the Shyok Suture. The arguments presented above bracket the age of the Shyok Suture between

102 Ma and 85–75 Ma, although Khan *et al.* (2009) suggested that the Shyok Suture was the last suture to form at *c.* 50 Ma.

One aspect of Kohistan that has been very difficult to reliably estimate is the original thickness of lithostratigraphical units. This is largely due to the structural complexity of the terrane, with a plethora of intrinsic shearing, faulting, thrusting and folding to unravel as well as the lack of diagnostic and widely correlatable subunits within any particular sequence. Detailed mapping has been undertaken only in relatively small areas and the mountainous and inaccessible terrain adds to the challenges. The few studies that addressed measurable estimates of unit thickness include those of Sullivan *et al.* (1993), Danishwar *et al.* (2001) and Petterson & Treloar 2004, whereas Coward *et al.* (1987) attempted a balanced cross-section approach to estimate shortening and unit thickness. Petterson & Treloar (2004) measured a 500 m section for the Yasin sediments, giving a minimum half-thickness, whereas data of Ivanac *et al.* (1956), Pudsey *et al.* (1985*a*, *b*), Pudsey (1986) and Robertson & Collins (2002), indicated a likely 2–3 km total thickness. Petterson & Treloar (2004) estimated a possible total thickness of 21 km for the Chalt Volcanic Group exposed in the Bagrot and Hunza valleys of which 66% is basalt to andesite sheets, 25% is volcaniclastic rocks and 9% is ignimbrites and rhyolites youngening predominantly northwards. There has undoubtedly been a high degree of thickening within the Chalt Volcanic Group caused by shearing, thrusting and isoclinal to tight folding. However, it is currently impossible to estimate precisely the degree of thickening, although Petterson & Treloar (2004) suggested that thickening was unlikely to have been more than a factor of 2–4, giving a possible stratigraphic thickness of 5–10 km for the Chalt Volcanic Group. Petterson & Treloar (2004) estimated a possible minimum thickness of 1.5–2 km for the Shamran/Teru volcanic rocks based on the height of mountains above the level of unconformity around Teru village. Danishwar *et al.* (2001) suggested a thickness of *c.* 3 km. Sullivan *et al.* 1993 estimated a total thickness of some 6 km for the Dir Group, comprising *c.* 3 km of turbiditic sandstones (the Baraul Banda Slates) and 3 km of Utror Volcanics (lower 2 km of volcaniclastic rocks and upper 1 km of rhyolites and ignimbrites). Estimates for the total thickness of the Jaglot Group (Khan *et al.* 2007) are particularly difficult because the unit has not, as yet, been properly defined, is made up of at least three sub-lithostratigraphic units, and is heavily intruded by the Kohistan batholith. It crops out over a thickness of *c.* 20–50 km but the great bulk of this will represent thickening as a result of batholith intrusion, folding (e.g. the Jaglot synform of Coward *et al.*

1982*a*, *b*), shearing and other structural thickening processes. This unit does form part of a long-lasting extensional basin that has produced the partly boninitic Hunza Volcanic Formation (Petterson *et al.* 1991*a*, *b*; Petterson & Windley 1992; Petterson & Treloar 2004; Bignold *et al.* 2006; Khan *et al.* 2007) and contains a dyke swarm interpreted as a back-arc spreading centre (Khan *et al.* 2007). The Jaglot Group has been subjected to intrusion by the exceptionally high-volume Chilas Complex (Treloar *et al.* 1996), which suggests it may be significantly thick (perhaps >4 and <8 km). The Chilas Complex is up to 40 km wide in outcrop, although Coward *et al.* (1982*a*, *b*, 1986) suggested that its real thickness is least half this value, as it forms a major antiformal structure, suggesting a minimum thickness of 15–20 km for the central part of the complex, wedging-out west and eastwards. The Kamila Amphibolites crop out over a thickness that varies between 10 and 45 km and are affected by very significant shearing and folding, with a crustal-scale shear zone situated within this unit (some 28 km wide accommodating crustal shortening, stacking and imbrication, Treloar *et al.* 1990; Dhuime *et al.* 2009; Fig. 4). A minimum thickness of *c.* 4–8 km is suggested for this unit, with high degrees of repetition and thickening being likely within the thickened shear zone. Although the Jijal Complex is up to 20 km wide at outcrop it too has undergone tectonic thickening. The Jijal Complex's sister unit, the Sapat Complex, cropping out around Babusar, has an estimated restored thickness (after structural balancing) of 7–8 km, which will be taken here as an approximate minimum thickness for the Jijal Complex (Jan *et al.* 1993; Khan *et al.* 1998; Dhuime *et al.* 2007; Garrido *et al.* 2007).

This basic analysis of possible thicknesses of single lithostratigraphic units indicates a total thickness of some 36–54 km. The Jijal Complex contains the transition zone between the crust and mantle (e.g. Garrido *et al.* 2006, 2007; Dhuime *et al.* 2007), which would suggest a possible original crustal thickness of some 32–50 km. This is rather thick for an intra-oceanic arc, with recent estimates of oceanic arc thickness ranging from 18 km for the Marianas arc (Dimalanta *et al.* 2002; Calvert *et al.* 2008) to between 20 and 30 km for arcs such as Izu–Bonin (16–22 km), the Marianas (22 km), the Aleutians (30 km), Tonga (20 km) and Vanuatu (26 km), suggesting that the lower estimate of *c.* 30 km is probably closer to the original Kohistan arc thickness.

Seismic crustal structure

Table 2 and Figure 6 summarize the key data published by Chroston & Simmons (1989) and Miller

Table 2. *Kohistan seismic velocity, crustal structure and density structure (after Miller & Christensen 1994)*

Lithostratigraphic unit	Crustal depth range (km)	V_p (km s^{-1})	Density (kg m^{-3})
Chalt Volcanic Group	0–3	4.3–6	2700–2900
Kohistan batholith	3–10	5–6.5	2500–3000
Chilas Complex	10–22	6–7	2900–3000
		(dunite >8)	3400
Kamila Amphibolites	22–36	7.2	2900–3000
Garnet granulites		7.8	3100–3200
Ultramafic Jijal Complex	36–45	8.0–8.4	3300
Upper Kohistan crust	0–10	6.2–6.4	2700–3000
Lower Kohistan crust	20–30	6.4–6.7	2900–3200
Average Kohistan crust	30–40	6.7 ± 0.05	3000–3100
Average arc	15–30	6.7	2900–3100

& Christensen (1994). Those researchers demonstrated that Kohistan is realistically representative of theoretical arc-like or Cordilleran crust as well as measured crust from other parts of the world (Fig. 6). Seismic P-wave velocities vary between 4.3 km s^{-1} and mantle-like velocities greater than 8.1 km s^{-1}. The generalized seismic structure of Miller & Christensen (1994) gives a P-wave velocity model crustal structure for Kohistan of c. 0–3 km, <5.7 km s^{-1}; 3–13 km, 5.7–6.4 km s^{-1}; 13–16 km, 6.4–6.8 km s^{-1}; 16–42 km, 6.8–7.8 km s^{-1}; and 42–50 km, >7.8 km s^{-1}. The mantle is indicated at c. 42 km depth (which coincides well with the maximum possible crustal thickness argument given in the section above). In comparison with other arcs and orogenic belts, Kohistan has a thinner upper crust (V_p < 6.4 km s^{-1}), a thinner mid-crustal unit (V_p 6.4–6.8 km s^{-1}) and a very thick lower crust (V_p 6.8–7.8 km s^{-1}). There are also differences between Kohistan crustal structure and those for Recent orogenic belts which have small volumes of crust with a V_p > 6.8 km s^{-1}) continental rifts (thick upper crust, thin mid–lower crust) and cratonic shields (small thickness of lower crust, thick middle crust, thick upper crust). An average Kohistan crustal V_p of 6.7 ± 0.05 km s^{-1} is identical to that for average arc or Cordilleran-type crust worldwide. A key characteristic of Kohistan crust is the abundance of basic to ultrabasic lower crustal units, which gives it an overall mafic character when compared with average continental crust. Figure 7 presents a highly instructive section through the Talkeetna obducted arc terrane in Alaska (Clift et al. 2005; Greene et al. 2006). Talkeetna is interpreted as representing some 30 km of arc crust with the lower and middle crust between depths of 12 and 30 km being dominated by gabbroic gabbronorites, the uppermost 5–7 km comprising volcanic rocks and an intermediate c. 7–12 km layer formed from intermediate–felsic to mafic plutonic rocks.

Geochemical, petrological and isotopic composition

A summary of over three decades of petrological, geochemical and isotopic data are presented in Tables 3–5, together with key references from which this section is drawn. The initial observation is the relatively low range in isotopic composition throughout the terrane. Kohistan has isotopic compositional ranges of $^{87}Sr/^{86}Sr_i$ = 0.70243–0.70559, $^{143}Nd/^{144}Nd_i$ = 0.5127–0.51303, εNd = +0.1 to +8.6, $^{206}Pb/^{204}Pb$ = 17.973–18.595; $^{207}Pb/^{204}Pb$ = 15.518–15.64 and $^{208}Pb/^{204}Pb$ = 38.048–38.829. No workers consider that the general isotopic composition of Kohistan reflects large crustal inputs; in other words, the great bulk of Kohistan is mantle-derived, primary, new additional material to the crust. There are arguments for addition of sediments into the source region and variable degrees of Kohistan crustal melting and even localized influences of Indian plate fluids win young granite sheets close to Nanga Parbat (e.g. George et al. 1993; Clift et al. 2002). However, it is worth noting that Kohistan largely represents juvenile crust derived by mantle melting. This represents a primary addition to the continental crust of some 0.9 × 10^6 km^3 or c. 1/60 the size of the Ontong–Java Plateau (or Alaska), the world's largest ocean plateau (e.g. Neal et al. 1997), or similar to the net volume addition to the crust of arcs such as the Aleutians, the Marianas and Izu–Bonin (Dimalanta et al. 2002). Thus Pakistani Kohistan represents a significant single addition to Earth's crust during the Phanerozoic.

Table 5 is after a table of Miller & Christensen (1994) demonstrating that the bulk major element composition of Kohistan is very similar to that of the world's only other obducted and well-exposed full crustal section, the Talkeetna terrane in Alaska (Pearcy et al. 1990; Greene et al. 2006). Both

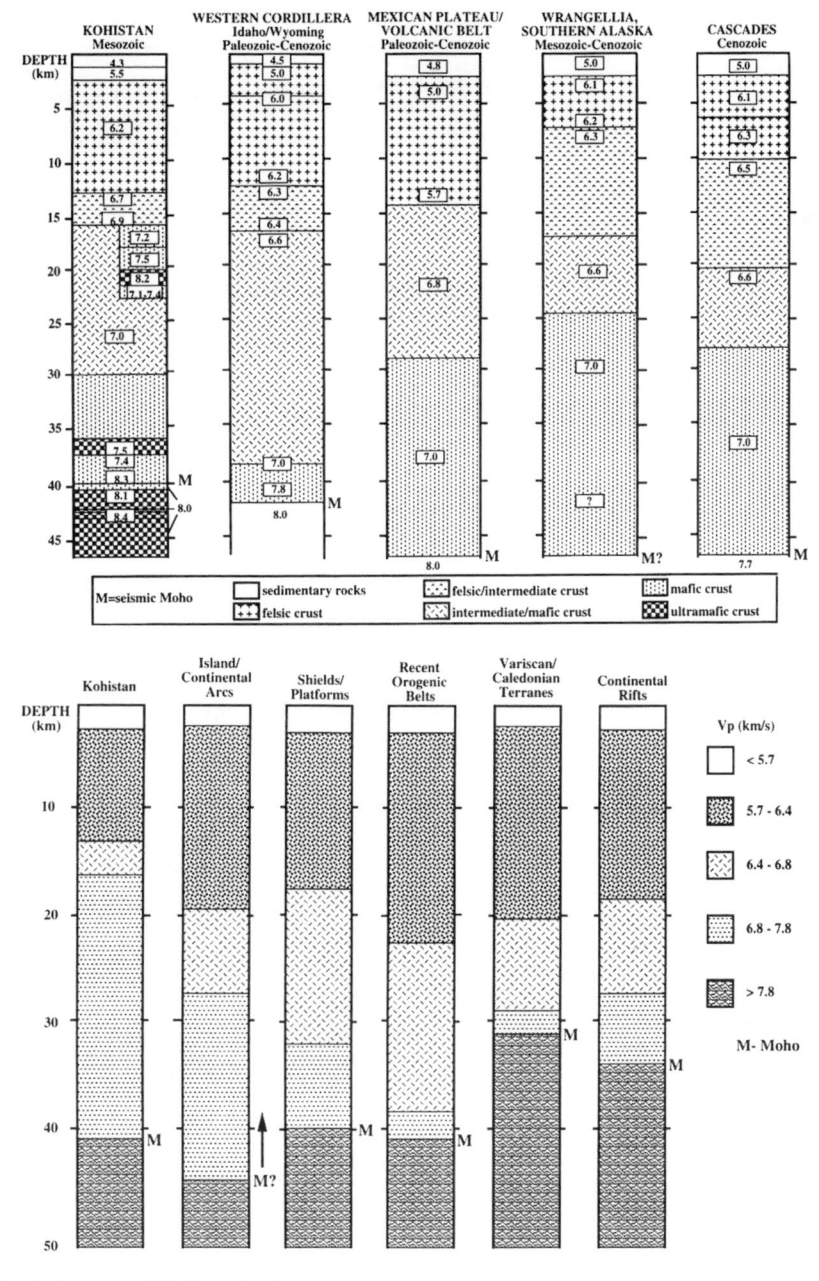

Fig. 6. Seismic Structure of Kohistan in relation to other types of crust [Note the dominance of intermediate–higher-velocity crust and thick layers of mafic–ultramafic crust within Kohistan (from Miller & Christensen 1994, acknowledgements American Geophysical Union).]

terranes are basic rather than intermediate with a SiO_2 content of only 51%, high levels of FeO_T, MgO and CaO, and relatively low levels of Na_2O and K_2O when compared with the most widely accepted compositional estimates of average continental crust.

Figures 8–10 illustrate the predominant trace element signature of the Jijal Complex, the Kamila Amphibolites and the various volcanic units of Kohistan. The general picture is one of enrichments in low field strength elements (LFSE; e.g. Rb, Ba, Sr, K, etc.) and light rare earth elements (LREE;

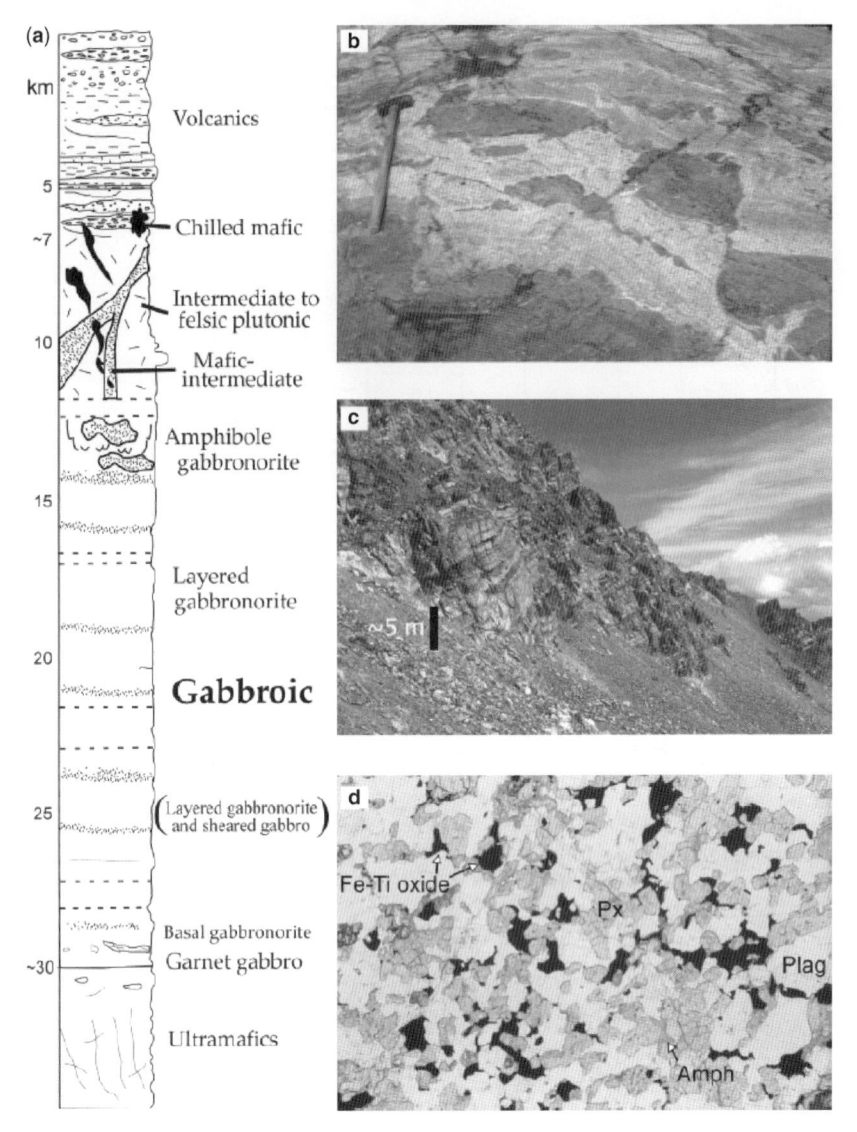

Fig. 7. (a) A crustal cross-section through the obducted Talkeetna arc terrane, which has remarkable similarities to the composition of Kohistan, including the presence of thick mafic middle–lower crust (Greene *et al.* 2009, acknowledgements Oxford University Press). Parts (**b–d**) illustrate details of the upper central and lower parts of the section, respectively.

e.g. La, Ce, Nd, etc.) relative to high field strength elements (HFSE; e.g. Y, Nb, Ti, Zr, etc.) and heavy rare earth elements (HREE; e.g. Tm, Yb, Lu), and features such as negative Nb anomalies. such trace element characteristics are widely interpreted as being typical of magmas produced at subduction zones. There are exceptions to this rule. Parts of some units have a geochemistry more typical of oceanic environments (e.g. mid-ocean basalt); an example here is about 30% of the Kamila amphibolites so far analysed (e.g. Khan *et al.* 1993). Others, however, are candidates for back-arc environments (e.g. part of the Hunza Formation, Petterson & Treloar 2004). There is also evidence for crustal melting, the best example being a component of the granite sheets that crop out close to Nanga Parbat (e.g. Petterson & Windley 1991; George *et al.* 1993). Petterson & Windley (1991) also suggested that the *c.* 100 Ma Matum–Das type trondhjemites that have a diagnostic low-K, flat REE geochemistry could be formed through melting of primitive crust such as

Table 3. *Summary of key geochemical, petrological and isotopic parameters*

Geological unit	Key isotopic parameters	Key geochemical–petrological parameters	References
Teru/Shamran Volcanics	$^{87}Sr/^{86}Sr_i$: 0.70243–0.7048819 $^{143}Nd/^{144}Nd_i$ 0.51281–0.51289 εNd: 3.29–5.42 $^{206}Pb/^{204}Pb$: 18.15–18.70 $^{207}Pb/^{204}Pb$: 15.54985–15.87 $^{208}Pb/^{204}Pb$: 37.99–38.82989	Highly evolved mature arc–Andean margin geochemical signatures (e.g. LREE/HREE $\gg$ 1). Dominated by ignimbrite, rhyolite, volcanic breccias and silicic volcaniclastic rocks.	Danishwar *et al.* 2001 Khan *et al.* 2004, 2009 Petterson & Treloar 2004 Bignold *et al.* 2006
Dir Group		Highly evolved mature arc–Andean margin geochemical signatures (e.g. LREE/HREE $\gg$ 1). Dominated by ignimbrite, rhyolite, volcanic breccias and silicic volcaniclastic rocks. Lower subaqueous turbidite unit	Sullivan *et al.* 1993 Treloar *et al.* 1996
Yasin sediments		Mixed sequence of sandstones, siltstones, mudstones, turbidites, carbonates, volcaniclastic rock, basalt–andesite lavas.	Pudsey 1985 Pudsey *et al.* 1986 Robertson & Collins 2002 Petterson & Treloar 2004
Chalt Volcanic Group	$^{87}Sr/^{86}Sr_i$: 0.70408–0.70559 $^{143}Nd/^{144}Nd_i$: 0.51296–0.51303 εNd: +6.24– +8.01 $^{206}Pb/^{204}Pb$: 18.091–18.345 $^{207}Pb/^{204}Pb$: 15.518–15.569 $^{208}Pb/^{204}Pb$: 38.048–38.340	Hunza Fm: boninite-like high–Mg and low-Mg basalt–basaltic andesite dominated, with andesites, rhyolites and minor thin ignimbrites. Ghizar Fm: low-Mg, mainly arc-like, basalt–andesite dominated, large areas tuff-dominated, proximal and distal facies, one proximal Strombolian–Vulcanian centre	Petterson & Windley 1991 Petterson *et al.* 1991a, b Treloar *et al.* 1996 Khan *et al.* 1997 Bignold & Treloar 2003 Petterson & Treloar 2004 Bignold *et al.* 2006
Kohistan batholith	$^{87}Sr/^{86}Sr$: 0.7032–0.7054 $^{143}Nd/^{144}Nd$: 0.51261–0.51293 εNd: +0.1 – +6.91 $^{206}Pb/^{204}Pb$: 18.52–18.72 $^{207}Pb/^{204}Pb$: 15.53–15.64 $^{208}Pb/^{204}Pb$: 38.48–38.95	Bimodal stage 1 plutonic rocks: arc-like (LREE/HREE $\gg$ 1) gabbroic diorites, SiO_2 45–60% and low-K high–SiO_2 (>70%) trondhjemites (LREE/HREE = 1); Stage 2: arc–Andean margin-like (LREE/HREE $\gg$ 1) medium–K gabbros to granites; Stage 3: high–SiO_2 granite and leucogranite sheets with Andean and crustal melt affinities	Petterson & Windley 1985 George *et al.* 1993 Petterson *et al.* 1993 Khan *et al.* 1997
Jaglot Sedimentary Group	$^{87}Sr/^{86}Sr_i$: 0.70323–0.7045 (0.70791) $^{143}Nd/^{144}Nd_i$: 0.51230–0.51286 (0.51230) εNd: (−6.93) +4.0 to +6.44 $^{206}Pb/^{204}Pb$: 18.192–18.504 $^{207}Pb/^{204}Pb$: 15.532–15.64 $^{208}Pb/^{204}Pb$: 38.182–38.651	Meta–sediment (psammites, pelites, semipelites, turbidites) dominated, with local sequences of arc-basalt–basaltic andesite (LREE/HREE > 1) and MORB–back-arc affinities (LREE/HREE < 1)	Treloar *et al.* 1996 Bignold & Treloar 2003 Bignold *et al.* 2006 Khan *et al.* 2007

(*Continued*)

Table 3. Continued

Geological unit	Key isotopic parameters	Key geochemical–petrological parameters	References
Chilas Complex	$^{87}Sr/^{86}Sr_i$: 0.70396–0.70414 $^{143}Nd/^{144}Nd_i$: 0.51274–0.51280 εNd: +2.9–+8.4 $^{206}Pb/^{204}Pb$: 18.526–18.574 $^{207}Pb/^{204}Pb$: 15.623–15.640 $^{208}Pb/^{204}Pb$: 38.671–38.783	85% 'gabbronorite association', homogeneous mafic-dominated, arc-like (LREE/HREE > 1); remainder dominated by 'ultramafic association' of dunites, peridotites, pyroxenites, etc.; arc–like magma chemistry	Khan et al. 1989, 1993, 1997 Jagoutz et al. 2006
Kamila Amphibolites	$^{87}Sr/^{86}Sr_i$: 0.70363–0.70467 $^{143}Nd/^{144}Nd_i$: 0.51274–0.51302 εNd: +2.03–7.40 $^{206}Pb/^{204}Pb$: 17.973–18.469 $^{207}Pb/^{204}Pb$: 15.456–15.611 $^{208}Pb/^{204}Pb$: 37.951–38.644	Predominantly basic–intermediate volcanic and plutonic rocks with smaller volume of acid material. Two main geochemical types: (1) MORB–like (LREE/HREE < 1); (2) arc-liKe (LREE/HREE > 1). Dhuime et al. (2009) recognized five compositional types and a separate intrusive granite phase	Jan 1988 Treloar et al. 1990, 1996 Khan et al. 1993, 1997 Bignold & Treloar 2003 Bignold et al. 2006 Dhuime et al. 2009
Jijal Complex	$^{87}Sr/^{86}Sr_i$: 0.70355–0.70500 $^{143}Nd/^{144}Nd$: 0.51271–0.51293 εNd: (100 Ma): +5.8–+8.6 $^{206}Pb/^{204}Pb$: 18.144–18.595 $^{207}Pb/^{204}Pb$: 15.49–15.577 $^{208}Pb/^{204}Pb$: 38.216–38.799	Lower ultramafic and upper mafic section. Probable exposure of petrological and seismic Moho with sequence. Part mantle, part lower crust. Arc affinities	Khan et al. 1997 Dhuime et al. 2007 Garrido et al. 2007

Table 4. *Whole–Kohistan isotopic composition range*

Parameter	Range
$^{87}Sr/^{86}Sr_i$	0.70243–0.70559 (0.70791)
$^{143}Nd/^{144}Nd_i$	0.51230–0.51303 (51330)
εNd	+0.1 (−6.93) − +8.6
$^{206}Pb/^{204}Pb$	17.973–18.70
$^{207}Pb/^{204}Pb$	15.518–15.87
$^{208}Pb/^{204}Pb$	38.048–38.95

References are given in Table 3; parentheses indicate one rather extreme sample for main compositional field.

the Kamila Amphibolites. Dhuime *et al.* (2007) (Fig. 9) presented multi-element trace element plots of the Jijal Complex (dunites, wehrlites, websterites, clinopyroxenites, garnet-rich and garnet-poor clinopyroxenites, hornblendites and hornblende gabbronorites). The basal ultramafic rocks (dunites, wehrlites and clinopyroxenites) show high Mg numbers and highly depleted REE patterns, no marked HFSE anomalies and high LFSE enrichments. The overlying mafic rocks display lower Mg numbers, enriched REE patterns, and marked HFSE anomalies. The ultramafic and mafic rocks are interpreted to have originated from different sources. The ultramafic rocks sampled the initiation stages of subduction and produced boninitic-type magmas whereas the mafic magmas have been influenced by more mature arc-subduction processes. Dhuime *et al.* (2009, and Fig. 9a, b) has recognized two distinct geochemical suites and five geochemically distinct compositional units within the Kamila Amphibolites that record increasing subduction inputs with time, granulite-grade metamorphism at 105–99 Ma and arc-root melting to produce intrusive granites. Kamila Amphibolites magmas evolve from a LREE- and LFSE-depleted magma to a highly evolved LREE- and LFSE-enriched high Ce/Y magma, reflecting the increasing influence of subduction-related inputs. The Chilas Complex is dominated (85%) by a relatively slightly LREE- and LFSE-enriched gabbronorite and a depleted ultramafic unit, probably representing a single large-volume melt event perhaps related to a sub-arc plume. The overlying volcanic units (Fig. 11) record typical arc-like evolved volcanism and penecontemporaneous boninitic style eruptions.

Figure 12 is a summary of key trends with time for $^{87}Sr/^{86}Sr_i$, SiO_2, Zr_N/Nb_N and Ce_N/Yb_N (N denotes that the values have been normalized relative to chondrite and/or mid-ocean ridge basalt (MORB): Ce and Yb relative to chondrite; Zr and Nb relative to MORB). In general, the composition of Kohistan units becomes progressively more evolved with time in terms of increasing SiO_2, $^{87}Sr/^{86}Sr_i$ and LFSE/HFSE ratios as represented by Ce/Yb and Zr/Nb. As the terrane ages, the proportion of andesite and rhyodacite increases, culminating in the later granitic stages of the Kohistan batholith and the rhyodactic cover volcanic units (Dir Group and Shamran/Teru volcanic rocks). The general increase in $^{87}Sr/^{86}Sr_i$ reflects an increasing subduction influence with time, a gradual Kohistan crustal evolution as the crust ages and differentiates (higher Rb/Sr crust producing radiogenic Sr with time) and, possibly, in the latest stages the involvement of older, non-Kohistan basement. Increasing SiO_2 with time reflects the maturing and thickening arc or Andean margin as crustal melts are tapped and/or magmas are forced to fractionate further as they pass through thicker crust. The penecontemporaneity of magmas with different characteristics (i.e. high-SiO_2, high

Table 5. *Bulk Crustal composition of Kohistan compared with estimates of continental crust (after Miller & Christensen 1994)*

Composition (wt%)	Kohistan	Average crust (Taylor & McLennan 1981)	Average crust (Weaver & Tarney 1984)	Average crust (Clarke 1924)	Talkeetna obducted arc terrane (Pearcy *et al.* 1990)
SiO_2	51.0	58.0	63.2	59.0	51.1
TiO_2	1.1	0.8	0.6	1.0	0.7
Al_2O_3	15.9	18.0	16.1	15.2	15.0
FeO_T	10.3	7.5	4.9	6.8	9.5
MgO	8.8	3.5	2.8	3.5	11.2
CaO	9.5	7.5	4.7	5.1	9.2
MnO	0.2	0.1	0.1	0.1	0.2
Na_2O	2.6	3.5	4.2	3.7	2.5
K_2O	0.6	1.5	2.1	3.1	0.5
P_2O_5	0.2		0.2	0.3	0.1

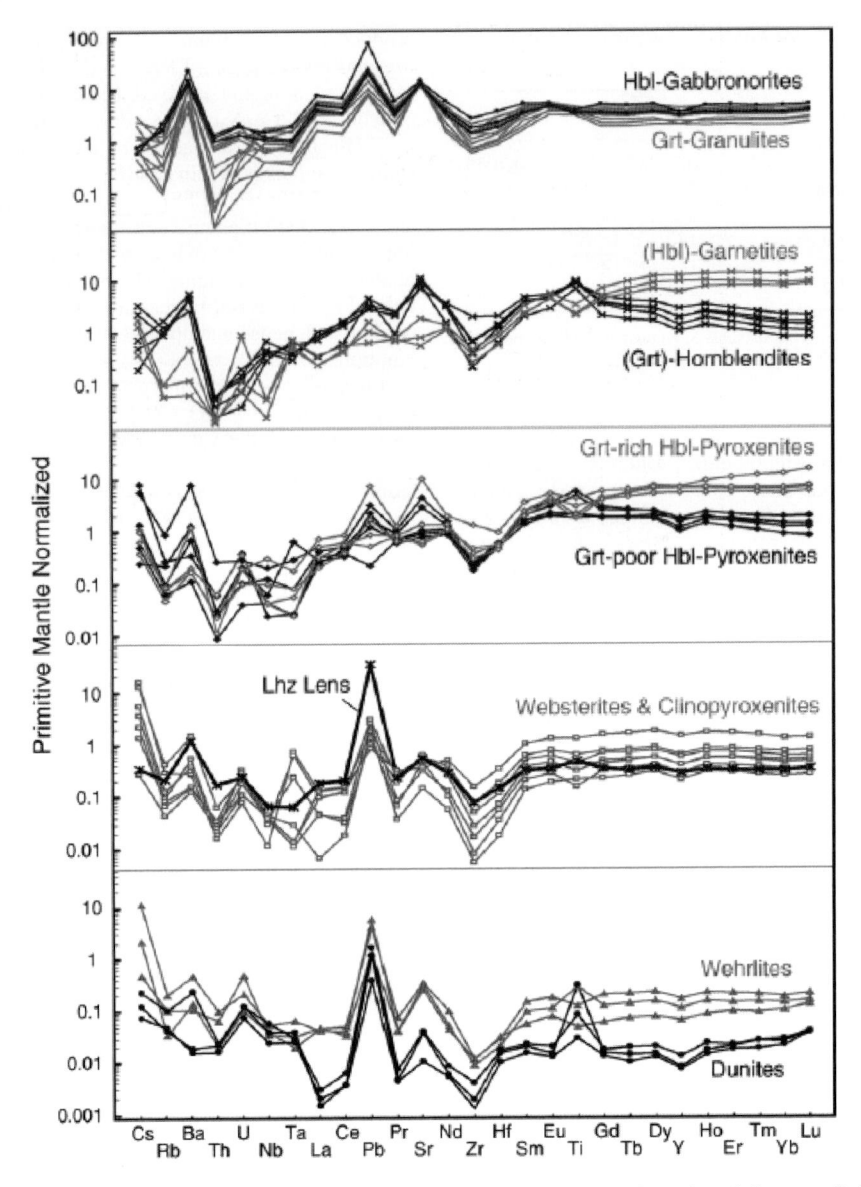

Fig. 8. REE and multi-element composition of the Jijal Complex (Dhuime *et al.* 2007, acknowledgements Oxford University Press).

Ce/Yb, low Zr/Nb with low SiO$_2$, low Ce/Y, high Zr/Nb) suggests that at any one time a range of source compositions were being tapped, a common theme in many researchers' work (e.g. Bignold *et al.* 2006). In broad terms, comparisons with modern arcs suggest that most Kohistan units are similar to Izu–Bonin, Tonga and the Marianas, with the more evolved parts of the Kohistan batholith resembling mature or continental arcs such as Japan or the Aegean (Clift *et al.* 2002).

Section 2: Discussion

Character of the Kohistan Arc?

Chroston & Simmons (1989) compared Kohistan with a range of arcs in terms of crustal seismic structure (e.g. Lesser and Greater Antilles, Japan, New Ireland, New Britain, Bali, Sunda, Aleutians, Kermadoc, Izu–Marian as and South Sandwich). Mature arcs are 25–30 km thick, with two (albeit

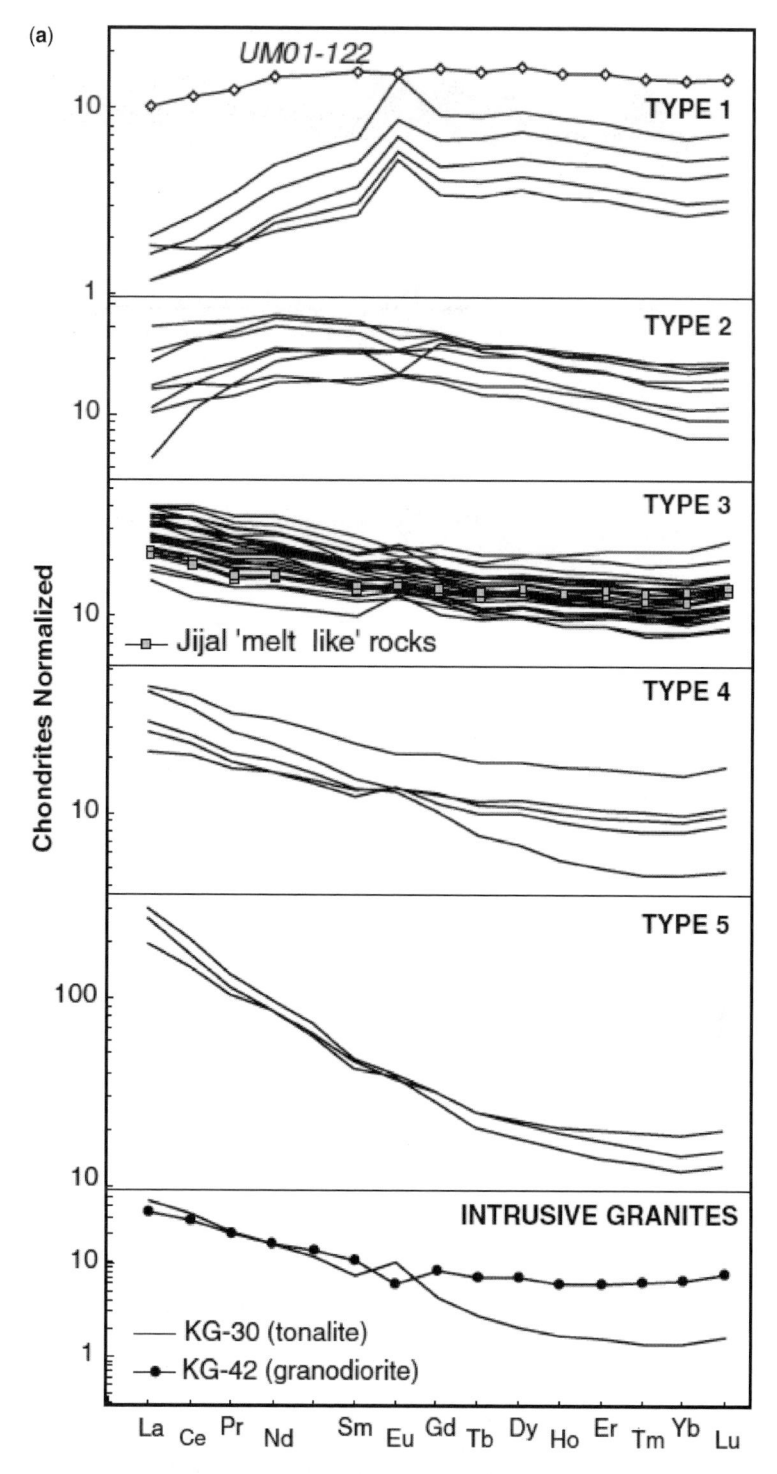

Fig. 9. REE and multi-element composition of the Kamila Amphibolites (Dhuime *et al.* 2009, acknowledgements Oxford University Press). Patterns referred to as types 1–5: type 1, gabbros; type 2, ultramafic–gabbro rocks; type 3, ultramafic–diorite rocks with variable Mg numbers; type 4, more differentiated compositions; type 5, differentiated gabbros with Mg number between 42 and 48. The diagrams record increasing inputs from subduction processes with time.

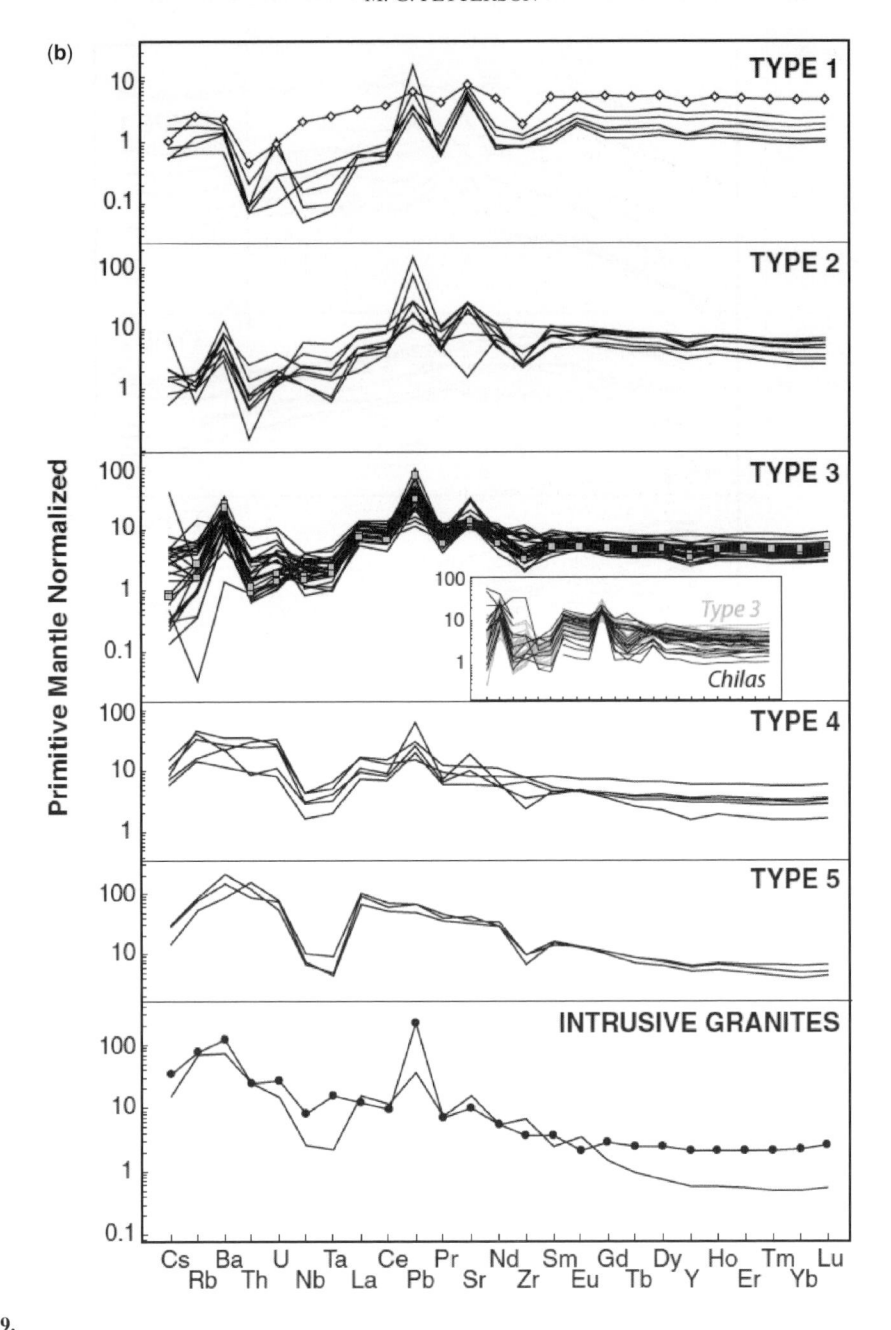

Fig. 9.

variable) main velocity layers: a lower crust with V_p between 6.6 and 7.1 km s^{-1} (average 6.9 km s^{-1} with thicknesses varying between 10 and 20 km) and upper crust with V_p 5–6.7 km s^{-1} (uppermost crust may have values as low as 1.5–3 km s^{-1}) some 8–15 km thick. In detail, arc structure varies considerably; an immature basic arc such as the South Sandwich Islands has a crustal thickness of c. 20 km with an upper layer, 5 km thick, with $V_p = 2$–5.3 km s^{-1} and a lower layer, 15 km thick, with $V_p = 6.9$–7.5 km s^{-1}, whereas a mature arc such as Japan is c. 35 km thick with

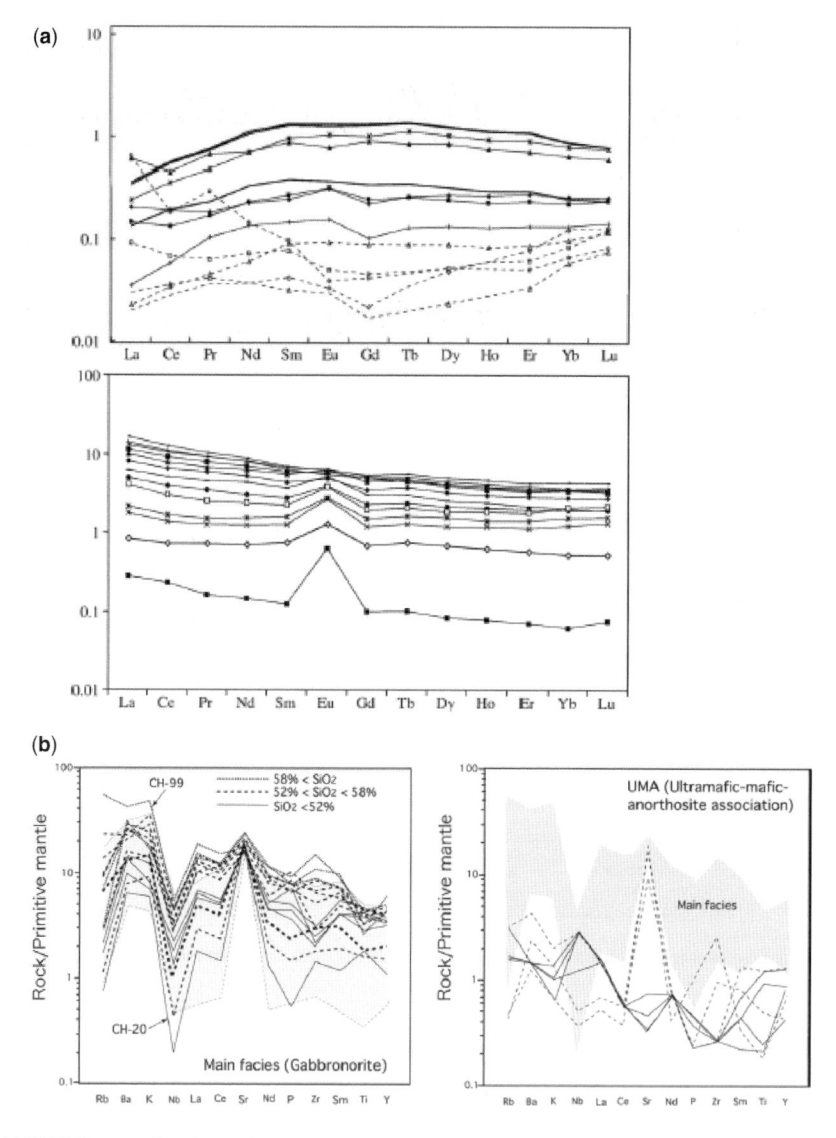

Fig. 10. (a) REE diagrams for ultramafic rocks and gabbro-norites, and (b) multi-element plots for gabbro-norites and ultramafic rocks from the Chilas Complex (*a*, Jagoutz *et al.* 2009, acknowledgements Elsevier; *b*, Takahashi *et al.* 2007, acknowledgements Elsevier).

comprising an upper 20 km with V_p 5.5–6 km s^{-1} and a lower 20 km with $V_p = 6.6$–6.8 km s^{-1}. Kohistan compares most closely, in terms of seismic structure, to arcs such as the Lesser and Greater Antilles, Japan and New Ireland, all of which have c. 35 km thick crust. In general, in comparison with those arcs, Kohistan has material with much higher V_p (up to 7.8 km s^{-1}) probably as a result of the high proportion of granulite material.

Kohistan volcanic evolution

The earliest part of the Kohistan extrusive story is recorded by the Kamila Amphibolites. Khan *et al.* (1993) and others identified two key types of volcanic rocks within the Kamila Volcanic sequence that changed the way this unit was viewed. A minor part of the Kamila sequence has geochemical characteristics more akin to main ocean rather than arc whereas the bulk of the

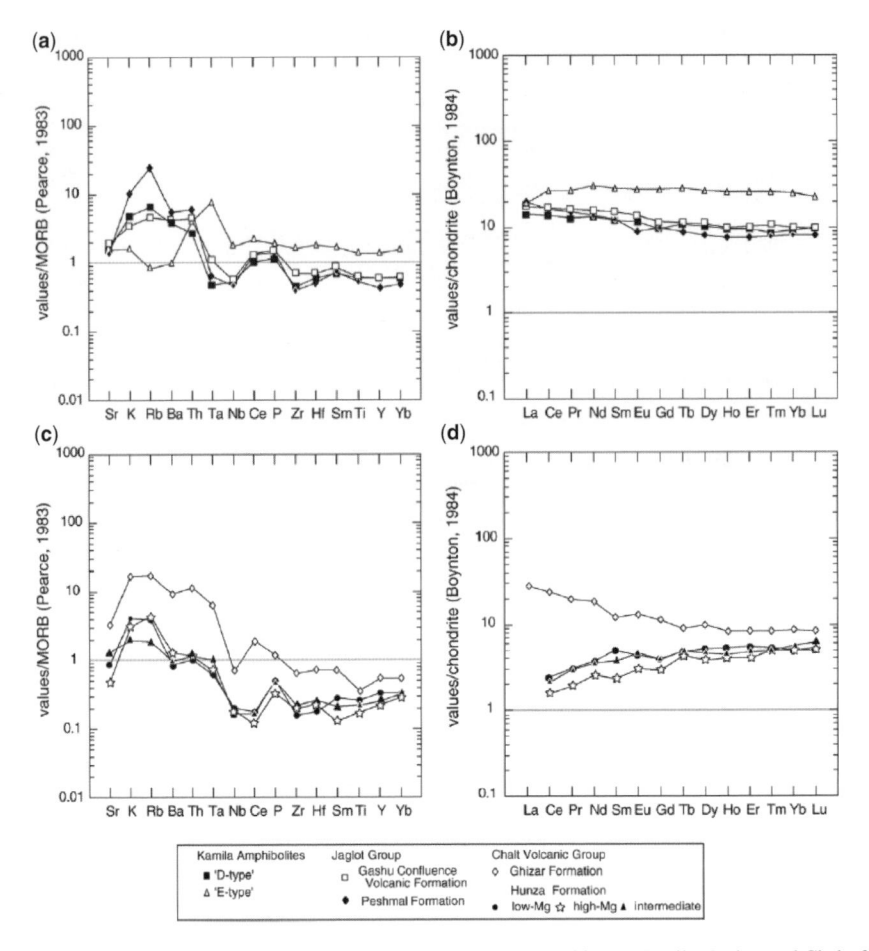

Fig. 11. Typical trace element patterns from metavolcanic sequences of Kohistan (Kamila, Jaglot, and Chalt; from Bignold *et al.* 2006, acknowledgements Elsevier).

Kamila Volcanic sequence represents arc material. Treloar *et al.* (1996) suggested that this reflects arc growth at an intra-oceanic site with arc volcanic rocks being emplaced upon, within, and interdigitating with oceanic basalts. Kamila volcanism was a mixture of quietly and explosively extrusive, subaqueous, basaltic–andesitic activity accompanied by plutonic emplacement and pyroclastic activity. Dhuime *et al.* (2009) demonstrated an evolving source and increasing subduction influence within the Kamila Amphibolites with time. As the Kamila volcanic phase waned, the arc underwent a period of extension, forming the Jaglot Group basin, which culminated in the largely subaqueous Hunza Formation boninitic, basaltic andesitic lava dominated with later pyroclastic and rhyolitic volcanism (Fig. 13a). Synchronous with the Hunza Formation was the largely subaerial Strombolian–Vulcanian proximal to distal explosive eruptions of basaltic andesite and andesite with local lava-dominated centres as represented by the Ghizar Formation (Fig. 13). Plutonism became important as the earliest stages of the batholith intruded into the Chalt Volcanics. Finally volcanism waned and siliciclastic and turbiditic sedimentation took over, with occasional lavas, volcaniclastic eruptions and periods of lower-energy carbonate deposition (recorded in the Yasin sediments).

The latest stages of volcanism in Kohistan record highly silicic ignimbritic and pyroclastic explosive volcanism as exemplified by the Shamran/Teru Volcanic Group in NW Kohistan (Fig. 13b–d). This highly evolved stage of volcanism is interpreted as a late to post-collisional stage of volcanism with a thickened mature arc crust.

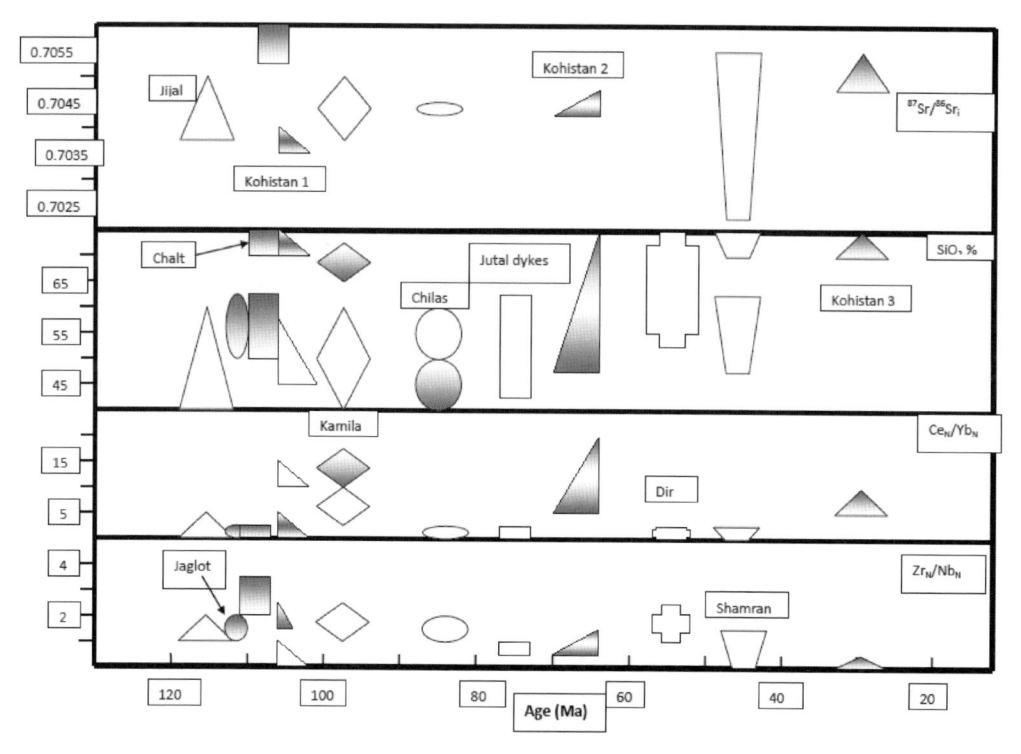

Fig. 12. Time–geochemical plot for the key lithostratigraphic units within Kohistan. Noteworth, features are the the overall increase in $^{87}Sr/^{86}Sr_i$ and SiO_2, and decrease in Zr_N/Nb_N with time, reflecting increasing subduction-related components with or without the involvement of older crust in later times, and the coexistence at various times of high SiO_2 and high Ce_N/Yb_N rocks with low SiO_2 and low Ce_N/Yb_N rocks, indicating the presence of a range of sources at any one time.

Chilas and Jijal palaeoenvironments

The two mafic to ultramafic complexes have, perhaps, been the most difficult units to interpret within a holistic model of Kohistan. Coward et al. (1982b) suggested that Chilas and Jijal were the same body connected at depth and had been caught up in complex deformation events. Age data now clearly show that this is not possible, as the most recent age for the original emplacement of Jijal is 117 or 118 Ma with a c. 95 Ma age for granulite-facies metamorphism (see Table 1) whereas the Chilas Complex has been dated on numerous occasions at around 85 Ma (Table 1 and references therein). Most workers (e.g. Jan 1988; Jan et al. 1993; Yamamoto 1993; Khan et al. 1998; Yamamoto & Yoshini 1998; Yoshino & Okudaira 2004; Garrido et al. 2006; Dhuime et al. 2007) have agreed that the Jijal Complex and related units represent part mantle, part crust. Where the petrological–seismic Moho is actually situated is a matter of debate, but it most probably coincides with an absence of plagioclase, a predominant upper granulite-facies metamorphic grade, and $V_p > 8$ km s^{-1}. A literature consensus (see above) concerning the original environment of the Jijal Complex indicates an origin at depths of 30–50 km, within a suprasubduction-zone environment affected by partial melting, granulite-grade metamorphism, magma mixing, magma–subsolidus crystalline rock interaction, fluid–rock interaction, and cumulate activity. Magmatic-dominated processes were dominant at c. 117 or 118 Ma and metamorphic-dominated processes at c. 95 Ma. It is the very root of an island arc forming before any subsequent Andean and/or continent collision stages.

The Chilas Complex is more enigmatic and is a class of its own, from a Kohistan perspective, being such a large-volume mafic intrusion comparable in size with other global large-volume eruptions. Until Treloar et al. (1996) published their paper the Chilas Complex was widely thought to be part of the earliest story of Kohistan, together with the Jijal Complex. However, it is apparent that the Chilas Complex has intrusive relationships with both the Kamila Amphibolites to the south and the Jaglot Group to the north, and that much

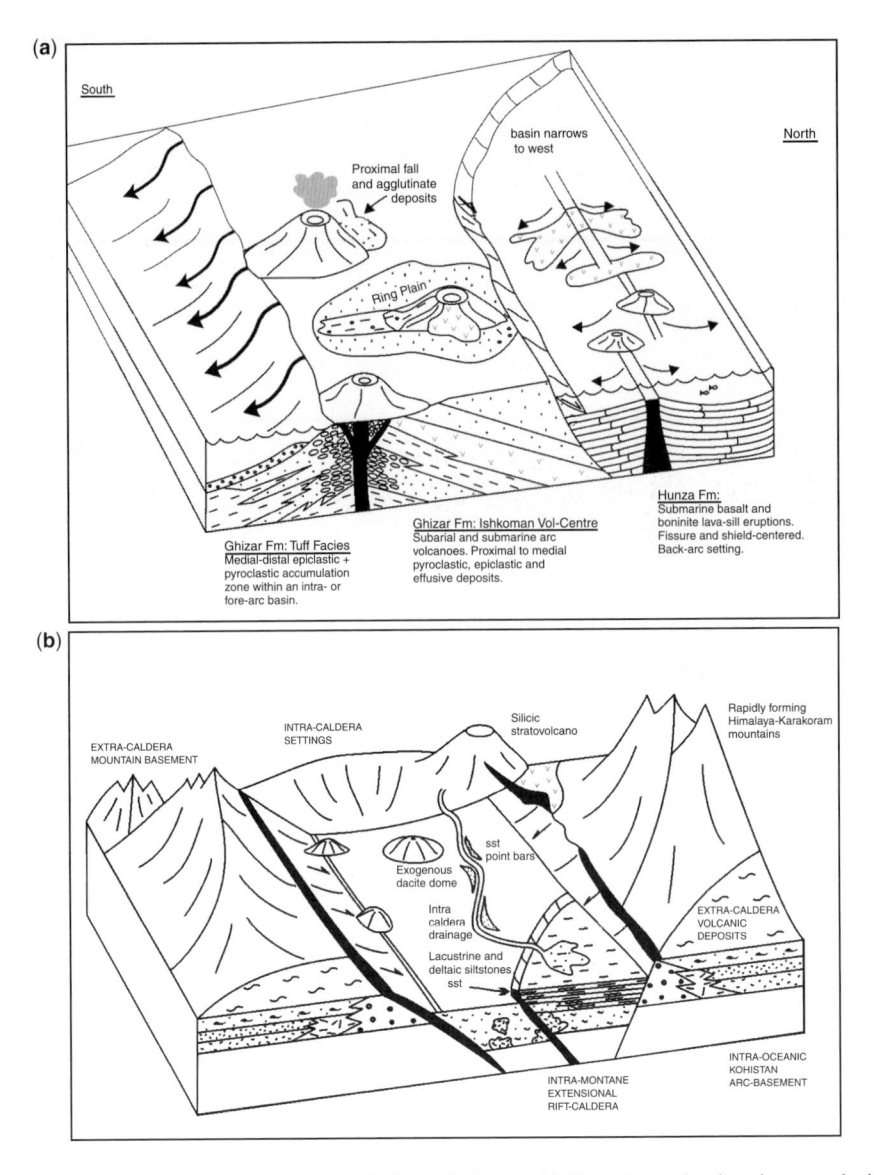

Fig. 13. (**a**) Palaeogeographical setting of the Chalt Volcanic Group with Hunza Fm volcanic units erupted within back-arc basins and the Ghizar Fm erupted subaerially within a stratovolcanic environment (from Petterson & Treloar 2004, acknowledgements Elsevier). (**b**) Palaeogeographical setting of the Shamran/Teru Volcanics indicating silicic volcanic eruptions within an intramontane basin (from Petterson & Treloar 2004, acknowledgements Elsevier). (**c**) Geological sketch map of the Shamran/Teru Volcanic Group and its unconformable relationship with the Chalt–Ghizar Volcanic Group and Kohistan batholith (from Petterson & Treloar 2004, acknowledgements Elsevier). (**d**) Typical lithologies from the Shamran/Teru Volcanic Group exhibiting highly silicic pyroclastic and volcaniclastic textures (from Petterson & Treloar 2004, acknowledgements Elsevier).

of it exhibits a lower scale of penetrative deformation relative to other Kohistan basement units. Additionally, its age (c. 85 Ma) is significantly younger than that of the Jijal Complex, granulite-facies metamorphism in the Kamila and Jijal units (c. 95 Ma) and the oldest parts of the batholith (c. 102 Ma). The Chilas Complex intruded as a huge magmatic body after much of Kohistan was already formed and during the later stages of the first deformation period, perhaps as the result of a plume within an extensional back-arc to intracontinental environment. Some workers (e.g.

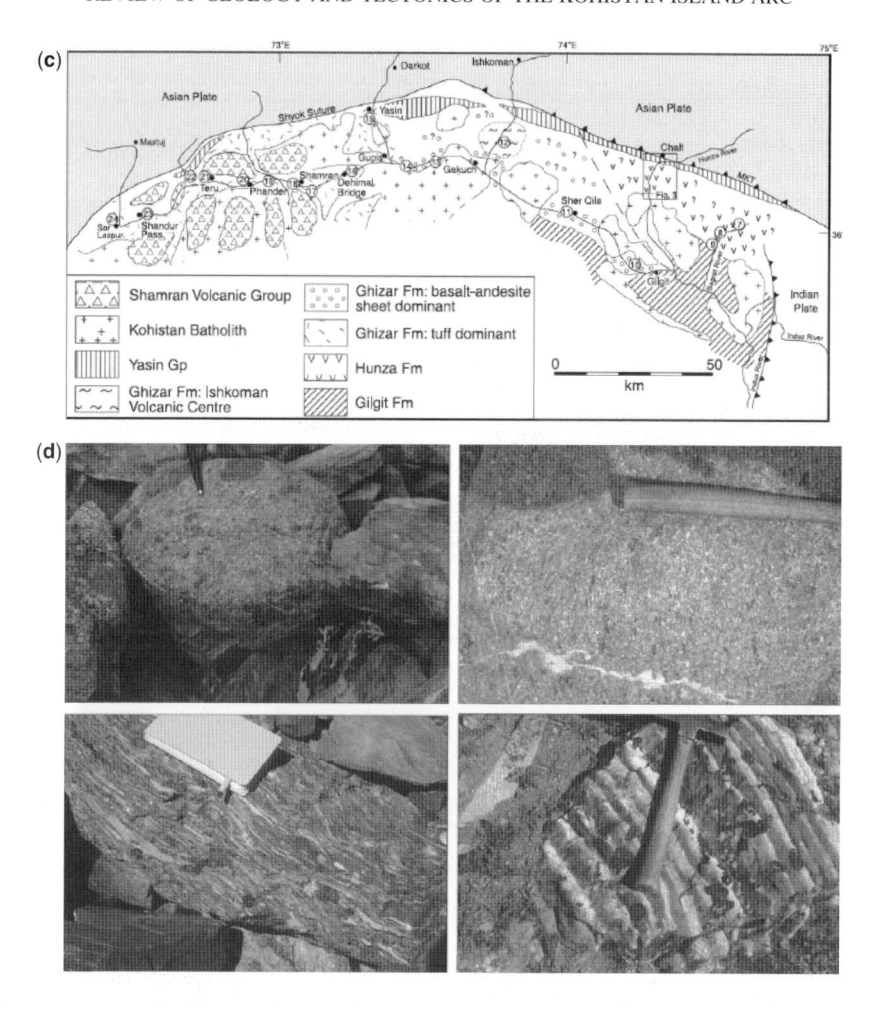

Fig. 13.

Burg et al. 1998; Jagoutz et al. 2006) interpreted the ultramafic association exposed in the deeply eroded parts of the Complex, close to Nanga Parbat, as mid-crustal mantle diapirs that fed the Chilas Complex within kilometre-scale mid-crustal channels (e.g. Burg et al. 1998, 2005).

The intrusion of such a large body of c. 1100–1200 °C basic magma at mid-crustal levels together with the input of latent heat of crystallization as the plutonic body solidified had a phenomenal impact on the temperature gradient of Kohistan at c. 85 Ma. This has led to researchers attempting to model the impact of the emplacement of the Chilas Complex alongside other parameters to investigate the thermal structure of Kohistan (e.g. Yoshino & Okudaira 2004; Yoshino et al. 1998). The Kamila Amphibolites are at their lowest grade in the centre of the unit and approach granulite grade towards both the Jijal and Chilas Complexes, respectively below and above the Kamila rocks. Both the Jijal and Chilas Complexes are essentially at granulite facies: The Jijal Complex achieved granulite grade at c. 95 Ma and the Chilas Complex as it intruded. Southern Kohistan cooled through 500 °C at c. 80 Ma. In broad terms the Chilas Complex would have formed a mid-crustal 'hot zone' in addition to the already warm-hot lower crust–mantle Jijal Complex, with the Kamila Amphibolites sandwiched between the two hot layers and the upper crust absorbing this high thermal anomaly through conduction, localized melting and fluid convection. It is interesting to note that this anomalously high geothermal gradient that existed at c. 85 Ma had cooled to below 500 °C within 5 Ma, suggesting that there were some rapid heat-advecting processes taking place within the Kohistan crust, an important topic for further research.

The Rheological Importance of the Jaglot Group and its importance in the evolution of Kohistan

The Jaglot Group represents a significant accumulation of relatively low-density volcano-sedimentary material that was highly deformed and locally melted during the first deformation stage. The density and rheological contrast between the Kamila Amphibolites and the uppermost crust as represented by the earliest batholith and Chalt–Yasin units made an appropriate crustal neutral-buoyancy level for the roof of Chilas to halt, fractionate and solidify, producing the current very thick lower crustal seismic character of Kohistan.

The Jaglot Group also served an extremely important role in the development of the arc stage of Kohistan. Coward et al. (1982a, b, 1986) recognized major synforms within the Jaglot Group with a half-wavelength of some 50 km. It is possible that the synformal structure reflects the original depocentre character of the Jaglot Group. Following the extrusion of the Kamila Volcanics, Kohistan experienced a protracted phase of extension that accommodated a thick clastic-dominated sedimentary sequence into which was extruded or intruded a number of local basalt–andesite centres (e.g. the Thelichi Formation) and dyke swarms interpreted as a back-arc feeder complex (Khan et al. 2007), and allowed for the formation of a number of highly extended, high-temperature back-arc and possible intra-arc basins that allowed the ingress of boninites and high-Mg andesites within the Chalt Volcanic Group.

Tectonic and crustal evolution of Kohistan

Ideas relating to the tectonic evolution of Kohistan fall into two camps. All researchers agree on the intra-oceanic origin for the arc but then divergence of views occurs. Some researchers argue for a model that involves initial collision between Eurasia and Kohistan followed by collision with India, whereas others argue for a 'deep-south' provenance of Kohistan in the southern hemisphere with collision first with India and later with Eurasia. This section will review the arguments for each of these models. Figure 14a–i shows illustrations of a range of researchers tectonic models.

Earliest stages of Kohistan evolution

Both models agree that the early arc stage of Kohistan was the main stage in terms of crustal evolution. Dhuime et al. (2009) suggested that subduction initiation began around 117 Ma at a boundary between older and younger oceanic crust, possibly at a transform fault, with the older, denser crust subducting beneath the younger crust. From 117 to c. 105 Ma subduction processes generated the Jijal Complex and the earlier arc volcanic rocks. Between 105 and 99 Ma (Fig. 14a–c) there was a period of gabbroic magmatic underplating, granulite-facies metamorphism and the increasing generation of felsic batholithic and volcanic rocks. Magmatic activity decreased between c. 95 and 85 Ma, although this period ended with the intrusion of the massive Chilas Complex. Garrido et al. (2006) similarly described a model whereby the early arc matured as a result of: (1) intrusions into pre-existing oceanic crust; (2) further intrusions into the middle-arc crust and intra-arc crustal differentiations; (3) fractionation of andesitic melts within closed systems, which generated a range of amphibole-bearing rocks; (4) younger shallow intrusions, which depressed old intrusions into the roots of Kohistan; (5) deep crustal hornblende plutonic rocks undergoing dehydration melting (when Kohistan attained a thickness of c. 25–30 km), leading to garnet–pyroxene-bearing restites and felsic magmas, which also further differentiated the more mature arc crust, producing an essentially andesitic middle crust. A series of workers (e.g. Petterson & Windley 1985, 1992; Coward et al. 1986; Pudsey 1986; Treloar et al. 1996; Bignold et al. 2003, 2006; Petterson & Treloar 2004) have documented the evolution of the upper parts of the intra-oceanic arc. Volcanic sequences formed within a series of depocentres, some dominated by high-Mg andesitic style magmatism (e.g. Chalt Volcanic Group) and others by more typical lower-Mg andesite to rhyolite style volcanism (e.g. Ghizar Formation, Jaglot Volcanic Group). Volcanic rocks were overlain by carbonates and arc clastic turbiditic sediments. These upper crustal units were intruded by a bimodal sequence of trondhjemites and gabbro–diorites.

Model 1: initial collision with Eurasia and northerly dipping subduction beneath Kohistan

Arc stage and collision with Eurasia. This model has been promoted or quoted as a preferred tectonic model by workers such as Tahirkheli (1979), Tahirkheli & Jan (1979), Coward et al. (1982a, b, 1986), Petterson & Windley (1985, 1986, 1991, 1992), Pudsey et al. (1985a, b, 1986), Pudsey (1986), Coward et al. (1987), Treloar et al. (1989, 1990, 1996), Petterson et al. (1991a, b, 1993), Searle (1991), Sullivan et al. (1993), Kazmi & Jan (1997), Burg et al. (1998), Searle et al. (1999), Clift et al. (2002), Robertson & Collins (2002), Schaltegger et al. (2002), Bignold & Treloar

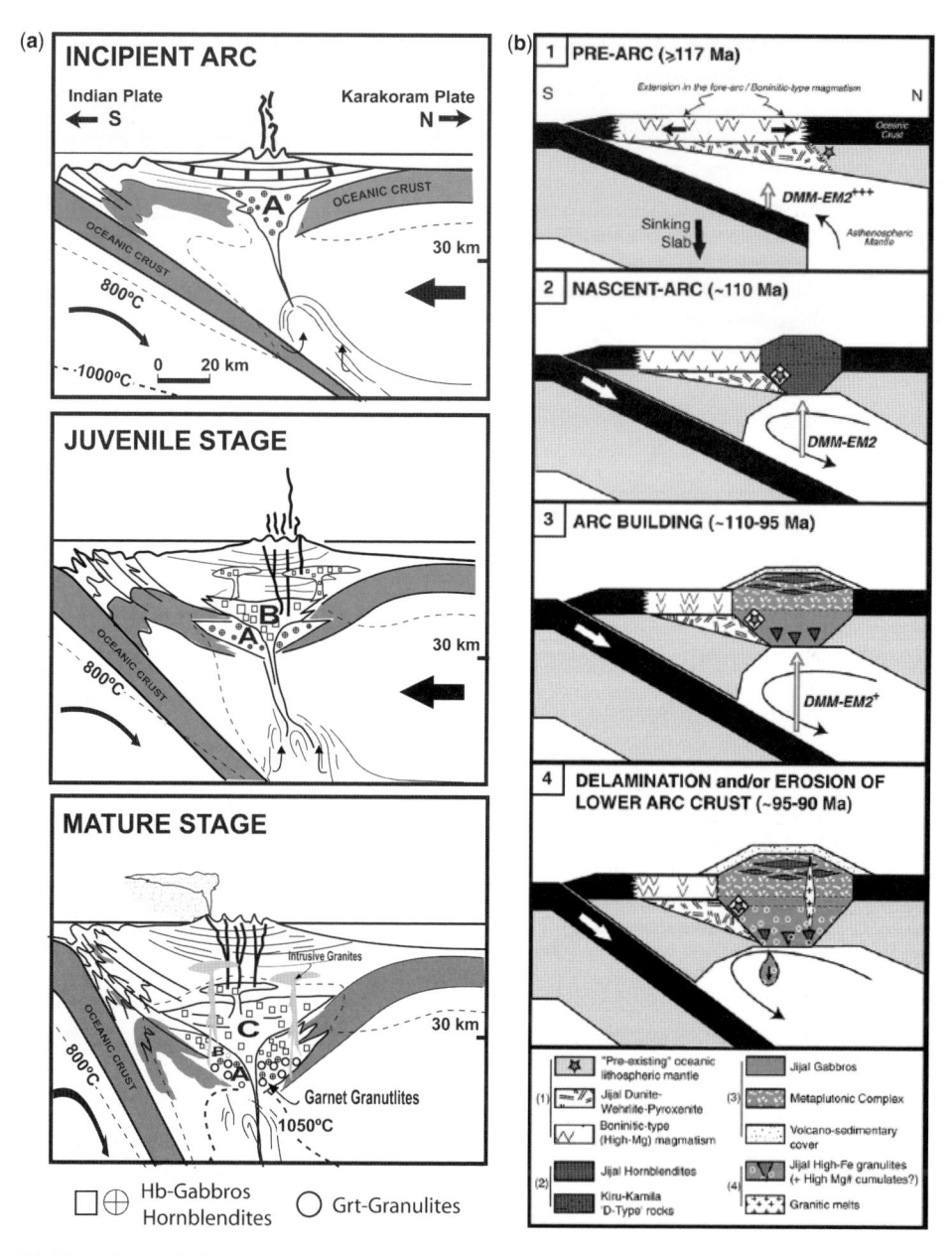

Fig. 14. Tectonic models for the evolution of Kohistan. (**a**) Early to mature arc growth stages showing the gradual thickening and underplating of Kohistan above an intra-oceanic subduction zone (Garrido *et al.* 2006, acknowledgements Oxford University Press). (**b**) Arc development from 117 to 90 Ma showing northward-directed subduction and early arc magmatic growth (Dhuime *et al.* 2007, acknowledgements Oxford University Press).

(2003), Petterson & Treloar (2004), Garrido *et al.* (2006, 2007) and Yin (2006). This list of references emphasizses the longevity of and wide support gained for this model (Fig. 14d–h).

The model begins with a period of intra-oceanic subduction that involved the Tethys oceanic plate subducting northwards beneath the southern margin of Kohistan. Kohistan was separated from Eurasia at this time by a relatively small ocean, possibly a back-arc ocean. A key difference from the competing model at this stage of tectonic development is the assertion that Kohistan evolved close

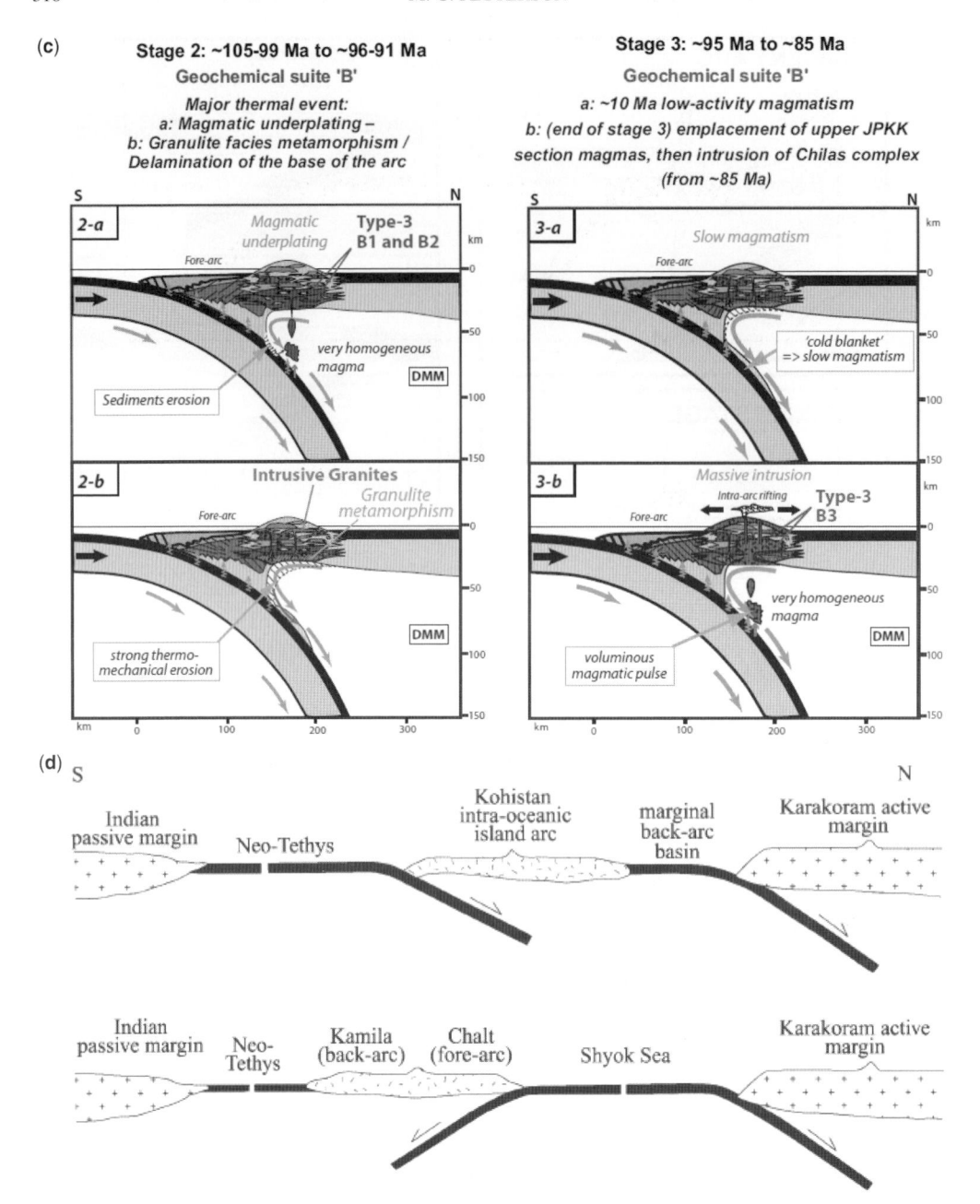

Fig. 14. (**c**) Arc development from 105 to 85 Ma showing northward-directed subduction, magmatic underplating, granulite-facies metamorphism, and the intrusion of the Chilas Complex (Dhuime et al. 2009, acknowledgements Oxford University Press). (**d**) (1) Model 1 as defined in the text with northward-directed subduction under Kohistan and a marginal basin between Kohistan and the Karakoram–Eurasian margin that subsequently closed at c. 84 Ma. (2) Elements of model 2 as discussed in the text with a Helmahera-style double-subduction style of tectonics beneath Kohistan and Eurasia (after Khan et al. 1997). (Bignold & Treloar 2003, acknowledgments Geological Society of London).

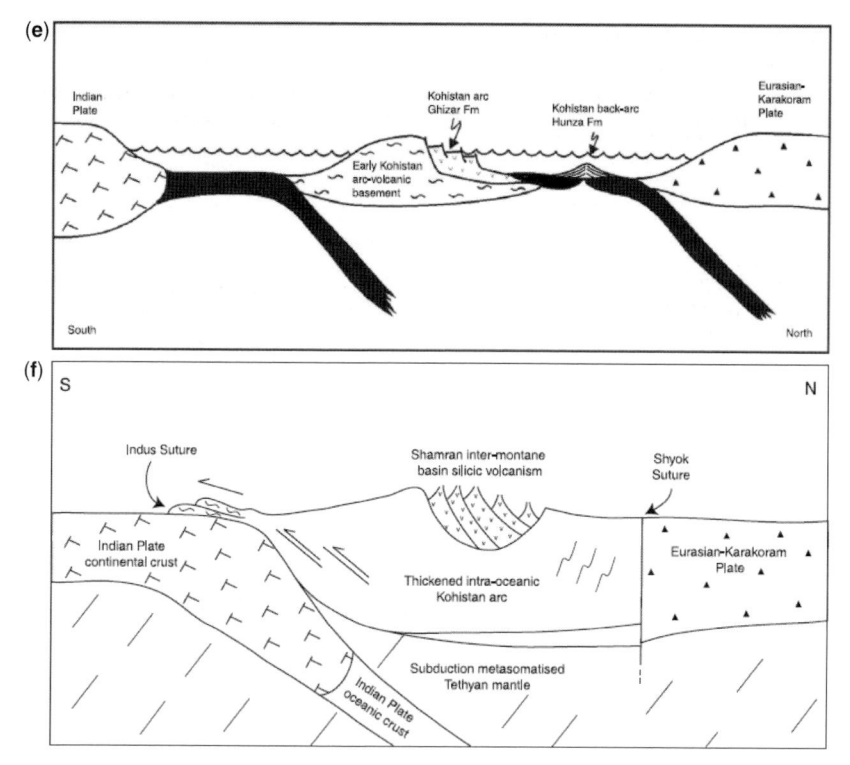

Fig. 14. (**e**) The High-Mg Hunza Formation Chalt Volcanic Group erupted within a back-arc basin with penecontemporaneous eruption of the low-Mg Ghizar Formation within an intra-arc tectonic setting. Both are situated above a northward-directed subduction zone (from Petterson & Treloar 2004, acknowledgements Elsevier). (**f**) The Shamran/Teru Volcanic Group erupted within a post-collisional tectonic setting after Kohistan had collided with both Eurasia and India (from Petterson & Treloar 2004, acknowledgements Elsevier).

to the equator or within the northern hemisphere. At around 85 Ma Kohistan collided with Eurasia and was transformed into an Andean continental margin. Finally India collided with an amalgamated Eurasia–Kohistan terrane during the Himalayan orogeny.

The first collision with Eurasia produced: (1) crustal-scale folding and shearing in the deeper ductile zones with dense faulting and thrusting in the shallower brittle zones; (2) an arc-wide strong penetrative rock foliation; (3) significant crustal thickening; (4) an enhanced geothermal gradient such that the lowermost crust experienced granulite-facies metamorphism; (5) a range of late kinematic vectors that ranged from vertical to strike-slip and horizontal, indicating that in the later stages of collision at least this was not a simple orthogonal compressional event (e.g. Pudsey *et al.* 1986; Khan & Coward 1991; Robertson & Collins 2002).

The key evidence that has led to the development of the early collision model is the existence of an early, Kohistan-wide penetrative fabric that is, demonstrably cut by dated later events, and the observation that the penetrative fabrics are linked to movements along the Northern or Shyok Suture, indicating an early major collisional event linked to the suturing of Kohistan with Asia. The clearest evidence for the age of the suture remains the presence of undeformed dolerite dykes dated at 75 Ma cutting deformed *c.* 102 Ma trondhjemites between Gilgit and the Northern Suture (Petterson & Windley 1992). Treloar *et al.* (1996) suggested that the age of the Chilas Complex (85 Ma) gives a closer indication of the age of suturing as it cuts highly deformed Kamila and Jaglot amphibolite–volcanic–metasedimentary sequences, although itself exhibiting only weak deformation. Clift *et al.* (2002) suggested, on the basis of structural and stratigraphic evidence in Ladakh, that the age of the Dras–Kohistan arc with Eurasia is Turonian–Santonian (*c.* 84–94 Ma). Heuberger *et al.* (2007) quoted new U–Pb age and Hf isotope data from the Karakoram and Kohistan terranes demonstrating that: (1) plutonic activity in Kohistan extended back to at least 112 Ma; (2) the magmatic and tectonic history of the Northern or Shyok Suture did not end at 85 or 75 Ma but continued into the Eocene.

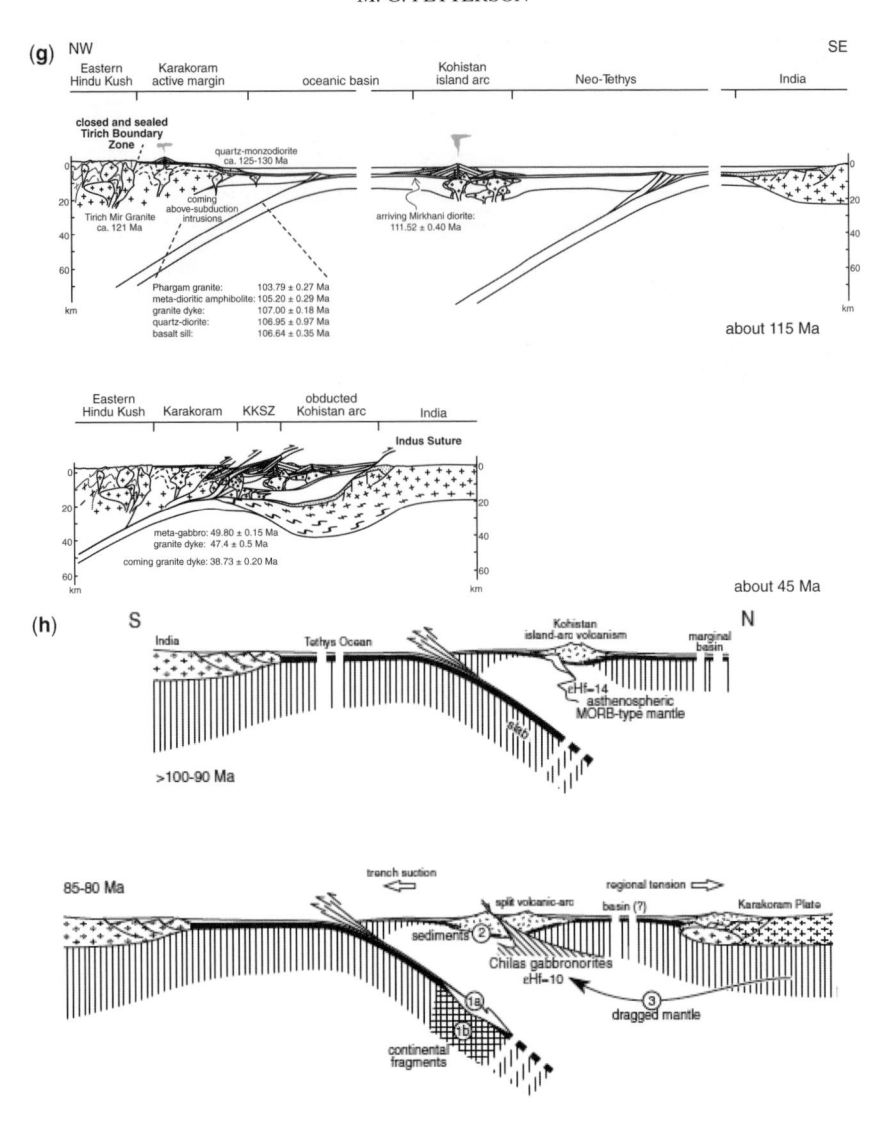

Fig. 14. (**g**) Pre- and post-collisional tectonic setting for Kohistan between India and Eurasia (Heuberger *et al.* 2007, acknowledgements *Swiss Journal of Geosciences*). (**h**) Kohistan is situated above a north-directed intra-oceanic subduction zone. The arc split at *c.* 85 Ma and upwelling mantle plumes intruded the Chilas Complex. (Schaltegger *et al.* 2002, acknowledgements Blackwell Science.)

Andean to post-collisional period. From *c.* 90–85 Ma to *c.* 55–45 Ma (Petterson & Windley 1985; Beck *et al.* 1995) Kohistan was within an Andean environment at the edge of the Asian continent facing southwards across a major subduction zone towards India (Fig. 14d, e). This period was one of crustal consolidation with the intrusion of medium–large volumes of gabbroic–granitic plutons, the uplift of western Kohistan in particular being followed by extrusion of silicic volcanic deposits within intramontane basins (Petterson & Windley 1985, 1992; Petterson & Treloar 2004; Fig. 13) and an increasing acid/basic crustal ratio with time.

The Andean to post-collisional volcanic episode of Kohistan as represented by the Shamran/Teru and Dir Group remains an area for further study. A number of workers (Sullivan *et al.* 1993; Shah & Shervais 1999; Danishwar *et al.* 2001; Khan *et al.* 2004, 2009; Petterson & Treloar 2004) have studied various aspects of these units. The Utror Volcanics of the Dir Group are more accessible than the Shamran/Teru Volcanics and have been described well by in Sullivan *et al.* (1993). This

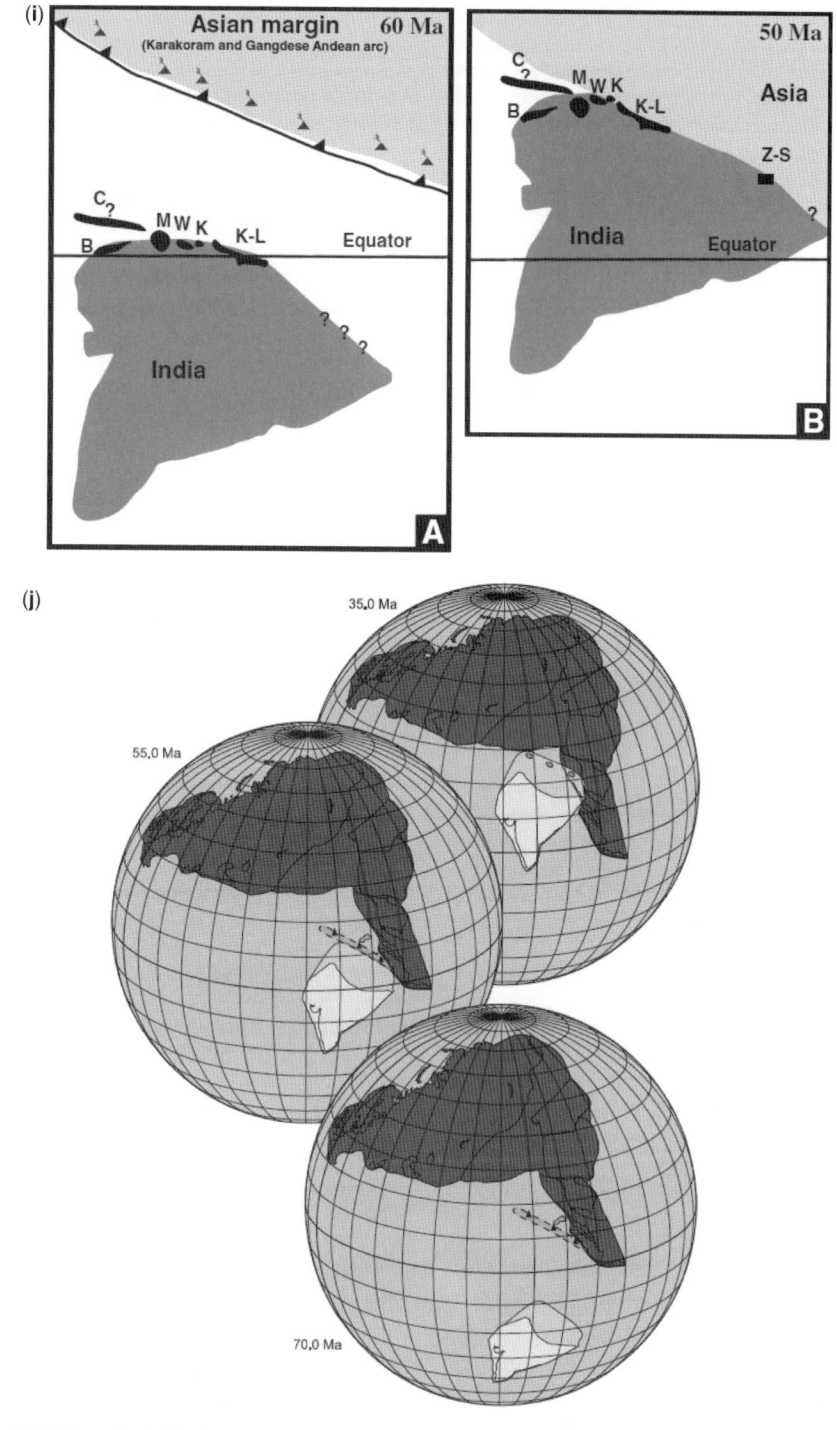

Fig. 14. (i) Kohistan–Ladakh (K–L) collided first with India at c. 61 Ma along with other intra-oceanic terranes, and subsequently collided with Asia at c. 50 Ma (Khan *et al.* 2009, acknowledgements Geological Society of America). (j) Global context of India moving northward from 70–35 Ma and colliding with Asia. Aitchison *et al.* (2007) favoured a later, c. 34 Ma collision between India and Eurasia. Many workers have dated this at c. 50 Ma and some would like the collision to be as old as 70 Ma. (Aitchison *et al.* 2007, acknowledgements American Geophysical Union.) (Terrane abbreviations: B, Bela ophiolite; C, Chagi arc; K, Khost ophiolite; K–L, Kohistan–Ladakh; M, Muslimbagh ophiolite; W, Waziristan ophiolite; Z–S, Zapur Shan Range).

sequence is dominated by coarse and fine volcaniclastic rocks as well as ignimbrites and rhyolites. Trace element multi-element diagrams are more evolved than those for earlier volcanic units, with high LFSE/LILE ratios typical of mature arcs and Andean margins. Exposures described by Danishwar et al. (2001) at lower levels of the Ghizar valley around Teru village could not be found by Petterson & Treloar (2004) although the geochemistry of the rocks is highly evolved and stratigraphic sections include high proportions of ignimbrite and rhyolite. Cross-sections drafted by Danishwar et al. (2001) suggest interdigitation of the Shamran/Teru unit with basement Kohistan units. Petterson & Treloar (2004) described the Shamran/Teru sequence as a unit resting unconformably upon the older Kohistan basement (Chalt Volcanic Group and Kohistan batholith), with the plane of unconformity being mapped high above the valley floor (Fig. 13). In a similar vein, Shah & Shervais (1999) described high-Mg and basalt–andesite-dominated rocks around Dir that they assigned to the Utror Volcanics, which, however, appear to be more closely related to the Chalt Volcanic Group in terms of geochemical composition and lithology. This remains a confusing area that warrants further investigation and the mapping of undeniable 'younger' volcanic sequences resting unconformably upon older basement sequences. What is clear is that a series of approximately Paleocene–Oligocene silicic and undeformed volcanic rocks with highly evolved mature arc–continental margin geochemical affinities are present in central and western areas of north Kohistan and these were erupted in an Andean margin to collisional setting.

Oligocene granite and leucogranite sheets close to Nanga Parbat have geochemical and isotopic signatures suggesting that they represent the melting of young crustal material without significant input of old Indian plate crust (e.g. Petterson & Windley 1985, 1991; George et al. 1993; Petterson et al. 1993). This area of Kohistan has experienced extremely rapid exhumation in recent times related to the formation of the Nanga Parbat syntaxis, and fluids from the Indian plate may have interacted with Kohistan granite magmas (George et al. 1993). Crustal thickening during post-collisional times may have been the major cause for intra-Kohistan crustal melting.

Model 2: southward subduction beneath Kohistan and/or collision first with India

This model has been proposed and/or supported or quoted as the preferred model by researchers such as Bard (1983), Reuber (1986), Khan et al. (1997, 2009), Corfield et al. (1999, 2001), Ziabrev et al. (2004) and Yin (2006) (see Fig. 14d, i).

Khan et al. (1997), argued that a reduction in the subduction component from north to south across Kohistan, as represented by 48–59% SiO_2 volcanic–plutonic rocks from Chalt, Chilas and Kamila, indicates a subduction polarity contrary to the more widely proposed northern subduction beneath Kohistan. They prefer a southward-directed subduction with subduction reversal occurring after Kohistan collided with Eurasia at 85 Ma. Khan et al. (1997) also argued from an isotopic compositional perspective that Kohistan was situated within the southern hemisphere and entrained enriched DUPAL-type mantle material in its source region, typical of south of the equator.

Khan et al. (2009) cited palaeomagnetic data from Kohistan as indicating an equatorial palaeolatitude (e.g. Zaman & Torri 1999). These data, together with the data of Khan et al. (1997) described above and those presented by Garrido et al. (2007) for the Jijal Complex indicating enriched EMII–DUPAL-type isotopic source, and new data from the Teru/Solidus Shamran volcanic rocks, they argued, preclude a close palaeogeographical proximity of Kohistan to Eurasia (possibly situated some 3000 km to the north in the Cretaceous). Instead, these data suggest a close palaeogeographical proximity to India. Furthermore, if only U–Pb zircon geochronological data are considered, Khan et al. (2009) argued that calc-alkaline volcanism ended by 61 Ma in Kohistan, which dates the collision between India and Kohistan, the first collision. Kohistan did not experience an Andean-style tectonic period; the Teru/Shamran volcanic rocks represent the final calc-alkaline magmatism in Kohistan. India plus Kohistan then collided with Eurasia at around 50 Ma. The suture was stitched by two plutons from northern Kohistan (with ages of 47 Ma and 41 Ma respectively) that indicate Eurasian inputs to their magmatic source. Khan et al. (2009) cited a range of structural-dominated research that estimated timings of ophiolite obduction from the Spontang Ophiolite, Ladakh, the Waziristan region and the Yarlung–Tsangpo Suture zone in Tibet (see above) to support the India-first collisional hypothesis, suggesting that these data are more compatible with a c. 61 Ma India–Kohistan collision.

Collision with India

The collision of India with Kohistan, or India–Kohistan with Eurasia, dependent on tectonic models, is generally bracketed at around 55–45 Ma (Klootwijk & Conaghan 1979; Patriat & Achache 1984; Beck et al. 1995; Hodges 2000) and involved the closure of the Tethys Ocean, and the continental 'subduction' of India at a rate of c. 4–5 cm a^{-1}. However, as with many major issues in geosciences, some workers disagree with the timing of this collision, preferring ages as young as c. 34 Ma (Aitchison et al. 2007) or as old as c. 70 Ma (e.g. Yin 2006); readers are directed to Aitchison et al. (2007) for further discussion. It is probable that the Indian plate had not fully underplated northern Kohistan until the Miocene. Beaumont et al. (2004) presented India as completely underplating the Lhasa terrane and downwelling beneath the Qiantang terrane in Tibet at present. The main result of the Indian collision on Kohistan was to rotate, shear and thrust the terrane to such an extent that it was turned into a vertical position, with lower crust to the south and upper crust to the north. Southern Kohistan was imbricated into a series of north-dipping thrusts and blueschists formed (at c. 44–46 Ma) as upper crust was taken down the India–Kohistan contact zone only to rebound later (e.g. Treloar et al. 2003). Kohistan–Ladakh was later split almost in two with the rise of Nanga Parbat at the apex of a diverging thrust system. Kohistan remains seismically active today with the continuing collision.

Future tests of tectonic models

This section demonstrates that in spite of decades of research in Kohistan there remains much to be learned, and hypotheses and models need to be further developed, refined or invented. The most controversial debate revolves around the tectonic model that best explains the geotectonic development of Kohistan. The majority of researchers have quoted model 1, although model 2 was only recently published by Khan et al. (2009) and will influence present and future researchers. Many workers (e.g. Aitchison et al. 2007) have pointed to the paucity of reliable palaeomagnetic data, especially data that can be unambiguously interpreted in spite of deformation and structural complexities. Aitchison et al. (2007) pointed out that the palaeo-wander curve for many parts of Asia is particularly poorly known. Khan et al. (2009) have chosen to interpret palaeomagnetic data to indicate the remoteness, in Cretaceous times, of Kohistan with respect to Asia. This review suggests that the generation of new high-resolution palaeomagnetic data from Kohistan, Karakoram and surrounding terranes will be essential to further elucidating the tectonic history of Kohistan. The existence of a DUPAL–EMII enriched mantle signature has been used by a number of workers to indicate a definitive southern hemisphere palaeo-position for Kohistan in the Cretaceous, asserting that present-day enriched oceanic mantle source domains are present in the southern Indian Ocean. Other workers (e.g. Bignold & Treloar 2003, and references therein) have suggested that the DUPAL signature is non-unique and can be formed through oceanic sedimentary inputs and/or subduction-zone fluids to the source regions of arc magmas. The Indian ocean DUPAL signature may represent ultimate recycling of oceanic sediment and subduction zones in the deep mantle. The Cretaceous Indian ocean mantle may not even have possessed a DUPAL signature. Thus it is questionable whether or not a DUPAL-type geochemistry can be cited as evidence for a non-unique source or a non-unique palaeogeographical position. Debates have raged in the literature for decades about the timing of emplacement of ophiolites around the Himalayan and other orogens, what exactly obduction of a particular ophiolite represents, how far the ophiolite has moved from its original position, and what exactly is the nature and age of the substrate beneath a particular ophiolite. It is timely for further investigation and synthesis of the timing and nature of ophiolites around the edge of Tethys, and examination of patterns in the data. Such a study could reveal new insights and tests of tectonic models but will require significant fieldwork and collection or generation of structural or geochronological data, and may never yield unambiguous conclusions. Table 1 presents age data from 118 Ma to 26 Ma for a range of Kohistan igneous rocks. The table includes a range of dating methods but limits itself to U–Pb, Rb–Sr whole-rock, Sm–Nd and Ar–Ar. U–Pb zircon data are the most reliable and precise data at present and one day the whole of Kohistan and all rock types will be covered by such data. It is, however, possibly premature to rule out non-U–Pb age dating data, particularly at an arbitrary level that may suit certain arguments, and also particularly when U–Pb data are not necessarily discrediting other dating systems; more accurately, they are refining and making more precise the context of previously published geochronological data. The literature indicates a large volume of 60 to c. <30 Ma magmatism across Kohistan that is, readily explained through an Andean-type model but less well explained as a passive margin on the northern edge of India in post-61 Ma times. The Northern–Shyok Suture remains poorly understood. One test of tectonic models is to evaluate and quantitatively assess the earliest movements on this suture and to generate structural–chronological data that document the main events related to this suture from its first formation (Possibly at c. 84 Ma or 50 Ma depending on the model) to its most recent movement across Kohistan and beyond. Finally, the author remains unconvinced that the Shamran/Teru volcanic sequence has been properly or fully documented and assessed in terms of basic geological mapping, the relationship of this unit to other units, the volcanology of the sequence, the identification of possible eruptive centres and modelling of eruptive mechanisms, and its range in geochemical composition and age. This remains an area for significant and potentially tectonic model-testing research.

Crustal accretion

Kohistan is a world-class natural laboratory in which to study crustal accretion processes. The most likely order of emplacement of units in time (from oldest to youngest) is: (1) oceanic-type Kamila Amphibolites; (2) arc-type Kamila Amphibolites; (3) Jijal Complex lower crustal–mantle arc-root; (4) Jaglot Group sediments and volcanic rocks; (5) Chalt volcanic rocks,; (6) Kohistan batholith stage 1 plus Yasin sediments; (7) Chilas Complex; (8) Kohistan batholith stage 2; (9) Dir Group and Shamran/Teru volcanic rocks; (10) Kohistan batholith stage 3.

The basement volcano-sedimentary rocks (units 1, 2, 4, 5 and 6) represent a long period of arc growth (Possibly Jurassic or early Cretaceous to Late Cretaceous or some 60–70 Ma of time) dominated by extrusion of volcanic deposits (lavas, intrusions, volcaniclastic deposits) perhaps representing a cumulative thickness of c. 15 to >20 km, although not necessarily all in vertical continuity (probably rather in a series of basins). Around 117 or 118 Ma part of the subcrustal mantle started to accrete to the base of the sub-arc crust and form lower-crustal intrusive complexes. The thick volcanic sequences would also have been fed from intra-crustal magma chambers that subsequently froze to become part of the thickening Kohistan crust. Stage 1 gabbro–diorites intruded as bulk transfers from the mantle whereas intra-arc amphibolite melting produced rocks such as the c. 102 Ma Matum Das trondhjemite. By c. 95 Ma the crust had attained sufficient thickness (>30–35 km) to create conditions for granulite-facies metamorphism in the lower parts of the arc. The Eurasian collision very significantly deformed and thickened the crust, which was then intruded by the c. 85 Ma Chilas Complex, which added another c. 5–20 km of material to the lower crust. Andean crustal growth added stage 2 batholith units and extrusive volcanic units (c. 3 km thick) but also led to uplift and erosion in parts. Kohistan then collided with India, rotated, partially melted and was completely underplated by India.

The author would like to express his gratitude for working with so many interesting and knowledgeable people over the years in relation to Kohistan, in particular B. Windley, A. Khan, Q. Jan, P. Treloar, I. Mian, M. Coward and C. Pudsey, and for the kindness of the Kohistani people over many field campaigns.

References

AITCHISON, J. C., ALI, J. R. & DAVIS, M. A. 2007. When and where did India and Asia collide? *Journal of Geophysical Research*, **112**, B05424, doi:10.1029/2006JB004706.

ANCZKIEWICZ, R. & VANCE, D. 2000. Isotopic constraints on the evolution of metamorphic conditions in the Jijal–Patan Complex and Kamila Belt of the Kohistan Arc, Pakistan Himalaya. *In*: KHAN, M. A., TRELOAR, P. J., SEARLE, M. P. & JAN, M. Q. (eds) *Tectonics of the Nanga Parbat Syntaxis and the Western Himalaya*. Geological Society, London, Special Publications, **170**, 321–331.

BARD, J. P. 1983. Metamorphism of an obducted island arc: example of the Kohistan sequence (Pakistan) in the Himalayan collided range. *Earth and Planetary Science Letters*, **65**, 133–144.

BARD, J. P., MULUSKI, H., MATTE, P. & PROUST, F. 1980. The Kohistan sequence: crust and mantle of an obducted island arc. *Special Issue, Geological Bulletin University of Peshawar*, **13**, 87–94.

BARKER, F. & GRANTZ, A. 1982. Talkeetna Formation in the southeastern Talkeetna Mountains, southern Alaska: an early Jurassic andesitic intraoceanic island arc. *Geological Society of America, Abstract with Programs*, **14**, 147.

BEAUMONT, C., JAMIESON, R. A., NGUYEN, M. H. & MEVDEV, S. 2004. Crustal channel flows: 1. Numerical models with applications to the tectonics of the Himalayan–Tibetan orogen. *Journal of Geophysical Research*, **109**, B06406, doi:10.1029/2003JB002809.

BECK, R. A., BURBANK, D. W. *ET AL*. 1995. Stratigraphic evidence for an early collision between northwest India and Asia. *Nature*, **373**, 55–57.

BIGNOLD, S. M. & TRELOAR, P. J. 2003. Northward subduction of the Indian Plate beneath the Kohistan island arc, Pakistan Himalaya: new evidence from isotopic data. *Journal of the Geological Society, London*, **160**, 377–384.

BIGNOLD, S. M., TRELOAR, P. J. & PETFORD, N. 2006. Changing sources of magma generation beneath intra-oceanic island arcs: an insight from the juvenile Kohistan island arc, Pakistan Himalaya. *Chemical Geology*, **233**, 46–74.

BURG, J. P., BODINIER, J.-L., CHAUDRY, S., HUSSAIN, S. & DAWOOD, H. 1998. Infra-arc mantle–crust transition and intra-arc mantle diapirs in the Kohistan Complex (Pakistan Himalaya): petro-structural evidence. *Terra Nova*, **10**, 74–80.

BURG, J.-P., ARBARET, L., CHAUDHRY, M. N., DAWOOD, H., HUSSAIN, S. & ZEILINGER, G. 2005. Shear strain localization from the upper mantle to the middle crust of the Kohistan Arc (Pakistan). *In*: BRUHA, D. & BURLINI, L. (eds) *High-strain Zones: Structure and Physical Properties*. Geological Society, London, Special Publications, **245**, 25–38.

CALVERT, A. J., KLEMPERER, S. L., TAKAHASHI, N. & KERR, B. C. 2008. Three-dimensional crustal structure of the Mariana island arc from seismic tomography. *Journal of Geophysical Research* **113**, B01406.

CASNEDI, R., DESIO, A., FORCELLA, F., NICOLETTI, M. & PETRUCCIANI, C. 1978. Absolute age of some granitoid rocks between Hindu Raj and Gilgit river (W. Karakorum). *Rendiconti della Accademia Nazionale dei Lincei*, **64**, 204–210.

CHROSTON, P. N. & SIMMONS, G. 1989. Seismic velocities from the Kohistan volcanic arc, northern Pakistan. *Journal of the Geological Society, London*, **146**, 971–979.

CLIFT, P. D., HANNIGAN, R., BLUSZTAJN, J. & DRAUT, A. E. 2002. Geochemical evolution of the Dras–Kohistan Arc during collision with Eurasia: evidence from the Ladakh Himalaya, India. *Island Arc*, **11**, 255–273.

CLIFT, P. D., PAVLIS, T., DEBARR, S. M., DRAUT, A. E., RIOUX, M. & KELEMEN, P. B. 2005. Subduction erosion of the Jurassic Talkeetna–Bonanza arc and the Mesozoic accretionary tectonics of western North America. *Geology*, **33**, 881–884.

CORFIELD, R. I., SEARLE, M. P. & GREEN, O. R. 1999. Photang thrust sheet: an accretionary complex structurally below the Spontang ophiolite constraining timing and tectonic environment of ophiolite obduction. *Journal of the Geological Society, London*, **156**, 1031–1044.

CORFIELD, R. I., SEARLE, M. P. & PEDERSEN, R. B. 2001. Tectonic setting, origin and obduction history of the Spontang ophiolite, Ladakh Himalaya, NW India. *Journal of Geology*, **109**, 715–736.

COWARD, M. P., JAN, M. Q., REX, D., TARNEY, J., THIRLWALL, M. & WINDLEY, B. F. 1982a. Geotectonic framework of the Himalaya of N Pakistan. *Journal of the Geological Society, London*, **139**, 299–308.

COWARD, M. P., JAN, M. Q., REX, D., TARNEY, J., THIRLWALL, M. & WINDLEY, B. F. 1982b. Structural evolution of a crustal section in the western Himalaya. *Nature*, **295**, 22–24.

COWARD, M. P., WINDLEY, B. F. ET AL. 1986. Collision tectonics in NW Himalayas. *In*: COWARD, M. P. & RIES, A. C. (eds) *Collision Tectonics*. Geological Society, London, Special Publications, **19**, 203–219.

COWARD, M. P., BUTLER, R. W. H., KHAN, M. A. & KNIPE, R. J. 1987. The tectonic history of Kohistan and its implications for Himalayan structure. *Journal of the Geological Society, London*, **144**, 377–391.

DANISHWAR, S., STERN, R. J. & KHAN, M. A. 2001. Field relations and structural constraints for the Teru volcanic formation, northern Kohistan Terrane, Pakistani Himalayas. *Journal of Asian Earth Sciences*, **19**, 683–695.

DEBARI, S. M. & COLEMAN, R. G. 1989. Examination of the deep levels of an island-arc – evidence from the Tonsina ultramafic–mafic assemblage, Tonsina, Alaska. *Journal of Geophysical Research*, **94**, 4373–4391.

DESIO, A. 1959. Cretaceous beds between Karakorum and Hindu Kush ranges (Central Asia). *Rivista Italiana di Paleontologiae Stratigrafia*, **65**, 221–229.

DESIO, A. 1963a. Review of the geologic 'Formations' of the western Karakoram (Central Asia). *Rivista Italiana di Paleontologiae Stratigrafia*, **69**, 475–501.

DESIO, A. 1963b. Review of the geologic 'Formations' of the western Karakoram (Central Asia). *Rivista Italiana di Paleontologiae Stratigrafia*, **69**, 475–501.

DESIO, A. 1964. *Geological Tentative Map of the western Karakorum, 1:500 000*. Istituto di Geologia, Milan.

DESIO, A. 1966. The Devonian sequence in Mastuj valley (Chitral, NW Pakistan). *Rivista Italiana di Paleontologiae Stratigrafia*, **72**, 293–320.

DESIO, A. 1974. Karakorum Mountains. *In*: SPENCER, A. M. (ed.) *Mesozoic–Cenozoic Orogenic Belts: Data for Orogenic Studies*. Geological Society, London, Special Publications, **4**, 255–266.

DESIO, A. 1975. Some geological notes and problems on the Chitral, (NW Pakistan). *Rendiconti della Accademia Nazionale dei Lincei*, **58**, 1–7.

DESIO, A. 1977. The occurrence of blueschists between the Middle Indus and Swat valleys as an evidence of subduction (N Pakistan). *Rendiconti della Accademia Nazionale dei Lincei, Series*, 8, **62**, 1–9.

DESIO, A. 1980. *Geology of the Shaksgam Valley*. E. J. Brill, Leiden.

DESIO, A. & ZANETTIN, B. 1970. *Geology of the Baltoro Basin*. Vol. 2, E. J. Brill, Leiden.

DESIO, A. & MARTINA, E. 1972. Geology of the upper Hunza valley, Karakorum, W. Pakistan. *Bollettino della Societa Geologica Italiana*, **91**, 283–314.

DESIO, A., TONGIORGI, E. & FERRARA, G. 1964. On the geological age of some granites of the Karakorum, Hindu Kush and Badakhshan (Central Asia). *Report of 22nd International Geological Congress*, **11**, 479–496.

DESIO, A., SILVA, I. P. & RONCHETTI, C. R. 1977. On the Cretaceous outcrop in the Chumarkhan and Laspur valleys, Gilgit–Chitral, NW Pakistan. *Rivista Italiana di Paleontologia*, **83**, 561–574.

DHUIME, B., BOSCH, D. ET AL. 2007. Multistage evolution of the Jijal ultramafic–mafic complex (Kohistan, N Pakistan): implications for building the roots of island arcs. *Earth and Planetary Science Letters*, **261**, 179–200.

DHUIME, B., BOSCH, D., GARRIDO, C. J., BODINIER, J.-L, BRUGUIER, O., HUSSAIN, S. S. & DAWOOD, H. 2009. Geochemical architecture of the lower- to middle-crustal section of a paleo-island arc (Kohistan Complex, Jijal–Kamila area, northern Pakistan): implications for the evolution of an oceanic subduction zone. *Journal of Petrology*, **50**, 531–569, doi:10.1093/petrology/egp010.

DIMALANTA, C., TAIRA, A., YUMUL, G. P. JR, TOKUYAMA, H. & MOCHIZUKI, K. 2002. New rates of western Pacific island arc magmatism from seismic and gravity data. *Earth and Planetary Science Letters*, **202**, 105–115.

GANSSER, A. 1964. *Geology of the Himalayas*. Wiley Interscience, London.

GANSSER, A. 1979. Map of ophiolitic belts of Himalayan and Tibetan region. *In: International Geological* drift of the Indian Plate. *Chemical Geology*, **182**, 139–178.

SCHALTEGGER, U., ZEILINGER, G., FRANK, M. & BURG, J. P. 2002. Multiple mantle sources during island arc magmatism: U–Pb and Hf isotopic evidence from the Kohistan arc complex, Pakistan. *Terra Nova*, **14**, 461–468.

SEARLE, M. P. 1991. *Geology and Tectonics of the Karakoram Mountains*. Wiley, Chichester.

SEARLE, M. P., KHAN, M. A., FRASER, J. E., GOUGH, S. J. & JAN, M. Q. 1999. The tectonic evolution of the Kohistan–Karakoram collision belt along the Karkoram highway transect, North Pakistan. *Tectonics*, **18**, 929–949.

SHAH, M. T. & SHERVAIS, J. W. 1999. The Dir–Utror metavolcanic sequence, Kohistan arc terrane, northern Pakistan. *Journal of Asian Earth Sciences*, **17**, 459–476.

SULLIVAN, M. A., WINDLEY, B. I., SAUNDERS, A. D., HAYNES, J. R. & REX, D. C. 1993. A palaeogeographic reconstruction of the Dir Group: evidence from magmatic arc migration within Kohistan, N. Pakistan. *In*: TRELOAR, P. J. & SEARLE, M. P. (eds) *Himalayan Tectonics*. Geological Society, London, Special Publications, **74**, 139–160.

TAHIRKHELI, R. A. K. 1979. Geology of Kohistan and adjoining Eurasian and Indo-Pakistan continents, Pakistan. *Special Issue, Geological Bulletin, University of Peshawar*, **11**, 1–30.

TAHIRKHELI, R. A. K. 1982. Geology of Hindu Kush, Himalayas and Karakoram in Pakistan. *Special Issue, Geological Bulletin, University of Peshawar*, **15**.

TAHIRKHELI, R. A. K. & JAN, M. Q. 1979. A preliminary geological map of Kohistan and the adjoining areas, N Pakistan. *Geological Bulletin, University of Peshawar*, **11**.

TAHIRKHELI, R. A. K., MATTAUER, M., PROUST, F. & TAPPONNIER, P. 1979. The India–Eurasia suture zone in northern Pakistan: synthesis and interpretation of recent data at plate scale. *In*: FARAH, A. & DE JONG, K. A. (eds) *Geodynamics of Pakistan*. Geological Survey of Pakistan, Quetta, 125–130.

TAKAHASHI, Y., MIKOSHIBA, M. U., TAKAHASHI, Y., KAUSAR, A. B., KHAN, T. & KUBO, K. 2007. Geochemical modelling of the Chilas Complex in the Kohistan Terrane, northern Pakistan. *Journal of Asian Earth Sciences*, **29**, 336–349.

TAYLOR, S. R. & McLENNAN, S. M. 1981. The composition and evolution of the continental crust: rare earth element evidence from sedimentary rocks. *Philosophical Transactions of the Royal Society of London, Series A*, **301**, 381–399.

TRELOAR, P. J., REX, D. C. ET AL. 1989. K–Ar and Ar–Ar geochronology of the Himalayan collision in NW Pakistan: constraints on the timing of suturing, deformation, metamorphism and uplift. *Tectonics*, **8**, 881–909.

TRELOAR, P. J., BRODIE, K. H. ET AL. 1990. The evolution of the Kamila Shear Zone, Kohistan, Pakistan. *In*: SALISBURY, M. H. & FOUNTAIN, D. M. (eds) *Exposed Cross-sections of the Continental Crust*. NATO ASI Series, **C317**, 175–214.

TRELOAR, P., PETTERSON, M. G., JAN, M. Q. & SULLIVAN, M. A. 1996. A re-evaluation of the stratigraphy and evolution of the Kohistan arc sequences, Pakistan Himalaya: implications for magmatic and tectonic arc-building processes. *Journal of the Geological Society, London*, **153**, 681–693.

TRELOAR, P. J., O'BRIEN, P. J., PARRISH, R. R. & KHAN, M. A. 2003. Exhumation of early Tertiary, coesite-bearing eclogites from the Pakistan Himalaya. *Journal of the Geological Society, London*, **160**, 367–376.

WADIA, D. N. 1932. Notes on the geology of Nanga Parbat, Mt Diamir, and adjoining parts of Chilas, Gilgit district, Kashmir. *Geological Survey of India Records*, **66**, 212–234.

WEAVER, B. L. & TARNEY, J. 1984. Major and trace element composition of the continental lithosphere. *In*: POLLACK, H. N. & MURTHY, V. R. (eds) *Structure and Evolution of the Continental Lithosphere*. Pergamon, New York, 39–68.

YAMAMOTO, H. 1993. Contrasting metamorphic $P-T$ time paths of the Kohistan granulites and tectonics of the western Himalayas. *Journal of the Geological Society, London*, **150**, 843–856.

YAMAMOTO, H & NAKAMURA, E. 1996. Sm–Nd dating of garnet granulites from the Kohistan Complex, northern Pakistan. *Journal of the Geological Society, London*, **153**, 965–969.

YAMAMOTO, H. & NAKAMURA, E. 2000. Timing of magmatic and metamorphic events in the Jijal Complex of the Kohistan Arc deduced from Sm–Nd dating of mafic granulites. *In*: KHAN, M. A., TRELOAR, P. J., SEARLE, M. P. & JAN, M. Q. (eds) *Tectonics of the Nanga Parbat Syntaxis and the Western Himalaya*. Geological Society, London, Special Publications, **170**, 313–319.

YAMAMOTO, H. & YOSHINO, T. 1998. Superposition of replacements in the mafic granulites of the Jijal complex of the Kohistan arc, northern Pakistan: dehydration and rehydration within deep arc crust. *Lithos*, **43**, 219–234.

YAMAMOTO, H., NAKAMURA, E., KANEKO, Y. & KAUSAR, A. B. 2005. U–Pb zircon dating of regional deformation in the lower crust of the Kohistan Arc. *International Geology Review*, **47**, 1035–1047.

YIN, A. 2006. Cenozoic evolution of the Himalayan orogen as constrained by along-strike variation of structural geometry, exhumation history, and foreland sedimentation. *Earth-Science Reviews*, **76**, 1–131.

YOSHINO, T. & OKUDAIRA, T. 2004. Crustal growth by magmatic accretion constrained by metamorphic $P-T$ paths and thermal models of the Kohistan Arc, NW Himalayas. *Journal of Petrology*, **45**, 2287–2302, doi:2210.1093/petrology/egh2056.

YOSHINO, T., YAMAMOTO, H., OKUDAIRA, T. & TORIUMI, M. 1998. Crustal thickening of the lower crust of the Kohistan arc (N Pakistan) deduced from Al zoning in clinopyroxene and plagioclase. *Journal of Metamorphic Geology*, **16**, 729–748.

ZAMAN, H. & TORRI, M. 1999. Palaeomagnetic study of Cretaceous red beds from the eastern Hindu Kush ranges, northern Pakistan: palaeoreconstruction of the Kohistan–Karakoram composite unit before the

India–Asia collision. *Geophysical Journal International*, **136**, 719–738.

ZEITLER, P. K. 1985. Cooling history of the NW Himalaya. *Pakistan Tectonics*, **4**, 127–151.

ZEITLER, P. K. & CHAMBERLAIN, C. P. 1991. Petrogenetic and tectonic significance of young leucogranites from the NW Himalaya, Pakistan. *Tectonics*, **10**, 729–741.

ZIABREV, S. V., AITCHISON, J. C., ABRAJEVITCH, A. V., BADENGZHU DAVIS, A. M. & LUO, H. 2004. Bainang terrane, Yarlung–Tsangpo suture, southern Tibet (Xizang, China): a record of intra-NeoTethyan subduction–accretion processes preserved on the roof of the world. *Journal of Geology*, **161**, 523–538.

Roles of strike-slip faults during continental deformation: examples from the active Arabia–Eurasia collision

MARK B. ALLEN

Department of Earth Sciences, University of Durham, Durham DH1 3LE, UK
(e-mail: m.b.allen@durham.ac.uk)

Abstract: This paper concerns the kinematics of active strike-slip faults in the Arabia–Eurasia collision zone, and how they accommodate plate convergence. Several roles are discernible: *(1) collision zone boundaries*, the left-lateral Dead Sea Fault System and right-lateral faults in eastern Iran form the western and eastern boundaries of the collision zone; *(2) tectonic escape structures*, the North and East Anatolian faults transport intervening crust westwards, out of the path of the Arabia; *(3) strain partitioning*, right-lateral slip on the Zagros Main Recent Fault and NW–SE-striking thrusts to its SW produce north–south convergence, parallel to the plate vector; left-lateral slip along the Alborz range and thrusts across it produce oblique left-lateral shortening; *(4) shortening arrays*, arrays of strike-slip faults (e.g. Kopeh Dagh and eastern Iranian faults) rotate about vertical axes, producing north–south shortening without crustal thickening; *(5) transfer zones*, fold trends and earthquake slip vectors change orientation across strike-slip faults in the Zagros, suggesting that these faults allow for changes in thrust transport along strike in the orogen. These different roles emphasize the complex behaviour of continental crust, and the advantages of studying active tectonics rather than ancient examples.

This paper reviews active strike-slip faults from the Arabia–Eurasia collision zone (Figs 1–3), to summarize the different ways in which such faults help achieve plate convergence during continent–continent collision. This is an important issue for two reasons. The first is that it is part of the more general problem of how faults in the upper crust collectively produce the velocity fields required by plate motions. The second is that strike-slip faults are common features in the geological record of the continents, but it is not always easy to determine why such faulting took place. Active tectonics provides data and constraints not available in ancient settings, principally through studies of decadal to millennial slip vectors [via global positioning system (GPS) and seismicity studies] and through use of the landscape to deduce deformation patterns. The approach is to use case studies from different regions to make general conclusions about the way in which the strike-slip faults in the upper crust behave during continental deformation. This paper is not intended to provide a systematic account of every active strike-slip fault in SW Asia, nor does it dwell on the many other aspects of the collision. Other papers (e.g. Mann 2007) have synthesized the structures associated with continental strike-slip faults, regardless of their origins; such material is not repeated here.

Active slip rates and finite offsets are known for many of the strike-slip faults, and in some cases there are data for the timing of onset. Therefore it is possible to compare the patterns of short-term and long-term deformation in the collision zone, and by implication in continental crust in general. Strike-slip faults are easier to work with in this respect than thrusts or normal faults, where the overall shortening or extension may be poorly constrained through lack of subsurface data.

Continental collision zones are excellent places in which to study continental deformation processes in general because of the widespread and highly variable nature of the deformation that takes place. Although collision by definition implies plate convergence, this can be accommodated in a tremendous variety of ways by combinations of compressional, strike-slip and even extensional structures (Dewey *et al.* 1986). Faulting is the main way in which strain is accomplished within the brittle upper crust, therefore the kinematics of fault zones are revealing about overall strain. However, there are few active continental collision zones in the world, compared with active subduction zone boundaries for example. One is the Arabia–Eurasia collision, part way along the network of Cenozoic orogenic belts between the Pyrenees and SE Asia known collectively as the Alpine–Himalayan system. Following a geological overview of the collision, later sections focus on single faults and groups of faults, to show how their kinematics fit into the overall plate convergence. Figure 3 is a summary map of the main active strike-slip faults in the Arabia–Eurasia collision, but also highlights the generic roles outlined in this paper, namely: collision zone boundaries (Dead Sea Fault System, eastern Iranian faults); tectonic escape structures (North and East Anatolian faults; NAF

From: KUSKY, T. M., ZHAI, M.-G. & XIAO, W. (eds) *The Evolving Continents: Understanding Processes of Continental Growth.* Geological Society, London, Special Publications, **338**, 329–344.
DOI: 10.1144/SP338.15   0305-8719/10/$15.00 © The Geological Society of London 2010.

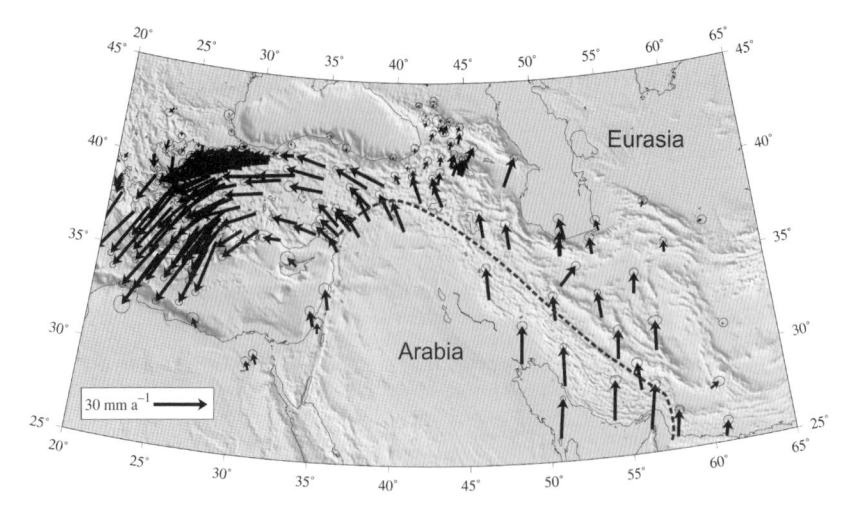

Fig. 1. GPS-derived velocity field of the Arabia–Eurasia collision, with respect to stable Eurasia. The dashed line is the Bitlis–Zagros suture. Compiled from McClusky et al. (2000) and Vernant et al. (2004a).

and EAF); strain partitioning elements (Main Recent Fault of the Zagros; Mosha Fault in the Alborz); shortening arrays (Kopeh Dagh); transfer and tear faults (Sangavar Fault).

Geological background

Collision between Arabia and Eurasia initially took place along the Bitlis–Zagros suture, which curves through SE Turkey before running NW–SE through southern Iran (Fig. 1). The plate boundaries were a passive continental margin on the northern side of the Arabian plate and an active continental margin along southern Eurasia (Şengör et al. 1988; Beydoun et al. 1992).

The plate-scale present-day convergence between Arabia and Eurasia is well understood: GPS studies show that roughly 18 ± 2 mm a^{-1} north–south convergence takes place between the

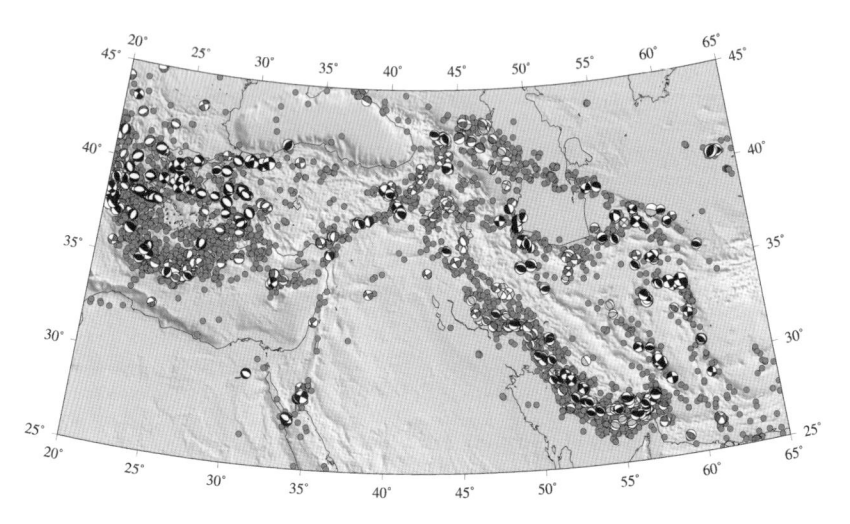

Fig. 2. Seismicity of the Arabia–Eurasia collision zone (from Allen et al. 2006). Small circles are epicentres from the catalogue of Engdahl et al. (1998). Focal mechanisms are from the following sources. Black, waveform modelled, from Jackson (2001) and references therein, with additional events from Talebian et al. (2004) and Walker et al. (2005). Dark grey, best double-couple CMT solutions from the Harvard catalogue (http://www.seismology/harvard.edu/CMTsearch.html) for earthquakes with depth ≤35 km, Mw ≥5.5 and double-couple component ≥70%, in the interval 1977–2002. Light Grey, first motion solutions from Jackson & McKenzie (1984). Earthquakes deeper than 35 km associated with the subduction zones in the Makran, South Caspian and Hellenic Trench have been omitted.

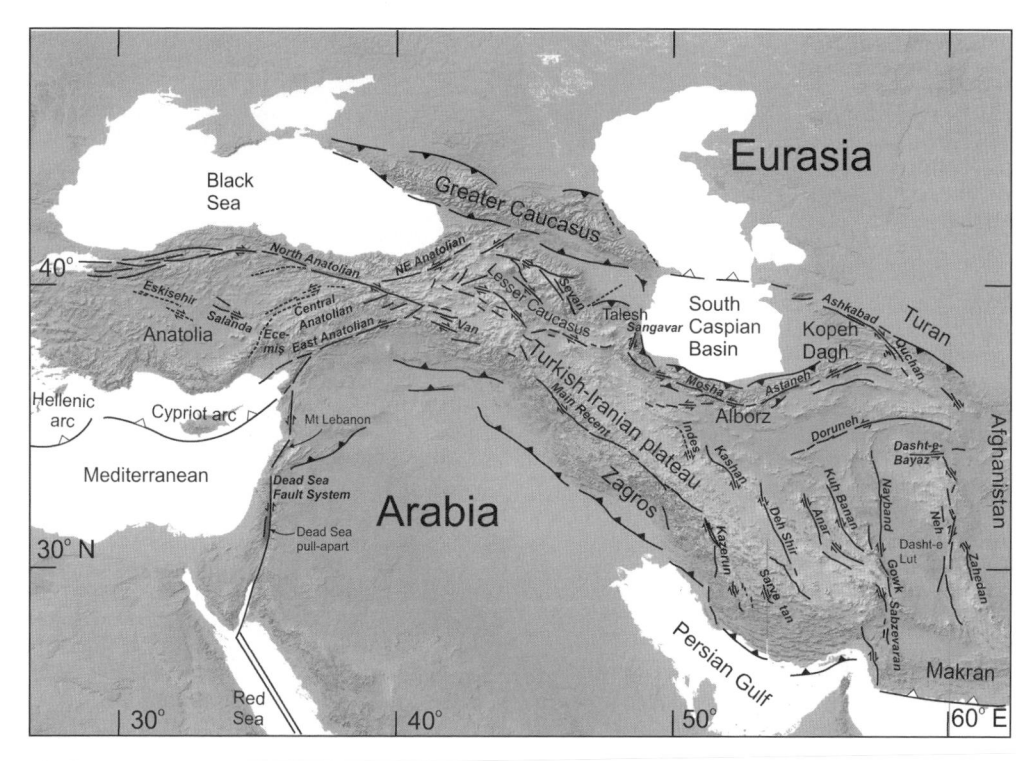

Fig. 3. Major active strike-slip fault zones within the Arabia–Eurasia collision zone. Derived from Allen *et al.* (2006) (Iran), Copley & Jackson (2006) (NW Iran), Allen *et al.* (2003) (northern Iran), Bozkurt (2001) (central and NW Turkey) and Kocyigit *et al.* (2001) (eastern Turkey). Activity on strike-slip faults in much of Anatolia is debated (e.g. Kocyigit & Beyhan 1998; Westaway 1999), so that the Eskisehir and Central Anatolian faults are marked by dashed lines, and others shown by Bozkurt (2001) and Kocyigit *et al.* (2001) are not shown at all. The Salanda Fault is in the vicinity of a strike-slip earthquake of 1938 (Jackson & McKenzie 1984), and so is more confidently assigned as active. Barbed lines show active thrust fronts, schematically. Thrust zones are typically harder to map as precisely, because many of the active thrusts are blind. White barbs are subduction zones at the margins of the South Caspian Basin and Makran and along the Cypriot and Hellenic arcs. Red Sea oceanic spreading is shown schematically by the double line.

stable interiors of Arabia and Eurasia at longitude 48°E (Fig. 1; McClusky *et al.* 2000). Convergence velocities increase and azimuths swing anticlockwise west to east along the collision zone, with a rotation pole in the NE Africa–eastern Mediterranean region (McClusky *et al.* 2003) and velocities *c.* 10 mm a^{-1} higher in eastern Iran than the western side of the collision. GPS and seismicity studies together show that deformation is concentrated between the Persian Gulf and the north side of the Greater Caucasus and Kopeh Dagh ranges; there is a good correlation between the limits of seismicity and topographic fronts (Fig. 2). However, deformation is not distributed evenly within these northern and southern limits. Seismogenic thrusting, and hence plate convergence achieved by crustal shortening and thickening, is concentrated at present in areas below the 1 km topographic contour (Talebian & Jackson 2004). This is mainly within the lower parts of the Zagros and Alborz–Caucasus regions at the southern and northern sides of the collision respectively (Fig. 3). The intervening region has lower relief, elevations commonly over 1.5 km and is known as the Turkish–Iranian plateau. GPS data from within the collision zone reveal that little active internal shortening takes place within this plateau (*c.* 2 mm a^{-1} or less; Vernant *et al.* 2004*a*), and large areas are aseismic. It is not totally quiescent: Late Cenozoic volcanic rocks occur in discrete fields across it (Pearce *et al.* 1990; Keskin *et al.* 1998; Kheirkhah *et al.* 2009), and strike-slip faults are locally associated with historical earthquakes, indicating at least some tectonic activity (Copley & Jackson 2006). Another area of low internal deformation at present is the South Caspian Basin, north of the Alborz (Fig. 3). Para-oceanic basement to this basin is in the early stages of subducting under the northern and possibly western basin margins (Mangino & Priestley 1998; Jackson et al. 2002). This basement is detached from folds within the thick sedimentary cover; these folds are not typically associated with major seismicity, indicating that the basement behaves as a rigid block, presumably because of unusually strong basement.

The western margin of the collision zone is sharply defined along the Dead Sea Fault System, which allows the largely stable interior of Arabia to move northwards with respect to the eastern Mediterranean. This basement to the latter area is not well known as it is buried beneath a thick sedimentary cover, including salt. It is probably underlain by highly thinned continental or even oceanic crust (de Voogd et al. 1992). West of a triple junction at the northern end of the Dead Sea Fault System, subduction of eastern Mediterranean basement takes places along the Cypriot and Hellenic arcs. Collision has not yet taken place in these regions, and north of the Hellenic arc the Aegean crust is rapidly extending. This extensional province merges eastwards in onshore Turkey, into the crust of Anatolia. Here there is little active internal deformation, but wholesale westwards transport between the North and East Anatolian faults (McKenzie 1972). The eastern side of the collision roughly coincides with the political boundary of Iran and Afghanistan; the latter is part of stable Eurasia, in the context of the active deformation field. There is active subduction of Indian plate oceanic lithosphere under the Makran (Regard et al. 2005).

Less is known about the earlier evolution of the collision zone. Even the onset of collision is debated, with recent estimates ranging from Late Eocene (c. 35 Ma) to mid–late Miocene (12–10 Ma) (McQuarrie et al. 2003; Vincent et al. 2005; Guest et al. 2006a; Verdel et al. 2007). Allen & Armstrong (2008) proposed that there was evidence from many localities on both sides of the original suture for Late Eocene (c. 35 Ma) deformation, uplift or changing sedimentation patterns, and that this was the true time of initial collision. This debate on the collision timing highlights how difficult it can be to interpret geological data from ancient settings. It arises in part because we can never have an overview for past times across the entire orogen, in the way that remote sensing, seismicity and GPS all provide for the active tectonics. Therefore data from one region for initial rock uplift, say, can be treated as though it is representative of the entire collision zone. This is misguided, given how the present-day tectonics show the wide variety of deformation, and quiescence, that takes place at any one time.

As a general point, there is no systematic difference in the depths of the strike-slip and thrust earthquakes in the various regions of the collision zone, such as the Alborz and Zagros ranges (Fig. 3). They are typically up to c. 15–20 km; that is, within the crystalline basement of the crust (Jackson et al. 2002; Talebian & Jackson 2004). This indicates that the strike-slip deformation described in this paper is 'thick-skinned' in structural geology terms.

Collision zone boundaries

Reduced to its simplest, the Arabia–Eurasia collision represents roughly north–south convergence between a promontory (Arabia) and a much broader continental mass (Eurasia). Figure 4 is a schematic illustration that highlights the main elements of the collision, and illustrates the role of strike-slip faults and the boundaries of deformation. The real locations of these structures are shown in Figure 3. The pre-collision position of the Arabian and Eurasian plate margins is not precisely known, but the north–south convergence vector requires hundreds of kilometres of northward motion of the stable interior of Arabia with respect to stable Eurasia, over tens of millions of years (McQuarrie et al. 2003; Allen & Armstrong 2008). Therefore it is unsurprising that the northern and southern limits to deformation are marked by thrusting (Fig. 2), allowing for plate convergence via crustal thickening, whereas the western and eastern limits are strike-slip fault zones, allowing the Arabian plate to move past adjacent crust. A crucial difference between the strike-slip faults on the western and eastern margins of the collision is that the former, the Dead Sea Fault System, decouples Arabia from the eastern Mediterranean, but both regions were part of the combined African–Arabian plate before collision. In the case of the east Iranian faults, the great majority of the region involved was part of Eurasia before the initial collision.

Deformation is sharply focused along the c. 1000 km long, left-lateral Dead Sea Fault System (Garfunkel 1981; Fig. 3), except for local splays at releasing and restraining bends such as the Dead Sea pull-apart basin (Manspeizer 1985) and the Mount Lebanon range. The southern end of the fault links into the active extension within the Red Sea: debate continues as to the interaction of extension in this region and initial collision on the northern side of the Arabian plate (Jolivet & Faccenna 2000; McQuarrie et al. 2003). The northern end links into the folds and thrust belts in southeastern Anatolia and the Zagros. Total offset across the fault is c. 105 km south of the Dead Sea (Quennell 1958), and this is fully observed in an offset dyke swarm dated at 22–18 Ma (Eyal et al. 1981). Active and late Quaternary slip rate estimates are variable, at 2–8 mm a^{-1} (e.g. Klinger et al. 2000), although more recent studies have indicated

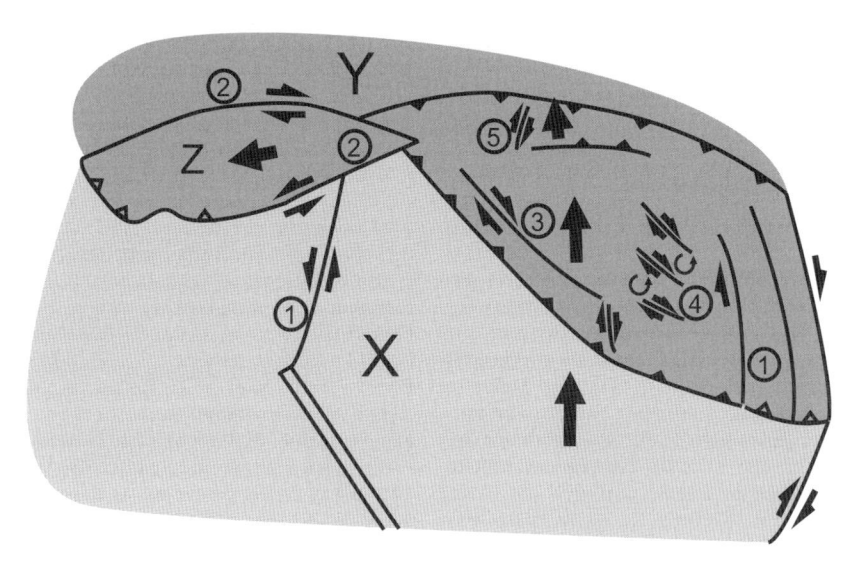

Fig. 4. Schematized kinematics of a continent–continent collision between plates X and Y, modelled after the Arabia–Eurasia collision and showing westward tectonic escape of block Z (i.e. Anatolia) and lateral strike-slip faults at the western and eastern boundary zones. Lines with filled triangles indicate thrusts at the margins of the collision zone; those with open triangles indicate adjacent subduction zones. Bold black arrows indicate velocities with respect to the stable interior of block Y, with length proportional to velocity. The five roles of strike-slip faults described in this paper are highlighted as follows: (1) collision zone boundaries, either diffuse or focused; (2) tectonic escape structures; (3) strain partitioning elements; (4) shortening arrays with vertical axis rotations; (5) transfer zones.

values of c. 5 mm a^{-1} (Ferry et al. 2007; Gomez et al. 2007). This velocity would need c. 20 Ma to achieve the full offset, consistent with the age of the offset dykes, but inconsistent with the fault having operated at this slip rate since the proposed Late Eocene start of collision.

North–south right-lateral faulting in eastern Iran forms the eastern boundary to the collision zone (Figs 1–3). Oceanic subduction takes place under the Makran region, such that right-lateral faults in the extreme SE of Iran juxtapose the easternmost Zagros (originating on the Arabian passive margin) with the accretionary prism to the east (Regard et al. 2005; Bayer et al. 2006). Further north, right-lateral faults to the east (Neh and Zahedan) and west (Nayband and Gowk) of the inert Dasht-e-Lut have a total offset estimated by Walker & Jackson (2004) as c. 80 km. A difference between the eastern and western margins to the collision zone is that in the west there is only one active fault system, whereas in eastern Iran there are at least two active, parallel fault systems, and possibly several more. There is little doubt that the Nayband and Gowk faults and the Neh and Zahedan faults take up most of the slip between Iran and Afghanistan (Walker & Jackson 2004; Walker et al. 2009), but the Deh Shir, Anar and Kuh Bahnan faults are also active (Meyer et al. 2006; Meyer & Le Dortz 2007), plausibly slip at 1–2 mm a^{-1} in the Holocene, and so may contribute part of the overall shear. A more fundamental problem is why deformation is so focused at the western collision margin but distributed in the east. The reason may be the distinct contrast in crustal type at the western side, where the Arabian crust was juxtaposed with para-oceanic basement to the eastern Mediterranean long before initial collision, when both regions formed part of the passive margin at the northern side of the African–Arabian plate. In eastern Iran and Afghanistan there is a mosaic of similar Gondwana-derived basement blocks (Şengör et al. 1988). Those blocks east of the Arabian indentor are not being deformed by the Arabia–Eurasia collision, but there is no sharp contrast within this crust as there is in the west.

The GPS-derived right-lateral shear between eastern Iran and Afghanistan is c. 16 mm a^{-1} (Vernant et al. 2004a). This requires only 5 Ma to achieve the total observed offset along the Neh–Zahedan and Nayband–Gowk faults. Given that all estimates of the initial collision put it much earlier than 5 Ma, something else accomplished right-lateral shear at the eastern side of the collision. The obvious explanation is that the region must contain faults that are now inactive, or only weakly active. The Deh Shir, Anar and Kuh-e Bahnan faults may have contributed relatively more to the boundary shear in the past, regardless of their precise present contribution. There may be further structures within the deserts of eastern Iran as yet unquantified or unrecognized.

Tectonic escape structures

The Arabia–Eurasia collision zone contains the first recognized example of so-called escape tectonics, in the case of Anatolian crust between the North and East Anatolian faults (McKenzie 1972). These are active right- and left-lateral faults respectively, and act to transport intervening crust westwards, largely without internal deformation (Fig. 1). Figure 4 reduces the kinematics to their simplest. The left-lateral East Anatolian Fault is the boundary between Arabia and Anatolia (Fig. 3), and runs for c. 400 km SW of its intersection with the North Anatolian Fault at Karliova, at c. 39.5°N, 41°E. There are several strands to the fault zone, with localized pull-apart basins and push-up zones (Lyberis et al. 1992; Westaway 1994). The GPS-derived slip rate is 9 ± 1 mm a^{-1} (McClusky et al. 2000) and needs to operate for only c. 3 Ma to achieve the geological offset of 27–33 km (Westaway & Arger 1996; Westaway et al. 2006), constrained by offset geological markers. This is in good agreement with the age of initial offset as late Pliocene (c. 3 Ma) or younger (Şaroğlu et al. 1992; Westaway & Arger 2001), based on the offset of volcanic rocks of this age.

The right-lateral North Anatolian Fault (NAF) achieves the slip between Eurasian and Anatolian crust for >1200 km (Figs 1 & 2), at a GPS-derived slip rate of 24 ± 1 mm a^{-1} (McClusky et al. 2000). The western end of the fault splits where it enters the north Aegean and passes into the extensional deformation in that region. Roughly 80–85 km is emerging as a consensus figure for the total offset of most of the length of the fault zone, based on combinations of geological and drainage offsets (Seymen 1975; Westaway 1994; Armijo et al. 1999). Distributed strike-slip and/or extension took place in the mid- or late Miocene, before the establishment of the present fault trace in some regions (e.g. Barka & Hancock 1984; Tüysüz et al. 1998; Coskun 2000; Şengör et al. 2005). There is no consensus on a precise age for the start of motion on the NAF, despite several estimates of c. 5 Ma (see Bozkurt 2001). The GPS-derived slip rate (24 ± 1 mm a^{-1}) achieves the total offset of 80–85 km in only c. 3.5 Ma, less than most geological estimates for the fault age. It seems that (1) the slip rate is higher now than in the past (but this is uncertain), and (2) the fault has not been active since the start of collision (this is more definite).

As in Dead Sea Fault System, the narrowness of both the NAF and East Anatolia Fault (EAF) and the sharp velocity contrasts across them resemble plate boundaries, as utilized as long ago as McKenzie (1972) in his vector calculations. However, this is a nearly instantaneous picture, and it is striking that both faults are young with respect to the overall collision zone, and need only a few million years at their present slip rates to achieve their total offset. In the case of the EAF, other faults may have played similar kinematic roles in the past. Other (inactive?) left-lateral faults have been identified in eastern Turkey, such as the Malatya–Ovacik Fault (Westaway & Arger 2001), with c. 29 km offset between 3 and 5 Ma, and the Ecemiş Fault (Jaffey & Robertson 2001), with c. 60 km offset, mainly between the Late Eocene and Miocene. Activity on the Central Anatolian Fault (Kocyigit & Beyhan 1998) is disputed (Westaway 1999). However, as the triple junction at the eastern end of NAF and EAF should migrate west with time, it is difficult to see how any of these inactive left-lateral faults in eastern Anatolia were the precise equivalent of the modern EAF.

Elements in strain partitioning

Plate boundaries are rarely orthogonal to plate vectors (Woodcock 1986). This fact underlies the origins of many continental strike-slip faults, not only in collision zones. Accommodation of north–south convergence by east–west-trending faults would be likely in idealized, isotropic crust, but has not happened in the heterogeneous crust of both Arabia and Eurasia. The suture zone trends NW–SE for much of its length (mainly within Iran), at roughly 45° to the plate convergence vector. Pre-collision structural fabrics commonly lie parallel to the suture within both plates (e.g. Sarkarinejad et al. 2008). The pattern of active faulting in the Zagros strongly suggests that pre-collision normal faults in the Arabian passive margin are now active as thrusts. Conclusive evidence for single fault reactivation is rarely available, largely because of a thick sediment carapace over blind thrusts, but most folds and thrusts in the NW Zagros trend NW–SE, parallel to both the suture and the trend of pre-collision sediment isopachs (Beydoun et al. 1992). The resultant NE–SW shortening is therefore oblique to the north–south plate convergence, and cannot achieve it on its own. The answer is the combination of this thrusting with adjacent strike-slip faulting, in an example of so-called strain partitioning (Fig. 4).

Along the NE side of the Zagros, loosely along the line of the original suture, there is a right-lateral strike-slip fault, the Main Recent Fault (MRF) (Talebian & Jackson 2002; Fig. 3). Offset along the MRF is c. 50 km (Talebian & Jackson 2002).

Shortening across the widest structural unit in the Zagros, the Simple Folded Zone, is similar in magnitude (Blanc et al. 2003; McQuarrie 2004). Combining the two estimates suggests c. 70 km of north–south convergence across the Zagros, by applying Pythagoras' rule (Fig. 5a). This is valid only if the strains took place at the same time. It is clear that shortening across the Zagros is active, and focused on lower elevations (<1 km) in the Simple Folded Zone. Vernant et al. (2004a) estimated 6.5 ± 2 mm a^{-1} north–south convergence at longitude c. 51°E, in their GPS survey of Iran. Likewise, both seismicity and GPS data indicate right-lateral slip along the Main Recent Fault, and the difference in slip vector azimuths between the Main Recent Fault and the Simple Folded Zone emphasizes the effectiveness of partitioning. However, the active slip rates do not fit a Pythagorean triangle as neatly as the total displacements, because GPS-derived slip along the MRF is only

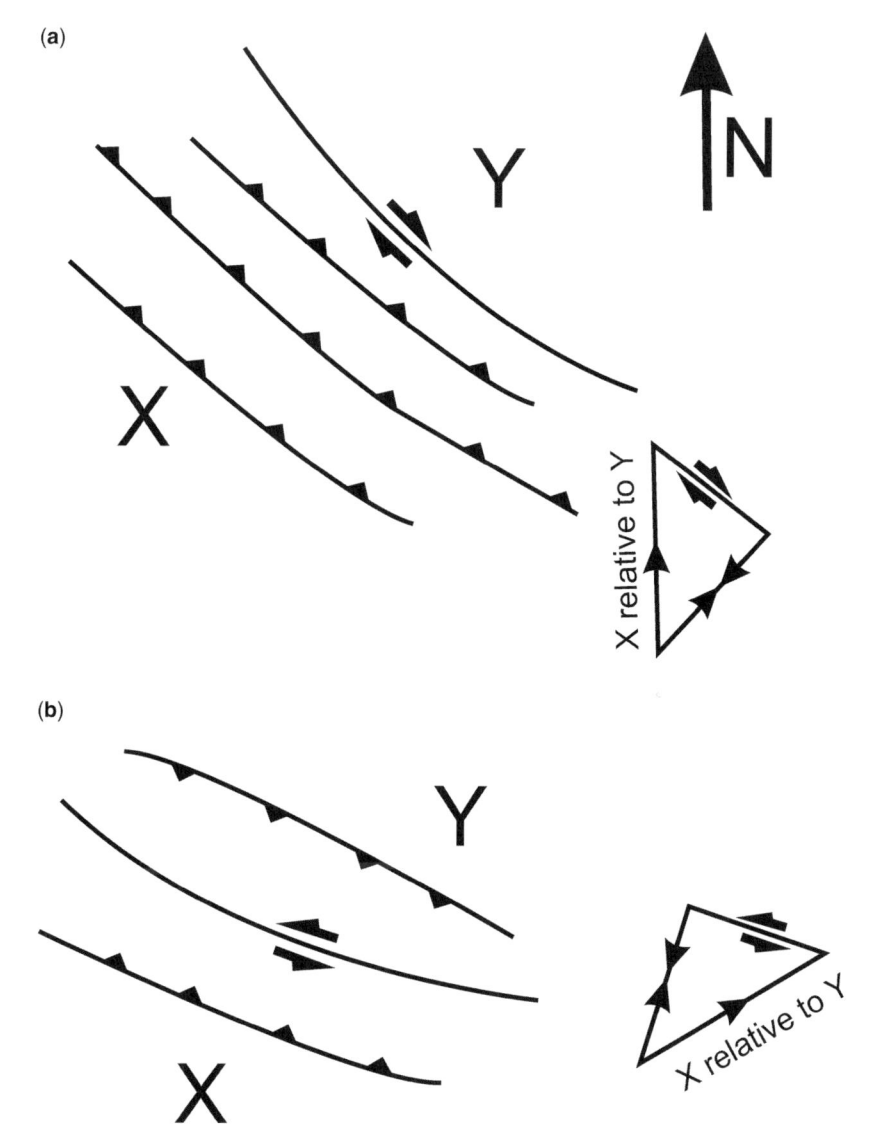

Fig. 5. The concept of strain partitioning. (**a**) Combined slip on the strike-slip fault and shortening across the adjacent thrust belt produces net convergence oblique to the fault trends; northward motion of block X with respect to Y. This scenario is similar to the NW Zagros Simple Folded Zone. (**b**) Strain partitioning where the strike-slip fault system lies within the interior of the thrust zone. This geometry is similar to the Alborz mountains.

3 ± 2 mm a^{-1} (Vernant et al. 2004a). This is less than the expected ≥ 10 mm a^{-1}, if the onset of slip was ≤ 5 Ma (Talebian & Jackson 2004). A further complication is that the Simple Folded Zone deformation may have begun earlier than 5 Ma, as suggested by syn-fold deposition at c. 8 Ma near the Zagros foreland (Homke et al. 2004).

Another example of strain partitioning in the active collision zone is from the Alborz mountains of northern Iran (Jackson et al. 2002; Allen et al. 2003; Guest et al. 2006b). This range lies between the Turkish–Iranian plateau to the south and the South Caspian Basin to the north (Fig. 3). It is actively thrusting to both the north and south, and cut by range-parallel left-lateral strike-slip faults with offsets in the order of several tens of kilometres (Mosha, Astaneh; Fig. 6) (Allen et al. 2003; Ritz et al. 2006; Hollingsworth et al. 2008). These are apparently segmented along strike, and at least locally more than one parallel fault segment is active, such as the Damghan Fault south of the longer Astaneh Fault. The resultant oblique motion across the range allows for westward motion of the rigid South Caspian basement with respect to Iran. As in the Zagros, the variation in earthquake slip vector azimuths helps make the case for effective strain partitioning (Jackson et al. 2002). Thus in contrast to the Zagros example, the strike-slip component of oblique shortening takes place predominantly within the thrust belt (Figs 5b & 6). Vernant et al. (2004b) determined the north–south shortening rate across the Alborz as 5 ± 2 mm a^{-1} and the left-lateral shear as 4 ± 2 mm a^{-1}, from a GPS study. Ritz et al. (2006) identified an extensional component on some of the left-lateral faults, which they suggested represented a Quaternary reorganization of the deformation.

There is evidence for older, but probably late Cenozoic, right-lateral faulting along parts of the range (Axen et al. 2001; Allen et al. 2003; Guest et al. 2006b; Zanchi et al. 2006). Thus at least part of the Alborz strike-slip system shows evidence of rapid reversal of its sense of motion, possibly within the last few million years. Given that the folding within the South Caspian cover succession is only a few million years old at most (Devlin et al. 1999) and that the overall westward motion of the South Caspian basement is very young (Jackson et al. 2002), the present fault configuration may be as recent as the Quaternary (Ritz et al. 2006). In contrast, Hollingsworth et al. (2008) showed that present slip rates in the eastern Alborz require c. 10 Ma to achieve the total offset, suggesting that the present kinematics go back further in time.

The combination of left-lateral faulting along the Alborz and right-lateral faulting along the Zagros has attracted repeated interest over the years,

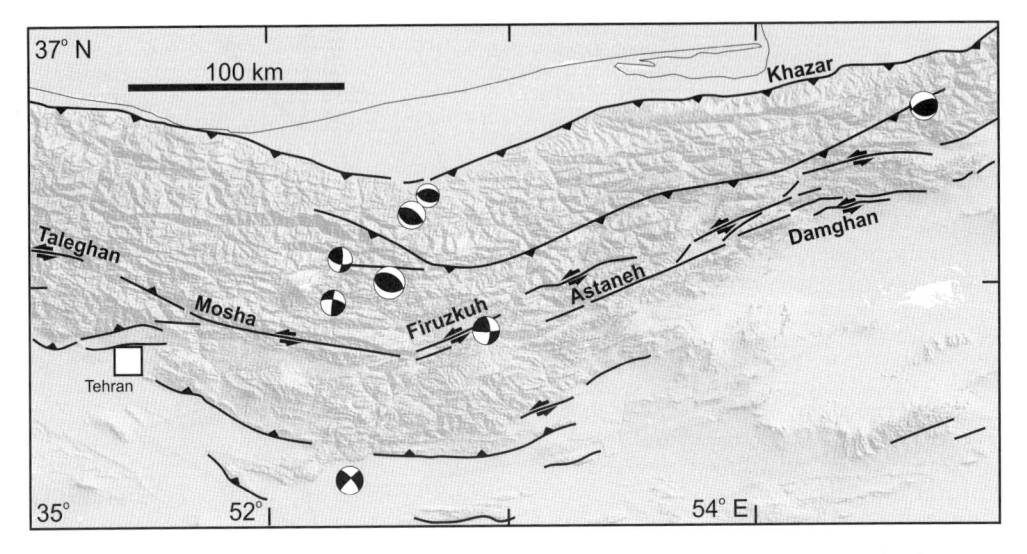

Fig. 6. Active faults in the Alborz between 51° and 55°E. Left-lateral faulting occurs within the range interior, principally on the Taleghan, Mosha, Firuzkuh and Astaneh faults, which collectively form a segmented fault system. Thrusting takes place on inward-dipping faults at both the northern and southern margins of the range. The continuity of the Khazar Fault may be an artefact of Caspian lake highstands bevelling southwards against the bedrock of the range: the thrust is blind. Map derived from Allen et al. (2003), Ritz et al. (2006), Hollingsworth et al. (2008) and analysis of Shuttle Radar Topography Mission (SRTM) digital topography; focal mechanisms from Jackson et al. (2002) and Tatar et al. (2007).

promoting the idea of eastward escape of Iranian crust out of the collision zone, in an apparent mirror image to the westward transport of Anatolian crust (McKenzie 1972; Axen et al. 2001; Bachmanov et al. 2004). Both seismicity data (Jackson et al. 1995) and the GPS-derived velocity field (Vernant et al. 2004a) show that this is not the case (Fig. 1), and that the strike-slip faults parallel to each range help accommodate oblique convergence across them (Allen et al. 2006). It the case of the Zagros, the resultant convergence is parallel to the regional plate vector. The Alborz strike-slip relates to the South Caspian basement moving as a rigid block within the collision zone, at a high angle to the overall plate convergence vector. This case study is a warning for all interpretations of escape tectonics in ancient orogens, where seismicity data and GPS-velocity fields are not feasible and the regional plate kinematics is not known: it is possible that such settings represent the strike-slip component of strain partitioning as outlined here. It should be feasible to distinguish between real and illusory escape tectonics, given that an essential component of strain partitioning is an adjacent zone of contemporary thrusting. In Anatolia, the neotectonic strike-slip faulting postdates previous thrusting and thickening.

Jackson (1992) noted that pure dip-slip thrusting in the Greater Caucasus took place on slip vectors oriented clockwise of the overall convergence vector at this longitude. The overall convergence vector is achieved by combining this shortening in the Greater Caucasus with right-lateral strike-slip faulting to the south, within the Lesser Caucasus and the interior of the Turkish–Iranian plateau. This is most active in a WNW–ENE-trending swarm of right-lateral faults including Van (Fig. 3). Copley & Jackson (2006) also found that an array of NW–SE right-lateral strike-slip faults accommodate a NW–SE velocity gradient of NE-directed velocity; these faults are located between the Van and Sevan faults. An aspect of this right-lateral shear within the Turkish–Iranian plateau (south of the Greater Caucasus) is that it is distributed across many faults, rather than focused on one main structure, which is the case to the west and SE in the NAF and Main Recent Fault respectively. In part this may be because of the presence of linear pre-Cenozoic sutures in the latter areas, available for reactivation. But it also relates to the way strain is partitioned across a much wider area than either the Zagros or Alborz, with the shortening component in the Greater Caucasus located north of the strike-slip faults (Jackson 1992). The strike-slip fault system is constantly transported northwards by the shortening in the Greater Caucasus, in a way that does not happen in either the Alborz or Zagros.

Shortening arrays

Escape tectonics is one scenario where continental shortening takes place without crustal thickening. Strike-slip faults can achieve crustal shortening in another way, via arrays of en echelon faults rotating about vertical axes as they slip (Fig. 4). The situation has parallels with the behaviour of normal faults in rift zones; in the latter case the faults rotate about horizontal axes as they slip and thin and extend the crust. In the strike-slip setting the net result is shortening across the fault zone and lengthening along it. Such fault arrays have recently been recognized in several places within the Arabia–Eurasia collision zone, mainly by James Jackson and colleagues.

The Kopeh Dagh in northeastern Iran lies on the northern side of the collision, between the Turkish–Iranian plateau to the south and the undeformed crust of the Turan platform to the north (Fig. 3). Its structure is dominated by arcuate but broadly NW–SE-trending folds and thrusts, which deform and expose Mesozoic and Early Cenozoic strata at current exposure levels. The right-lateral and range-parallel Ashkabad Fault lies along the northeastern margin of the range, trending WNW–ESE, such that the combination of slip along this fault and shortening–thickening across the range is another example of strain partitioning in the collision zone (Lyberis & Manby 1999). However, the folds and thrusts are offset by an en echelon array of right-lateral faults that strike NNW–SSE or NW–SE (Hollingsworth et al. 2006), such as the Quchan Fault. Palaeomagnetic data are not available to quantify tectonic rotations, but the folds of Mesozoic strata can be traced across the fault zones and the rotations thereby quantified. Knowing the rotations and the present dimensions of the fault arrays allows the total north–south shortening achieved by these faults to be estimated as c. 60 km (Hollingsworth et al. 2006). The geometry of such a fault array is shown schematically in Figure 7. GPS data (Vernant et al. 2004a) put the total north–south convergence across the Kopeh Dagh as c. 7 mm a^{-1}. As there are no detailed estimates for crustal shortening via thrusting and thickening, it is difficult to compare geodetic and long-term deformation rates across the range.

A similar fault array exists south of the Kopeh Dagh (Fig. 3), at the northern end of the north–south right-lateral structures within eastern Iran, where these faults die out and are replaced by left-lateral faults that appear to be rotating clockwise about vertical axes (Dasht-e Bayaz and Doruneh; Jackson & McKenzie 1984; Walker & Jackson 2004). The slip along the Deh Shir, Anar and Kuh Bahnan faults further south again (Fig. 3) may be another example of this behaviour fault array is active and allows for shortening within the tip of the Arabian promontory (Copley & Jackson 2006). Other right-lateral faults trend NNE–SSW or NW–SE across central Iran (e.g. Kashan, Indes). There is limited seismicity on some of these (Fig. 2), but little indication that they contribute much to the overall strain pattern at present.

Deformation in the Greater Caucasus represents the northern component of the collision zone at present. Initial uplift in the range may be as old as Late Eocene (Vincent et al. 2007), such that this range carries a longer record of compressional deformation than most parts of the collision zone. Attention has focused on range-parallel thrusts, held responsible for a present-day convergence rate of c. 10 mm a^{-1} across it (Reilinger et al. 2006). However, there are oblique features within or close to the Greater Caucasus that look like fault zones at high angles to the overall structural trend. In particular, several folds terminate along NW–SE lines, just inland of the Caspian shoreline (Fig. 3). Other structural breaks have the same orientation in the same region. No offsets are identifiable in the exposed geology, so that it is uncertain what these trends mean.

Transfer zones and tear faults

A textbook explanation for strike-slip faults within zones of compressional deformation is that they link along-strike sections of the thrusts, either where the latter die out laterally and strain needs to be relayed to another structure, or because it would be mechanically unfeasible to move the thrust sheets if they were too long. Such strike-slip faults are known as tear faults, or transfer faults. They have not been highlighted within the active fold and thrust belts of the Arabia–Eurasia collision. In part this may relate to the blind nature of many thrusts within the Zagros, Alborz, Caucasus and Kopeh Dagh: thrust earthquakes do not typically rupture to the surface through the thick sedimentary cover of these ranges. (This is in contrast to many of the longer strike-slip faults, where earthquake magnitudes can be higher, and surface ruptures are common for the larger events.)

Transfer zones are present on larger scales, although there is potential overlap with some of the other kinematic roles defined in this paper (Fig. 4). The Zagros Simple Folded Zone is cut by NNW–SSE- or NE–SE-trending right-lateral faults such as Kazerun and Sabz Pushan (Fig. 8). These have offsets of a few to a few tens of kilometres. Higher estimates, based on range-wide structural and geomorphological correlations (Berberian 1995) are not confirmed by local

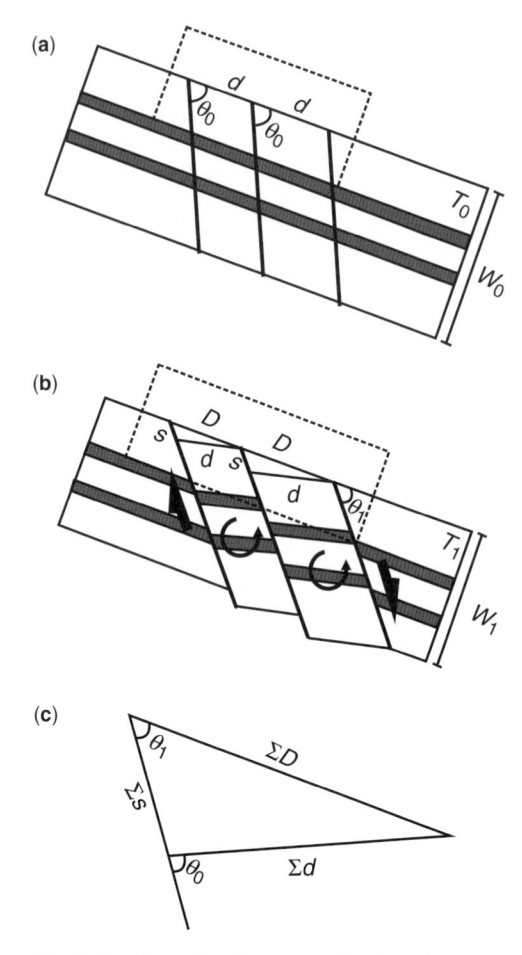

Fig. 7. Rotating strike-slip arrays acting to produce shortening and along-strike elongation (from Hollingsworth et al. 2006), as seen in the Kopeh Dagh. (**a**) Fault blocks have initial width d and angle θ_0 with the deformation zone boundary, across a zone of width W_0. Grey bands represent fold trends, which act as strain markers as the faults and fault blocks are offset and rotated. (**b**) Offset and fault block rotation produces new boundary length D, and angle θ_1, across a width W_1. (**c**) If all fault block rotations are of the same amount, the geometry simplifies to a single triangle with lengths ΣD, Σd and Σs. Measurement of ΣD, Σs, θ_0 and θ_1 allows the original length of the deforming boundary (Σd) to be calculated using the cosine rule.

(Walker & Jackson 2004) and not simply related to the eastern margin of the collision zone (Meyer & Le Dortz 2007). This explanation has the advantage that such faults are well within the interior of Iran, and so seem poorly located to contribute to shear resulting from the contrast with Afghanistan beyond the collision zone. At the far NW of Iran and in easternmost Turkey a similar right-lateral

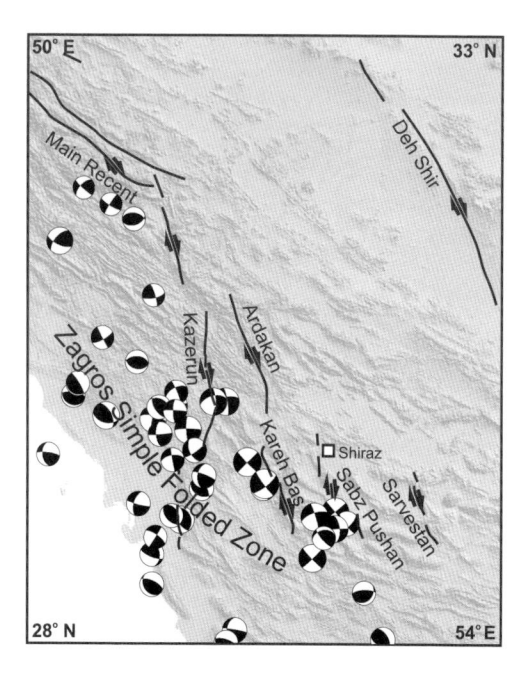

Fig. 8. Active strike-slip faults in the Central Zagros. Several segmented right-lateral faults fan out from the southeastern end of the Main Recent Fault. Fault locations derived from Authemayou *et al.* (2006) and analysis of SRTM imagery. Focal mechanisms for thrust and strike-slip events in the region are from the Harvard and USGS catalogues (http://neic.usgs.gov/neis/sopar/) for earthquakes with Mw ≥ 5 and double-couple component ≥70%, in the interval 1986–2005.

studies (Authemayou *et al.* 2006). Talebian & Jackson (2004) related these faults to the strike-slip deformation present along the MRF, and the need for lengthening along the Simple Folded Zone as a result of this slip. This is the same style of behaviour as the rotating fault arrays described in the previous section. However, predicted anticlockwise rotations have not been detected palaeomagnetically (Aubourg *et al.* 2008). Blanc *et al.* (2003) noted that the strain partitioning in the NW Zagros does not occur in the east, where folds and thrusts are aligned roughly east–west, orthogonal to the convergence vector, with no strike-slip equivalent to the motion of the MRF. The strike-slip faults within the Simple Folded Zone act to link the zones of strain partitioning and no strain partitioning; single folds cut by the strike-slip faults also change orientation across them, becoming more east–west further east.

Another scale of transfer behaviour occurs at the western side of the Alborz, where the north–south right-lateral Sangavar Fault (Berberian & Yeats 1999) links the Alborz to the folds and thrusts in the Talesh (Talysh) range to the north (Fig. 9). The arcuate and highly 3D nature of the structure in this part of the collision zone relates to the rigid basement of the South Caspian Basin, which underthrusts the Talesh to its west on very gently dipping thrusts (Jackson *et al.* 2002). This is superimposed on a component of the regional north–south convergence, such that the overall kinematics appears highly variable in this region (Masson *et al.* 2006), despite the remarkable consistency in the velocity field with respect to Eurasia (Fig. 9). Deformation at the SE corner of the collision zone is similarly complex, where the eastern Zagros abuts the Makran accretionary prism (Regard *et al.* 2005; Bayer *et al.* 2006).

Discussion

The examples described above demonstrate the different roles that strike-slip faults can play in one timeframe of one collision zone. Some generalizations are possible. Strike-slip faults form the boundaries of major deformation zones, where these involve translation rather than convergence or extension. Strain partitioning involves strike-slip faults acting in concert with adjacent, parallel thrusts to achieve the overall convergence vector required by far-field conditions. 'Far field' mainly means the overall plate convergence zone, but can be rigid blocks moving within it, such as the South Caspian basement. Such partitioning produces the potential for the misinterpretation of strike-slip faults as tectonic escape structures. Tectonic escape is the valid interpretation for the NAF and EAF, where independent estimates of the regional velocity field confirm the westward transport of Anatolia with respect to both Arabia and Eurasia. This is not the case for central Iran, where strike-slip faults along the Alborz and Zagros ranges work with parallel thrusts to produce oblique convergence across each range. Geoscientists typically think of thrusts as the predominant structures in orogens, with mountain building as the result. En echelon right-lateral strike-slip faults within Iran show the potential for rotating arrays to achieve plate convergence, without crustal thickening. Such arrays are found both within areas of active thickening (Zagros, Kopeh Dagh, and, possibly, the Greater Caucasus), but also within the Turkish–Iranian plateau, where crustal thickening has ceased (Allen *et al.* 2004). In the latter case, the strike-slip mechanism for convergence has the advantage that it does not require work against gravity, which is important in areas of thickened and/or elevated crust where buoyancy forces oppose crustal thickening. A textbook explanation for strike-slip faults within fold and thrust belts is that they link single thrusts, and

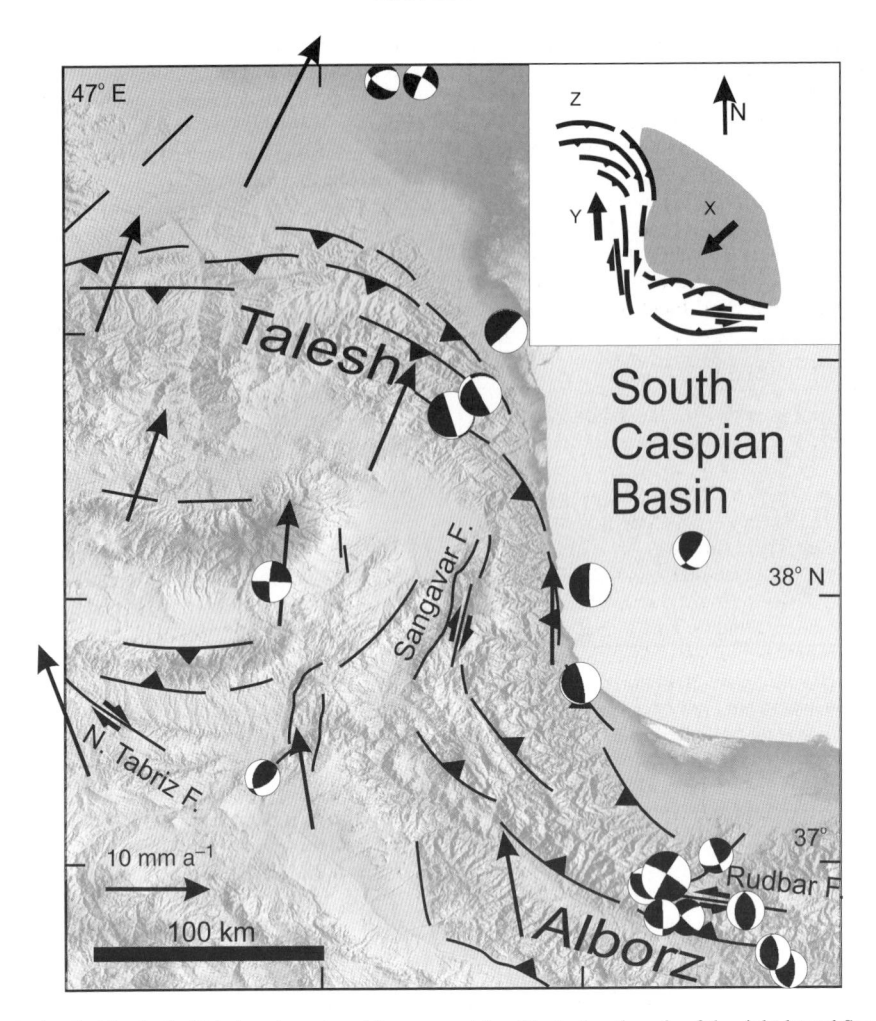

Fig. 9. Active faulting in the Talesh and western Alborz mountains, illustrating the role of the right-lateral Sangavar Fault as a transfer fault between the regions. Focal mechanisms from Jackson *et al.* (2002), with three additional events from the Harvard and USGS catalogues (http://neic.usgs.gov/neis/sopar/) for earthquakes with Mw ≥ 5 and double-couple component ≥70%, in the interval 2002–2007. Arrows show GPS-derived velocities with respect to Eurasia, from Masson *et al.* (2006). These do not change markedly across the region, despite the wide variation in fault strikes and focal mechanisms. The inset shows a schematic transfer zone between two thrust belts, modelled on the junction of the Talesh and Alborz ranges. Deformation not only wraps around the rigid basement of block X, but also has to accommodate its motion independent of the north–south convergence of larger regions Y and Z. This produces highly arcuate and complex fault geometries, which are unlikely to be stable over long periods.

ensure the continuity of strain across large regions. Such features have not been emphasized to date within the Arabia–Eurasia collision zone, but this may be because many thrusts in actively thickening areas are blind. Larger transfer zones exist, linking entire fold and thrust belts such as the western Alborz and southern Talesh (Fig. 9).

In Woodcock's (1986) review of strike-slip faults at plate boundaries, all of the faults described in this paper would be included in the type 'Indent-linked strike-slip fault', with the exception of the collision zone boundary faults, which partly equate to the 'Boundary transform' type. The kinematics of the faults within the Arabia–Eurasia collision, and interpretations on the roles they play in plate convergence, permit a more specific analysis. The five categories listed here (collision zone boundaries, tectonic escape structures, strain partitioning elements, shortening arrays and transfer zones; Fig. 3) are not meant to be rigid. No doubt future studies will allow further refinement. The different kinematic roles are not necessarily mutually exclusive. Strike-slip faults in the Zagros link the western and eastern parts of this fold and thrust belt, but also contribute a small amount of shortening across the range (Fig. 8).

In recent years there has been a debate as to whether continental deformation is best described by continuum models (where the emphasis is on the smoothness of the velocity field; England & Molnar 2005), or a rigid block model (where the role of single fault zones is paramount, and a quasi-plate-tectonic approach to the kinematics is valid; Thatcher 2007). The Arabia–Eurasia collision has been involved in this debate, because of the availability of GPS and seismicity data on its deformation. Reilinger et al. (2006) modelled the behaviour of the collision zone as a series of blocks, which collectively satisfied the overall velocity field. This approach involved reducing regions as broad and complex as the Zagros (200–300 km width) to a single boundary. Liu & Bird (2008) performed a finite-element analysis of active deformation between eastern Anatolia and Burma, modelling geodetic data, geological fault slip rates and seismic moment tensor orientations. They showed that throughout the entire collision zone deformation was distributed, with only a few embedded rigid blocks, such as the South Caspian and Black Sea basins. These have para-oceanic basement distinct from the surrounding continental crust. The derived anelastic strain rate (0.7% Ma^{-1}) across the collision zone, apart from these rare blocks, is inconsistent with a rigid microplate model.

The two approaches outlined above produce radically different results. Each is correct in the technical sense that the data are properly handled in the framework of the model parameters. As Thatcher (2007) noted, the transition between the two end-member behaviours is blurred: as fault number increases, block size decreases. The important question is, which is the more realistic model of continental behaviour, given the way faulting is distributed across the continental crust in the active examples we have available for study? In this context it is not only the number of fault zones within the Arabia–Eurasia collision that is notable, but also their ability to rotate, reverse, accelerate or die within geologically short length and time scales. Such mobility indicates that a distributed model is the more useful way of understanding the deformation, rather than reduction to a small number of rigid microplates. Most of this review has focused on active or at least late Quaternary deformation, because of the wealth of data available for fault slip rates on these time scales. However, a satisfactory description of how deformation occurs within the continents may appear only when we have enough data on the pre-neotectonic kinematics. To apply the phrase Brian Windley has made famous, their behaviour cannot be summarized by a snapshot; the key lies in how the continents evolve.

It is a pleasure to acknowledge and thank Brian Windley for his support and guidance over the last two decades. I am also grateful to the Geological Survey of Iran and the Geology Institute, Azerbaijan Academy of Sciences, for their collaborations on the Arabia–Eurasia collision. The data and ideas reviewed in this paper draw heavily on numerous conversations over the years with J. Jackson and R. Walker, and their insightful papers on the active tectonics of Iran.

References

ALLEN, M., JACKSON, J. & WALKER, R. 2004. Late Cenozoic reorganization of the Arabia–Eurasia collision and the comparison of short-term and long-term deformation rates. *Tectonics*, **23**, pTC2008, doi:10.1029/2003TC001530.

ALLEN, M. B. & ARMSTRONG, H. A. 2008. Arabia–Eurasia collision and the forcing of mid Cenozoic global cooling. *Palaeogeography, Palaeoclimatology, Palaeoecology*, **265**, 52–58.

ALLEN, M. B., GHASSEMI, M. R., SHAHRABI, M. & QORASHI, M. 2003. Accommodation of late Cenozoic oblique shortening in the Alborz range, northern Iran. *Journal of Structural Geology*, **25**, 659–672.

ALLEN, M. B., WALKER, R., JACKSON, J., BLANC, E. J.-P., TALEBIAN, M. & GHASSEMI, M. R. 2006. Contrasting styles of convergence in the Arabia–Eurasia collision: why escape tectonics does not occur in Iran. *In*: DILEK, Y. & PAVLIDES, S. (eds) *Postcollisional Tectonics and Magmatism in the Mediterranean Region and Asia*. Geological Society of America, Special Papers, **409**, 579–589.

ARMIJO, R., MEYER, B., HUBERT, A. & BARKA, A. 1999. Westward propagation of the North Anatolian fault into the northern Aegean: timing and kinematics. *Geology*, **27**, 267–270.

AUBOURG, C., SMITH, B., BAKHTARI, H. R., GUYA, N. & ESHRAGHI, A. 2008. Tertiary block rotations in the Fars Arc (Zagros, Iran). *Geophysical Journal International*, **173**, 659–673.

AUTHEMAYOU, C., CHARDON, D., BELLIER, O., MALEKZADEH, Z., SHABANIAN, E. & ABBASSI, M. R. 2006. Late Cenozoic partitioning of oblique plate convergence in the Zagros fold-and-thrust belt (Iran). *Tectonics*, **25**, TC3002, doi:10.1029/2005tc001860.

AXEN, G. J., LAM, P. S., GROVE, M., STOCKLI, D. F. & HASSANZADEH, J. 2001. Exhumation of the west–central Alborz Mountains, Iran, Caspian subsidence, and collision-related tectonics. *Geology*, **29**, 559–562.

BACHMANOV, D. M., TRIFONOV, V. G. ET AL. 2004. Active faults in the Zagros and central Iran. *Tectonophysics*, **380**, 221–241.

BARKA, A. A. & HANCOCK, P. L. 1984. Neotectonic deformation patterns in the convex-northwards arc of the North Anatolian fault. *In*: DIXON, J. E. &

ROBERTSON, A. H. F. (eds) *The Geological Evolution of the Eastern Mediterranean*. Geological Society, London, Special Publications, **17**, 763–773.

BAYER, R., CHERY, J. ET AL. 2006. Active deformation in Zagros–Makran transition zone inferred from GPS measurements. *Geophysical Journal International*, **165**, 373–381.

BERBERIAN, M. 1995. Master 'blind' thrust faults hidden under the Zagros folds: active basement tectonics and surface morphotectonics. *Tectonophysics*, **241**, 193–224.

BERBERIAN, M. & YEATS, R. S. 1999. Patterns of historical earthquake rupture in the Iranian plateau. *Bulletin of the Seismological Society of America*, **89**, 120–139.

BEYDOUN, Z. R., HUGHES CLARKE, M. W. & STONELEY, R. 1992. Petroleum in the Zagros Basin: a late Tertiary foreland basin overprinted onto the outer edge of a vast hydrocarbon-rich Paleozoic–Mesozoic passive-margin shelf. *In*: MACQUEEN, R. & LECKIE, D. (eds) *Foreland Basins and Foldbelts*. AAPG Memoirs, **55**, 309–339.

BLANC, E. J.-P., ALLEN, M. B., INGER, S. & HASSANI, H. 2003. Structural styles in the Zagros Simple Folded Zone, Iran. *Journal of the Geological Society, London*, **160**, 400–412.

BOZKURT, E. 2001. Neotectonics of Turkey – a synthesis. *Geodinamica Acta*, **14**, 3–30.

COPLEY, A. & JACKSON, J. 2006. Active tectonics of the Turkish–Iranian Plateau. *Tectonics*, **25**, doi:10.1029/2005TC001096.

COSKUN, B. 2000. North Anatolian Fault–Saros Gulf relationships and their relevance to hydrocarbon exploration, northern Aegean Sea, Turkey. *Marine and Petroleum Geology*, **17**, 751–772.

DEVLIN, W., COGSWELL, J. ET AL. 1999. South Caspian Basin: young, cool, and full of promise. *GSA Today*, **9**, 1–9.

DE VOOGD, B., TRUFFERT, C., CHAMOTROOKE, N., HUCHON, P., LALLEMANT, S. & LEPICHON, X. 1992. 2-ship deep seismic-soundings in the basins of the eastern Mediterranean Sea (Pasiphae cruise). *Geophysical Journal International*, **109**, 536–552.

DEWEY, J. F., HEMPTON, M. R., KIDD, W. S. F., SAROGLU, F. & ŞENGÖR, A. M. C. 1986. Shortening of continental lithosphere: the neotectonics of Eastern Anatolia – a young collision zone. *In*: COWARD, M. & RIES, A. (eds) *Collision Tectonics*. Geological Society, London, Special Publications, **19**, 3–36.

ENGDAHL, E. R., VAN DER HILST, R. & BULAND, R. 1998. Global teleseismic earthquake relocation with improved travel times and procedures for depth determination. *Bulletin of the Seismological Society of America*, **88**, 722–743.

ENGLAND, P. & MOLNAR, P. 2005. Late Quaternary to decadal velocity fields in Asia. *Journal of Geophysical Research – Solid Earth*, **110**, doi:10.1029/2004JB003541.

EYAL, M., EYAL, Y., BARTOV, Y. & STEINITZ, G. 1981. The tectonic development of the western margin of the Gulf of Elat, Aqaba, rift. *Tectonophysics*, **80**, 39–66.

FERRY, M., MEGHRAOUI, M., ABOU KARAKI, N., AL-TAJ, M., AMOUSH, H., AL-DHAISAT, S. & BARJOUS, M. 2007. A 48-kyr-long slip rate history for the Jordan Valley segment of the Dead Sea Fault. *Earth and Planetary Science Letters*, **260**, 394–406.

GARFUNKEL, Z. 1981. Internal structure of the Dead Sea leaky transform (rift) in relation to plate kinematics. *Tectonophysics*, **80**, 81–108.

GOMEZ, F., KARAM, G. ET AL. 2007. Global Positioning System measurements of strain accumulation and slip transfer through the restraining bend along the Dead Sea fault system in Lebanon. *Geophysical Journal International*, **168**, 1021–1028.

GUEST, B., STOCKLI, D. F., GROVE, M., AXEN, G. J., LAM, P. S. & HASSANZADEH, J. 2006a. Thermal histories from the central Alborz Mountains, northern Iran: implications for the spatial and temporal distribution of deformation in northern Iran. *Geological Society of America Bulletin*, **118**, 1507–1521.

GUEST, B., AXEN, G. J., LAM, P. S. & HASSANZADEH, J. 2006b. Late Cenozoic shortening in the west–central Alborz Mountains, northern Iran, by combined conjugate strike-slip and thin-skinned deformation. *Geosphere*, **2**, 35–52.

HOLLINGSWORTH, J., JACKSON, J., WALKER, R., GHEITANCHI, M. R. & BOLOURCHI, M. J. 2006. Strike-slip faulting, rotation, and along-strike elongation in the Kopeh Dagh mountains, NE Iran. *Geophysical Journal International*, **166**, 1161–1177.

HOLLINGSWORTH, J., JACKSON, J., WALKER, R. & NAZARI, H. 2008. Extrusion tectonics and subduction in the eastern South Caspian region since 10 Ma. *Geology*, **36**, 763–766.

HOMKE, S., VERGÉS, J., GARCÉS, M., EMAMI, H. & KARPUZ, R. 2004. Magnetostratigraphy of Miocene–Pliocene Zagros foreland deposits in the front of the Push-e Kush Arc (Lurestan Province, Iran). *Earth and Planetary Science Letters*, **225**, 397–410.

JACKSON, J. 1992. Partitioning of strike-slip and convergent motion between Eurasia and Arabia in eastern Turkey and the Caucasus. *Journal of Geophysical Research*, **97**, 12471–12479.

JACKSON, J. 2001. Living with earthquakes: know your faults. *Journal of Earthquake Engineering*, **5**, 5–123.

JACKSON, J. & MCKENZIE, D. 1984. Active tectonics of the Alpine–Himalayan belt between western Turkey and Pakistan. *Geophysical Journal of the Royal Astronomical Society*, **77**, 185–264.

JACKSON, J., HAINES, A. J. & HOLT, W. E. 1995. The accommodation of Arabia–Eurasia plate convergence in Iran. *Journal of Geophysical Research*, **100**, 15 205–15 209.

JACKSON, J., PRIESTLEY, K., ALLEN, M. & BERBERIAN, M. 2002. Active tectonics of the South Caspian Basin. *Geophysical Journal International*, **148**, 214–245.

JAFFEY, N. & ROBERTSON, A. H. F. 2001. New sedimentological and structural data from the Ecemis Fault Zone, southern Turkey: implications for its timing and offset and the Cenozoic tectonic escape of Anatolia. *Journal of the Geological Society, London*, **158**, 367–378.

JOLIVET, L. & FACCENNA, C. 2000. Mediterranean extension and the Africa–Eurasia collision. *Tectonics*, **19**, 1095–1106.

KESKIN, M., PEARCE, J. A. & MITCHELL, J. G. 1998. Volcano-stratigraphy and geochemistry of collision-related volcanism on the Erzurum–Kars Plateau, northeastern Turkey. *Journal of Volcanology and Geothermal Research*, **85**, 355–404.

KHEIRKHAH, M., ALLEN, M. B. & EMAMI, M. 2009. Quaternary syn-collision magmatism from the Iran/Turkey borderlands. *Journal of Volcanology and Geothermal Research*, **182**, 1–12.

KLINGER, Y., AVOUAC, J. P., ABOU KARAKI, N., DORBATH, L., BOURLES, D. & REYSS, J. L. 2000. Slip rate on the Dead Sea transform fault in northern Araba valley (Jordan). *Geophysical Journal International*, **142**, 755–768.

KOCYIGIT, A. & BEYHAN, A. 1998. A new intracontinental transcurrent structure: the Central Anatolian Fault Zone, Turkey. *Tectonophysics*, **284**, 317–336.

KOCYIGIT, A., YILMAZ, A., ADAMIA, S. & KULOSHVILI, S. 2001. Neotectonics of East Anatolian Plateau (Turkey) and Lesser Caucasus: implication for transition from thrusting to strike-slip faulting. *Geodinamica Acta*, **14**, 177–195.

LIU, Z. & BIRD, P. 2008. Kinematic modelling of neotectonics in the Persia–Tibet–Burma orogen. *Geophysical Journal International*, **172**, 779–797.

LYBERIS, N. & MANBY, G. 1999. Oblique to orthogonal convergence across the Turan Block in the Post-Miocene. *AAPG Bulletin*, **83**, 1135–1160.

LYBERIS, N., YURUR, T., CHOROWICZ, J., KASAPOGLU, E. & GUNDOGDU, N. 1992. The East Anatolian Fault: an oblique collisional belt. *Tectonophysics*, **204**, 1–15.

MANGINO, S. & PRIESTLEY, K. 1998. The crustal structure of the southern Caspian region. *Geophysical Journal International*, **133**, 630–648.

MANN, P. 2007. Global catalogue, classififcation and tectonic origins of restraining and releasing bends on active and ancient strike-slip fault systems. *In:* CUNNINGHAM, W. D. & MANN, P. (eds) *Tectonics of Strike-slip Restraining and Releasing Bends*. Geological Society, London, Special Publications, **290**, 13–142.

MANSPEIZER, W. 1985. The Dead Sea Rift: impact of climate and tectonism on Pleistocene and Holocene sedimentation. *In:* BIDDLE, K. & CHRISTIE-BLICK, N. (eds) *Strike-slip Deformation, Basin Formation and Sedimentation*. Society of Economic Paleontolgists and Mineralogists, Special Publications, **37**, 143–158.

MASSON, F., DJAMOUR, Y. *ET AL.* 2006. Extension in NW Iran driven by the motion of the south Caspian basin. *Earth and Planetary Science Letters*, **252**, 180–188.

MCCLUSKY, S., BALASSANIAN, S. *ET AL.* 2000. Global Positioning System constraints on plate kinematics and dynamics in the eastern Mediterranean and Caucasus. *Journal of Geophysical Research*, **105**, 5695–5719.

MCCLUSKY, S., REILINGER, R., MAHMOUD, S., BEN SARI, D. & TEALEB, A. 2003. GPS constraints on Africa (Nubia) and Arabia plate motions. *Geophysical Journal International*, **155**, 126–138.

MCKENZIE, D. P. 1972. Active tectonics of the Mediterranean region. *Geophysical Journal of the Royal Astronomical Society*, **30**, 109–185.

MCQUARRIE, N. 2004. Crustal scale geometry of the Zagros fold–thrust belt, Iran. *Journal of Structural Geology*, **26**, 519–535.

MCQUARRIE, N., STOCK, J. M., VERDEL, C. & WERNICKE, B. 2003. Cenozoic evolution of Neotethys and implications for the causes of plate motions. *Geophysical Research Letters*, **30**, 2036, doi:10.1029/2003GL017992.

MEYER, B. & LE DORTZ, K. 2007. Strike-slip kinematics in Central and Eastern Iran: estimating fault slip-rates averaged over the Holocene. *Tectonics*, **26**, Tc5009, doi:10.1029/2006tc002073.

MEYER, B., MOUTHEREAU, F., LACOMBE, O. & AGARD, P. 2006. Evidence of Quaternary activity along the Deshir Fault: implication for the Tertiary tectonics of central Iran. *Geophysical Journal International*, **164**, 192–201.

PEARCE, J. A., BENDER, J. F. *ET AL.* 1990. Genesis of collision volcanism in eastern Anatolia, Turkey. *Journal of Volcanology and Geothermal Research*, **44**, 189–229.

QUENNELL, A. M. 1958. The structure and evolution of the Dead Sea rift. *Quarterly Journal of the Geological Society, London*, **64**, 1–24.

REGARD, V., BELLIER, O. *ET AL.* 2005. Cumulative right-lateral fault slip rate across the Zagros–Makran transfer zone: role of the Minab–Zendan fault system in accommodating Arabia–Eurasia convergence in southeast Iran. *Geophysical Journal International*, **162**, 177–203.

REILINGER, R., MCCLUSKY, S. *ET AL.* 2006. GPS constraints on continental deformation in the Africa–Arabia–Eurasia continental collision zone and implications for the dynamics of plate interactions. *Journal of Geophysical Research – Solid Earth*, **111**, B05411, doi:10.1029/2005JB004051.

RITZ, J. F., NAZARI, H., GHASSEMI, A., SALAMATI, R., SHAFEI, A., SOLAYMANI, S. & VERNANT, P. 2006. Active transtension inside central Alborz: a new insight into northern Iran–southern Caspian geodynamics. *Geology*, **34**, 477–480.

ŞAROĞLU, F., EMRE, O. & KUSCU, I. 1992. The East Anatolian Fault of Turkey. *Annales Tectonicae*, **6**, 99–125.

SARKARINEJAD, K., FAGHIH, A. & GRASERNANN, B. 2008. Transpressional deformations within the Sanandaj–Sirjan metamorphic belt (Zagros Mountains, Iran). *Journal of Structural Geology*, **30**, 818–826.

ŞENGÖR, A. M. C., ALTINER, D., CIN, A., USTAOMER, T. & HSU, K. J. 1988. Origin and assembly of the Tethyside orogenic collage at the expense of Gondwana Land. *In:* AUDLEY-CHARLES, M. G. & HALLAM, A. (eds) *Gondwana and Tethys*. Geological Society, London, Special Publications, **37**, 119–181.

ŞENGÖR, A. M. C., TUYSUZ, O. *ET AL.* 2005. The North Anatolian Fault: a new look. *Annual Review of Earth and Planetary Sciences*, **33**, 37–112.

SEYMEN, I. 1975. *Tectonic characteristics of the North Anatolian Fault zone in the Kelkit Valley segment*. Istanbul Teknik Universitesi Maden Fakultesi Yayinlari, Istanbul.

TALEBIAN, M. & JACKSON, J. 2002. Offset on the Main Recent Fault of NW Iran and implications for the late Cenozoic tectonics of the Arabia–Eurasia collision zone. *Geophysical Journal International*, **150**, 422–439.

TALEBIAN, M. & JACKSON, J. 2004. A reappraisal of earthquake focal mechanisms and active shortening in the Zagros mountains of Iran. *Geophysical Journal International*, **156**, 506–526.

Tatar, M., Jackson, J., Hatzfeld, D. & Bergman, E. 2007. The 2004 May 28 Baladeh earthquake (M-w 6.2) in the Alborz, Iran: overthrusting the South Caspian Basin margin, partitioning of oblique convergence and the seismic hazard of Tehran. *Geophysical Journal International*, **170**, 249–261.

Thatcher, W. 2007. Microplate model for the present-day deformation of Tibet. *Journal of Geophysical Research – Solid Earth*, **112**, B01401, doi:10.1029/2005jb004244.

Tuysuz, O., Barka, A. & Yigitbas, E. 1998. Geology of the Saros graben and its implications for the evolution of the North Anatolian fault in the Ganos–Saros region, northwestern Turkey. *Tectonophysics*, **293**, 105–126.

Verdel, C., Wernicke, B. P., Ramezani, J., Hassanzadeh, J., Renne, P. R. & Spell, T. L. 2007. Geology and thermochronology of Tertiary Cordilleran-style metamorphic core complexes in the Saghand region of central Iran. *Geological Society of America Bulletin*, **119**, 961–977.

Vernant, P., Nilforoushan, F. et al. 2004a. Contemporary crustal deformation and plate kinematics in Middle East constrained by GPS measurements in Iran and northern Iran. *Geophysical Journal International*, **157**, 381–398.

Vernant, P., Nilforoushan, F. et al. 2004b. Deciphering oblique shortening of central Alborz in Iran using geodetic data. *Earth and Planetary Science Letters*, **223**, 177–185.

Vincent, S. J., Allen, M. B., Ismail-Zadeh, A. D., Flecker, R., Foland, K. A. & Simmons, M. D. 2005. Insights from the Talysh of Azerbaijan into the Paleogene evolution of the South Caspian region. *Geological Society of America Bulletin*, **117**, 1513–1533.

Vincent, S. J., Morton, A. C., Carter, A., Gibbs, S. & Barabadze, T. G. 2007. Oligocene uplift of the Western Greater Caucasus: an effect of initial Arabia–Eurasia collision. *Terra Nova*, **19**, 160–166.

Walker, R. & Jackson, J. 2004. Active tectonics and late Cenozoic strain distribution in central and eastern Iran. *Tectonics*, **23**, TC5010; doi:10.1029/2003TC001529.

Walker, R. T., Bergman, E., Jackson, J., Ghorashi, M. & Talebian, M. 2005. The 2002 June 22 Changureh (Avaj) earthquake in Qazvin province, northwest Iran: epicentral relocation, source parameters, surface deformation and geomorphology. *Geophysical Journal International*, **160**, 707–720.

Walker, R. T., Gans, P., Allen, M. B., Jackson, J., Khatib, M., Marsh, N. & Zarrinkoub, M. 2009. Late Cenozoic volcanism and rates of active faulting in eastern Iran. *Geophysical Journal International*, **177**, 783–805.

Westaway, R. 1994. Present-day kinematics of the Middle-East and Eastern Mediterranean. *Journal of Geophysical Research*, **99**, 12 071–12 090.

Westaway, R. 1999. Comment on 'A new intracontinental transcurrent structure: the Central Anatolian Fault Zone, Turkey' by A. Kocyigit and A. Beyhan. *Tectonophysics*, **314**, 469–479.

Westaway, R. & Arger, J. 1996. The Golbasi basin, southeastern Turkey: a complex discontinuity in a major strike-slip fault zone. *Journal of the Geological Society, London*, **153**, 729–743.

Westaway, R. & Arger, J. 2001. Kinematics of the Malatya–Ovacik fault zone. *Geodinamica Acta*, **14**, 103–131.

Westaway, R., Demir, T., Seyrek, A. & Beck, A. 2006. Kinematics of active left-lateral faulting in SE Turkey from offset Pleistocene river gorges: improved constraint on the rate and history of relative motion between the Turkish and Arabian plates. *Journal of the Geological Society, London*, **163**, 149–164.

Woodcock, N. H. 1986. The role of strike-slip fault systems at plate boundaries. *Philosophical Transactions of the Royal Society of London, Series A*, **317**, 13–29.

Zanchi, A., Berra, F., Mattei, M., Ghassemi, M. R. & Sabouri, J. 2006. Inversion tectonics in central Alborz, Iran. *Journal of Structural Geology*, **28**, 2023–2037.

Was Late Cretaceous–Paleocene obduction of ophiolite complexes the primary cause of crustal thickening and regional metamorphism in the Pakistan Himalaya?

M. P. SEARLE[1] & P. J. TRELOAR[2]

[1]*Department of Earth Sciences, Oxford University, Parks Road, Oxford OX1 3PR, UK*

[2]*Centre for Earth and Environmental Science Research, Kingston University, Penrhyn Road, Kingston-upon-Thames KT1 2EE, UK*

*Corresponding author (e-mail: mike.searle@earth.ox.ac.uk)

Abstract: Regional metamorphic rocks in the Pakistan Himalaya include both UHP coesite eclogite-facies and MP/T kyanite–sillimanite-grade Barrovian metamorphic rocks. Age data show that peak metamorphism of both was c. 47 Ma. $^{40}Ar-^{39}Ar$ hornblende cooling ages date post-peak metamorphic cooling of both through 500 °C by 40 Ma, some 20 Ma earlier than for metamorphic rocks in the central and eastern Himalaya. Typically these ages have been explained by obduction of the Kohistan arc onto the Indian plate at about 50 Ma and India–Asia collision. We suggest instead that the earlier metamorphic and cooling ages of the Pakistani Barrovian metamorphic sequence could be partially explained by Late Cretaceous to Early Paleocene crustal thickening linked to obduction of an ophiolite thrust sheet onto the leading edge of the Indian plate, similar to the Spontang Ophiolite in Ladakh. Heating following on from this Paleocene crustal thickening explains peak Barrovian metamorphism within 5–10 Ma of subsequent obduction of Kohistan. Remnants of the ophiolite sheet, and underlying Tethyan sediments, are preserved in NW India and in western Pakistan but not in northern Pakistan. Tectonic erosion removed all cover sequences (including the ophiolites) from the Indian plate basement.

The Kohistan arc developed during the Cretaceous as an intra-oceanic island arc within the Neo-Tethyan ocean that lay between continental India and the Asian plate margin. On its northern side the arc is joined to the southern margin of the Asian plate along the Shyok Suture (Fig. 1). On its southern side the arc is thrust onto the leading edge of the Indian plate along the Main Mantle Thrust zone. Suturing of the arc to Asia was generally thought to have occurred at between 102 and 75 Ma (Petterson & Windley 1985, 1992; Treloar *et al.* 1989*a*) with recent isotopic data from Heuberger *et al.* (2007) and Jagoutz *et al.* (2009) leaning towards the younger end of that range. Parrish *et al.* (2006) dated zircons and allanites from coesite-bearing eclogites along the leading edge of the Indian plate at 46.4 ± 0.1 Ma, an age that they interpreted as dating peak ultrahigh-pressure (UHP) metamorphism. However, similar eclogites have been recorded within Permian–Mesozoic shelf carbonate rocks structurally beneath the Semail Ophiolite along the leading edge of the Arabian plate in Oman (Searle *et al.* 2004), a region that has certainly not yet undergone continent–continent collision. This structural setting suggests that UHP metamorphic ages from the Himalaya cannot be used to interpret conclusive evidence for the age of suturing or continental collision.

An important issue is the timing of amphibolite-grade metamorphism in the Indian plate of northern Pakistan. Here, cooling through 500 °C after peak metamorphism at kyanite- and sillimanite-bearing conditions had taken place by 40 Ma (Treloar & Rex 1990*a, b*; Chamberlain *et al.* 1991; Tonarini *et al.* 1993). Regional metamorphism is usually modelled as taking tens of millions of years to accomplish (England & Thompson 1984). Could peak Barrovian style regional metamorphic temperatures in the NW Himalaya have been attained within a few million years of collision? Treloar (1997) explained this through arguing that the major heat source was dissipative shear heating on the footwall of the Main Mantle Thrust as Kohistan was thrust southward onto the Indian plate, and that a thin, inverted metamorphic sequence was subsequently thickened by imbrication on the hanging wall of later thrusts. However, the thickness of the metamorphic rocks in Pakistan, even in single thrust slices, is still far greater than that in metamorphic soles along the base of ophiolite complexes (e.g. Searle & Cox 1999, 2002).

One potential solution to the metamorphic timing dilemma may be that heating of the leading

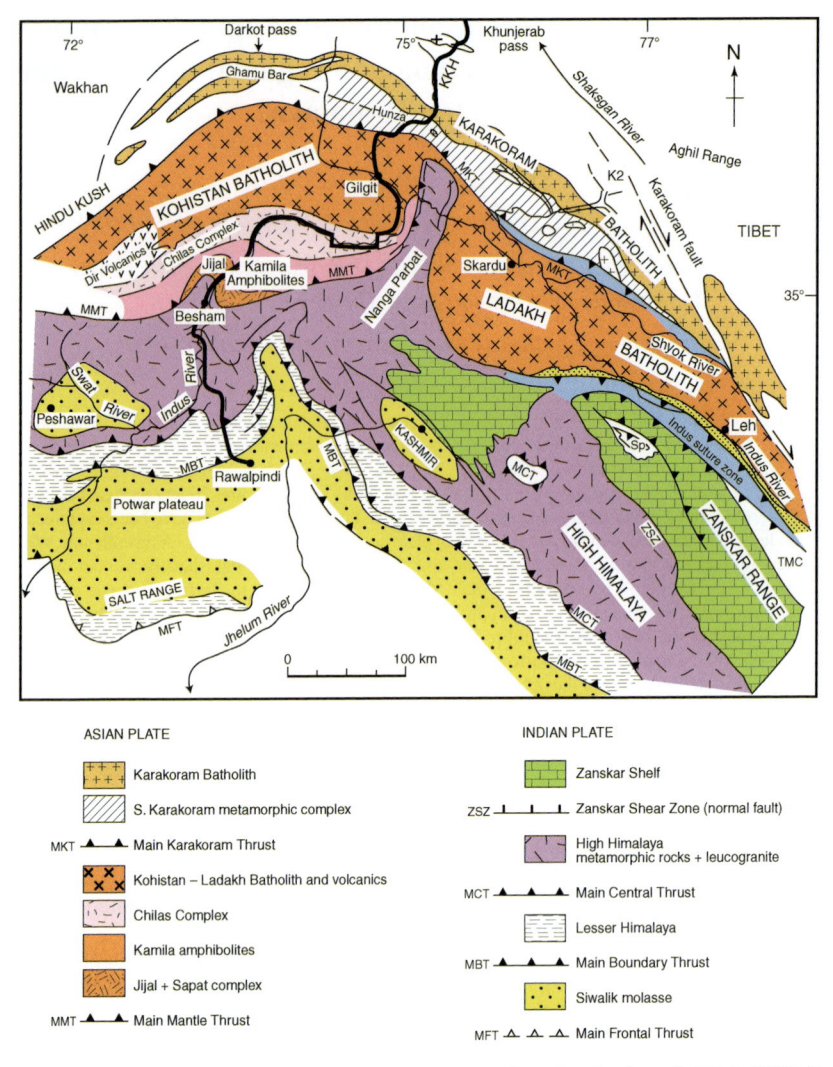

Fig. 1. Geological map of the Kohistan–Ladakh western Himalaya region, after Searle *et al.* (1999). KKH, Karakoram Highway.

edge of the Indian plate started as early as 65 Ma as a result of ophiolite emplacement onto the northern continental margin of India–Pakistan. Although not preserved in northern Pakistan there is abundant evidence for ophiolite obduction onto the Indian plate margin in the Paleocene, including the Bela, Khost and Waziristan ophiolites of western Pakistan, the Spontang Ophiolite in Ladakh and the Yamdrock mélange-associated ophiolites along southern Tibet (Fig. 2). This includes precise age data from the Spontang Ophiolite in Ladakh (Corfield *et al.* 2001; Pedersen *et al.* 2001) and faunal evidence from the Khost and Waziristan ophiolite sequences in western Pakistan and eastern Afghanistan (see Treloar & Izatt 1993, for a review).

In this paper we take a critical look at the data that constrain the tectonic evolution of the NW Himalaya, in particular the major time lines in the evolution of the Kohistan arc. Is it possible that the arc collided with India before Asia as suggested by Khan *et al.* (2008)? Could Paleocene ophiolite obduction explain the earlier crustal thickening and metamorphism in the Pakistan sector of the Himalaya and the older Palaeogene metamorphism and cooling ages in the Indian plate rocks of north Pakistan? We first summarize the geology of the Kohistan arc in Pakistan, the Ladakh batholith and Indus Suture zone in Ladakh and then propose a new tectonic evolution model for the western Himalaya.

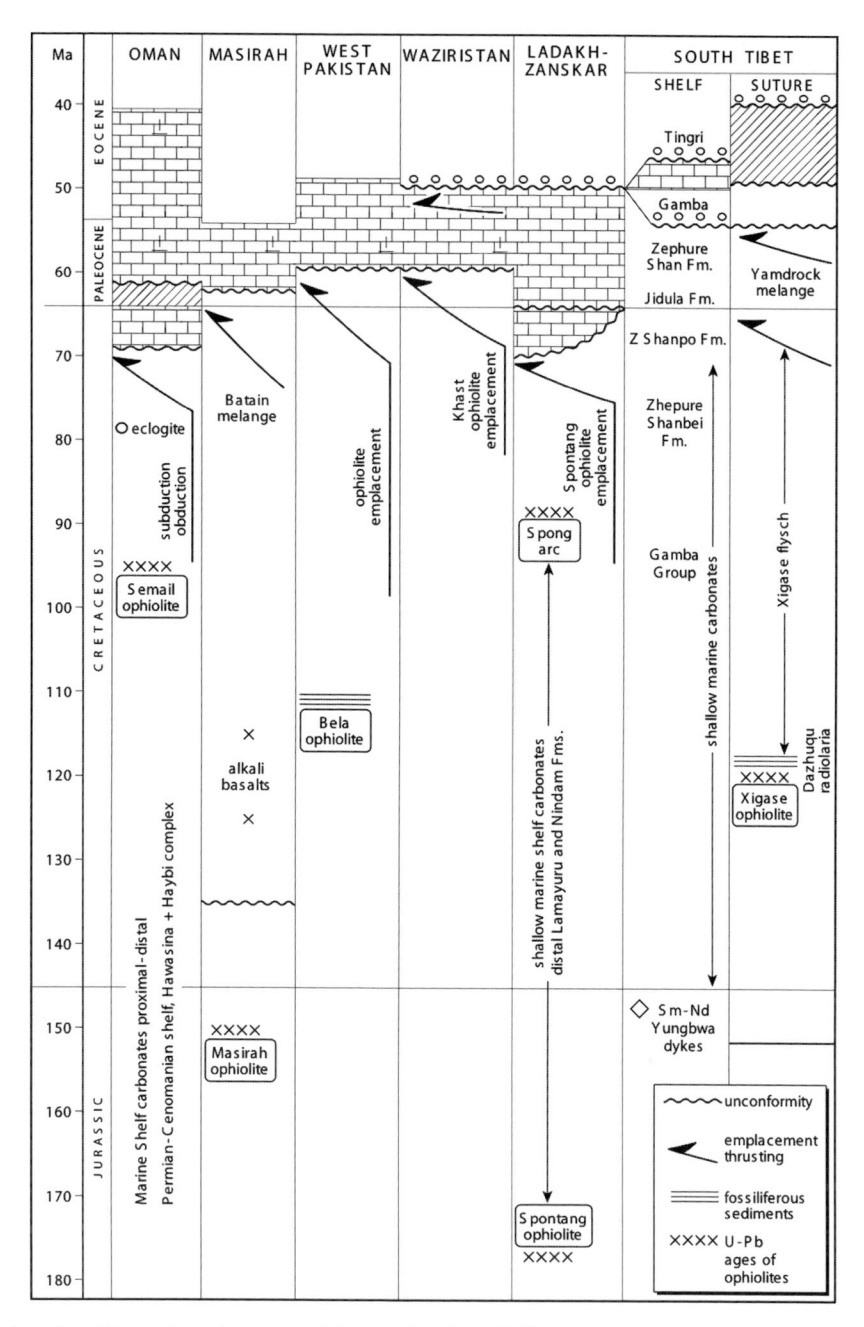

Fig. 2. Jurassic to Eocene time chart summarizing age data for ophiolite complexes and associated sediments from Oman, Pakistan, Ladakh and south Tibet, after Green *et al.* (2008).

The Kohistan arc

The Kohistan arc provides a well-exposed and complete section through a Mesozoic intra-oceanic island arc sequence sandwiched between the Indian plate to the south and the Karakoram (Asian) plate to the north (Tahirkheli *et al.* 1979; Coward *et al.* 1982, 1986; Pettersen & Windley 1985, 1991, 1992; Khan *et al.* 1989, 1997; Treloar *et al.* 1996; Pettersen & Treloar 2004; Burg *et al.* 1998, 2006; Jagoutz *et al.* 2006, 2007, 2009). Some key geochronological data for Kohistan are

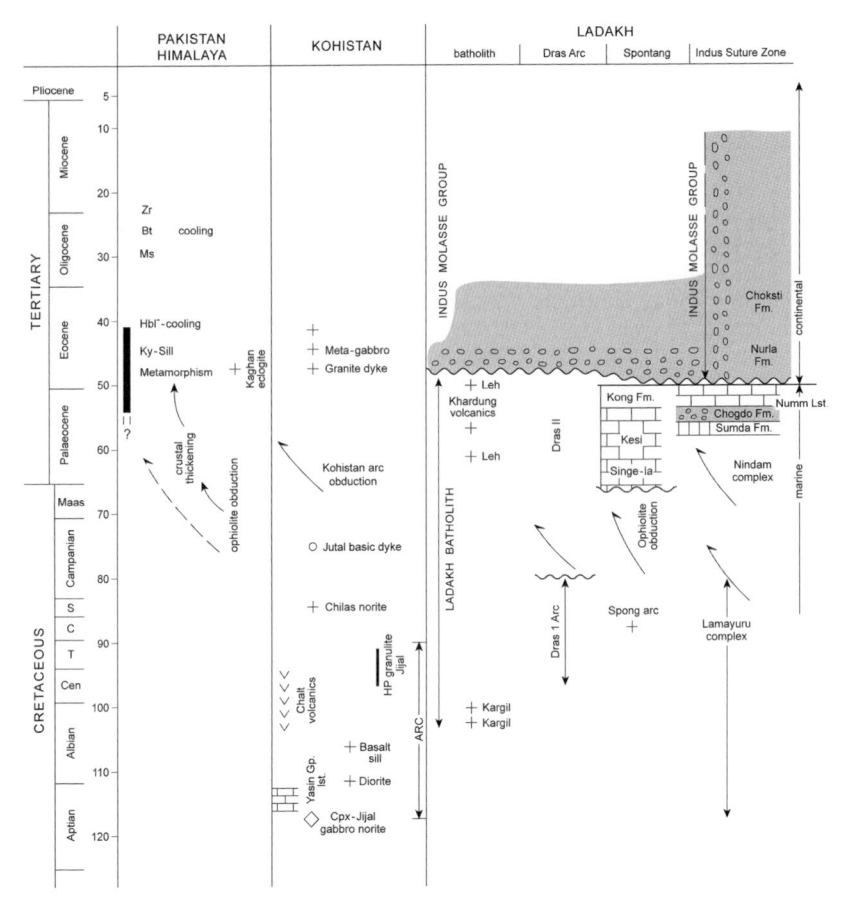

Fig. 3. Summary time chart showing age constraints for the Pakistan Himalaya, Kohistan and Ladakh regions. Sources of data are Treloar & Rex (1990a, b) and Treloar et al. 1989a–c) for Pakistan Himalayan metamorphism; Parrish et al. (2006) for Kaghan eclogites; Pettersen & Windley (1985, 1991, 1992), Yamamoto & Nakamura (1996), Schaltegger et al. (2002, 2003), Dhuime et al. (2007), Heuberger et al. (2007) and Jagoutz et al. (2009) for Kohistan; Reuber (1989) for the Dras arc; Honegger et al. (1982), Schärer et al. (1984), Weinberg & Dunlap (2000) and Weinberg et al. (2000) for the Ladakh batholith; Pedersen et al. (2001) for the Spontang Ophiolite; and Garzanti et al. (1987), Garzanti & van Haver (1988) Green et al. (2008) for stratigraphic ages of Tertiary sedimentary rocks. Timing of major structural events is mainly from Searle (1986) and Searle et al. (1988, 1997).

plotted in Figure 3. Mid-ocean ridge basalt (MORB) type oceanic rocks were intruded by ultramafic rocks of the Jijal Complex. Dated at about 117 ± 7 Ma these include a dunite–wehrlite–websterite ultramafic series that passes upwards into a series of mafic garnet granulites and metagabbros (Jan & Howie 1981; Yamamoto & Nakamura 1996; Garrido et al. 2006, 2007; Dhuime et al. 2007, 2009). The garnet granulites have been interpreted by Garrido et al. (2006) as being the result of partial melting of hornblende-bearing two-pyroxene gabbros in the lower crust at pressures of c. 10 kbar.

The Jijal Complex is flanked to the north by a sequence of strongly sheared volcanic rocks of the Kamila Amphibolite belt intruded by gabbros and hornblende pegmatites. The rocks dip moderately towards the north, with stretching lineations and shear criteria suggesting a top-side-north sense of shear (Coward et al. 1987; Treloar et al. 1990, 1991). Volcanic rocks within the belt are divided into two: E-type low-Ti MORB-type sea-floor rocks with clear pillow lava sequences and D-type, high-Ti units with a clear arc-related chemical signature (Khan et al. 1993; Treloar et al. 1996; Bignold & Treloar 2003; Bignold et al. 2006). The former represent the substrate on which the arc was built, the latter represent the arc building environment.

The Kamila amphibolites are intruded on their northern margin by gabbro norites of the Chilas Complex. The most robust age data for the Chilas Complex suggest that it was emplaced at about 85 Ma (Schaltegger et al. 2002). The Chilas Complex is a dominantly gabbro-norite body, although with ultramafic units that could represent either direct mantle-derived magmas (Jagoutz et al. 2006, 2007) or magma chamber differentiation (Khan et al. 1989). This large body is interpreted as having been emplaced during arc rifting (Treloar et al. 1996; Burg et al. 2006). Burg et al. (2006) suggested that the magmatic fabrics, initially reported by Treloar et al. (1996), show that the body is composed of juxtaposed plutonic bodies emplaced along south-dipping listric faults. Burg et al. (2006) suggested that 'the northward dip of the southern side of the Kohistan [arc] is inherited from late Cretaceous rotation of crustal blocks in the hanging wall of the listric faults'.

On its northern margin the Chilas Complex includes screens of volcaniclastic rocks of the Gilgit Formation. The Gilgit Formation is part of a sequence of volcanic and sedimentary rocks, largely metamorphosed to amphibolite facies, which make up most of the northern part of the arc. Clastic sediments are overlain by volcanic rocks of the Chalt Volcanic group (Coward et al. 1982). On the basis of chemistry Bignold et al. (2006) and Petterson & Treloar (2004) divided the Chalt Volcanic Group into two. The Ghizar Valley Formation has geochemistry typical of intra-arc volcanic sequences. By contrast the Hunza Valley Formation contains basalts and andesites with boninitic chemistries and strongly depleted light rare earth element (LREE) signatures. Bignold et al. (2006) interpreted these as back-arc rocks derived from the melting of an already depleted mantle source. The Chalt Volcanic Group is overlain by apparently unmetamorphosed limestones of Albian–Cenomanian age containing *Orbitolina* sp. foraminifers (Pudsey 1986).

There are two other units that help to constrain the tectonomagmatic history of the Kohistan arc. The first is late-stage, unmetamorphosed felsic to mafic extrusive rocks. The second is a series of, generally felsic, granitoid plutons. The late-stage extrusive rocks can be divided into two: the Utror Volcanic Sequence, which crops out in the Dir and Swat valleys of western central Kohistan (Sullivan et al. 1993), and the Shamran Volcanic Sequence, which crops out in northern Kohistan (Sullivan et al. 1993; Danishwar et al. 2001; Khan et al. 2004, 2008; Petterson & Treloar 2004). The bimodal, dominantly rhyolitic, Utror Volcanic Sequence, overlies the volcaniclastic Baraul Banda Slate Formation. Sullivan et al. (1993) dated the latter on faunal criteria at c. 60 Ma, which is consistent with a hornblende Ar–Ar date of c. 55 Ma from the Utror Volcanic Sequence rocks (Treloar et al. 1989a). However, both ages can now be questioned on the basis of robust U–Pb zircon ages of 72.3 ± 5.1 and 70.3 ± 5.0 Ma from granites that apparently intrude the Utror Volcanic Sequence (Jagoutz et al. 2009). The Shamran Volcanic Sequence was defined by Sullivan et al. (1993), and the name was used also by Treloar et al. (1996) and Petterson & Treloar (2004). However, Danishwar et al. (2001) used the name Teru Volcanic Formation instead and continued to do so later (Khan et al. 2004, 2008). We use the term Teru/Shamran Volcanic Formation here to ensure no confusion. Despite the bias in the analytical data presented by Khan et al. (2008) the Teru/Shamran sequence is a bimodal one with rhyolite dominant above basalt to basaltic-andesite (Petterson & Treloar 2004). Sullivan et al. (1993) interpreted both the Utror and Teru/Shamran sequences as being of the same age. However, Khan et al (2008) reported three zircon ages with a combined age of 64.8 ± 1.0 Ma for the Teru/Shamran Volcanic Sequence at about 63 Ma. Both volcanic sequences suggest a continuing contribution of calc-alkaline, mantle wedge derived magmas to the growing arc to at least 65 Ma.

Kohistan batholith

The Kohistan–Ladakh batholith is part of the Trans-Himalayan calc-alkaline batholith, which extends 2500 km from NW Pakistan across northern Kohistan and Ladakh and south Tibet to the eastern Himalayan syntaxis and probably further around the syntaxis into Burma. It is an Andean-type ('I-type'), subduction-related batholith that formed above the north-dipping oceanic subduction zone along the southern margin of Asia prior to the collision of India c. 50 Ma ago. The batholith is variably called the Kohistan batholith in Pakistan, the Ladakh batholith in India and the Gangdese batholith in Tibet. The Kohistan batholith is dominated by granodioritic, dioritic and quartz-dioritic plutons, although early stages of the batholith include tonalites and some gabbros. Petterson & Windley (1985) divided these into deformed Stage 1 plutons and undeformed Stage 2 plutons. They dated the Matum Das pluton using Rb–Sr methods at 102 ± 12 Ma. However, the validity of this date, from a pluton subsequently metamorphosed to amphibolites facies, is questionable, especially as Schaltegger et al. (2003) reported a zircon age of 153 Ma from it. Jagoutz et al. (2009) dated four calc-alkaline granitoid intrusions from western Kohistan and reported $^{206}Pb/^{238}U$ ages ranging between 75.1 ± 4.5 and 42.1 ± 4.4 Ma. Those workers concluded that Kohistan granitoids originated from mantle-derived melts and evolved through amphibole-dominated fractionation and intra-crustal assimilation. A tholeiitic metagabbro from Kohistan, derived from partial melting of a depleted mantle MORB-type source and dated at 50 Ma (Heuberger et al. 2007) may be the youngest tholeiitic component. This metagabbro has been intruded by a crustal-derived granite dated at 47 Ma (Heuberger et al. 2007), implying that suturing was complete by then.

The Kohistan Arc and batholith seems to have undergone three distinct phases during its evolution (Fig. 3). An intra-oceanic arc phase (c. 118–90 Ma) was followed by a phase of arc rifting and emplacement of the giant Chilas gabbro-norite body (c. 85 Ma) and finally by a phase of calc-alkaline batholith intrusion (c. 75–47 Ma). The last stage of magmatism is represented by granite sheets emplaced near the Indus valley confluence east of Gilgit. Although these Indus confluence dykes have been dated by Rb/Sr at c. 26 Ma (Petterson & Windley 1985; George et al. 1993) there are as yet no robust U–Pb data on this magmatic phase. The two-mica granite sheets of the Indus confluence are more probably related to crustal thickening and lower crustal melting rather than to calc-alkaline subduction-related magmatism.

Ladakh batholith

The Ladakh batholith comprises dominantly biotite granite, tonalite and biotite + hornblende granite, but also has more primitive components including gabbros, norites and diorites exposed around Kargil and in the Deosai plateau region of NE Pakistan. Volcanic eruptive phases include andesites, trachyandesites and rhyolitic ignimbrites along the Ladakh Range NW of Leh (Raz & Honegger 1989).

U–Pb zircon ages from the batholith range from 103 ± 3 Ma (Honegger et al. 1982) and 101 ± 2 Ma (Schärer et al. 1984) to around 50 Ma, the age of India–Asia collision (Green et al. 2008). Magmatic intrusion of the granite near Leh occurred between 70 and 50 Ma, with the last main pulse of magma at c. 49.8 ± 0.8 Ma (Weinberg & Dunlap 2000). Minor dyke intrusion occurred at 46 ± 1 Ma and all magmatic activity ended essentially once the Indian plate had collided and oceanic subduction beneath the Asian margin ceased. The youngest phase of magmatism in the Ladakh batholith is a suite of leucogranite dykes that invade the granite at Chumathang, east of Leh, and at the Indus–Gilgit River confluence in northern Pakistan.

Geochemistry and isotope chemistry show that the Ladakh batholith was mantle derived with almost no crustal influence, consistent with its I-type subduction-related origin (Weinberg & Dunlap 2000). $^{87}Sr/^{86}Sr$ ratios are 0.7033–0.7053 and ε_{Nd} values close to zero indicate recent separation from the mantle and a short 60–70 Ma. crustal residence time. Batholith growth is thought to have occurred by the progressive addition of basaltic magmas to the base of the Asian crust and successive remelting of earlier magmatic intrusions to produce more evolved melts of essentially andesitic composition. This process is also widely thought to explain early Archaean crustal growth.

The Khardung volcanic rocks crop out along the northern side of the Ladakh range. These andesitic volcanic rocks are the eruptive component of the Ladakh granites, and have U–Pb ages of 60.5–67.4 Ma (Dunlap & Wysoczanski 2002). These rocks also include rhyolite pyroclastic flows and coarse volcaniclastic rocks. The northern margin of the Khardung Volcanic Group is the steep, north-dipping, south-verging Khalsar Thrust and Shyok Suture, which separates the Ladakh ranges to the south from the Karakoram terrane to the north. The Shyok Suture in Ladakh is problematic because there is no evidence of ophiolitic rocks in the Nubra valley or the southern Saltoro hills (Weinberg et al. 2000).

Indus Suture Zone

The Indus Suture Zone (ISZ) extends around the northern margin of the Indian plate from NW Pakistan (Waziristan–Kohistan Main Mantle Thrust; MMT) through Ladakh and south Tibet to the Eastern Himalayan syntaxis, thence south through the Indo-Burman ranges to the Andaman Sea. The ISZ closed during the Early Eocene (c. 50 Ma), the age of the youngest marine sediments preserved along the suture and along the northern margin of the Indian plate (Searle et al. 1997; Green et al. 2008). The rocks within the Indus Suture Zone consist of the following.

(1) Mesozoic distal shelf-slope (Lamayuru Complex) to Tethyan ocean basin (Karamba Complex) sediments that are time-equivalent to the Mesozoic passive shelf margin rocks of the Zanskar range to the south (Robertson & Sharp 1988; Searle et al. 1988; Gaetani & Garzanti 1991; Robertson & Degnan 1993).

(2) Permian and Triassic exotic limestone blocks associated with alkaline volcanic rocks, interpreted as off-axis, within-plate ocean island seamounts within the distal part of Tethys away from the continental margin.

(3) Late Jurassic–Cretaceous Dras (–Kohistan) island arc sequence, a complex intra-oceanic island arc comprising basaltic andesites, andesites, dacites and rhyolites, together with forearc

sediments shed off the arc (Nindam Formation; Reuber 1989; Robertson & Degnan 1994).

(4) Mélanges, sometimes ophiolitic mélanges, broken formations or disrupted thrust sheets, consisting of blocks of radiolarian cherts, passive margin sedimentary rocks, alkaline volcanic rocks, rare glaucophane-bearing blueschsists (Sapi-la schists), and ophiolitic rocks, usually serpentinized harzburgites and dunites (Robertson 2000).

(5) Post-collisional Indus Molasse Group sedimentary rocks, mainly conglomerates (Hemis conglomerates), shales and sandstones formed in an intramontane basin along the Indus Suture Zone and accumulating erosional debris mostly from the north (Ladakh granites and andesites) but also from the suture zone (radiolarian cherts, exotic limestones, ophiolitic peridotites), and less commonly from the southern margin of India (Zanskar shelf carbonates; Garzanti & van Haver 1988; Searle *et al.* 1988, 1990).

Complex deformation involves (1) early, Late Cretaceous, pre-Middle Paleocene south- or SW-vergent thrusting and folding associated with obduction of the Spontang Ophiolite and underlying Tethyan thrust sheets onto the Indian passive margin; (2) post-collisional folding and thrusting following the India–Asia collision and closing of Tethys (c. 50 Ma); (3) Late Tertiary north- or NE-directed back-thrusting (Colchen *et al.* 1986; Searle 1986; Garzanti *et al.* 1987; Searle *et al.* 1988, 1997; Steck 2003).

Himalayan metamorphism

The age of peak kyanite (c. 35–30 Ma) and peak sillimanite (c. 24–18 Ma) grade metamorphism is fairly consistent along the length of the Indian, Nepal, Sikkim and Bhutan Himalaya to the east of Pakistan (see reviews by Hodges 2000; Godin *et al.* 2006; Searle *et al.* 2007). However, in Pakistan peak amphibolite-facies metamorphism in NW Pakistan must predate 40 Ma (the age of cooling back through the hornblende blocking temperature of 500 °C). Few U–Pb age constraints exist for the Pakistan Himalaya, unlike the main Himalayan ranges to the east in India and Nepal. The metamorphic peak in Pakistan has been dated at about 47 Ma. Smith *et al.* (1994) dated zircons by U–Pb in basement gneisses from near Naran (Fig. 1) at 47 ± 3 Ma, and Foster *et al.* (2002) dated garnets from north of Naran at 46.3 ± 1.7 Ma by the Sm–Nd method. These ages are similar to the U–Pb zircon and allanite ages obtained by Parrish *et al.* (2006) from the Kaghan valley coesite eclogites. Cooling through 500 °C after this early Tertiary metamorphism was before c. 40 Ma (Treloar & Rex 1990a, b; Chamberlain *et al.* 1991; Tonarini *et al.* 1993). Although the end of volcanism in the arc and the age of UHP metamorphism on the Indian plate of northern Pakistan are both consistent with the start of subduction of continental crust of India beneath Kohistan at about 50–55 Ma, the age of peak Barrovian metamorphism might imply that the thermal structure of the leading edge of continental India was affected by an earlier pre-continental collision event.

Late Cretaceous–Paleocene deformation in the Tethyan Himalaya

An important, if often unremarked feature of the north Indian margin is that it was affected by a significant Paleocene to Early Eocene ophiolite emplacement event (Searle *et al.* 1988, 1997; Corfield *et al.* 2001). The Spontang Ophiolite in Ladakh is a complete ophiolite including upper mantle harzburgite and a full crustal sequence that includes gabbros, diorites, sheeted dykes and pillow lavas (Searle 1986; Searle *et al.* 1988; Corfield & Searle 2000; Corfield *et al.* 2001). U–Pb ages of zircons from plagiogranites date the ophiolite crust as Jurassic at 177 ± 1 Ma, but a dacite–andesite arc sequence built on top of the ophiolite was dated at 88 ± 5 Ma (Pedersen *et al.* 2001). The Spontang Ophiolite with the Spong Arc on top was obducted southward onto the Mesozoic passive margin of India during the Late Cretaceous together with underlying Photang and Lamayuru Complex thrust sheets of Tethyan oceanic rocks (Searle 1986; Searle *et al.* 1988, 1997; Corfield *et al.* 1999, 2005). Structural mapping across Ladakh has shown that a major regional unconformity exists beneath the Paleocene–Eocene shallow marine carbonates (Figs 4 & 5). The cliff sections along the Zanskar River clearly show that the Mesozoic sedimentary rocks are intensely deformed with isoclinal folds, and numerous thrusts (Searle 1986; Searle *et al.* 1988, 1997; Corfield & Searle 1999). Above these rocks allochthonous Lamayuru Complex deep-water sedimentary rocks have been thrust above the Mesozoic shelf carbonates. The Lamayuru Complex rocks underlie the Spongtang Ophiolite and are unconformably overlain by the Paleocene–Eocene shallow marine limestones (Fig. 5). The latter show gentle south-vergent folds with far less shortening than the deeper Mesozoic shelf carbonates (Searle *et al.* 1999). Thus a major crustal shortening and thickening event must have occurred around the Late Cretaceous–Paleocene boundary, an event that we correlate with the timing of obduction of the Spontang Ophiolite onto the previous passive continental margin of India.

Although there is no ophiolite complex exposed on the northern margin of the Indian plate in

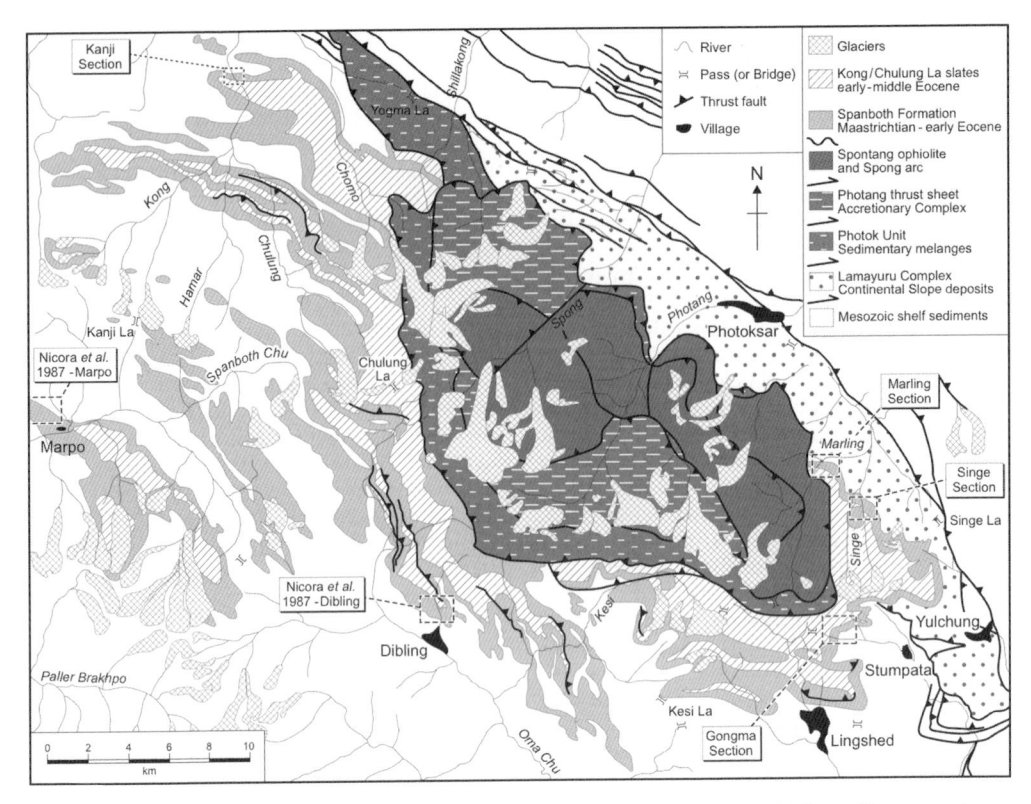

Fig. 4. Geological map of Ladakh after Corfield & Searle (2000) showing the Spontang Ophiolite and Lamayuru thrust sheets overlying the Zanskar shelf carbonates. The location of the photograph in Figure 5 is across the Zanskar river gorge view NW towards Lingshed (bottom right of map) Nicora *et al.* (1987).

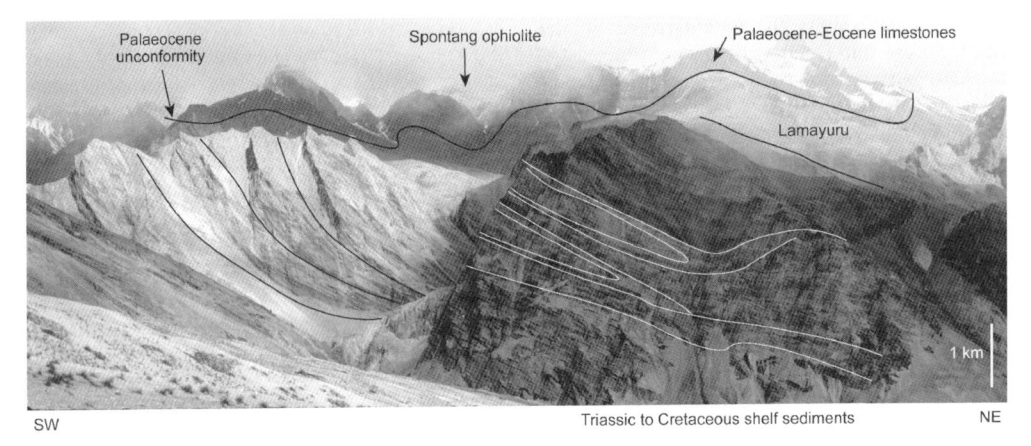

Fig. 5. Photograph of the Zanskar river gorge in Ladakh (see Fig. 4 for location). Mesozoic shelf carbonates showing tight to isoclinal folding and thrusting are overlain by allochthonous Lamayuru Complex deep-sea sedimentary rocks thrust towards the south beneath the Spontang Ophiolite during the Late Cretaceous obduction event. The Spontang Ophiolite is in the far distance. All Mesozoic units are unconformably overlain by Paleocene–Eocene limestones showing south-vergent folds with only a minor amount of internal shortening.

Pakistan, the western margin of the plate is structurally overlain by a number of ophiolite complexes. These include the Khost, Waziristan, Zhob, Bela and Muslim Bagh ophiolites. All of these include Maastrichtian rocks in the tectonic mélange and are overlain by Early to Middle Eocene neo-autochthonous sediments suggesting that widespread ophiolite obduction-related deformation occurred during the Paleocene to Early Eocene (Mattauer et al. 1978; Ahmad & Abbas 1979; Alleman 1979; Tapponnier et al. 1981). It is unclear as to whether the Dargai klippe, which crops out to the south of the Kohistan arc, is part of this sequence of ophiolite units emplaced during the Maastrichtian or part of the forearc to the Kohistan arc emplaced during the early Tertiary.

We see the potential here to develop a direct analogy with the Oman ophiolite in terms of pre-continental collision ophiolite emplacement and metamorphism. The Oman Mountains preserve a complete ophiolite sequence including over 15–20 km thickness of upper mantle harzburgite and dunite with 4–6 km of crustal gabbros, sheeted dykes and pillow lavas (e.g. Searle & Cox 1999). The entire sequence has been obducted onto the formerly passive northern continental margin of Arabia during the Late Cretaceous. The SW foreland region of the Oman Mountains shows a classic thin-skinned fold–thrust belt but the deeper structural levels in the hinterland show extensive HP metamorphism (eclogite-, blueschist- and carpholite-grade metamorphism and thick-skinned thrust sheets involving the passive margin and the basement (e.g. Searle & Cox 1999, 2002; Searle et al. 2004). The HP metamorphism and major deformation involving folding and thrusting in Oman was clearly related to the latest stages of ophiolite obduction, and not to any continent–continent collision, an event that has not yet occurred in Oman.

We suggest here that ophiolite obduction onto the northern margin of continental India during the Late Cretaceous–Paleocene could have resulted in significant crustal thickening prior to any continental collision as seen in Oman and Ladakh. This event could have resulted in increased pressure and temperature in the Indian plate crust. Hence, once the major India–Asia collision followed the closing of Neo-Tethys about 50 Ma (Green et al. 2008), the crust would already have had an elevated thermal gradient, thus explaining the short time scale between collision and peak amphibolite-facies, Barrovian-style regional metamorphism in Pakistan.

Tectonic summary

We propose the following tectonic scenario for the western Himalaya in Pakistan and Ladakh.

(1) Kohistan arc formation

The Kohistan island arc terrane shows three distinct phases of evolution from (a) an initial intra-oceanic arc building phase (c. 117 to c. 90 Ma) with subduction-related magmatism and metamorphism (Jijal Complex) and calc-alkaline volcanism (Chalt Group) through (b) an arc rifting phase when intrusion of the Chilas gabbro-norite occurred (c. 85 Ma) to (c) a calc-alkaline batholith phase when numerous granitoid intrusions were emplaced across northern Kohistan (c. 75–47 Ma). Calc-alkaline volcanism continued until about 65 Ma, although granitoid emplacement continued until about 40 Ma. However, the lack of exposed extrusive rocks younger than 65 Ma might simply be a result of high erosion rates through the Tertiary, which have now exposed the batholithic roots of the arc. This suggests that the mantle wedge remained active until about 40 Ma and ended around the time of India–Asia collision, when the oceanic slab was cut off.

(2) Ophiolite formation

The Spontang Ophiolite formed at a mid-ocean ridge spreading centre during the Mid-Jurassic (U–Pb zircon age $177 + 1$ Ma; Pedersen et al. 2001) with a later andesite–dacite volcanic arc (the Spong arc) formed above a north-dipping subduction zone (U–Pb zircon age $88 + 5$ Ma; Corfield et al. 2001; Pedersen et al. 2001). The latter age is interpreted as dating initiation of subduction that led to the southward obduction of the ophiolite complex onto the previously passive northern continental margin of India. The Spong arc is a separate arc from the more northerly Dras arc, which is correlated with Kohistan. This suggests that during the Late Cretaceous there must have been two northerly dipping subduction zones and two arc complexes located to the north of India (Corfield et al. 2001, fig. 14).

(3) Ophiolite obduction

Late Cretaceous to Paleocene obduction of the Tethyan ophiolite complexes southward over the passive margin of the Indian–Pakistan plate led to regional crustal shortening and thickening. Mesozoic continental shelf rocks beneath the Spontang Ophiolite in Ladakh are known to have undergone widespread regional folding and thrusting prior to the mid-Paleocene (Searle 1986; Searle et al. 1988, 1997). We propose that a similar large ophiolite slab was emplaced onto the Pakistan sector of the Himalaya, resulting in similar Late Cretaceous to Paleocene crustal thickening. At this stage there must still have been continuing oceanic subduction beneath the Kohistan arc as calc-alkaline magmatism continued until c. 50 Ma. By analogy with Oman, the early stage of ophiolite obduction was concomitant with subduction of the Tethyan oceanic crust that was attached to the Indian passive margin. The later stage of obduction involved the attempted subduction of the thinned continental crust of the leading margin of India. This led to the deep subduction of continental crust, eventual slab (eclogitic root) break-off and the rapid rebound of the subducted continental rocks back up the same subduction zone.

(4) India–Asia collision

The closure of Neo-Tethys along the Indus–Tsangpo Suture Zone is well constrained in Ladakh by the age of the youngest marine sedimentary rock formations both along the suture zone and in the Zanskar range, the northern continental margin of the Indian plate. Shallow marine limestones deposited during the planktonic foraminifera zone (P8; 50.5 Ma, Early Eocene) are the youngest marine deposits (Green et al. 2008). These rocks are overlain by continental clastic rocks and conglomerates that contain boulders and cobbles eroded from the southern margin of Asia (Ladakh–Gangdese granites). In Pakistan there is no trace of the suture zone rocks seen in Ladakh and so we have no direct evidence for the timing of collision between India and Kohistan. Also missing in Pakistan is the 4 km thick succession of Phanerozoic unmetamorphosed shelf sediments of the northern margin of India that are exposed across the Zanskar range (Corfield & Searle 2000). We suspect that these rocks would have been present prior to the ophiolite subduction event and the subsequent thrusting of the Kohistan arc onto the leading edge of the Indian plate. As with the missing ophiolite sequence of northern Pakistan it is fair to ask where these rocks are now. Thrusting of the >30 km thick Kohistan arc onto the leading edge of India decoupled the cover rocks from the basement (Treloar et al. 1989b, 1991) stacking them up in a south-vergent thrust wedge (Coward & Butler 1985; Coward et al. 1987). These rocks, and the ophiolites that structurally overlay them, have subsequently been eroded.

(5) UHP eclogite metamorphism

Subduction of the thinned continental margin of the Indian plate to >100 km depth resulted in coesite eclogite UHP metamorphism at c. 27 kbar (O'Brien et al. 2001; Treloar et al. 2003), at c. 47 Ma (Parrish et al. 2006). Tertiary eclogites are known from two localities, the Kaghan region of northern Pakistan and the Tso Morari region of Ladakh. In both regions eclogite bodies (usually boudinaged sills or dykes) are widespread regionally and are enclosed in later sillimanite- and kyanite-grade gneisses (Tso Morari) or Proterozoic orthogneisses (Kaghan). Eclogite metamorphism could be related to either the ophiolite obduction event (as in Oman; Searle et al. 2004), which predated India–Asia collision, or the earliest stage of the continent–continent collision process following the Early Eocene (c. 50 Ma) closure of Tethys (Green et al. 2008), which involved thrusting of the Kohistan arc onto the Indian plate margin.

(6) Kyanite–sillimanite metamorphism

Rapid exhumation of the Kaghan eclogites through the mantle from 27 kbar depths up to the base of the crust at c. 47 Ma is shown by U–Pb dating of zircon, allanite and titanite in the Kaghan eclogites (Parrish et al. 2001). Eclogites were thrust onto kyanite–sillimanite-grade regional metamorphic rocks formed at 7–12 kbar (Treloar 1997; Foster et al. 2002) along the thickened continental margin. Peak amphibolite-facies regional kyanite–sillimanite metamorphism in Pakistan is dated at about 47 Ma (Smith et al. 1994; Foster et al. 2002) with subsequent cooling through the Ar–Ar hornblende blocking temperature by 40 Ma. Crustal thickening along the Indian plate could have lasted c. 25 Ma from the later phase of Kohistan arc obduction onto the Indian plate margin to c. 47–45 Ma, a time scale similar to that recorded beneath the Oman ophiolite (Searle et al. 2004; Searle 2007).

(7) Eocene exhumation and cooling

By c. 40 Ma Greater Himalayan metamorphic rocks in the Swat and Kaghan thrust sheets in Pakistan had cooled below 500 °C and post-metamorphic south-vergent thrusts were active, placing high-grade rocks onto low-grade rocks (Treloar et al. 1989c, d). Basement gneisses in the Besham block were unaffected by Himalayan metamorphism (Treloar 1997).

(8) Miocene thrusting and low-angle normal faulting (Main Mantle Thrust)

Widespread Miocene sillimanite-grade metamorphism, migmatization and melting, so widespread along the Indian and Nepalese Himalaya, are notably absent in Pakistan. Large Miocene leucogranites common everywhere along the central and eastern Himalaya are also notably absent in Pakistan. There is therefore no real evidence to suggest that 'channel flow' processes were operative in this far western part of the Himalaya (Treloar

1997), unlike the main Himalayan ranges in India, Nepal, Bhutan and south Tibet (e.g. Grujic et al. 1996, 2002; Searle et al. 1997, 2003, 2006; Searle & Szulc 2005). Instead, in Pakistan, south-vergent thrust tectonics dominated, with thrusts generally propagating towards the foreland in the direction of transport. At the rear of the Himalayan metamorphic pile, low-angle normal faulting along the Main Mantle Thrust (MMT) dropped the Kohistan arc down to the north against high-grade metamorphic rocks of the Kaghan and upper Swat thrust sheets (Vince & Treloar 1996; Treloar 1997). In reality the MMT probably acted as a 'passive roof fault' with southward displacement and exhumation of Himalayan footwall rocks relative to a fixed Kohistan hanging wall. In this respect we propose that the MMT acted in a similar fashion to the South Tibetan Detachment during Miocene southward motion of Greater Himalayan rocks along the central and eastern Himalaya. Total displacement across these faults must be measurable in terms of tens to hundreds of kilometres. On their hanging wall are rocks from the base of the >30 km thick Kohistan arc. On their footwall are a variety of metamorphic rocks that range from greenschist to amphibolite facies. The uppermost rocks in the footwall sequence (Tethyan sediments and obducted ophiolites) have all been subsequently eroded.

Conclusions

Indian plate rocks of the Pakistan Himalaya contain kyanite- and sillimanite-grade metamorphic rocks similar to those in the rest of the Himalayan chain although with several key differences. In Pakistan, the widespread migmatites and crustal melt leucogranites seen in the Indian and Nepalese Himalaya are missing. There is no inverted metamorphic sequence along the hanging wall of the Main Central Thrust or condensed sequence of right-way-up isograds beneath the South Tibetan Detachment low-angle fault as seen elsewhere along the Himalaya. There is also an important difference in timing of metamorphism and cooling in the Pakistan Himalaya compared with the main range in India and Nepal. In Pakistan, peak metamorphism was earlier (>47 Ma) than in India-Nepal, where peak temperatures were attained between 23 and 16 Ma. Because final closure of Tethys and India-Asia collision is well constrained at c. 50 Ma, some crustal thickening and regional metamorphism must predate continental collision. The early timing of regional metamorphism in the Pakistan Himalaya is consistent with an earlier crustal thickening event and associated thermal relaxation.

We suggest here that southward obduction of a large ophiolite sheet, similar to the Spontang Ophiolite in Ladakh, occurred also in northern Pakistan, resulting in thickening of the Indian plate margin during the Late Cretaceous–Paleocene. Regional crustal thickening, as evidenced by the large-scale recumbent folds and thrusts in the Mesozoic shelf carbonates beneath the Spontang Ophiolite in Ladakh, would have meant that crustal heating in the Indian plate could have started at about 65 Ma.

Following ophiolite obduction, oceanic subduction to the north of India continued beneath Kohistan until c. 50 Ma. This postdates the last recorded extrusive volcanism at 65 Ma but predates cessation of late-stage calc-alkaline pluton emplacement into the arc, which continued until c. 42 Ma (Jagoutz et al. 2009). This means that two north-dipping subduction zones are required, one beneath the Spontang Ophiolite and the other, more northerly, one beneath the Dras–Kohistan arc. We speculate that >10–20 km of erosion has removed the ophiolite and upper crustal sedimentary cover (equivalent to the Zanskar zone in northern india) in Pakistan. Additionally, north-vergent 'extension' along the Main Mantle Thrust shear zone has dropped the Kohistan arc down to the north, juxtaposing the lower parts of the arc (Jijal Complex) directly against mid–lower crust metamorphic rocks of the Pakistan Himalaya. Like the South Tibetan Detachment, the 'extension' along the MMT was relative, not absolute, with extensional fabrics being a result of southward expulsion of the footwall rocks relative to the stationary hanging-wall Kohistan arc rocks.

Ophiolite obduction provides a mechanism to explain early (pre-continent-continent collision) deformation and early metamorphism in Pakistan compared with the rest of the Himalaya. The Kohistan arc collided with the already deformed and partially metamorphosed north Indian margin probably c. 60–50 Ma (Searle et al. 1999) when the Indian plate jammed against the subduction zone, cutting off the source for calc-alkaline granites. Crustal shortening and thickening propagated southward across the Pakistan Himalaya, and during exhumation of the kyanite–sillimanite-grade rocks towards the south beneath the Kohistan arc extensional fabrics along the Main Mantle Thrust overprinted earlier obduction-related fabrics.

We are very grateful to Brian Windley for his support and guidance over many years and for introducing us to the amazing geology of northern Pakistan. We would also like to thank our Pakistani colleagues from the University of Peshawar, Northwest Frontier province, notably R. K. Tahirkheli, Q. Jan and A. Khan. Our work has been funded mainly by the Natural Environment Research Council, UK and the Royal Society.

References

AHMAD, Z. & ABBAS, S. G. 1979. The Muslim Bagh Ophiolites. *In*: FARAH, A. & DEJONG, K. A. (eds) *Geodynamics of Pakistan*. Geological Survey of Pakistan, Quetta, 243–249.

ALLEMAN, F. 1979. Time of emplacement of the Zhob Valley ophiolites and Bela ophiolites, Baluchistan. *In*: FARAH, A. & DEJONG, K. A. (eds) *Geodynamics of Pakistan*. Geological Survey of Pakistan, Quetta, 215–242.

BIGNOLD, S. M. & TRELOAR, P. J. 2003. Northward subduction of the Indian Plate beneath the Kohistan island arc, Pakistan Himalaya: new evidence from isotopic data. *Journal of the Geological Society, London*, **160**, 377–384.

BIGNOLD, S. M., TRELOAR, P. J. & PETFORD, N. 2006. Changing sources of magma generation beneath intra-oceanic island arcs: An insight from the juvenile Kohistan island arc, Pakistan Himalaya. *Chemical Geology*, **233**, 46–74.

BURG, J.-P., BODINIER, J. L., CHAUDHRY, S., HUSSAIN, S. & DAWOOD, H. 1998. Infra-arc mantle–crust transition and intra-arc mantle diapirs in the Kohistan complex (Pakistani Himalaya): Petrostructural evidence. *Terra Nova*, **10**, 74–80.

BURG, J.-P., JAGOUTZ, O., DAWOOD, H. & HUSSAIN, S. 2006. Precollision tilt of crustal blocks in rifted island arcs: structural evidence from the Kohistan arc. *Tectonics*, **25**, doi:10.1029/2005TC001835.

CHAMBERLAIN, C. P., ZEITLER, P. K. & ERIKSON, E. 1991. Constraints on the tectonic evolution of the northwestern Himalaya from geochronologic and petrologic studies of Babusar Pass. *Journal of Geology*, **99**, 829–849.

COLCHEN, M., MASCLE, G. & VAN HAVER, T. 1986. Some aspects of collision tectonics in the Indus suture zone, Ladakh. *In*: COWARD, M. P. & RIES, A. (eds) *Collision Tectonics*. Geological Society, London, Special Publications, **19**, 173–184.

CORFIELD, R. I. & SEARLE, M. P. 2000. Crustal shortening estimates across the north Indian continental margin, Ladakh, NW India. *In*: KHAN, M. A., TRELOAR, P. J., SEARLE, M. P. & JAN, M. Q. (eds) *Tectonics of the Nanga Parbat Syntaxis and the Western Himalaya*. Geological Society, London, Special Publications, **170**, 395–410.

CORFIELD, R. I., SEARLE, M. P. & GREEN, O. R. 1999. Photang thrust sheet—an accretionary complex structurally below the Spontang ophiolite constraining timing and tectonic environment of ophiolite obduction, Ladakh Himalaya. *Journal of the Geological Society, London*, **156**, 1031–1044.

CORFIELD, R. I., SEARLE, M. P. & PEDERSON, R. B. 2001. Tectonic setting, origin and obduction history of the Spontang Ophiolite, Ladakh Himalaya, NW India. *Journal of Geology*, **109**, 715–736.

CORFIELD, R. I., WATTS, A. B. & SEARLE, M. P. 2005. Subsidence history of the North Indian continental margin, Zanskar–Ladakh Himalaya, NW India. *Journal of the Geological Society, London*, **162**, 135–146.

COWARD, M. P. & BUTLER, R. W. H. 1985. Thrust tectonics and the deep structure of the Pakistan Himalaya. *Geology*, **13**, 417–420.

COWARD, M. P., JAN, M. Q., REX, D. C., TARNEY, J., THIRLWALL, M. F. & WINDLEY, B. F. 1982. Geotectonic framework of the Himalaya of northern Pakistan. *Journal of the Geological Society, London*, **139**, 299–308.

COWARD, M. P., WINDLEY, B. F. ET AL. 1986. Collision tectonics in the NW Himalayas. *In*: COWARD, M. P. & RIES, A. (eds) *Collision Tectonics*. Geological Society, London, Special Publications, **19**, 205–221.

COWARD, M. P., BUTLER, R. W. H., KHAN, M. A. & KNIPE, R. J. 1987. The tectonic history of Kohistan and its implications for Himalayan structure. *Journal of the Geological Society, London*, **144**, 377–392.

DANISHWAR, S., STERN, R. J. & KHAN, M. A. 2001. Field relations and structural constraints for the Teru volcanic formation, northern Kohistan Terrane, Pakistani Himalayas. *Journal of Asian Earth Sciences*, **19**, 683–695.

DHUIME, B., BOSCH, D., BODINIER, J.-L., GARRIDO, C. J., BRUGUIER, O., HUSSAIN, S. S. & DAWOOD, H. 2007. Multistage evolution of the Jijal ultramafic–mafic complex (Kohistan N. Pakistan): implications for building the roots of island arcs. *Earth and Planetary Science Letters*, **261**, 179–200.

DHUIME, B., BOSCH, D., GARRIDO, C. J., BODINIER, J.-L., BRUGUIER, O., HUSSAIN, S. S. & DAWOOD, H. 2009. Geochemical architecture of the lower to middle crustal section of a paleo-island arc (Kohistan Complex, Jial-Kamila Area, Northern Pakistan): implications for the evolution of an oceanic subduction zone. *Journal of Petrology*, **50**, 531–569.

DUNLAP, W. J. & WYSOCZANSKI, R. 2002. Thermal evidence for early Cretaceous metamorphism in the Shyok suture zone and age of the Khardung volcanic rocks, Ladakh, India. *Journal of Asian Earth Sciences*, **20**, 481–490.

ENGLAND, P. C. & THOMPSON, A. B. 1984. Pressure–temperature–time paths of regional metamorphism. 1: Heat transfer during the evolution of regions of thickened crust. *Journal of Petrology*, **25**, 894–925.

FOSTER, G. L., VANCE, D., ARGLES, T. W. & HARRIS, N. B. W. 2002. The Tertiary collision-related thermal history of the NW Himalaya. *Journal of Metamorphic Geology*, **20**, 827–844.

GAETANI, M. & GARZANTI, E. 1991. Multicyclic history of the northern India continental margin (North-western Himalaya). *AAPG Bulletin*, **75**, 1427–1446.

GARRIDO, C. J., BODINIER, J.-L. ET AL. 2006. Petrogenesis of mafic garnet granulite in the lower crust of the Kohistan paleo-arc complex (N. Pakistan): Implications for intracrustal differentiation of island arcs and generation of continental crust: *Journal of Petrology*, **47**, 1873–1914.

GARRIDO, C. J., DHUIME, B., BOSCH, D., BRUGUIRE, O., HUSSAIN, D. & BURG, J.-P. 2007. Origin of the island arc Moho transition zone via melt–rock reaction and its implications for intracrustal differentiation of island arcs: Evidence from the Jijal complex (Kohistan complex, northern Pakistan). *Geology*, **35**, 683–687.

GARZANTI, E. & VAN HAVER, T. 1988. The Indus clastics: forearc basin sedimentation in the Ladakh Himalaya (India). *Sedimentary Geology*, **59**, 237–249.

GARZANTI, E., BAUD, A. & MASCLE, G. 1987. Sedimentary record of the northward flight of India and its collision with Eurasia (Ladakh Himalaya, India). *Geodinamica Acta*, **1**, 297–312.

GEORGE, M. T., BUTLER, R. W. H. & HARRIS, N. W. B. 1993. The tectonic implications of contrasting granite magmatism between the Kohistan island arc and the Nanga Parbat–Haramosh massif. *In*: TRELOAR, P. J. & SEARLE, M. P. (eds) *Himalayan Tectonics*. Geological Society, London, Special Publications, **74**, 173–191.

GODIN, L., GRUJIC, D., LAW, R. D. & SEARLE, M. P. 2006. Channel flow, ductile extrusion and exhumation in continental collision zones: an introduction. *In*: LAW, R. D., SEARLE, M. P. & GODIN, L. (eds) *Channel Flow, Ductile Extrusion and Exhumation in Continental Collision Zones*. Geological Society, London, Special Publications, **268**, 1–23.

GREEN, O. R., SEARLE, M. P., CORFIELD, R. I. & CORFIELD, R. M. 2008. Cretaceous–Tertiary carbonate evolution and the age of the India–Asia collision along the Ladakh Himalaya (Northwest India). *Journal of Geology*, **116**, doi:10.1086/588831.

GRUJIC, D., CASEY, M., DAVIDSON, C., HOLLISTER, L., KUNDIG, K., PAVLIS, T. & SCHMID, S. 1996. Ductile extrusion of the Higher Himalayan crystalline in Bhutan: evidence from quartz microfabrics. *Tectonophysics*, **260**, 21–43.

GRUJIC, D., HOLLISTER, L. & PARRISH, R. R. 2002. Himalayan metamorphic sequence as a orogenic channel: insight from Bhutan. *Earth and Planetary Science Letters*, **198**, 177–191.

HEUBERGER, S., SCHALTEGGER, U. ET AL. 2007. Age and isotopic constraints on magmatism along the Karakoram–Kohistan suture zone, NW Pakistan: evidence for subduction and continued convergence after the India–Asia collision. *Swiss Journal of Geoscience*, **100**, 85–107.

HODGES, K. V. 2000. Tectonics of the Himalaya and southern Tibet from two perspectives. *Geological Society of America Bulletin*, **112**, 324–350.

HONEGGER, K., DIETRICH, V., FRANK, W., GANSSER, A., THONI, M. & TROMMSDORF, V. 1982. Magmatism and metamorphism in the Ladakh Himalayas (the Indus–Tsangpo suture zone). *Earth and Planetary Science Letters*, **60**, 253–292.

JAGOUTZ, O., MÜNTENER, O., BURG, J.-P., ULMER, P. & JAGOUTZ, E. 2006. Lower continental crust formation through focused flow in km-scale melt conduits: the zoned ultramafic bodies of the Chilas Complex in the Kohistan island arc (NW Pakistan). *Earth and Planetary Science Letters*, **242**, 320–342.

JAGOUTZ, O., MUNTENER, O., ULMER, P., PETTKE, T., BURG, J.-P., DAWOOD, H. & HUSSAIN, S. 2007. Petrology and mineral chemistry of lower crustal intrusions: the Chilas complex, Kohistan (NW Pakistan). *Journal of Petrology*, **48**, 1895–1953.

JAGOUTZ, O. E., BURG, J.-P., HUSSAINS, S., DAWOOD, H., PETTKE, T., IIZUKA, T. & MARUYAMA, S. 2009. Construction of the granitoid crust of an island arc part 1: geochronological and geochemical constraints from the plutonic Kohistan (NW Pakistan). *Contributions to Mineralogy and Petrology*, doi:10.1007/s00410-009-0408-3.

JAN, M. Q. & HOWIE, R. A. 1981. The mineralogy and geochemistry of the metamorphosed basic and ultrabasic rocks of the Jijal complex, Kohistan, NW Pakistan. *Journal of Petrology*, **22**, 85–126.

KHAN, M. A., JAN, M. Q., WINDLEY, B. F., TARNEY, J. & THIRLWALL, M. F. 1989. The Chilas mafic–ultramafic igneous complex; the root of the Kohistan island arc in the Himalaya of northern Pakistan. *In*: MALINCONICO, L. M. & LILLIE, R. J. (eds) *Tectonics of the Western Himalayas*. Geological Society of America, Special Papers, **232**, 75–94.

KHAN, M. A., JAN, M. Q. & WEAVER, B. L. 1993. Evolution of the lower arc crust in Kohistan, N. Pakistan: temporal arc magmatism through early, mature and intra-arc rift stages. *In*: TRELOAR, P. J. & SEARLE, M. P. (eds), *Himalayan Tectonics*. Geological Society, London, Special Publications, **74**, 123–138.

KHAN, M. A., STERN, R. J., GRIBBLE, R. F. & WINDLEY, B. F. 1997. Geochemical and isotopic constraints on subduction polarity, magma sources, and palaeogeography of the Kohistan intra-oceanic arc, northern Pakistan Himalaya. *Journal of the Geological Society, London*, **154**, 935–946.

KHAN, S. D., STERN, R. J., MANTON, W. I., COPELAND, P., KIMURA, J. I. & KHAN, M. A. 2004. Age, geochemical and Sr–Nd–Pb isotopic constraints for mantle source characteristics and petrogenesis of Teru Volcanic Formation, northern Kohistan terrane, Pakistan: *Tectonophysics*, **393**, 263–280.

KHAN, S. D., WALKER, D. J., HALL, S. A., BURKE, K. C., SHAH, M. T. & STOCKLI, L. 2008. Did the Kohistan–Ladakh island arc collide first with India? *Geological Society of America Bulletin*, doi:10.1130/B26348.1.

MATTAUER, M., PROUST, F., TAPPONIER, P. & CASSAIGNEAU, C. 1978. Ophiolites, obductions et tectonique globale dans l'est de l'Afghanistan. *Comptes Rendus de l'Académie des Sciences, Série D*, **287**, 983–985.

NICORA, A., GARZANTI, E. & FOIS, E. 1987. Evolution of the Tethys Himalaya continental shelf during Maastrichtian to Palaeocene (Zanskar, India). *Rivista Italiana di Paleontologia e Stratigraphifia*, **92**, 439–496.

O'BRIEN, P. J., ZOTOV, N., LAW, R. D., KHAN, M. A. & JAN, M. Q. 2001. Coesite in Himalayan eclogite and implications for models of India–Asia collision. *Geology*, **29**, 435–438.

PARRISH, R. R., GOUGH, S. J., SEARLE, M. P. & WATERS, D. J. 2006. Plate velocity exhumation of ultrahigh-pressure eclogites in the Pakistan Himalaya. *Geology*, **34**, 989–992.

PEDERSEN, R., SEARLE, M. P. & CORFIELD, R. I. 2001. U–Pb zircon ages from the Spontang Ophiolite, Ladakh Himalaya. *Journal of the Geological Society, London*, **158**, 513–520.

PETTERSEN, M. G. & TRELOAR, P. J. 2004. Volcanostratigraphy of arc volcanic sequences in the Kohistan arc, north Pakistan: Volcanism within island arc, backarc-basin, and intracontinental tectonic settings. *Journal of Volcanology and Geothermal Research*, **130**, 147–178.

PETTERSEN, M. G. & WINDLEY, B. F. 1985. Rb–Sr dating of the Kohistan arc batholith in the Trans Himalaya of N. Pakistan and tectonic implications. *Earth and Planetary Science Letters*, **74**, 54–75.

PETTERSEN, M. G. & WINDLEY, B. F. 1991. Changing source regions of magmas and crustal growth in the Trans-Himalayas: Evidence from the Chalt volcanics and Kohistan batholith, Kohistan, N. Pakistan. *Earth and Planetary Science Letters*, **102**, 326–346.

PETTERSEN, M. G. & WINDLEY, B. F. 1992. Field relations, geochemistry and petrogenesis of the Cretaceous basaltic Jutal dykes, Kohistan, northern Pakistan. *Journal of the Geological Society, London*, **149**, 107–114.

PUDSEY, C. J. 1986. The Northern Suture, Pakistan: margin of a Cretaceous island arc. *Geological Magazine*, **123**, 405–423.

RAZ, U. & HONEGGER, K. 1989. Magmatic and tectonic evolution of the Ladakh Block from field studies. *Tectonophysics*, **161**, 107–118.

REUBER, I. 1989. The Dras arc: two successive volcanic events on eroded oceanic crust. *Tectonophysics*, **161**, 93–106.

ROBERTSON, A. H. F. 2000. Formation of mélanges in the Indus Suture Zone, Ladakh Himalaya by successive subduction-related collisional and post-collisional processes during Late Mesozoic–Late Tertiary time. *In*: KHAN, M. A., TRELOAR, P. J., SEARLE, M. P. & JAN, M. Q. (eds) *Tectonics of the Nanga Parbat Syntaxis and the Western Himalaya*. Geological Society, London, Special Publications, **170**, 395–410.

ROBERTSON, A. H. F. & DEGNAN, P. J. 1993. Sedimentology and tectonic implications of the Lamayuru Complex: deep-water facies of the Indian passive margin, Indus Suture Zone, Ladakh Himalaya. *In*: TRELOAR, P. J. & SEARLE, M. P. (eds) *Himalayan Tectonics*. Geological Society, London, Special Publications, **74**, 299–321.

ROBERTSON, A. H. F. & DEGNAN, P. 1994. The Dras Arc complex: lithofacies and reconstruction of a Late Cretaceous oceanic volcanic arc in the Indus Suture Zone, Ladakh Himalaya. *Sedimentary Geology*, **92**, 117–145.

ROBERTSON, A. & SHARP, I. 1988. Mesozoic deep-water slope/rise sedimentation and volcanism along the North Indian passive margin: Evidence from the Karamba complex, Indus suture zone (western Ladakh Himalaya). *Journal of Asian Earth Sciences*, **16**, 195–215.

SCHALTEGGER, U., ZEILINGER, G., FRANK, M. & BURG, J.-P. 2002. Multiple mantle sources during island arc magmatism: U–Pb and Hf isotope evidence from the Kohistan arc complex, Pakistan. *Terra Nova*, **14**, 461–468.

SCHALTEGGER, U., FRANK, M. & BURG, J.-P. 2003. A 120 million years record of magmatism and crustal melting in the Kohistan batholith. *Geophysical Research Abstracts*, EGU–AGU–EUG Joint Assembly 5, EAE03-A-08307.

SCHÄRER, U., HAMET, J. & ALLÈGRE, C. J. 1984. The Transhimalaya (Gangdese) plutonism in the Ladakh region: a U–Pb and Rb–Sr study. *Earth and Planetary Science Letters*, **67**, 327–339.

SEARLE, M. P. 1986. Structural evolution and sequence of thrusting in the High Himalayan, Tibetan–Tethys and Indus suture zones of Zanskar and Ladakh, Western Himalaya. *Journal of Structural Geology*, **8**, 923–936.

SEARLE, M. P. 2007. Diagnostic features and processes in the construction and evolution of Oman-, Zagros-, Himalayan-, Karakoram-, and Tibetan-type orogenic belts. *In*: HATCHER, R. D., CARLSON, M. P., MCBRIDE, J. H. & CATALAN, J. R. (eds) *4-D Framework of Continental Crust*. Geological Society of America, Memoirs, **200**, 41–61, doi:10.1130/207.1200(04).

SEARLE, M. P. & COX, J. S. 1999. Tectonic setting, origin and obduction of the Oman Ophiolite. *Geological Society of America Bulletin*, **111**, 104–122.

SEARLE, M. P. & COX, J. S. 2002. Subduction zone metamorphism during formation and emplacement of the Semail Ophiolite in the Oman Mountains. *Geological Magazine*, **139**, 241–255.

SEARLE, M. P. & SZULC, A. G. 2005. Channel flow and ductile extrusion of the High Himalayan slab, Kangchenjunga–Darjeeling profile, Sikkim Himalaya. *Journal of Asian Earth Sciences*, **25**, 173–185.

SEARLE, M. P., COOPER, D. J. W. & REX, A. J. 1988. Collision tectonics of the Ladakh–Zanskar Himalaya. *Philosophical Transactions of the Royal Society of London, Series A*, **326**, 117–150.

SEARLE, M. P., PICKERING, K. T. & COOPER, D. J. 1990. Restoration and evolution of the intermontane Indus molasse basin, Ladakh Himalaya, India. *Tectonophysics*, **174**, 301–314.

SEARLE, M. P., CORFIELD, R. I., STEPHENSON, B. J. & MCCARRON, J. 1997. Structure of the north Indian continental margin in the Ladakh–Zanskar Himalayas: implications for the timing of obduction of the Spontang ophiolite, India–Asia collision and deformation events in the Himalaya. *Geological Magazine*, **134**, 297–316.

SEARLE, M. P., KHAN, M. A., FRASER, J. E., GOUGH, S. J. & JAN, M. Q. 1999. The tectonic evolution of the Kohistan–Karakoram collision belt along the Karakoram Highway transect, North Pakistan. *Tectonics*, **18**, 929–949.

SEARLE, M. P., SIMPSON, R. L., LAW, R. D., PARRISH, R. R. & WATERS, D. J. 2003. The structural geometry, metamorphic and magmatic evolution of the Everest massif, High Himalaya of Nepal–south Tibet. *Journal of the Geological Society, London*, **160**, 345–366.

SEARLE, M. P., WARREN, C. J., WATERS, D. J. & PARRISH, R. R. 2004. Structural evolution, metamorphism and restoration of the Arabian continental margin, Saih Hatat region, Oman Mountains. *Journal of Structural Geology*, **26**, 451–473.

SEARLE, M. P., LAW, R. D., JESSUP, M. & SIMPSON, R. L. 2006. Crustal structure and evolution of the Greater Himalaya in Nepal–South Tibet: implications for channel flow and ductile extrusion of the middle crust. *In*: LAW, R. D., SEARLE, M. P. & GODIN, L. (eds) *Channel Flow, Ductile Extrusion and Exhumation in Continental Collision Zones*. Geological Society, London, Special Publications, **268**, 355–378.

SEARLE, M. P., STEPHENSON, B. J., WALKER, J. D. & WALKER, C. B. 2007. Restoration of the Greater Himalaya in Zanskar–Lahoul: implications for metamorphic protoliths, thrust and normal faulting, and

channel flow in the Western Himalaya. *Episodes*, **30**, 242–257.
SMITH, H. A., CHAMBERLAIN, C. P. & ZEITLER, P. K. 1994. Timing and duration of Himalayan metamorphism within the Indian Plate, Northwest Himalaya, Pakistan. *Journal of Geology*, **102**, 493–508.
STECK, A. 2003. Geology of the NW Indian Himalaya. *Eclogue Geologicae Helvetiae*, **96**, 147–196.
SULLIVAN, M. A., WINDLEY, B. F., SAUNDERS, A. D., HAYNES, J. R. & REX, D. C. 1993. A palaeogeographic reconstruction of the Dir Group: Evidence for magmatic arc migration within Kohistan, N. Pakistan. *In*: TRELOAR, P. J. & SEARLE, M. P. (eds) *Himalayan Tectonics*. Geological Society, London, Special Publications, **74**, 139–160.
TAHIRKHELI, R. A. K., MATTAUER, M., PROUST, F. & TAPPONNIER, P. 1979. The India–Eurasia suture zone in northern Pakistan: Synthesis and interpretation of data at plate scale: *In*: FARAH, A. & DEJONG, K. A. (eds) *Geodynamics of Pakistan*. Geological Survey of Pakistan, Quetta, 125–130.
TAPPONNIER, P., MATTAUER, M., PROUST, F. & CASSAIBNEAU, C. 1981. Mesozoic ophiolites, sutures and large scale tectonic movements in Afghanistan. *Earth and Planetary Science Letters*, **52**, 355–371.
TONARINI, S., VILLA, I., OBERLI, F., MEIER, M., SPENCER, D. A., POGNANTE, U. & RAMSAY, J. G. 1993. Eocene age of eclogite metamorphism in the Pakistan Himalaya: implications for India–Eurasian collision. *Terra Nova*, **5**, 13–20.
TRELOAR, P. J. 1997. Thermal controls on early-Tertiary, short-lived, rapid regional metamorphism in the NW Himalaya, Pakistan. *Tectonophysics*, **273**, 77–104.
TRELOAR, P. J. & IZATT, C. N. 1993. Tectonics of the Himalayan collision zone between the Indian plate and the Afghan Block: a synthesis. *In*: TRELOAR, P. J. & SEARLE, M. P. (eds) *Himalayan Tectonics*. Geological Society, London, Special Publications, **74**, 69–87.
TRELOAR, P. J. & REX, D. C. 1990a. Cooling, uplift and exhumation rates in the crystalline thrust stack of the North Indian Plate, west of the Nanga Parbat syntaxis. *Tectonophysics*, **180**, 323–349.
TRELOAR, P. J. & REX, D. C. 1990b. Post-metamorphic cooling history of the Indian Plate crystalline thrust stack, Pakistan Himalaya. *Journal of the Geological Society, London*, **147**, 735–738.
TRELOAR, P. J., REX, D. C. ET AL. 1989a. K–Ar and Ar–Ar geochronology of the Himalayan collision on NW Pakistan: constraints on the timing of suturing, deformation, metamorphism and uplift. *Tectonics*, **8**, 881–909.
TRELOAR, P. J., COWARD, M. P., WILLIAMS, M. P. & KHAN, M. A. 1989b. Basement–cover imbrication south of the Main Mantle Thrust, North Pakistan. *In*: MALINCONICO, L. M. & LILLIE, R. J. (eds) *Tectonics of the Western Himalayas*. Geological Society of America, Special Papers, **232**, 137–152.
TRELOAR, P. J., BROUGHTON, R. D., WILLIAMS, M. P., COWARD, M. P. & WINDLEY, B. F. 1989c. Deformation, metamorphism and imbrication of the Indian Plate, south of the Main Mantle Thrust, North Pakistan. *Journal of Metamorphic Geology*, **7**, 111–125.
TRELOAR, P. J., WILLIAMS, M. P. & COWARD, M. P. 1989d. Metamorphism and crustal stacking in the north Indian Plate, North Pakistan. *Tectonophysics*, **165**, 167–184.
TRELOAR, P. J., BRODIE, K. H., COWARD, M. P., JAN, M. Q., KNIPE, R. J., REX, D. C. & WILLIAMS, M. P. 1990. The evolution of the Kamila shear zone, Kohistan, Pakistan. *In*: SALISBURY, M. H. & FOUNTAIN, D. M. (eds) *Exposed Cross-sections of the Continental Crust*. Kluwer, Dordrecht, 175–214.
TRELOAR, P. J., COWARD, M. P., CHAMBERS, A. F., IZATT, C. N. & JACKSON, K. C. 1991. Thrust geometries, interferences and rotations in the northwest Himalaya. *In*: McCLAY, K. R. (ed.) *Thrust Tectonics*. Chapman & Hall, London, 325–342.
TRELOAR, P. J., PETTERSON, M. G., JAN, M. Q. & SULLIVAN, M. A. 1996. A re-evaluation of the stratigraphy and evolution of the Kohistan arc sequence, Pakistan Himalaya: implications for magmatic and tectonic arc-building processes. *Journal of the Geological Society, London*, **153**, 681–693.
TRELOAR, P. J., O'BRIEN, P. J., PARRISH, R. R. & KHAN, M. A. 2003. Exhumation of early Tertiary coesite-bearing eclogites from the Pakistan Himalaya. *Journal of the Geological Society, London*, **160**, 367–376.
VINCE, K. J. & TRELOAR, P. J. 1996. Miocene, north-vergent extensional displacements along the Main Mantle Thrust, NW Himalaya, Pakistan. *Journal of the Geological Society, London*, **153**, 677–680.
WEINBERG, R. F. & DUNLAP, W. J. 2000. Growth and deformation of the Ladakh batholith, NW Himalayas: implications for timing of continental collision and origin of calc-alkaline batholiths. *Journal of Geology*, **108**, 303–320.
WEINBERG, R. F., DUNLAP, W. J. & WHITEHOUSE, M. J. 2000. New field, structural and geochronological data from the Shyok and Nubra valleys, northern Ladakh: linking Kohistan to Tibet. *In*: KHAN, M. A., TRELOAR, P. J., SEARLE, M. P. & JAN, M. Q. (eds) *Tectonics of the Nanga Parbat Syntaxis and the Western Himalaya*. Geological Society, London, Special Publications, **170**, 253–275.
YAMAMOTO, H. & NAKAMURA, E. 1996. Sm–Nd dating of garnet granulites from the Kohistan complex, northern Pakistan. *Journal of the Geological Society, London*, **153**, 965–969.

Tectonic setting and structural evolution of the Late Cenozoic Gobi Altai orogen

DICKSON CUNNINGHAM

University of Leicester, Department of Geology, Leicester LE17RH, UK
(e-mail: wdc2@le.ac.uk)

Abstract: The Gobi Altai is an intraplate, intracontinental transpressional orogen in southern Mongolia that formed in the Late Cenozoic as a distant response to the Indo-Eurasia collision. The modern range formed within crust constructed by successive terrane accretion and ocean suturing events and widespread granite plutonism throughout the Palaeozoic. Modern reactivation of the Gobi Altai crust and the kinematics of Quaternary faults are fundamentally controlled by Palaeozoic basement structural trends, the location of rigid Precambrian blocks, orientation of SH_{max} and possible thermal weakening of the lower crust as a result of an extensive history of Mesozoic–Cenozoic basaltic volcanism in the region, and the presence of thermally elevated asthenosphere under the Hangay Dome to the north. Modern mountain building processes in the Gobi Altai typically involve reactivation of NW–SE-striking basement structures in thrust mode and development of linking east–west left-lateral strike-slip faults that crosscut basement structures within an overall left-lateral transpressional regime. Restraining bends, other transpressional ridges and thrust basement blocks are the main range type, but are discontinuously distributed and separated by internally drained basins filling with modern alluvial deposits. Unlike a contractional thrust belt, there is no orogenic foreland or hinterland, and thrusts are both NE and SW directed with no evidence for a basal décollement. Normal faults related to widespread Cretaceous rifting in the region appear to be unfavourably oriented for Late Cenozoic reactivation despite widespread topographic inversion of Cretaceous basin sequences. Because the Gobi Altai is an actively developing youthful mountain range in an arid region with low erosion rates, it provides an excellent opportunity to study the way in which a continental interior reactivates as a result a distant continental collision. In addition, it offers important insights into how other more advanced intracontinental transpressional orogens may have developed during earlier stages of their evolution.

This paper presents a review of the geological evolution of southern Mongolia, focusing on the development of the Late Cenozoic Gobi Altai orogen (Fig. 1). The Gobi Altai is an actively forming intraplate and intracontinental mountain range that forms part of the complex deformation field of Central Asia north of Tibet. Continuing deformation in the Gobi Altai is believed to be driven by NE-directed compressional stress derived from the Indo-Eurasia collision 2000 km to the south (Tapponnier & Molnar 1979; Zoback 1992). The range is c. 900 km long from east to west and between 250 and 350 km wide from north to south. Highest elevations are nearly 4000 m (Ih Bogd) whereas adjacent basins are typically between 1000 and 1500 m in elevation (Fig. 1). The Gobi Altai occurs within a larger region that is sometimes referred to as the Mongolian Plateau because of its regionally elevated nature (Windley & Allen 1993). Although generally regarded as a distinctively separate mountain belt, the Gobi Altai is structurally and topographically linked to the Altai to the NW, the easternmost Tien Shan to the west and the Beishan to the SW (Fig. 1). Because it is an arid region with excellent exposure of both basement and cover sequences, the Gobi Altai affords an excellent opportunity to study the way in which a continental interior region reactivates in response to a distant continental collision.

Geological evolution

To understand the modern evolution of the Gobi Altai orogen, it is important to consider the geological history of the region and the crustal preconditions that existed prior to Late Cenozoic crustal reactivation. Without appreciation of this crustal context, it is impossible to gain a full understanding of the distribution and kinematics of faults that have reactivated the region and accommodated mountain building.

Precambrian–Palaeozoic history

The Gobi Altai crust is part of the Central Asian Orogenic Belt (CAOB), a terrane collage that records Palaeozoic closure of the Palaeo-Asian ocean (Fig. 2; Badarch *et al.* 2002). Spirited debate has revolved around competing continental

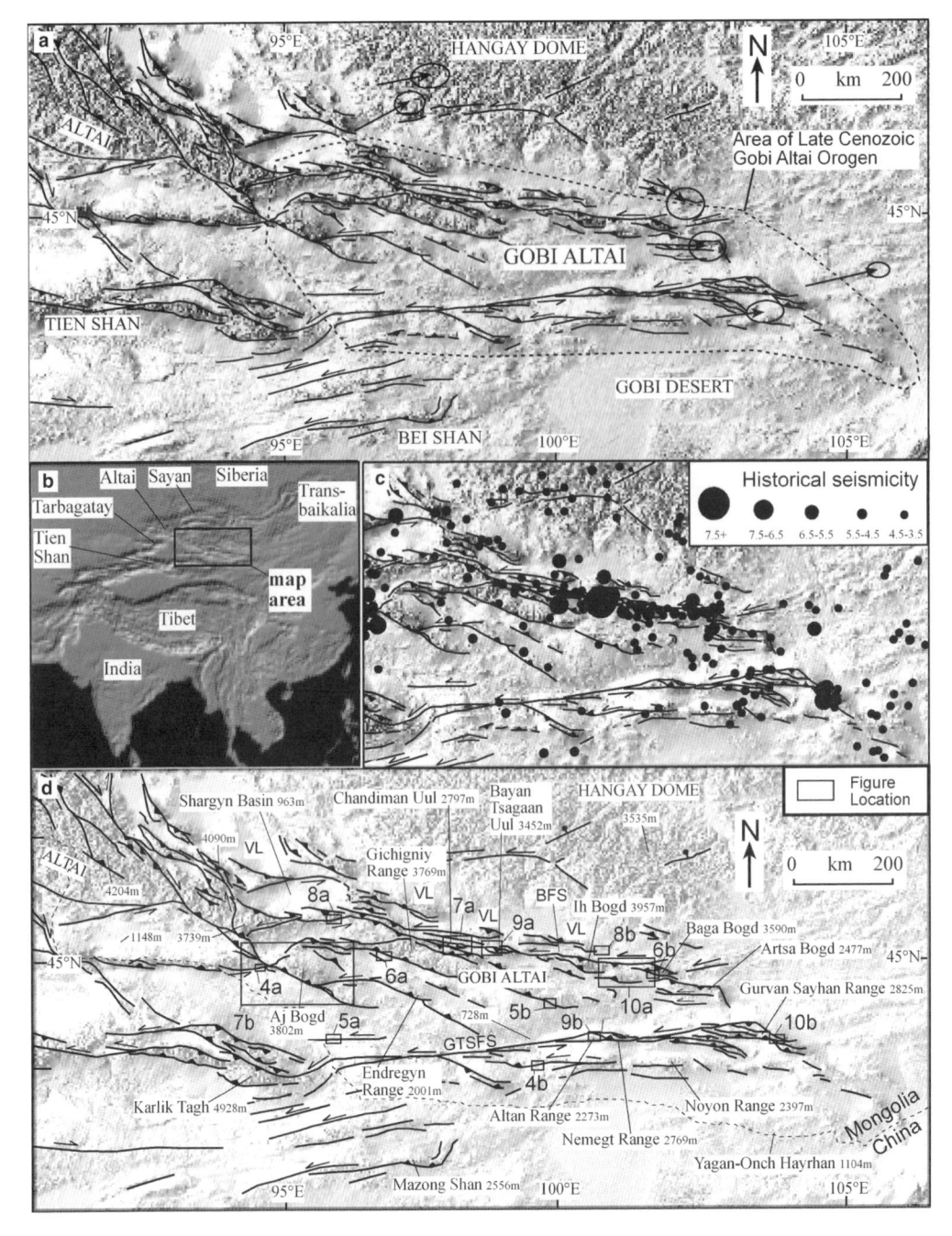

Fig. 1. The Gobi Altai orogen in Central Asia: (**a**) Quaternary fault map; (**b**) surrounding orogenic belts and location of (a), (c) and (d); (**c**) historical seismicity in Gobi Altai region from Adiya *et al.* (2003); (**d**) map showing areas mentioned in text and location of subsequent figures. BFS, Bogd Fault System; GTSFS, Gobi–Tien Shan Fault System; VL, Valley of Lakes.

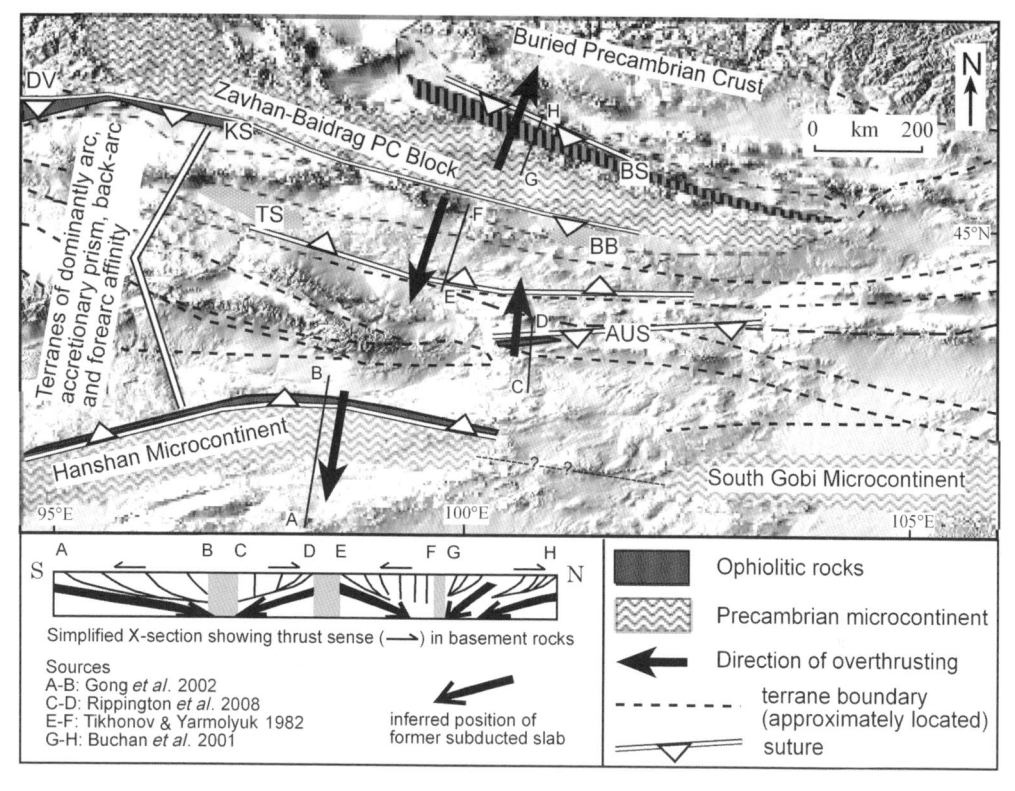

Fig. 2. Location of Precambrian blocks, Palaeozoic terrane boundaries, ophiolitic belts and suture zones in the Gobi Altai region. Direction of overthrusting at suture zones is indicated. BS, Bayanhongor Suture; KS, Khantairshir Suture; DV, Dariv ophiolite; TS, Tseel Block; BB, Baga Bogd; AUS, Altan Uul Suture. Terrane boundaries from Badarch *et al.* (2002). Evidence for Precambrian crust in the Tseel terrane and at Baga Bogd from Demoux *et al.* (2008).

growth models for the region between proponents of a single long-lived subduction–accretion system (e.g. Sengor *et al.* 1993) and those who favour an archipelago- or Indonesian-type model of multiple subduction zones and terrane collisions (Filippova *et al.* 2001; Windley *et al.* 2007, and references therein). Precambrian crustal fragments are exposed in the southern and western Hangay region (Archaean Baidrag Massif; Mitrofanov *et al.* 1985; Kozakov *et al.* 1997, 1999; Kovalenko *et al.* 2004), northeastern Gobi Altai (Mesoproterozoic–Neoproterozoic Baga Bogd massif; Demoux *et al.* 2008) and the extreme SE Gobi Altai region (Hutag Uul Terrane, also called the south Gobi Microcontinent; Badarch *et al.* 2002; Yarmolyuk *et al.* 2005). Nd isotopic data from metaigneous and metavolcanic rocks in the Gichigeniy Range (Tseel terrane of Badarch *et al.* 2002) also suggest that there is ancient continental crust (at least 1.5 Ga) at depth there (Fig. 2; Helo *et al.* 2006). These ancient blocks are separated by various terranes of mainly island arc, accretionary complex and ophiolitic affinity. Island arc magmatism is well documented in the northern Gobi Altai in the Cambrian–Silurian, and the timing of arc magmatism youngs progressively southward (Sengor *et al.* 1993) with Devonian and Carboniferous arcs well documented in southernmost Mongolia (Sengor *et al.* 1993; Lamb & Badarch 1997; Lamb *et al.* 2008). In addition, various unmetamorphosed to low–medium-grade metamorphosed sedimentary terranes interpreted to be back-arc or forearc assemblages exist between arc terranes, but typically were also intruded by arc-related intrusions (Badarch *et al.* 2002; Helo *et al.* 2006). Throughout the Palaeozoic period of terrane docking, and amalgamation, the construction of continental crust in the Gobi Altai region was enhanced by numerous periods of magmatism including voluminous and widespread post-tectonic granites that typically have juvenile characteristics implying juvenile sources (Jahn *et al.* 2000, 2001; Helo *et al.* 2006). Windley *et al.* (2007) proposed that in addition to normal arc magmatism, ridge subduction was an episodic event that provided an additional heat source that contributed to widespread granite plutonism during the multiple stages of ocean closure and terrane accretion in the Gobi Altai region.

At least four ophiolite-bearing sutures are recognized in specific locations in the Gobi Altai region, although their regional continuity is not yet documented: (1) a latest Precambrian–Cambrian WNW–ESE-striking suture in the Bayanhongor region along the south flank of the Hangay Dome north of the Baidrag Block (Fig. 2; Buchan et al. 2001); (2) a latest Precambrian–Cambrian east–west-striking suture in the Khantaishir–Dariv region SW of the Zavhan–Baidrag Block (Dijkstra et al. 2006); (3) an east–west-striking Carboniferous suture in the Nemegt region (Rippington et al. 2008); (4) the Permian Solonker suture, which occurs in the SE Mongolia–China border region and can be correlated westward into the approximately east–west strike belts of the Beishan immediately south of the Gobi Altai (Hsu et al. 1992; Xiao et al. 2003). In addition, Precambrian rocks in Baga Bogd correlate better geochronologically with Precambrian basement in the South Gobi microcontinent than the nearby Baidrag Block (Fig. 2; Demoux et al. 2008), suggesting that a separate cryptic suture separates the northeasternmost Gobi Altai basement from the Baidrag basement exposed in the southern Hangay region. This boundary is marked by a south-dipping, north-directed Palaeozoic fold and thrust belt in the Valley of Lakes previously termed the Lake Fold Zone by Dergunov (1989) and Zorin et al. (1993) and probably links along-strike with the Khantaishir ophiolite to the west (Fig. 2). Finally, other ophiolitic rocks have been reported from some arc and accretionary terranes (e.g. Gurvan Sayhan, Zoolen terranes of Badarch et al. 2002), but are more likely to be components of mélange complexes than true collisional suture zones.

The sense of former subduction polarity at each of these proposed suture zones is unproven and controversial, partly because of the paucity of detailed structural transects that might reveal the original sense of overthrusting and obduction at each former convergent boundary, and partly because existing geochronological data are insufficient to define separate arc terranes in time and space. However, at the Bayanhongor suture, a NE-directed thrust stack suggests that subduction was SSW-dipping (Buchan et al. 2001), whereas in the Gichigeniy Nuruu (Fig. 1d) Tikhonov & Yarmolyuk (1982) published a crustal profile that showed south-directed overthrust rocks including possible ophiolite fragments suggesting north-dipping subduction (compare the Trans-Altai fold zone of Zorin et al. 1993). Similarly, Lamb & Badarch (2001) suggested that Devonian–Carboniferous subduction in the southernmost Gobi Altai region was north-dipping and linked along-strike with coeval north-dipping subduction beneath the easternmost Tien Shan to the west (Xiao et al. 2004), and north-dipping subduction beneath SE Mongolia to the east, a region now famous for its Devonian–Carboniferous arc-hosted porphyry copper deposits (Lamb & Cox 1998; Blight et al. 2008). Rippington et al. (2008) presented structural evidence for southward subduction and northward ophiolite obduction and thrust transport in the Nemegt and Altan ranges in the southern Gobi Altai which may link along-strike to the east where Carboniferous back-arc basin closure as a result of southward subduction has been proposed in the Gurvan Sayhan region based on the identification of mélange units in the north of the range (Lamb & Badarch 2001). Finally, the end-Permian Solonker suture marks the closure of the final Palaeo-Asian ocean seaway and both north- and south-dipping subduction zones have been interpreted there because of the presence of two separate accretionary wedge complexes with opposing structural vergence (Xiao et al. 2003). Late Permian continental clastic sediments are widely deposited in southern–central Mongolia and overlap older marine deposits, indicating widespread terrestrialization in the Gobi Altai region by that time, although large Permian marine basins still existed in extreme SE Mongolia and in the Hangay region (Manankov et al. 2006).

The progressive north–south accretion of terranes south of the Precambrian blocks under the southern Hangay region has produced a strong crustal fabric throughout the Gobi Altai region consisting of terrane boundary faults, intra-terrane faults, metamorphic fabrics and sedimentary strike belts. This fundamental basement grain is WNW–ESE-striking in the western Gobi Altai, east–west-striking in the central Gobi Altai and ENE–WSW-striking in the easternmost Gobi Altai (Fig. 2).

Mesozoic history

Following terminal closure of the Palaeo-Asian ocean and consolidation of the CAOB in the Permian, southern Mongolia continued to be affected by regional tectonic events throughout much of the Mesozoic era. The absence of significant strata of Triassic–early Jurassic age in almost all of the Gobi Altai region except the extreme south (Tomurtogoo 1999) suggests that it was a period dominated by erosion in a mountainous landscape (Traynor & Sladen 1995). During the early–mid-Jurassic a major contractional event with significant thrust displacements occurred in southernmost Mongolia and in the Beishan (Fig. 1; Zheng et al. 1996). Mid–late Jurassic regional-scale folding and foreland basin style sedimentation also occurred in the Noyon Range in the southernmost Gobi Altai

(Fig. 1d), and fission-track data from the Mongolia–China border region record mid-Jurassic cooling that has been attributed to a major shortening event with rapid uplift and erosion (Fig. 1; Hendrix et al. 1996; Zheng et al. 1996; Dumitru & Hendrix 2001). Recently, Lamb et al. (2008) suggested Jurassic contractional deformation in the Gichigeniy Range (Fig. 1d) following Permian–Triassic terrane amalgamation, based on the observation that Jurassic sedimentary rocks are only mildly deformed and tilted, in contrast to nearby Permian strata, which are strongly cleaved and folded. Lamb et al. (1999) also suggested that two regionally significant NE–SW-striking left-lateral faults that cut across the central and southern Gobi Altai were active during the Triassic–mid-Cretaceous to account for apparent offsets in older Palaeozoic strike belts. In much of southern and eastern Mongolia, Late Triassic–Early Jurassic post-orogenic granitoids were rapidly exhumed and brought to the surface by an estimated 5 km or more of erosion between the early and mid-Jurassic (Traynor & Sladen 1995). This major erosion event is marked by a regional unconformity within Jurassic successions throughout the Gobi region and was recognized long ago (e.g. Berkey & Morris 1924a; Shuvalov 1969).

The driving force for early–mid-Jurassic crustal shortening in the Gobi Altai region is uncertain. The closure of the Mongol–Okhotsk ocean in the early–mid-Jurassic (Zorin 1999) in the region NE of the Gobi Altai may have generated a compressive stress field that reactivated the surrounding crust including the Gobi Altai region. However, the former western limit of the Mongol–Okhotsk ocean is uncertain, because a Jurassic suture has not yet been identified in the south-southeastern Hangay region. Indeed, the true southwestern terminus of the Mongol–Okhotsk ocean and the regional extent of deformation associated with its final closure is one of the outstanding unresolved problems in Mongolian tectonics. Other proposed drivers for early–mid-Jurassic contractional reactivation of the Gobi Altai region are distant collisional events that occurred along the southern margin of Asia, especially the Triassic–Jurassic collisions of the Qiangtang Block and Lhasa Block in central Tibet (Dewey et al. 1988; Yin & Harrison 2000).

In the Late Jurassic–early Cretaceous, much of Central Asia experienced continental rifting suggesting a change in the regional stress regime from compression to tension (Traynor & Sladen 1995; Graham et al. 2001; Davis et al. 2002). Southern–central and southeastern Mongolian basins are dominated by thick non-marine Jurassic–Cretaceous clastic sequences including widespread lacustrine facies that are potentially important petroleum source rocks (Sladen & Traynor 2000).

Late Jurassic–Early Cretaceous extensional rift development is well documented in SE Mongolia, where some basins are hydrocarbon producing and seismic profiles reveal the deeper basin architecture (Graham et al. 2001; Johnson 2004). Early Cretaceous crustal extension including metamorphic core complex development is also well documented south of the southeasternmost Gobi Altai along the Chinese–Mongolian border region in the Yagan–Onch Hayrhan core complex (Fig. 1d; Webb et al. 1999) and in the western Nemegt and Altan ranges (Cunningham et al. 2009). Cretaceous normal faults are also exposed in the Ih Bogd massif (author's unpublished data). Berkey & Morris (1927) also documented differential fault block movements in the NE Gobi Altai and dragged sedimentary sections that provide strong evidence for Cretaceous normal faulting in the Artsa Bogd region (Fig. 1d). Traynor & Sladen (1995) suggested that all of central and eastern Mongolia was characterized by rift development during the mid-Jurassic–Cretaceous, although they did not document normal faults of that age. Likewise, numerous papers on the Cretaceous stratigraphy and dinosaur palaeontology of the Gobi Altai region have suggested that Mesozoic Gobi basins were extensional grabens, but these papers lack normal fault documentation that might confirm the researchers' genetic interpretations for Cretaceous basin development (e.g. Jerzykiewicz & Russell 1991; Sladen & Traynor 2000).

Regional Late Jurassic–Early Cretaceous extension and metamorphic core complex development was coeval with collapse and extension of the early Mesozoic Mongol–Okhotsk contractional orogen in southern Siberia and NE Mongolia (Zorin 1999; Donskaya et al. 2008) and in the Yinshan–Yanshan contractional orogen in inner Mongolia (Davis et al. 1998, 2002; Graham et al. 2001). Increasingly, it appears that a huge and probably contiguous basin and range extensional province existed throughout much of western, southern and eastern Mongolia, Transbaikalia, and adjacent regions in north and NE China during the Cretaceous (Zorin 1999; Howard et al. 2003; Meng et al. 2003; Yang et al. 2007; Cunningham et al. 2009).

A brief period of transpressional inversion at the end of the early Cretaceous is suggested by seismic data in SE Mongolia (Graham et al. 2001; Johnson 2004), but has not been documented in the Gobi Altai region. In the Nemegt basin (Fig. 1d), normal faults cut the youngest Cretaceous strata (Maastrichtian), suggesting that crustal extension persisted throughout the Cretaceous in the southern Gobi Altai (author's unpublished data). No published seismic images exist for the basins of the Gobi Altai region, and because of strong Late Cenozoic crustal reactivation and alluvial sedimentation much of the older mid-Jurassic–Cretaceous basin history is overprinted, buried and poorly preserved at the surface (Cunningham et al. 2009).

The driving force for the widespread mid-Jurassic–Cretaceous extensional event in the region is controversial. However, the change to an extensional regime following contractional tectonism associated with final closure of the Palaeo-Asian ocean in the Permian and Triassic–Jurassic closure of the Palaeo-Pacific ocean along the Mongol–Okhotsk suture belt (Zorin 1999) suggests a major change in the regional stress field. Cessation of subduction presumably led to Palaeo-Pacific slab dropoff (Van der Voo et al. 1999; Meng 2003) removing the main driving force for plate convergence and contractional deformation in the CAOB. A change in regional stress regime, possible upwelling of hot asthenosphere to fill the space occupied by a detached and sinking slab(s), and eastward-directed slab rollback along the Pacific margin may have all promoted upper-plate extension in eastern central Asia.

Cenozoic reactivation of the Gobi Altai region

From the previous sections, it can be appreciated that the modern Gobi Altai orogen was not constructed on a blank crustal template, but rather on a complex assemblage of diverse crustal blocks that were repeatedly collided, rifted, intruded, metamorphosed, uplifted and eroded through much of the Phanerozoic. This complex history imparted widespread structural weaknesses in the crust that preconditioned it for subsequent reactivation. The Gobi Altai is one of a number of Palaeozoic terrane collages in central Asia that lacks old rigid cratonic lithosphere and is therefore relatively weak and more susceptible to reactivation by far-field compressive stresses (Westaway 1995).

Seismicity and geodetic data

The earthquake history of the Gobi Altai region (Fig. 1c) has been reviewed by Khil'ko et al. (1985), Baljinnyam et al. (1993), Kurushin et al. (1999) and Bayasgalan et al. (2005), and the reader is referred to those publications for a thorough discussion on the subject. During the last century, Mongolia has been one of the most seismically active intracontinental regions on Earth (Baljinnyam et al. 1993) with three $M > 8$ earthquakes recorded within the country's borders. This apparent clustering of large events in a short period of time in an intracontinental setting has been cited as a possible example of remote triggering of successive earthquakes through stress transfer over large distances between faults (Pollitz et al. 2003). Historical earthquakes recorded in the region are typically left-lateral strike-slip or left-lateral oblique-slip thrust events on c. east–west- to WNW–ESE-striking faults (Baljinnyam et al. 1993). The 1957 $M = 8.3$ earthquake in the northern Gobi Altai ruptured the Bogd Fault System (BFS; also known as the Gurvan Bogd, North Gobi Altai, and Gobi Altai Fault System) over a 250 km length producing spectacular surface scarps (Florensov & Solonenko 1965; Kurushin et al. 1999). This event involved a complex rupture process on multiple faults with different slip senses and, in some locations, no obvious surface connectivity (Bayarsayhan et al. 1996); although faults that moved during the 1957 event have not always moved coevally during previous events (Prentice et al. 2002). The BFS continues westward from the 1957 rupture where strain appears partitioned between several parallel strike-slip faults (Mushkin et al. 2004a). Older surface scarps along the system occur at Bayan Tsagaan Uul and Chandiman Uul (Fig. 1d; Khil'ko et al. 1985; Baljinnyam et al. 1993). Huge landslides have been reported in two areas of the northern Gobi Altai associated with palaeo-ruptures of the BFS and its western splays (Philip & Ritz 1999; Mushkin et al. 2004b). Other smaller landslides occurred during the 1957 M 8.3 earthquake (Florensov & Solonenko 1963). Elsewhere in the region, earthquakes in the southern and southwestern Gobi Altai are less common; however, a significant number of small events have been recorded from the Gurvan Sayhan region and surrounding areas directly south and SE (Fig. 1c; Khil'ko et al. 1985).

Regional faulting

The Gobi Altai is characterized by active sinistral transpression accommodated by an east–west array of left-lateral strike-slip faults linking with WNW–ESE-striking thrust faults over a north–south width of 250–350 km (Fig. 1). Within this array, deformation is diffusely distributed (Fig. 1). The northern boundary to the Gobi Altai is topographically expressed by the Valley of Lakes and the regional south-sloping, southern flank of the Hangay Dome, which is characterized by limited strike-slip and normal faulting (Cunningham 2001; Walker et al. 2007, 2008). The eastern and southeastern boundaries of the Gobi Altai are less well defined, as the range dies out as a series of small ridges in the Gobi Desert. However, low areas to the east and SE may be tectonically active at slow strain rates where Cenozoic faults have reportedly accommodated predominantly strike-slip displacements, instead of relief generating dip-slip movements (Webb & Johnson 2006).

Left-lateral strike-slip displacements

Left-lateral strike-slip and oblique-slip faults are east–west- to ENE–WSW-trending and typically crosscut older NW–SE-striking basement structures (Figs 3–5). The left-lateral faults commonly link with NW–SE-striking thrusts and are found as far north as the southern flank of the Hangay Dome (Walker *et al.* 2008), and southward all the way to the Beishan and northernmost Tibet. Thus they define a parallel array within relatively low terrain between the Hangay Dome and Tibet previously defined as the Gobi corridor of left-lateral transpression (Cunningham *et al.* 1996a; Cunningham 1998). Approximate spacing between major left-lateral faults in the eastern Gobi Altai and in the Beishan region is of the order of 30–50 km.

These left-lateral faults accommodate the east-northeastward displacement of the Gobi Altai and Beishan crust relative to the stable Siberian craton to the north similar to the Altyn Tagh and Kun Lun faults in northern Tibet, except that the faults in Mongolia are more numerous and each has accommodated a small component of the total left-lateral strain. Nevertheless, within the Gobi Altai array, the Gobi–Tien Shan Fault System (GTSFS) and BFS are clearly the longest and therefore dominant left-lateral faults in the region, with the greatest amounts of displacement (Figs 1d & 3a). Offset aeromagnetic anomalies and Palaeozoic strike-belts across the GTSFS suggest 30–40 km of total left-lateral slip (Cunningham *et al.* 2003a). The amount of total left-lateral displacement across the BFS is undetermined, although its continuously

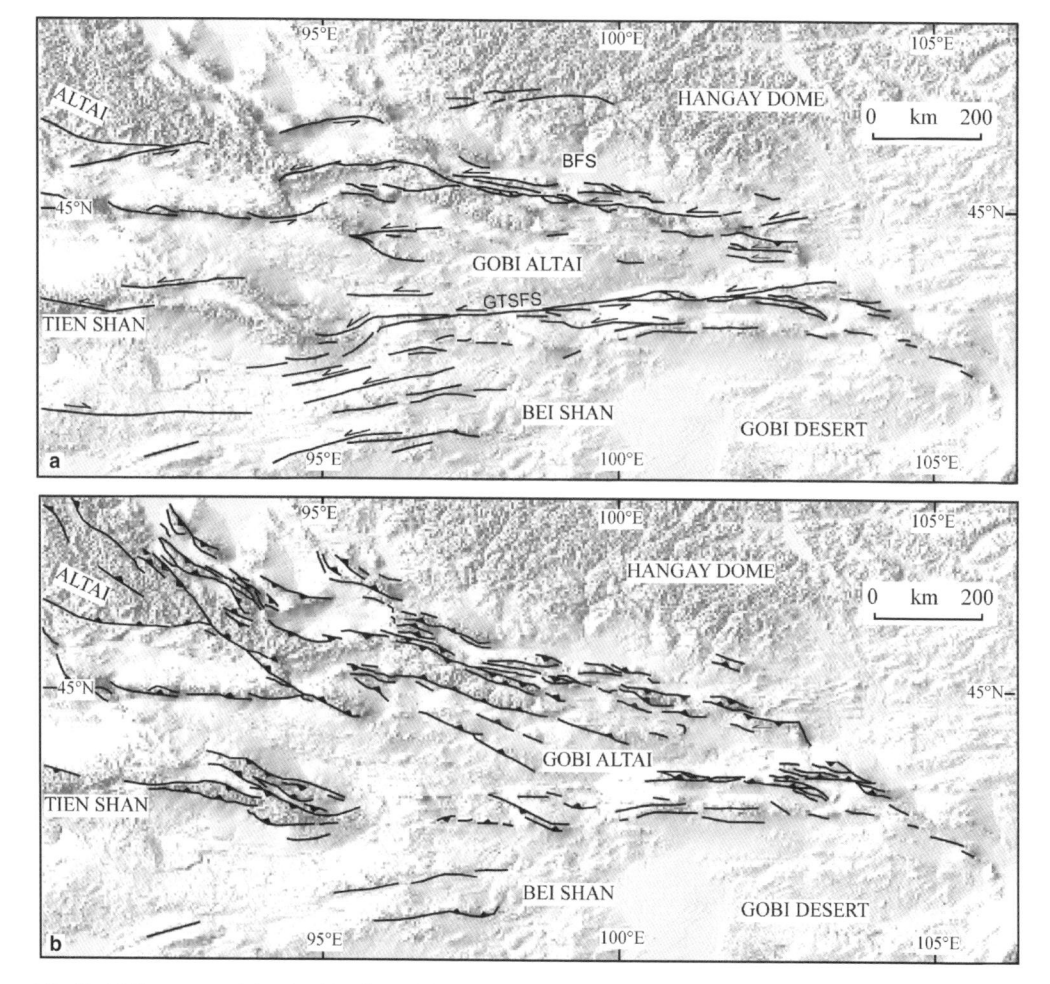

Fig. 3. (**a**) Quaternary sinistral strike-slip and oblique sinistral strike-slip faults in the Gobi Altai region. GPs vectors taken from Calais *et al.* (2003). BFS, Bogd Fault System; GTSFS, Gobi–Tien Shan Fault System. (**b**) Quaternary thrust faults in the Gobi Altai region.

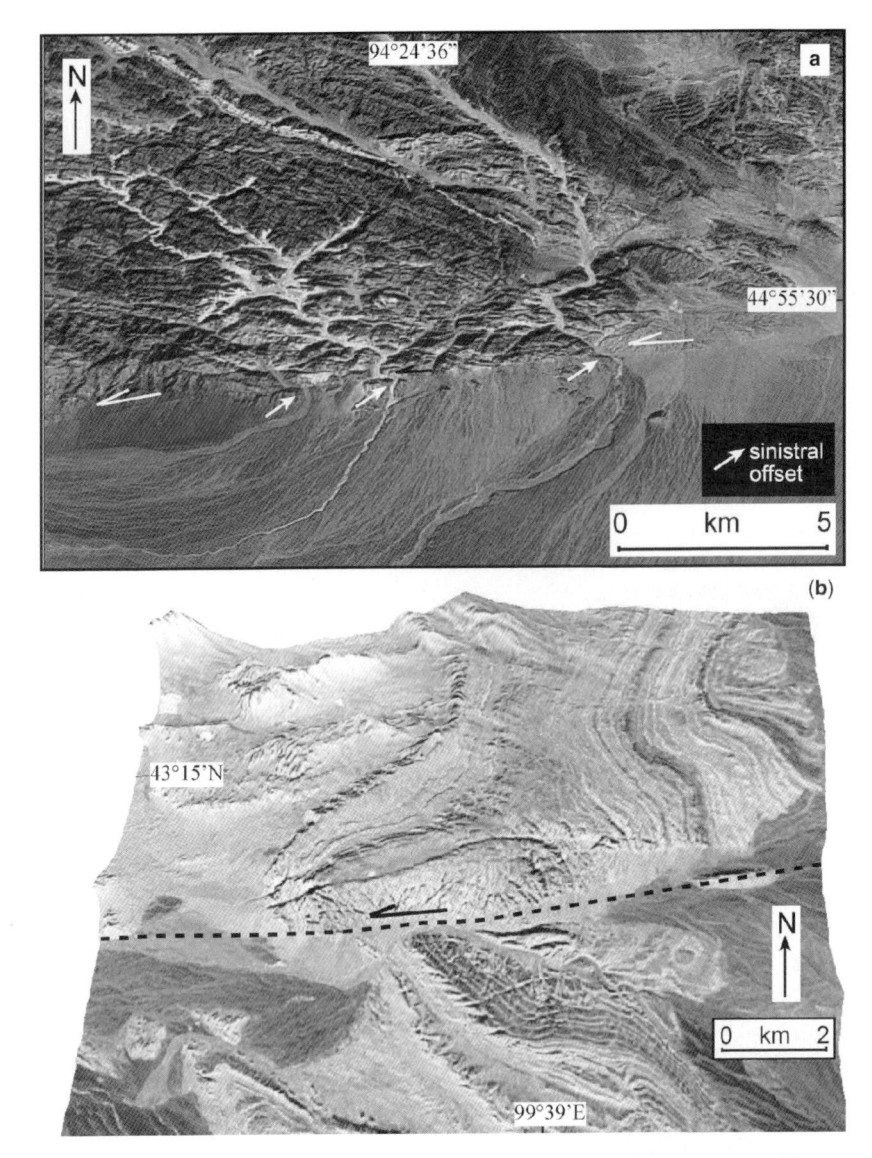

Fig. 4. Two examples of Quaternary left-lateral strike-slip faults in the Gobi Altai that cut across older structural trends: (**a**) Left-lateral stream offsets west of Aj Bogd; (**b**) left-laterally offset Cretaceous sedimentary rocks SW of Altan Uul. (See Fig. 1d for locations.)

traceable length of at least 550 km suggests that it has accommodated at least 15 km of offset using a conservative total displacement (TD)/length (L) scaling relation of $TD = 0.03L$ (Cowie & Scholz 1992).

Thrust displacements

Almost all thrust faults responsible for late Cenozoic uplift of the Gobi Altai mountains strike NW–SE to east–west and occur along mountain fronts that have low mountain front sinuosity (Roberts & Cunningham 2008). They typically are bordered by alluvial fan complexes and dip into the ranges they bound. Most are parallel to the pre-existing basement grain including metamorphic fabrics, older faults and terrane boundaries. In many cases, the frontal alluvial fan complexes are also cut by thrusts indicating continued outward growth of the range (Fig. 6). Many thrusts have

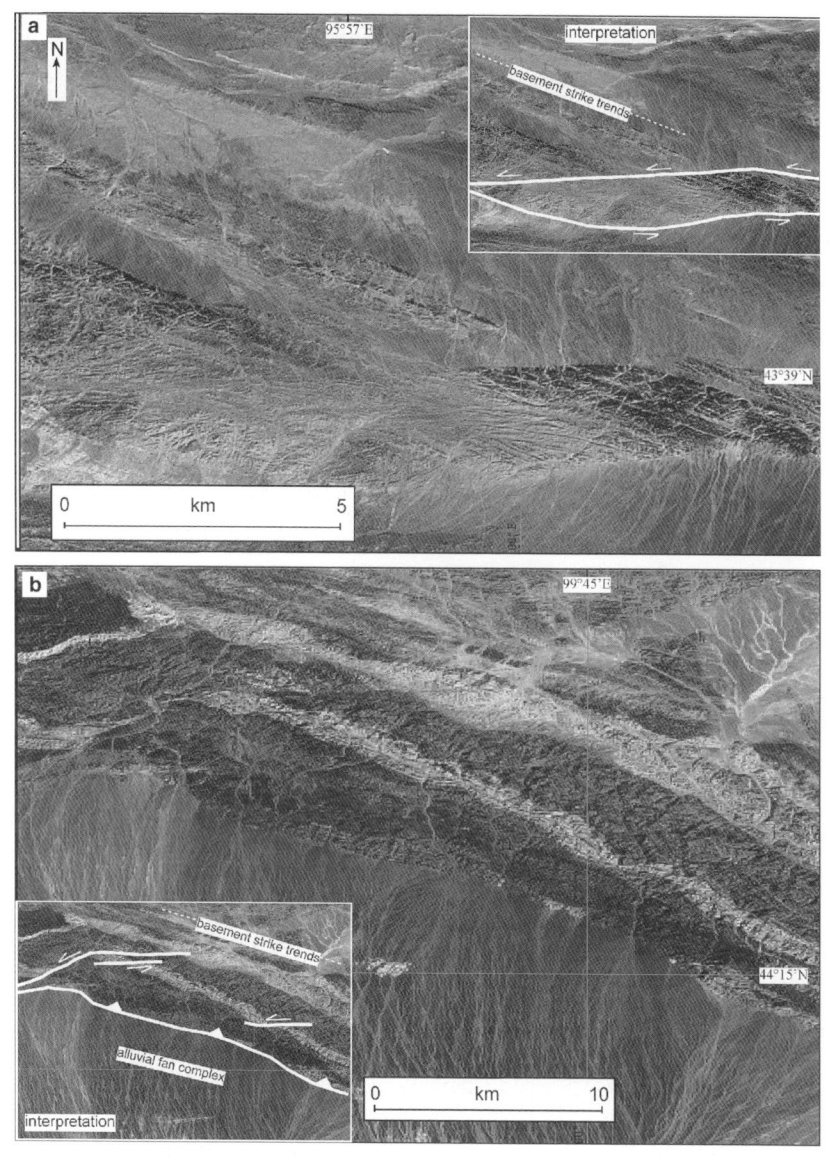

Fig. 5. (a) Sinistral strike-slip fault cutting across Silurian–Devonian strike-belt in southwestern Gobi Altai; (b) typical example of NW–SE-striking thrust that reactivates NW–SE-trending basement fabric in central Gobi Altai (SW Gichigniy Range front). It should be noted that small east–west sinistral strike-slip faults cut across fabric instead of reactivating it. (See Fig. 1d for locations.)

oblique-slip histories, especially near their tips where they link with strike-slip faults or where they change their strikes to more east–west orientations. Thrust faults in the region do not define a foreland style fold and thrust belt; single ranges have their own unique dominant transport direction (NE or SW; Fig 3b) and wide basins between ranges suggest that there is no basal décollement into which all Gobi Altai thrust faults root. Instead, it is suspected that thrusts root into vertical strike-slip faults as positive flower structures or are simply single basement-cored uplifts that reactivate pre-existing weak fabrics and faults. This style of deformation has also been proposed for the Altai and other intracontinental, intraplate transpressional belts (Cunningham *et al.* 1996*b*; Cunningham 2005).

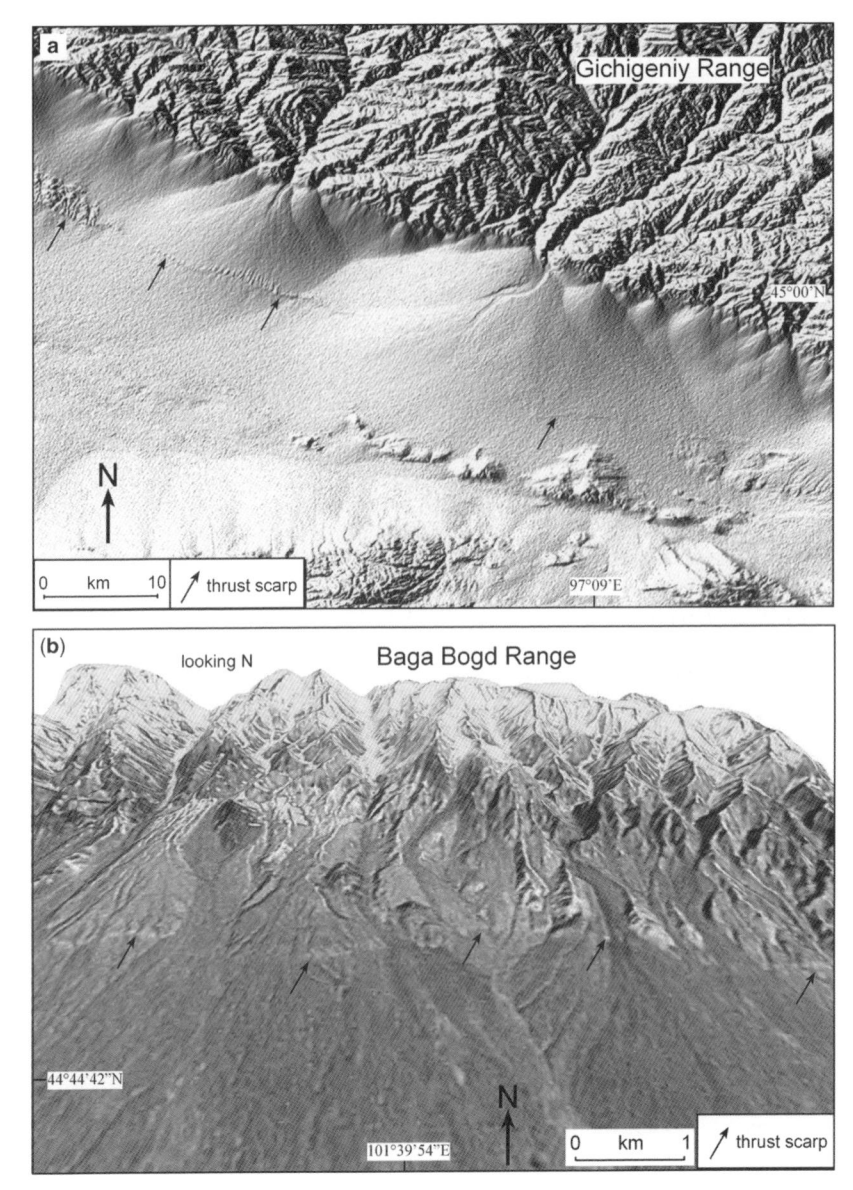

Fig. 6. Two examples of Quaternary range-bounding thrust faults in the Gobi Altai. It should be noted in both cases that thrusting has stepped out into the bordering alluvial fan complex where a degraded thrust scarp is visible. (See Fig. 1d for locations.)

Deformation rates as constrained by thermochronological, palaeoseismological and geodetic data

Most attempts to calculate Quaternary deformation rates in the Gobi Altai come from the Ih Bogd region, where the 1957 BFS scarps cut dateable landforms and where the greatest amount of topographic uplift in the region is believed to have occurred. Modelling of apatite fission-track data from Ih Bogd and Baga Bogd indicates that uplift began $c.\ 5 \pm 3$ Ma, at rates between 0.25 and 1 mm a^{-1} (Vassallo *et al.* 2007*a*). This age was extrapolated by Vassallo *et al.* to the entire Gobi Altai, suggesting that the modern orogen began forming only in the late Miocene. This conclusion is challenged by the presence of widespread Oligocene and Miocene clastic sediments in the northern

Gobi Altai region (Devyatkin 1981; Devyatkin & Badamgarav 1993), which suggests that eroding topography was present at that time, perhaps as a result of early stages of orogenic uplift. However, much of the older Cenozoic sediment in the northern Gobi Altai could also have been externally derived and transported southward from the Hangay region, as documented for Miocene successions from the Valley of Lakes 40 km north of Ih Bogd and Baga Bogd by Hock et al. (1999).

Vassallo et al. (2007b) used ^{10}Be dating of granite boulders on uplifted terraces in northern Ih Bogd to calculate an uplift rate during the late Pleistocene–Holocene of 0.1 mm a^{-1}. They also concluded that the range has a stairstep morphology as a result of outward growth of thrust faults through time, with the frontal thrust active since 600 ka. Vassallo et al. (2005) also calculated an uplift rate for the NW Artsa Bogd range front of 0.13 ± 0.01 mm a^{-1} during the last 160 ka.

Late Quaternary rates of left-lateral slip along the BFS segment 20 km NW of Ih Bogd were determined by Ritz et al. (1995) to be c. 1.2 mm a^{-1} based on ^{10}Be dating of offset alluvial surfaces. This rate is consistent with a 1–10 mm a^{-1} left-lateral slip rate determined previously by Baljinnyam et al. (1993). More recently, Vassallo et al. (2005) calculated a rate of 0.95 ± 0.29 mm a^{-1} for left-lateral slip along the BFS NW of Ih Bogd during the last 250 ka. It is estimated that great earthquakes occur along the BFS every 3–5 ka (Ritz et al. 2003; Prentice et al. 2002).

GPS data by Calais et al. (2003) reveal that the eastern Gobi Altai region is moving eastward relative to a fixed Siberia at rates between 3 and 5 mm a^{-1}. These data include one velocity measurement from a point east of the Gurvan Sayhan range (Fig. 1) with 4.8 mm a^{-1} of eastward motion, suggesting that eastward displacements continue in the lowlands east of the Gobi Altai (Fig. 1a). Calais et al. (2003) suggested that c. 15% of the overall India–Eurasia convergence (c. 10 mm a^{-1}) is accommodated in the huge region north of the Tien Shan and south of stable Eurasia including the Altai and Gobi Altai regions. This convergence is partly accommodated by parallel left-lateral strike-slip faults in the Gobi Altai, and by dextral transpressional deformation and probable anti-clockwise vertical axis rotations in the Altai (see Bayasgalan et al. 2005, fig. 9). Calais et al. (2003) also suggested that dynamic models for continuing deformation in Mongolia and adjacent areas of China and Russia that consider only stresses derived from the Indo-Eurasia collision may be incomplete and the effects of shear traction forces derived from retreating subduction zones along the Pacific margin of northeastern Asia may contribute to the force balance. Finally, updated world stress map data for the Gobi Altai region, as determined from focal mechanism solutions, indicate that SH$_{max}$ is northeasterly trending between 045 and 070 (Heidbach et al. 2008).

Restraining bends and other transpressional uplifts

The linkage of east–west left-lateral faults with NW–SE-striking thrust faults throughout the Gobi Altai has led to the development of a variety of restraining bend uplifts (see review by Cunningham 2007), which comprise most of the highest mountains in the region including Ih Bogd, Baga Bogd, Bayan Tsagaan Uul, Chandiman Uul, Aj Bogd, Gurvan Sayhan and Nemegt Uul (Fig. 7). These restraining bends occur in three possible fault settings: at single fault bends (e.g. Ih Bogd), at fault stopovers where parallel strike-slip faults join across the bend (e.g. Nemegt Range), or at horsetail splay fault termination zones (e.g. Gurvan Sayhan Range; Cunningham 2007). Typically, restraining bends are positive flower structures, although the degree of fault asymmetry varies, and most have the strike-slip component focused on one side of the range and different amounts of thrusting on either side, causing an overall tilt to the range. With time, restraining bends typically broaden through development of thrust forebergs with linking strike-slip faults in the bounding foreland (Bayasgalan et al. 1999a); the fault geometry that develops is best described as a transpressional duplex (see Cunningham et al. 2003b; Cunningham 2005). As restraining bends grow outwards and along strike they may link with other ranges and lose their topographic individuality, thus obscuring their origin as a transpressional uplift at a strike-slip fault bend or stepover (Cunningham et al. 1996a). This process appears to be occurring along the BFS today, where Baga Bogd, Ih Bogd and Bayan Tsagaan Uul are nearly linked up along-strike and new forebergs in their forelands attest to the continuing widening of these ranges (Cunningham 1998, 2007; Bayasgalan et al. 1999a).

Within the Gobi Altai, there are also topographically significant NW–SE-striking thrust ridges that are dominantly contractional rather than transpressional in origin, because they lack significant strike-slip faults. The most obvious examples are the Gichigniy and Endregiyn Nuruu (Fig. 1d), which are thrust ridges ≥300 km long with only minor left-lateral strike-slip faults cutting through their interiors or bounding their southeastern ends.

Transtensional deformation

The Gobi Altai contains very limited evidence for transtensional deformation. This is most certainly

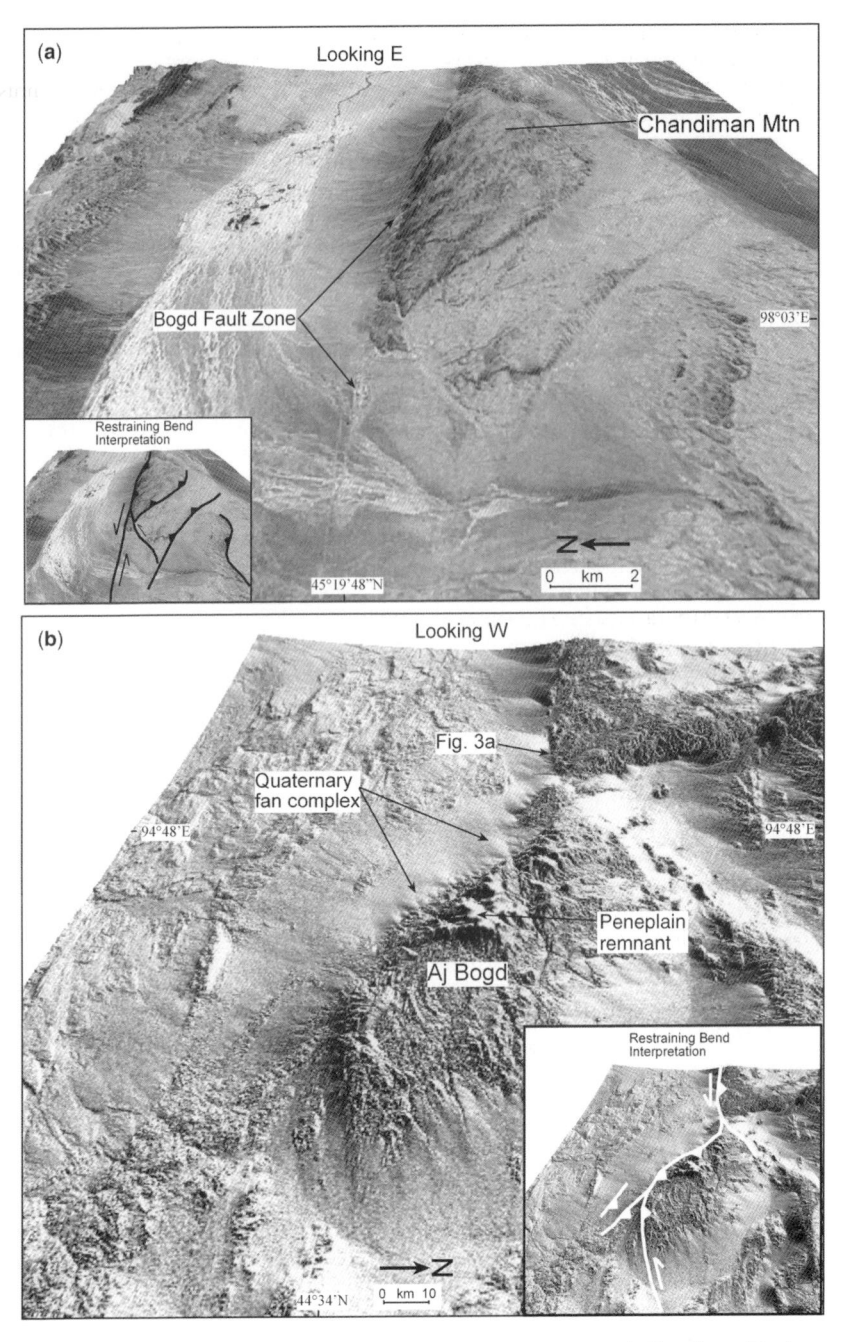

Fig. 7. Two examples of typical restraining bend mountain ranges in the Gobi Altai. Chandiman Mtn is a small restraining bend mountain range along the western trace of the BFS. Aj Bogd is an isolated massif with a peneplain remnant on its summit that was uplifted at a right step along an unnamed left-lateral strike-slip fault system in the extreme western Gobi Altai. Cunningham (2007) has provided further discussion of these bends and other examples in the Gobi Altai. (See Fig. 1d for locations.)

because east–west left-lateral strike-slip faults are almost everywhere right stepping, because linking thrust faults reactivate the predominant WNW–ESE-striking basement structures. A left-stepping and therefore releasing bend fault array for an east–west left-lateral fault system would require that faults that link strike-slip segments would have to strike NE–SW, cutting directly across the prevailing NW–SE-trending basement grain, which is mechanically difficult and less likely.

Only a handful of Late Cenozoic transtensional basins have been identified in the region. Two small pull-apart basins are present south of the Shargyn Basin in the extreme westernmost Gobi Altai (Fig. 8a) In addition, the Shargyn Basin itself is also unusually low [963 m above sea level (masl)] compared with surrounding regions (Fig. 1d), suggesting that a component of extensional deformation may have led to its development (although bounding faults with extensional components of displacement have not yet been documented). In addition, immediately SW of the Nemegt–Altan Ranges along the GTSFS, a topographically low trough along the trace of the fault system at c. 43°30′N, 100°E suggests that a transtensional basin may exist there, although dip-slip fault components have not yet been documented in that region and younger sediments obscure the deeper fault architecture. Finally, an actively developing small transtensional graben was identified in the Valley of Lakes at the NE curving tips of two east–west left-lateral strike-slip faults north of Ih Bogd (Fig. 8b). This is the only known transtensional feature in the Valley of Lakes, which is not believed to be an area characterized by Cenozoic extensional faulting, but is instead a remnant low area collecting sediment shed from the Hangay Dome and northernmost ranges of the Gobi Altai.

Tectonic geomorphology

The entire Gobi Altai region has a basin and range topography, although the modern range is not characterized by extensional faulting typical of other basin and range provinces (e.g. western North America). Instead, discrete thrust-bound ridges and restraining bends linked with strike-slip faults are separated by basinal lows containing clastic infill that are not typically down-dropped blocks, but simply remnant low areas relative to surrounding uplifts. Many ranges have low mountain front sinuosity and are flanked by actively forming alluvial fan complexes, suggesting that more faults may have been active in the Quaternary than has previously been appreciated, with sobering implications for those wanting to extrapolate slip rate calculations into the pre-Holocene (Roberts &

Cunningham 2008). In addition, Quaternary alluvial fans in the region typically occur along NW–SE-trending mountain fronts bounded by thrust faults (Figs 5b, 6 & 7), thus the fans are a mappable proxy for those faults with the greatest amount of relief-generating throw.

The region lacks a through-going drainage network and most basins are internally drained. There are no perennial rivers, although there are small permanent streams that exit from some of the higher ranges. The mountains typically are cut by steep canyons, and summit regions lack geomorphological evidence of former glaciations except for a few possible (but unproven) small cirques on Ih Bogd and Baga Bogd (Berkey & Morris 1924b; author's personal observations). Sediment eroded from actively uplifting ranges is either stored in frontal fan complexes or deposited more distally in basin centres as fine-grained alluvial deposits. Ephemeral lake deposits, evaporites and aeolian deposits fill some basin centres. Sediment leaves the region only by aeolian deflation. However, it has been estimated that up to 50% of eroded sediment has been removed from the Gobi Altai region by wind and has been redeposited to the SE in the Chinese loess region (Traynor & Sladen 1995). Because rainfall is of the order of $50-250$ mm a^{-1} (Anonymous 1990) and there is no removal of eroded sediment by any major river systems, the visible landscape strongly expresses the active tectonics without the obscuring influence of a strong climate and erosional signal, typical of more humid regions. Therefore, satellite imagery and high-resolution digital topographic data are particularly useful for tectonic analysis in the Gobi Altai region (e.g. Tapponnier & Molnar 1979; Roberts & Cunningham 2008).

The youthful ranges of the Gobi Altai commonly preserve a remnant summit peneplain interpreted as a palaeo-erosion surface (Devyatkin 1974). The surface is widespread, although it is differentially dissected and only partially preserved in many areas (Fig. 9). Berkey & Morris (1924b) first described the Gobi Altai peneplains and noted that Early Cretaceous sediments were deposited above the peneplain surface, therefore suggesting that the surface is pre-Early Cretaceous. Peneplain remants are common features of Gobi Altai mountains, with obvious examples on Ih Bogd, Baga Bogd, Aj Bogd, Bayan Tsagaan Uul, Gichigeniy Nuruu, Altan Uul, Gurvan Sayhan, and numerous ranges south and east of the Shargyn Basin area. In addition, similar summit peneplains of apparently Mesozoic age are well known from the Altai, eastern Tien Shan, Hangay Dome, Tarbagatay and Sayan ranges (Fig. 1b; Devyatkin 1974; Shahgedanova et al. 2002) and clearly predate the late Cenozoic reactivation of these regions. Many Gobi Altai ranges

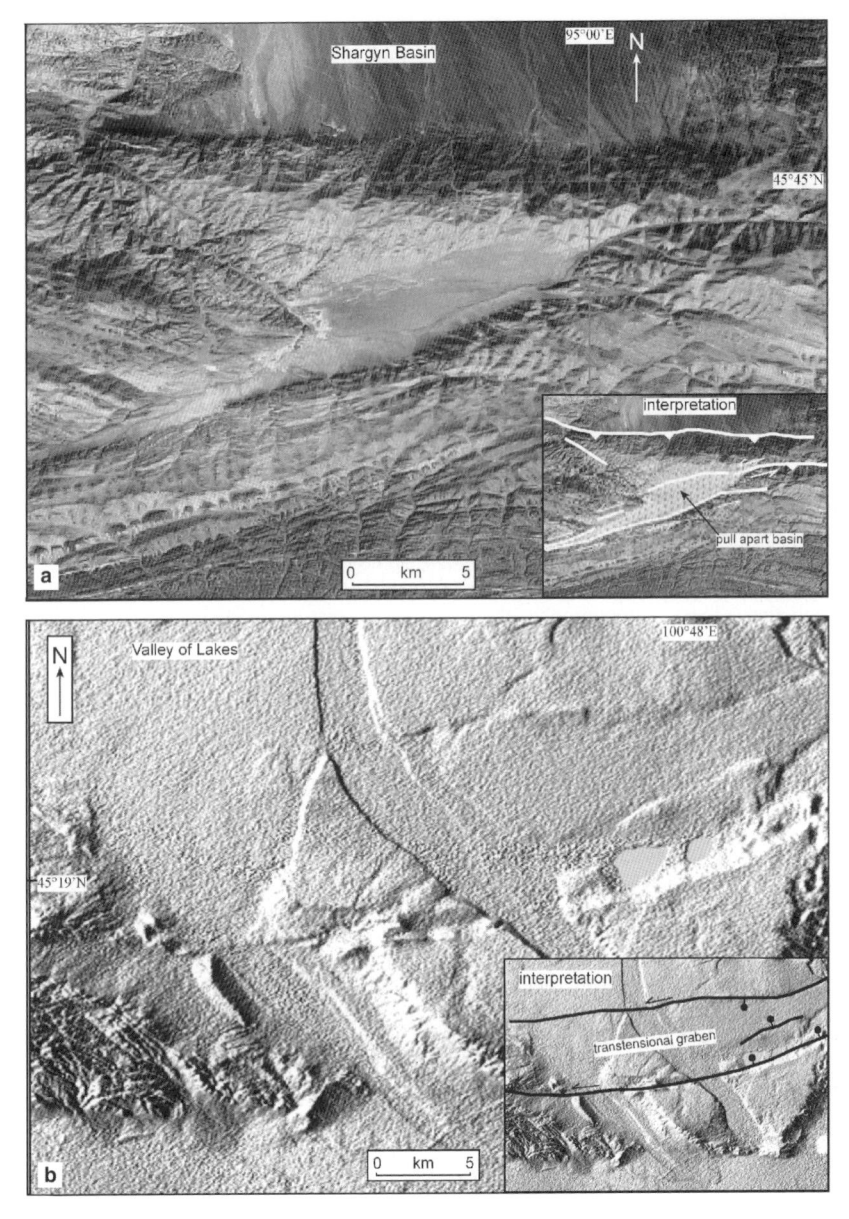

Fig. 8. Two examples of Late Cenozoic transtensional basin development in the Gobi Altai: (**a**) small sinistral pull-apart basin within range directly south of Shargyn Basin, northwesternmost Gobi Altai; (**b**) small transtensional graben in Valley of Lakes north of Ih Bogd. (See Fig. 1d for locations.)

also preserve a relatively smooth grass-covered internal upland that is not peneplained, but is a mature landscape with subdued slopes that is not yet redissected by younger drainages. This landscape has been described from summit areas of Nemegt Uul (Owen *et al.* 1999) and Baga Bogd (Berkey & Morris 1927) and is evident on other summits directly west of Ih Bogd (author's personal observations; Fig. 1d). Jolivet *et al.* (2007) attempted to date the Ih Bogd summit peneplain remnant (and also one Altai and one Sayan summit surface) using apatite fission-track samples and demonstrated a rapid Jurassic cooling event, which they interpreted as a period of erosional exhumation responsible for creation of the smooth erosion surface. Jolivet *et al.* (2007) suggested that the preservation of the surface

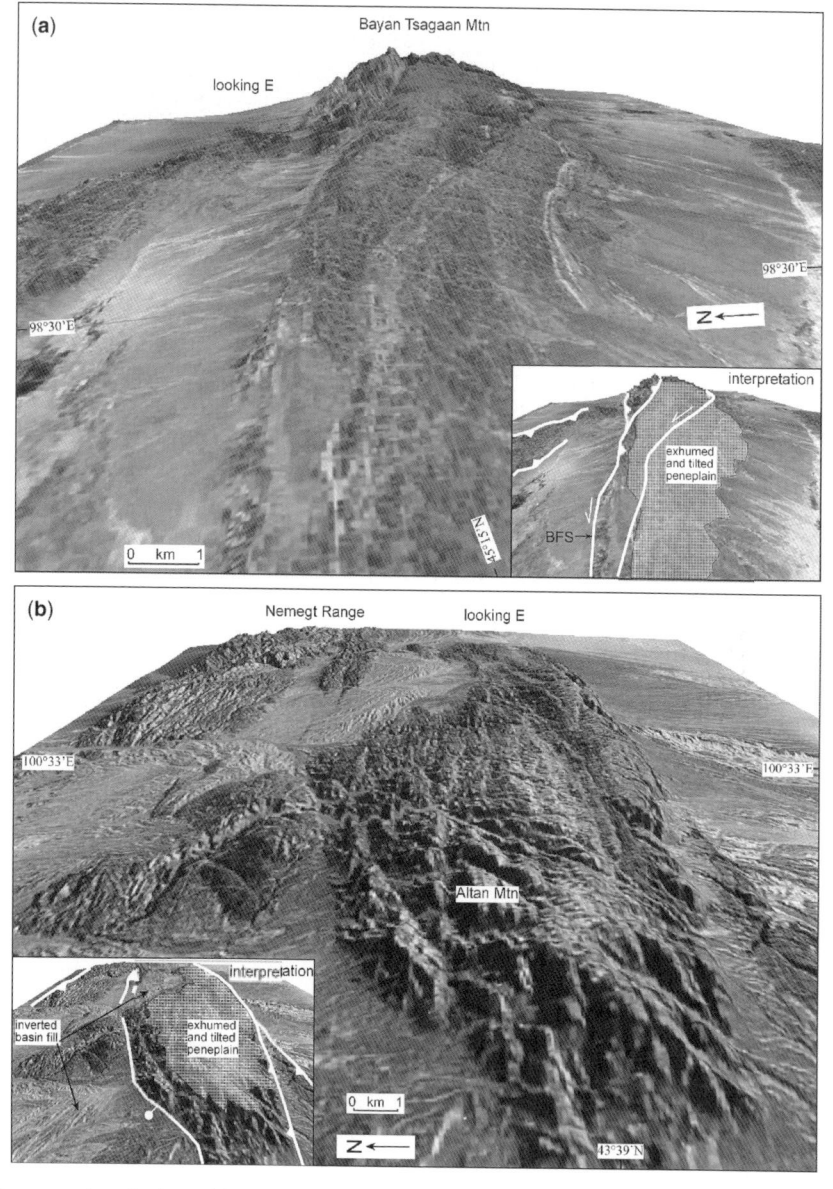

Fig. 9. Two examples of exhumed Jurassic–Cretaceous peneplains in the Gobi Altai. Both preserved peneplain remnants are exhumed within thrust-bound restraining bends. Their tilts provide an indication of which thrust faults have accommodated the most Quaternary displacement. BFS, Bogd Fault System. (See Fig. 1d for locations.)

rules out any major deformation events in the Gobi Altai region since the Jurassic and prior to 8 Ma. However, with timing constrained from one summit only, and clear evidence for Cretaceous extensional rifting in the Gobi Altai region (e.g. Berkey & Morris 1927; Traynor & Sladen 1995; Cunningham et al. 2009), their conclusions appear overstated.

The summit peneplain provides a useful datum for quantifying the amount of vertical uplift relative to surrounding base-level. In addition, if the dip of bounding thrust faults is known, then the amount of shortening can be cautiously determined by restoring the peneplain to its pre-thrust base-level elevation (Cunningham et al. 2003a). Peneplain tilt is also a useful indicator of overall range tilt and the location within the range of maximum uplift. When present on summits of restraining bends that have outward-directed thrusts on both sides, a tilted peneplain may also provide an indication of the side of the range where thrusting has been dominant (Cunningham et al. 2003b; Cunningham 2007).

Fault reactivation and inversion of Cretaceous sedimentary basin areas

Throughout the Gobi Altai region, basins between ranges are dominantly filled with Jurassic–Cretaceous clastic sediments. Younger Neogene–Recent alluvial, lacustrine and aeolian deposits are less volumetrically significant. In addition, older Jurassic–Cretaceous basin blocks are now topographically inverted and perched at high elevations within younger transpressional mountain ranges. These inverted basins are typically eroding and provide a sediment source for younger alluvial fan complexes at the basin margins (Fig. 10). The presence of topographically elevated Cretaceous basin deposits might suggest that Cretaceous normal faults that initially created the accommodation space for the basin fill have been reactivated in thrust mode, thereby elevating the basin fill within the modern mountain ranges. However, except for a few examples in the Nemegt range in the southern Gobi Altai (author's unpublished data), normal fault reactivation has not yet been demonstrated to be an important process in the generation of the modern Gobi Altai orogen. This is probably because Cretaceous normal faults in southern Mongolia (and adjacent regions of China; see Meng 2003) are ENE–WSW-striking and nearly parallel to the regional SH_{max}. Thus the Cretaceous normal faults are unfavourably oriented for Late Cenozoic reactivation and are instead passively uplifted in blocks raised by NW–SE-striking thrust faults that have reactivated older NW–SE-striking Palaeozoic basement structures (Fig. 11).

Volcanism

The Gobi Altai contains fairly widespread and scattered outcrops and small plateaux of Jurassic–Cretaceous and Cenozoic lavas, especially in the Ih Bogd, Baga Bogd and Artsa Bogd areas (Fig. 1d; Whitford-Stark 1987; Yarmolyuk et al. 1995; Barry 1999; Hankard et al. 2008; Van Hinsbergen et al. 2008). There is also a notable concentration of Cretaceous volcanic necks south of Artsa Bogd (Barry 1999). Yarmolyuk et al. (1995) reviewed the timing and extent of volcanism in the Hangay and Gobi Altai regions and concluded that there has been nearly continuous volcanism since 160 Ma, with by far the greatest outpourings in the northern Gobi Altai during the Late Jurassic–earliest Cretaceous, synchronous with purported widespread crustal extension in southern and eastern Mongolia (Traynor & Sladen 1995). Cenozoic lavas are also numerous in the northern Gobi Altai and Valley of Lakes, where significant Eocene, early Oligocene and Miocene lava outpourings have been dated by K–Ar (Devyatkin & Smelov 1979) or Ar/Ar (Barry & Kent 1998; Höck et al. 1999; Barry et al. 2003; van Hinsbergen et al. 2008). Based on analysis of teleseismic travel times, gravity data and regional topography, Petit et al. (2008) suggested that the domed topography under the Hangay region is supported by buoyant asthenosphere and that less dense and hotter upper mantle also extends beneath the northern Gobi Altai region. In addition, Vergnolle et al. (2003) reviewed seismological and petrological evidence to determine the viscosity of the upper mantle and lower crust in central and western Mongolia, and concluded that both are weakened thermally, most probably by the presence of partial melt in the uppermost mantle and possibly lower crust. This accords with Bayasgalan et al.'s (2005) rheological study of the Mongolian lithosphere using earthquake source parameters and gravity data, in which they determined the seismic thickness in the Gobi Altai to be less than 20 km and concluded that the strength of the lithosphere may reside in the upper crust. Barry & Kent (1998) suggested that the upper mantle in the northern Gobi Altai and southern Hangay regions was previously metasomatically enriched (hydrated) by earlier subducted slabs, which would have contributed to partial melting in the presence of hot buoyant asthenosphere. Thus, the existence of an anomalously hot uppermost mantle under southern Hangay and the northern Gobi Altai region since at least the Late Jurassic may have caused lower crustal weakening, thereby also predisposing the crust to Late Cenozoic reactivation.

Discussion

The modern Gobi Altai is an intracontinental transpressional orogen, a unique class of orogen characterized by oblique deformation in an intraplate setting (Cunningham 2005). Other examples of intracontinental transpressional orogens in Asia include the Altai, easternmost Tien Shan, Sayan, Tarbagatay and Stanovoy belts. These belts occur along the perimeter of the Indo-Eurasia deformation field and between rigid cratonic blocks within central Asia. Intracontinental transpressional orogens have topographic, geomorphological and structural elements that differ from classic contractional thrust belts (Tables 1 & 2). Because they form in a continental interior setting, they lack the typical architecture of a telescoped continental margin (e.g. Himalayas), contracted arc and/or back-arc (e.g.

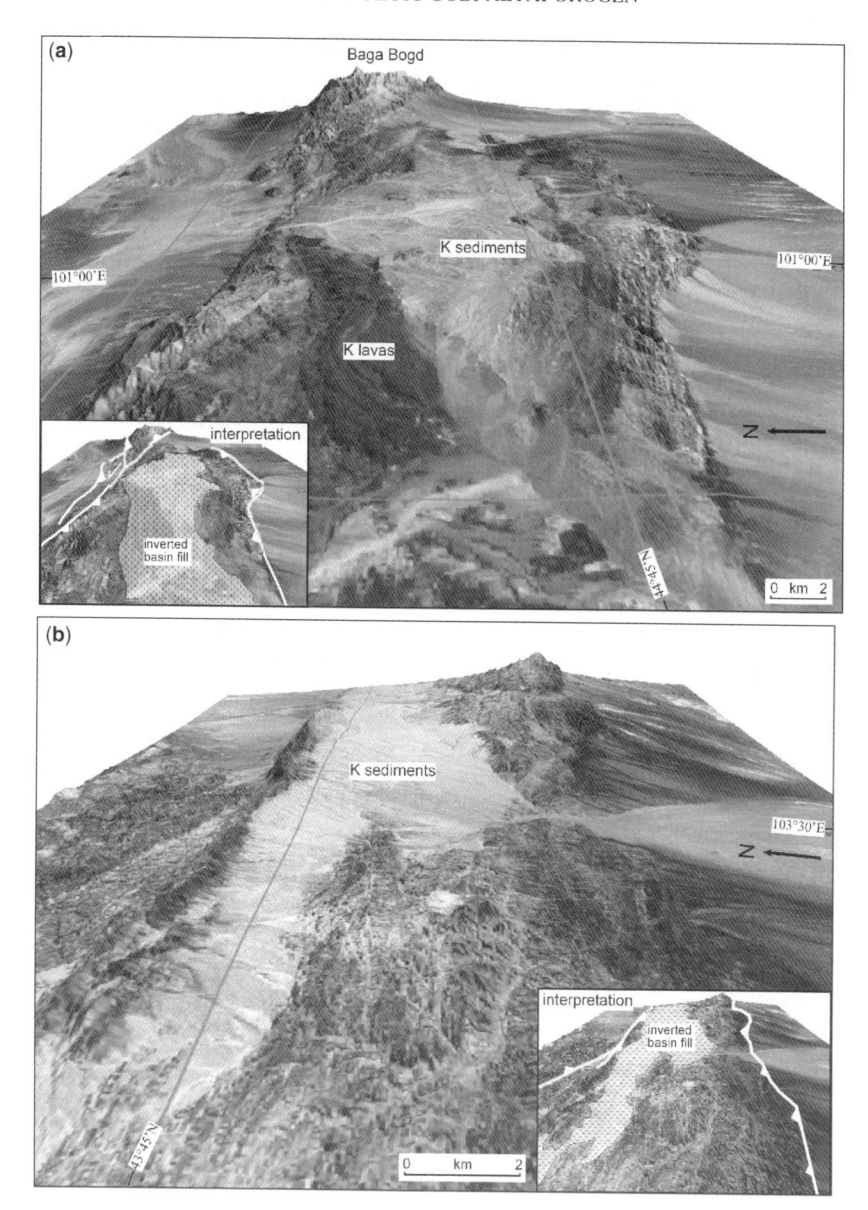

Fig. 10. Two examples of topographically inverted Cretaceous basin sequences now elevated and eroding within thrust-bound ranges in the Gobi Altai. (See Fig. 1d for locations.)

Andes) shortened accretionary wedge (e.g. Makran) or inverted rift (e.g. Pyrenees). Instead, an intracontinental transpressional orogen is characterized by a basin and range topography containing a diffuse belt of linked thrusts and strike-slip faults that generate pure thrust ridges, restraining bends, and other transpressional ridges, many of which are flower structures in cross-section. As the shortening component of deformation increases, parallel flower structures may grow, overlap and coalesce, thus obscuring their separate origins as uplifts that nucleated along a strike-slip fault bend or stepover. In addition, intermontane basins may close by inversion and/or overthrusting from one side or both sides (half-ramp and full-ramp basins; see Cobbold *et al.* 1993). Isolated high massifs may nucleate along major strike-slip systems anywhere within the orogen; thus there is no simple topographic gradient from a low foreland to a higher orogenic hinterland as is typical in purely contractional orogens.

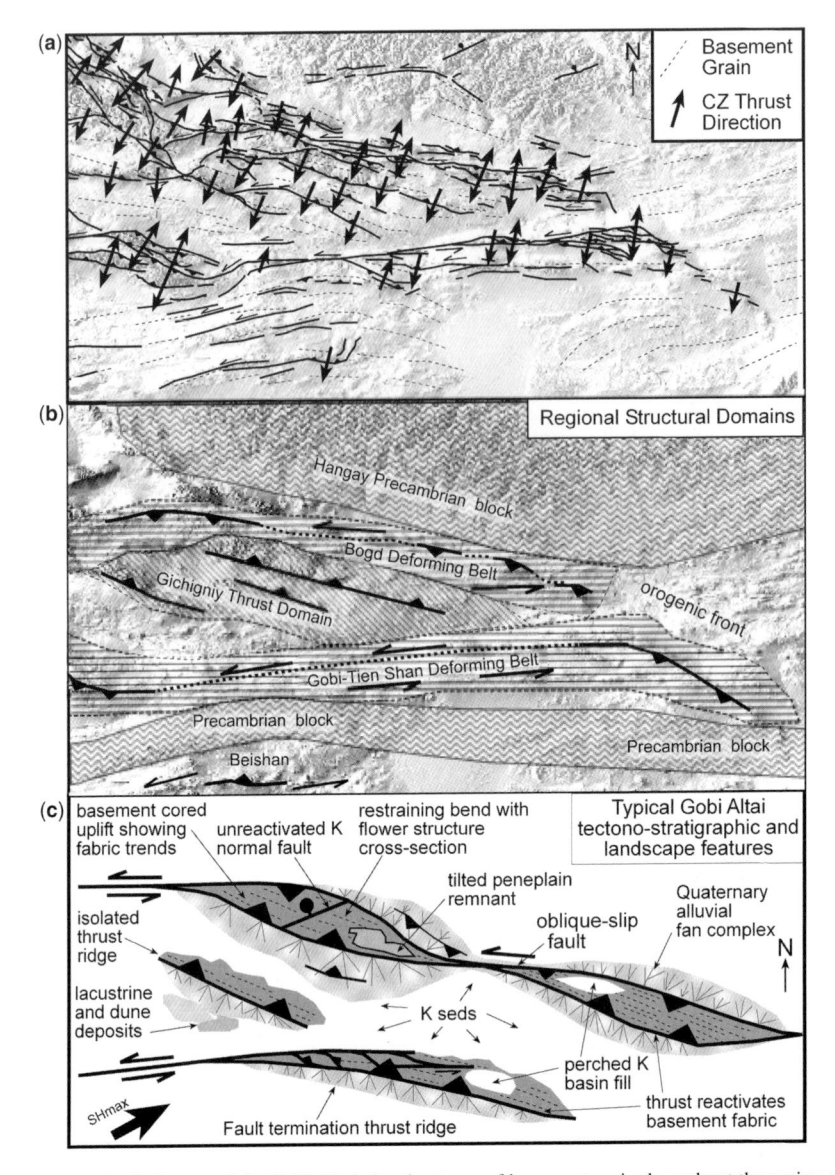

Fig. 11. (a) Quaternary fault map of the Gobi Altai showing trace of basement grain throughout the region and directions of Quaternary thrust movement. Noteworthy features are the close parallelism between Quaternary thrusts and older basement fabric, and the inconsistent directions of thrusting to both NE and SW throughout the region. (b) Simplified subdivision of Gobi Altai into three discrete deforming domains. The Gobi Altai is sandwiched between Precambrian basement blocks under the Hangay Dome and southernmost Gobi region. The Bogd and Gobi–Tien Shan deforming belts are characterized by important left-lateral strike-slip displacements and thrusting along linking ridges and at terminal restraining bends. In contrast, the Gichigniy belt is dominated by long thrust ridges and fewer east–west strike-slip faults. (c) Schematic diagram showing typical tectonic and sedimentary features of the Late Cenozoic Gobi Altai oblique deformation belt.

Intracontinental transpressional orogens typically lack consistent structural vergence and bilateral thrusting within single ranges is common (Fig. 11a). Because range-bounding thrusts are observed to link with strike-slip faults that enter the range along-strike, it is likely that the root structures for many thrusts are steep–vertical strike-slip faults and not shallow-dipping basal décollements; that is, there is no regional thrust wedge architecture. Intracontinental transpressional orogens contain similar

Table 1. *Major Phanerozoic events in the Gobi Altai region*

Time period	Deformation event/geological setting	Kinematics	Possible tectonic driving force	Key references
Miocene–Present	Intracontinental, intraplate crustal reactivation in sinistral transpressional mode, mountain building, clastic sedimentation, limited volcanism in northernmost Gobi Altai	NE–SW crustal contraction, ENE–WSW to east–west sinistral strike-slip displacements	Indo-Eurasia collision, NE SH_{max} owing to India's continued indentation, possible influence of eastward retreating Pacific margin subduction zones	Tapponnier & Molnar 1979; Baljinnyam *et al.* 1993; Cunningham *et al.* 1996*a*; Cunningham 1998, 2007; Bayasgalan *et al.* 1999; Calais *et al.* 2003; Vassallo *et al.* 2007*a* Berkey & Morris 1924*a, b*; Traynor & Sladen 1995; Höck *et al.* 1999
Paleocene–Miocene	Tectonic quiescence, erosion, limited sedimentation, minor basaltic volcanism in northernmost Gobi Altai	No known crustal movements during this period	Stable continental interior, distant effects of Indo-Eurasia collision not yet established	
Late Jurassic–Cretaceous	Regional crustal extension and development of diffuse continental rift province, basaltic magmatism, clastic sedimentation, development of basin and range landscape	NNW–SSE crustal extension	Terminal closure of Mongol–Okhotsk ocean, final slab dropoff, removal of slab pull driving force for crustal convergence in Central Asia, crustal relaxation and upwelling aesthenosphere in southern Hangay, northern Gobi Altai region	Traynor & Sladen 1995; Sladen & Traynor 2000; Graham *et al.* 2001; Davis *et al.* 2002; Meng *et al.* 2003; Johnson 2004; Cunningham *et al.* 2009
Early–Mid-Jurassic	Intracontinental contractional deformation, significant erosion and exhumation of deeper crust	North–south shortening	Terminal closure of Mongol–Okhotsk seaway to north and NE, collisional events to the south in Tibet	Traynor & Sladen 1995; Hendrix *et al.* 1996; Zheng *et al.* 1996; Zorin *et al.* 1999; Dumitru & Hendrix 2001; Lamb *et al.* 2008
Late Permian–Early Jurassic	Overall tectonic quiescence, widespread post-orogenic erosion and granitic plutonism, foreland sedimentation in southernmost Gobi Altai related to contractional deformation in northern China	Limited NE–SW sinistral strike-slip faulting, with possible minor contractional faulting in extreme southern Gobi Altai	Consolidation of accreted crustal components, isostatic adjustments, widespread erosion	Traynor & Sladen 1995; Hendrix *et al.* 1996; Lamb *et al.* 1999; Manankov *et al.* 2006
Cambrian–Permian	Amalgamation of Altaids terrane collage involving at least four separate ocean closure events involving both north- and south-dipping subduction zones, widespread arc and post-collisional magmatism, terrestrialization and eventual cessation of marine sedimentation	North–south contractional orogenesis, minor strike-slip displacements	Plate convergence, terane accretion and amalgamation, and terminal closure of Palaeo-Asian ocean between older Precambrian cratons and blocks	Sengor *et al.* 1993; Lamb & Badarch 1997; Jahn *et al.* 2000; Fillipova *et al.* 2001; Badarch *et al.* 2002; Helo *et al.* 2006; Windley *et al.* 2007, and references therein

Table 2. *A comparison between thrust belt orogens and intracontinental transpressional orogens*

	Thrust belt orogens	Intracontinental intraplate transpressional orogens
Topography	Lowest in foreland, highest in hinterland towards rear of thrust wedge, mountainous landscape, jagged summits, no preserved summit peneplain, antecedent rivers and integrated drainage systems leaving range	Highest ranges are typically structural culminations along transpressional fault systems and can be in centre or margins of orogen, basin and range landscape, summit peneplain preservation, internally drained basins, lack of integrated river systems
Fault architecture	Thrust wedge with internal duplex and imbricate geometries, basal décollement; strike-slip faults rare, except cross-strike tear faults	No regional thrust wedge, single restraining bends and flower structures common and their coalescence may lead to elongate thrust ridges with bilateral thrusting, isolated thrust blocks common, faults may root into vertical structures, not low-angle; strike-slip and oblique-slip faults common, often in parallel arrays
Transport direction	Thrusts dominantly foreland directed; folds dominantly foreland vergent	Inconsistent thrust propagation directions and fold vergence
Range elements	Thrust stacks, duplexes, imbricate fans, antiformal stacks	Various types of restraining bends (terminal, double, partial), thrust ridges linked to strike-slip faults, triangular block uplifts between conjugate strike-slip faults, single thrust ridges, and thrusted 'thick-skinned' basement blocks
Basin elements	Foreland basins, piggyback basins	Half and full ramp basins, open-sided thrust basins, pull-apart and strike-slip basins, unreactivated remnant low areas, no regional foreland basins
Crustal setting prior to orogeny	Continental margin, back-arc basin, rift basin, mechanically weak continental interior settings	Continental interior, terrane collages, mechanically weak crustal blocks between Precambrian cratons
Reactivation elements	Dominantly normal faults and possibly strike-slip faults	Any favourably oriented faults, fabrics, strike belts
Driving force	Plate boundary convergence, slab pull, continental collision	Intraplate compressive stress derived from distant convergent boundary or collision, gravitational stored energy

orogenic and basinal elements to transpressional continental transform boundaries except that they lack a single major fault system (e.g. San Andreas, North Anatolian Fault) and instead of transferring interplate motion to another plate boundary, they are dead-end zones where intraplate strain is terminally accommodated (Cunningham 2005).

In the Gobi Altai, sinistral transpression is the prevailing modern deformation regime because NE-directed SH_{max} (045–070) acts obliquely on crust with a regionally well-developed and penetrative WNW–ESE-striking fabric (090–110). Orthogonal components of the NE-directed compressive stress acting on the WNW–ESE-striking basement fabric promote structural reactivation in thrust or oblique-thrust mode. However, the non-orthogonal relationship between SH_{max} and the basement structural grain leaves a strike-slip vector component that is accommodated throughout the region on approximately east–west sinistral strike-slip faults that typically cut across older structures. The BFS and GTSFS are the most obvious examples of cross-cutting faults (Fig. 1), but many other strike-slip faults with shorter lengths (Figs 1, 4 & 5) cut the older basement fabrics and thus may be entirely Late Cenozoic in origin. The Gobi Altai tapers to the east in terms of its north–south width and average elevations (Fig. 1), with apparently decreasing amounts of thrust shortening in the easternmost ranges; this may be due to the diminished angle between SH_{max} and the east–west structural grain there, which promotes strike-slip motions more than relief-generating thrust displacements.

The angle between SH_{max} and the east–west strike-slip faults is everywhere >20° in the Gobi Altai and true strain partitioning appears to be limited in the region (Dewey et al. 1998), with oblique-slip thrusting and oblique-slip sinistral displacements expressed in outcrop (e.g. Cunningham et al. 1996a) and in focal mechanism solutions of modern earthquakes (Adiya et al. 2003; Bayasgalan et al. 2005). Likewise, vertical axis rotations associated with left-slip on the BFS have been proposed in the northern Gobi Altai (Bayasgalan et al. 1999b); however, palaeomagnetic data from lava flows in the Baga Bogd, Artsa Bogd and Ih Bogd regions indicate that any Neogene–Recent vertical axis rotations there were limited to less than 10° (van Hinsbergen et al. 2008).

The modern Gobi Altai lacks significant transtensional deformation, and paired restraining and releasing bends have not been identified as occurs in other strike-slip deforming belts, except possibly the Nemegt–Altan Uul restraining bend and the lowland to the SW. Likewise, conjugate strike-slip faulting occurs at the extreme NW end of the Gobi Altai, where it links with the Altai around the Shargyn Basin (Fig. 1; Tapponnier & Molnar 1979); however, conjugate strike-slip faulting is not generally observed elsewhere in the Gobi Altai.

The north–south distribution of reactivated crust in the Gobi Altai region is limited by the Precambrian Baidrag Block in the north and by the Hanshan–South Gobi microcontinent in the south (Fig. 2). Thus like other regions of central Asia, the distribution of older Precambrian blocks has strongly influenced the location of crustal reactivation (see Westaway 1995). This is presumably because these ancient blocks are thicker (colder?), and rheologically stronger than intervening Palaeozoic terrane collages and have thus resisted internal deformation. In addition, sporadic basaltic volcanism in the region during the last 160 Ma (Devyatkin & Smelov 1980), with lava and xenolith compositions suggesting anomalously elevated mantle temperatures at depth (Stosch et al. 1995; Ionov 2002; Barry et al. 2003), may have caused thermal weakening of the crust and preconditioned the Gobi Altai lithosphere to reactivation. Analogue models of orogens forming in mechanically weak crust caught between rigid blocks like the Gobi Altai ('vice-like orogens' of Cruden et al. 2006) suggest that upper crustal shortening may be balanced by lateral extrusion of lower crust and lithosphere mantle. This has been previously proposed for extruding regions of Tibet (Clark & Royden 2000) and the Altai region (Cunningham 2005), and may also apply to the Gobi Altai, where existing seismic anisotropy data indicate ENE to ESE fast directions (Gao et al. 1994; Barruol et al. 2008). Eastward-directed lower crustal–upper mantle flow or sublithospheric flow is kinematically compatible with east–west left-lateral strike-slip faulting in the brittle crust above and may thus be a major driver of the seismogenic deformation (Cunningham et al. 1997; Cunningham 1998; Barruol et al. 2008).

Although deformation is diffusely distributed throughout the Gobi Altai region (Fig. 1a), three major deforming belts appear to dominate the Quaternary deformation field (Fig. 11b): (1) the left-lateral transpressional ranges along the BFS; (2) the Gichigeniy and Endregyn thrust ridges; (3) the GTSFS and transpressional ranges that are forming along it, and at its termination zones. The easternmost Tien Shan is structurally linked to the southwestern and southernmost Gobi Altai ranges and thus it could be argued on tectonic grounds that the Tien Shan is actively propagating into southern Mongolia and structurally terminates there, instead of in China. If the growth of the eastern Tien Shan has been diachronous and eastward propagating during the late Cenozoic, then the young ranges and faults along the GTSFS in the SW and southernmost Gobi Altai may be good analogues for earlier stages of mountain building in the Tien Shan further west. Because the Gobi Altai is a discontinuous mountain belt, it provides useful examples of the orogenic architecture and tectonic geomorphology of an intracontinental transpressional orogen in an early stage of development, thus providing lessons for understanding the earlier stages of more advanced intracontinental transpressional orogens in other regions.

Conclusions

(1) The Gobi Altai is underlain by a Palaeozoic terrane collage consisting of arc, forearc, back-arc and accretionary complexes that accreted during consolidation of the CAOB during the Cambrian–Permian. At least four ophiolite-bearing sutures are present in the region and structural data suggest that the polarity of subduction was both north and south directed. Faults, metamorphic fabrics and sedimentary strike belts that formed during Palaeozoic terrane amalgamation represent the primary structural grain in the Gobi Altai region.

(2) During the early–mid-Jurassic, the region experienced a contractional orogenic event, which may have been caused by the Mongol–Okhotsk ocean closure or terrane collisions in Tibet. This was subsequently followed by Late Jurassic–Cretaceous regional extension, which is expressed by rift basins, thick clastic infill and widespread basaltic volcanism.

(3) During the late Cenozoic, Gobi Altai crust was reactivated by NE-directed stress as a result of

the Indo-Eurasia collision. The angular relationship between NE-directed SH_{max} and the WNW–ESE to east–west basement fabric promoted thrust reactivation of the older fabric and east–west strike-slip faulting within an overall sinistral transpression regime. Precambrian blocks that bound the orogen to the north and south appear to limit the region of Late Cenozoic mountain building. Reactivation of Gobi Altai crust may have been facilitated by thermal weakening of the lower crust and lithospheric mantle as a result of 160 Ma of episodic volcanism in the region since the Late Jurassic. Normal fault reactivation appears to be an uncommon mountain building process in the region, probably because major Late Jurassic–Cretaceous rift-bounding faults are NE–SW-trending and unfavourably oriented for reactivation by the modern NE-directed SH_{max}.

(4) Quaternary–Recent tectonic activity in the Gobi Altai is indicated by earthquakes, palaeoseismic surface ruptures, and sharply defined thrust fronts bordered by active alluvial fan complexes. The BFS and GTSFS are the dominant deforming transpressional belts whereas the Gichigeniy Nuruu is dominated by SW-directed thrusting. GPS data indicate the eastern Gobi Altai crust is moving eastwards between 3 and 5 mm a^{-1}. Left-lateral strike-slip rates along the BFS are of the order of 1 mm a^{-1}. Fission-track data from Ih Bogd suggest that the northern Gobi Altai began forming no later than 8 Ma, although more thermochronological and sedimentological data are needed throughout the region to better define the onset of the modern orogeny.

(5) The Gobi Altai is an intracontinental, intraplate transpressional orogen, a class of orogen that is relatively common within and along the perimeter of the Indo-Eurasia deformation field, but that is under-appreciated as an orogen type by many geologists. Intracontinental transpressional orogens are characterized by a basin and range topography with range and basin architectures and geomorphological characteristics that differ from typical plate boundary contractional orogens and continental transform boundaries. Specifically, deformation is diffusely distributed and terminally accommodated along thrusts, and strike-slip and oblique-slip faults, which link in complex ways to produce restraining bends and thrust ridges separated by internally drained basins. There is no orogenic foreland or hinterland.

(6) The Gobi Altai (and to some extent the southernmost Hangay Dome) accommodates eastward crustal displacements along an array of east–west sinistral strike-slip faults, similar to the Beishan and Tibetan regions to the south (Tapponnier et al. 1982). Thus in addition to NE–SW crustal shortening in the region, a component of eastward tectonic extrusion relative to stable Siberia is accommodated in southern Mongolia.

(7) Because the Gobi Altai is an actively developing youthful mountain range in an arid region with low erosion rates, it provides an excellent opportunity to study the way in which a continental interior reactivates as a result of a distant continental collision. In addition, it offers important insights into how other more advanced intracontinental transpressional orogens may have developed during earlier stages of their evolution. Tectonic problems of general interest that are especially suitable for investigation in the Gobi Altai include the spatial and temporal development of linked fault networks; controls on the extent of structural reactivation and basin inversion; crustal factors that determine nucleation sites for mountain uplift; flower structure growth and restraining bend coalescence; geometric relations between reactivated basement structures, the distribution and kinematics of active faults, SH_{max} and the geodetically derived velocity field; and the manner in which upper crustal structures terminally accommodate intraplate strain by non-partitioned oblique deformation.

I am very grateful to Brian Windley for all of his support with my research and academic career over the last 15 years. Brian has been an excellent collaborator and good friend, and his wise professional advice has always been appreciated. Most of all, his unwavering enthusiasm for fieldwork and tectonics research has been truly inspirational.

References

ADIYA, M., ANKHTSETSEG, D. ET AL. 2003. *One Century of Seismicity in Mongolia (1900–2000)*. Research Centre of Astronomy & Geophysics of the Mongolian Academy of Sciences (RCAG; Ulaan Baatar), Mongolia; Laboratoire de Teledetection et Risque Sismique, Bruyeres le Chatel, France.

ANONYMOUS 1990. *Mongolian National Atlas*. Joint Publication of the Mongolian Academy of Sciences, Ulaan Baatar, and Russian Academy of Sciences, Moscow.

BADARCH, G., CUNNINGHAM, W. D. & WINDLEY, B. F. 2002. A new terrane subdivision for Mongolia: implications for the Phanerozoic crustal growth of Central Asia. *Journal of Asian Earth Sciences*, **21**, 87–110.

BALJINNYAM, I., BAYASGALAN, A. ET AL. 1993. *Ruptures of Major Earthquakes and Active Deformation in Mongolia and its Surroundings*. Geological Society of America, Memoirs, **181**.

BARRUOL, G., DESCHAMPS, A. ET AL. 2008. Upper mantle flow beneath and around the Hangay dome, Central Mongolia. *Earth and Planetary Science Letters*, **274**, 221–233.

BARRY, T. L. 1999. *Origins of Cenozoic Basalts in Mongolia: A Chemical and Isotope Study*. PhD thesis, University of Leicester.

BARRY, T. L. & KENT, R. W. 1998. Cenozoic magmatism in Mongolia and the origin of central and east Asian basalts. *In*: FLOWER, M., CHUNG, S.-L., LO, C.-H. & LEE, T.-Y. (eds) *Mantle Dynamics and Plate Interactions in East Asia*. American Geophysical Union Monograph, Geodynamics Series, **27**, 347–364.

BARRY, T. L., SAUNDERS, A. D., KEMPTON, P. D., WINDLEY, B. F., PRINGLE, M. S., DORJNAMJAA, D. & SAANDAR, S. 2003. Petrogenesis of Cenozoic basalts from Mongolia: evidence for the role of asthenospheric versus metasomatized lithospheric mantle sources. *Journal of Petrology*, **44**, 55–91.

BAYARSAYHAN, C., BAYASGALAN, A., ENHTUVSHIN, B., HUDNUT, K. W., KURUSHIN, R. A., MOLNAR, P. & OLZIYBAT, M. 1996. 1957 Gobi-Altay, Mongolia, earthquake as a prototype for southern California's most devastating earthquake. *Geology*, **24**, 579–582.

BAYASGALAN, A., JACKSON, J., RITZ, J. F. & CARRETIER, S. 1999a. 'Forebergs', flower structures, and the development of large intracontinental strike-slip faults: The Gurvan Bogd fault system in Mongolia. *Journal of Structural Geology*, **21**, 1285–1302.

BAYASGALAN, A., JACKSON, J., RITZ, J. F. & CARRETIER, S. 1999b. Field examples of strike-slip fault terminations in Mongolia and their tectonic significance. *Tectonics*, **18**, 394–411.

BAYASGALAN, A., JACKSON, J. & MCKENZIE, D. 2005. Lithosphere rheology and active tectonics in Mongolia: relations between earthquake source parameters, gravity and GPS measurements. *Geophysical Journal International*, **163**, 1151–1179.

BERKEY, C. P. & MORRIS, F. K. 1924a. Basin structures in Mongolia. *Bulletin of the American Museum of Natural History*, **LI**, 103–127.

BERKEY, C. P. & MORRIS, F. K. 1924b. The peneplanes of Mongolia. *American Museum Novitates*, **136**, 1–11.

BERKEY, C. P. & MORRIS, F. K. 1927. *Geology of Mongolia: a reconnaissance report based on the investigations of the years 1922–1923*. American Museum of Natural History, New York.

BLIGHT, J. H. S., CUNNINGHAM, D. & PETTERSON, M. G. 2008. Crustal evolution of the Saykhandulaan Inlier, Mongolia: implications for Palaeozoic arc magmatism, polyphase deformation and terrane accretion in the Southeast Gobi Mineral Belt. *Journal of Asian Earth Sciences*, **32**, 142–164.

BUCHAN, C., CUNNINGHAM, D., WINDLEY, B. F. & TOMURHUU, D. 2001. Stuctural and lithological characteristics of the Bayankhongor ophiolite zone, Central Mongolia. *Journal of the Geological Society, London*, **158**, 445–460.

CALAIS, E., VERGNOLLE, M., SAN'KOV, V., LUKHNEV, A., MIROSHNITCHENKO, A., AMARJARGAL, S. & DEVERCHERE, J. 2003. GPS measurements of crustal deformation in the Baikal–Mongolia area (1994–2002): implications for current kinematics of Asia. *Journal of Geophysical Research B: Solid Earth*, **108**, ETG 14-1–ETG 14-13.

CLARK, M. K. & ROYDEN, L. H. 2000. Topographic ooze: building the eastern margin of Tibet by lower crustal flow. *Geology*, **28**, 703–706.

COBBOLD, P. R., DAVY, P. ET AL. 1993. Sedimentary basins and crustal thickening. *Sedimentary Geology*, **86**, 77–89.

COWIE, P. A. & SCHOLZ, C. H. 1992. Displacement-length scaling relationship for faults: data synthesis and discussion. *Journal of Structural Geology*, **14**, 1149–1156.

CRUDEN, A. R., NASSERI, M. H. B. & PYSKLYWEC, R. 2006. Surface topography and internal strain variation in wide hot orogens from three-dimensional analogue and two-dimensional numerical vice models. *In*: BUITER, S. J. H. & SCHREURS, G. (eds) *Analogue and Numerical Modelling of Crustal-Scale Processes*. Geological Society, London, Special Publications, **253**, 79–104.

CUNNINGHAM, W. D. 1998. Lithospheric controls on late Cenozoic construction of the Mongolian Altai. *Tectonics*, **17**, 891–902.

CUNNINGHAM, W. D. 2001. Cenozoic normal faulting and regional doming in the southern Hangay region, Central Mongolia: implications for the origin of the Baikal rift province. *Tectonophysics*, **331**, 389–481.

CUNNINGHAM, D. 2005. Active intracontinental transpressional mountain building in the Mongolian Altai: defining a new class of orogen. *Earth and Planetary Science Letters*, **240**, 436–444.

CUNNINGHAM, D. 2007. Structural and topographic characteristics of restraining bend mountain ranges of the Altai, Gobi Altai and easternmost Tien Shan. *In*: CUNNINGHAM, W. D. & MANN, P. (eds) *Tectonics of Strike-Slip Restraining and Releasing Bends*. Geological Society, London, Special Publications, **290**, 219–237.

CUNNINGHAM, W. D., WINDLEY, B. F., DORJNAMJAA, D., BADAMGAROV, J. & SAANDAR, M. 1996a. Late Cenozoic transpression in southwestern Mongolia and the Gobi Altai–Tien Shan connection. *Earth and Planetary Science Letters*, **140**, 67–81.

CUNNINGHAM, W. D., WINDLEY, B. F., DORJNAMJAA, D., BADAMGAROV, G. & SAANDAR, M. 1996b. A structural transect across the Mongolian Western Altai: active transpressional mountain building in central Asia. *Tectonics*, **15**, 142–156.

CUNNINGHAM, W. D., WINDLEY, B. F., OWEN, L. A., BARRY, T., DORJNAMJAA, D. & BADAMGARAV, J. 1997. Geometry and style of partitioned deformation within a late Cenozoic transpressional zone in the eastern Gobi Altai Mountains, Mongolia. *Tectonophysics*, **277**, 285–306.

CUNNINGHAM, D., OWEN, L. A., SNEE, L. W. & LI, J. 2003a. Structural framework of a major intracontinental orogenic termination zone: the easternmost Tien Shan, China. *Journal of the Geological Society, London*, **160**, 575–590.

CUNNINGHAM, D., DAVIES, S. & BADARCH, G. 2003b. Crustal architecture and active growth of the Sutai Range, western Mongolia: a major intracontinental, intraplate restraining bend. *Journal of Geodynamics*, **36**, 169–191.

CUNNINGHAM, D., DAVIES, S. & MCLEAN, D. 2009. Exhumation of a Cretaceous rift complex within a Late Cenozoic restraining bend, southern Mongolia: implications of the crustal evolution of the Gobi Altai region. *Journal of the Geological Society, London*, **166**, 1–13.

DAVIS, G. A., CONG, W., ZHENG, Y., ZHANG, J.-J., ZHANG, C.-H. & GEHRELS, G. E. 1998. The enigmatic Yinshan fold-and-thrust belt of northern China: new views on its intraplate contractional styles. *Geology*, **26**, 43–46.

DAVIS, G. A., DARBY, B. J., ZHENG, Y. & SPELL, T. L. 2002. Geometric and temporal evolution of an extensional detachment fault, Hohhot metamorphic core complex, Inner Mongolia, China. *Geology*, **30**, 1003–1006.

DEMOUX, A., KRONER, A., LIU, D. & BADARCH, G. 2008. Precambrian crystalline basement in southern Mongolia as revealed by SHRIMP zircon dating. *International Journal of Earth Sciences*, **98**, 1–16.

DERGUNOV, A. B. 1989. *The Caledonides of the Central Asia, Transactions, 437*, Nauka, Moscow.

DEVYATKIN, E. V. 1974. Structures and formational complexes of the Cenozoic activated stage. In: *Tectonics of the Mongolian People's Republic*. Nauka, Moscow, 182–195 [in Russian].

DEVYATKIN, E. V. 1981. The Cenozoic of Inner Asia (Stratigraphy, geochronology and correlation). In: NIKIFOROVA, K. V. (ed.) *The Joint Soviet–Mongolian Scientific-Research Geological Expedition, Transactions 27*. Nauka, Moscow [in Russian].

DEVYATKIN, E. V. & BADAMGARAV, D. 1993. Geological essay on Paleogene and Neogene deposits of the Valley of Lakes and Prealtai depressions. In: BARSBOLD, R. & AKHMETIEV, M. A. (eds) *International Geological Correlation Program, Project 326, Oligocene–Miocene Transitions in the Northern Hemisphere, Excursion Guidebook Mongolia: Oligocene–Miocene Boundary in Mongolia*. Joint Publication of the Mongolian Academy of Sciences, Ulaan Baatar, and Russian Academy of Sciences, Moscow, 7–27.

DEVYATKIN, E. V. & SMELOV, S. B. 1980. Position of basalts in the Cenozoic sedimentary sequence of Mongolia. *International Geology Review*, **22**, 307–317.

DEWEY, J. F., SHACKLETON, R. M., CHENGFA, C. & YIYIN, S. 1988. The tectonic evolution of the Tibetan Plateau. *Philosophical Transactions of the Royal Society of London, Series A*, **327**, 379–413.

DEWEY, J. F., HOLDSWORTH, R. E. & STRACHAN, R. A. 1998. Transpression and transtension zones. In: HOLDSWORTH, R. E. & STRACHAN, R. A. (eds) *Continental Transpressional and Transtensional Tectonics*. Geological Society, London, Special Publications, **135**, 1–14.

DIJKSTRA, A. H., BROUWER, F. M., CUNNINGHAM, W. D., BUCHAN, C., BADARCH, G. & MASON, P. R. D. 2006. Late Neoproterozoic proto-arc ocean crust in the Dariv Range, Western Mongolia: a supra-subduction zone end-member ophiolite. *Journal of the Geological Society, London*, **163**, 363–373.

DONSKAYA, T. V., WINDLEY, B. F. ET AL. 2008. Age and evolution of late Mesozoic metamorphic core complexes in southern Siberia and northern Mongolia. *Journal of the Geological Society, London*, **165**, 405–421.

DUMITRU, T. A. & HENDRIX, M. S. 2001. Fission-track constraints on Jurassic folding and thrusting in southern Mongolia and their relationship to the Beishan thrust of northern China. In: HENDRIX, M. S. & DAVIS, G. A. (eds) *Paleozoic and Mesozoic Tectonic Evolution of Central Asia: From Continental Assembly to Intracontinental Deformation*. Geological Society of America, Memoirs, **194**, 215–229.

FILIPPOVA, I. B., BUSH, V. A. & DIDENKO, A. N. 2001. Middle Paleozoic subduction belts: the leading factor in the formation of the Central Asian fold-and-thrust belt. *Russian Journal of Earth Sciences*, **3**, 405–426.

FLORENSOV, N. A. & SOLONENKO, V. P. (eds) 1963. *The Gobi-Altai Earthquake*. Akademiya Nauk USSR, Moscow [in Russian; English translation, 1965, US Department of Commerce, Washington, DC].

GAO, S., DEVIS, P. M. ET AL. 1994. Seismic anisotropy and mantle flow beneath the Baikal rift zone. *Nature*, **371**, 149–151.

GONG, Q., LIU, M., LI, H., LIANG, M. & DAI, W. 2002. The type and basic characteristics of the Beishan orogenic belt, Gansu. *Northwestern Geology*, **35**, 29–34 [in Chinese].

GRAHAM, S. A., HENDRIX, M. S. ET AL. 2001. Sedimentary record and tectonic implications of Mesozoic rifting in southeast Mongolia. *Geological Society of America Bulletin*, **113**, 1560–1579.

HANKARD, F., COGNE, J. P., LAGROIX, F., QUIDELLEUR, X., KRAVCHINSKY, V. A., BAYASGALAN, A. & LKHAGVADORJ, P. 2008. Palaeomagnetic results from Palaeocene basalts from Mongolia reveal no inclination shallowing at 60 Ma in Central Asia. *Geophysical Journal International*, **172**, 87–102.

HEIDBACH, O., TINGAY, M., BARTH, A., REINECKER, J., KURFEß, D. & MÜLLER, B. 2008. The 2008 release of the World Stress Map. Available online at: www.world-stress-map.org.

HELO, C., HEGNER, E., KRÖNER, A., BADARCH, G., TOMURTOGOO, O., WINDLEY, B. F. & DULSKI, P. 2006. Geochemical signature of Paleozoic accretionary complexes of the Central Asian Orogenic Belt in South Mongolia: constraints on arc environments and crustal growth. *Chemical Geology*, **227**, 236–257.

HENDRIX, M. S., GRAHAM, S. A., AMORY, J. Y. & BADARCH, G. 1996. Noyon Uul syncline, southern Mongolia: Lower Mesozoic sedimentary record of the tectonic amalgamation of central Asia. *Geological Society of America, Bulletin*, **108**, 1256–1274.

HÖCK, V., DAXNER-HÖCK, G. & SCHMID, H. P. 1999. Oligocene–Miocene sediments, fossils and basalts from the Valley of Lakes (Central Mongolia) – an integrated study. *Mitteilungen der Österreichischen Geologischer Gesellschaft*, **90**, 83–125.

HOWARD, J., CUNNINGHAM, D., DAVIES, S., DIJKSTRA, A. & BADARCH, G. 2003. Stratigraphic and structural evolution of the Dzereg Basin, Mongolia. *Basin Research*, **15**, 45–72.

HSU, K. J., YAO, Y., LI, J. & WANG, Q. 1992. Geology of the Beishan Mountains and the tectonic evolution of northwest China. *Eclogae Geologicae Helvetiae*, **85**, 213–225.

IONOV, D. 2002. Mantle structure and rifting processes in the Baikal–Mongolia region: geophysical data and evidence from xenoliths in volcanic rocks. *Tectonophysics*, **351**, 41–60.

JAHN, B. M., WU, F. & CHEN, B. 2000. Massive granitoid generation in Central Asia: Nd isotope evidence and implication for continental growth in the Phanerozoic. *Episodes*, **23**, 82–92.

Jahn, B. M., Wu, F. Y. & Chen, B. 2001. Growth of Asia in the Phanerozoic – Nd isotopic evidence. *Gondwana Research*, **4**, 640–642.

Jerzykiewicz, T. & Russell, D. A. 1991. Late Mesozoic stratigraphy and vertebrates of the Gobi Basin. *Cretaceous Research*, **12**, 345–377.

Johnson, C. L. 2004. Polyphase evolution of the East Gobi basin: sedimentary and structural records of Mesozoic–Cenozoic intraplate deformation in Mongolia. *Basin Research*, **16**, 79–99.

Jolivet, M., Ritz, J. F. et al. 2007. Mongolian summits: an uplifted, flat, old but still preserved erosion surface. *Geology*, **35**, 871–874.

Khil'ko, S. D., Kurushin, R. A., Kochetkov, V. M., Baljinnyam, I. & Monkoo, D. 1985. Earthquakes and the bases for seismic zoning of Mongolia. In: *Transactions 41, The Joint Soviet–Mongolian Scientific Geological Research Expedition*. Nauka, Moscow.

Kovalenko, V. I., Yarmolyuk, V. V., Kovach, V. P., Kotov, A. B., Kozakov, I. K., Salnikova, E. B. & Larin, A. M. 2004. Isotope provinces, mechanisms of generation and sources of the continental crust in the Central Asian mobile belt: geological and isotopic evidence. *Journal of Asian Earth Sciences*, **23**, 605–627.

Kozakov, I. K., Kotov, A. B., Kovach, V. P. & Sal'nikova, E. B. 1997. Crustal growth in the geologic evolution of the Baidarik block, Central Mongolia: evidence from Sm–Nd isotopic systematics. *Petrology*, **5**, 201–207.

Kozakov, I. K., Kotov, A. B. et al. 1999. Metamorphic age of crystalline complexes of the Tuva–Mongolia Massif: the U–Pb geochronology of granitoids. *Petrology*, **7**, 177–191.

Kurushin, R. A., Bayasgalan, A. et al. 1999. *The Surface Rupture of the 1957 Gobi-Altay, Mongolia, Earthquake*. Geological Society of America, Special Papers, **320**.

Lamb, M. A. & Badarch, G. 1997. Paleozoic sedimentary basins and volcanic-arc systems of southern Mongolia: new stratigraphic and sedimentologic constraints. *International Geology Review*, **39**, 542–576.

Lamb, M. A. & Badarch, G. 2001. Paleozoic sedimentary basins and volcanic arc systems of southern Mongolia: new geochemical and petrographic constraints. In: Hendrix, M. S. & Davis, G. A. (eds) *Paleozoic and Mesozoic Tectonic Evolution of Central Asia: From Continental Assembly to Intracontinental Deformation*. Geological Society of America, Memoirs, **194**, 117–149.

Lamb, M. A. & Cox, D. 1998. New ^{40}Ar/^{39}Ar age data and implications for porphyry copper deposits of Mongolia. *Economic Geology*, **93**, 524–529.

Lamb, M. A., Hanson, A. D., Graham, S. A., Badarch, G. & Webb, L. E. 1999. Left-lateral sense offset of Upper Proterozoic to Paleozoic features across the Gobi Onon, Tost, and Zuunbayan faults in southern Mongolia and implications for other central Asian faults. *Earth and Planetary Science Letters*, **173**, 183–194.

Lamb, M. A., Badarch, G., Navratil, T. & Poier, R. 2008. Structural and geochronologic data from the Shin Jinst area, eastern Gobi Altai, Mongolia: implications for Phanerozoic intracontinental deformation in Asia. *Tectonophysics*, **451**, 312–330.

Manankov, I. N., Shi, G. R. & Shen, S. 2006. An overview of Permian marine stratigraphy and biostratigraphy of Mongolia. *Journal of Asian Earth Sciences*, **26**, 294–303.

Meng, Q. R. 2003. What drove late Mesozoic extension of the northern China–Mongolia tract? *Tectonophysics*, **369**, 155–174.

Meng, Q. R., Hu, J. M., Jin, J. Q., Zhang, Y. & Xu, D. F. 2003. Tectonics of the late Mesozoic wide extensional basin system in the China–Mongolia border region. *Basin Research*, **15**, 397–416.

Mitrofanov, F. P., Bibikova, Y. V., Gracheva, Y., Kozakov, I. K., Sumin, L. V. & Shuleshko, I. K. 1985. Isotope age of 'grey' tonalites and gneisses of the Archean in Caledonian structures of central Mongolia. *Doklady Akademii Nauk SSSR*, **284**, 670–675.

Mushkin, A., Bayasgalan, A. & Gillespie, A. 2004a. Constraints on lateral offsets along the Gobi-Altay fault system, southwestern Mongolia. *Geological Society of America, Abstracts with Programs*, **36**, 136.

Mushkin, A., Javkhlanbold, D., Bayasgalan, A. & Gillespie, A. 2004b. Large paleo landslides along the western part of the Gobi-Altay fault system in southwestern Mongolia. *EOS Transactions, American Geophysical Union*, T11C-1280 AGU.

Owen, L. A., Cunningham, W. D., Windley, B. F., Badamgarov, J. & Dorjnamjaa, D. 1999. The landscape evolution of Nemegt Uul: a late Cenozoic transpressional uplift in the Gobi Altai, southern Mongolia. In: Smith, B. J., Whalley, W. B. & Warke, P. A. (eds) *Uplift, Erosion and Stability: Perspectives on Long-term Landscape Development*. Geological Society, London, Special Publications, **162**, 201–218.

Petit, C., Tiberi, C., Deschamps, A. & Deverchere, J. 2008. Teleseismic traveltimes, topography and the lithospheric structure across central Mongolia. *Geophysical Research Letters, Geophysical Research Letters*, **35**, L11301, doi: 10.1029/2008GL033993.

Philip, H. & Ritz, J. F. 1999. Gigantic paleolandslide associated with active faulting along the Bogd fault (Gobi-Altay, Mongolia). *Geology*, **27**, 211–214.

Pollitz, F., Vergnolle, M. & Calais, E. 2003. Fault interaction and stress triggering of twentieth century earthquakes in Mongolia. *Journal of Geophysical Research B: Solid Earth*, **108**, ETG 16-1–ETG 16-14.

Prentice, C. S., Kendrick, K., Berryman, K., Bayasgalan, A., Ritz, J. F. & Spencer, J. Q. 2002. Prehistoric ruptures of the Gurvan Bulag fault, Gobi Altay, Mongolia. *Journal of Geophysical Research B: Solid Earth*, **107**, ESE 1-1–1-18.

Rippington, S., Cunningham, D. & England, R. 2008. Structure and petrology of the Altan Uul ophiolite: new evidence for a Late Carboniferous suture in the Gobi Altai, southern Mongolia. *Journal of the Geological Society, London*, **165**, 711–723.

Ritz, J. F., Brown, E. T. et al. 1995. Slip rates along active faults estimated with cosmic-ray-exposure dates: application to the Bogd fault, Gobi-Altai, Mongolia. *Geology*, **23**, 1019–1022.

RITZ, J. F., BOURLÈS, D. ET AL. 2003. Late Pleistocene to Holocene slip rates for the Gurvan Bulag thrust fault (Gobi-Altay, Mongolia) estimated with ^{10}Be dates. *Journal of Geophysical Research B: Solid Earth*, **108**, ETG 8-1–8-16.

ROBERTS, N. & CUNNINGHAM, D. 2008. Automated alluvial fan discrimination, Quaternary fault identification, and the distribution of tectonically reactivated crust in the Gobi Altai region, southern Mongolia. *International Journal of Remote Sensing*, **29**, 6957–6969.

SENGOR, A. M. C., NATAL'IN, B. A. & BURTMAN, V. S. 1993. Evolution of the Altaid tectonic collage and Palaeozoic crustal growth in Eurasia. *Nature*, **364**, 299–307.

SHAHGEDANOVA, M., MIKHAILOV, N., LARIN, S. & BREDIKHIN, A. 2002. The mountains of southern Siberia. *In*: SHAHGEDANOVA, M. (ed.) *The Physical Geography of Northern Eurasia*. Oxford University Press, Oxford, 314–349.

SHUVALOV, V. F. 1969. Continental red beds of the upper Jurassic of Mongolia. *Doklady Akademii Nauk SSSR*, **189**, 1088–1091.

SLADEN, C. & TRAYNOR, J. J. 2000. Lakes during the evolution of Mongolia. *In*: GIERLOWSKI-KORDESCH, E. H. & KELTS, K. R. (eds) *Lake Basins through Space and Time*. American Association of Petroleum Geologists, Studies in Geology, **46**, 35–57.

STOSCH, H.-G., IONOV, D. A., PUCHTEL, I. S., GALER, S. J. G. & SHARPOURI, A. 1995. Lower crustal xenoliths from Mongolia and their bearing on the nature of the deep crust beneath central Asia. *Lithos*, **136**, 227–242.

TAPPONNIER, P. & MOLNAR, P. 1979. Active faulting and Cenozoic tectonics of the Tien Shan, Mongolia, and Baykal regions. *Journal of Geophysical Research*, **84**, 3425–3459.

TAPPONNIER, P., PELTZER, G., LE DAIN, A. Y., ARMIJO, R. & COBBOLD, P. 1982. Propagating extrusion tectonics in Asia: new insights from simple experiments with plasticine. *Geology*, **10**, 611–616.

TIKHONOV, V. I. & YARMOLYUK, V. V. 1982. Structure and evolution of the Gobi Altai, southern boundary structure of the Mongolian Caledonides. *Geotectonics*, **16**, 266–273.

TOMURTOGOO, O. 1999. *Geological Map of Mongolia, 1:1 000 000*. Mongolian Academy of Science, Institute of Geology and Mineral Resources, Ulaan Baatar.

TRAYNOR, J. J. & SLADEN, C. 1995. Tectonic and stratigraphic evolution of the Mongolian People's Republic and its influence on hydrocarbon geology and potential. *Marine and Petroleum Geology*, **12**, 35–52.

VAN DER VOO, R., SPAKMAN, W. & BIJWAARD, H. 1999. Mesozoic subducted slabs under Siberia. *Nature*, **397**, 246–249.

VAN HINSBERGEN, D. J. J., STRAATHOF, G. B., KUIPER, K. F., CUNNINGHAM, W. D. & WIJBRANS, J. 2008. No vertical axis rotations during Neogene transpressional orogeny in the NE Gobi Altai: coinciding Mongolian and Eurasian early Cretaceous apparent polar wander paths. *Geophysical Journal International*, **173**, 105–126.

VASSALLO, R., RITZ, J. F., BRAUCHER, R. & CARRETIER, S. 2005. Dating faulted alluvial fans with cosmogenic ^{10}Be in the Gurvan Bogd mountain range (Gobi-Altay, Mongolia): climatic and tectonic implications. *Terra Nova*, **17**, 278–285.

VASSALLO, R., JOLIVET, M. ET AL. 2007a. Uplift age and rates of the Gurvan Bogd system (Gobi-Altay) by apatite fission track analysis. *Earth and Planetary Science Letters*, **259**, 333–346.

VASSALLO, R., RITZ, J. F. ET AL. 2007b. Transpressional tectonics and stream terraces of the Gobi-Altay, Mongolia. *Tectonics*, **26**, TC5013, doi: 10.1029/2006TC002081.

VERGNOLLE, M., POLLITZ, F. & CALAIS, E. 2003. Constraints on the viscosity of the continental crust and mantle from GPS measurements and postseismic deformation models in western Mongolia. *Journal of Geophysical Research B: Solid Earth*, **108**, ETG 15-1–ETG 15-14.

WALKER, R. T., NISSEN, E., MOLOR, E. & BAYASGALAN, A. 2007. Reinterpretation of the active faulting in central Mongolia. *Geology*, **35**, 759–762.

WALKER, R. T., MOLOR, E., FOX, M. & BAYASGALAN, A. 2008. Active tectonics of an apparently aseismic region: distributed active strike-slip faulting in the Hangay Mountains of central Mongolia. *Geophysical Journal International*, **174**, 1121–1137.

WEBB, L. E. & JOHNSON, C. L. 2006. Tertiary strike-slip faulting in southeastern Mongolia and implications for Asian tectonics. *Earth and Planetary Science Letters*, **241**, 323–335.

WEBB, L. E., GRAHAM, S. A., JOHNSON, C. L., BADARCH, G. & HENDRIX, M. S. 1999. Occurrence, age, and implications of the Yagan–Onch Hayrhan metamorphic core complex, southern Mongolia. *Geology*, **27**, 143–146.

WESTAWAY, R. 1995. Crustal volume balance during the India–Eurasia collision and altitude of the Tibetan plateau: a working hypothesis. *Journal of Geophysical Research*, **100**, 15 173–15 192.

WHITFORD-STARK, J. L. 1987. *A Survey of Cenozoic Volcanism on Mainland Asia*. Geological Society of America, Special Paper, **213**.

WINDLEY, B. F. & ALLEN, M. B. 1993. Mongolian Plateau; evidence for a late Cenozoic mantle plume under Central Asia. *Geology*, **21**, 295–298.

WINDLEY, B. F., ALEXEIEV, D., XIAO, W., KRÖNER, A. & BADARCH, G. 2007. Tectonic models for accretion of the Central Asian Orogenic Belt. *Journal of the Geological Society, London*, **164**, 31–47.

XIAO, W., WINDLEY, B. F., HAO, J. & ZHAI, M. 2003. Accretion leading to collision and the Permian Solonker suture, Inner Mongolia, China: termination of the central Asian orogenic belt. *Tectonics*, **22**, 1069, doi: 10.1029/2002TC001484.

XIAO, W., WINDLEY, B. F., BADARCH, G., SUN, S., LI, J., QIN, K. & WANG, Z. 2004. Palaeozoic accretionary and convergent tectonics of the southern Altaids: implications for the growth of Central Asia. *Journal of the Geological Society, London*, **161**, 339–342.

YANG, J. H., WU, F. Y. ET AL. 2007. Rapid exhumation and cooling of the Liaonan metamorphic core complex: inferences from $^{40}Ar/^{39}Ar$ thermochronology and implications for Late Mesozoic extension in the eastern North China Craton. *Geological Society of America Bulletin*, **119**, 1405–1414.

YARMOLYUK, V. V., KOVALENKO, V. I., IVANOV, V. G. & SAMOYLOV, V. S. 1995. Dynamics and magmatism of Late Mesozoic–Cenozoic mantle hot spot, southern Khangai (Mongolia). *Geotectonics (English Translation Edition)*, **28**, 391–407.

YARMOLYUK, V. V., KOVALENDO, V. I. ET AL. 2005. U–Pb age of syn- and post-metamorphic granitoids of South Mongolia: evidence for the presence of Grenvillides in the Central Asian foldbelt. *Doklady Earth Sciences*, **404**, 986–990.

YIN, A. & HARRISON, T. M. 2000. Geologic evolution of the Himalayan–Tibetan orogen. *Annual Review of Earth and Planetary Sciences*, **28**, 211–280.

ZHENG, Y., ZHANG, Q. ET AL. 1996. Great Jurassic thrust sheets in Beishan (North Mountains) – Gobi areas of China and southern Mongolia. *Journal of Structural Geology*, **18**, 1111–1126.

ZOBACK, M. L. 1992. First- and second-order patterns of stress in the lithosphere – the World Stress Map project. *Journal of Geophysical Research*, **97**, 11703–11728.

ZORIN, Y. A. 1999. Geodynamics of the western part of the Mongolia–Okhotsk collisional belt, Trans-Baikal region (Russia) and Mongolia. *Tectonophysics*, **306**, 33–56.

ZORIN, Y. A., BELICHENKO, V. G. ET AL. 1993. The South Siberia–Central Mongolia transect. *Tectonophysics*, **225**, 361–378.

Landscape development of the Himalayan–Tibetan orogen: a review

LEWIS A. OWEN

Department of Geology, University of Cincinnati, Cincinnati, OH 45221, USA
(e-mail: Lewis.Owen@uc.edu)

Abstract: The Himalayan–Tibetan orogen provides one of the best natural laboratories in which to examine the nature and dynamics of landscape development within continent–continent collision zones. Many new tectonic–climatic–geomorphological theories and models have emerged and/or have been greatly influenced as a consequence of the study of the region and the quest to understand its geomorphological development. These include models of the interactions between tectonics, climate and surface processes, notably, the influence of climate on surface uplift by denudational unloading; the limiting of topography by glaciation (the glacial buzz-saw model); localized uplift at syntaxes by enhanced fluvial and glacial erosion that, in turn, weaken the lithosphere, enhancing surface uplift and exhumation (the tectonic aneurysm model); climate-driven out-of-sequence thrusting and crustal channel flow; glacial damming leading to differential erosion and uplift; paraglaciation; and the influence of extreme events such as earthquakes, landslides, and floods as major formative processes. The development of new technologies, including satellite remote sensing and global positioning systems, and analytical methods such as numerical dating is now allowing these theories and models to be tested and will inevitably lead to new paradigms.

The Himalaya and Tibet provide one of the best natural laboratories in which to examine the nature and dynamics of landscape development within continent–continent collision zones. Not surprisingly, many new geomorphological theories and models have emerged as the consequence of the study of the region and the quest to understand its geomorphological development. This paper will review the main studies that have been undertaken that have led to new insights into how mountain belts develop, which underlie the new paradigms that landscape development in mountains is a consequence of the interactions between tectonics, climate and Earth surface processes. The studies will be broadly discussed chronologically to help understand how the different theories and models have evolved.

Early exploration

The earliest studies in the Himalayan–Tibetan orogen were undertaken during the late 19th and early 20th centuries as western explorer–naturalists travelled the region and documented both its cultural and scientific characteristics (e.g. Shaw 1871; Drew 1875). Many of the early explorations probably had ulterior motives, being part of the 'Great Game', the political intrigue between British India and Russia (Keay 1996). Much emphasis was placed on surveying the region, and some of these surveys resulted in profound discoveries, such as the gravity surveys by Airy (1855) and Pratt (1859) that resulted in the realization that the Himalaya and other mountains had deep, low-density roots.

Probably the most notable geomorphological study was undertaken by Drew (1873, 1875) in the Indus valley, who noted numerous glacial, lacustrine and alluvial fan landforms. In his 1873 paper, Drew was the first to use the term 'alluvial fan' (Fig. 1). Drew's work focused on describing the morphology of the landforms, but little attention was given to their sedimentology. Nevertheless, he discussed their likely origins, influenced by the earlier work of Surell (1851) and Haast (1879) in the Alps and Canterbury Plains of New Zealand. In addition, Drew (1873, 1875) highlighted and emphasized the important evidence for past glaciation throughout Kashmir. Later geomorphological studies focused on examining the glacial geological record and included the notable works of Dainelli (1922, 1934, 1935), Norin (1925), Trinkler (1930), Misch (1935), de Terra & Paterson (1939) and Paffen et al. (1956) for regions throughout the western end of the Himalayan–Tibetan orogen.

Klute (1930) provided the first comprehensive map of the extent of glaciation for the Last Glacial period for the entire Himalayan–Tibetan region. At the same time as these studies of landforms and glacial geology were being undertaken, bedrock and structural geology were being examined by

Fig. 1. Views of an alluvial fan at Tigarc in the Nubra valley drawn by Drew (1875). (**a**) View from the mountains behind Charasa; (**b**) a profile view looking up the valley.

numerous geologists, most notably the Swiss geologist Augusto Gansser. Much of Gansser's work was published in his seminal book *Geology of the Himalaya* (Gansser 1964). However, these 'hard rock' studies had to await the development of plate-tectonic theory in the late 1960s and 1970s before a solid framework was available to appreciate their significance and to provide a foundation for tectonic, geomorphological, and landscape development studies.

Ice Age studies

Early exploration of the Himalaya was concurrent with the development of glacial geology and the realization that the world had experienced extensive glaciation in the recent past. Many observations were made on glacial geology and attempts were made to reconstruct former ice extent throughout the Himalaya. These early studies of glaciation were driven by a desire to correlate Himalayan glacial successions with those recognized in Europe, notably the seminal work of Penck & Brunkner (1909) in the Alps, who argued that four glaciations characterized the Quaternary. As a consequence, numerous researchers assigned their glacial geological evidence to four glaciations throughout areas of Tibet and the Himalaya (e.g. de Terra & Paterson 1939; Porter 1970; Zhang & Shi 1980).

The first truly modern comprehensive studies of glaciation were undertaken by von Wissmann (1959) and Frenzel (1960), who used the observations of the earliest explorers to construct a regional synthesis of the glaciation of Tibet and the Himalaya. They both suggested that during glacial times, ice caps expanded and extensive valley glacial systems developed throughout much of the Himalaya, Pamir, Kunlun, and Qilian Shan. This view was questioned by Kuhle (1985, 1986, 1987, 1988a, b, 1990a, b, 1991, 1993, 1995), who argued for an extensive ice sheet covering most of the Tibetan Plateau during the Last Glacial. Kuhle's Tibetan ice sheet hypothesis was important throughout the 1980s and 1990s in driving much of the study of the Quaternary glacial geology of Tibet and adjacent regions. Subsequently, however, numerous studies have disproven the existence an extensive ice sheet on Tibet (Derbyshire 1987; Zheng 1989; Burbank & Kang 1991; Derbyshire et al. 1991; Shi 1992; Hövermann et al. 1993a, b; Lehmkuhl 1995, 1997, 1998; Rutter 1995; Lehmkuhl et al. 1998; Zheng & Rutter 1998; Schäfer et al. 2002; Owen et al. 2003a). In particular, as Owen et al. (2008a) and Seong et al. (2008) pointed out, the differences in interpretation between Kuhle (1985, 1986, 1987, 1988a, b, 1990a, b, 1991, 1993, 1995) and other researchers is a consequence of their differing interpretation of landforms and sediments, the misuse of equilibrium-line altitudes for determining former ice extent, and poor chronological control. It is now generally accepted that a large ice sheet did not cover the Tibetan Plateau, at least not during the past few glacial cycles. Figure 2 illustrates different reconstructions for the extent of glaciation through the Himalaya and Tibet, with the reconstruction of Shi (1992) and Li et al. (1991) now being the generally accepted view of the former extent of glaciation.

With the development of marine oxygen isotope stratigraphy for the Quaternary during the 1980s researchers began to appreciate the complexity of glaciation in the Himalaya and Tibet specifically, that there were many more than four glaciations during the Quaternary Ice Age. However, defining the timing of glaciation in the Himalayan–Tibetan orogen was hindered by the lack of numerical ages on moraines and associated landforms. This was mainly due to the scarcity of organic matter necessary for radiocarbon dating, the standard dating technique for Quaternary landforms and sediments at that time. Röthlisberger & Geyh (1985), however, were able to obtain enough organic material from moraines in the Himalaya and Karakoram to determine 68 radiocarbon ages to define some 10 glacial advances in the Late Quaternary. Subsequent studies, however, have provided only a few additional radiocarbon ages on moraines (e.g. Derbyshire et al. 1991; Lehmkuhl 1997).

The development of optically stimulated luminescence (OSL) and terrestrial cosmogenic nuclide (TCN) surface exposure dating since the late 1980s have allowed workers to date glacial and associated landforms that were previously updateable. Furthermore, the techniques have allowed landforms older than >30 ka (the limit of standard radiocarbon dating) to be defined, and in some cases, to be taken back to many hundreds of thousands years. These method have resulted in a plethora of ages on moraines throughout the Himalaya and Tibet (see summary by Owen et al. 2008a).

Owen et al. (2008a) highlighted the limitations of OSL and TCN methods by re-evaluating published glacial chronologies throughout the Himalaya and Tibet. These limits, for example, include geological factors and uncertainties associated with modelling the appropriate production rates for TCN dating and variation in dose rates for OSL dating. The summary provided by Owen et al. (2008a) supported the view that expanded ice caps and extensive valley glacier systems existed throughout the Himalaya and Tibet during the late Quaternary, but it is not yet possible to determine whether or not the timing of the extent of maximum glaciation was synchronous throughout the entire region. The data do show considerable variations in the extent of glaciation from one region to the next during a glaciation. Glaciers throughout monsoon-influenced Tibet, the Himalaya, and the Transhimalaya are probably synchronous both with climate change resulting from oscillations in the South Asian monsoon and with Northern Hemisphere cooling cycles. In contrast, glaciers in Pamir in the far western regions of the Himalayan–Tibetan orogen advanced asynchronously to regions that are monsoon-influenced and appear to be mainly in phase with the Northern Hemisphere

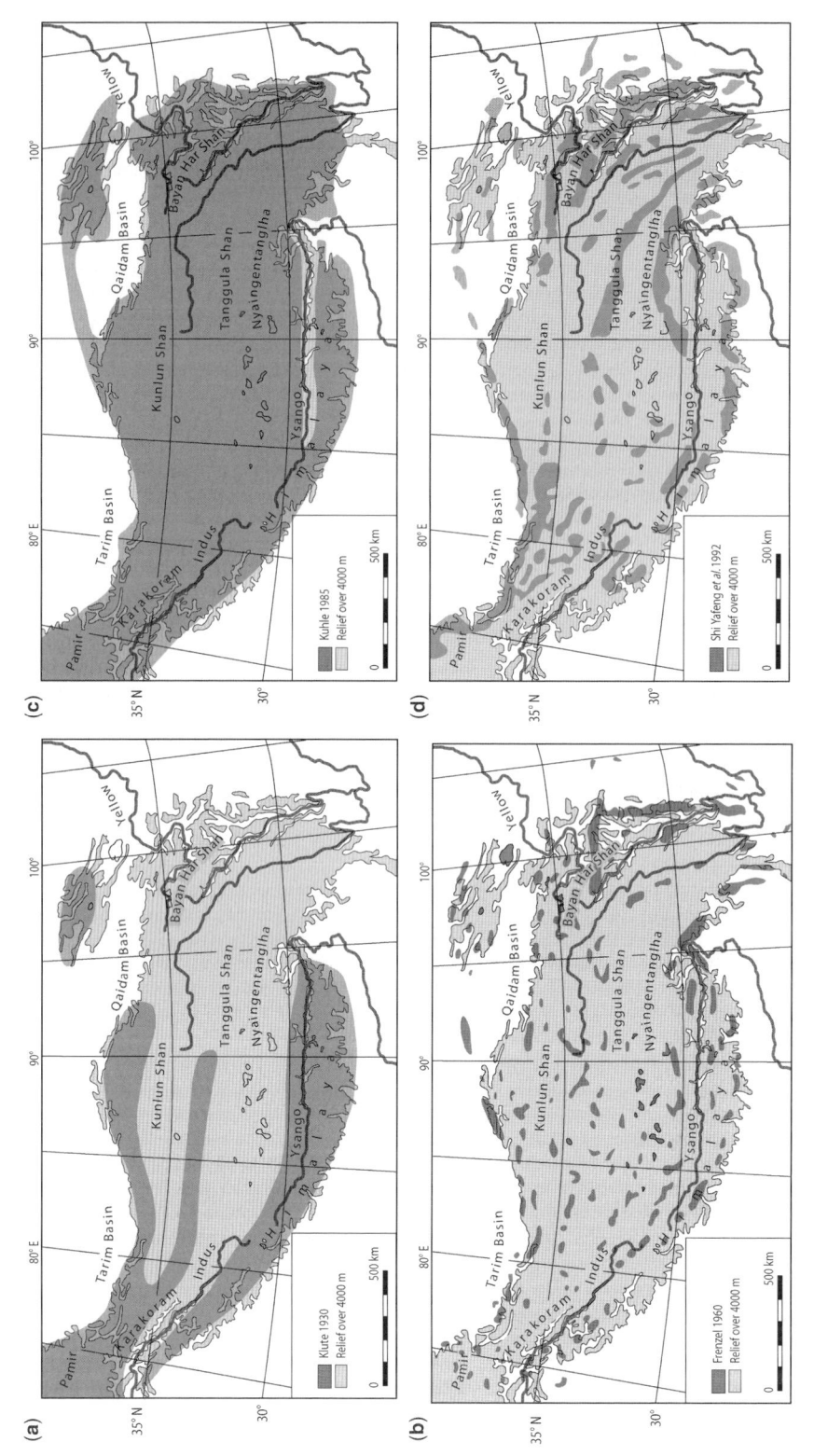

Fig. 2. Selected reconstructions for the extent of glaciation across Tibet and the bordering mountains for the maximum extent of glaciation during the Last Glacial (after Owen et al. 2008a). Light grey, relief over 4000 m above sea level; dark grey, areas considered glaciated. (**a**) Klute's (1930) reconstruction based on a temperature depression of 4 °C, with a shift of climatic zones to the south and an intensification of atmospheric circulation such that precipitation increased towards the dry areas of central Asia. (**b**) Frenzel's (1960) reconstruction based on the detailed work of Wissmann (1959), who evaluated the observations of the earliest explorers. (**c**) Kuhle's (1985) reconstruction based on field observations and extrapolation of large equilibrium-line altitude depressions (>1000 m) from the margins of Tibet into the interior regions. (**d**) Reconstruction of Shi (1992) and Li et al. (1991) based on detailed field mapping of glacial and associated landforms and sediments.

cooling cycles. Owen et al. (2008a) also pointed out that broad patterns of local and regional variability based on equilibrium-line altitudes have yet to be fully assessed, but have the potential to help define changes in climatic gradients over time. Clearly, accurate reconstructions of the former extent and timing of glaciation is important for assessing and quantifying the tectonic–climate–landscape models.

Paraglaciation

In the 1970s, studies on Late Quaternary glaciofluvial deposits on Baffen Island and in British Columbia resulting in the realization that rates of sedimentation increase significantly during times of deglaciation (Ryder 1971a, b; Church & Ryder 1972). This culminated in the concept of paraglaciation, which argues that nonglacial processes are directly conditioned by glaciation and that landscape changes are greatest during times of deglaciation (Ryder 1971a, b; Church & Ryder 1972; and reviewed by Ballantyne 2002, 2004). The term paraglacial was applied to describe the processes that operate and the landforms that are produced during deglaciation as the landscape readjusts to new climatic and environmental conditions. Paraglacial time was considered to be the period when paraglacial processes are dominant, mainly fluvial erosion and resedimentation, and mass movement. Paraglacial processes are particularly important in the transfer and resedimentation of glacial and proglacial sediments within and beyond the high-mountain landscapes, thus helping to contribute to the net denudation of high mountains.

When valley fill successions, river terraces, and alluvial fans began to be studied in the Karakoram during late 1980s and early 1990s, the concept of paraglaciation was applied to help understand their origin (Owen 1989; Derbyshire & Owen 1990; Owen & Derbyshire 1993). Subsequently, many studies have argued for the importance of paraglaciation in the landscape development of other regions of the Himalayan–Tibetan orogen (Owen & Sharma 1996; Owen et al. 1995, 2002, 2006; Yang et al. 2002; Barnard et al. 2004a, b, 2006a, b; Seong et al. 2009a, b). Figure 3 provides an example to illustrate how the relative composition of landforms within Himalayan valleys that were deglaciated at different times changes from moraine-dominated to terrace- and fan-dominated with time.

Glacial lake outburst floods (GLOFs) constitute one of the most dramatic manifestations of paraglaciation. They result from rapid drainage of moraine-dammed or supraglacial lakes and in recent years have attracted much attention because of the potential catastrophic impact they may have on settlements throughout the Himalaya. Their frequency may probably increase as glaciers continue to melt and retreat owing to human-induced climate change (Reynolds 2000; Richards & Reynolds 2000; Benn et al. 2001). Evidence for past GLOFs is abundant throughout the Himalaya and includes mega-ripples, giant boulder clusters, scour pools and large terraces (Coxon et al. 1996; Barnard et al. 2006a; Seong et al. 2009a).

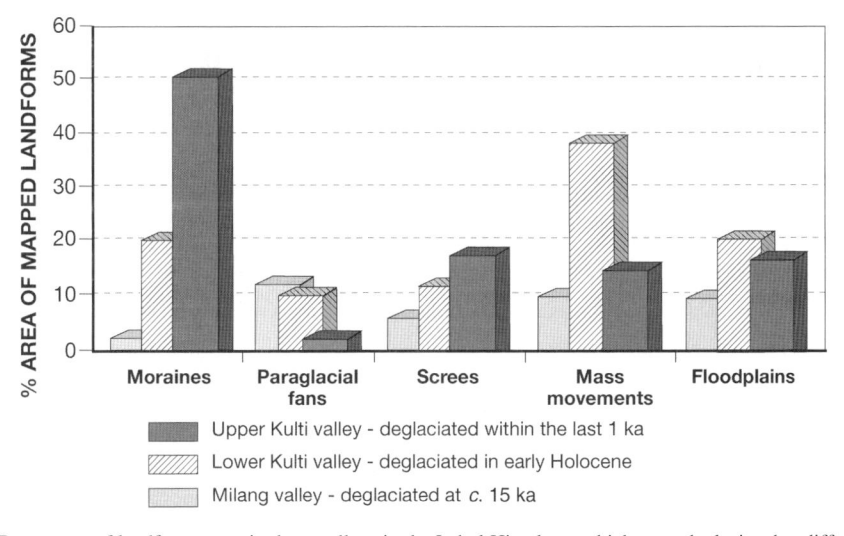

Fig. 3. Percentage of landform types in three valleys in the Lahul Himalaya, which were deglaciated at different times (after Owen et al. 1995). The data show an increase in paraglacial fans, screes and terraces, and a decrease of moraines with increasing age since deglaciation.

It is now generally accepted that landscapes in the Himalayan–Tibetan orogen have been continuously readjusting to changing climatic and environmental conditions associated with the high frequency of the oscillations of glaciers (on millennial time scales) throughout the late Quaternary.

Plate tectonics and landscape development

The Himalaya and Tibet received much attention as plate-tectonic theory developed in the late 1960s and early 1970s, but logistical and political access to the region was somewhat limited, hindering study of the region. In the late 1970s, a series of new roads were constructed to traverse the Himalaya and Tibet, permitting relatively easy access to the region (Owen 1996a). Furthermore, China began to strengthen its international research collaboration, allowing more access to Tibet and adjacent regions. It was not until the early 1980s, however, that researchers actually began to describe tectonic landforms and interpret their origin.

Much of the early work on tectonic geomorphology was stimulated by the realization that a significant component of the collision of the Indian and Asian continental lithospheric plates might be accommodated by lateral extrusion of Tibet along major strike-slip faults (Tapponnier & Molnar 1976; Fig. 4). Faulted landforms, notably faulted moraines, were examined to help estimate displacement rates along the major strike-slip faults (Tapponnier & Molnar 1977; Molnar & Tapponnier 1978; Armijo et al. 1986).

As the structure of the Himalaya and Tibet began to be mapped and examined by integrating thermochronological and geobarometric methods, areas of differential uplift were identified. Most notable was focused uplift around the Himalayan western and eastern syntaxes, particularly Nanga Parbat (Zeitler 1985). This resulted in the initial examination of the tectonic geomorphology of these zones of focused uplift (e.g. Owen 1988a; Shroder et al. 1989). It soon became apparent that relating landforms and deformed sediments to tectonic processes was not simple. Many of the Himalayan landforms are polygenetic in origin (Owen 1989) and therefore any

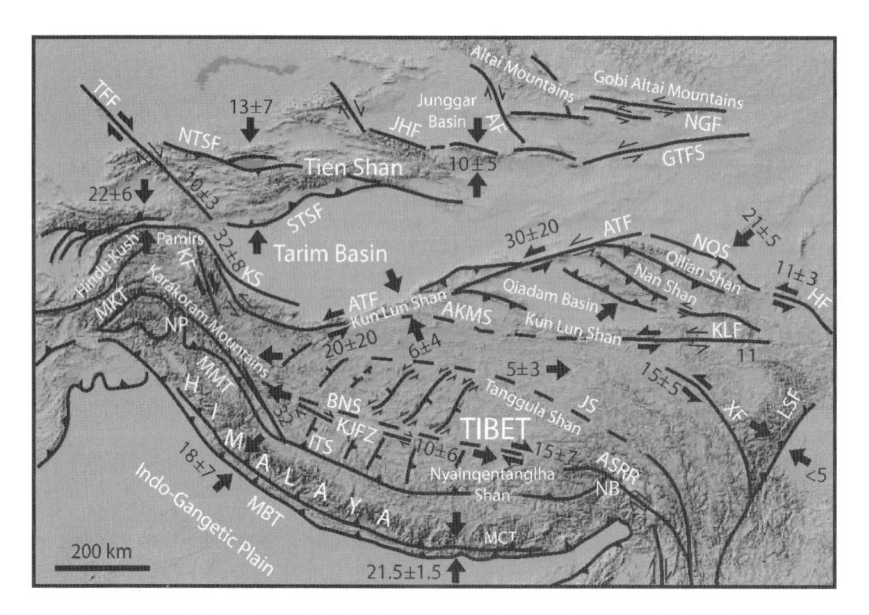

Fig. 4. Digital elevation model of Tibet and the bordering mountains showing major faults and sutures (adapted from Owen 2004, compiled from Searle 1991; Cunningham et al. 1996; Chung et al. 1998; Yin et al. 1999; Yin & Harrison 2000; Blisniuk et al. 2001; Hurtado et al. 2001). Estimates of Late Quaternary strike-slip, convergent and extensional rates are shown in mm a^{-1} (after Larson et al. 1999; Tapponnier et al. 2001). AKMS, Ayimaqin–Kunlun–Mutztagh Suture; ASRR, Ailao Shan–Red River Shear Zone; AF, Altai Fault; ATF, Altyn Tagh Fault; BNS, Bangong Nujiang Suture; GTFS, Gobi–Tien Shan Fault System; HF, Haiyuan fault; ITS, Indus Tsangpo Suture; JHF, Junggar Hegen Fault; JS, Jinsha Suture; KF, Karakoram Fault; KJFZ, Karakoram Jiali Fault Zone; KLF, Kunlun Fault; KS, Kudi Suture; LSF, Longmen Shan Fault; MBT, Main Boundary Thrust; MCT, Main Central Thrust; MKT, Main Karakoram Thrust (Shyok Suture Zone); MMT, Main Mantle Thrust; NB, Namche Barwa; NGF, North Gobi Fault; NP, Nanga Parbat; NQS, North Qilian Suture; NTSF, North Tien Shan Fault; STSF, South Tien Shan Fault; TFF, Talus–Fergana Fault; XF, Xianshuihe Fault.

tectonic influences could not easily be elucidated by simple study of landforms. In addition, glacial and mass movement processes in Himalayan–Tibetan environments produce deformation structures that are easily misinterpreted as tectonic, such as the glaciotectonized deposits throughout the Skardu Basin and landslides along the middle Indus valley in northern Pakistan (Owen 1988a, b; Owen & Derbyshire 1988).

Over the last two decades new mechanical and thermomechanical models to explain the tectonic evolution of Himalayan–Tibetan orogen have been proposed. These have evolved from an orogenic wedge model (notably that of Burchfiel et al. 1992), to pervasive ductile flow, indicated by folded isograds within a tectonic wedge (Grujic et al. 1996), to channel flow of the middle crust (Jamieson et al. 2004), with variants of mid-crustal flow based on numerical modelling (Beaumont et al. 2004). Harris (2007) described the history and essential components of these models and summarized the current view, which proposes movement of a low-viscosity crustal layer in response to topographic loading. This potential mechanism results in the eastward flow of the Asian lower crust, causing the peripheral growth of the Tibetan Plateau and the southward flow of the Indian middle crust and its extrusion along the Himalayan topographic front. Harris (2007) stressed that thermomechanical models for channel flow link extrusion to focused orographic precipitation at the surface. These models are exciting because they recognize the links between tectonics, climate and surface processes.

In addition to the tectonic models for the Himalaya, numerous models have been proposed for the evolution of the high elevations in Tibet. Tapponnier et al. (2001) summarized two end-members: (1) continuous thickening and widespread viscous flow of the crust and mantle beneath the entire plateau; (2) time-dependent, localized shear between coherent lithosphere blocks. Tapponnier et al. (2001) favoured the latter and argued for dominant growth of the Tibetan Plateau toward the east and NE. This in turn has resulted in the development of new mountain ranges and basins, with the basins becoming progressively filled to form flat, high plains.

Interactions between climate, glacial erosion, uplift and topography

During the late 1980s and 1990s, much attention was focused on the potential role of uplift of the Tibetan–Himalayan orogen in driving Cenozoic global cooling and the growth of large continental ice sheets during the Late Tertiary and Quaternary (Raymo et al. 1988; Ruddiman & Kutzbach 1989, 1991; Prell & Kutzbach 1992; Raymo & Ruddiman 1992). The uplift of the orogen was thought to have altered atmospheric circulation and to have changed the concentrations of gases in the atmosphere as a result of changes in biogeochemical cycles associated with the increased weathering of newly exposed rock surfaces, which, in turn, caused global cooling. Controversy ensued, however, over defining the timing and magnitude of Tibetan–Himalayan uplift and the complex feedbacks involved in the biogeochemical cycles and climate (Molnar & England 1990; Dupont-Nivet et al. 2008; Garzione 2008). In a broad sense, however, it is generally accepted that uplift of the orogen probably played a significant role, in addition to many other factors, in Cenozoic cooling (Ruddiman 1997; Owen 2006). In particular, the uplift of the orogen probably helped initiate the south Asian monsoon at c. 8 Ma (Burbank et al. 1993), which has had a profound impact on the dynamics of surface processes in Central Asia.

The high elevation of the Himalayan–Tibetan orogen makes it the most glaciated region outside of the polar realm. Furthermore, the widespread evidence of more extensive glaciation in the past made it inevitable that glaciers were considered important agents in the denudation that resulted in long-term lowering of the orogen and limited the orogen's average elevation (Broecker & Denton 1990). This view was challenged, however, by Molnar & England (1990), who suggested that glacial erosion generating local relief in the cores of mountain ranges might actually initiate tectonic uplift because of erosional unloading of the crust in the cores of ranges (Fig. 5). Furthermore, Molnar & England (1990) suggested that the onset of the Quaternary Ice Age and enhanced glaciation led to accelerated uplift of Himalayan and Tibetan peaks as a consequence of great glacial and associated erosion. Therefore, in the Molnar & England (1990) view there is potential for a significant positive feedback loop in which glacial erosion leads to uplift, which, in turn, results in larger glaciers and more erosion and hence more peak uplift.

Molnar & England's (1990) paper stimulated much debate and similar connections have been suggested elsewhere (e.g. Small & Anderson 1998), but whether localized erosion can actually result in mountain peak uplift remains controversial (Gilchrist & Summerfield 1991).

In contrast, Brozovic et al. (1997) argued that topographic data from Nanga Parbat show that glaciers are rapidly eroding topography above the equilibrium-line altitude (ELA) exerting an altitudinal limit on mountain height. This has become known as the glacial buzz-saw model and has subsequently been applied to understanding the

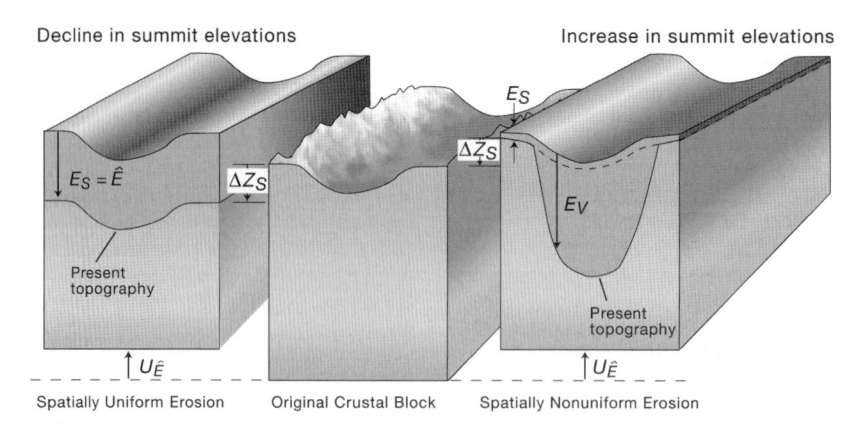

Fig. 5. Model of relief generation via glacial erosion for the case of mountain erosion after Small & Anderson (1998; modified from Molnar & England 1990). Centre panel represents initial condition. When erosion (shaded area) is spatially uniform ($E_S = \hat{E}$, where E_S is spatially uniform erosion and $\hat{E}$ spatially averaged erosion), the sum of erosionally-driven rock uplift ($U_{\hat{E}}$) and summit erosion (E_S) results in lower summit elevations (ΔZ_S) (left panel). When erosion is spatially variable (right panel), changes in summit elevation are positive because rock uplift is greater than summit erosion [$\Delta Z_S = E_S + U_{\hat{E}}(x,y)$]. Rock uplift is the same in each case because $\hat{E}$, which drives rock uplift, is equal (shaded areas are the same size). The present geophysical relief ($\hat{R}$) is the mean elevation difference between a smooth surface connecting the highest points in the landscape (dashed line) and the current topography. It represents the average of valley erosion (E_V) minus summit erosion calculated at each point in the landscape, including summit flats.

topography in other mountains, including the Cascade mountains in Washington State and the Chugach Range in Alaska (Meigs & Sauber 2000; Montgomery *et al.* 2001; Mitchell & Montgomery 2006; Spotila *et al.* 2004) Figure 6 shows the relationship between mountain topography, latitude, and present ELAs.

Other studies, however, such as Whipple *et al.* (1999), argued that there are geomorphological limits to climate-induced increases in topographic relief and that neither fluvial nor glacial erosion is likely to induce significant isostatic peak uplift. Whipple *et al.* (1999) provided a quantitative overview and constraints, including empirical evidence to define the effects of a transition from fluvial to glacial erosion. Whipple *et al.* (1999) showed that in almost all non-glaciated landscapes an increase in erosivity of the fluvial system is anticipated to result in a reduction both in trunk stream and tributary valley relief. When coupled with the constraint that hillslope relief rapidly attains a maximum in active orogens, this observation implies that ridge to valley bottom relief will actually decrease under these conditions. Whipple *et al.* (1999) suggested that relief increase is possible only if a given climate change induces a decrease in erosivity along headwater channel segments in concert with a simultaneous increase in erosivity downstream. An onset of glaciation increases hillslope relief, valley widening, and the formation of hanging valleys and overdeepenings, and relief reduction over short wavelengths. In contrast, over long wavelengths there is a reduction in relief along trunk and tributary valley profiles. If the upper reaches have thin cold-based glaciers, ridges and peaks may be protected from erosion and glacier valley profiles may become more concave, adding potentially to overall glacial relief. However, for warm-based glaciers Whipple *et al.* (1999) argued that relief production associated with each of the various glacial relief production mechanisms scales with ice thickness. They also argued that relief production is limited to several hundred metres and it is unlikely that climate change would induce significant amounts of isostatic relief production.

This ability of glaciers and rivers to accomplish rapid erosion underlies a proposed explanation for localized rapid rock uplift in the syntaxes of the Himalaya, a phenomenon that has been dubbed the 'tectonic aneurysm' (Zeitler *et al.* 2001; Koons *et al.* 2002). The aneurysm model argues that the dynamic interactions of focused erosion, topographic stresses, peak uplift, rapid exhumation, thermal weakening of the lithosphere, and deformation lead to localized feedbacks between erosion, deformation, and uplift (Fig. 7). Significant debate exists, however, over the importance of mountain glaciers versus other processes, such as fluvial erosion and mass movement, in shaping mountain landscapes and driving the tectonic aneurysm (Hallet *et al.* 1996).

The discovery in the 1980s of the South Tibetan Detachment fault near the crest of the Himalaya and

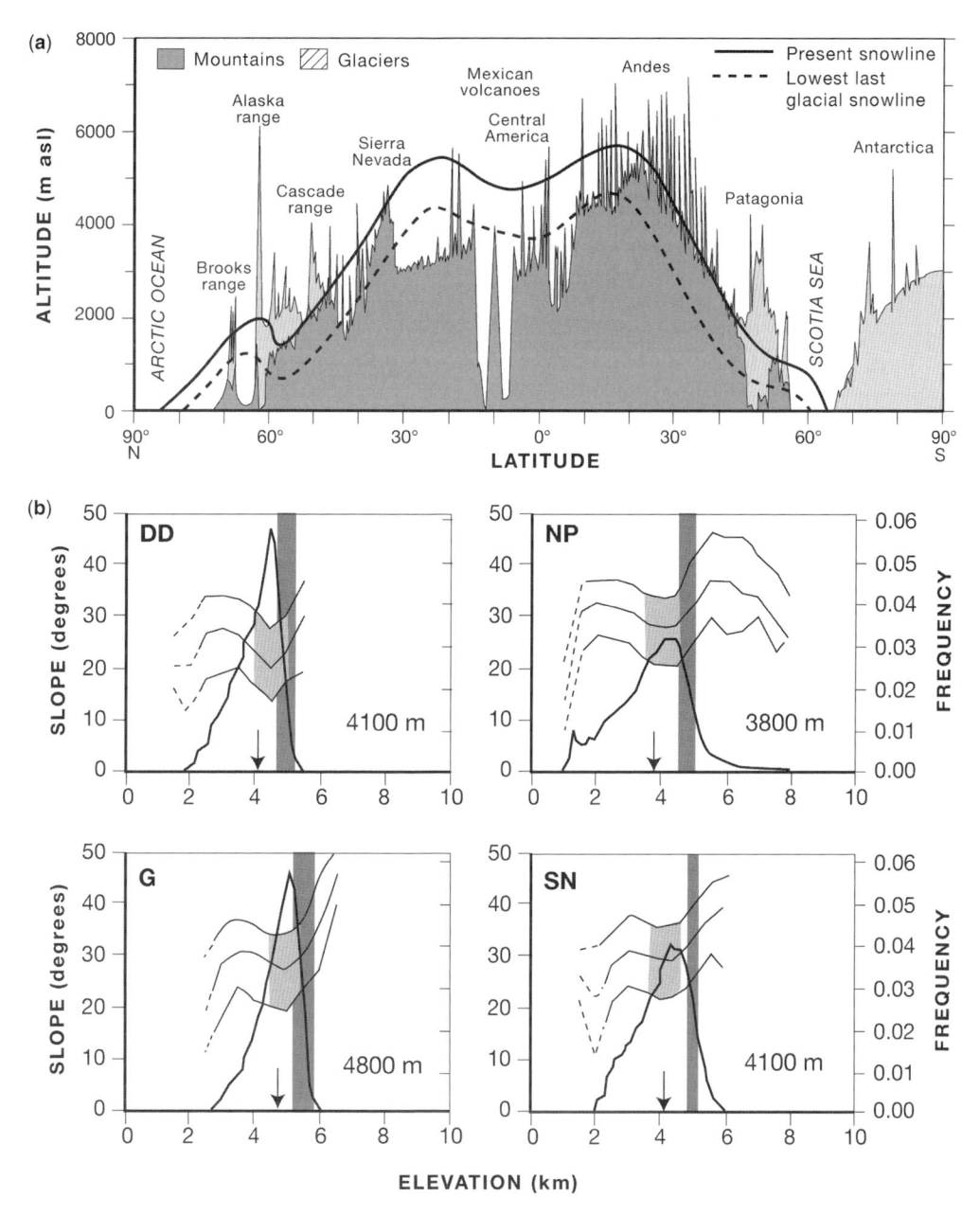

Fig. 6. Glacial buzz-saw model. (**a**) The present and minimum snowline elevations during the Last Glacial plotted along a topographic transect along the cordillera of North and South America illustrating the positive relationship between topography and snowline elevation (after Broecker & Denton 1990; Skinner *et al.* 2004). (**b**) Frequency distribution of altitude and slope angles for selected physiographic areas within the western syntaxis of the Himalaya [DD, dissected portion of Deosai Plateau; NP, Nanga Parbat; G, Ghujerab mountains (northern Karakoram); SN, area north of Skardu] after Brozovic *et al.* (1997). The arrows on the horizontal axis indicate the mean elevation for each region; values are shown inside the graphs. The dark vertical lines running from top to bottom in the graphs show the range of snowline altitudes for each region. The three subhorizontal grey lines show the 25th (lower line), 50th and 75th (upper line) percentile of slope distribution as a function of altitude for each region. The light grey shaded areas on the slope distribution curves highlight the regions of moderate slopes that generally coincide with the modal elevations. These graphs illustrate the strong relationship between snowline elevation and topography, suggesting that glaciation places limits on topography.

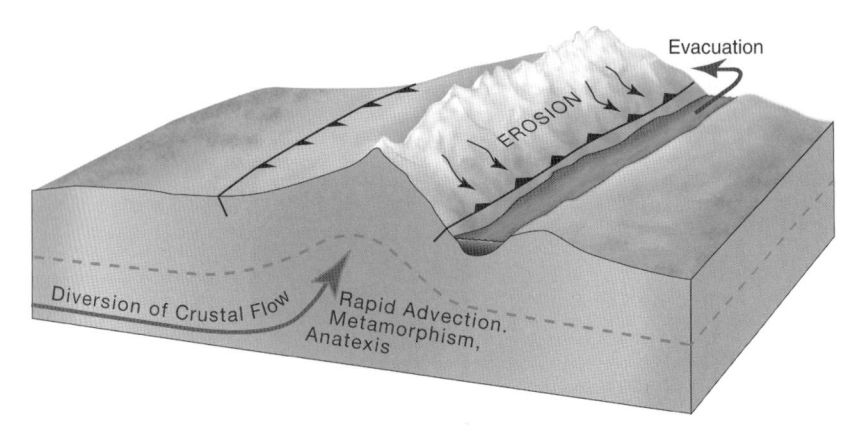

Fig. 7. Schematic representation illustrating the dynamics of a tectonic aneurysm, shown at a mature stage based on Nanga Parbat (after Zeitler *et al.* 2001).

the realization that it defined northward normal displacement of Tibet stimulated the development of the extrusive flow model (sometimes referred to as the extrusion model; Burchfiel *et al.* 1992; Fig. 8). The extrusive flow model theorizes southward flow of ductile middle and lower crust that continuously replenishes the Himalayan range front, transporting material from between the Main Frontal Thrust and the South Tibetan fault system, especially the rocks bounded by the South Tibetan Detachment fault system above and the Main Central (sole) Thrust below (Hodges *et al.* 2001; Harris 2007). This southward extruding zone is thought to represent the ductile lower crustal

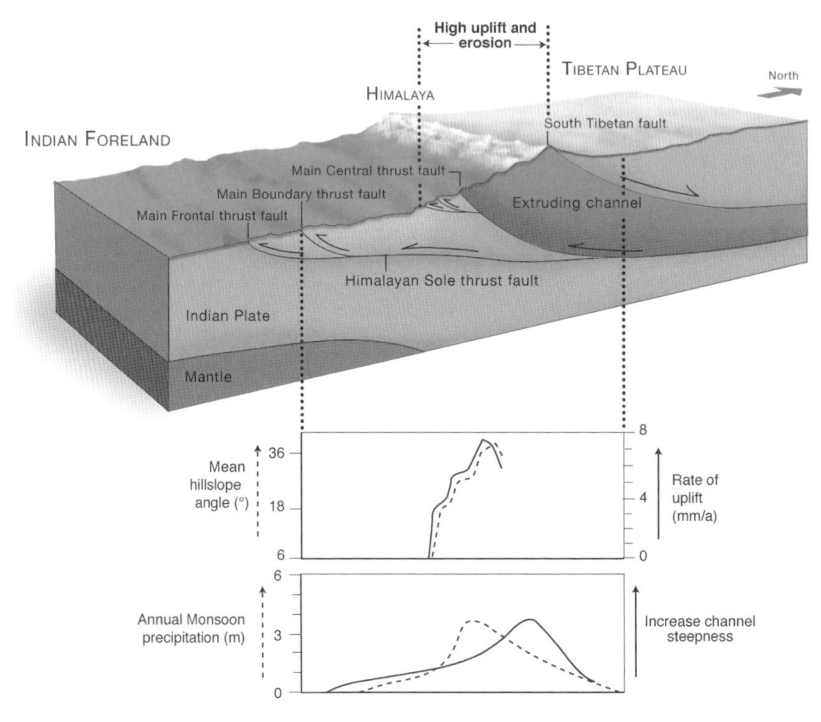

Fig. 8. Schematic section across the Himalaya and southern Tibet illustrating channel flow and extrusion of the channel, and the variation of mean hillslope angle, rate of uplift, annual monsoon precipitation and relative stream channel steepness (adapted from Hodges 2006).

channel of Tibet that has ground its way to the surface since the Miocene (Harris 2007). Theoretical models by Beaumont et al. (2001) argued that channel flow is maintained by erosion of the Himalayan front. This view was enhanced by geomorphological studies that suggested a feedback between erosion and extrusive flow (Wobus et al. 2003; Hodges et al. 2004).

Examining the Marsyandi River valley, which that traverses a region of extremely high exhumation between the South Tibetan Detachment and the Main Central Thrust, and where there is very high precipitation, Burbank et al. (2003) argued that there is no direct precipitation–erosion linkage because erosion rates are constant based on apatite fission-track ages that show no systematic trend. Rather, they suggested that additional factors that influence river incision rates such as channel width and sediment concentrations must compensate for differences in precipitation across the region. Moreover, spatially constant erosion is a response to uniform, upward tectonic transport of Greater Himalayan rock above a crustal ramp.

In contrast, on the basis of geological mapping of the Marsyandi valley in central Nepal, Hodges et al. (2004) showed that a zone of recent faulting is coincident with an abrupt change in river gradient, which is thought to mark the transition from rapid uplift of the Higher Himalaya to a region of slower uplift to the south, and probably reflects active thrusting at the topographic front. Hodges et al. (2004) suggested that the zone of active thrusting is coincident with a zone of intense monsoon precipitation, probably indicating a positive feedback between focused erosion and deformation at the front of the Higher Himalayan ranges.

Korup & Montgomery (2008) suggested a new twist on the landscape development of the Himalayan–Tibetan orogen in their study of the geomorphology of Namche Barwa. In theory, rivers in this region should progressively erode towards their heads, resulting in the steady degradation of the plateau's margins, and that over time knickpoints should migrate upstream and become less apparent as erosion progresses. Recently, however, geological evidence in the Namche Barwa region has shown that the Tsangpo and its tributary streams that traverse Namche Barwa have probably been stable, remaining at the same positions over at least the last million years, effectively preserving the margin of the Tibetan Plateau (Finnegan et al. 2008). Various theories have been invoked to help explain the preservation of the margin of the Himalayan–Tibetan orogen, including theories of differential rock uplift matching erosion and other mechanisms, such as landsliding that armours valley floors and helps protect them from river erosion (Lavé & Avouac 2001; Ouimet et al. 2007).

Korup & Montgomery (2008) highlighted that broad stream valleys with substantial terraces and thick valley fill sediments are present upstream of moraines that mark where glaciers and their moraines dammed the Tsangpo and its tributaries. In contrast, stream channels are more confined and are usually entrenched into bedrock downstream of the moraines. They argued that the glaciers and/or the moraines impounded the streams to produce large lakes that fill with sediment, increasing river sedimentation upstream (Fig. 9). These sediments then essentially protected the upstream valley from erosion as stream power decreased in the newly established broad valleys. The sediments become incised to form terraces, with little overall valley lowering. In contrast, the stream channels are confined as they cut through the moraines, resulting in increased stream power and leading to enhanced erosion and entrenchment into bedrock

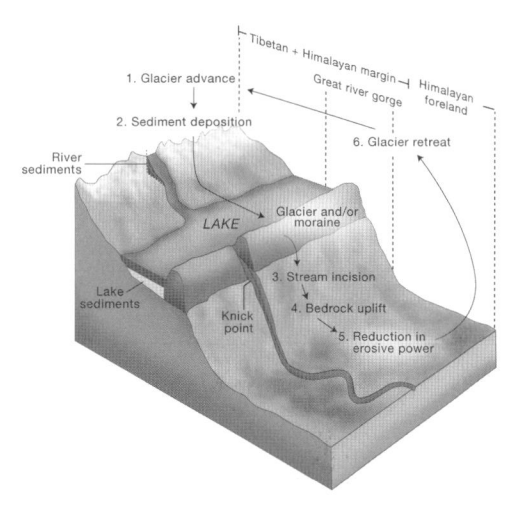

Fig. 9. Possible mechanism for the preservation of the edge of SE Tibet as suggested by Korup & Montgomery (2008) and adapted from Owen (2008) with the following as the sequence of events: 1, glacier advance dams a river; 2, the deposition of sediments in the resulting upstream lake and stream both protects the valley floor from erosion and reduces the stream's erosive power; 3, streams draining through the glacier and/or moraine are confined, so stream flow is strong and downstream erosion increases; 4, stream incision into the bedrock essentially weakens the crust and enhances bedrock uplift to the surface; 5, as streams travel farther from the knick point, smaller gradients reduce their erosive power; 6, the glacier retreats, but the events are repeated numerous times as glaciers advance and retreat in response to climate change. The repetition of this sequence of events maintains equilibrium between erosion and bedrock uplift, essentially preserving the topography of the margin of SE Tibet.

downstream of the moraine. This, in turn, enhanced bedrock uplift in the gorge sections, essentially preserving the margin of Tibet.

All these models highlight an exciting trend in the Earth sciences linking endogenetic and exogenic processes, and highlighting the complex feedbacks between these processes that are reflected in the landscapes that develop.

Extreme events

The relative roles of high-magnitude–low-frequency and low-magnitude–high-frequency events in landscape development have been long debated (Brunsden & Jones 1984). Defining what constitutes a high-magnitude–low-frequency event is difficult for the Himalayan–Tibetan orogen where geomorphological processes are often an order of magnitude greater in size and effect than most other terrestrial settings, and the definition depends on the temporal and spatial framework that is being considered. However, a catastrophic flood that destroys extensive farmland, highways and villages, a giant landslide displacing $>1 \times 10^6$ m^3 of displaced debris, and an earthquake of magnitude 7 or higher clearly constitute high magnitude events. For clarity, this type of event will be referred to as an extreme event.

Probably the most evident extreme geomorphological events are landslides triggered by large earthquakes, yet there have not been many studies of earthquake-triggered landslides in the Himalayan–Tibetan orogen because of the relatively low occurrence of large earthquakes over the hundred years since records have been kept. However, the 1991 and 1999 Garhwal, the 2001 Kokoxili, the 2005 Kashmir, and the 2008 Wenchuan (Sichuan) earthquakes have provided opportunities to study the effects of earthquake-triggered landslides on landscape development (Owen *et al.* 1996; Barnard *et al.* 2001; Van der Woerd *et al.* 2004; Liu & Kusky 2008).

During the Garhwal earthquakes, landsliding comprised mainly rock and debris avalanches that were concentrated along the lower stretches of the valley slopes. Owen *et al.* (1996) and Barnard *et al.* (2001) calculated the equivalent net lowering (denudation) of the landscape by mapping and measuring the amount of sediment produced and moved during and shortly after the earthquakes, and concluded that this was small compared with earthquake-induced landsliding in other mountainous regions. Furthermore, Owen *et al.* (1996) and Barnard *et al.* (2001) were able to show that long-term denudation rates and sediment flux resulting from human and monsoon activity in the region were far more significant than earthquake-induced mass movements and associated processes. These studies, therefore, questioned the relative importance of earthquake-induced landforms and processes in landscape evolution of the Himalaya.

In contrast, landslides triggered by the 2005 Kashmir earthquake were much more numerous,

Fig. 10. The Hattian Bala landslide triggered by the 2005 Kashmir earthquake. The landslide buried four villages resulting in >450 fatalities, blocked two streams and formed two lakes.

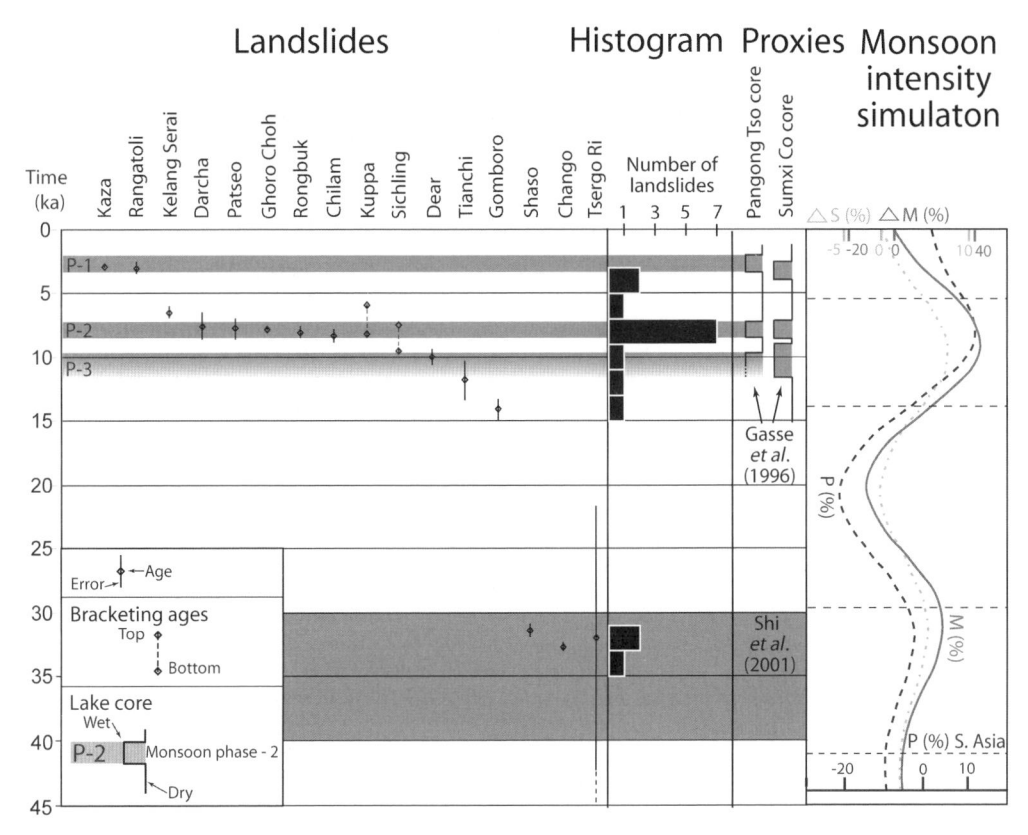

Fig. 11. The relationship between landslides of known age and curves of monsoon intensity and precipitation (after Dortch et al. 2009). The thin grey intervals indicate enhanced monsoon phases and subsequent increased precipitation based on lake core data taken from Gasse et al. (1996). The histogram uses a 2 ka bin width for landslide occurrence. Dashed horizontal lines on right side show maximum age ranges of rock avalanche clusters crossing the modelled monsoon intensity curve. Thick grey bar is period of enhanced monsoon precipitation between 30 and 40 ka inferred from lake terraces, pollen, lake cores and ice cores by Li et al. (1991). The three curves in the proxy data section show the simulated monsoon pressure index (ΔM percentage, continuous line) for the Indian Ocean, simulated changes in precipitation (P percentage, thick black dashed line) in southern Asia, and variations in Northern Hemisphere solar radiation (ΔS percentage, thin grey dashed line) after Prell & Kutzbach (1987).

occurring throughout a region of >7500 km², but were highly concentrated, associated with six geomorphological–geological–anthropogenic settings (Kamp et al. 2008; Owen et al. 2008b). Owen et al. (2008b) estimated that several thousand landslides were triggered, mainly rock falls and debris falls, although translational rock and debris slides also occurred. In addition, a debris avalanche (the Hattian Bala sturzstrom) comprising >80 × 10⁶ m³ occurred at Hattian, which blocked streams and created two lakes (Fig. 10). The 2008 Wenchuan earthquake triggered numerous large landslides, including one that buried >700 people (Liu & Kusky 2008; US Geological Survey 2008), and probably many thousands of smaller landslides, yet their full impact has still to be assessed.

Landsliding triggered by the 14 November 2001 magnitude 7.9 Kokoxili earthquake has not been fully assessed, but Van der Woerd et al. (2004) were able to map the occurrence of several giant ice avalanches (each involving >1 × 10⁶ m³ of ice and snow) initiated by slope failure from ice caps as a result of strong ground motion. These ice avalanches transported little rock debris, and Van der Woerd et al. (2004) concluded that it is unlikely that ice avalanches are very important in contributing to the landscape development in Tibet. Van der Woerd et al. (2004) argued, however, that given the appropriate geological and climatic conditions, ice avalanching may be an important process in the landscape evolution of high mountainous terrains.

In summary, the importance of earthquake-triggered landslides for landscape development has not been thoroughly assessed and is likely to be an important topic of future research. However, non-earthquake-triggered landslides are pervasive throughout much of the Himalaya and are triggered by undercutting of slopes by fluvial and glacial erosion. In monsoon areas, heavy seasonal rainfall enhances landsliding. There are numerous large or giant landslides, which comprise $>1 \times 10^6$ m^3 of debris, throughout the orogen. Many of these are ancient. Recent studies are suggesting that their formation is temporally clustered and their movement is associated with times of enhanced monsoon, such as during the early Holocene, when higher groundwater levels enhance failures (Bookhagen et al. 2005; Dortch et al. 2009; Fig. 11).

Landsliding can also initiate extreme flooding events. This occurs when landslides dam drainages and create lakes that may drain catastrophically if the landslide dam fails. The Indus valley provides several infamous examples of such extreme events. The most notable was the Indus flood of 1841, which was the result of the drainage of a >60 km long lake in the Indus and Gilgit valleys, which were blocked as the result of the earthquake-triggered collapse of the Lichar Spur on Nanga Parbat in 1840. The flood waters advanced >400 km out of the Himalaya and devastated a Sikh army that was camped on the Chach Plain near Attock (Mason 1929).

Owen (1996b) provided other examples of landslide-dammed lakes that failed catastrophically. The geomorphological consequences of such failures have not been systematically assessed and quantified, but clearly they result in major landscape changes involving large-scale erosion and resedimentation.

The Garhwal, Kashmir and Wenchuan earthquakes have focused much interest on determining the frequency and magnitude of large and great earthquakes along the Himalaya and in Tibet (e.g. Wesnousky et al. 1999; Bilham et al. 2001; Kumar et al. 2001, 2006; Feldl & Bilham 2006; Burchfiel et al. 2008; Kondo et al. 2008). These studies show that great earthquakes are not uncommon in the Himalaya and are probably very important for landscape development.

Future trends

Predicting the advances of future research trends in any discipline is challenging. The major paradigm shifts and new technological developments applied to understanding landscape development of the Himalaya–Tibetan orogen over the last few decades illustrate how difficult it would have been to have predicted where the discipline would have progressed just a few decades ago. Probably one of the greatest challenges for us is to examine the complex variability between regions within the Himalayan–Tibetan orogen, and to quantify the timing, rates and magnitude of landscape development on various time scales ranging from 10^0 to 10^6 years. The development and application of new technologies, including satellite remote sensing (e.g. studies such as that by Bookhagen & Burbank 2006) and global positioning systems (e.g. studies such as those by Chen et al. 2004; Jade et al. 2004), plus analytical methods such as numerical dating, will allow many of the new theories and models to be tested.

Conclusion

The continuous study of the Himalayan–Tibetan orogen over the last century has resulted in exciting new paradigms on the nature and dynamics of landscape development within continent–continent collision zones. Of particular note is the realization that complex links, interactions and feedbacks exist between tectonics, climate and landscape development. New models based on these complex relationships include those on the influence of climate on rock uplift by denudational unloading of the crust; the limiting of topography by glaciation (the glacial buzz-saw model); localized uplift at syntaxes by enhanced fluvial and glacial erosion that, in turn, weaken the lithosphere and so enhance bedrock uplift (the tectonic aneurysm model); climate-driven out-of-sequence thrusting and focused erosion driving extrusion of ductile crustal channels; and glacial damming leading to differential erosion and uplift. Improved mapping and dating of landforms will inevitably allow us to test these models and will probably result in the development of new paradigms.

This paper was written to express my sincere and profound thanks to Brian Windley for supervising me as a doctoral student, introducing me to the Himalayan–Tibetan orogen, and for his continued encouragement throughout my career. Many thanks go to T. Phillips for drafting most of the figures included in this paper, and C. Dietsch and D. Nash for comments on the manuscript. Thanks go to T. Kusky and N. Pinter for their constructive and extremely useful comments on this paper.

References

AIRY, G. B. 1855. On the computation of the effect of the attraction of the mountain masses disturbing the apparent astronomical latitude of stations in geodetic surveys. *Philosophical Transactions of the Royal Society of London*, **145**, 101–104.

ARMIJO, R., TAPPONNIER, P., MERCIER, J. L. & HANZ, T.-L. 1986. Quaternary extension in southern Tibet: field observations and tectonic implications. *Journal of Geophysical Research*, **91**, 13 803–13 872.

BALLANTYNE, C. K. 2002. Paraglacial geomorphology. *Quaternary Science Reviews*, **21**, 1935–2017.

BALLANTYNE, C. K. 2004. Paraglacial landsystems. *In*: EVANS, D. J. (ed.) *Glacial Landsystems*. Edward Arnold, London, 432–461.

BARNARD, P. L., OWEN, L. A., SHARMA, M. C. & FINKEL, R. C. 2001. Natural and human-induced landsliding in the Garwhal Himalaya of Northern India. *Geomorphology*, **40**, 21–35.

BARNARD, P. L., OWEN, L. A. & FINKEL, R. C. 2004a. Style and timing of glacial and paraglacial sedimentation in a monsoonal influenced high Himalayan environment, the upper Bhagirathi Valley, Garhwal Himalaya. *Sedimentary Geology*, **165**, 199–221.

BARNARD, P. L., OWEN, L. A., SHARMA, M. C. & FINKEL, R. C. 2004b. Late Quaternary (Holocene) landscape evolution of a monsoon-influenced high Himalayan valley, Gori Ganga, Nanda Devi, NE Garhwal. *Geomorphology*, **61**, 91–110.

BARNARD, P. L., OWEN, L. A. & FINKEL, R. C. 2006a. Quaternary fans and terraces in the Khumbu Himalaya, south of Mt. Everest: their characteristics, age and formation. *Journal of the Geological Society, London*, **163**, 383–400.

BARNARD, P. L., OWEN, L. A., FINKEL, R. C. & ASAHI, K. 2006b. Landscape response to deglaciation in a high relief, monsoon-influenced alpine environment, Langtang Himal, Nepal. *Quaternary Science Reviews*, **25**, 2162–2176.

BEAUMONT, C., JAMIESON, R. A., HGUYEN, M. H. & LEE, B. 2001. Himalayan tectonics explained by extrusion of a low-viscosity crustal channel coupled to focused surface denudation. *Nature*, **414**, 738–742.

BEAUMONT, C., JAMIESON, R. A., NGUYEN, M. H. & MEDVEDEV, S. 2004. Crustal channel flows: 1. Numerical models with applications to the tectonics of the Himalayan–Tibetan orogen. *Journal of Geophysical Research*, **109**, B06406.

BENN, D. I., WISEMAN, S. & HANDS, K. 2001. Growth and drainage of supraglacial lakes on the debris-mantled Ngozumpa Glacier, Khumbu Himal. *Journal of Glaciology*, **47**, 626–638.

BILHAM, R., GAUR, V. K. & MOLNAR, P. 2001. Himalayan seismic hazard. *Science*, **293**, 1442–1444.

BLISNIUK, P. M., HACKER, B. R., GLODNY, J. ET AL. 2001. Normal faulting in central Tibet since at least 13.5 Myr ago. *Nature*, **412**, 628–632.

BOOKHAGEN, B. & BURBANK, D. W. 2006. Topography, relief, and TRMM-derived rainfall variations along the Himalaya. *Geophysical Research Letters*, **33**, L08405.

BOOKHAGEN, B., THIEDE, R. C. & STRECKER, M. R. 2005. Late Quaternary intensified monsoon phases control landscape evolution in the northwest Himalaya. *Geology*, **33**, 149–152.

BROECKER, W. S. & DENTON, G. H. 1990. What drives glacial cycles? *Scientific American*, January, 49–56.

BROZOVIC, N., BURBANK, D. W. & MEIGS, A. J. 1997. Climatic limits on landscape development in the northwestern Himalaya. *Science*, **276**, 571–574.

BRUNSDEN, D. & JONES, D. K. C. 1984. The geomorphology of high magnitude–low frequency events in the Karakoram Mountains. *In*: MILLER, K. (ed.) *International Karakoram Project*. Cambridge University Press, Cambridge, 343–388.

BURBANK, D. W. & KANG, J. C. 1991. Relative dating of Quaternary moraines, Rongbuk Valley, Mount Everest, Tibet: implications for an ice sheet on the Tibetan Plateau. *Quaternary Research*, **36**, 1–18.

BURBANK, D. W., DERRY, L. & FRANCE-LANORD, C. 1993. Lower Himalayan detrital sediment delivery despite an intensified monsoon at 8 Ma. *Nature*, **364**, 48–50.

BURBANK, D. W., BLYTHE, A. E. ET AL. 2003. Decoupling of erosion and precipitation in the Himalayas. *Nature*, **426**, 652–655.

BURCHFIEL, B. C., CHEN, Z., HODGES, K. V., LIU, Y., ROYDEN, L. H., DENG, C. & XU, J. 1992. The South Tibetan detachment system, Himalayan orogen: extension contemporaneous with and parallel to shortening in a collisional mountain belt. *In*: BURCHFIEL, B. C., CHEN, Z. & HODGES, K. V. (eds) *The South Tibetan Detachment System, Himalayan Orogeny: Extension Contemporaneous with and Parallel to Shortening in a Collisional Mountain Belt*. Geological Society of America Special Paper, **269**, 1–41.

BURCHFIEL, B. C., ROYDEN, L. H. ET AL. 2008. A geological and geophysical context for the Wenchuan earthquake of 12 May 2008, Sichuan, People's Republic of China. *GSA Today*, **18**, 4–11.

CHEN, Q., FREYMUELLER, J. T., YANG, Z., XU, C., JIANG, W., WANG, Q. & LIU, J. 2004. Spatially variable extension in southern Tibet based on GPS measurements. *Journal of Geophysical Research*, **109**, B09401.

CHUNG, S. L., LO, C. H. ET AL. 1998. Diachronous uplift of the Tibet Plateau starting 40 Myr ago. *Nature*, **394**, 769–773.

CHURCH, M. A. & RYDER, J. M. 1972. Paraglacial sedimentation: a consideration of fluvial processes conditioned by glaciation. *Geological Society of America Bulletin*, **83**, 3059–3071.

COXON, P., OWEN, L. A. & MITCHELL, W. A. 1996. A late Quaternary catastrophic flood in the Lahul Himalayas. *Journal of Quaternary Science*, **11**, 495–510.

CUNNINGHAM, W. D., WINDLEY, B. F., DORJNAMJAA, D., BADAMGAROV, G. & SAANDER, M. 1996. Late Cenozoic transpression in southwestern Mongolia and the Gobi Altai–Tien Shan connection. *Earth and Planetary Science Letters*, **140**, 67–82.

DAINELLI, G. 1922. *Spedizione Italiana de Filippi nell' Himalaia, Caracorum e Turchestan Cinese (1913–1914). Ser. II, 3*. Zanichelli, Bologna.

DAINELLI, G. 1934. *Spedizione Italiana de Filippi nell' Himalaia, Caracorum e Turchestan Cinese (1913–1914). Ser. II, 3*. Zanichelli, Bologna.

DAINELLI, G. 1935. *Spedizione Italiana de Filippi nell' Himalaia, Caracorum e Turchestan Cinese (1913–1914). Ser. II, 2*. Zanichelli, Bologna.

DERBYSHIRE, E. 1987. A history of the glacial stratigraphy in China. *Quaternary Science Reviews*, **6**, 301–314.

DERBYSHIRE, E. & OWEN, L. A. 1990. Quaternary alluvial fans in the Karakoram Mountains. *In*: RACHOCKI, A. H. & CHURCH, M. (eds) *Alluvial Fans: A Field Approach*. Wiley, Chichester, 27–53.

DERBYSHIRE, E., SHI, Y., LI, J., ZHENG, B., LI, S. & WANG, J. 1991. Quaternary glaciation of Tibet: the geological evidence. *Quaternary Science Reviews*, **10**, 485–510.

DE TERRA, H. & PATERSON, T. T. 1939. *Studies on the Ice Age in India and associated human cultures*. Carnegie Institute of Washington Publications, **493**.

DORTCH, J., OWEN, L. A., HANEBERG, W. C., CAFFEE, M. W., DIETSCH, C. & KAMP, U. 2009. Nature and timing of large landslides in the Himalaya and Transhimalaya of northern India. *Quaternary Science Reviews*, **28**, 1037–1056.

DREW, F. 1873. Alluvial and lacustrine deposits and glacial records of the upper Indus basin; Part 1, Alluvial deposits. *Geological Society of London Quarterly Journal*, **29**, 449–471.

DREW, F. 1875. *The Jummoo and Kashmir Territories: A Geographical Account*. Indus Publications, Karachi [reprinted 1980].

DUPONT-NIVET, G., HOORN, C. & KONERT, M. 2008. Tibetan uplift prior to the Eocene–Oligocene climate transition: evidence from pollen analysis of the Xining Basin. *Geology*, **36**, 987–990.

FELDL, N. & BILHAM, R. 2006. Great Himalayan earthquakes and the Tibetan plateau. *Nature*, **444**, 165–170.

FINNEGAN, N. J., HALLET, B., MONTGOMERY, D. R., ZEITLER, P. K., STONE, J. O., ANDERS, A. M. & LIU, Y. 2008. Coupling of rock uplift and river incision in the Namche Barwa–Gyala Peri massif, Tibet. *Geological Society of America Bulletin*, **120**, 142–155.

FRENZEL, B. 1960. Die Vegetations- und Landschaftszonen Nordeurasiens während der letzten Eiszeit und während der Postglazialen Warmezeit. *Akademie der Wissenschaften und der Literatur in Mainz, Abhandlungen der Mathematisch-Naturwissenschaftlichen Klasse*, **13**, 937–1099.

GANSSER, A. 1964. *Geology of the Himalaya*. Wiley-Interscience, London.

GARZIONE, C. N. 2008. Surface uplift of Tibet and Cenozoic global cooling. *Geology*, **36**, 1003–1004.

GASSE, F., FONTES, J. C., VAN CAMPO, E. & WEI, K. 1996. Holocene environmental changes in Bangong Co basin (Western Tibet). Part 4: discussion and conclusions. *Paleogeography, Paleoclimatology, Paleoecology*, **120**, 79–92.

GILCHRIST, A. R. & SUMMERFIELD, M. A. 1991. Denudation, isostasy and landscape evolution. *Earth Surface Processes and Landforms*, **16**, 555–562.

GRUJIC, D., CASEY, M., DAVIDSON, C., HOLLISTER, L. S., KÜNDIG, R., PAVLIS, T. & SCHMID, S. 1996. Ductile extrusion of the Himalayan Crystalline in Bhutan: evidence from quartz microfabrics. *Tectonophysics*, **260**, 21–43.

HAAST, J. 1879. *Geology of the Provinces of Canterbury and Westland, New Zealand*. Times office, Christchurch.

HALLET, B., HUNTER, L. & BOGEN, J. 1996. Rates of erosion and sediment evacuation by glaciers: a review of field data and their implications. *Global Planetary Change*, **12**, 213–235.

HARRIS, N. 2007. Channel flow and the Himalayan–Tibetan orogen: a critical review. *Journal of the Geological Society, London*, **164**, 511–523.

HODGES, K. 2006. Climate and the evolution of mountains. *Scientific American*, August, 72–79.

HODGES, K. V., HURTADO, J. M. & WHIPPLE, K. X. 2001. Southward extrusion of Tibetan crust and its effects on Himalayan tectonics. *Tectonics*, **20**, 799–809.

HODGES, K. V., WOBUS, C., RUHL, K., SCHILDGEN, T. & WHIPPLE, K. 2004. Quaternary deformation, river steepening, and heavy precipitation at the front of the Higher Himalayan Ranges. *Earth and Planetary Science Letters*, **220**, 379–389.

HÖVERMANN, J., LEHMKUHL, F. & PÖRTGE, K.-H. 1993a. Pleistocene glaciations in Eastern and Central Tibet – Preliminary results of the Chinese–German joint expeditions. *Zeitscrift für Geomorphologie*, **92**, 85–96.

HÖVERMANN, J., LEHMKUHL, F. & SÜSSENBERGER, H. 1993b. Neue Befunde zur Paläoklimatologie Nordafrikas und Zentralasiens. *Abhandlungen der Braunschweigischen Wissenschaftlichen Gesellschaft*, **43**, 127–150.

HURTADO, J. M., HODGES, K. V. & WHIPPLE, K. X. 2001. Neotectonics of the Thakkhola graben and implications for recent activity on the South Tibetan fault system in the central Himalaya. *Geological Society of America Bulletin*, **113**, 222–240.

JADE, S., BHATT, B. C. ET AL. 2004. GPS measurements from the Ladakh Himalaya, India: preliminary tests of plate-like or continuous deformation in Tibet. *GSA Bulletin*, **116**, 1385–1391.

JAMIESON, R. A., BEAUMONT, C., MEDVEDEV, S. & NGUYEN, M. H. 2004. Crustal channel flows: 2. Numerical models with implications for metamorphism in the Himalayan–Tibetan orogen. *Journal of Geophysical Research*, **109**, B06407.

KAMP, U., GROWLEY, B. J., KHATTAK, G. A. & OWEN, L. A. 2008. GIS-based landslide susceptibility mapping for the 2005 Kashmir Earthquake Region. *Geomorphology*, **101**, 631–642.

KEAY, J. 1996. *Explorers of the Western Himalayas, 1820–95: 'When Men and Mountains Meet' and 'Gilgit Game'*. John Murray, London.

KLUTE, F. 1930. Verschiebung der Klimagebiete der letzten Eiszeit. *Petermanns Mitteilungen Ergänzungsheft*, **209**, 166–182.

KONDO, H., NAKATA, T. ET AL. 2008. Long recurrence interval of faulting beyond the 2005 Kashmir earthquake around the northwestern margin of the Indo-Asian collision zone. *Geology*, **36**, 731–734.

KOONS, P. O., ZEITLER, P. K., CHAMBERLAIN, C. P., CRAW, D. & MELTZER, A. S. 2002. Mechanical links between erosion and metamorphism in Nanga Parbat, Pakistan Himalaya. *American Journal Science*, **302**, 749–773.

KORUP, O. & MONTGOMERY, D. R. 2008. Tibetan plateau river incision inhibited by glacial stabilization of the Tsangpo gorge. *Nature*, **455**, 786–789.

KUHLE, M. 1985. *Ein subtropisches Inlandeis als Eiszeitauslöser, Südtibet un Mt. Everest expedition 1984*. Georgia Augusta, Nachrichten aus der Universität Gottingen, May, 1–17.

KUHLE, M. 1986. The upper limit of glaciation in the Himalayas. *GeoJournal*, **13**, 331–346.

KUHLE, M. 1987. The problem of a Pleistocene inland glaciation of the northeastern Qinghai–Xizang Plateau. *In*: HÖVERMANN, J. & WANG, W. (eds

Reports of the Qinghai–Xizang (Tibet) Plateau. Science Press, Beijing, 250–315.

KUHLE, M. 1988a. Geomorphological findings on the built-up of Pleistocene glaciation in Southern Tibet and on the problem of inland ice. *GeoJournal,* **17,** 457–512.

KUHLE, M. 1988b. Topography as a fundamental element of glacial systems. *GeoJournal,* **17,** 545–568.

KUHLE, M. 1990a. The cold deserts of high Asia (Tibet and contiguous mountains). *GeoJournal,* **20,** 319–323.

KUHLE, M. 1990b. Ice marginal ramps and alluvial fans in semiarid mountains: convergence and difference. *In:* RACHOCKI, A. H. & CHURCH, M. (eds) *Alluvial Fans: A Field Approach.* Wiley, Chichester, 55–68.

KUHLE, M. 1991. Observations supporting the Pleistocene inland glaciation of High Asia. *GeoJournal,* **25,** 131–231.

KUHLE, M. 1993. A short report of the Tibet excursion 14-A, Part of the XIII INQUA Congress 1991 in Beijing. *GeoJournal,* **29,** 426–427.

KUHLE, M. 1995. Glacial isostatic uplift of Tibet as a consequence of a former ice sheet. *GeoJournal,* **37,** 431–449.

KUMAR, S., WESNOUSKY, S. G., ROCKWELL, T. K., RAGONA, D., THAKUR, V. C. & SEITZ, G. G. 2001. Earthquake recurrence and rupture dynamics of Himalayan Frontal Thrust, India. *Science,* **294,** 2328–2331.

KUMAR, S., WESNOUSKY, S. G., ROCKWELL, T. K., BRIGGS, R. W., THAKUR, V. C. & JAYANGONDAPERUMAL, R. 2006. Paleoseismic evidence of great surface-rupture earthquakes along the Indian Himalaya. *Journal of Geophysical Research,* **111,** B003309.

LARSON, K., BURGMANN, R., BILLHAM, R. & FREYMUELLER, J. T. 1999. Kinematics of the India–Eurasian collision zone from GPS measurements. *Journal of Geophysical Research,* **104,** 1077–1094.

LAVÉ, J. & AVOUAC, J. P. 2001. Fluvial incision and tectonic uplift across the Himalayas of Central Nepal. *Journal of Geophysical Research,* **106,** 26 561–26 592.

LEHMKUHL, F. 1995. Geomorphologische Untersuchungen zum Klima des Holozäns und Jungpleistozäns Osttibets. *Göttinger Geographische Abhandlungen,* **102,** 1–184.

LEHMKUHL, F. 1997. Late Pleistocene, Late-glacial and Holocene glacier advances on the Tibetan Plateau. *Quaternary International,* **38–39,** 77–83.

LEHMKUHL, F. 1998. Extent and spatial distribution of Pleistocene glaciations in Eastern Tibet. *Quaternary International,* **45–46,** 123–134.

LEHMKUHL, F., OWEN, L. A. & DERBYSHIRE, E. 1998. Late Quaternary glacial history of northeastern Tibet. *Quaternary Proceedings,* **6,** 121–142.

LI, B., LI, J. & CUI, Z. (eds) 1991. Quaternary glacial distribution map of Qinghai-Xizang (Tibet) Plateau 1:3,000,000. SHI, Y. (Scientific Advisor), Quaternary Glacier, and Environment Research Center, Lanzhou University.

LIU, J. G. & KUSKY, T. M. 2008. After the quake: a first-hand report on an international field excursion to investigate the aftermath of the China earthquake. *Earth Magazine,* October 2008, 48–51.

MASON, K. 1929. Indus floods and Shyok glaciers. *Himalayan Journal,* **1,** 10–29.

MEIGS, A. & SAUBER, J. 2000. Southern Alaska as an example of the long-term consequences of mountain building under the influence of glaciers. *Quaternary Science Reviews,* **19,** 1543–1562.

MISCH, P. 1935. *Ein gefalteter junger Sandstein im Nordwest-Himalaya und sein Gefüge. Festschrift zum 60 Geburstag von Hans Stille.* Verlag, Stuttgart.

MITCHELL, S. G. & MONTGOMERY, D. R. 2006. Influence of a glacial buzzsaw on the height and morphology of the Cascade Range in central Washington State, USA. *Quaternary Research,* **65,** 96–107.

MOLNAR, P. & ENGLAND, P. 1990. Late Cenozoic uplift of mountain ranges and global climatic change: chicken or egg? *Nature,* **46,** 29–34.

MOLNAR, P. & TAPPONNIER, P. 1978. Active tectonics of Tibet. *Journal of Geophysical Research,* **83,** 5361–5375.

MONTGOMERY, D. R., BALCO, G. & WILLETT, S. D. 2001. Climate, tectonics and the morphology of the Andes. *Geology,* **29,** 579–582.

NORIN, E. 1925. Preliminary notes on the Late Quaternary glaciation of the North Western Himalaya. *Geografiska Annaler,* **7,** 165–194.

OUIMET, W. B., WHIPPLE, K. X., ROYDEN, L. H., SUN, Z. & CHEN, Z. 2007. The influence of large landslides on river incision in a transient landscape: eastern margin of the Tibetan Plateau (Sichuan, China). *Geological Society of America Bulletin,* **119,** 1462–1476.

OWEN, L. A. 1988a. Wet-sediment deformation of Quaternary and Recent sediments in the Skardu Basin, Karakoram Mountains, Pakistan. *In:* CROOTS, D. (ed.) *Glaciotectonics.* Balkema, Rotterdam, 123–147.

OWEN, L. A. 1988b. Neotectonics and glacial deformation in the Karakoram Mountains and Nanga Parbat Himalaya. *Tectonophysics,* **163,** 227–265.

OWEN, L. A. 1989. Terraces, uplift and climate in the Karakoram Mountains, Northern Pakistan: Karakoram intermontane basin evolution. *Zeitschrift für Geomorphologie,* **76,** 117–146.

OWEN, L. A. 1996a. High roads high risks. *Geographical Magazine,* **118,** 12–15.

OWEN, L. A. 1996b. Quaternary lacustrine deposits in a high-energy semi-arid mountain environment, Karakoram Mountains, northern Pakistan. *Journal of Quaternary Science,* **11,** 461–483.

OWEN, L. A. 2004. Cenozoic evolution of global mountain systems. *In:* OWENS, P. N. & SLAYMAKER, O. (eds) *Mountain Geomorphology.* Edward Arnold, London, 33–58.

OWEN, L. A. 2006. Tectonic uplift and continental configurations. *In:* SCOTT, E. (ed.) *Encyclopedia of Quaternary Science Vol. 2.* Elsevier, Oxford, 1011–1016.

OWEN, L. A. 2008. How Tibet might keep its edge. *Nature,* **455,** 748–749.

OWEN, L. A. & DERBYSHIRE, E. 1988. Glacially deformed diamictons in the Karakoram Mountains, Northern Pakistan. *In:* CROOTS, D. (ed.) *Glaciotectonics.* Balkema, Rotterdam, 149–176.

OWEN, L. A. & DERBYSHIRE, E. 1993. Quaternary and Holocene intermontane basin sedimentation in the Karakoram Mountains. *In:* SHRODER, J. F. (ed.) *Himalaya to the Sea: Geology, Geomorphology and the Quaternary.* Routledge, London, 108–131.

OWEN, L. A. & SHARMA, M. C. 1998. Rates and magnitudes of paraglacial fan formation in the Garhwal Himalaya: implications for landscape evolution. *Geomorphology*, **26**, 171–184.

OWEN, L. A., BENN, D. I. ET AL. 1995. The geomorphology and landscape evolution of the Lahul Himalaya, Northern India. *Zeitschrift für Geomorphologie*, **39**, 145–174.

OWEN, L. A. & SHARMA, M. 1996. Landscape modification and geomorphological consequences of the 20 October 1991 earthquake and the July–August 1992 monsoon in the Garhwal Himalaya. *Zeitschrift für Geomorphologie*, **103**, 359–372.

OWEN, L. A., KAMP, U., SPENCER, J. Q. & HASERODT, K. 2002. Timing and style of Late Quaternary glaciation in the eastern Hindu Kush, Chitral, northern Pakistan: a review and revision of the glacial chronology based on new optically stimulated luminescence dating. *Quaternary International*, **97–98**, 41–56.

OWEN, L. A., FINKEL, R. C., MA, H., SPENCER, J. Q., DERBYSHIRE, E., BARNARD, P. L. & CAFFEE, M. W. 2003. Timing and style of Late Quaternary glaciations in NE Tibet. *Geological Society of America Bulletin*, **115**, 1356–1364.

OWEN, L. A., FINKEL, R. C., MA, H. & BARNARD, P. L. 2006. Late Quaternary landscape evolution in the Kunlun Mountains and Qaidam Basin, Northern Tibet: a framework for examining the links between glaciation, lake level changes and alluvial fan formation. *Quaternary International*, **154–155**, 73–86.

OWEN, L. A., CAFFEE, M. W., FINKEL, R. C. & SEONG, B. Y. 2008a. Quaternary glaciations of the Himalayan–Tibetan orogen. *Journal of Quaternary Science*, **23**, 513–532.

OWEN, L. A., KAMP, U., KHATTAK, G. A., HARP, E. L., KEEFER, D. K. & BAUER, M. A. 2008b. Landslides triggered by the October 8, 2005, Kashmir Earthquake. *Geomorphology*, **94**, 1–9.

PAFFEN, K. H., PILLEWIZER, W. & SCHNEIDE, H. J. 1956. Forschungen im Hunza–Karakoram. *Erdkunde*, **10**, 1–33.

PENCK, A. & BRUCKNER, E. 1909. *Die Alpen im Eiszeitalter*. Tauchnitz, Leipzig.

PORTER, S. C. 1970. Quaternary glacial record in the Swat Kohistan, West Pakistan. *Geological Society of America Bulletin*, **81**, 1421–1446.

PRATT, J. H. 1859. On the deflection of the plumb line in India. *Philosophical Transactions of the Royal Society of London*, **149**, 745–796.

PRELL, W. L. & KUTZBACH, J. F. 1987. Monsoon variability over the past 150 000 yr. *Journal of Geophysical Research*, **92**, 8411–8425.

PRELL, W. L. & KUTZBACH, J. F. 1992. Sensitivity of the Indian monsoon to forcing parameters and implications for its evolution. *Nature*, **360**, 647–652.

RAYMO, M. E. & RUDDIMAN, W. F. 1992. Tectonic forcing of late Cenozoic climate. *Nature*, **359**, 117–122.

RAYMO, M. E., RUDDIMAN, W. F. & FROELICH, P. N. 1988. Influence of late Cenozoic mountain building on ocean geochemical cycles. *Geology*, **16**, 649–653.

REYNOLDS, J. M. 2000. On the formation of supraglacial lakes on debris-covered glaciers. *In*: NAKAWO, M., RAYMOND, C. F. & FOUNTAIN, A. (eds) *Debris-Covered Glaciers*. International Association of Hydrological Sciences Publication, **264**, 153–161.

RICHARDSON, S. D. & REYNOLDS, J. M. 2000. An overview of glacial hazards in the Himalayas. *Quaternary International*, **65–66**, 31–47.

RÖTHLISBERGER, F. & GEYH, M. 1985. Glacier variations in Himalayas and Karakoram. *Zeitschrift für Gletscherkunde und Glazialgeologie*, **21**, 237–249.

RUDDIMAN, W. F. (ed.) 1997. *Tectonic Uplift and Climate Change*. Plenum, New York.

RUDDIMAN, W. F. & KUTZBACH, J. E. 1989. Forcing of Late Cenozoic Northern Hemisphere climate by plateau uplift in southern Asia and the America west. *Journal of Geophysical Research*, **94**, 409–427.

RUDDIMAN, W. F. & KUTZBACH, J. E. 1991. Plateau uplift and climate change. *Scientific American*, **264**, 42–51.

RUTTER, N. W. 1995. Problematic ice sheets. *Quaternary International*, **28**, 19–37.

RYDER, J. M. 1971a. Some aspects of the morphology of paraglacial alluvial fans in south–central British Columbia. *Canadian Journal of Earth Sciences*, **8**, 1252–1264.

RYDER, J. M. 1971b. The stratigraphy and morphology of paraglacial alluvial fans in south–central British Columbia. *Canadian Journal of Earth Sciences*, **8**, 279–298.

SCHÄFER, J. M., TSCHUDI, S. ET AL. 2002. The limited influence of glaciation in Tibet on global climate over the past 170 000 yr. *Earth and Planetary Science Letters*, **194**, 287–297.

SEARLE, M. P. 1991. *Geology and Tectonics of the Karakoram Mountains*. Wiley, Chichester.

SEONG, Y. B., OWEN, L. A. ET AL. 2008. Reply to comments by Matthias Kuhle on 'Quaternary glacial history of the central Karakoram'. *Quaternary Science Reviews*, **27**, 1656–1658.

SEONG, Y. B., BISHOP, M. P. ET AL. 2009a. Landforms and landscape evolution in the Skardu, Shigar and Braldu Valleys, Central Karakoram. *Geomorphology*, **103**, 251–267.

SEONG, Y. B., OWEN, L. A., YI, C., FINKEL, R. C. & SCHOENBOHM, L. 2009b. Geomorphology of anomalously high glaciated mountains at the northwestern end of Tibet: Muztag Ata and Kongur Shan. *Geomorphology*, **103**, 227–250.

SHAW, R. B. 1871. *Visits to High Tartary, Yarkard and Kashgar*. Oxford University Press, Hong Kong [reprinted 1984].

SHI, Y. 1992. Glaciers and glacial geomorphology in China. *Zeitschrift für Geomorphologie*, **86**, 19–35.

SHRODER, J. F., KHAN, M. S., LAWRENCE, R. D., MADIN, I. P. & HIGGINS, S. M. 1989. Quaternary glacial chronology and neotectonics in the Himalaya of northern Pakistan. *In*: MALINCONICO, L. L. & LILLIE, R. J. (eds) *Tectonics of Western Himalayas*. Geological Society of America, Special Papers, **232**, 275–294.

SKINNER, B. J., PORTER, S. C. & PARK, J. 2004. *Dynamic Earth: An Introduction to Physical Geology*. Wiley, Chichester.

SMALL, E. E. & ANDERSON, R. S. 1998. Pleistocene relief production in Laramide Mountain Ranges, western U.S. *Geology*, **26**, 123–126.

SPOTILA, J. A., BUSHER, J. T., MEIGS, A. J. & REINERS, P. W. 2004. Long-term glacier erosion of active mountain belts: example of the Chugach–St. Elias Range, Alaska. *Geology*, **32**, 501–504.

SURELL, A. 1851. *Etude sur les Torrents des Hautes Alpes*. Dunod, Paris.

TAPPONNIER, P. & MOLNAR, P. 1976. Slip-line theory and large-scale continental tectonics. *Nature*, **264**, 319–324.

TAPPONNIER, P. & MOLNAR, P. 1977. Active faulting and tectonics in China. *Journal of Geophysical Research*, **82**, 2095–2930.

TAPPONNIER, P., XU, Z., ROGER, F., MEYER, B., ARNAUD, N., WITTLINGER, G. & YANG, J. 2001. Oblique stepwise rise and growth of the Tibet plateau. *Science*, **294**, 1671–1677.

TRINKLER, E. 1930. The Ice-Age on the Tibetan Plateau and in the adjacent region. *Geography Journal*, **75**, 225–232.

US GEOLOGICAL SURVEY 2008. Earthquake Hazard program: Magnitude 7.9 – Eastern Sichuan, China. World Wide Web Address: http://earthquake.usgs.gov/eqcenter/eqinthenews/2008/us2008ryan/#summary.

VAN DER WOERD, J., OWEN, L. A., TAPPONNIER, P., XU, X., KERVYN, F., FINKEL, R. C. & BARNARD, P. L. 2004. Giant, M~8 earthquake-triggered, ice avalanches in the eastern Kunlun Shan (Northern Tibet): characteristics, nature and dynamics. *Geological Society of America Bulletin*, **116**, 394–406.

VON WISSMANN, H. 1959. Die heutige Vergletscherung und Schneegrenze in Hochasien mit Hinweisen auf die Vergletscherung der letzten Eiszeit. *Akademie der Wissenschaften und der Literatur in Mainz, Abhandlungen der Mathematisch-Naturwissenschaftlichen Klasse*, **14**, 121–123.

WESNOUSKY, S. G., KUMAR, S., MOHINDRA, R. & THAKUR, V. C. 1999. Holocene slip rate of the Himalayan Frontal Thrust of India, observations near Dehra Dun. *Tectonics*, **18**, 967–976.

WHIPPLE, K. X., KIRBY, E. & BROCKLEHURST, S. H. 1999. Geomorphic limits to climate-induced increases in topographic relief. *Nature*, **401**, 39–43.

WOBUS, C. W., HODGES, K. V. & WHIPPLE, K. X. 2003. Has focused denudation sustained active thrusting at the Himalayan topographic front? *Geology*, **31**, 861–864.

YANG, X., ZHU, Z., JAEKEL, D., OWEN, L. A. & HAN, J. 2002. Late Quaternary palaeoenvironmental change and landscape evolution along the Keriya River, Xingjiang, China: the relationship between high mountain glaciation and landscape evolution in foreland desert regions. *Quaternary International*, **97–98**, 155–166.

YIN, A. & HARRISON, T. M. 2000. Geologic evolution of the Himalayan–Tibetan orogen. *Annual Reviews of Earth and Planetary Sciences*, **28**, 211–280.

YIN, A., KAPP, P. A. ET AL. 1999. Significant late Neogene east–west extension in northern Tibet. *Geology*, **27**, 787–790.

ZEITLER, P. K. 1985. Cooling history of the NW Himalaya. *Tectonics*, **4**, 127–151.

ZEITLER, P. K., MELTZER, A. S. ET AL. 2001. Erosion, Himalayan geodynamics and the geomorphology of metamorphism. *GSA Today*, **11**, 4–8

ZHANG, X. & SHI, Y. 1980. *Changes of the Batura Glacier in the Quaternary and recent times*. Professional Papers on the Batura Glacier, Karakoram Mountains. Science Press, Beijing, 173–190.

ZHENG, B. 1989. Controversy regarding the existence of a large ice sheet on the Qinghai–Xizang (Tibetan) Plateau during the Quaternary Period. *Quaternary Research*, **32**, 121–123.

ZHENG, B. & RUTTER, N. 1998. On the problem of Quaternary glaciations, and the extent and patterns of Pleistocene ice cover in the Qinghai–Xizang (Tibet) plateau. *Quaternary International*, **45–46**, 109–122.

Index

Note: Figures are shown in *italic* font and tables in **bold**

Aasiaat domain 217, 229–232
accretionary orogen 1, 2, 79–85, 90, *132*
   circum-Pacific 35–45
   Mona Complex 55–72
accretionary wedge 215, *229*
   Gobi Altai orogen 364, 377
adakite 279, 280
African rift valley 91, *99*
Alaska-type arc *36, 38*, 39–41, 43
Alborz strike-slip system *336*, 340
Aleutian arc 22, *24*, 29
alluvial fan 389, *390*, 393
   faulted 368, *370*, 371, 373
Alpine–Himalaya system 329
Altaid Tectonic Collage 117
Altaids (Central Asian Orogenic Belt) 35, 36, 42–44, *45*
Amasia 82
Ampandandrava, diopsidites 157
amphibole 199
   mineral analyses **203**, 205–206
amphibolite-facies, Nagssugtoqidian 217–221
Anatolian faults 329, *331*, 332, 334, 337, 339
Andean margin 305, 317–318, 320, 321, 322
Andean-type arcs 7–10, 211, 349, 172–173
Andrahomana pegamatite **144, 148**, 152–154, 157
Anglesey, ocean plate stratigraphy 1–2, 61–63, *67*, 70
Anglesey–Lleyn 55–72
anorthositic complex 3–4, 197–211, 254, 272
   crystallization history 206
   melt composition 207–208
   mineral analyses **201, 202, 203**
   petrogenesis 209
Anshan complex 236–237
Antarctica, phlogopites 157
anti-continent/crust 93, 95, 96
anti-plate tectonics 78
apatite, Khanda Block 122
Arabia–Eurasia collision 329–341
arc systems 1, 36–40
   accretion rate 181
   circum-Pacific 35–45
   development *302, 316*, 322
   magma composition 208–209
   transition with continental margins 40–42
   trench deposits 228–229
arc–arc collision 42, 85–86
arc–back-arc associations 240
Archaean convection 96
Archaean craton, China 235–255
Archaean geotherms *94*, 96, 97
Archaean mantle processes, Greenland 3, 197–211
Archaean orogen 95, 96–98
Archaean terrane, Brazil 263
Archaean–Proterozoic boundary 244–245
arrested orogens 90–92, 110
Asia, tectonic map *43*
atmospheric circulation and tectonics 395

back-arc basin 1, 81, 102
   intra-oceanic arc 9, 10, *11*, 14–17
   magma *20*, 21, 30, 349
   type 37, 40, 41
   upper mantle 21–26
back-scattered electron imaging 179
banded iron formation 4
   Brazil 265, 274
   China 236, *238*, 241
   geochemistry 224–226
   Greenland 217, *220*, 228
Barrovian hydration events 87
basement uplift, North China craton 250–251, 255
basin and range, Gobi Altai 365, 377, 382
batholiths 82, 289, 291–292
Benioff thrust/plate 83, 85, 102, 103, *104*
biochemical cycles 395
Bitlis–Zagros suture *330*
blueschist 38, 82, 83, 95, 98, 102, 351
   Khanda Block 130–131, 133–134
Blueschist Unit 57, 58, 61, 63, *67, 69*, 72
Bogd Fault system *362, 367*, 370, 371, *375*, 380–382
boninites 17, 21, 29
boron-fluid evolution 2, 139–157
   composition and transport 140, 156–157
Bouguer anomaly, megamullion 101, 102
Brazil, diopsidite 156
Brazil, Precambrian terranes 4, 263–282

calc-alkaline magmatism, Kohistan arc 349–350, 353–355
Caldeirão shear belt **269**, 277–278
caldera, submarine 14
Canada, diopsidites 156
Caraíba complex **268**, 270–274, 278–281
carbonate 79, *80*, 119, 133
carbonatites 106, 253
cathodoluminescence 179, 181, *182*
Cenozoic cooling 395
Cenozoic deformation, Pakistan 351–353
Cenozoic orogenic belt 329
Central Asian Orogenic Belt 2, 117–134, 181, 187
   Japanese terranes 127–128, 133
   Jiamusi Block 130–133
   Khanka Block *118*, 119–121
   Mongolia 361–382
Chalt Volcanic Group 292, **303**, 310, *317*, 322, 349
chemistry *see* geochemistry
chert 55–60, *64*, 69, 70
Chilas Complex 291, 293, 295, 305, *309*, 349
   age and origin 311–314, *316*, 322
   deformation and metamorphism 297–299
CHIME analyses 127, 128
chromite 198–199, 267
   mineral analyses 200, **201**, 204–205
chromitite 3–4, 197–211
   analytical methods 199–200
   petrography 198–199
chrondrites *307*

INDEX

climate change 6, 391, 393–397, 402
CO$_2$ vapour flux 3, *11, 99*
Coedana granite 57, 58, 72
coesite 80, 83, 84, 95, 108, 345
collision tectonics 5
    Arabia–Eurasia 329–341
    Archaean 242
    Brazil 265, 279–280
    circum-Pacific 35–45
    continent–continent 100–101
    Eurasia 314–321, *330–331*
    Europe–Africa 101
    India–Asia 354, 355
    Indo-Eurasia 361, 365, 371, 382
    Madagascar 139
    Kohistan 287
collision type orogens 79–82, 83, 85–87, 90
Columbia supercontinent 188, 253, 255
concealed orogens 90, *91*, 92
continental crust, growth 29–30, 263
continental evolution 1–6, 236
continental reactivation, geochronology 188
convection and subduction 102–105
convergence rate 330–332, 335, 338
cooling, dating 184, 354
copper ore 246, 364
cordierite 274
Cordillera-type arc 35, *36*, *37*, 39, 40–41, 43
core-mantle boundary 78, 90, *91*, 108
cratonization, China 239–245
crust 183, *185*, 322
    composition 172, *305*, 314
    generation 210–211
    growth 29, 237–239
    intra-oceanic arc 21–26
    seismic structure 299–301, 366, 308–309
crustal thickness 86, 90, 98
    Arabia–Eurasia 339, 346, 355
    intra-oceanic arc 9, *12*, 13, 14, 20, 22, 26, 39
    Kohistan 299, 308, 322
Curaçá orogen 264–267

dating techniques 179
Dead Sea Fault System 329, 332
Deep Sea Drilling Programme 71
deformation, Quaternary 370
dehydration reaction 102, 108
delamination *11*, 57, 71, *315*
density and seismic velocity *300*
diamond 80, 84, 95, 182
diopside 143, **144**, 145
diopsidites 140, 142–156
Dir Group 292, 297, **303,** 318, 322
dunite 22, 24, 289
duplex *56*, 57, 60, 61, 72
dyke swarm, mafic
    Brazil 265
    China 2–5, 163–174, 242–244, 251–252
    Kohistan 314

earthquakes 102, *103*, 108
    Arabia–Eurasia 331, 332, 336
    Gobi Altai 366, 376, 381, 382
    Himalayan–Tibetan orogen 400–402

eclogite 80, 95, 98, *107*, 108
    metamorphism 354
    regressed 247–249
equilibrium-line altitude 395–396
Equutiit Killiat schist 219, 221–224, 228
    geochemistry **222–223**, *224, 225*, 226
    geochronology 226–228
    metamorphosed rocks 228
Erguna Block 118
escape structures *333*, 334, 337, 339
evaporite 373
exhumation of orogens 82–85, 354
    rate of 84–85
extant orogens 90, *91*, 92
extension 79
    and dyke swarms 163
    Mesozoic 365–366, 376
extrusive flow model 398–399

Fiskenæsset anorthositic complex 3, 197–211
fission track data, Gobi Altai 364, 370, 374, 382, 399
flower structure 369, 371, 377
fluid circulation 105–106, 109–110
forearc, intra-oceanic arc systems *11*, 12–13, *26, 28*
    development 26–29, 37, 81
    upper mantle and crust 21–26, 30

gabbro norite 349, 350, 353
Gangdese batholith 349
Garhwal earthquake 400, 402
Gavião Block *266*, 267, 280
geobarometry 2
geochemistry *20*, 349, 350
    Brazil **268–271**, *272–276*
    chromitites 197–211
    intra-oceanic arc systems 17–21
    mafic dyke swarm 165, **166**, 168–172, 174
    Nagssugtoqidian orogen **222–223**, *224*, 224–226
    North China craton 239–241, 243–244, 246–247, 253–254
    Tranomaro Belt 145–149, 153
geochronology 3, 179–189
    Brazil 264–265, 267–279
    dating techniques 179
    Kohistan arc 347–348
    Nagssugtoqidian orogen 226–228
    North China craton 236–2251, 255
geomorphology and tectonics 4, 373–376
geophysical relief 396
geothermal gradient 24, *26*, 83, 84, 95, 102
    Arab–Eurasia 313, 317
    Kohistan arc 353
geothermometry 2, *88*, 89, 93–108
ghost orogens 2, 86, 90, *91*, 93, 110
glacial buzz-saw model 395, *397*
glacial erosion and uplift 395–397
glacial lake outburst floods 393, 402
glacial record 389, 391–394, 395
gneiss protolith 184–187
gneiss, Brazil 263–273, 281
gneiss, west Greenland 210
Gobi Altai orogen 4–5, 361–382
    basement 361–364
    Cenozoic reactivation 361, 366

deformation 370–373
events **379–380**
faulting 366–370, 376
geomorphology 373–376
Mesozoic basins 364–366
volcanism 363, 376
Gobi–Tien Shan Fault system *362, 367*, 380, 381
Godzilla mullion 102
Gondwana 82, 83, 131, 133, 188, 289, 333
Gondwana, borosilicates 2, 155
grandidierite 143–145, 150, 152–154, 156, 157
granite–greenstone terrane 98
granitoid 123, **269**, 278–279
granulite 79, 87, 93, 98, 155
　Brazil 267, **268–269**, 272, 278, 279, 281
　Khanda Block 118, 123, 130, 131, 133
　Kohistan 294–295, 305, 313
　North China craton 247–251
gravity survey 389
Greenland, anorthositic complex 3, 4, 197–211
Greenland, grandidierite 157
Greenland, Precambrian 197–211, 213–232
greenstone belt
　Brazil 263–267, 270–277
　North China craton 236, 239–240, 242, 254
Guguan volcano *18*
Gulf of California, oceanic crust 41
Gwna Group 57–59, 61, *68*, 70, 72

harzburgite 4, 21, 22, 24, 29, 30, 95
　chemistry 12, 208, 209
　Khanka 121
　Kohistan 289, 351
Hattian Bala landslide *400*, 401
heavy minerals 184, 226
Hebei terrane 240–242
Heilongjiang Complex 118, 119, 121–123
　metamorphism 130–131, 133–134
Hida Block 127, 128, 133
hillslope angle, Himalayas–Tibet *398*
Himalayan tectonics *100*, 298, 329, 346
　review 287–322
Himalayan–Tibetan orogen, landscape development 389–402
　climate and topography 6, 395–400
　extreme events 400–402
　glaciations 391–393
　literature review 389–390
　paraglaciation 393–394
　plate tectonics 394–395
hot plate subduction 82
hotspot 13, 55, *56*
Hunde Ejland mineralization 221, **222–223**
hydrocarbon source rocks 365
hydrothermal mineralization 13

ice age study 391–393
ice avalanches 401
ICP–MS analysis 123, **128–129**
igneous province, North China craton 170, 172
Indian ocean, hot orogen 101–102
Indus Suture Zone 350–351
intercontinental transpressional orogen 376–382

intra-oceanic arc systems (IOAS) 7, 345, 350
　definition 8–10
　early development 26–30
　magma 13, 17–21
　upper mantle and crust 21–26
　sediments 10, 14–17, *18*
　structure 10–14
　trace elements 19–20, 21
inversion, 376, *377*
iron quadrangle 264
island arc 35–45, 363
　see also under Kohistan
isostatic relief 396
isotope geochemistry 180–181, 184, 187, 363
　parameters 300, **303–305**
　plate tectonics 188–189
Itabuna orogen 264–267
Izu arc *12*, 14, 21
Izu–Bonin–Mariana arc 7, 29, 38, 86

Jacobina group 277, 280
Jaglot Group 291, 299, **303**, 310, 314, 322
Japan, age of orogens 81–82
Japan, cold orogen 102–105
Japan-type arc *36, 37*, 38–41, 43–44
Jiamusi Block 118–119, *132*
Jijal Complex 289, 293–297, 299, 301, 305, *306*
　age and origin 311–314, 322, 348
juvenile crust 5, 9, *27*, 29–30
　growth 180–182
　Khanda Block 117
　North China craton 237
juvenile magma 276, 363

Kamila Amphibolites 289, 291, *295–297*, 299, 348, 349
　age 293, **294**, 322
　geochemistry 301, **304**, 305, 307–313
Kashmir earthquake 400, 402
Kazakhstan 42, 43
Kermadec arc 14
Khanda Block, China 2, 117–134
　age 121, 123, **124–125**
　geological setting 120
　and Japanese terranes 127–128, 133–134
　and Jiamusi terrane 130–133
khondalite series 130, 245, 247, 249–250
kimberlite 95, 98, 106, 189
Kohistan batholith 289, 291–293, **303**, 322, 349–350
Kohistan island arc 4–6, 24, 287–322
　collision 287, 314–322
　crustal structure 289, 299–302, 306, 308–309, 322
　geochemistry 300–306, *307–311*, 321
　geochronology 293, **294**, 347–349
　igneous rocks 289–293, *295–298*, 309–310
　lithostratigraphy 289, 292
　metamorphism 292, 293–299, 314, 317
　palaeogeography *312–313*
　review of literature 288
　seismic crustal structure 299–300, *301*, 306, 308–309
　structure *290*, 293–294
　tectonics 289, *290*, 298, 314–321, 353
Kokoxili earthquake 401
komatiite 239, 240, 242, 264
Kopeh Dagh fault array 337

kornerupine 155
kyanite–sillimanite metamorphism 80, 351, 354

Ladakh batholith 349, 350
Lamayuru Complex 351, *352*
lamproites 106
landscape and tectonics 6, 329, 389–402
landslides 395, 400–402
laser-ablation inductively couple plasma spectrometry (LA–ICP–MS) 179
layered igneous intrustion 198, 206
lherzolite 21, 22, 30, 94, 95
Llanddwyn Island, stratigraphy 59–61, *64–66*, 70
Lleyn, stratigraphy 1–2, 61, 63, 66, *68*
loess 373
lost orogens/continents 78, 87, 93, 108
lunar anorthosite 197

Macquarie Island, hot orogen 102
Madagascar, Tranomaro Belt 2, 139–157
mafic *see under* ultramafic
magma composition 207–209
magmatic arc/front *11*, 13, 17–26, *28*
magmatism 204, 208, *209*, 349–355, 349, 363
　　North China craton 163–174, 242–244, 251–254
　　volume 321
Main Mantle Thrust 354–355
Mairi complex **268**
mantle 1, 12, 299
　　convection *11*, *96*
　　intra-oceanic arcs 21–26
　　primitive *306*, *308*, *309*
　　temperature 376, 381
　　wedge 10, *19*, *84*, 102, *103*, *109*
　　84, 102, *103*, *109*
　　xenoliths 78, 89, 92–94, 181
mantle-derived crust 300, 321
mantle-derived magma 204, 208, *209*, 350
Mariana Trench 7, 12, 14–17, *18*, 85
　　sediments 70, *71*
　　upper mantle and crust 22–26, 299, 306
Mariana-type arc *37*, 38, 40–41, 43
Mashan Complex 118, 130, 132
mechanical models, Himalayan–Tibetan orogen 395
mega-mullion 79, 101
mélange 69–70, 351, 364
metal deposits 44
metamorphism 83, 87, 90, 101, 106–108, *109*
　　analytical techniques 122
　　dating 182–184, 198
　　spreading ridge 221, 224
metamorphism, Brazil 263–282
metamorphism, Himalayan 345, 351, 353, 354
metamorphism, Khanda Block 117–134
metamorphism, North China craton 172, 235–236, 242, 245–253
　　high temperature–pressure 245–251
metamorphism, Tranomaro Belt 139–157
metamorphism, west Greenland 213–232
metasedimentary units 291
metasomatism, borosilicates 154–155, 157
metavolcanic rocks 217–218, 252–253
microcontinents 36, 44
　　Khanda Block 117–120, 127, 131
　　North China craton *239*, 244

microprobe *see* SHRIMP
mid ocean ridge basalt 21, 38, 78–80, 89, *93*, 94, 102
　　Greenland 198, *225*, 228
　　Kohistan 305, 348, 350
　　Mona Complex 55, 58, 60, 61
mine, phlogopite *142*, **144**
mineral age 179, 182–184
mineral chemistry, anorthositic complex 200–206
mineral deposits 246, 267, 364
mineralization, spreading ridge 221, **222–223**
mobile belts, Palaeoproterozoic 236, 245–251, 255
Moho 98, 101, 102, *104*
　　intra-oceanic arcs 21–23, *24*, 26, 30
molasse 351
Mona Complex 55–72
monazite in crustal dating 183, *185*
Mongolia, arc system 43–44
Mongolia, Central Asian orogenic belt 6, 187, 361–382
Mongol–Okhotsk ocean closure 381
monsoon 391, *398*, 399, 400–402
moraines 391, 394, 399, 400
MORB *see* mid ocean ridge basalt
mud volcano 82
Mundo Novo greenstone belt 267, **268**

Nagssugtoqidian orogen, Greenland 213–232
　　geochemistry **222–223**, *224*, 224–226
　　geochronology 217, 226–228
Nankai Trough 71, 72, 85
Naternaq supercrustal belt 217, *218*
Nb isotope data 1, 180
　　Khanda Block 18
　　Brazil **268–269**, 274, 276
Nefyn, ocean plate stratigraphy 63, *68, 69*
New Harbour Group 57–58, 72
North China craton 235–255
　　basement uplift 250–251, 255
　　cratonization 239–245
　　dyke swarms 163–174
　　geochemistry 239–241, 243–244, 246–247, 253–254
　　geochronology 236–239, 241–244, 246–247, 249, 251–252
　　magmatism 242–244, 251–254
　　metamorphism 235–236, 242, 245–251, 252–253
　　rifting 252–254

obduction, ophiolite 351, 353–354
Ocean Drilling Programme 70
ocean plate stratigraphy 97
　　Mona Complex 57–59, 72, 228
　　olistostrome-type 65, 72
　　rate of sedimentation 70
　　ridge-trench transition 59–61, 66, *69*
　　type section 56–57
ocean-floor–arc-trench deposits 215, *220*
oceanic core complex 101
olistolith, Khanka 120
olistostrome 70, 81, 85
　　accretionary complex 58, 65, *68, 69*, 72
Oman ophiolite/harzburgite *204*, 206, *209*, 353
Oman, diopsidite 156
ophiolite 12, 21, 29, 121, 206, 321, 364
　　age *347*
　　complex 4, 6, 345–355

optically stimulated luminescence (OSL) 391
ore deposits 13, 168, 246, 364
orogens *36, 91*
    accretionary-type 79–85
    age 90–98, 106–108
    characteristics 78–86
    classification 2, 86–98
    cold 98–99, 102–105, 106
    collision-type 79–82, 83, 85–87
    exhumation 82–85
    hot 99–102
    structure and size 81
    tectonic erosion 85
    Wilson cycle 81–82

Pacific Ocean, age of 81
Pakistan *see under* Kohistan
Palaeo-Asian ocean 44
    closure 361, 364, 366
palaeomagnetism 321, 381
Palaeoproterozoic terrane, Brazil 263
    palaeosutures *231*
    plate tectonic model 215–217
Pan African event 188
Papua New Guinea arc system 41
paraglaciation 393–394
passive continental margin 58, 79, 80, 81, 91
    Cenozoic 321, 351, 353
    Precambrian 232, 254
pegamatite 152–154, 217
peneplain remnants 373–376
peridotite 12, 22
petrography, chromitite 198–199
petrography, Tranomaro Belt 143–145, 153
phase relations 149–152, 153–154, 206
phlogopite 140, 142–144, 149, 150, 153, 154, 157
pillow lavas *64*, 69, *218, 220, 225*
plagioclase 199, **202**, 205
plate boundaries *9, 71, 99*, 330, 332, 334
plate tectonics 78, 215–217, 229–232
    initiation of 96–98, 188–189
    and landscape development 394–395
plume *71*, 78, *91, 93*, 99–108, 172
porphyry copper deposits 364
Precambrian basement, Brazil 263–282
Precambrian tectonics 3, 4, 263
    China 235–255
    Greenland 197–211, 213–232
    Mona Complex 55–72
protolith 184–187, 279
Pythagoras' rule 335

Qaqqarsiatsiaq, geochemistry **222–223**
Qeqertarssuatsiaq Island, anorthositic complex *198, 200*

radiocarbon dates 391
radiolaria 55, 56
radiometric age 57, 58, 179–189
    Khanka Block 121–131, 133–134
    Kohistan arc 349–350
    North China craton 165, **166**, 168
rapakivi granite 254
rare earth elements 349
    Brazil 267, 273, 275

Greenland 207, *208*, 224–226
Kohistan 301, 302, 305, *306–309*
North China craton 238, 239, 241, 243, 246, 253
    dykes 251
restraining bend 371, *372*, 373, *375*, 381
rheology 24, *26*, 314, 376
rifting 14, *15, 99*, 252–254, 353
Rinkian fold belt 213, 215, 229, *230*, 232
Rio Capin greenstone belt **268–269**, 274–275
Rio Itapicuru greenstone belt **269**, 275, 277, 281
Rodinia 187–188
roll-back, trench 9, 14
Russia, Aldan Shield 154, 156
Russia, Central Asian Orogenic Belt 118–122
rutile, crustal dating 183

Saharan Metacraton 188
Salvador orogen 264–267
São Francisco craton 263–282
    geochronology and geochemistry 264–265, 267–279
    metamorphic rocks 265–267
    tectonic model 264–265
sapphirine 154–155, 265, 277
Saúde complex **269**, 277
sea-floor spreading, rate of 14
seamount 55, *56*, 70, *71*
secondary ionization mass spectrometry (SIMS) 180, 183, 184
sedimentary basins *16*, 364–366
sedimentation *18*, 38, 79, 82
    rate of 14–16
seismic crustal structure *13*, 299–300, *301*, 306, 308–309
seismic profile, Mariana arc 17
seismic thickness 376
seismicity *330*, 338, *362*, 366
serendibite 143, **144, 145, 147**, *150*, 151–152, 154
serpentinite mud volcano 12, *13, 16*
Serrinha Block *266*, 267, **268,** 272, 274, 279–283
Shamran/Teru Volcanic Formation 292, 293, 299, **303**, 320–322, 349
    development of 310, *312–313, 317*, 318
shortening 365, 375, 381, 382
shortening arrays, Arabia–Eurasia 330, *333*, 337–338
SHRIMP (zircon sensitive high resolution ion microprobe) *180*, 183, *185, 186*, 189
    Brazil 267, 272–278
    Central Asia Orogenic Belt 121–123, **124–127, 129**
    Madagascar 140
    North China craton 236–237, 241, 243–244, 246, 249, 253–254
    dykes 165, 251
single-zircon thermal ionization mass spectrometry (TIMS) 179, 184
sinhalite 143, **144, 145, 147**, *150*, 152, 154
Sino-Korean craton *see* North China craton
skarn 154
slab break-off 40, 83, *84*, 85, 107–109, 354
    and related melts 280
slab descent and fluid flux 18, 29, 30
slab graveyards 78, 90
slab melting 182, 210
slab rollback 366
slab subduction *91, 97, 103, 107*, 108
slab, age of *93*, 94, 98

Songliao Block 118
South Stack Group 58, 72
spilite–keratophyre 169
spinel 22, 24, 94, 95
    Tranomaro belt 140, **144, 145**, 149, 153, 155
Spontang Ophiolite 351, *352*, 353, 355
spreading ridge, west Greenland 213–232
Sri Lanka, metasomatism 154–155
staurolite 221, 224, 226
strain partitioning, Arabia–Eurasia *333*, 334–337, 339
strike-slip faults 5, 366–368, *369, 372*
    Arabia–Eurasia 329–341
    rate of movement 329, *330*, 332, 334–336, 371, 382
subduction 314–320
    continental collision 332–333
    geotherms *93, 94*
    intra-oceanic arc system 26–29
    rate of 7, 61, 86, 92, 321
    see also under slab
subduction polarity 40, 57, 63, 70, 81, 364, 381
supercontinent 37, 263, 281
    Columbia 188, 253, 255
    Neoarchaean, 244
sutures, Palaeoproterozoic 230–232

Taihang–Lvliang mafic dyke swarm 165–167, *169*, **171**
Talkeetna arc 24, *26*
tear fault *see* transfer zones
tectonic aneurysm 396, *398*
tectonic erosion 12, 85, 181,
tectonic framework of orogens *36*
    cold 98–99, 102–105
    hot 99–102
tectonic geomorphology 394
tectonic model *42, 44*
    Kohistan *315, 318–319*, 321–322
    North China craton *173, 255*
    São Francisco craton 280, *281*
Tein Shan faults 4–5, *362, 367*, 380, 381
terraces, river 393, 399
terrane accretion, geochronology 187–188
terrestrial cosmogenic nuclide 391
Teru volcanic rocks 292
Tethyan ocean 82, 350
    closure 320, 321, 353, 354, 355
thermal history of orogens 79, 87–90, *93*
thermal relaxation 355
thermomechanical model *316*, 395
thick-skinned deformation 332
thrust fault 368–369, *370*
    Miocene 354–355
thrust ridges 371, 373
thrust zones *331*, 337, 338
thrusting and topography 399
thrusts, blind 334, 339
Tibet–Himalaya, hot orogen 100–101
Tibet, underplating 321
Tibetian Plateau, growth of 395, 399
Timor–Tanimbar belt 82–85, 86, 92
titanite in crustal dating 183–184
tonalite–trondhjemite–granodiorite
    batholiths 78, 79, *80*, 90, 97
    Brazil 263, 279
    gneiss 185, 197, 198, 210

topography and deformation 365, 370, 381, 382
trace elements 19–20, 21, 30
    Brazil **270–271**, 274, 279
    Greenland **223**, 224–225
    Kohistan 301, 302, 305, *310*, 320
    North China craton *168*, 238
Tranomaro Belt, Madagascar 139–157
    deformation history 139
    geology 140–142
    metamorphism 152, 154
transfer zones, Arabia–Eurasia 330, *333*, 338–339
transpression 366, 371, 382
transtensional deformation 371, 373, *374*
trench and intra-oceanic arc systems 10–12
triple junction 2, 332, 334

Ultima Pangaea 82
ultramafic–mafic rocks 349
    Kohistan 289, 291, 289, *296*, 305, *309*
    North China craton *164*, **166**
underplating 57, 61, 70, 82, 180
    Kohistan 314, 321
U–Pb zircon age 217, 226, **227**, *228*
    see also SHRIMP
uplift and topography 6, 399, 402
uplift, dating 184
uplift, rate of 255, 370–371, *398*

velocity *23–24*
Vohibola diopsidites *141*, 142–152, 156
volcanic magmatic arc 13–14
volcanic province, North China craton 167–170
    geochemistry/geochronology 244–246, 252–253
volcanic rocks, Kohistan 292–293, *295*, **303**, 309–310, 348, 349
volcanism 363, 376
    volume of eruption 170, 172

wedge extrusion 83, 84, 108
wehrlite 24, 94, 95
Wenchuan earthquake 401, 402
Wilson cycle 44, 81–82
Windley, B F, biography vii–viii, *ix*, 1
wrench tectonics 280

xenocryst 123, 181, 272
xenolith 22, 95, 106
    Archaean *88, 94*, 107, 108
    Khanda Block 121
    mantle 78, 89, 92, 93, 94
Xing'an Block 118
Xiong'er volcanic province 167–170, **171**, 253

Yasin volcano-sedimentary formation 292, **303**

Zagros faults 329, 334–337, *339*, 341
Zagros Simple Folded Zone 338, 339
Zimbabwe 2, 3
zircon *182, 183*
    analyses 122–129, 133, 140
    dating 179, *180*
zircon age, Khanda Block 121–129, 130–131
zircon sensitive high resolution ion microprobe
    see SHRIMP